Cell Engineering

Volume 10

Series Editor
Mohamed Al-Rubeai, University College Dublin, Belfield, Dublin 4, Ireland

Integrating advances in molecular biology into bioprocesses presents a continuous challenge to scientists and bioengineers. This series is conceived to help meet this challenge. It examines and assesses the feasibility of new approaches for the modification of cellular function such as gene expression, protein processing, secretion, glycosylation, immortalisation, proliferation, and apoptosis as well as the systematic study of the metabolic genotype-phenotype relationship. The series provides detailed coverage of the methodology for improving cellular properties of cells used in the production of biopharmaceuticals, gene and cell therapies and tissue engineering. It also seeks to explain the cellular mechanisms underlying *in vitro* physiological activity and productivity.

More information about this series at http://www.springer.com/series/5728

Ralf Pörtner
Editor

Cell Culture Engineering and Technology

In appreciation to Professor Mohamed Al-Rubeai

Springer

Editor
Ralf Pörtner
Institute of Bioprocess and Biosystems Engineering
Hamburg University of Technology
Hamburg, Germany

ISSN 1389-6946
Cell Engineering
ISBN 978-3-030-79870-3 ISBN 978-3-030-79871-0 (eBook)
https://doi.org/10.1007/978-3-030-79871-0

This Springer imprint is published by the registered company Springer Nature Switzerland AG
The registered company address is: Gewerbestrasse 11, 6330 Cham, Switzerland

Preface

Mammalian cells are used in industry as well as in research for a variety of applications. Examples are the production of monoclonal antibodies or proteins for diagnostic or therapeutic use, production of viral vaccines, as well as cultivation of tissue cells for artificial organs or for gene therapy. This volume is aimed to provide high-quality contributions reviewing cutting-edge progress in cell culture engineering and technology. It is dedicated to Prof. Dr. Mohamed Al-Rubeai, University College Dublin (Ireland), on the occasion of his 70th birthday.

Cell Engineering 9 – Animal Cell Culture, released in 2015, reviewed the state of the art with in-depth assessments that emphasize the practical aspects of efficient operation of cell culture techniques, including selection and characterization of cell lines, suitable types of bioreactor system, optimal physical and chemical environment and appropriate production mode, as well as the relevance of process analytical technology (PAT) in the biopharmaceutical industry.

This volume, *Cell Engineering 10 – Cell Culture Engineering and Technology* is dedicated towards the progress achieved within the last years. It starts with chapters on cell engineering, novel technologies, and the use of omics, including products from animal cells, use of remote control for therapeutic designer cells, mRNA-techniques, omic's data-driven design of biopharmaceutical production, media optimization, glycoengineering, and cell engineering platforms.

Topics with respect to development of cell-based technologies and therapeutics cover the power of metabolic understanding, techniques for manufacturing of different types of stem cells, as well as production processes related to gene therapy.

Aspects of cell culture process engineering, advanced bioreactor concepts, modelling, as well as monitoring and control are addressed by chapters on scale-up (and scale-down) techniques, use of continuous single-use systems, mathematical modelling and model-assisted design concepts for mammalian cell culture processes, as well as bioprocess monitoring and control, especially process analytical technology for improved process control.

In summary, the volume comprises contributions of researchers active in the field of cell culture development for the production of recombinant protein, cell line development, cell therapy, and gene therapy, with consideration of media development, process scale-up, reactor design, monitoring and control, and model-assisted strategies for process design. The knowledge and expertise of the authors cover disciplines like cell biology, engineering, biotechnology, and biomedical sciences. Inevitably, some omissions will occur in the text, but the authors have sought to avoid duplications by extensive cross-referencing to chapters in other volumes of this series and elsewhere. We hope the volume provides a useful compendium of techniques for scientists in industrial and research laboratories that use mammalian cells for biotechnology purposes.

I am grateful for the support of all the excellent contributors; the series editor Prof. Dr. Mohamed Al-Rubeai, University College Dublin (Ireland); the staff at Springer for their efficiency in preparing this volume; and Dr. Gonzalo Cordova for his help and patience.

Appreciation for Professor Mohamed Al-Rubeai

***Cell Engineering* (volume 10) is dedicated to Professor Mohamed Al-Rubeai on the occasion of his retirement after 45 years of scientific activities in academia.**

Mohamed Al-Rubeai is an outstanding professor of biochemical engineering who has dedicated his life to scientific research, education, charity, and human rights.

Professor Al-Rubeai is emeritus professor of biochemical engineering and founding director of The Cell Culture Engineering Research Centre in the University College Dublin School of Chemical and Bioprocess Engineering.

He obtained his PhD in genetics from Queen Mary College at the University of London in 1979. Following a brief stint at the University of Surrey, he joined the University of Birmingham in 1988, where he established himself as a world leader in cell culture technology with his work on the shear sensitivity of animal cells in bioreactors, and his efforts helped to establish the importance of flow cytometry for monitoring and controlling cell culture. He wrote his most important papers at this time, including showing that animal cells in *in vitro* culture die by apoptosis. He pioneered the development of cell engineering method to improve the robustness and viability of cell lines and to optimize culture productivity. In 2005, he moved to University College Dublin to take up the newly established Chair of Biochemical Engineering. His recruitment to UCD was a pivotal step in the establishment of the National Institute for Bioprocessing Research and Training (NIBRT) and preceded the development of a globally significant biopharmaceutical manufacturing industry in Ireland.

Professor Al-Rubeai has authored over 500 professional papers, reviews, chapters, articles, and patents and has a record of >17,000 citations. This impressive body of work reflects his broad interests in the production of biopharmaceuticals, mammalian cell culture, tissue engineering, stem cell bioprocessing, metabolic engineering, nano-biotechnology, and biomaterials. These achievements are supported by the esteem in which Professor Al-Rubeai is held within the global biotechnology community across both academia and industry, where his work is universally known and acknowledged. Professor Al-Rubeai is, without a doubt, one of the founding fathers of the UK biotechnology industry.

He is a Fellow of the Conway Institute of Biomolecular & Biomedical Research, and Fellow of the UK Society of Biology (FSB). In 2007, he won the Astellas Award and in 2008 was awarded the IChemE Donald Medal for outstanding services in biochemical engineering. The UK Research Assessment Exercise gave his research its highest score (5 stars) in its last two sessions in 1996 and 2001.

When I think of exceptional people I know, Professor Al-Rubeai is right at the front of the line. We celebrate his anniversary with a special volume of *Cell Engineering* which we recommend for everyone working in cell culture and production of biopharmaceuticals.

Hamburg, Germany Ralf Pörtner
May 2021

Contents

Part I
Biopharmaceutical Production

Biopharmaceuticals Produced from Cultivated Mammalian Cells

Rasoul Al-Majmaie, Darrin Kuystermans, and Mohamed Al-Rubeai

Abbreviations

ADCC	Antibody-dependent cellular cytotoxicity
ATF	Alternating tangential flow
BHK	Baby hamster kidney
CAGR	Compound annual growth rate
CD2	Cluster of differentiation 2
CD20	Cluster of differentiation 20
CD3	Cluster of differentiation 3
CD4	Cluster of differentiation 4
CDRs	Complementarity-determining regions
CHMP	Committee for Medicinal Products for Human Use
CHO	Chinese hamster ovary
CMO	Contract manufacturing organization
DHFR	Dihydrofolate reductase
EMA	European Medicines Agency
Fab	Antigen-binding fragment

This chapter is an extended update of the previously published chapter: Kuystermans D. and Al-Rubeai M. (2015) Biopharmaceutical Products from Animal Cell Culture. In: Cell Engineering, Animal Cell Culture, vol. 9, pp. 717–757.

R. Al-Majmaie · M. Al-Rubeai (✉)
School of Chemical and Bioprocess Engineering, University College Dublin, Dublin, Ireland
e-mail: rasoul.al-majmaie@ucdconnect.ie; m.al-rubeai@ucd.ie

D. Kuystermans
Sanford-Burnham Medical Research Institute at Lake Nona, Orlando, FL, USA
e-mail: dkuystermans@sanfordburnham.org

R. Pörtner (ed.), *Cell Culture Engineering and Technology*, Cell Engineering 10,
https://doi.org/10.1007/978-3-030-79871-0_1

Fc	Fragment crystallizable
FcRn	Neonatal Fc receptor
FDA	Food and Drug Administration
HAMA	Human anti-mouse antibody
HEK-293	Human embryonic kidney epithelial cells
hGH	Recombinant human growth hormone
IgG	Immunoglobulin G
kDa	Kilodaltons
LFA-3	Human leukocyte function antigen 3
mAb	Monoclonal antibody
NK	Natural killer cells
NS0	Murine myeloma cells
Sp2/0	Murine myeloma cells
TNF	Tumor necrosis factor
TNFR	Tumor necrosis factor receptor
VEGF	Vascular endothelial growth factor

1 Introduction

The term "biopharmaceuticals" here refers to protein, nucleic acid- and engineered cell-based pharmaceutical substances produced by genetic engineering other than extraction from a native biological source [1, 2]. Since the first biopharmaceutical (Humulin®) approved in 1982, the biopharmaceutical industry has seen incredible growth over the past three decades, with total annual revenue increasing from around $4.4 billion in 1990 to over $180 billion in 2017. The global biopharmaceutical market is estimated to be about $270 billion in 2019 and is further projected to reach close to $500 billion by 2025 at a compound annual growth rate (CAGR) of around 13% between 2018 and 2025, which doubles the growth rate of conventional pharmaceutical industry [3]. In the meantime, there has been a rapid advancement in the number of biopharmaceutical approvals. While only 29 such products were approved by the US Food and Drug Administration (FDA) prior to 1996, the number has increased by more than five folds to 145 over the past 5 years only from 2016 to 2020 [4–6].

In the early developmental stage of biopharmaceuticals, recombinant protein products (e.g., hormones and enzymes), known as replacement therapies, have been widely approved for treating patients with diseases caused by the deficiency of specific molecules. These recombinant protein biopharmaceuticals provided a therapeutic strategy for managing disease states by restoring or supplementing endogenous proteins [7, 8]. During this period, the first monoclonal antibody (mAb) therapeutic, muromonab-CD3 (Orthoclone OKT3®), was approved for the treatment of acute, glucocorticoid-resistant rejection of allogeneic renal, heart and liver transplants in 1986. This antibody-based drug targeting the CD3 receptor was manufactured through mouse ascites, with antibody-rich ascitic fluid harvested from peritoneal tumors in mice that were induced by injecting hybridoma cells into the peritoneum [9, 10]. Although muromonab-CD3 is not under consideration in this review due to the non-recombinant production process, it appears that the approval of the mAb has opened the gateway for the development of specifically designed biopharmaceuticals targeting specific mediators (e.g., genes, proteins or cells) in human diseases. The therapeutic strategy has greatly stimulated the development of biopharmaceuticals including recombinant protein, nucleic acid- and engineered cells-based products [4]. In particular, the area of recombinant mAb biopharmaceuticals constitutes the largest growing segment of the global biopharmaceutical market with exceeded US$98 billion in sales and a CAGR of 18.3% in 2017 [11].

Recombinant protein biopharmaceuticals hold sway over the biopharmaceutical industry from the beginning, although nucleic acid- and engineered cell-based biopharmaceuticals are slowing starting to gather momentum. From a manufacturing process perspective, the dominant position of recombinant protein biopharmaceuticals can be greatly attributed to a variety of recombinant protein expression systems that meet therapeutic needs. To date, recombinant protein biopharmaceuticals are produced by prokaryotic and eukaryotic expression systems including bacteria, fungi, insect cells, mammalian cells, transgenic animal and plant organisms [12, 13]. Among these, the mammalian cells-based expression system has indisputably become the predominant choice for the production of recombinant protein biopharmaceuticals based on a comprehensive consideration of various factors, such as their ability to perform post-translational modifications suitable for therapeutic requirements, protein yield and process operability [14, 15]. Over the last 5 years (2016–2020), more than 80% of recombinant protein biopharmaceuticals approved in US and European markets were produced via cultivated mammalian cells. In parallel, eight of the top ten best-selling drugs in 2018 (including both traditional pharmaceuticals and biopharmaceuticals) were recombinant proteins produced in cultivated mammalian cells [4, 16].

This chapter will give an overview of the biopharmaceutical products manufactured using cultivated mammalian cells in the US and EU markets and put more emphasis on the development of antibody related products including mAbs and Fc-fusion biologics. In addition, since 2005, biosimilars have been being approved for the purpose of reducing drug expenditures. Here we also briefly outline the current trend of biosimilars approved and the impact on the biopharmaceutical industry. The biopharmaceuticals produced via mammalian cell culture brought onto the EU and US market from 1987 till February of 2021 have also been tabulated for reference along with their respective platform host cell lines and indications.

2 Early Mammalian Cells-Based Biopharmaceuticals

The approval of Chinese hamster ovary (CHO) cells-derived tissue plasminogen activator (Activase®) for the treatment of myocardial infarctions in 1987 ushered

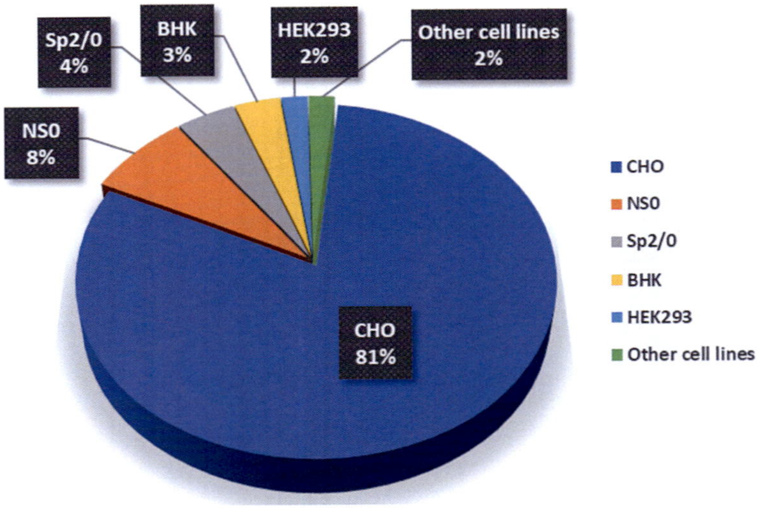

Fig. 1 Percentage of the type of mammalian cell lines used in commercial scale manufacturing of biopharmaceuticals from 1987 till February 2021

in the era of utilizing mammalian cell expression systems to produce biopharmaceuticals. Over the past decades, a group of mammalian cell expression systems employed for the production of biopharmaceuticals include murine myeloma lymphoblastoid type cells such as NS0 [17, 18] and Sp2/0 [19], CHO [20, 21], baby hamster kidney (BHK) [22–25], and human embryonic kidney epithelial cells (HEK-293) [26, 27]. Among these mammalian cells expression systems, CHO cells have become the predominant workhorse of the biopharmaceuticals industry: from 1987 to February 2021, 81% of approved biopharmaceuticals produced via cultivated mammalian cells were made using CHO cells (Fig. 1). The popularity of CHO cells can be ascribed in large part to the inherit attributes. The immense adaptability of CHO cells and their ease of genetic manipulations result in the uses of suspension culture model and various mutant CHO production cell lines, such as dihydrofolate reductase (DHFR) deficient cell lines [28, 29]. In addition, CHO cells are capable of undergoing complex post-translational modifications that meet therapeutic requirements. Another beneficial attribute is that CHO cells as non-human origin cells, are less vulnerable to contamination by human viruses than those cell lines of human origin [30].

The early mammalian cells-based production processes had very low yields of recombinant protein product, sometimes a 100-times less when compared to today's processes. The bottlenecks were due to the lack of optimization in both upstream and downstream processes. Regarding the upstream process, the concentrations of mammalian cells-based products were typically in the range of 1–100 mg/L in 1980s, a far cry from today's average of 2000–5000 mg/L using currently the most widely used fed batch process [31]. Over the last three decades the optimizations of protein yields have been carried out through the improvement in media development, cell line construction with improved selection methods, and the optimization of bioreactor designs and configurations [32–35]. Nowadays, with the use of an emerging perfusion technology utilizing alternating tangential flow (ATF) perfusion, average productivity in perfusion is 7.5 times greater than fed-batch along with improved product qualities through a side-by-side comparison of two mammalian cell culture models [36]. In terms of the early stage of the downstream process, the purification steps in many cases only gave yields below 20% which meant that the manufacturer had to increase the scale of the process to achieve sufficient products to meet the demand in the market. The improved downstream process technologies and designs have greatly increased the efficiency of the bioprocesses which allows to operate the manufacturing process at a smaller scale to satisfy market demand, and further develop flexible multi-product operating facilities [37, 38].

While the first round of mammalian cells-based biopharmaceuticals were mainly copies of recombinant clotting factors, cytokines, and enzymes, this tendency was changed in 1994 when the first approved and successfully marketed antibody drug produced via mammalian cell culture was manufactured by Centocor (now known as Janssen Biotech) with the trade name ReoPro® (Abciximab). Abciximab is a fragment antigen-binding (Fab) fragment of the chimeric human murine monoclonal antibody 7E3 which is designed to overcome the obstacles of murine based antibodies for human therapeutics since murine antibodies possess glycosylation patterns that are highly immunogenic to humans [39, 40] as can be seen with studies about muromonab-CD3 where 50% patients have experienced a potentially lethal human anti-mouse antibody (HAMA) response [41, 42]. Abciximab is specifically designed to retain the Fab fragment by the removal of the Fc fragment in order to reduce possible complement-activating and immunogenicity reactions caused by the Fc fragment [43, 44]. Years prior to Abciximab, Centocor had almost reached the brink of bankruptcy due to approval denial for an IgM antibody drug expressed from a Sp2/0 cell line in an industrial scale perfusion process. The biopharmaceutical, known as nebacumab (Centoxin®), was already approved in Netherlands, Britain, Germany, and France in 1991 where it was indicated for the treatment for Gram-negative sepsis, but soon after the FDA rejected the approval in the USA due to new clinical trial data shown that the treatment failed to reduce mortality which eventually led to the withdrawal of nebacumab from the market. The lessons learned and the bioprocesses developed for nebacumab allowed Centocor in partnership with Eli Lilly to develop the perfusion process and gain the marketing approval of ReoPro®, 8 years after the first antibody-based drug was introduced into the market [45]. Since the success of Abciximab, a great number of chimeric and humanized mAbs that have the ability to trigger effector functions in humans with longer circulatory half-life and decreased immunogenicity have been approved and come to the biopharmaceutical market.

3 Antibody Related Biopharmaceuticals

As shown in Table 1, a total of more than 207 mammalian cells-derived biopharmaceuticals were approved between 1987 and 2020 in the EU and USA. In the meantime, it is apparent that mAbs have become the essential driving force in the biopharmaceutical market as the fastest growing source of innovation and revenue. Four major antibody types that have been approved are murine antibody, chimeric antibody, humanized antibody and fully human antibody. Chimeric antibody is the first engineered antibody where the murine constant regions are replaced by human constant regions. The humanization process is then introduced with an engineered antibody where only the complementarity determining regions (CDRs) of variable regions from a non-human antibody are grafted onto human variable region frameworks. Fully human antibodies can be generated either in vivo by the immunization of genetically modified mice or in vitro by phage display technology [46, 47]. With the development of antibodies from murine origin to fully human origin, the concerns regarding immunogenicity, weak efficacy, and short serum half-life, have been reduced significantly. Currently only one murine mAb therapeutics is marketed, named Zevalin® (Ibritumomab tiuxetan). Ibritumomab tiuxetan was approved in 2002 as a conjugated IgG1 drug targeting CD20 for the treatment of low grade or transformed B cell non-Hodgkin's lymphoma, a type of lymphoproliferative disorder. The murine antibody is conjugated to the radioactive isotope yttrium 90 via the chelate tiuxetan and has a half-life of 1.25 days. It showed a lower immunogenicity, with below 8% of patients suffered from HAMA responses, considering that murine antibodies generally can cause HAMA responses in 50–100% of patients [48, 49].

Table 1 The biopharmaceutical drug approvals in the EU and USA between 1987 and February 2021 which are produced by mammalian cell culture are summarized in Table 1 below. The table starts from the most recent approvals. Data was collected from several sources (European Medicines Agency, 2014; United States Food and Drug Administration, 2014). The products withdrawn from the market are not taken into account. Those shaded in green are biosimilars

Product	Company	Target/Initial Therapeutic Indication	Product Category	Cell Line	First Approved	Region
Evkeeza	Regeneron Pharmaceuticals, Inc.	ANGPTL3 (angiopoietin-like 3) inhibitor for the treatment of adult and pediatric patients, aged 12 years and older, with homozygous familial hypercholesterolemia (HoFH).	Fully human monoclonal antibody	CHO	11-Feb-21	USA
2020						
Ebanga (ansuvimab-zykl)	Ridgeback Biotherapeutics	A Zaire ebolavirus glycoprotein (EBOV GP)-indicated for the treatment of infection caused by Zaire ebolavirus in adult and pediatric patients.	Fully human monoclonal antibody	CHO	21-Dec-20	USA
Riabni™ (rituximab-arrx)	Amgen, Inc.	A biosimilar of RITUXAN indicated for the treatment of non-Hodgkin's Lymphoma, Chronic Lymphocytic Leukemia, Granulomatosis with Polyangiitis (GPA) (Wegener's Granulomatosis) and Microscopic Polyangiitis (MPA) in adult patients.	Chimeric monoclonal antibody	CHO	17-Dec-20	USA
Margenza	MacroGenics Inc.	A HER2/neu receptor antagonist indicated, in combination with chemotherapy, for the treatment of adult patients with metastatic HER2 positive breast cancer.	Chimeric monoclonal antibody	CHO	16-Dec-20	USA
Danyelza® (naxitamab-gqgk)	Y-mABs Therapeutics, Inc	A GD2-binding monoclonal antibody indicated for the treatment of pediatric patients 1 year of age and older and adult patients with relapsed or refractory high-risk neuroblastoma in the bone or bone marrow.	Humanized monoclonal antibody	CHO	25-Nov-20	USA
Inmazeb™ (atoltivimab, maftivimab, and odesivimab-ebgn)	Regeneron Pharmaceuticals, Inc.	A combination of Zaire ebolavirus glycoprotein-directed human monoclonal antibodies (atoltivimab, maftivimab, and odesivimab), indicated for the treatment of infection caused by Zaire ebolavirus in adult and pediatric patients.	Fully human monoclonal antibody	CHO	14-Oct-20	USA

(continued)

Table 1 (continued)

Product	Company	Target/Initial Therapeutic Indication	Product Category	Cell Line	First Approved	Region
Enspryng™ (satralizumab-mwge)	Genentech, Inc.	An interleukin-6 (IL-6) receptor antagonist indicated for the treatment of neuromyelitis optica spectrum disorder (NMOSD) in adult patients who are anti-aquaporin-4 (AQP4) antibody positive.	Humanized monoclonal antibody	CHO	14-Aug-20	USA
Blenrep	GlaxoSmithKline	A B-cell maturation antigen (BCMA)-directed antibody and microtubule inhibitor conjugate indicated for the treatment of adult patients with relapsed or refractory multiple myeloma.	Humanized monoclonal antibody	CHO	5-Aug-20	USA, EU
Monjuvi® (tafasitamab-cxix)	MorphoSys US Inc.	A CD19-directed cytolytic antibody indicated for the treatment of adult patients with relapsed or refractory diffuse large B-cell lymphoma.	Humanized monoclonal antibody	CHO	31-Jul-20	USA
Hulio® (adalimumab-fkjp)	Mylan Pharmaceuticals Inc.	A biosimilar of Humira for all of Humira's approved indications.	Fully human monoclonal antibody	CHO	16-Sep-18	USA, EU
Phesgo (pertuzumab, trastuzumab, and hyaluronidase-zzxf)	Genentech, Inc.	A combination of pertuzumab and trastuzumab, HER2/neu receptor antagonists, and hyaluronidase, an endoglycosidase, indicated for the treatment of patients with HER2-positive, locally advanced, inflammatory, or early stage breast cancer.	Humanized monoclonal antibody	CHO	29-Jun-20	USA, EU
Uplizna™(inebilizumab-cdon)	Viela Bio	A CD19-directed cytolytic antibody indicated for the treatment of neuromyelitis optica spectrum disorder	Humanized monoclonal antibody	CHO	11-Jun-20	USA

Product	Company	Target/Initial Therapeutic Indication	Product Category	Cell Line	First Approved	Region
		(NMOSD) in adult patients who are anti-aquaporin-4 (AQP4) antibody positive.				
Darzalex Faspro™ (daratumumab and hyaluronidase-fihj)	Janssen Biotech, Inc.	A combination of daratumumab, a CD38-directed cytolytic antibody, and hyaluronidase, an endoglycosidase, for the treatment of adult patients with multiple myeloma.	Fully human monoclonal antibody	CHO	1-May-20	USA
Trodelvy™ (sacituzumab govitecan-hziy)	Immunomedics, Inc.	A Trop-2-directed antibody and topoisomerase inhibitor conjugate indicated for the treatment of adult patients with metastatic triple-negative breast cancer.	Humanized monoclonal antibody	NS0	22-Apr-20	USA
Sarclisa® (isatuximab-irfc)	sanofi-aventis U.S. LLC	A CD38-directed cytolytic antibody indicated, in combination with pomalidomide and dexamethasone, for the treatment of adult patients with multiple myeloma.	Chimeric monoclonal antibody	CHO	2-Mar-19	USA, EU
Tepezza (teprotumumab-trbw)	Horizon Therapeutics Ireland DAC	An insulin-like growth factor-1 receptor inhibitor indicated for the treatment of Thyroid Eye Disease.	Fully human monoclonal antibody	CHO	21-Jan-20	USA
		2019				
Enhertu® (fam-trastuzumab deruxtecan-nxki)	Daiichi Sankyo, Inc	A HER2-directed antibody and topoisomerase inhibitor conjugate indicated for the treatment of adult patients with unresectable or metastatic HER2-positive breast cancer.	Humanized monoclonal antibody	CHO	20-Dec-19	USA, EU
Padcev™ (enfortumab vedotin-ejfv)	Astellas Pharma US, Inc.	A Nectin-4-directed antibody and microtubule inhibitor conjugate indicated for the treatment of	Fully human monoclonal antibody	CHO	18-Dec-19	USA

(continued)

Table 1 (continued)

Product	Company	Target/Initial Therapeutic Indication	Product Category	Cell Line	First Approved	Region
		adult patients with locally advanced or metastatic urothelial cancer.				
Avsola (infliximab-axxq)	Amgen, Inc.	A biosimilar of Remicade for most of Remicade's approved indications.	Chimeric monoclonal antibody	CHO	6-Dec-19	USA
Adakveo® (crizanlizumab-tmca)	Novartis Pharmaceuticals Corporation	A selectin blocker indicated to reduce the frequency of vasoocclusive crises in adults and pediatric patients aged 16 years and older with sickle cell disease.	Humanized monoclonal antibody	CHO	15-Nov-19	USA, EU
Abrilada™ (adalimumab-afzb)	Pfizer, Inc.	A biosimilar of Humira for all of Humira's approved indications.	Humanized monoclonal antibody	CHO	15-Nov-19	USA, EU
Reblozyl® (luspatercept-aamt)	Celgene Corporation	An erythroid maturation agent indicated for the treatment of Anemia in adult patients with beta thalassemia who require regular red blood cell (RBC) transfusions.	Fc-fusion biologic	CHO	11-Aug-19	USA, EU
Ruxience™ (rituximab-pvvr)	Pfizer Ireland Pharmaceuticals	A biosimilar of Rituxan for most of Rituxan's approved indications.	Chimeric monoclonal antibody	CHO	23-Jul-19	USA, EU
Hadlima (adalimumab-bwwd)	Samsung Bioepis Co., Ltd.	A biosimilar of Humira for all of Humira's approved indications.	Humanized monoclonal antibody	A mammalian cell line unpub	23-Jul-19	USA, EU
Zirabev™ Bevacizumab-bvzr	Pfizer	A biosimilar of Avastin for metastatic colorectal cancer and HER2-negative metastatic breast cancer.	Humanized monoclonal antibody	CHO	28-Jun-19	USA

Product	Company	Target/Initial Therapeutic Indication	Product Category	Cell Line	First Approved	Region
Kaniinti™ Trastuzumab-anns	Amgen/Allergan	A biosimilar of Herceptin for treatment of HER2-positive breast cancer and gastric cancer.	Humanized monoclonal antibody	CHO	13-Jun-19	USA
Nucala	GlaxoSmithKline	An interleukin-5 antagonist monoclonal antibody (IgG1 kappa) indicated for add-on maintenance treatment of patients with severe asthma aged 12 years and older, and with an eosinophilic phenotype.	Humanized monoclonal antibody	CHO	6-Jun-19	USA
Eticovo™ Etanerceot-ykro	Samsung Bioepis	A biosimilar of Enbrel for the treatment of rheumatoid arthritis. ankylosing spondylitis, plaque psoriasis, psoriatic arthritis and polyarticular juvenile idiopathic arthritis.	Fc-fusion biologic	CHO	25-Apr-19	USA
Skyrizi™ Risankizumab-rzaa	AbbVie	An interleukin-23 antagonist for treatment of plaque psoriasis.	Humanized monoclonal antibody	CHO	23-Apr-19	USA
Evenity® Romosozumab	Amgen	Targeting and inhibiting the protein sclerostin, thereby preventing inhibition of bone formation by allowing Wnt to bind to LDL receptor-related proteins 5 and 6 for treatment osteoporotic fracture.	Humanized monoclonal antibody	CHO	9-Apr-19	USA
Trazimera™ Trastuzumab-qyyp	Pfizer	A biosimilar of Herceptin for treatment of HER2-positive breast cancer.	Humanized monoclonal antibody	CHO	11-Mar-19	EU
Herceptin Hylecta™ Trastuzumab and hyaluronidase	Genentech/Roche	A combination of trastuzumab and hyaluronidase for treatment of breast cancer.	Humanized monoclonal antibody/Enzyme	CHO	28-Feb-19	USA

(continued)

Table 1 (continued)

Product	Company	Target/Initial Therapeutic Indication	Product Category	Cell Line	First Approved	Region
Esperoct ® Turoctocog alfa pegol	Novo Nordisk	A recombinant analogue of human coagulation Factor VIII (FVIII) conjugated with a 40-kDa polyethylene glycol (PEG) molecule.	Clotting factor	CHO	19-Feb-19	BJ & USA
Ontruzant® Trastuzumab-dttb	Samsung Bioepis	A biosimilar of Herceptin for treatment of HER2-positive breast cancer.	Humanized monoclonal antibody	CHO	16-Jan-19	USA
2018						
Ultomiris® Ravulizumab-cwvz	Alexion Pharmaceuticals	Binding to the particular complement protein C5 with high affinity, thereby inhibiting its cleavage to C5a and C5b and preventing the generation of the terminal complement complex C5b9 for treatment of adult patients with paroxysmal nocturnal hernoglobinuria.	Humanized monoclonal antibody	CHO	21-Dec-18	BJ & USA
Herzuma® Trastuzumab-okrb	Celltrion/Teva	A biosimilar of Herceptin for all of Herceptin's approved indications.	Humanized monoclonal antibody	CHO	15-Dec-18	USA
Truxima® Rituximab-abbs	Celltrion	A biosimilar of Rituxan for a good number of Rituxan's indications.	Chimeric monoclonal antibody	CHO	28-Nov-18	USA
Gamifant® Emapalumab-lzsg	Novimmune SA/Sobi	Targeting interferon gamma for the treatment of hemophagocytic lymphohistiocytosis.	Fully human monoclonal antibody	CHO	20-Nov-18	BJ & USA
Hyrimoz® Adalimumab -adaz	Sandoz/Novartis	A biosimilar of Humira for all of Humira's approved indications.	Fully human monoclonal antibody	CHO	31-Oct-18	USA

Product	Company	Target/Initial Therapeutic Indication	Product Category	Cell Line	First Approved	Region
Libtayo® Cemiplimab-rwlc	Regeneron/Sanofi	Binding to the PD-1 receptor and blocks its interaction with PD-L1 and PD-L2 ligands, causing PD-1pathway-mediated inhibition of the immune response the treatment of patients with metastatic cutaneous squamous cell carcinoma (CSCC) or locally advanced CSCC.	Fully human monoclonal antibody	CHO	28-Sep-18	USA
Emgality® Galcanezumab-gnlm	Eli Lilly	Binding and antagonizing the calcitonin gene-related peptide receptor (CGRPR) as a means to prevent migraines.	Fully human monoclonal antibody	CHO	27-Sep-18	EU& USA
Ajovy® Fremanezumab-vfrm	Teva	Binding and antagonizing the calcitonin gene-related peptide receptor (CGRPR) as a means to prevent migraines.	Fully human monoclonal antibody	CHO	14-Sep-18	EU & USA
Jivi® Damoctocog alfa pegol	Bayer	A longer-acting Factor VIII therapy formulated with polyethylene glycol (PEG) to reduce the number of infusions necessary to prevent bleeds in patients diagnosed with Haemophilia A.	Clotting factor	BHK	30-Aug-18	USA
Takhzyro® Lanadelumab-flyo	Shire	A plasma kallikrein inhibitor for the treatment of types I and II hereditary angioedema.	Fully human monoclonal antibody	CHO	23-Aug-18	EU & USA
Poteligeo® Mogamulizumab-kpkc	Kyowa Kirin/Ultragenyx	Targeting CC chemokine receptor 4 (CCR4) for the treatment of Mycosis Fungoides (MF) and Sezary Syndrome (SS), the most common subtypes of cutaneous T-cell lymphoma.	Humanized monoclonal antibody	CHO	8-Aug-18	USA

(continued)

Table 1 (continued)

Product	Company	Target/Initial Therapeutic Indication	Product Category	Cell Line	First Approved	Region
Aimovig ® Erenumab-aooe	Amgen/Novartis	Binding and antagonizing the calcitonin gene-related peptide receptor (CGRPR) as a means to prevent migraines.	Fully human monoclonal antibody	CHO	17-May-18	EU & USA
Retacrit™ Epoetin Alfa-epbx	Hospira/ Pfizer	A biosimilar to Epogen for the treatment of all indications of the reference product.	Cytokine	CHO	17-May-18	USA
Andexxa® Andexanet alfa	Portola Pharmaceuticals	A recombinant human coagulation Factor Xa that promotes blood coagulation.	Clotting factor	CHO	3-May-18	USA
Crysvita® Burosumab-twza	Ultragenyx Pharmaceutical Inc.	Binds to excess fibroblast growth factor 23 for treatment of x-linked hypophosphatemia.	Fully human monoclonal antibody	CHO	17-Apr-18	EU & USA
Ilumya™ Tildrakizumab	Sun Pharmaceutical Industries Ltd.	Targeting interleukin 23 p19 that shows promise in the evaluation of treatment strategy in chronic plaque psoriasis.	Humanized monoclonal antibody	CHO	20-Mar-18	USA
Trogarzo ® Ibalizumab-uiyk	TaiMed Biologics	Binding to CD4 receptors on the surface of CD4-positive cells preventing HIV particle entry into the lymphocytes.	Humanized monoclonal antibody	NS0	6-Mar-18	USA
2017						
Ixifi™ Infliximab-qbtx	Pfizer	A tumor necrosis factor blocker and biosimilar to Remicade for treatment of the many indications approved for Remicade.	Chimeric monoclonal antibody	A mammalian cell line unpub	13-Dec-17	USA
Ogivri ® Trastuzumab-dkst	Mylan/Biocon	A biosimilar to Herceptin for the treatment of patients with breast or metastatic stomach cancer whose tumors overexpress the HER2 gene.	Humanized monoclonal antibody	CHO	1-Dec-17	USA

Product	Company	Target/Initial Therapeutic Indication	Product Category	Cell Line	First Approved	Region
Hemlibra® Emicizumab-kxwh	Genentech/Roche	Mimicking the function of the coagulation Factor VIII and it has the capacity to bind simultaneously to activated Factor IX and Factor X for treatment of hemophilia A with factor VIII inhibitors.	Humanized monoclonal antibody	CHO	16-Nov-17	USA
Mepsevii™ Vestronidase alfa-vjbk	Ultragenyx Pharmaceutical	A recombinant human lysosomal beta glucuronidase for treatment of inherited metabolic condition called mucopolysaccharidosis type VII.	Enzyme	CHO	15-Nov-17	USA
Fasenra® Benralizumab	Astra Zeneca	Specifically binds to the alpha chain of the interleukin 5 receptor (IL-5R) expressed on eosinophils and basophils for treatment of asthma.	Humanized monoclonal antibody	CHO	14-Nov-17	USA
Mvasi® Bevacizumab-awwb	Sanofi	A biosimilar to Avastin for the treatment of certain colorectal , lung, brain, kidney and cervical cancer.	Humanized monoclonal antibody	CHO	14-Sep-17	USA
Cyltezo® Adalimumab-adbm	Boehringer Ingelheim	An anti-TNF-α monoclonal antibody biosimilar to Humira for the treatment of various inflammatory diseases.	Fully human monoclonal antibody	CHO	25-Aug-17	EU & USA
Besponsa® Inotuzumab ozogamicin	Pfizer	An antibody-drug conjugate including CD22-targeting monoclonal antibody covalently attached to calicheamicin derivative, N-acetyl-gamma-calicheamicin dimethylhydrazine, which is a potent DNA-binding cytotoxic agent for treatment of relapsed or refractory B- cell precursor acute lymphoblastic leukaemia.	Humanized monoclonal antibody	CHO	17-Aug-17	EU & USA

(continued)

Table 1 (continued)

Product	Company	Target/Initial Therapeutic Indication	Product Category	Cell Line	First Approved	Region
Tremfya® Guselkumab	Janssen Biotech	Selectively blocks interleukin-23. IL-23 is an inflammatory cytokine that activates the CD4+ T-helper (Th17) cell pathway to mediate the inflammatory cascade that induces psoriatic plaque formation for treatment of plaque psoriasis.	Fully human monoclonal antibody	CHO	13-Jul-17	EU & USA
Rebinyn® nonacog beta pegol	Novo Nordisk	Recombinant Coagulation Factor IX for treatment of hemophilia B.	Clotting factor	CHO	31-May-17	EU & USA
Kevzara® Sarilumab	Sanofi/Regeneron	Binding to both membrane bound and soluble interleukin 6 (IL-6) receptor forms, thus blocking the cis- and trans- inflammatory signalling cascades of IL-6 for treatment of severely active rheumatoid arthritis.	Fully human monoclonal antibody	CHO	22-May-17	EU & USA
Tecentriq® Atezolizumab	Genentech/Roche	A programmed death ligand 1(PD-LI) blocker for treatment of advanced urothelial carcinoma; the treatment of metastatic non-small cell lung cancer (NSCLC); extensive-stage small cell lung cancer; and for use in combination with Abraxane for the treatment of metastatic triple- negative breast cancer.	Humanized monoclonal antibody	CHO	18-May-17	USA
Imfinzi® Durvalumab	MedImmkune/AstraZeneca	A programmed death ligand 1(PD-LI) blocker for treatment of urothelial carcinoma.	Fully human monoclonal antibody	CHO	1-May-17	USA
Brineura® Cerliponase alfa	BioMarin Pharmaceutical Inc.	An enzyme replacement treatment for a specific form of Batten disease.	Enzyme	CHO	27-Apr-17	EU & USA
Renflexis® Infliximab-abda	Merck	A tumor necrosis factor blocker and biosimilar of Remicade for the treatment of Remicade's approved indications.	Chimeric monoclonal antibody	CHO	21-Apr-17	USA

Product	Company	Target/Initial Therapeutic Indication	Product Category	Cell Line	First Approved	Region
Dupixent® Dupilumab	Sanofi/Regeneron	Binding to the interleukin-4 (IL-4) receptor, inhibiting the receptor signaling pathways for the treatment of moderate-to-severe eczema.	Fully human monoclonal antibody	CHO	28-Mar-17	EU & USA
Ocrevus® Ocrelizumab	Genentech/Roche	A CD20-directed cytolytic antibody indicated for the treatment of patients with relapsing or primary progressive forms of multiple sclerosis.	Humanized monoclonal antibody	CHO	28-Mar-17	USA
Bavencio®Avelumab	Merck	A programmed death ligand-1(PD-LI) blocking antibody indicated for the treatment of adults and pediatric patients 12 years and older with metastatic Merkel cell carcinoma.	Fully human monoclonal antibody	CHO	25-Mar-17	EU & USA
Siliq™ Brodalumab	Valeant Pharmaceuticals	Binding with high affinity to interleukin (IL)-17 receptor A, thereby inhibiting several pro-inflammatory cytokines from the IL-17 family for the treatment of moderate-to-severe plaque psoriasis.	Fully human monoclonal antibody	CHO	15-Feb-17	EU & USA
2016						
Zinplava™ Bezlotoxumab	Merck	Binding to Clostridium difficile toxin B and neutralizes its effects for reducing recurrence of Clostridium difficile infection.	Fully human monoclonal antibody	CHO	21-Oct-16	USA
Lartruvo® Olaratumab	Eli Lilly	Selectively binds the external domain of human platelet- derived growth factor receptor with high affinity and blocks ligand binding for treatment of advanced soft tissue sarcoma in combination with doxorubicin.	Fully human monoclonal antibody	NS0	19-Oct-16	EU & USA

(continued)

Table 1 (continued)

Product	Company	Target/Initial Therapeutic Indication	Product Category	Cell Line	First Approved	Region
Amjevita ® Adalimumab-atto	Amgen	A biosimilar of Humira for all of Humira's approved indications.	Fully human monoclonal antibody	CHO	23-Sep-16	USA
Erelzi™ Etanercept	Sandoz/Novartis	A biosimilar of Enbrel. Dimeric fusion protein consisting of the extracellular ligand-binding portion of human tumor necrosis factor receptor linked to the Fc portion of human IgG1.	Fc-fusion biologic	CHO	26-Aug-16	USA
Afstyla® Lonoctocog alfa	CSL Behring	B-domain-truncated rh coagulation factor VIII for treatment hemophilia A.	Clotting factor	CHO	25-May-16	USA
Cinqair® Reslizumab	Teva	Interleukin-5 (IL-5) antagonist for treatment of severe eosinophilic asthma.	Humanized monoclonal antibody	NS0	23-Mar-16	EU & USA
Taltz® Ixekizumab	Eli Lilly	Humanized immunoglobulin G subclass 4 (IgG4) monoclonal antibody (mAb) against interleukin-17A for treatment of adults with moderate-to-severe plaque psoriasis.	Humanized monoclonal antibody	CHO	22-Mar-16	EU & USA
Anthim® Obiltoxaximab	Elusys Therapeutics, Inc.	Chimeric IgG1 kappa monoclonal antibody (mAb) that binds the PA component of B for prevention and treatment of infection and death caused by anthrax toxin.	Chimeric monoclonal antibody	NS0	18-Mar-16	USA
Kovaltry® Octocog alfa	Rayer AG	Recombinant human Factor VIII for treatment of hemophilia.	Clotting factor	BHK	17-Mar-16	EU & USA
Idelvion® Albutrepenonacog alfa	CSL Behring	Recombinant fusion protein that links a recombinant coagulation factor IX with a recombinant human albumin for control and prevention of bleeding.	Clotting factor	CHO	4-Mar-16	EU & USA

Product	Company	Target/Initial Therapeutic Indication	Product Category	Cell Line	First Approved	Region
2015						
Vonvendi® von Willebrand factor	Baxalta U.S.	von Willebrand Factor for treatment of van Willebrand's disease.	Clotting factor	CHO	9-Dec-15	USA
Empliciti® Elotuzumab	Bristol Myers Squibb	Targeting SLAMF7,also known as Signalling Lymphocytic Activation Molecule Family member 7, a cell surface glycoprotein for the treatment of patients with multiple myeloma who have received 1-3 prior therapies.	Humanized monoclonal antibody	NS0	30-Nov-15	USA
Portrazza® Necitumumab	Eli Lilly	EGFR antagonist for treatment of metastatic squamous non-small cell lung cancer in combination with gemcitabine and cisplatin.	Fully human monoclonal antibody	NS0	24-Nov-15	USA
Darzalex® Daratumumab	Janssen Biotech, Inc.	Inducing apoptosis of cancer cells by targeting the CD38 epitope, which is highly expressed on haematological malignancies for treatment of multiple myeloma.	Fully human monoclonal antibody	CHO	16-Nov-15	USA
Nucala® Mepolizumab	GlaxoSmithKline	IL-5 antagonist monoclonal antibody for the treatment of asthma.	Humanized monoclonal antibody	CHO	4-Nov-15	EU & USA
Strensiq® Asfotase alfa	Alexion Pharmaceuticals	First-in-class bone-targeted enzyme replacement therapy designed to address the underlying cause of hypophosphatasia (HPP0-deficient alkaline phosphatase (ALP).	Enzyme	CHO	23-Oct-15	EU & USA
Praxbind® Idarucizumab	Boehringer Ingelheim	Humanized monoclonal antibody fragment (Fab) derived from an immunoglobulin G1 isotype molecule that binds to and inactivates the oral	Humanized monoclonal antibody	CHO	16-Oct-15	EU & USA

(continued)

Table 1 (continued)

Product	Company	Target/Initial Therapeutic Indication	Product Category	Cell Line	First Approved	Region
		anticoagulant dabigatran, thereby reversing its anticoagulant effect.				
Nuwiq® Simoctocog alfa	Octapharma USA	Recombinant B-domain deleted (BDD) rFVIII for treatment of hemophilia A.	Clotting factor	HEK293	16-Oct-15	EU
Repatha® Evolocumab	Amgen Inc.	Targeting proprotein convertase subtilisin kexin type 9(PCSK9) for treatment of high low-density lipoprotein (LDL) cholesterol levels.	Fully human monoclonal antibody	CHO	27-Aug-15	EU & USA
Praluent® Alirocumab	Sanofi/Regeneron	Targeting proprotein convertase subtilisin kexin type 9(PCSK9) for treatment of high low-density lipoprotein (LDL) cholesterol levels.	Fully human monoclonal antibody	CHO	24-Jul-15	EU & USA
Ixinity® Coagulation factor IX	Cangene/Emergent Biosolutions	Recombinant Coagulation Factor IX is a purified Factor IX glycoprotein for treatment of hemophilia B.	Clotting factor	CHO	29-Apr-15	USA
Unituxin® Dinutuximab	United Therapeutics Corp	Dinutuximab is an IgG1 monoclonal human/mouse chimeric antibody against GD2, a disialoganglioside expressed on tumors of neuroectodermal origin, including human neuroblastoma and melanoma, with highly restricted expression on normal tissues for the treatment of paediatric patients with high-risk neuroblastoma.	Chimeric monoclonal antibody	Sp2/0	10-Mar-15	USA

Product	Company	Target/Initial Therapeutic Indication	Product Category	Cell Line	First Approved	Region
Opdivo® Nivolumab	Bristol-Myers Squibb	Targeting the immune checkpoint programmed death receptor-1(PD-1).This molecule was produced entirely on mice and grafted onto human kappa and IgG4 Fc region with the mutation S228P for additional stability and reduced variability for treatment of patients with metastatic squamous non-small cell lung cancer with progression on or after platinum-based chemotherapy.	Humanized monoclonal antibody	CHO	4-Mar-15	USA
Cosentyx® Secukinumab	Novartis	An interleukin-17A (IL-17A) inhibitor for treatment of adults with moderate to severe plaque psoriasis.	Fully human monoclonal antibody	CHO	21-Jan-15	EU & USA
2014						
Opdivo® Nivolumab	Bristol-Myers Squibb	Targeting the immune checkpoint programmed death receptor-1 (PD-1). This molecule was produced entirely on mice and grafted onto human kappa and IgG4 Fc region with the mutation S228P for additional stability and reduced variability for treatment of advanced melanoma.	Humanized monoclonal antibody	CHO	22-Dec-14	USA
Blincyto® Blinatumomab	Amgen	A bispecific CD19-directed CD3 T-cell engager that binds to CD19 expressed on the surface of cells of B-lineage origin and CD3 expressed on the surface of T cells for treatment of Philadelphia chromosome-negative relapsed/refractory B-precursor acute lymphoblastic leukemia.	Bi-specific antibody	CHO	3-Dec-14	USA

(continued)

Table 1 (continued)

Product	Company	Target/Initial Therapeutic Indication	Product Category	Cell Line	First Approved	Region
Obizur® Susoctocog alfa	Baxter	A recombinant, B-domain deleted, porcine sequence antihaemophilic factor VIII (FVIII) product that has recently been approved for the treatment of bleeding episodes in adults with acquired haemophilia A.	Clotting factor	BHK	24-Oct-14	USA
Trulicity® Dulaglutide	Eli Lilly	A novel glucagon -like peptide-1 agonist (GLP-1) biologic drug consisting of a dipeptidyl peptidase-IV-protected GLP-1 analogue covalently linked to a human IgG4-Fc heavy chain by a small peptide linker for treatment of adult patients with adults with type 2 diabetes.	Fc-fusion biologic	CHO	18-Sep-14	USA
Keytruda® Pembrolizumab	Merck	Highly selective IgG4-kappa humanized monoclonal antibody against PD-1 receptor for treatment of patients with advanced melanoma not responding to other therapies.	Humanized monoclonal antibody	CHO	4-Sep-14	USA
Plegridy ® Peginterferon beta-1a	Biogen Idec	An interferon beta-1a to which a single, linear 20,000 Dalton (Da) methoxy poly(ethylene glycol)-O-2-methylpropionaldehyde (PEG) molecule is covalently attached to the alpha amino group of the N-terminal amino acid residue for the treatment of relapsing forms of multiple sclerosis.	Cytokine	CHO	15-Aug-14	EU & USA
Eloctate ® Efmoroctocog alfa	Swedish Orphan Biovitrum/Biogen Idec	First commercially available rFVIII-Fc fusion protein (rFVIIIFc) where the conjugated molecule of rFVIII to polyethylene glycol is covalently fused to the dimeric Fc domain of human immunoglobulin G1 for the treatment of hemophilia A.	Clotting factor	HEK293	6-Jun-14	USA

Product	Company	Target/Initial Therapeutic Indication	Product Category	Cell Line	First Approved	Region
Entyvio vedolizumab	Takeda Pharmaceuticals	Antibody that binds to integrin α4β7 blocking the α4β7 integrin resulting in gut-selective anti-inflammatory activity, for the treatment of ulcerative colitis and Crohn's disease.	Humanized monoclonal antibody	CHO	20-May-14	EU & USA
Cyramza™ Ramucirumab	Eli Lilly	Binds to VEGFR2 blocking the binding of vascular endothelial growth factor (VEGF) to VEGFR2 indicated for the treatment of patients with advanced or metastatic, gastric or gastro-esophageal junction adenocarcinoma.	Humanized monoclonal antibody	NS0	23-Apr-14	USA
Sylvant™ Siltuximab	Janssen Biotech (Johnson & Johnson)	Antibody that specifically binds to and neutralizes human IL-6 with high affinity for the treatment of adult patients with multicentric Castleman's disease (MCD).	Chimeric monoclonal antibody	CHO	23-Apr-14	USA
Alprolix™ Coagulation Factor IX	Biogen Idec.	Coagulation factor human IgG1 Fc fusion, which binds to the neonatal Fc receptor (FcRn) for treatment of hemophilia B.	Fc-fusion biologic	HEK293	28-Mar-14	USA
Vimizim™ elosulfase alfa	BioMarin Pharmaceutical Inc.	Recombinant N-acetylgalactosamine-6-sulfate sulfatase replacement therapy for treatment of Mucopolysaccharidosis type IVA (MPS IVA; Morquio A syndrome).	Enzyme	CHO	14-Feb-14	USA
2013						
Obinutuzumab ®(Gazyva GA101-CD20 mAb)	Genentech/Roche	Treatment of chronic lymphocytic leukemia (CLL) Obinutuzumab targets CD20 and kills B cells.	Humanized monoclonal antibody	CHO	1-Nov-13	USA

(continued)

Table 1 (continued)

Product	Company	Target/Initial Therapeutic Indication	Product Category	Cell Line	First Approved	Region
Novoeight® ,Antihemophilic Factor VIII	Novo Nordisk	Recombinant factor VIII for the treatment and prophylaxis of bleeding in patients with haemophilia A (congenital factor VIII deficiency).	Clotting factor	CHO	15-Oct-13	EU & USA
Ovaleap™ Follitropin	Teva Pharma B.V.	A biosimilar to Remicade, against tumour necrosis factor alpha (TNF- α) used to treat Crohn's disease, ankylosing spondylitis, alcerative Colitis, psoriatic arthritis, plaque psoriasis and rheumatoid arthritis.	Hormone	CHO	27-Sep-13	EU
Inflectra™ Infliximab	Hospira	A biosimilar to Remicade, against tumour necrosis factor alpha (TNF- α) used to treat Crohn's disease, ankylosing spondylitis, alcerative Colitis, psoriatic arthritis, plaque psoriasis and rheumatoid arthritis.	Chimeric monoclonal antibody	Sp2/0	10-Sep-13	EU
Remsima™ Infliximab	Celltrion	A biosimilar to Remicade, against tumour necrosis factor alpha (TNF- α) used to treat Crohn's disease, ankylosing spondylitis, alcerative Colitis, psoriatic arthritis, plaque psoriasis and rheumatoid arthritis.	Chimeric monoclonal antibody	Sp2/0	10-Sep-13	EU
Simponi® Aria™ Golimumab	Janssen Biotech (Johnson & Johnson)	Inhibits inflammatory response by suppressing tumor necrosis factor (TNF) as a treatment of moderately to severely active rheumatoid arthritis.	Fully human monoclonal antibody	CHO	18-Jul-13	USA
Rixubis®- (Coagulation Factor IX)	Baxter international Inc.	Recombinant Coagulation Factor IX for treatment of hemophilia B, indicated for the control and prevention of bleeding episodes (prophylaxis).	Clotting factor	CHO	27-Jun-13	USA

Product	Company	Target/Initial Therapeutic Indication	Product Category	Cell Line	First Approved	Region
Kadcyla ® Ado-Trastuzumab Emtansine, T-DM1	Genentech/Roche	Trastuzumab alone stops growth of cancer cells by binding to the HER2/neu receptor, whereas mertansine enters cells and destroys them by binding to tubulin.	Humanized monoclonal antibody	CHO	22-Feb-13	EU & USA
2012						
Abthrax ® Raxibacumab injection	GlaxoSmithKline	Monodonal antibody that neutralizes the toxins produced by the anthrax bacterium Bacillus anthracis intended for the prophylaxis and treatment of inhaled anthrax.	Fully human monoclonal antibody	NS0	14-Dec-12	USA
Zaltrap™ ziv-afliberceot	Regeneron/Sanofi Aventis	Fusion protein consisting IgG1 Fc fused with VEGF-binding portions from the extracellular domains of human VEGF receptors 1and 2 for the treatment of metastatic colorectal cancer as Zaltrao (aka Elyea).	Fc-fusion biologic	CHO	3-Aug-12	EU & USA
Perjeta® Pertuzumab	Genentech/ Roche	Indicated for the treatment of HER2-positive breast cancer, in combination with trastuzumab and docetaxel.	Humanized monoclonal antibody	CHO	4-Mar-12	EU & USA
2011						
Eylea™ Aflibercept	Regeneron/Bayer	Fusion protein with VEGF-binding portions from the extracellular domains of human VEGF receptors 1and 2 for the treatment of neovascular (wet) Age-related Macular Deaeneration (AMD) as Eylea (aka Zaltrap).	Fc-fusion biologic	CHO	18-Nov-11	EU & USA
Adcetris™ Brentuximab vedotin	Millennium Pharmaceuticals/Takeda	Antibody that which targets the cell-membrane protein CD30 linked to cathepsin cleavable linker for the treatment of Hodgkin lymphoma (HL) and systemic anaplastic large cell lymphoma (sALCL).	Chimeric monoclonal antibody	CHO	19-Aug-11	EU & USA

(continued)

Table 1 (continued)

Product	Company	Target/Initial Therapeutic Indication	Product Category	Cell Line	First Approved	Region
Nulojix ® Belatacept	Bristol-Myers Squibb	Fc fragment of a human IgG1 linked to the extracellular domain of CTLA-4 blocking the process of T-cell activation for the prevention of acute rejection in adult kidney transplants.	Fc-fusion biologic	CHO	15-Jun-11	EU & USA
Yervoy® Ipilimumab	Bristol-Myers Squibb	Activates the immune system by targeting CTLA-4 for the treatment of late-stage melanoma.	Fully human monoclonal antibody	CHO	25-Mar -11	EU & USA
Benlysta ® Belimumab	GlaxoSmithKline	Inhibits B-cell activating factor, also known as B-lymphocyte stimulator for treatment of adults with active, autoantibody-positive systemic lupus erythematosus.	Fully human monoclonal antibody	NS0	9-Mar-11	EU & USA
2010						
Xgeva® Denosumab	Amgen	Inhibits RANK ligand , which acts as the primary signal for bone removal for the treatment of bone loss due to cancer.	Fully human monoclonal antibody	CHO	18-Nov-10	EU & USA
Prolia® Denosumab	Amgen	Inhibits RANK ligand, which acts as the primary signal for bone removal for the treatment of osteoporosis, bone metastases, multiple myeloma, and tumors of the bone.	Fully human monoclonal antibody	CHO	1-Jun-10	EU & USA
Lumizyme® Alglucosidase alfa	Genzyme	Lysosomal glycogen-specific enzyme indicated for patients 8 years and older with late (non-infantile) onset Pompe disease (replaces Myozyme®).	Enzyme	CHO	24-May-10	USA
Vpriv® Velaglucerase alfa	Shire	VPRIV® is used as a long term treatment as an enzyme replacement in patients with Type I Gaucher disease.	Enzyme	Human Fibroblasts	26-Feb-10	EU & USA

Product	Company	Target/Initial Therapeutic Indication	Product Category	Cell Line	First Approved	Region
Elonva® Corifollitropin alfa	Merck Sharp & Dohme Limited	A modified rhFSH in which the carboxy-terminal peptide of the beta subunit of hCG is fused to the FSH beta chain indicated for the controlled stimulation of the ovaries.	Hormone	CHO	25-Jan-10	EU
Actemra ® Tocilizumab	Genentech/ Roche	Targets interleukin-6 receptor (IL-6R) for the treatment of rheumatoid arthritis (RA) and systemic juvenile idiopathic arthritis.	Humanized monoclonal antibody	CHO	8-Jan-10	USA
2009						
Eporatio® Epoetin theta	Ratiopharm GmbH (Biopoin® and Epon	Recombinant erythropoietin that stimulates the production of red blood cells from the bone marrow and is used to treat anaemia, chronic renal failure, and in adults with non- myeloid cancer.	Cytokine	CHO	29-Oct-09	EU
Arzerra ® Ofatumumab	GlaxoSmithKline	CD20 monodonal antibody for the treatment for chronic lymphocytic leukaemia in patients who have not responded to Campath (alemtuzumab) or fludarabine.	Fully human monoclonal antibody	NS0	26-Oct-09	EU & USA
Biopoin® Epoetin theta	Teva GmbH (Biopoin® and Eporatio® a	Recombinant erythropoietin that stimulates the production of red blood cells from the bone marrow and is used to treat anaemia, chronic renal failure, and in adults with non- myeloid cancer.	Cytokine	CHO	23-Oct-09	EU
Stelara™ Ustekinumab	Janssen Biotech (Johnson & Johnson)	Directed against interleukin 12 and interleukin 23, proteins that regulate the immune-mediated inflammatory disorders to decrease the inflammatory response.	Full human monoclonal antibody	CHO	16-Jan-09	EU & USA

(continued)

Table 1 (continued)

Product	Company	Target/Initial Therapeutic Indication	Product Category	Cell Line	First Approved	Region
Ilaris® Canakinumab	Novartis	Monoclonal antibody targeted at interleukin-1 beta for treatment of Cryopyrin Associated Periodic Syndrome (CAPS).	Fully human monoclonal antibody	Sp2/0	17-Jun-09	EU & USA
Simponi® Golimumab	Janssen Biotech (Johnson & Johnson)	Inhibits inflammatory response by suppressing tumor necrosis factor (TNF) as a treatment of moderately to severely active rheumatoid arthritis.	Fully human monoclonal antibody	CHO	24-Apr-09	EU & USA
RoActemra® Tocilizumab	Genentech/ Roche	Targets interleukin-6 receptor (IL-6R) for the treatment of rheumatoid arthritis (RA) and systemic juvenile idiopathic arthritis.	Humanized monoclonal antibody	CHO	16-Jan-09	EU
2008						
Arcalyst ® Rilonacept	Regeneron	For treatment of two Cryopyrin-Associated Periodic Syndromes (CAPS) disorders: Familial Cold Auto-Inflammatory Syndrome (FCAS) and Muckle-Wells Svndrome (MWS).	Fc-fusion biologic	CHO	27-Feb-08	USA
Xyntha® Factor VIII	Wyeth Pharmaceuticals/ Pfizer	Recombinant factor VIII (updated version of Refacto with no-animal origin raw materials used) for the treatment and prophylaxis of bleeding in patients with haemophilia A (congenital factor VIII deficiency).	Clotting factor	CHO	21-Feb-08	USA
Recothrom® Thrombin	ZymoGenetics, Inc.	Recombinant thrombin used in the prevention of minor bleeding during surgery.	Clotting factor	CHO	17-Jan-08	USA
2007						
Silapo® Epoetin-Zeta	Stada Arzneimittel AG	A biosimilar medicine of Eprex/Erypo, a recombinant erythropoietin that stimulates erythropoiesis and is used to treat anemia, chronic renal failure related to cancer chemotherapy.	Cytokine	CHO	18-Dec-07	EU

Product	Company	Target/Initial Therapeutic Indication	Product Category	Cell Line	First Approved	Region
Retacrit® Epoetin-Zeta	Hospira UK Limited	A biosimilar medicine of Eprex/ Erypo, a recombinant erythropoietin that stimulates erythropoiesis and is used to treat anemia , chronic renal failure related to cancer chemotherapy.	Cytokine	CHO	18-Dec-07	EU
Vectibix ® Panitumumab	Amgen	Monoclonal antibody specific to the epidermal growth factor receptor (EGFR)for the treatment of EGFR expressing metastatic colorectal cancer.	Fully human monoclonal antibody	CHO	3-Dec-07	EU
Abseamed ® Epoetin alfa	Medice Arzneimittel Putter GmbH & Co	A biosimilar medicine of Eprex/ Erypo, a recombinant erythropoietin that stimulates erythropoiesis and is used to treat anemia , chronic renal failure related to cancer chemotherapy.	Cytokine	CHO	28-Aug-07	EU
Binocrit ® Epoetin alfa	Sandoz GmbH	A biosimilar medicine of Eprex/Erypo, a recombinant erythropoietin that stimulates erythropoiesis and is used to treat anemia , chronic renal failure related to cancer chemotherapy.	Cytokine	CHO	28-Aug-07	EU
Epoetin alfa Hexal®	Hexal AG	A biosimilar medicine of Eprex/Erypo, a recombinant erythropoietin that stimulates erythropoiesis and is used to treat anemia, chronic renal failure related to cancer chemotherapy.	Cytokine	CHO	28-Aug-07	EU
Mircera® Methoxy polyethene glycol-epoetin beta	Roche	Chemically linked erythropoietin with methoxy polyethylene glycol butanoic acid stimulating erythropoiesis and is used to treat anemia associated with chronic kidney failure.	Cytokine	CHO	20-Jul-07	EU & USA

(continued)

Table 1 (continud)

Product	Company	Target/Initial Therapeutic Indication	Product Category	Cell Line	First Approved	Region
Pergoveris® Follitropin alfa/Lutropin alfa	Merck Serono Europe Ltd.	Recombinant hormone combination indicated to stimulate the development of follicles in the ovaries of infertile adults with low FSH and LH.	Hormone	CHO	25-Jun-07	EU
Orencia® Abatacept	Bristol-Myers Squibb	IgG1 Fc fusion to CTLA-4 binding with more avidity to CD80 (B7-1) than to CD86 (B7-2) for second-line treatment of rheumatoid arthritis in moderate to severe adult patients.	Fc-fusion biologic	CHO	21-May-07	EU
Soliris® Eculizumab	Alexion	Monoclonal antibody that is a terminal complement inhibitor and approved for the treatment of paroxysmal nocturnal hemoglobinuria (PNH).	Humanized monoclonal antibody	NS0	16-Mar-07	EU & USA
2006						
Vectibix ® Panitumumab	Amgen	Monoclonal antibody specific to the epidermal growth factor receptor (EGFR) for the treatment of EGFR expressing metastatic colorectal cancer.	Fully human monoclonal antibody	CHO	27-Sep-06	USA
Elaprase® Idursulfase	Shire	Recombinant lysosomal enzyme iduronate-2-sulfatase for the treatment of Hunter syndrome (mucopolysaccharidosis II· MPS II).	Enzyme	HT-1080	24-Jul-06	EU & USA
Tysabri® Natalizumab	Biogen Idec.	Monoclonal antibody against the cell adhesion molecule α 4-integrin for treatment of multiple sclerosis.	Humanized monoclonal antibody	NS0	27-Jun-06	EU
Myozyme® Alglucosidase alfa	Genzyme Corporation	Recombinant glucosidase for enzyme replacement therapy for the treatment of Pompe disease (Glycogen storage disease type II), a rare lysosomal storage disorder (LSD).	Enzyme	CHO	29-Mar-06	EU & USA

Product	Company	Target/Initial Therapeutic Indication	Product Category	Cell Line	First Approved	Region
Naglazyme® galsulfase	BioMarin	Recombinant N-acetylgalactosamine 4-sulfatase increasing the catabolism of glycosaminoglycans (GAG) is indicated for patients with mucopolysaccharidosis VI(MPS VI).	Enzyme	CHO	24-Jan-06	EU
		2005				
Orencia® Abatacept	Bristol-Myer Squibb	IgG1 Fc fusion to CTLA-4 binding with more avidity to CD80 (B7-1) than to CD86 (B7-2) for second-line treatment of rheumatoid arthritis in moderate to severe adult patients.	Fc-fusion biologic	CHO	26-Dec-05	USA
Hylenex® Cumulase® Hyaluronidase	Halozyme Therapeutics, Baxter Healthcare	Recombinant hyaluronidase for use as a "spreading agent" to enhance the delivery of local anaesthesia, contrast agents, and for subcutaneous fluid replacement (hypodermoclysis).	Enzyme	CHO	5-Dec-05	USA
Naglazyme® galsulfase	BioMarin	Recombinant N-acetylgalactosamine 4-sulfatase increasing the catabolism of glycosaminoglycans (GAG) is indicated for patients with mucopolysaccharidosis VI (MPS VI).	Enzyme	CHO	31-May-05	USA
		2004				
Tysabri® Natalizumab	Biogen Idec.	Monoclonal antibody against the cell adhesion molecule α 4-integrin for treatment of multiple sclerosis.	Humanized monoclonal antibody	NS0	24-Nov-04	USA
Luveris® LH	EMO Serano, Inc.	Luteinizing hormone (recombinant human LH) for the treatment of female infertility and it is indicated for use in combination with human follicle-stimulating hormone (Gonal-f®).	Hormone	CHO	24-May-04	USA

(continued)

Table 1 (continued)

Product	Company	Target/Initial Therapeutic Indication	Product Category	Cell Line	First Approved	Region
Avastin® Bevacizumab	Genentech/ Roche	An angiogenesis inhibitor (inhibits VEGF-A) where the drug slows the growth of new blood vessels to treat colorectal, lung, breast, glioblastoma, kidney and ovarian cancers.	Humanized monoclonal antibody	CHO	26-Feb-04	EU& USA
Erbitux® Cetuximab	Bristol-Myer Squibb, Eli Lilly	Mouse/human antibody for treatment of patients with epidermal growth factor receptor (EGFR)-expressing, metastatic colorectal cancer, head and neck cancer.	Chimeric monoclonal antibody	Sp2/0	12-Feb-04	EU& USA
		2003				
Advate ® Antihemophilic Factor	Baxter international Inc.	Recombinant factor VIII for the treatment and prophylaxis of bleeding in patients with haemophilia A (congenital factor VIII deficiency).	Clotting factor	CHO	25-Jul-03	EU& USA
Xolair® Omalizumab	Genentech Inc.	Targeting human immunoglobulin E (IgE) which treats the symptoms of asthma and chronic idiopathic urticaria by limiting the allergic response for treatment of moderate- to-severe allergic asthma.	Humanized monoclonal antibody	CHO	20-Jun-03	EU& USA
Aldurazyme® Laronidase	BioMarin	Human recombinant alpha-L-iduronidase for the treatment of mucopolysaccharidosis.	Enzyme	CHO	30-Apr-03	USA
		2002				
Humira ® Adalimumab	Abbott Laboratories (AbbVie)	Adalimumab binds to tumor necrosis factor-alpha (TNFα) and this TNFα inactivation has proven to be important in downregulating the inflammatory reactions.	Fully human monoclonal antibody	CHO	30-Dec-02	USA
Inductos® dibotermin alfa	Medtronic BioPharma B.V.	Acts on the bone structure since it is a recombinant version of bone morphogenetic protein 2 (BMP-2)	Cytokine	CHO	9-Sep-02	EU

Product	Company	Target/Initial Therapeutic Indication	Product Category	Cell Line	First Approved	Region
		stimulating bone formation to help heal fractures and aid in lower-back spine fusion surgery.				
InFUSE- Bone Graft/LT-CAGE™	Medtronic Sofamor Dane	Acts on the bone structure since it is a recombinant prophylaxis of bleeding in patients with haemophilia A version of bone morphogenetic protein 2 (BMP-2) (congenital factor VIII deficiency). stimulating bone formation to help heal fractures and aid in bone graft surgery.	Cytokine	CHO	2-Jul-02	USA
Zevalin® Ibritumomab tiuxetan	Spectrum Pharmaceuticals/Biogen Idec	Radioimmunotherapy treatment for relapsed or refractory, low grade or transformed B cell nonHodgkin's lymphoma, a lymphoproliferative disorder.	Murine monoclonal antibody	CHO	19-Feb-02	USA
2001						
Xigris® Drotrecogin Alpha	Eli Lilly	Recombinant form of human activated protein C that has anti-thrombotic, anti-inflammatory, and profibrinolytic properties for the treatment of sepsis (withdrawn).	Clotting factor	HEK293	21-Nov-01	USA
Aranesp® Darbepoetin alfa	Amgen	Recombinant erythropoietin to simulate erythropoiesis (increases red blood cell levels) and is used to treat anemia, commonly associated with chronic renal failure and cancer chemotherapy.	Cytokine	CHO	17-Sep-01	EU & USA
Fabrazyme® Agalsidase	Genzyme Corporation	Recombinant alpha-galactosidase A or alpha-GAL for enzyme replacement indicated for the treatment of Fabry disease.	Enzyme	CHO	3-Aug-01	EU & USA

(continued)

Table 1 (continued)

Product	Company	Target/Initial Therapeutic Indication	Product Category	Cell Line	First Approved	Region
Replagal® Agalsidase beta	Shire Human Genetic Therapies AB	Recombinant alpha-galactosidase A or alpha-GAL for enzyme replacement indicated for the treatment of Fabry disease.	Enzyme	CHO	3-Aug-01	EU
Campath®, Mabcampath® Alemtuzumab	Genzyme Corporation	Binds to CD52, on the surface of mature lymphocytes treatment of chronic lymphocytic leukemia (CLL), cutaneous T-cell lymphoma (CTCL) and T-cell lymphoma.	Humanized monoclonal antibody	CHO	7-May-01	USA
2000						
Ovidrel® Choriogonadotropin Alpha	EMO Serano, Inc.	Ovidrel (recombinant Chorionic Gonadotropin (Hcg) used as part of a treatment program for certain fertility problems in women and generally in combination with another hormone (FSH).	Hormone	CHO	20-Sep-00	EU & USA
Helixate FS®/ NexGen® Factor VIII octocog alfa	CSL Behring	A recombinant antihemophilic factor that is indicated for the control and prevention of bleeding episodes in patients with hemophilia A.	Clotting factor	BHK	4-Aug-00	EU
Kogenate Bayer® Factor VIII octocog alfa	Bayer Healthcare	A recombinant antihemophilic factor that is indicated for the control and prevention of bleeding episodes in patients with hemophilia A.	Clotting factor	BHK	4-Aug-00	EU
TNKase® Tenecteplase	Genentech/Roche	Recombinant fibrin-specific plasminogen activator indicated for use in the reduction of mortality associated with acute myocardial infarction (AMI).	Enzyme	CHO	2-Jun-00	USA
Enbrel® Etanercept	Pfizer	Tumor necrosis factor receptor2-immune globulin G1 Fc fusion protein for treatment of rheumatoid arthritis, psoriasis, and ankylosing spondylitis.	Fc-fusion biologic	CHO	3-Feb-00	EU
1999						

Product	Company	Target/Initial Therapeutic Indication	Product Category	Cell Line	First Approved	Region
Synagis® Palivizumab	MedImmune Inc.	Monoclonal antibody (IgG) directed against an epitope in the A antigenic site of the F protein of respiratory syncytial virus (RSV) for treatment of RSV infections.	Humanized monoclonal antibody	NS0	13-Aug-99	EU
ReFacto® Antihemophilic Factor	Wyeth Pharmaceuticals/ Pfizer	Recombinant factor VIII for the treatment and prophylaxis of bleeding in patients with haemophilia A (congenital factor VIII deficiency).	Clotting factor	CHO	13-Apr-99	EU & USA
1998						
Thyrogen® (thyrotropin alfa)	Genzyme Corporation	Recombinant form of human thyroid-stimulating hormone (TSH) which binds to TSH receptors used as a diagnostic tool for patients that may have thyroid cancer.	Hormone	CHO	30-Nov-98	EU & USA
Enbrel® Etanercept	Amgen	Tumor necrosis factor receptor2-immune globulin Gl Fc-fusion protein for treatment of rheumatoid arthritis, psoriasis, and ankylosing spondylitis.	Fc-fusion biologic	CHO	2-Nov-98	USA
Herceptin® Trastuzumab	Genentech/ Roche	A treatment that mainly interferes with the HER2/neu receptor to treat certain breast cancers.	Humanized monoclonal antibody	CHO	25-Sep-98	EU& USA
Remicade® Infliximab	Janssen Biotech (Johnson & Johnson)	Chimeric monoclonal antibody against tumour necrosis factor alpha (TNF-α) used to treat autoimmune diseases such as Crohn's disease and rheumatoid arthritis.	Chimeric monoclonal antibody	Sp2/0	24-Aug-98	EU & USA
Synagis® Palivizumab	MedImmune Inc.	Monoclonal antibody (IgG) directed against an epitope in the A antigenic site of the F protein of respiratory syncytial virus (RSV) for treatment of RSV infections.	Humanized monoclonal antibody	NS0	19-Jun-98	USA

(continued)

Table 1 (continued)

Product	Company	Target/Initial Therapeutic Indication	Product Category	Cell Line	First Approved	Region
Simulect® Basiliximab	Novartis	Mouse-human monoclonal antibody to the α chain (CD25) of the IL-2 receptor to prevent organ transplantation rejection.	Chimeric monoclonal antibody	Sp2/0	12-May-98	USA
Rebif® Interferon Beta-1a	Merck Serano Europe Ltd.	Recombinant interferon beta 1a for the treatment of relapsing forms of multiple sclerosis.	Cytokine	CHO	4-May-98	EU
1997						
Zenapax ® Daclizumab	Roche	First humanized antibody that binds to CD25, the alpha subunit of the IL-2 receptor of T cells for treatment of organ transplant rejection and investigated for treatment of multiple sclerosis.	Humanized monoclonal antibody	Sp2/0	10-Dec-97	USA
Rituxan® Rituximab	Genentech (Roche)/Biogen Idec	Monoclonal antibody against the protein CD20 found primarily on B cells allowing for the treatment of diseases with excessive or dysfunctional B cells.	Chimeric monoclonal antibody	CHO	26-Nov-97	EU & USA
Cerezyme® Imiglucerase	Genzyme Europe B.V .	Recombinant analogue of human β-glucocerebrosidase used in the treatment of Gaucher's disease, in which a fatty substance (lipid) accumulates in cells and certain organs.	Enzyme	CHO	17-Nov-97	EU
Follistim ® Follitropin-Beta	Merck & Co	Recombinant follicle-stimulating hormone for the treatment of infertility by stimulating ovaries to produce one or more eggs during each treatment.	Hormone	CHO	29-Sep-97	USA
Gonal-F® Follitropin-alfa	EMD Serono Inc.	Recombinant follicle-stimulating hormone for use in conjunction with assisted reproductive technologies, such as in vitro fertilization by the induction of ovulation.	Hormone	CHO	29-Sep-97	USA

Product	Company	Target/Initial Therapeutic Indication	Product Category	Cell Line	First Approved	Region
BeneFIX® Factor IX	Wyeth Pharmaceuticals/ Pfizer	For the control and prevention of hemorrhagic episodes in patients with hemophilia B (congenital factor IX deficiency and Christmas disease), including control and prevention of bleeding in surgical settings.	Clotting factor	CHO	11-Feb-97	EU & USA
		1996				
Cathflo®, Activase ® Actilyse® (Alteplase)	Genentech, Boehringer Ingelheim	Recombinant tissue plasminogen activator that is responsible for clot breakdown an indicated for treating heart attacks and acute massive pulmonary embolism such as in strokes.	Enzyme	CHO	18-Jun-96	EU & USA
NeoRecormon ® Epoetin-Beta	Roche	Recombinant erythropoietin .It stimulates erythropoiesis and is used to treat anemia, commonly associated with chronic renal failure and cancer chemotherapy.	Cytokine	CHO	14-Jun-96	USA
Avonex ® Interferon Beta-1a	Biogen Idec, Inc.	Recombinant interferon beta 1a is a drug in the interferon family used to treat multiple sclerosis (MS) by balancing the expression of pro- and anti-inflammatory agents in the brain.	Cytokine	CHO	17-May-96	EU & USA
Puregon® Follitropin beta	N.V. Organon	Recombinant follicle-stimulating hormone for the treatment of infertility by stimulating ovaries to produce one or more eggs during each treatment.	Hormone	CHO	3-May-96	EU
NovoSeven® Factor VIIa	Novo Nordisk	Treating or preventing bleeding episodes in certain patients with bleeding problems such as hemophilia A or B, acquired haemophilia or congenital FVII deficiency.	Clotting factor	CHO	23-Feb-96	EU & USA
		1995				

(continued)

Table 1 (continued)

Product	Company	Target/Initial Therapeutic Indication	Product Category	Cell Line	First Approved	Region
Gonal-F® Follitropin-alfa	Merck Serono Europe Ltd	Recombinant follicle-stimulating hormone for use in conjunction with assisted reproductive technologies, such as in vitro fertilization by the induction of ovulation.	Hormone	CHO	20-Oct-95	EU
1994						
Reopro® Abciximab	Janssen Biotech and Eli Lilly	Antibody fragment and a platelet aggregation inhibitor mainly used during and after coronary artery procedures like angioplasty to prevent thrombus (blood clots).	Chimeric monoclonal antibody	Sp2/0	14-Nov-94	EU & USA
Cerezyme® Imiglucerase	Genzyme Corporation	Recombinant analogue of human β-glucocerebrosidase use.	Enzyme	CHO	6-Jun-94	USA
1993						
Bioclate™ factor VIII	Aventis Behring	A recombinant antihemophilic factor that is indicated for the control and prevention of bleeding episodes in patients with hemophilia A.	Clotting factor	CHO	31-Dec-93	USA
Pulmozyme ® human deoxyribonuclease I (rhDNase)	Genentech/ Roche	Recombinant human deoxyribonuclease I (rhDNase) hydrolyzes the DNA present in sputum/mucus of cystic fibrosis patients and reduces viscosity in the lungs, promoting improved clearance of secretions.	Enzyme	CHO	30-Dec-93	USA
Kogenate FS® Factor VIII octocog alfa	Bayer Healthcare	A recombinant antihemophilic factor that is indicated for the control and prevention of bleeding episodes in patients with hemophilia A.	Clotting factor	BHK	28-Feb-93	USA
Helixate FS® Factor VIII octocog alfa	CSL Behring	A recombinant antihemophilic factor that is indicated for the control and prevention of bleeding episodes in patients with hemophilia A.	Clotting factor	BHK	28-Feb-93	USA

Product	Company	Target/Initial Therapeutic Indication	Product Category	Cell Line	First Approved	Region
1992						
Recombinate™ Antihemophilic Factor VIII	Baxter international Inc.	A recombinant antihemophilic factor that is indicated for the control and prevention of bleeding episodes in patients with haemophilia A.	Clotting factor	CHO	21-Dec-92	USA
1988						
Eprex® Erypo® (Epoetin-Alfa)	Janssen Biotech (Johnson & Johnson)	Recombinant erythropoietin that stimulates erythropoiesis and is used to treat anaemia, commonly associated with chronic renal failure and cancer chemotherapy.	Cytokine	CHO	4-Aug-88	EU
1987						
Activase ® (Alteplase)	Genentech	Recombinant tissue plasminogen activator that is responsible for clot breakdown an indicated for treating acute mycardial infarctions.	Enzyme	CHO	13-Nov-87	USA

The introduction of chimeric full-length antibodies as biopharmaceuticals started with Rituximab (Rituxan®) in 1997, Rituximab was first conceived and developed by IDEC Pharmaceutical Corporation (now known as Biogen Idec) as IDEC-C2B8 of the development name for the treatment of non-Hodgkin's lymphoma [50, 51]. As a chimeric antibody, Rituximab possesses the variable region fragments of a murine anti-human CD20 that is expressed on the surface of malignant and normal B cells, and the human IgG heavy and light chain constant regions [52]. The mAb therapeutic was eventually brought to market in collaboration with Genentech (now a subsidiary of Roche). Rituximab was designed with improved antibody-dependent cellular cytotoxicity (ADCC) with human effector cells and 22 days of long serum half-life due to a modified human Fc region [53]. While Rituximab has shown a low immunogenic such as 1.1% of patients, immunogenicity is generally a concern for chimeric antibodies such as basiliximab and infliximab both demonstrating immunogenicity in up to 44% and 37% of patients, respectively [49]. Apart from Rituximab, Roche also developed the first humanized mAb approved for marketing in 1997, known as Daclizumab (Zenapax®) for the treatment of organ transplant rejection. Antibody humanization through engineering an antibody with a reduced non-human amino acid sequence can reduce the immunogenicity of engineered antibodies. However, humanized antibodies still have immunogenicity risks due to various factors that can induce immunogenicity. In addition to the presence of amino acid sequences from non-human origin that can contribute to increased immunogenicity, there can be several other intrinsic and even extrinsic factors that may increase immunogenicity for mAb therapeutics. It is known that the carbohydrate side-chains attached via glycosylation has a major impact on immunogenicity of an antibody and plays a major intrinsic role as well as other post translational events that may modify the antibody sequence such as oxidation, non-enzymatic glycosylation, and deamination of the amino side chains [54, 55]. It has also been found that antibodies that target insoluble factors, such as cell surface markers, may pose a risk of increased immunogenicity to the patient. Another intrinsic factor is the presence of CD4 positive T helper epitopes that can lead to an immune response depending on the amino acid sequence [47]. Apart from patients' immunological status and the effects of co-medication [47, 56], extrinsic factors may cause immunogenicity due to the composition of the antibody drugs manufacturers formulate. Some formulations may be able to cause increased immunogenicity issues due to the presence of adjuvant-like contaminants and aggregates [57, 58]. In spite of the concern on immunogenicity, humanization of a non-human antibody has still been the preferred choice for the mAb therapeutics.

Adalimumab, the first fully human antibody, was selected via phage display of the human variable heavy and light chain sequences, but it is also possible to produce fully human antibodies from an engineered mouse via a process known as XenoMouse technology. With XenoMouse technology, the immunoglobulin genes within the transgenic mouse are of human origin [59, 60] making the possibility of natural in vivo affinity maturation of the sequences which may contribute to a further reduction in immunogenicity. The first therapeutic mAb to be approved for marketing that utilized the XenoMouse technology was panitumumab (Vectibix®), in 2006 [61]. Panitumumab has a very low immunogenicity of 3–4%, due to

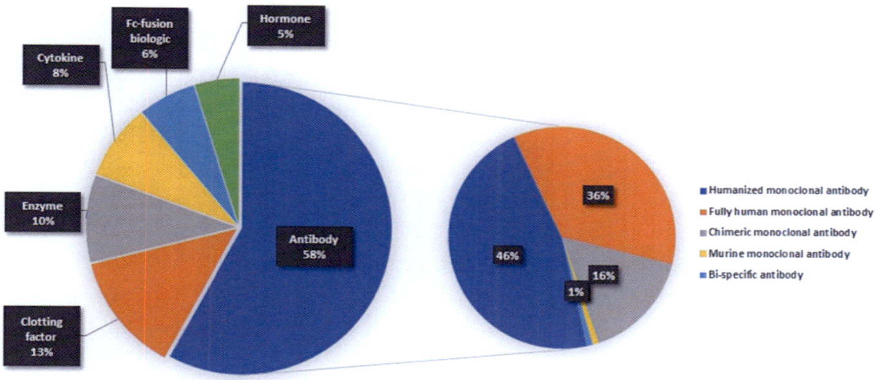

Fig. 2 The variety of mammalian cells-based products approved in the EU and USA, from the year 1987 till February 2021, revealing that antibodies take 58% of the approved mammalian cells-based biologics on the market. Humanized and fully human antibodies make up the majority of the antibodies approved

the antibody development strategy employed, thus fully human derived antibodies can contain no murine sequences, unlike humanized antibodies, but immune responses can still occur. Thus, the development of fully human antibodies are not a guarantee of non-immunogenicity, but it is possible that with further development steps immunogenicity of engineered mAbs can be reduced or even eliminated by a combination of CDR-sequence engineering, optimized cell culture bioprocess development strategies, and formulation engineering to help fine tune the intrinsic and extrinsic factors that can reduce immunogenicity.

Biopharmaceuticals produced by cultivated mammalian cells include hormones, enzymes, cytokines, clotting factors, Fc-fusion proteins and antibodies. The majority of mammalian-cells based biopharmaceuticals approved from 1987 to February 2021, are antibodies which account for about 58% in the biopharmaceutical market (Fig. 2). Among them, three major types including chimeric, humanized and fully human monoclonal antibodies constitute about 98% of antibody products. Surprisedly, based on the statistical analysis in Fig. 2, far more humanized antibodies have come to the market compared to fully human antibodies, although fully human antibodies are considered a lower risk for inducing immune responses in humans compared to humanized antibodies.

What has also been observed is that serum half-life can vary greatly with humanized and fully human antibody drugs compared to natural antibodies such as IgG which has a mean half-life of 25–32 days [62]. These engineered therapeutic antibodies have a serum half-life that varies greatly from a low of 7.5 days to a range similar to natural antibodies [53]. Varying serum half-life can also be the result of variations in post translational processing of these recombinant antibodies with the use of non-human originating cell lines including the culture conditions during manufacturing as we know that glycans also influence immunogenicity and efficacy [63].

The serum half-life of mAb's is usually high compared to other recombinant proteins due the neonatal Fc receptor of IgG (FcRn). The FcRn is a MHC Class I like molecule that binds to the CH2–CH3 hinge region of IgG which starts a process that ultimately protects IgG from degradation thereby promoting the extended half-life of this class of antibody in the serum [64, 65]. In further detail, IgG is bound to the Fc receptor of a cell within an acidic endosome that is destined to be internalized via pinocytosis, the IgG can be recycled to the cell surface and released back into a neutral pH environment preventing the faith of lysosomal degradation that unbound proteins face when taken in by the endosome. This recycling can extend the serum half- life of IgG [66], although, further studies are required since studies have shown that an increase in binding affinity of an engineered IgG molecule to the FcRn is not proportional to half-life [67]. One study demonstrated this with variants of mAb drug, Herceptin™, from Genentech with 3 and 12 fold higher binding affinities for the FcRn that still had similar half-life compared to Herceptin at the end [68]. Currently, more than 20 glyco-engineered mAbs, with enhanced ADCC, are being evaluated in clinical studies. Two of these mAbs have already been approved, mogamulizumab (Poteligeo®) on March 30th, 2012 for marketing in Japan, an antibody developed exclusively by Kyowa Hakko Kirin, and obinutuzumab (Gazyva®), approved on November 1st, 2013 in the USA (see Table 1), confirming the success of this approach. The glyco-engineered Fc region of obinutuzumab has a bisected, complex, non-fucosylated oligosaccharides attached to asparagine 297, that enhances the binding affinity to FcγRIII an Fc receptor [69]. The glycol-engineering of obinutuzumab has significantly improved the efficacy over earlier therapeutic molecules such as rituximab and earlier developed mAb in B-cell malignancies.

Since the Fc region of antibody binds to the FcRn to confer longer circulatory half-life there has been great success with the use of this natural molecular process to engineer proteins that can take advantage of this. In 1998, Enbrel® (etanercept) was the first CHO cell culture produced Fc fusion biologic, to gain marketing approval by the FDA. Enbrel®, a recombinant human soluble tumor necrosis factor (TNF) receptor able to bind and inactivate soluble and cell bound TNF and lymphotoxin competing with the cellular TNF receptors for the treatment of rheumatoid arthritis and has been one of the most successful biopharmaceuticals on the market with global sales reaching $8.4 billion in 2013 just behind Humira® of $10.7 billion as two of the most successful drugs the biopharmaceutical industry has ever developed. Enbrel® consists of an intracellular portion of the human p75 TNFR linked to the Fc portion of IgG1 to form a dimeric protein. The benefit of the Fc fusion bestows the etanercept molecule with an extended median half-life of 4.8 days, together with a high binding affinity this contributes to Enbrel® overall effectiveness as an arthritis drug compared to others on the market at the time of its approval [70]. A CHO cell line is used as the host for expression of the 150kDa dimeric etanercept molecule. Enbrel® is part of a class of biologics that work by inhibiting the binding of TNF such as adalimumab (Humira®), golimumab (Simponi®, Simponi ARIA®), and infliximab (Remicade®), this allows these biologics to suppress the cascade of reactions that lead to an inflammatory response within the body that can actually destroy joint tissue as is characteristic with rheumatoid arthritis.

After market approval and release of Enbrel®, several other Fc fusion molecules were approved in the EU and USA that required a mammalian cell culture process in order to produce their complex fusion molecules. In 2003, a second mammalian cell expressed fusion product was approved by the name of Amevive® (alefacept) [71, 72] which utilized the Fc portion for apoptosis induction apart from boosting half-life. This 91.4 kDa protein has a Fc region of IgG1 linked to human leukocyte function antigen 3 (LFA-3) that can bind, with high affinity, to CD2, a functionally important and widely distributed T lymphocyte surface glycoprotein. Upon human LFA-3/IgG1 fusion protein administration the LFA-3 binds to CD2 inhibiting T-cell activation and proliferation. The Fc portion extends the circulatory half-life to 11.25 days [73] in addition to interacting with the FcγRIII receptor on the surface of NK cells which results in NK induced apoptosis of T-lymphocytes [74]. This overall effect suppresses the immune system and can be used in the treatment of psoriasis, a skin condition that causes skin redness and irritation.

Two of the Fc-fusion biologics are not solely immunosuppressive, and these are Elyea® and Alprolix™. The drug Eylea® is an anti-angiogenic used in the treatment of neovascular age-related macular degeneration, an eye disease due to blood vessels leaking fluid into the macula. Zaltrap® (named ziv-aflibercept for distinction) is the same drug but approved as an anti-cancer agent for treatment of metastatic colorectal cancer. Aflibercept is a vascular endothelial growth factor (VEGF) trap that consists of an Fc region fused with the VEGF-binding portions from the extracellular domains of human VEGF receptors 1 and 2. The VEGF trap has a highly variable half-life of 1.7–7.4 days depending on the dosage. Alprolix™, which was approved for marketing on the 28th of March 2014, is the only Fc-fusion protein produced in HEK-293 cells that has been approved by the FDA and also the first fusion drug for the treatment of hemophilia B. The Fc fusion increased the half-life of the drug to 3.6 days, a considerable increase over the other coagulation factor IX drugs, Rixubis® and BeneFIX®. Rixubis® was introduced in 2013 with a half-life more than 3 times less than Alprolix®, at 26 hours, and BeneFIX® was approved in 1997 with a maximum half-life of 24 h. The coagulation factor IX drugs activate the coagulation pathway to ultimately convert prothrombin to thrombin which converts fibrinogen to fibrin so that a clot can be formed for the treatment of bleeding episodes.

The future outlook of mAb drugs is bright, with the start of this decade having close to 300 mAb's in various stages of clinical development. Of these mAb's approximately 150 new monoclonal antibodies are in development for the area of oncology treatments and close to 70 mAb's are in clinical development for treatment of inflammatory and autoimmune diseases with the rest are for indications that include metabolic disorders, cardiovascular disorders, central nervous system disorders, infectious diseases, and transplant rejection. Currently, mAbs are the strongest growing segment of the pharmaceutical market and is expected to further grow at a fast pace along with subcategories such as fusion protein drugs that use antibody components to carry-out their function.

4 Biosimilars or Follow-On Biologics

As competition drives the price of the generic products down in the pharmaceutical sector, the impact of biosimilars would be expected to have a similar affect. Once a patent expires generic drugs can be legally produced, although loopholes can exist in areas or countries where the patent is not enforceable, or the patent can be proven invalid. While small molecule formulated pharmaceuticals can have exact copies made that can pass the regulatory framework, biopharmaceutical manufactured products such as, recombinant proteins can have a high degree of molecular complexity that includes the post translational modification which are all affected by the manufacturing process. The term biosimilar, or follow-on biologic, was introduced by the regulatory authorities in the EU and USA to imply that the newly introduced product would be similar to the original biologic but might not be an identical molecular copy of the parent biologic. The FDA tends to use the term follow-on biologics while in the EU biosimilar is used by the EMA [75, 76]. In order to prove a biological entity is a biosimilar to the regulatory authorities, data has to be compiled through clinical, animal and analytical studies where the results must indicate that the biological entity reproduces the same clinical results as the parent drug. The time and costs associated with mammalian biopharmaceutical development and manufacturing of biosimilars will be a far greater investment for a pharmaceutical compared to what is required for small-molecule generics as it is estimated that the development time for a biosimilar recombinant proteins could range from 5–8 years compared to the 1–2 years for a generic small molecule [77, 78]. This is due to the complexity of large glycosylated molecule and the process development required. For example, the host cell will tend to go through the process of cell line selection and development to create a suitable host for the target bioprocess to produce the biosimilar, which can take months to years depending on the biological product. This is in addition to the development of the commercial scale manufacturing process that requires strict quality controls and process monitoring of all upstream and downstream processes till final formulation and product testing.

In 2006, the first biosimilar was approved in both the EU and the USA under the trade name Omnitrope™. This biosimilar was an E.coli expressed 22.1 kDa recombinant growth hormone (hGH) identical to the native protein consisting of a 191 amino acid single chain polypeptide manufactured by Sandoz for the treatment of growth hormone deficiencies [75]. This approval was due to the regulatory framework that was established since 2005 by the EMA and established legislation in the USA [79, 80]. The EU has further build on their regulatory framework that by 2010 draft guidelines were established for mAb biosimilars leading to a final version of the guidelines completed by the EMA's Committee for Medicinal Products for

Human Use (CHMP) that included IgG1 Fc-fusion protein biosimilars in the scope of mAb biosimilars [81]. The regulatory framework in the EU allowed the EMA to approve several biosimilars that are recombinant biopharmaceutical proteins produced in mammalian cell culture (see Table 1) beginning with a biosimilar for recombinant erythropoietin (Epoetin alfa). The first five Epoetin alfa biosimilars were approved in 2007 each marketed by Hexal AG, Sandoz GmbH, Medice Arzneimittel Pütter GmbH & Co. KG, Hospira UK Limited, and Stada Arzneimittel AG under the trade names; Epoetin alfa Hexal®, Binocrit®, Abseamed®, Retacrit®, and Silapo® respectively. This was a milestone accomplishment for a getting biosimilars of a glycosylated protein onto the market and helped set the stage for mAbs. In June 2013, Celltrion and Hospira received permission from the EMA's CHMP to market their biosimilars to Johnson & Johnson's Remicade® (infliximab) under the trade names Remsima® and Inflectra® respectively (see Table 1 for further details). At the time of writing this review, a total of 30 biosimilars produced in mammalian cells have been approved in the USA and EU which account for about 14% of the mammalian cells-based biopharmaceuticals market.

The impact of mammalian cell produced biologics is becoming the major contributor to pharmaceutical industry growth pipelines and the existence of a regulatory framework for biosimilars has meant further increases in mammalian cell culture capacity. BioProcess Technology Consultants, Inc. have given an interesting analysis of the global mammalian cell culture capacity with currently, the existence of one contract manufacturing organization (CMO) (Lonza), one excess capacity company acting as both product manufacturer and CMO (Boehringer Ingelheim), and 10 product companies with an installed capacity greater than 100,000L each. These companies are Roche, Johnson & Johnson, Amgen, Pfizer, Sanofi-Aventis, Novartis, Eli Lilly, Biogen Idec, Bristol-Myers Squibb, and Celltrion. Samsung Biologics being an additional CMO to join the list by 2017 [82]. Celltrion, Samsung BioLogics, and Innovent Biologics are examples of companies outside of the USA and Europe increasing capacity due to growing interests in mammalian cell culture biopharmaceutical manufacturing and the biosimilar market. This growth in worldwide capacity is an approximate 57% increase in capacity since 2010.

5 Conclusion

As the pharmaceutical market has demonstrated over the last decade, the mammalian cell derived biologicals market has continued to thrive and drive major growth in the pharmaceutical industry. The ability to provide post translational modifications and the continued need for monoclonal antibody therapies and the rise of Fc-fusion protein therapies have given mammalian expression systems a dominant advantage over other expression systems for the next generation of engineered biopharmaceuticals. With the introduction of disposable technologies, the improvement of cell culture processes, cell line selection and development strategies giving higher titers and specific productivities, including the establishment of EU and US regulatory pathways to bring biosimilars to the market, an infusion of growth has occurred at a rate currently faster than any other pharmaceutical sector. This growth is exemplified by the increase in global manufacturing capacity, including a substantial increase in the construction of Asian GMP mammalian culture facilities over the last few years. The introduction of new players in the biopharmaceutical industry, alongside the arrival of biosimilars promising lowered healthcare expenses of animal cell derived biopharmaceuticals, allows us to make the prediction that the next decade of mammalian bioprocesses and their biological products will continue to grow at a remarkable pace in innovation and discovery providing an increased affordability of these biological medicines.

References

1. Rader RA (2008) (Re)defining biopharmaceutical. Nat Biotechnol 26(7):743–751. https://doi.org/10.1038/nbt0708-743
2. Walsh G (2002) Biopharmaceuticals and biotechnology medicines: an issue of nomenclature. Eur J Pharm Sci 15(2):135–138. https://doi.org/10.1016/S0928-0987(01)00222-6
3. Owczarek B, Gerszberg A (2019) A brief reminder of systems of production and chromatography-based recovery of recombinant protein. Biopharmaceuticals 2019:4216060. https://doi.org/10.1155/2019/4216060
4. Walsh G (2018) Biopharmaceutical benchmarks 2018. Nat Biotechnol 36(12):1136–1145. https://doi.org/10.1038/nbt.4305
5. Administration UFaD (2021) Purple Book Database of Licensed Biological Products. https://purplebooksearch.fda.gov/
6. Wilson AW, Neumann PJ (2012) The cost-effectiveness of biopharmaceuticals: a look at the evidence. MAbs 4(2):281–288. https://doi.org/10.4161/mabs.4.2.18812
7. Hughes B, Hann LE (2007) The production of biopharmaceuticals. In: Boehncke W-H, Radeke HH (eds) Biologics in general medicine. Springer, Berlin/Heidelberg, pp 59–66. https://doi.org/10.1007/978-3-540-29018-6_7
8. Sanchez-Garcia L, Martín L, Mangues R, Ferrer-Miralles N, Vázquez E, Villaverde A (2016) Recombinant pharmaceuticals from microbial cells: a 2015 update. Microb Cell Factories 15(1):33. https://doi.org/10.1186/s12934-016-0437-3
9. Jackson LR, Trudel LJ, Fox JG, Lipman NS (1999) Monoclonal antibody production in murine ascites. I. Clinical and pathologic features. Lab Anim Sci 49(1):70–80
10. McGuill MW, Rowan AN (1989) Refinement of monoclonal antibody production and animal well-being. ILAR J 31(1):7–11. https://doi.org/10.1093/ilar.31.1.7
11. Grilo AL, Mantalaris A (2019) The increasingly human and profitable monoclonal antibody market. Trends Biotechnol 37(1):9–16. https://doi.org/10.1016/j.tibtech.2018.05.014
12. Kesik-Brodacka M (2018) Progress in biopharmaceutical development. Biotechnol Appl Biochem 65(3):306–322. https://doi.org/10.1002/bab.1617
13. Schmidt FR (2004) Recombinant expression systems in the pharmaceutical industry. Appl Microbiol Biotechnol 65(4):363–372. https://doi.org/10.1007/s00253-004-1656-9
14. Warnock JN, Al-Rubeai M (2006) Bioreactor systems for the production of biopharmaceuticals from animal cells. Biotechnol Appl Biochem 45(Pt 1):1–12. https://doi.org/10.1042/ba20050233
15. Wurm FM (2004) Production of recombinant protein therapeutics in cultivated mammalian cells. Nat Biotechnol 22(11):1393–1398. https://doi.org/10.1038/nbt1026

16. Lindsley CW (2019) Predictions and statistics for the best-selling drugs globally and in the United States in 2018 and a look forward to 2024 projections. ACS Chem Neurosci 10(3):1115–1115. https://doi.org/10.1021/acschemneuro.9b00112
17. Bebbington CR, Renner G, Thomson S, King D, Abrams D, Yarranton GT (1992) High-level expression of a recombinant antibody from myeloma cells using a glutamine synthetase gene as an amplifiable selectable marker. Bio/technology (Nat Publ Company) 10(2):169–175. https://doi.org/10.1038/nbt0292-169
18. Barnes LM, Bentley CM, Dickson AJ (2001) Characterization of the stability of recombinant protein production in the GS-NS0 expression system. Biotechnol Bioeng 73(4):261–270. https://doi.org/10.1002/bit.1059
19. Shulman M, Wilde CD, Köhler G (1978) A better cell line for making hybridomas secreting specific antibodies. Nature 276(5685):269–270. https://doi.org/10.1038/276269a0
20. Cockett MI, Bebbington CR, Yarranton GT (1990) High level expression of tissue inhibitor of metalloproteinases in Chinese hamster ovary cells using glutamine synthetase gene amplification. Bio/technology (Nat Publ Company) 8(7):662–667. https://doi.org/10.1038/nbt0790-662
21. Milbrandt JD, Azizkhan JC, Hamlin JL (1983) Amplification of a cloned Chinese hamster dihydrofolate reductase gene after transfer into a dihydrofolate reductase-deficient cell line. Mol Cell Biol 3(7):1274–1282. https://doi.org/10.1128/mcb.3.7.1274
22. Carvalhal AV, Moreira JL, Carrondo MJ (2001) Strategies to modulate BHK cell proliferation by the regulation of IRF-1 expression. J Biotechnol 92(1):47–59. https://doi.org/10.1016/s0168-1656(01)00365-0
23. Christie A, Butler M (1999) The adaptation of BHK cells to a non-ammoniagenic glutamate-based culture medium. Biotechnol Bioeng 64(3):298–309
24. Geserick C, Bonarius HP, Kongerslev L, Hauser H, Mueller PP (2000) Enhanced productivity during controlled proliferation of BHK cells in continuously perfused bioreactors. Biotechnol Bioeng 69(3):266–274. https://doi.org/10.1002/1097-0290(20000805)69:3<266::aid-bit4>3.0.co;2-z
25. Kirchhoff S, Kröger A, Cruz H, Tümmler M, Schaper F, Köster M, Hauser H (1996) Regulation of cell growth by IRF-1 in BHK-21 cells. Cytotechnology 22(1–3):147–156
26. Baldi L, Muller N, Picasso S, Jacquet R, Girard P, Thanh HP, Derow E, Wurm FM (2005) Transient gene expression in suspension HEK-293 cells: application to large-scale protein production. Biotechnol Prog 21(1):148–153. https://doi.org/10.1021/bp049830x
27. Schlaeger EJ, Christensen K (1999) Transient gene expression in mammalian cells grown in serum-free suspension culture. Cytotechnology 30(1–3):71–83. https://doi.org/10.1023/a:1008000327766
28. Jayapal KP, Wlaschin KF, Hu W, Yap MG (2007) Recombinant protein therapeutics from CHO cells-20 years and counting. Chem Eng Prog 103(10):40
29. Urlaub G, Chasin LA (1980) Isolation of Chinese hamster cell mutants deficient in dihydrofolate reductase activity. Proc Natl Acad Sci 77(7):4216–4220
30. Wiebe M, Becker F, Lazar R, May L, Casto B, Semense M, Fautz C (1989) A multifaceted approach to assure that recombinant tPA is free of adventitious virus
31. Huang YM, Hu W, Rustandi E, Chang K, Yusuf-Makagiansar H, Ryll T (2010) Maximizing productivity of CHO cell-based fed-batch culture using chemically defined media conditions and typical manufacturing equipment. Biotechnol Prog 26(5):1400–1410
32. Browne SM, Al-Rubeai M (2007) Selection methods for high-producing mammalian cell lines. Trends Biotechnol 25(9):425–432
33. Chon JH, Zarbis-Papastoitsis G (2011) Advances in the production and downstream processing of antibodies. New Biotechnol 28(5):458–463
34. Kelley B (2009) Industrialization of mAb production technology: the bioprocessing industry at a crossroads. In: MAbs, vol 5. Taylor & Francis, pp 443–452
35. Kuystermans D, Al-Rubeai M (2011) Bioreactor systems for producing antibody from mammalian cells. Antibody Expression Prod:25–52

36. Walther J, Lu J, Hollenbach M, Yu M, Hwang C, McLarty J, Brower K (2019) Perfusion cell culture decreases process and product heterogeneity in a head-to-head comparison with fed-batch. Biotechnol J 14(2):1700733
37. Low D, O'Leary R, Pujar NS (2007) Future of antibody purification. J Chromatogr B 848(1):48–63
38. Shukla AA, Hubbard B, Tressel T, Guhan S, Low D (2007) Downstream processing of monoclonal antibodies – application of platform approaches. J Chromatogr B 848(1):28–39
39. Butler M (2005) Animal cell cultures: recent achievements and perspectives in the production of biopharmaceuticals. Appl Microbiol Biotechnol 68(3):283–291
40. Jenkins N, Parekh RB, James DC (1996) Getting the glycosylation right: implications for the biotechnology industry. Nat Biotechnol 14(8):975–981
41. Niaudet P, Jean G, Broyer M, Chatenoud L (1993) Anti-OKT3 response following prophylactic treatment in paediatric kidney transplant recipients. Pediatr Nephrol 7(3):263–267
42. Richards J, Auger J, Peace D, Gale D, Michel J, Koons A, Haverty T, Zivin R, Jolliffe L, Bluestone JA (1999) Phase I evaluation of humanized OKT3: toxicity and immunomodulatory effects of hOKT3γ4. Cancer Res 59(9):2096–2101
43. Knight DM, Wagner C, Jordan R, McAleer MF, DeRita R, Fass DN, Coller BS, Weisman HF, Ghrayeb J (1995) The immunogenicity of the 7E3 murine monoclonal Fab antibody fragment variable region is dramatically reduced in humans by substitution of human for murine constant regions. Mol Immunol 32(16):1271–1281
44. Tam SH, Sassoli PM, Jordan RE, Nakada MT (1998) Abciximab (ReoPro, chimeric 7E3 Fab) demonstrates equivalent affinity and functional blockade of glycoprotein IIb/IIIa and αvβ3 integrins. Circulation 98(11):1085–1091
45. Marks L (2012) The birth pangs of monoclonal antibody therapeutics: the failure and legacy of Centoxin. In: MAbs, vol 3. Taylor & Francis, pp 403–412
46. Almagro JC, Daniels-Wells TR, Perez-Tapia SM, Penichet ML (2018) Progress and challenges in the design and clinical development of antibodies for cancer therapy. Front Immunol 8:1751
47. Harding FA, Stickler MM, Razo J, DuBridge R (2010) The immunogenicity of humanized and fully human antibodies: residual immunogenicity resides in the CDR regions. In: MAbs, vol 3. Taylor & Francis, pp 256–265
48. Hwang WYK, Foote J (2005) Immunogenicity of engineered antibodies. Methods 36(1):3–10
49. Levêque D, Wisniewski S, Jehl F (2005) Pharmacokinetics of therapeutic monoclonal antibodies used in oncology. Anticancer Res 25(3C):2327–2343
50. Grillo-López AJ (2000) Rituximab: an insider's historical perspective. Semin Oncol 6(Suppl 12):9–16
51. Maloney DG, Grillo-López AJ, White CA, Bodkin D, Schilder RJ, Neidhart JA, Janakiraman N, Foon KA, Liles T-M, Dallaire BK (1997) IDEC-C2B8 (Rituximab) anti-CD20 monoclonal antibody therapy in patients with relapsed low-grade non-Hodgkin's lymphoma. Blood 90(6):2188–2195
52. Silverman GJ, Weisman S (2003) Rituximab therapy and autoimmune disorders: prospects for anti–B cell therapy. Arthritis Rheum 48(6):1484–1492
53. Yoon S, Kim Y-S, Shim H, Chung J (2010) Current perspectives on therapeutic antibodies. Biotechnol Bioprocess Eng 15(5):709–715
54. Arnold JN, Wormald MR, Sim RB, Rudd PM, Dwek RA (2007) The impact of glycosylation on the biological function and structure of human immunoglobulins. Annu Rev Immunol 25:21–50
55. Sheeley DM, Merrill BM, Taylor LC (1997) Characterization of monoclonal antibody glycosylation: comparison of expression systems and identification of terminal α-linked galactose. Anal Biochem 247(1):102–110
56. Hendrickson JE, Desmarets M, Deshpande SS, Chadwick TE, Hillyer CD, Roback JD, Zimring JC (2006) Recipient inflammation affects the frequency and magnitude of immunization to transfused red blood cells. Transfusion 46(9):1526–1536
57. Shire SJ (2009) Formulation and manufacturability of biologics. Curr Opin Biotechnol 20(6):708–714

58. Rosenberg AS (2006) Effects of protein aggregates: an immunologic perspective. AAPS J 8(3):E501–E507
59. Lonberg N, Taylor LD, Harding FA, Trounstine M, Higgins KM, Schramm SR, Kuo C-C, Mashayekh R, Wymore K, McCabe JG (1994) Antigen-specific human antibodies from mice comprising four distinct genetic modifications. Nature 368(6474):856–859
60. Green LL (1999) Antibody engineering via genetic engineering of the mouse: XenoMouse strains are a vehicle for the facile generation of therapeutic human monoclonal antibodies. J Immunol Methods 231(1-2):11–23
61. Jakobovits A, Amado RG, Yang X, Roskos L, Schwab G (2007) From XenoMouse technology to panitumumab, the first fully human antibody product from transgenic mice. Nat Biotechnol 25(10):1134–1143
62. Maarschalk-Ellerbroek L, Hoepelman I, Ellerbroek P (2011) Immunoglobulin treatment in primary antibody deficiency. Int J Antimicrob Agents 37(5):396–404
63. Ghaderi D, Zhang M, Hurtado-Ziola N, Varki A (2012) Production platforms for biotherapeutic glycoproteins. Occurrence, impact, and challenges of non-human sialylation. Biotechnol Genet Eng Rev 28(1):147–176. https://doi.org/10.5661/bger-28-147
64. Simister NE, Mostov KE (1989) An Fc receptor structurally related to MHC class I antigens. Nature 337(6203):184–187. https://doi.org/10.1038/337184a0
65. Kuo TT, Baker K, Yoshida M, Qiao SW, Aveson VG, Lencer WI, Blumberg RS (2010) Neonatal Fc receptor: from immunity to therapeutics. J Clin Immunol 30(6):777–789. https://doi.org/10.1007/s10875-010-9468-4
66. Rodewald R (1976) pH-dependent binding of immunoglobulins to intestinal cells of the neonatal rat. J Cell Biol 71(2):666–669. https://doi.org/10.1083/jcb.71.2.666
67. Roopenian DC, Akilesh S (2007) FcRn: the neonatal Fc receptor comes of age. Nat Rev Immunol 7(9):715–725. https://doi.org/10.1038/nri2155
68. Petkova SB, Akilesh S, Sproule TJ, Christianson GJ, Al Khabbaz H, Brown AC, Presta LG, Meng YG, Roopenian DC (2006) Enhanced half-life of genetically engineered human IgG1 antibodies in a humanized FcRn mouse model: potential application in humorally mediated autoimmune disease. Int Immunol 18(12):1759–1769. https://doi.org/10.1093/intimm/dxl110
69. Mössner E, Brünker P, Moser S, Püntener U, Schmidt C, Herter S, Grau R, Gerdes C, Nopora A, van Puijenbroek E, Ferrara C, Sondermann P, Jäger C, Strein P, Fertig G, Friess T, Schüll C, Bauer S, Dal Porto J, Del Nagro C, Dabbagh K, Dyer MJS, Poppema S, Klein C, Umaña P (2010) Increasing the efficacy of CD20 antibody therapy through the engineering of a new type II anti-CD20 antibody with enhanced direct and immune effector cell–mediated B-cell cytotoxicity. Blood 115(22):4393–4402. https://doi.org/10.1182/blood-2009-06-225979
70. Mohler KM, Torrance DS, Smith CA, Goodwin RG, Stremler KE, Fung VP, Madani H, Widmer MB (1993) Soluble tumor necrosis factor (TNF) receptors are effective therapeutic agents in lethal endotoxemia and function simultaneously as both TNF carriers and TNF antagonists. J Immunol 151(3):1548–1561
71. Krueger GG, Callis KP (2003) Development and use of alefacept to treat psoriasis. J Am Acad Dermatol 49(2 Supplement):87–97. https://doi.org/10.1016/mjd.2003.552
72. Krueger GG, Ellis CN (2003) Alefacept therapy produces remission for patients with chronic plaque psoriasis. Br J Dermatol 148(4):784–788. https://doi.org/10.1046/j.1365-2133.2003.05239.x
73. Kimchi-Sarfaty C, Schiller T, Hamasaki-Katagiri N, Khan MA, Yanover C, Sauna ZE (2013) Building better drugs: developing and regulating engineered therapeutic proteins. Trends Pharmacol Sci 34(10):534–548. https://doi.org/10.1016/j.tips.2013.08.005
74. Majeau GR, Meier W, Jimmo B, Kioussis D, Hochman PS (1994) Mechanism of lymphocyte function-associated molecule 3-Ig fusion proteins inhibition of T cell responses. Structure/function analysis in vitro and in human CD2 transgenic mice. J Immunol 152(6):2753–2767
75. European Medicine Agency (2014) EMA. http://www.ema.europa.eu/ema/. Accessed 5 Apr 2014
76. Food and Drug Administration (2014) FDA. http://www.fda.gov/. Accessed 5 Apr 2014

77. Lanthier M, Behrman R, Nardinelli C (2008) Economic issues with follow-on protein products. Nat Rev Drug Discov 7(9):733–737
78. Grabowski H, Cockburn I, Long G (2006) The market for follow-on biologics: how will it evolve? Health Aff 25(5):1291–1301. https://doi.org/10.1377/hlthaff.25.5.1291
79. Woodcock J, Griffin J, Behrman R, Cherney B, Crescenzi T, Fraser B, Hixon D, Joneckis C, Kozlowski S, Rosenberg A, Schrager L, Shacter E, Temple R, Webber K, Winkle H (2007) The FDA's assessment of follow-on protein products: a historical perspective. Nat Rev Drug Discov 6(6):437–442
80. Schneider CK, Kalinke U (2008) Toward biosimilar monoclonal antibodies. Nat Biotechnol 26(9):985–990
81. Schneider CK, Vleminckx C, Gravanis I, Ehmann F, Trouvin J-H, Weise M, Thirstrup S (2012) Setting the stage for biosimilar monoclonal antibodies. Nat Biotechnol 30(12):1179–1185. https://doi.org/10.1038/nbt.2447
82. Ecker DM, Ransohoff TC (2014) Mammalian Cell Culture Capacity for Biopharmaceutical Manufacturing. In: Zhou W, Kantardjieff A (eds) Mammalian cell cultures for biologics manufacturing, Advances in biochemical engineering/biotechnology, vol 139. Springer, Berlin/Heidelberg, pp 185–225. https://doi.org/10.1007/10_2013_215

Part II
Cell Engineering, Novel Technologies and the Use of Omics

Remote Control of Mammalian Therapeutic Designer Cells

Maysam Mansouri and Martin Fussenegger

Abbreviations

CAR	chimeric antigen receptor
ChR2	channelrhodopsin-2
CRISPR	clustered regularly interspaced short palindromic repeats
ER	endoplasmic reticulum
FUS	focused ultrasound
Gal4	transcription factor for galactose gene
HEK	human embryonic kidney
hGLP1	human glucagon-like peptide-1
HSP	heat-shock promoters
IL-1Ra	interleukin-1 receptor antagonist
mActRIIB^{ECD}-hFc	activin type IIB, receptor ligand trap protein
MRI	magnetic resonance imaging
NFAT	nuclear factor of activated T-cells
NIR	near infrared
PAC	photoactivated adenylyl cyclase
PhyB-PIF	phytochrome B-phytochrome-interacting factor
RBS	ribosome binding site
RF	radio frequency
scFV	single-chain variable fragment
synNotch	synthetic Notch

M. Mansouri
Department of Biosystems Science and Engineering, ETH Zurich, Basel, Switzerland

M. Fussenegger (✉)
Department of Biosystems Science and Engineering, ETH Zurich, Basel, Switzerland

Faculty of Science, University of Basel, Basel, Switzerland
e-mail: fussenegger@bsse.ethz.ch

R. Pörtner (ed.), *Cell Culture Engineering and Technology*, Cell Engineering 10,
https://doi.org/10.1007/978-3-030-79871-0_2

TetO	Tet operator
TetR	Tet repressor
TRPV1	thermosensitive ion channels
UAS	upstream activator sequence
UCNP	upconverting nanoparticles
UV	ultraviolet
VP16	transcription factor
VP64	transcription factor
VPR	VP64-p65-Rta
μLEDs	micro LEDs

1 Introduction

Synthetic biology, the science of engineering complex biological systems to create useful entities, harnesses the capabilities of genetic circuits to program novel or improved biological functions in a predictable and controllable way [1, 2]. A synthetic cell system can be developed through either "top-down" or "bottom-up" strategies. Top-down approaches include manipulations of the genetic or protein content of the engineered cells, and in such cases the modified cell is still closely related to its biological ancestor [3]. Alternatively, bottom-up strategies aim to build artificial systems from scratch in a cell-free environment by reconstituting functional modules from natural or synthetic molecular building blocks [4].

Synthetic biology-inspired cell therapy strategies often rely on top-down strategies employing genetically engineered cells, called designer cells, to sense a user-defined input signal, process it, and respond appropriately with a customizable therapeutic output (Fig. 1) [5]. These programmable man-made therapeutic biomachines can be organized into three groups based on their origins. The first group consists of augmented naturally committed cells, e.g., immune cells equipped with synthetic genetic circuits enabling these cells to combat diseases that natural immune cells normally cannot handle [6]. The second group consists of designer stem cells carrying, for example, synthetic lineage-control networks that enable them to be differentiated into therapeutic cells in a robust and reliable way [7, 8]. The third group consists of artificial designer cells, including malleable cells upon which natural-cell characteristics can be conferred. This last group is exemplified by human embryonic kidney (HEK) cells, which were engineered to mimic β cell function by sensing high levels of blood glucose and producing recombinant insulin in response [9].

Control of cellular behavior is key for cell engineering to alter the overall activity or dosage of a therapeutic output, and careful regulation is essential [10]. Currently, two types of control systems based on closed-loop and open-loop strategies are available for obtaining precisely timed induction when required, while leaving the cells inactive at other times [11]. In a closed-loop system, a disease-upregulated biomarker triggers a genetic circuit in designer cells, leading to expression of a

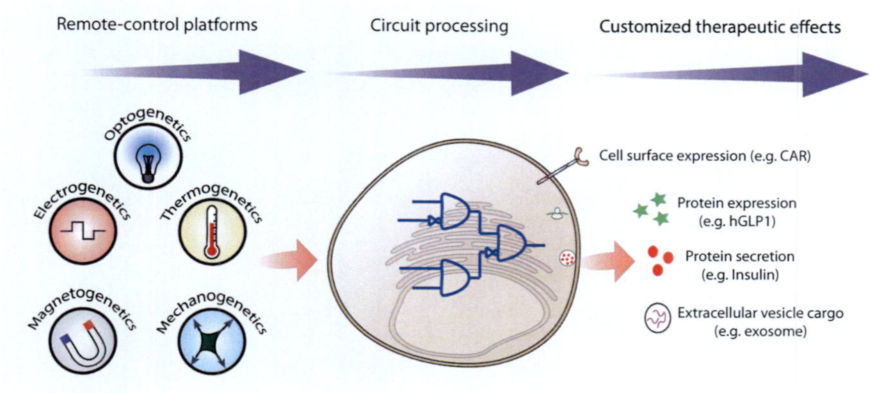

Fig. 1 Remote control of mammalian designer cells. Traceless trigger platforms can wirelessly induce engineered cells (left) and activate a synthetic genetic circuit (middle) to produce a customized therapeutic function (right)

therapeutic protein(s) that in turn causes the level of the biomarker to drop, down-regulating the system. In contrast, open-loop systems do not have negative feedback implemented, and they often rely on a user-defined exotic inducer, which activates designer cells to produce therapeutic agents. Although closed-loop systems provide a self-sufficient strategy for disease treatment, open-loop systems offer more flexible on-demand control over therapeutic effects.

Programming designer cells with open loop systems through external signals mainly relies on chemical and physical cues [12]. Chemical cues include organic and inorganic compounds, stimulatory peptides and odorants. Despite high induction potential and ease of use, potential clinical applications of chemical inducers are often limited by side effects, bioavailability or pharmacodynamics [2]. Chemical inducers can also diffuse freely and may cause side effects due to off-target activity elsewhere in the body. Additionally, it can be hard to remove the inducer quickly and reliably if required [13, 14]. Conversely, traceless physical inducers can provide a robust, efficient and precise way to wirelessly control cellular behaviors at a desired time and place. Approaches based on physical inducers also reduce the likelihood of cross-reactivity and off-target effects, and additionally avoid the need for invasive access to the tissue or targeted organ.

In this review, we first introduce available remote-control techniques, including optogenetics, magnetogenetics, mechanogenetics, thermogenetics and electrogenetics, for programming therapeutic designer cells (Fig. 2). The main advantages of each approach, as well as its bottlenecks, are discussed. We also describe some genetic circuits programmed by traceless physical inducers e.g., light, magnetic field, mechanical force, heat and electrical pulses. Finally, we summarize recent applications and advances in the wireless control of designer cells to express therapeutic proteins.

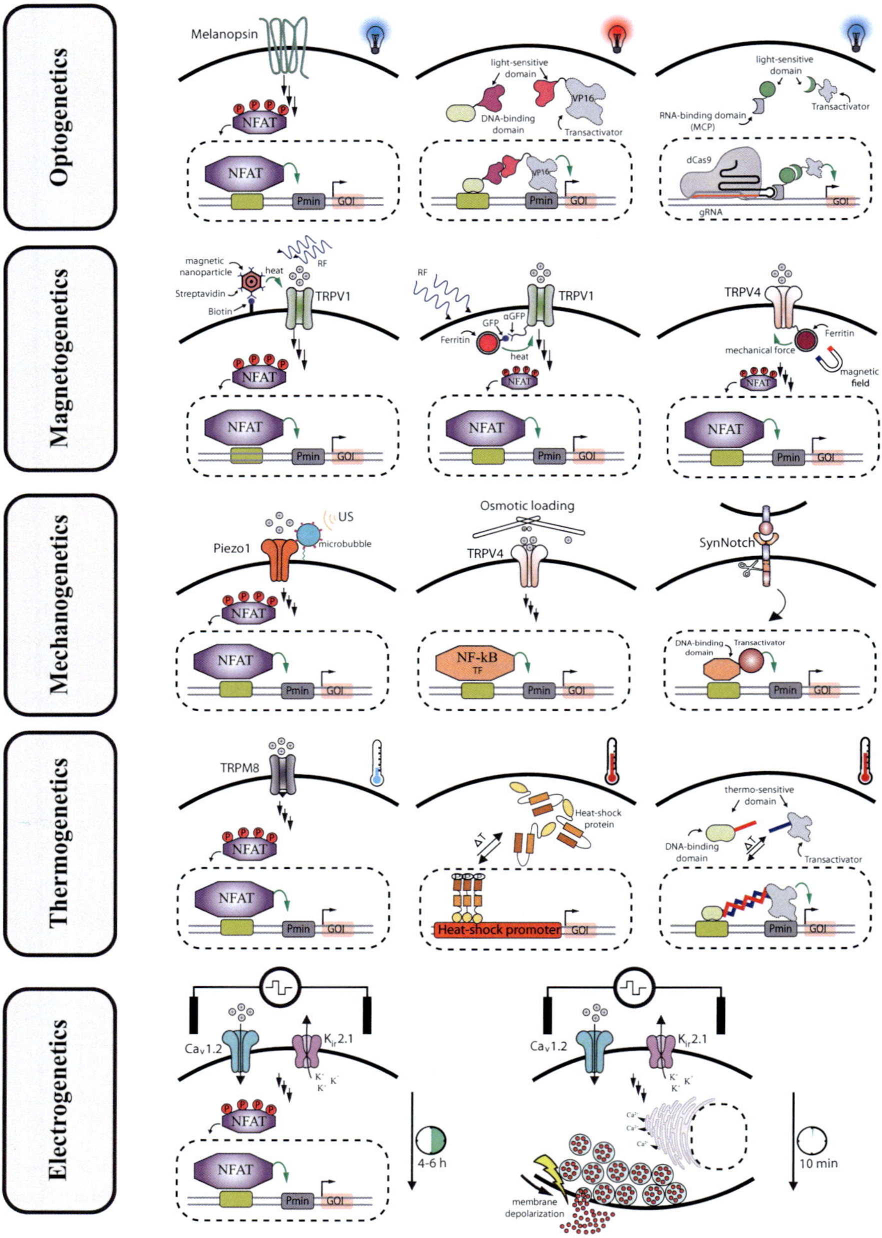

Fig. 2 Remote-controlled genetic circuits in engineered designer cells for production of therapeutic agents

Optogenetics; light can serve as a traceless trigger to activate endogenous signaling pathways (left), orthogonal synthetic circuits (middle), or targeted dCas9 to express the gene of interest

2 Optogenetics

Engineered light-controllable mammalian cells were first established in neurobiology in order to manipulate the membrane potential of neurons [15]. However, this combination of light and genetic sciences, called optogenetics, was soon taken up by synthetic biologists and applied to other cell types [16], because light has unprecedented features that makes it an excellent inducer for remote-control of therapeutic gene expression in designer cells [17]; First, it provides very high resolution, enabling spatiotemporally precise activation of the target cells. Second, we can precisely regulate the gene expression level by adjusting the light intensity or exposure time (tunability). Third, induction of gene expression by light is conveniently reversible (reversibility).

Nowadays, many different optogenetic systems are available for on-demand therapeutic gene expression in mammalian cells [18]. These systems contain light-responsive receptors or photo-switchable proteins and work in different wavelength regions, ranging from ultraviolet (UV) to near-infrared (NIR). This provides great flexibility to build a system according to the user's needs. In general, optogenetic systems that are active in the red and NIR regions have higher tissue penetration (important for controlling cells in deeper tissues) and lower cytotoxicity compared to systems that are activated by UV or blue light [19]. In addition, different light-responsive proteins with minimum cross-reactivity can be multiplexed in the same experiment to increase the level of complexity (e.g., to control the expression of multiple genes in parallel) or improve the level of safety (e.g., by incorporating logic gates) [20].

Here, we divide available optogenetic systems into two groups based on their use for therapeutic cell engineering: systems that trigger endogenous signaling pathways and orthogonal optogenetic systems.

Fig. 2 (continued)
Magnetogenetics; magnetic nanoparticles convert magnetic field energy to heat (left and middle) or mechanical force (right) and trigger endogenous signaling pathways by initiating calcium influx into the cytoplasm
Mechanogenetics; ultrasound (left), compression (middle) or stretching (left) can harness either endogenous calcium signaling (left and middle) or orthogonal transcription factors to produce on-demand therapeutic agents
Thermogenetics; hypothermal (left) or hyperthermal (middle and right) stimulation can serve as inducers to program designer cells
Electrogenetics; voltage-gated circuits can induce engineered cells to express desired transgenes within 4–6 h upon electrostimulation (left). In a fast-release system, changes in voltage potential trigger the rapid secretion of pre-formed granular vesicles containing therapeutic protein (right) with a very fast response time (within 10 min) after electrical stimulation

2.1 *Light-Responsive Genetic Circuits Triggering Endogenous Signaling Pathways*

Examples of this group include the use of a natural [21] or chimeric [22] light-inducible receptor to activate endogenous signaling pathways upon illumination. Activation of the pathway leads to a cascade of biological events, including increases in the levels of intracellular second messengers, activation of transcription factors, or changes in cellular metabolism. Optogenetics can be used in three ways to control cellular behaviour. First, light-activated endogenous transcription factors can directly initiate transcription of the desired transgene from a synthetic promoter. For example, it was shown that activation of a blue light-sensitive opsin called melanopsin in HEK cells can be rewired to a synthetic expression unit to produce the clinically licensed human glucagon-like peptide-1 (hGLP1) [21]. Activated melanopsin receptors ultimately increase intracellular calcium levels through calcium mobilization from the endoplasmic reticulum (ER), followed by calcium influx through calcium ion channels. Raising cytoplasmic calcium levels in turn induces dephosphorylation of a transcription factor called NFAT (nuclear factor of activated T-cells). Dephosphorylated NFAT translocates into the nucleus and initiates hGLP1 transcription from a synthetic reporter containing NFAT-responsive elements. Cell implants harboring this genetic circuit improved blood glucose levels and associated symptoms in mice with experimental type-2 diabetes. This melanopsin-controlled gene expression system was also used for the traceless production of recombinant products in bioreactors. This method is of particular interest to the pharmaceutical industry, because it can help to reduce the need for sophisticated and expensive downstream processing to remove impurities derived from chemical inducers. Second, light can be used to control a master transcription factor (master regulator) regulating the activation or suppression of multiple other downstream cell modulators. Here, regulation of gene expression takes place at genomic loci instead of in synthetic expression units as introduced earlier. To give an example, early activation of the endogenous RAF/MEK/ERK pathway by a blue-light mediated optogenetic system, called optoRAF1, can induce the expression of glial markers in neural progenitor cells and promote astrocytogenesis [23]. Third, light can also activate endogenous pathways to trigger a non-transcriptional event in therapeutic cells. For example, β cells are specialized to produce insulin and store it in granular vesicles within the cell. These β cells release insulin in response to a high level of blood glucose through fusion of pre-formed insulin-containing granules with the membrane, releasing insulin into the blood stream [24]. Optogenetically engineered β cells have been developed to provide on-demand insulin secretion in response to light [25]. Such systems are often based on regulation of a second messenger (e.g., calcium or cAMP). For example, expression of a light-sensitive adenylyl cyclase (bacterial PAC), that converts ATP to cAMP upon blue light illumination, is able to boost the level of cAMP within β cell lines, resulting in insulin secretion 2 h after induction. This optogenetically based rapid-release system was shown to ameliorate type-1 diabetes in an experimental mouse model upon exposure to blue light [26]. Similarly, ectopic expression of bacterial channelrhodopsin-2 (ChR2) in pancreatic β cells was used to increase intracellular calcium concentration and ultimately release accumulated insulin granules in response to blue light [27].

2.1.1 Orthogonal Light-Inducible Gene Expression Systems

Optogenetic systems of this type are designed to minimize cross-talk between light-activated components and endogenous signaling pathways. Orthogonal systems are especially beneficial in cases when either the integrity of endogenous signaling is crucial for functionality (e.g., in stem cell biology) or when by-products of cell activation should be avoided [28]. Orthogonal optogenetic systems in mammalian designer cells often rely on ectopically expressed light-sensitive domains from other kingdoms of life that either dimerize or monomerize in response to light [29]. Already a variety of orthogonal optogenetic systems based on different designs and topologies have been developed [18]. One common approach is based on a topology blueprint of the pioneering-tetracycline-repressible gene regulation system [30]. Here, a light-sensitive domain is fused to a DNA-binding domain (e.g., TetR or Gal4), while the light-dimerizing counterpart is fused to a transcriptional activator (e.g., VPR, VP16, VP64). Co-expression of these chimeric proteins along with synthetic promoter unit containing DNA-binding responsive elements (e.g., TetO or UAS) leads to transcription of a gene of interest upon light exposure in ON-type systems [31].

Alternatively, opto-CRISPR systems have been designed to induce expression of a gene of interest that can be either located on a synthetic expression unit or in the host genome. These systems are mainly based on light-inducible reconstitution of two-split dCas9 moieties [32] or binding of a transactivator (e.g., p65) to dCas9 through light-sensitive domains [33].

Although light has been successfully used for traceless spatiotemporal remote-control of cellular behavior and precise expression dosing of therapeutic transgenes in a variety of in vitro and in vivo studies [15, 21, 34–37], there are still some limitations that need to be considered before engineered cells can be applied for biomedical therapies. For example, blue-light-based optogenetic applications often suffer from high cytotoxicity [38, 39], low tissue penetration [40], and unwanted upregulation of endogenous genes [41]. In addition, some optogenetic systems need complex chromophores for activation (e.g., PhyB-PIF system) [42]. However, the optogenetics community has developed some solutions to these issues. To enable activation of blue-light-dependent systems even in deep tissues, upconverting nanoparticles (UCNPs) can be used [43]. These nanoparticles collect energy from more than one photon and emit a single photon with a shorter wavelength. A straightforward approach to circumvent the tissue barrier is to directly equip the implant with its own light source that can be powered wirelessly via electromagnetic induction. With the advent of very small and efficient micro LEDs (μLEDs), this strategy is becoming increasingly feasible [44]. Also, NIR-activated optogenetic systems with better tissue penetration and compatible chromophores [20] can be used for clinical applications. All in all, optogenetic systems now provide a precise and reliable remote-control platform for therapeutic designer cells.

3 Magnetogenetics

Magnetogenetics is a physical technique that uses magnetic stimuli to alter cellular behavior. Recent examples include the use of magnetic fields to activate magnetic nanoparticles targeted to receptors in the plasma membrane of engineered cells [45]. Subsequently, the activated receptors initiate a cascade of biological events, which can be harnessed for therapeutic gene expression or other cellular outputs. Magnetogenetics provides a wireless and facile approach to remotely program engineered cells based on the intensity and polarity of externally applied magnetic fields [46]. In contrast to optogenetics systems, which are often restricted by the limited tissue penetration of light, magnetic fields can freely pass through tissues regardless of their composition and size.

The magnetic particles used in magnetogenetics are often either chemical nanoparticles or genetically encoded nanoparticles. These particles convert energy received from magnetic fields to a variety of physical stimuli, such as thermal relaxation, resulting in heating up of the surroundings, or strong mechanical forces [47]. However the plausibility of the proposed mechanisms especially those related to genetically encoded magnetic nanoparticles was called into question on the basis of the underlying physics [48]. Nevertheless, researchers have reported successful activation of endogenous or ectopically expressed thermo-/mechano-sensitive receptors by a magnetic field to initiate downstream signaling pathways.

Here, we will illustrate three different strategies that have been used to program engineered mammalian cells by means of magnetogenetics. First, HEK cells were engineered to co-express thermosensitive ion channels (TRPV1) as well as a biotin-conjugated receptor on the cell surface [49]. Targeting these cells with streptavidin-labeled magnetic nanoparticles ($MnFe_2O_4$) led to localized heating when the cells were subjected to a radio-frequency (RF) magnetic signal. This local heating opens TRPV1 channels, resulting in calcium influx into the cell cytoplasm. Second, Stanley and colleagues have developed a similar magneto-thermo strategy using activated magnetic nanoparticles to stimulate TRPV1 channels and trigger calcium influx [50]. However, in their work a genetically encoded ferritin nanoparticle was coupled to the thermosensitive TRPV1 channel. They also demonstrated that the rise in local temperature produced by magnetic fields at radio-wave frequency (465 kHz, 23–32 mT) can be rewired to the synthesis and release of insulin from a synthetic calcium-responsive expression unit in engineered HEK cells. Third, another magnetogenetic strategy was based on a single chimeric construct (called Magneto2.0) consisting of the mechanosensitive TRVP4 channel genetically fused to a ferritin protein complex [51]. Application of a strong static magnetic field (~50 mT) exerted a mechanical force on the ferritin particles, which subsequently activated TRVP4 channels, resulting in calcium influx into the cells.

Even though magnetogenetics can provide a wireless induction platform to program cellular behavior, major issues remain. First, the underlying mechanisms of some of the strategies are not fully understood, limiting the feasibility of an evidence-based design approach. Additionally, most of the experiments need a sophisticated infrastructure and expensive magnetic field generators. This technique also seems to be less efficient than other well-established systems such as optogenetics.

4 Mechanogenetics

Mechanogenetics is based on sensing mechanical perturbations in the plasma membrane of engineered cells and using them to induce customized responses [52]. Mechanically engineered cells can be built by the introduction of mechanosensitive ion channels or receptors. Mechanogenetics provides a safe, non-invasive and remotely controllable approach with high spatiotemporal resolution, and circumvents existing issues of optogenetics and magnetogenetics, such as limited tissue penetration and low spatial resolution. Mechanical cues are included (but not limited to) acoustically mediated shear stress, compression and stretching [53].

Acoustic mechanogenetics employs ultrasound waves to control designer cells [54]. For example, mammalian cells were engineered to ectopically express the mechanosensitive Piezo1 ion channel, which can trigger calcium influx and related downstream transcription factors (e.g., NFAT) upon low-frequency ultrasound stimulation [55]. This system was implemented to remotely and precisely induce expression of anti-CD19 chimeric antigen receptors (CARs) on T cells in tumor sites. Here, microbubbles, that had to be present in the vicinity of the T cells, were used to amplify the ultrasound signal and activate Piezo1 channels. More recently, an acoustogenetic approach based on focused ultrasound (FUS) as a stimulus for activation of engineered cells has been developed. FUS can deliver energy in a safe and non-invasive way into tissues to a depth of centimeters. Wu and colleagues created a FUS-controllable CAR T cell therapy technology that enables T cells to express CARs from a temperature-sensitive heat-shock promoter and successfully eradicated targeted tumors [56]. Here, application of MRI-guided FUS raises the temperature at the tumor site, leading to activation of designer T cells by heat-induced transcription of the CAR gene at a precisely controlled location and time.

Alternatively, Nims and colleagues have engineered cells to drive expression of therapeutic anti-inflammatory cytokine interleukin-1 receptor antagonist (IL-1Ra) in response to mechanical compression [57]. To do this, they created a synthetic genetic circuit composed of a mechanosensitive TRPV4 channel on chondrocytes, rewired to a synthetic promoter driving expression of the IL-1Ra gene. TRVP4 is an osmotically sensitive cation channel, which allows calcium to leak into the cells upon activation and ultimately leads to NF-kB signaling. Indeed, activated transcription factors downstream of TRPV4 were able to initiate expression of therapeutic IL-1Ra.

Another example of signal transduction via mechanosensation between engineered therapeutic cells is provided by synNotch receptors on the surface of immune T cells. This system was designed to enable programmable cell-to-cell communication at the site of contact. synNotch follows a lift-and-cut (stretch) model, as described for the wild-type Notch receptor [58]. In the native context, two neighboring cells can communicate through physical contact between the Notch receptor and a ligand, expressed on opposing cell surfaces. When the ligand engages the receptor, the resulting mechanical force is thought to expose a hidden protease site, initiating a multi-step process leading to release of the intracellular transcriptional domain, which translocate to the nucleus and initiates transcription of relevant genes. In the synthetic Notch (synNotch) setup, the ligand and ligand-binding domain, as well as the transcription factor, are customized for mechanically induced gene expression [59]. For example, T cells equipped with synNotch carrying an scFV against HER2 (α-HER2) were targeted to HER2-expressing cancer cells. In response to binding of α-HER2 to HER2, engineered T cells would either express a CAR receptor to recognize other antigens on cancer cells [60] or secrete immune mediators to stimulate the immune system at the site of the tumor [61].

Overall, mechanogenetics can provide reliable remote-control systems if programmed with precise devices, such as ultrasound. However, like magnetogenetics, this strategy often relies on sophisticated infrastructure, which can restrict its applicability.

5 Thermogenetics

In thermogenetics, temperature change serves as input signal to activate genetic circuits in designer cells. Temperature change can be induced in a non-invasive way by laser stimulation, even in deep tissues [62]. Since we have already described magneto/mechano-thermogenetics, we will focus in this section on genetic systems that are directly controlled by low temperature (cooling, hypothermia) or high temperature (heating, hyperthermia), regardless of the method of induction.

In a blueprint for a hypothermal gene switch, therapeutic insulin as well as activin type IIB, a receptor ligand trap protein (mActRIIB^{ECD}-hFc), were expressed in designer HEK cells in response to a cool environment (15–18 °C) or a cooling compound (e.g., menthol) [63]. Here, cells were engineered to express the transient receptor potential melastatin 8 (TRPM8) ion channel to sense the temperature threshold. At low temperatures, the hTRPM8-activated signaling pathway is rerouted to a synthetic promoter containing NFAT-binding elements. Implanted designer cells containing this hypothermal gene switch could alleviate type-1 diabetes and muscle atrophy in different experimental mouse models upon stimulation with cooling compounds.

Most hyperthermal systems are activated by means of mild heating (~39–42 °C). These systems often rely on temperature-gated channel/receptors that control endogenous pathways to alter the expression levels of selected genes. Temperature-gated channels are mainly from the TRPV family, and were described above in the magneto-/mechano-genetics sections. Three strategies are available for transcriptional/translational-based hyperthermal techniques to induce gene expression in designer cells. The first employs synthetic heat-shock promoters (HSP). Raising the temperature normally induces cellular stress, resulting in activation of related transcription factors, such as heat-shock proteins. Those activated heat-shock proteins can be harnessed to initiate transcription of transgenes from a synthetic heat-shock promoter. For example, Miller and colleagues constructed a thermal promoter from the heat shock protein 70B′ (called HSPA6 promoter). This promoter was activated in T cells by photothermal induction, reaching fold changes of up to 200 [64]. The second strategy employs thermo-sensitive domains that dimerize/monomerize upon heating. A chimeric thermosensitive-domain fused to orthogonal transcription factors can up-regulate or suppress gene of interest [65]. The third strategy is based on engineered biological structures that undergo conformational changes on heating. An example of such a thermo-responsive structure is provided by an RNA thermometer which consists of a secondary RNA structure that protects the ribosome binding site (RBS) of the controlled gene until heating causes the structure to unfold, thereby exposing the RBS. Subsequently, the ribosome can initiate translation of mRNA and produce the desired protein [66].

Although thermogenetics can modulate cellular functions non-invasively, even in deep tissues, it does induce cellular stress within engineered cells, and the consequences of this may need to be considered before clinical use.

6 Electrogenetics

Electrogenetics is a relatively new technique that uses electrical pulses to directly program the behavior of designer cells. We have recently developed electrogenetic methodology in mammalian cells to achieve expression and secretion of desired proteins, including therapeutics, in response to electrical pulses generated by a wearable electronic device. To this end, electro designer cells were engineered to ectopically express a voltage-gated circuit including a voltage-gated calcium channel ($Ca_V1.2$) coupled to an inwardly rectifying potassium channel ($K_{ir}2.1$). Upon application of a pulsed electrical field, $Ca_V1.2$ channels mediate an influx of calcium into the cell. The change in intracellular calcium concentration was first rewired to express a transgene from a synthetic expression unit containing binding sites for activated NFAT transcription factor. This transcription-based electrogenetics system induces transgene expression in a time frame of ~4–6 h after electrostimulation. To develop electrically responsive designer cells with faster kinetics, a similar voltage-gated genetic circuit was introduced into a pancreatic β cell line, termed ${}_{\text{Electro}}\beta$ cells. In ${}_{\text{Electro}}\beta$ cells, insulin is produced and accumulates in vesicular granules. These granules release insulin to their surroundings only upon electrostimulation, irrespective of the glucose concentration in the vicinity.

Engineered $_{\text{Electro}}\beta$ cells producing various therapeutic proteins in response to electrical signals could be valuable tools for treating diseases that require therapeutic proteins for short-term homeostasis (e.g., type-1 diabetes). Indeed, $_{\text{Electro}}\beta$ cells encased in a wearable electronic device that was subcutaneously implanted in an alloxan-induced type-1 diabetic mouse model rapidly reversed hyperglycemia when subjected to electrostimulation.

Overall, electrogenetics enables reliable wireless control of engineered cells by means of electrostimulation. Nevertheless, electrostimulation devices require sophisticated design techniques, and this may hamper their progress towards clinical applications.

7 Concluding Remarks

Engineering of therapeutic cells to produce therapeutic agents on demand is already moving beyond proof-of-concept. Engineered therapeutic cells can be customized to sense signals, process them, and respond in a controllable, programmable and predictable way. A toolbox of traceless inducers, including light, magnetic and mechanical forces, heat and electrical stimulation, is available for non-invasive, safe, and reliable physical stimulation to remotely control engineered cells. Despite certain limitations of these physical inducers, all of them and their related genetic platforms have been successfully employed in clinical research. We think future work will be focused on pushing the boundaries of these platforms to resolve current limitations, and also on developing wireless control systems with higher level of safety, precision and efficiency. Carefully chosen combinations of these platforms may have the potential to create multi-tasking cells in which each function is separately programmable by a different wireless control system.

Acknowledgement Work in the laboratory of M.F. is financially supported in part through a European Research Council advanced grant (ElectroGene, no. 785800) and in part by the National Centre of Competence in Research (NCCR) for Molecular Systems Engineering. The authors would like to thank Tobias Strittmatter for helpful discussions and critical comments on the paper. The authors apologize to all colleagues whose important papers were not cited due to space limitations.

Competing Interest Declaration The authors declare no competing interests.

References

1. Xie M, Fussenegger M (2015) Mammalian designer cells: engineering principles and biomedical applications. Biotechnol J 10:10051018. https://doi.org/10.1002/biot.201400642
2. Xie M, Fussenegger M (2018) Designing cell function: assembly of synthetic gene circuits for cell biology applications. Nat Rev Mol Cell Biol 19:507–525

3. Göpfrich K, Platzman I, Spatz JP (2018) Mastering complexity: towards bottom-up construction of multifunctional eukaryotic synthetic cells. Trends Biotechnol 36:938–951
4. Ausländer S, Ausländer D, Fussenegger M (2017) Synthetic biology-the synthesis of biology. Angew Chem Int Ed 56:6396–6419. https://doi.org/10.1002/anie.201609229
5. Scheller L, Fussenegger M (2019) From synthetic biology to human therapy: engineered mammalian cells. Elsevier Ltd, London
6. Roybal KT, Lim WA (2017) Synthetic immunology: hacking immune cells to expand their therapeutic capabilities. Annu Rev Immunol 35:229–253
7. Saxena P, Bojar D, Zulewski H, Fussenegger M (2017) Generation of glucose-sensitive insulin-secreting beta-like cells from human embryonic stem cells by incorporating a synthetic lineage-control network. J Biotechnol 259:39–45. https://doi.org/10.1016/j.jbiotec.2017.07.018
8. Saxena P, Heng BC, Bai P et al (2016) A programmable synthetic lineage-control network that differentiates human IPSCs into glucose-sensitive insulin-secreting beta-like cells. Nat Commun 7:1–14. https://doi.org/10.1038/ncomms11247
9. Xie M, Ye H, Wang H et al (2016) β-Cell-mimetic designer cells provide closed-loop glycemic control. Science 354:1296–1301. https://doi.org/10.1126/science.aaf4006
10. Lim WA, June CH (2017) The Principles of Engineering Immune Cells to Treat Cancer. Cell 168:724–740. https://doi.org/10.1016/J.CELL.2017.01.016
11. Wang Y, Wang M, Dong K, Ye H (2018) Engineering mammalian designer cells for the treatment of metabolic diseases. Biotechnol J 13:1700160. https://doi.org/10.1002/biot.201700160
12. Xie M, Haellman V, Fussenegger M (2016) Synthetic biology – application-oriented cell engineering. Curr Opin Biotechnol 40:139–148
13. Müller K, Naumann S, Weber W, Zurbriggen MD (2015) Optogenetics for gene expression in mammalian cells. Biol Chem 396:145–152. https://doi.org/10.1515/HSZ-2014-0199
14. Beyer HM, Naumann S, Weber W, Radziwill G (2015) Optogenetic control of signaling in mammalian cells. Biotechnol J 10:273–283
15. Boyden ES, Zhang F, Bamberg E et al (2005) Millisecond-timescale, genetically targeted optical control of neural activity. Nat Neurosci 8:1263–1268. https://doi.org/10.1038/nn1525
16. Krueger D, Izquierdo E, Viswanathan R et al (2019) Principles and applications of optogenetics in developmental biology. Development 146:dev175067. https://doi.org/10.1242/dev.175067
17. Repina NA, Rosenbloom A, Mukherjee A et al (2017) At light speed: advances in optogenetic systems for regulating cell signaling and behavior. Annu Rev Chem Biomol Eng 8:13–39
18. Kolar K, Knobloch C, Stork H et al (2018) OptoBase: a web platform for molecular optogenetics. ACS Synth Biol 7:1825–1828. https://doi.org/10.1021/acssynbio.8b00120
19. Mansouri M, Strittmatter T, Fussenegger M (2019) Light-controlled mammalian cells and their therapeutic applications in synthetic biology. Adv Sci 6:1800952
20. Redchuk TA, Omelina ES, Chernov KG, Verkhusha VV (2017) Near-infrared optogenetic pair for protein regulation and spectral multiplexing. Nat Chem Biol 13:633–639. https://doi.org/10.1038/nchembio.2343
21. Ye H, Baba MDE, Peng RW, Fussenegger M (2011) A synthetic optogenetic transcription device enhances blood-glucose homeostasis in mice. Science (80-) 332:1565–1568. https://doi.org/10.1126/science.1203535
22. Kainrath S, Stadler M, Reichhart E et al (2017) Green-light-induced inactivation of receptor signaling using cobalamin-binding domains. Angew Chem Int Ed 56:4608–4611. https://doi.org/10.1002/anie.201611998
23. Su Y, Huang X, Huang Z et al (2020) Early but not delayed optogenetic RAF activation promotes astrocytogenesis in mouse neural progenitors. J Mol Biol 432:4358–4368. https://doi.org/10.1016/j.jmb.2020.06.020
24. Straub SG, Sharp GWG (2002) Glucose-stimulated signaling pathways in biphasic insulin secretion. Diabetes Metab Res Rev 18:451–463
25. Frank JA, Broichhagen J, Yushchenko DA et al (2018) Optical tools for understanding the complexity of β-cell signalling and insulin release. Nat Rev Endocrinol 14:721–737
26. Zhang F, Tzanakakis ES (2019) Amelioration of diabetes in a murine model upon transplantation of pancreatic β-cells with optogenetic control of cyclic adenosine monophosphate. ACS Synth Biol 8:2248–2255. https://doi.org/10.1021/acssynbio.9b00262

27. Reinbothe TM, Safi F, Axelsson AS et al (2014) Optogenetic control of insulin secretion in intact pancreatic islets with β-cell-specific expression of Channelrhodopsin-2. Islets 6:e28095. https://doi.org/10.4161/isl.28095
28. Mukherjee A, Repina NA, Schaffer DV, Kane RS (2017) Optogenetic tools for cell biological applications. J Thorac Dis 9:4867–4870
29. Yamada M, Nagasaki SC, Ozawa T, Imayoshi I (2020) Light-mediated control of gene expression in mammalian cells. Neurosci Res 152:66–77
30. Gossen M, Freundlieb S, Bender G et al (1995) Transcriptional activation by tetracyclines in mammalian cells. Science (80-) 268:1766–1769. https://doi.org/10.1126/science.7792603
31. Redchuk TA, Kaberniuk AA, Verkhusha VV (2018) Near-infrared light–controlled systems for gene transcription regulation, protein targeting and spectral multiplexing. Nat Protoc 13:1121–1136. https://doi.org/10.1038/nprot.2018.022
32. Nihongaki Y, Kawano F, Nakajima T, Sato M (2015) Photoactivatable CRISPR-Cas9 for optogenetic genome editing. Nat Biotechnol 33:755–760. https://doi.org/10.1038/nbt.3245
33. Nihongaki Y, Yamamoto S, Kawano F et al (2015) CRISPR-Cas9-based photoactivatable transcription system. Chem Biol 22:169–174. https://doi.org/10.1016/J.CHEMBIOL.2014.12.011
34. Chow BY, Boyden ES (2013) Optogenetics and translational medicine. Sci Transl Med 5:177ps5-177ps5. https://doi.org/10.1126/scitranslmed.3003101
35. Huang Z, Wu Y, Allen ME et al (2020) Engineering light-controllable CAR T cells for cancer immunotherapy. Sci Adv 6:eaay9209. https://doi.org/10.1126/sciadv.aay9209
36. Kim T, Folcher M, Baba MD-E, Fussenegger M (2015) A synthetic erectile optogenetic stimulator enabling blue-light-inducible penile erection. Angew Chem Int Ed 54:5933–5938. https://doi.org/10.1002/anie.201412204
37. Shao J, Xue S, Yu G et al (2017) Smartphone-controlled optogenetically engineered cells enable semiautomatic glucose homeostasis in diabetic mice. Sci Transl Med 9:eaal2298. https://doi.org/10.1126/scitranslmed.aal2298
38. Waldchen S, Lehmann J, Klein T et al (2015) Light-induced cell damage in live-cell super-resolution microscopy. Sci Rep 5:15348. https://doi.org/10.1038/srep15348
39. Stockley JH, Evans K, Matthey M et al (2017) Surpassing light-induced cell damage in vitro with novel cell culture media. Sci Rep 7:1–11. https://doi.org/10.1038/s41598-017-00829-x
40. Ruggiero E, Alonso-De Castro S, Habtemariam A, Salassa L (2016) Upconverting nanoparticles for the near infrared photoactivation of transition metal complexes: new opportunities and challenges in medicinal inorganic photochemistry. Dalton Trans 45:13012–13020. https://doi.org/10.1039/c6dt01428c
41. Tyssowski KM, Gray JM (2019) Blue light increases neuronal activity-regulated gene expression in the absence of optogenetic proteins. eNeuro 6:ENEURO.0085-19.2019. https://doi.org/10.1523/ENEURO.0085-19.2019
42. Müller K, Engesser R, Metzger S et al (2013) A red/far-red light-responsive bi-stable toggle switch to control gene expression in mammalian cells. Nucleic Acids Res 41:e77. https://doi.org/10.1093/nar/gkt002
43. Wen S, Zhou J, Zheng K et al (2018) Advances in highly doped upconversion nanoparticles. Nat Commun 9:1–12. https://doi.org/10.1038/s41467-018-04813-5
44. Scharf R, Tsunematsu T, McAlinden N et al (2016) Depth-specific optogenetic control in vivo with a scalable, high-density μLED neural probe. Sci Rep 6:28381. https://doi.org/10.1038/srep28381
45. Long X, Ye J, Zhao D, Zhang SJ (2015) Magnetogenetics: remote non-invasive magnetic activation of neuronal activity with a magnetoreceptor. Sci Bull 60:2107–2119. https://doi.org/10.1007/s11434-015-0902-0
46. Christiansen MG, Hornslien W, Schuerle PS, Schuerle S (2020) A possible inductive mechanism for magnetogenetics. bioRxiv 2020.07.16.207126. https://doi.org/10.1101/2020.07.16.207126
47. Nimpf S, Keays DA (2017) Is magnetogenetics the new optogenetics? EMBO J 36:1643–1646. https://doi.org/10.15252/embj.201797177

48. Meister M (2016) Physical limits to magnetogenetics. elife 5:e17210. https://doi.org/10.7554/eLife.17210
49. Huang H, Delikanli S, Zeng H et al (2010) Remote control of ion channels and neurons through magnetic-field heating of nanoparticles. Nat Nanotechnol 5:602–606. https://doi.org/10.1038/nnano.2010.125
50. Stanley SA, Gagner JE, Damanpour S et al (2012) Radio-wave heating of iron oxide nanoparticles can regulate plasma glucose in mice. Science (80-) 336:604–608. https://doi.org/10.1126/science.1216753
51. Wheeler MA, Smith CJ, Ottolini M et al (2016) Genetically targeted magnetic control of the nervous system. Nat Neurosci 19:756–761. https://doi.org/10.1038/nn.4265
52. Zhu L, Wu Y, Yoon CW, Wang Y (2020) Mechanogenetics for cellular engineering and cancer immunotherapy. Curr Opin Biotechnol 66:88–94
53. Dufort CC, Paszek MJ, Weaver VM (2011) Balancing forces: architectural control of mechanotransduction. Nat Rev Mol Cell Biol 12:308–319
54. Pan Y, Yoon S, Zhu L, Wang Y (2018) Acoustic mechanogenetics. Curr Opin Biomed Eng 7:64–70
55. Pan Y, Yoon S, Sun J et al (2018) Mechanogenetics for the remote and noninvasive control of cancer immunotherapy. Proc Natl Acad Sci U S A 115:992–997. https://doi.org/10.1073/pnas.1714900115
56. Wu Y, Liu Y, Huang Z et al (2020) Acoustogenetic control of CAR T cells via focused ultrasound. bioRxiv 2020.02.18.955005. https://doi.org/10.1101/2020.02.18.955005
57. Nims RJ, Pferdehirt L, Ho NB et al (2020) A synthetic mechanogenetic gene circuit for autonomous drug delivery in engineered tissues. bioRxiv 2020.04.29.069294. https://doi.org/10.1101/2020.04.29.069294
58. Gordon WR, Vardar-Ulu D, Histen G et al (2007) Structural basis for autoinhibition of notch. Nat Struct Mol Biol 14:295–300. https://doi.org/10.1038/nsmb1227
59. Morsut L, Roybal KTKT, Xiong X et al (2016) Engineering customized cell sensing and response behaviors using synthetic notch receptors. Cell 164:780–791. https://doi.org/10.1016/j.cell.2016.01.012
60. Roybal KT, Rupp LJ, Morsut L et al (2016) Precision tumor recognition by T cells with combinatorial antigen-sensing circuits. Cell 164:770–779. https://doi.org/10.1016/j.cell.2016.01.011
61. Roybal KT, Williams JZ, Morsut L et al (2016) Engineering T cells with customized therapeutic response programs using synthetic notch receptors. Cell 167:419. https://doi.org/10.1016/j.cell.2016.09.011
62. Gamboa L, Zamat AH, Kwong GA (2020) Synthetic immunity by remote control. Theranostics 10:3652–3667
63. Bai P, Liu Y, Xue S et al (2019) A fully human transgene switch to regulate therapeutic protein production by cooling sensation. Nat Med 25:1266–1273
64. Miller IC, Gamboa Castro M, Maenza J et al (2018) Remote control of mammalian cells with heat-triggered gene switches and photothermal pulse trains. ACS Synth Biol 7:1167–1173. https://doi.org/10.1021/acssynbio.7b00455
65. Piraner DI, Wu Y, Shapiro MG (2019) Modular thermal control of protein dimerization. ACS Synth Biol 8:2256–2262. https://doi.org/10.1021/acssynbio.9b00275
66. Sen S, Apurva D, Satija R et al (2017) Design of a toolbox of RNA thermometers. ACS Synth Biol 6:1461–1470. https://doi.org/10.1021/acssynbio.6b00301

Next Generation Cell Engineering Using microRNAs

Florian Klingler, Nadja Raab, Nikolas Zeh, and Kerstin Otte

Abbreviations

Ago	Argonaute
NGS	Next generation sequencing
CHO	Chinese hamster ovary
VCD	viable cell density
miRNA	microRNAs
mAb	monoclonal antibody (mAb)
Cdkn1b	cyclin-dependent kinase inhibitor p27
SEAP	secreted alkaline phosphatase
HIPK1	Homeodomain-interacting protein kinase 1
ATF6β	activating transcription factor 6 beta
UPR	unfolded protein response
HDAC	histone deacetylase
VPA	valproic acid
CAP	Cevec's Amniocyte Production
mitosRNAs	mitochondrial RNAs
CerS2	Ceramide Synthase 2
Tbc1D20	Rab1 GAP Tbc domain family member 20
BCL2L11	Bcl-2-like protein 11

Author Contribution: Florian Klingler, Nadja Raab and Nikolas Zeh contributed equally with all other contributors.

F. Klingler · N. Raab · N. Zeh · K. Otte (✉)
Institute of Applied Biotechnology, Biberach University of Applied Sciences, Biberach, Germany
e-mail: klingler@hochschule-bc.de; raab@hochschule-bc.de; zeh@hochschule-bc.de; otte@hochschule-bc.de

R. Pörtner (ed.), *Cell Culture Engineering and Technology*, Cell Engineering 10,
https://doi.org/10.1007/978-3-030-79871-0_3

PTEN	Phosphatase and tensin homolog
FUT8	fucosyltransferase 8
ADCC	antibody-dependent cellular cytotoxicity
CMP-Neu5Ac	cytidine monophospho-N-acetylneuraminic acid
GSL	glycosphingolipid
UGCG	GSL UDP-glucose ceramide glucosyltransferase
amiRNA	artificial miRNA
MREs	miRNA response elements
UTR	untranslated region
GOI	gene of interest
TRE	tetracycline-responsive element
KD	Knock-Down
KO	Knock-Out
ZFNs	zinc finger nucleases
TALENs	Transcription activator-like effector nucleases
CRISPR/Cas9	Clustered Regularly Interspaced Short Palindromic Repeats/ CRISPR-associated proteins
shRNA	short hairpin RNA
Usp14	Ubiquitin carboxyl-terminal hydrolase 14

1 Introduction

1.1 CHO Cells Are the Preferred Expression Systems to Meet the Rising Demand for Biopharmaceuticals

The long lasting success of mammalian expression systems for the production of biopharmaceutical products is based on the fact that a majority of biopharmaceutical drugs require post-translational modifications which cannot be provided by bacterial expression systems. The most widely used mammalian cell-based systems are Chinese hamster ovary (CHO) cell lines [1] due to several advantages including their robust growth in chemically defined, serum-free media in suspension culture, resistance to human pathogenic viruses and expression of recombinant proteins with human-like post-translational modifications [2]. In addition several robust selection systems are available, to easily integrate and express recombinant genes in CHO cells. Commonly used industrial CHO cell lines include the CHO-K1, the CHO-DG44, and the CHO-DXB11 also referred to as CHO-DUKX cell line. In recent years, the productivity of mammalian-based processes has been drastically improved and the produced recombinant proteins can reach concentrations of 5–10 g/L [2, 3]. However, to facilitate these high product titers and to overcome other drawbacks like undesired growth behavior or product quality issues, major efforts in cell line engineering and process optimization are required. The main goals of cell line engineering are therefore the enhancement of growth characteristics and recombinant protein expression, the repression of cell death and the modulation of post-translational modifications.

1.2 MicroRNAs as Novel Option for Production Cell Line Engineering

Cell engineering has traditionally used the expression modulation of protein coding genes displaying positive or negative influence on bioprocess relevant parameters as viable cell density (VCD), growth, titer or specific productivity [4, 5]. However, a novel and powerful tool for cell line engineering has emerged in the recent past. By modulating the expression of small non coding RNAs, including microRNAs (miRNA), the cellular intrinsic mechanism of RNA induced silencing of gene expression is used for the purpose of cell engineering.

MicroRNAs are small non coding RNA molecules with a length of ~22 nt encoded in the cellular genome. After transcription, the pri-miRNAs are further processed and exported to the cytoplasm, where they are processed to form a miRNA duplex [6, 7]. A protein of the Argonaute (Ago) then binds to the miRNA duplex and incorporates the mature, single-strand miRNA into the RNA protein complex RISC [8, 9], which regulates the activity of the specific target mRNAs and thereby causes downregulation of target gene expression by inhibition of mRNA translation or mRNA cleavage.

MiRNAs are highly interesting targets for cell engineering in the bioprocess context due to several specific molecular characteristics. As miRNAs can act as regulators of gene expression in many cellular pathways like growth, cell survival and protein synthesis, they fulfill potentially interesting cellular functions which could be exploited for bioprocess development [10]. Furthermore, miRNAs are able to regulate multiple targets in the same or different pathways and can therefore be used as modulators for complex cellular phenotypes. Finally, miRNAs have no coding potential for proteins and therefore do not burden the translation machinery in a recombinant protein expressing cell line.

All in all cell line engineering using miRNAs has the potential to use their molecular influence on complex cellular mechanisms to create more beneficial cellular phenotypes for improvement of recombinant protein production in CHO cells.

2 Identification of Process Relevant miRNAs

2.1 Discovery and Annotation of miRNAs in CHO Cells

A major step towards the utilization of miRNAs as tools for CHO cell engineering was the initial identification of miRNA expression in a variety of CHO cells. Using this information, the subsequent analysis of differential expression under a variety of culture and bioprocessing conditions could then shed light on their potential functional impact on bioprocess relevant parameters.

The first identified miRNA in CHO was cgr-miR-21 described by Gammell et al. in 2007 [11], which was previously linked to the regulation of cell growth and apoptosis in human cells [12]. By using microarrays with human, mouse and rat miRNA probes and qPCR assays, differential expression of miR-21 was demonstrated in a biphasic cultivation process after temperature shift in a CHO-K1 cell line [11].

The knowledge about miRNAs expressed in CHO cell lines was further increased by a study of Johnson et al. using six cDNA libraries of small RNAs isolated from four CHO cell lines including CHO-K1 as well as CHO-DG44 cell lines cultured under different conditions including temperature shift, butyrate treatment and sampling at different cultivation phases. Next generation sequencing (NGS) and subsequent homology comparison with other species identified 350 CHO expressed miRNAs showing a high level of conservation between CHO and other mammals. This study also emphasized the importance to develop CHO specific miRNA microarrays and qRT-PCR assays for further exploration of CHO expressed miRNAs [13].

2.2 *Differential Expression Analysis of miRNAs in CHO Cells During Bioprocessing*

To further improve the annotation of CHO miRNAs and identify novel miRNA targets for cell line engineering, Hackl et al. [14] performed NGS experiments of six different biotechnologically relevant CHO cell lines including CHO-K1 as well as CHO-DUXB11 cell lines and analysed the resulting transcriptome. By using known miRNA hairpin sequences to map the sequencing data, which allowed for a more precise annotation of conserved miRNAs, this study identified 387 mature CHO miRNAs including 235 conserved as well as 11 novel miRNA genes. Subsequent analysis of differential expression indicated that miRNA expression was highly responsive to respective culture conditions. In addition, 18 miRNAs were regulated by adaption of CHO cells to suspension and serum-free growth, including well characterized miRNAs such as miR-31, miR-221-3p and miR-92a involved in cell proliferation [15], apoptosis [16], tumor development and aging [17–19].

An additional study analysed the influence of culture conditions on differential miRNA expression under fed-batch process conditions [20]. MiRNA expression profiles were shown to change after an induced temperature shift, a common process strategy to enhance recombinant protein production. 89 miRNAs were identified to be differentially expressed in different CHO-DG44 cell lines and cultivation phases. 19 miRNAs were subsequently validated to influence process relevant parameters like apoptosis, necrosis, productivity or proliferation underlining the relevance of miRNAs for cellular reactions to changes in culturing and process environment.

In addition to focusing on culture conditions, several studies were performed investigating differences of miRNA expression between recombinant producing

CHO cells and their parental cell lines to uncover miRNA influence on transgene expression.

Lin et al. [21] performed a study profiling the expression of conserved miRNAs in monoclonal antibody (mAb) expressing CHO cell lines and their parental hosts including CHO-DG44 and CHO-K1 using miRNA microarrays for human, mouse and rat. Further validation of 16 miRNAs by quantitative RT-PCR identified miR-221 and miR-222 as significantly downregulated in several mAb-producing CHO-DG44 cell lines, which were reported to target cyclin-dependent kinase inhibitor p27 (Cdkn1b) and therefore regulating cell cycle progression [22]. As overexpression of p27 was shown to have a beneficial effect on productivity of recombinant proteins and culture longevity by induction of G1 cell cycle arrest [23, 24], miR-221/222 were suggested as potential engineering targets. Furthermore three miRNAs (miR-19a, miR-17, miR106b) belonging miR-17-92 cluster were found to be upregulated in antibody producing cell lines [21], whereof miR-19 and miR-17 were previously reported to enhance proliferation and cell cycle progression [25].

Yet another study focused on the differential miRNA transcriptome between producing and non-producing cell lines using NGS methodology [26]. Here, 190 miRNAs were identified to be expressed in CHO cells and a number greater 80 % of these miRNAs showed differential expression in CHO-K1 and two recombinant protein producing CHO cell lines. Also Maccani et al. confirmed that miRNA expression profiles differ between producing and non-producing cells but could also differ between CHO cells producing different products [27].

A further step forwards the analysis of CHO expression profiles was the combination of transcriptomic and proteomic experiments to allow for a better correlation between miRNA expression and their functional target regulation on mRNA and protein level. Clark et al. [28] combined miRNA, mRNA and protein expression analyses to analyze the miRNA influence on cellular growth rate. Using microarray analysis of mRNA, qPCR for miRNA and quantitative LC-MS/MS for protein expression profiles from several mAb expressing cell lines covering a variety of different growth behaviors, 51 miRNAs were identified to possibly correlate with increased growth rate, whereof 35 miRNAs were upregulated and 16 miRNAs were downregulated. Finally, the availability of a CHO K1 genomic sequence published by Xu et al. in 2011 [29] further enhanced the potential of miRNA expression studies, since it provided the possibility to identify novel CHO specific miRNAs by mapping to the genomic sequence.

All the above discussed studies have contributed to the annotation of 353 mature miRNA sequences for the hamster in the current miRBase Version 22 (www.mirbase.org). However, this number is still low in comparisons to other organism like human with 2693 mature miRNA sequence entries and mouse with 2013 sequence entries. Considering these numbers there is still the potential to identify new endogenous expressed hamster or CHO miRNAs with potential use as tools for miRNA based cell line engineering.

2.3 *The Use of Transient Screening Techniques to Identify Process Relevant miRNAs*

While expression analysis of miRNAs only reveals indirect information on their potential functionality within CHO cells, functional screenings with a defined phenotypic readout may allow for creating direct new insights into pathway regulation initiated by specific miRNAs. Using functional miRNAs screening techniques, not only single miRNA targets but their impact on complex cellular pathways can be analyzed. In addition, a screening approach is not limited to endogenous expressed miRNAs but offers the chance to test miRNAs from another species, which could possibly also trigger cellular effects in CHO. Such functional RNAi screens can either utilize synthetic miRNA mimics or antagomiRs for a transient screening approach preferably for screening of large mimic libraries or can be performed by stable expressing miRNAs using expression plasmids with a smaller number of candidates.

A stable screening method was developed by Jadhav et al. [30]. As model cell line a recombinant Epo-Fc producing CHO cell line was used. For method setup and protocol optimization four known CHO miRNAs were used, cgr-miR-17, cgr-miR-221, cgr-miR-21 and cgr-miR-210. These four miRNAs were cloned into small hairpin vectors containing a GFP marker cassette. Upon stable overexpression mR-17 and miR-221 were shown to increase growth rate while specific productivity was not affected, resulting in increased final titers. Stable overexpression of miR-21 and miR-210 showed a decrease in specific productivity.

To identify miRNAs that increase the productivity of CHO cells expressing an IgG1 antibody, Strotbek et al. performed a functional transient screen using a human mimic library containing 879 miRNAs. Using this approach nine miRNAs were found to improve productivity of CHO cells. The results were validated using another CHO cell line producing human serum albumin, therefore demonstrating that the miRNAs act in a product independent manner. Further two miRNAs namely miR-557 and miR-1287 were identified by the transient screening to influence viable cell density and specific productivity. Subsequent stable overexpression in the IgG producing cell line resulted in increased specific productivity of the cells. Since both miRNAs were shown to be not endogenously expressed by CHO cells these data highlight the value of screening experiments to identify novel miRNAs with a beneficial regulative function within CHO cells [31].

Another transient high-content miRNA mimic screen in CHO-DG44 cells was reported by Fischer et al. in 2014 [32]. By transient transfection of miRNA mimics from a murine mimic library containing 1139 miRNAs, their influence on cellular phenotypes including growth, viability, cell death and titer were analyzed. A proportion of 21% of these were identified to increase cell-specific transgene productivity of secreted alkaline phosphatase (SEAP), whereas 5% of the miRNAs were demonstrated to accelerate cell proliferation. Further data evaluation highlighted the miR-30 family as potential tool for increasing titer and bioprocess performance. By stable overexpression of miR-30a twofold increase of transgene productivity and maximum cell density in comparison to the parental cell line was achieved. All above discussed studies are summarized in Table 1.

The discovery of miRNAs to influence process relevant parameters in CHO is still a recent and highly valuable research topic. By further increasing insights into the miRNA regulation mechanisms within CHO cells new miRNA based engineering approaches will arise and overcome today´s limitations of the CHO cell systems to further improve bioprocess performance.

3 Cell Engineering Using Overexpression of miRNAs

3.1 Advantages of miRNA Overexpression for Production Cell Line Engineering

MiRNAs are an interesting tool for the modification of industrial relevant CHO cell lines as they modulate serval bioprocess related pathways [10]. For cell line engineering, the most commonly used approach is stable overexpression of specific miRNAs to down-regulate unfavorable target gene expression and promote positive phenotypes for bioproduction [33–36]. Here, the most common aim is to improve CHO cell characteristics like productivity, product quality, secretion, metabolism, cell growth or viability during the production process [4, 37]. Technically, miRNAs can easily be overexpressed by stable genomic integration of precursor miRNA sequences. This overexpression of non-coding RNAs does not result in a large additional translational burden for the production cell line and thus does not waste cellular energy required for the biosynthesis of the product or cell growth [38]. In addition, one miRNA can regulate several target genes, which can lead to additive effects on one phenotype or affect several positive phenotypes at once. Using this characteristic, overexpression of one miRNA could enable several benefits and therefore eliminate the need for multiple, expensive adjustments on the host cell line [4]. Finally, miRNA overexpression can also be applied for fine-tuning to reduce the expression of negative genes, which may be lethal if knocked out [39].

3.2 Methodology for Overexpression of Noncoding miRNAs

For the overexpression of miRNAs a variety of methods are available. Most commonly, a single miRNA is overexpressed to regulate an individual, predefined pathway or gene. Here, the vectors pSP65 or pMSCV have successfully been used to overexpress e.g. miRNA-34a or miRNA-135a in CHO cells [40, 41]. It is also possible to multiplex the expression of miRNAs to regulate several different pathways or multiple genes of a single pathway simultaneously by using the vector pCMV-Tag4 [42]. For this purpose, multi-hairpin constructs were cloned

Table 1 Overview of all miRNA expression profiling and screening studies conducted in CHO cells

Methods	Cell line	Experimental Setup	Outcome	Publications
Microarray, qPCR	CHO-K1	Comparison of miRNA expression in biphasic cultivation process	Differential expression of cgr-miR-21 after temperature shift was shown	Gammell et al. (2007)
NGS	CHO-K1 CHO-DG44	Sequencing of six cDNA libraries of small RNAs isolated from various culture conditions	Identification of 350 CHO expressed miRNAs	Johnson et al. (2011)
NGS	CHO-K1 CHO-DUXB11	Transcriptome analysis of six CHO cell lines; differential expression analysis upon adaption to suspension growth	Identification of 387 mature CHO miRNAs including 11 novel miRNA genes; Differential expression of 18 miRNAs under suspension growth	Hackl et al. (2011)
NGS	CHO-DG44	Differential miRNA expression under fed-batch cultivation and temperature-shift	Identification of 19 Differentially expressed miRNAs influencing relevant process parameters	Stiefel et al. (2016)
Microarray, qPCR	CHO-DG44 CHO-K1	miRNA expression profiling in mAb producing cell lines	Identification of miR-221/222 as potential engineering targets; Upregulated expression of miR-19a/17/106b in mAb expressing cell lines	Lin et al. (2011)
NGS	CHO-K1	Differential expression analysis producing vs. non-producing cells	Identification of 190 CHO expressed miRNAs with 80% of these showing differential expression	S. Hammond et al. (2012)
Microarray, qPCR	CHO-DUXB11	Differential expression analysis of parental vs. producing cell lines under steady-state conditions	Identification of differential miRNA expression between parental and producing cell lines but also between CHO cells expressing different products	Maccani et al. (2014)

(continued)

Table 1 (continued)

Methods	Cell line	Experimental Setup	Outcome	Publications
Microarray, qPCR, quantitative LC-MS/MS	Several mAb expressing CHO cells	Correlation of mRNA, miRNA and protein expression profiles	Identification differential regulation of 51 growth rate correlated miRNAs	Clarke et al. (2012)
Stable miRNA expression screen	Epo-Fc producing CHO cell line	Stable expression of miR-17, miR-221, miR-21, miR-210	miR-17 and miR-221 increased growth rate miR-21 and miR-210 decreased specific productivity	Jadhav et al. (2012)
Transient Mimic screen	IgG1 expressing CHO cell line	Human mimic library (879 mimics)	miR-557 and miR-1287 increased specific productivity	Strotbeck et al. (2013)
Transient Mimic Screen	SEAP expressing CHO-DG44 cell line	Murine mimic library (1139 mimics)	miR-30a increased productivity and growth	Fischer et al. (2014)

intronically into the plasmid flanked by a splicing donor and splicing acceptor site leading to a simultaneous expression of up to five miRNAs [42]. Furthermore, the possibility to combine miRNA overexpression with the expression of a gene of interest is a very attractive option to add up synergistic effects and generate an optimized cellular phenotype. This strategy was published by Greber et al., 2007 using the pDG1 vector [43], where a miRNA is intronically encoded between two exons of a gene of interest guaranteeing comparable expression levels of the miRNA and the overexpressed mRNA.

3.3 *MiRNA Overexpression Can Enhance Cellular Productivity*

Up to date, there are several studies where miRNA overexpression has successfully been applied to optimize CHO cell expression systems (Table 2). The predominantly optimized phenotype of CHO cells is the specific productivity resulting in increased volumetric titer. A well-described example is miRNA-22, which is known to contribute to increased recombinant protein expression. After Xiao et al. described the productivity-enhancing effect of miRNA-22, also Inwood et al. achieved a 2.4 fold increase of specific productivity by overexpression of miRNA-22 and additionally described the downregulated target of miRNA-22 Homeodomain-interacting protein kinase 1 (HIPK1) [44–46]. When focusing on difficult-to-express proteins, the overexpression of miRNA-577 increased the specific productivity of CHO cells and a significant titer increase could be measured while product quality was maintained

Table 2 Summary of overexpressed miRNAs for the improvement of biotechnologically relevant cell lines

miRNA	Phenotypic effect	Molecular mechanism	Cell line	Publications
miRNA-22	productivity	HIPK1 involved in cellular stress and apoptotic pathways	HEK293	Xiao et al. (2015) and Inwood et al. (2018, 2020)
miRNA-557	productivity	May targets genes involved in ribosome formation inhibition	CHO	Strotbek et al. (2013) and Fischer et al. (2017)
miRNA-1287	productivity	Leads to a more balanced unfolded protein response by targeting GRP78	CHO	Strotbek et al. (2013) and Pieper et al. (2017)
miRNA19b	productivity	Represses PTEN resulting in reduced apoptosis	CHO	Loh et al. (2014)
miRNA-2861	productivity	Inhibits HDAC5 and enhances GOI transcription	CHO	Fischer et al. (2015)
mitosRNA-1978	productivity	Targets the ER proteins CerS2 and Tbc1D20 and supports vesicular trafficking	CHO	Pieper et al. (2017)
miRNA-483	productivity	May inhibit PDK4, KANK4 and MAPK3 for enhanced proliferation and optimized Glucose metabolism	CHO	Emmerling et al. (2016)
miRNA-143	productivity	Represses MAPK7 with a potential shift of cellular proliferation to protein production	CHO	Schoellhorn et al. (2017)
miRNA-26a	productivity	Inhibits histone deacetylases	HEK293	Meyer et al. (2017)
miRNA-136	productivity	Targets UPS17L for prolonged stationary growth phase	CAP	Weis et al. (2018)
miRNA-3074	productivity	unknown	CAP	Weis et al. (2018)
miRNA-30a	growth	unknown	CHO	Fischer et al. (2014)

(continued)

Table 2 (continued)

miRNA	Phenotypic effect	Molecular mechanism	Cell line	Publications
miRNA-18b	growth	unknown	CHO	Bort et al. (2012)
miRNA-20a+b	growth	Repress BIM and inhibit apoptosis	CHO	Bort et al. (2012)
miRNA-17	productivitygrowth	Targets TBC1D2 leading to enhanced membrane trafficking and protein secretion	CHO	Jadhav et al. (2014), Loh et al. (2014), and Bort et al. (2012)
miRNA-92a	productivitygrowth	Targets ITGA5 and promotes angiogenesis and BIM to inhibit apoptosis	CHO	Loh et al. (2014, 2017) and Bort et al. (2012)
miRNA-7	productivitygrowth	Regulates α-synuclein, Pak1 and IGF-1R	HEK293CHO	Koh et al. (2009) and Barron et al. (2011)
miRNA-221/222	productivitygrowth	unkown	CHO	Klanert et al. (2016)
miRNA-16	productivitygrowth	Inhibits CCNE1 and induces proliferation	HEK293	Koh et al. (2009)

[31, 47]. Also, the study by Schoellhorn et al., focused on intensified production of difficult-to-express proteins and could achieve a 30 % increased productivity using miRNA-143 overexpression. Here MAPK7 was identified as a target of miR-143 and its downregulation by siRNAs as well increased productivity [35]. Another example is miRNA-1287, which regulates the gene activating transcription factor 6 beta (ATF6β). Overexpression of miRNA-1287 reduced the unfolded protein response (UPR) and led to an improved phenotype in the context of viability and productivity [31, 39]. Several studies analyzed the potential of the miRNA-17-92 family for cell line engineering. Overexpression of the cluster members miRNA-17, miRNA-18, miRNA-19a+b and miRNA-92a was published along with improved industrial cell phenotypes like specific productivity or growth [14, 30, 44, 48–50]. These studies showed that by regulating BCL2, JAK1 and CCND1 in case of miRNA-17 or INSIG1 in case of miRNA-92a the expression of recombinant proteins could be improved with comparable or even increased cell proliferation [30, 48–50]. Another interesting example is miRNA-2861, where in contrast to the unspecific histone deacetylase (HDAC) inhibitor valproic acid (VPA), this miRNA was able to inhibit HDAC5 specifically and thereby improving the titer of around 20 % without influencing product quality [51]. In contrast, the effect of the comparable HDAC inhibitor miRNA-26a could be further enhanced with VPA. While the use of this miRNA alone resulted in an increased titer of ~50

%, an additional supply of VPA generated a hyperproductive phenotype with a 2-fold higher protein yield [52]. Focusing on miRNAs which are mainly expressed under certain conditions, Emmerling et al. could show that miRNA-483 was up-regulated during mild hypothermia and its overexpression led to an up to 4-fold increase in specific productivity, while a knock-down of miRNA-483 resulted in a 50 % reduced protein expression [53]. In this case, the modulation of a specific miRNA helped to provide an understanding of the increased specific productivity during mild hypothermia and served as a next-generation cell line engineering tool to constantly optimize productivity.

Furthermore, miRNAs have also been successfully used in alternative recombinant expression systems such as Cevec's Amniocyte Production (CAP) cells to increase specific productivity. As reported by Weis et al., both miRNA-136 and miRNA-3074 when overexpressed in CAP cells induced a desired phenotype with increased product quantity [36]. A further very interesting approach for next-generation cell line development is the use of noncoding mitochondrial RNAs (mitosRNAs) as published by Pieper et al. Through the overexpression of these functionally with miRNAs related RNAs, a downregulation of the two endoplasmic reticulum proteins Ceramide Synthase 2 (CerS2) and the Rab1 GAP Tbc domain family member 20 (Tbc1D20) could be achieved leading to an increase in specific productivity of 60 % [54].

3.4 Increasing Cell Proliferation and Growth Capacity by miRNA Overexpression

Besides increasing specific productivity or titer, the induction of proliferation is also of great industrial interest in order to obtain the same or additional amount of recombinant protein in less time. Also for this purpose, certain miRNAs have been identified which exert a positive effect on this phenotype. In this context, Fischer et al., reported on the potential of members of the miRNA-30 family to increase growth in combination with elevated titer following overexpression [32]. Especially miRNA-30a revealed a significant positive effect on the maximal viable cell density with simultaneously increased titer after transient transfection as well as after stable overexpression [32]. Similar results were obtained with overexpression of the miRNA-17-92 family, which have been identified in CHO-K1 cells in the context of growth promotion [55]. The members of the miRNA-17-92 family (miRNA-17, miRNA-18b, miRNA-20a+b, miRNA-92) control growth by regulating the cell cycle inhibitor p21 and prolonging cell longevity by downregulation of Bcl-2-like protein 11 (BCL2L11) and Phosphatase and tensin homolog (PTEN), resulting in increased VCD [55]. Furthermore, miRNA-221 and miRNA-222, which are encoded in close chromosomal proximity, show positive effects on the growth of CHO cells [56]. While these "oncomiRs" are used in diagnostic as tumor markers, they represent an interesting alternative for the engineering of CHO cells, which could permanently promote the proliferation of industrially used mammalian cells [57, 58].

3.5 Synthetic Biology Applying RNAi Enables Fine-Tuning of Production Cell Growth

Besides the simple overexpression of miRNAs to regulate genes and optimize mammalian cells, a novel and promising approach is the use of synthetic biology, where short non-coding RNAs are exploited as regulatory elements. Especially endogenous or artificial miRNAs are used for the purpose to align the expression of multiple vectors and to ensure the controlled expression of certain genes [59]. Using miRNA response elements (MREs) placed in the untranslated region (UTR) of the GOI, the respective miRNAs ensure via an ON/OFF principle that the correct amount of the GOI is expressed at the right time and balance for the production cell to meet ideal conditions [60]. In CHO cells, two studies applied synthetic biology using RNAi to improve the production cell phenotype. Malphettes et al. established a two-level transcription-translations control circuit allowing the expression of GOI to be precisely fine-tuned [61]. The system is based on a marcolide inducible RNAi expression vector and a tetracycline inducible vector for the expression of a GOI. In addition, the miRNA can bind in the UTR of the transcript of the GOI and regulate it negatively. As a result, the operator can increase the expression of a GOI to a defined amount during a certain part of the cultivation using tetracycline administration and at a later time inhibit the expression by macrolide addition [61]. Using the two-level control system, a growth-promoting gene can be overexpressed in the exponential phase in order to reach a higher cell density earlier and subsequently release additional translation capacities for the production of recombinant proteins in the stationary phase using RNAi. A more complex system using synthetic biology via miRNA overexpression was established by Lillacci et al. [62]. This system is based on two vectors, where one vector expresses the synthetic tetracycline transactivator tTA-Advanced, which is able to bind to the tetracycline-responsive element (TRE) of the second vector. Thus, the expression of a GOI can be induced with simultaneous expression of a miRNA intronically encoded on the vector. Lillacci et al. developed a toolbox of different vectors, where an intronically encoded miRNA can bind either in the UTR of the transactivator, the GOI or in both UTRs in order to regulate the expression of the GOI to different extents. In addition, the use of the synthetic tetracycline transactivator enables the operator to further limit the expression of the GOI externally to a minimal level using doxycycline [62]. This RNAi-based system for regulated gene expression can be used to express growth or production-promoting genes to a low extent without significantly increasing the expression burden for the production cell, potentially leading to favorable phenotypes. All above discussed studies are summarized in table 2.

Taken together, the diverse examples for miRNA overexpression or use as synthetic biology tool to optimize the phenotype of cells for recombinant protein expression demonstrates the potential of miRNAs in cell line development. Especially the upcoming approach of synthetic biology, which uses miRNAs as negative regulators of biological cycles underlines their image as new tools for next-generation cell line development.

4 Cell Engineering Using Knock-Down or Knock-Out of miRNAs

In order to improve bioprocess relevant parameters in CHO production cell lines, several strategies have been developed aiming at the depletion of miRNAs displaying detrimental influence on CHO cell production performance. These loss-of-function experiments can be categorized as "Knock-Down" (KD) or "Knock-Out" (KO) approaches depending on their functional mechanism and the degree of down-regulation they seek to achieve. In KD attempts, also known as gene silencing, the original miRNA encoding gene remains unchanged whereas endogenous miRNA-levels are reduced to a certain extent. In contrast, KO approaches aim at the complete loss of the miRNA encoding gene, which is achieved by its genomic deletion through the use of endonucleases like e.g. zinc finger nucleases (ZFNs), Transcription activator-like effector nucleases (TALENs) or Clustered Regularly Interspaced Short Palindromic Repeats/ CRISPR-associated proteins (CRISPR/Cas9) [63–66].

4.1 miRNA Knockdown Studies Using antagomiR Inhibitors Enhance CHO Production Cells

Initial KD attempts to silence miRNA function took advantage of the canonical Watson-Crick base pairing capacity between complementary RNA molecules. In 2004 Meister et al. and Hutvágner et al. used nuclease-resistant antisense 2′-O-methyl–modified oligoribonucleotides to efficiently knock down target miRNAs in HeLa cells and *C. elegans* independently of one another [67, 68]. The 2′-O-methyl-modification of used antisense oligoribonucleotides hereby mediates resistance against degradation by nucleases and in parallel accelerates stable and fast binding of complementary single-stranded RNA molecules [69, 70]. However, there are no reports that this KD approach was ever tested in CHO cells. In contrast, an advancement of these antisense oligonucleotides termed "antagomiRs" is a prominent transient miRNA KD method also applied in CHO cells. In addition to the 2'-O-methyl group, antagomiRs possess a partially modified phosphorothioate

RNA backbone which is additionally coupled to a cholesterol residue [71, 72]. The conjugation of RNA and cholesterol enables a more efficient delivery and systemic *in vivo* activity of the miRNA inhibitor [73, 74].

In 2011 Barron et al. identified several differentially expressed miRNAs upon a temperature shift from 37 °C to 28 °C–33 °C during batch cultivation of CHO cells. Among those, miR-7 was downregulated to the greatest extent and subsequent transient overexpression of miR-7 led to a complete inhibition of proliferation. However, KD experiments using transfection of miR-7 antagomiRs showed no effect on cell growth at all [33].

In contrast Fischer et al. were able to identify miR-30 family during a high content miRNA screen using SEAP secreting CHO cells to have a positive impact on cell proliferation and productivity. The beneficial influence of miR-30 family on CHO production cells was further supported by transient transfection experiments using miR-30 family antagomiRs that did not increase viable cell density or productivity [32, 75].

As for Fischer et al. Emmerling et al. and Loh et al. also performed antagomiR control transfections in order to demonstrate the beneficial effect if the investigated miRNA is present at higher levels. Thus, Emmerling et al. observed a significant reduction of mAb productivity in CHO cells as well as reduced viral particle production of HeLa cells after transfection of miR-483 antagomiRs [53].

Loh et al. observed a correlation of miR-92a overexpression and specific productivity of CHO cells [49]. Consequently, they implemented miR-92a mimic and antagomiR transfection experiments to investigate the regulation of the potential target gene *insig1* of miR-92a [50].

A different study work focused on the involvement of miRNAs in apoptosis. Druz et al. identified differentially expressed miRNAs in CHO cells as a consequence of apoptosis provoked by nutrient depletion. Using microarray analysis the authors identified miR-466h to be upregulated during the apoptotic phase and concluded that miR-466h induces apoptosis indirectly by inhibition of anti-apoptotic genes. In subsequent transient KD experiments with antagomiR-466h a reduced Caspase-3/7 activity was observed in parallel to elevated cell viability [76].

To investigate if this beneficial effect was also applicable over the whole process time Druz et al. engineered a stable miR-466h KD cell line using a vector based miRNA inhibition system based on short hairpin RNA (shRNA) constructs [77, 78]. The engineered stable anti-miR-466h SEAP expressing CHO cell line showed an improvement in viability and viable cell density along with delayed Caspase-3/7 activity. As a result the final increase in specific productivity accounted for 11 % whereby SEAP titer was even raised by 43 %, showing the potential of using miRNAs for CHO cell line engineering [78].

4.2 Use of miRNA Sponge Knockdown Approaches Raise Viability and Proliferation

A further miRNA KD strategy apart from those stated earlier is termed "miRNA sponges". Sponges are artificial transcripts which possess multiple tandem binding sites for the miRNA-of-interest. These molecules mimic the natural binding of miRNAs to their complementary target sites in the UTRs of mRNAs. By recombinant expression of sponge constructs endogenous miRNA levels are reduced [79] and the same degree of downregulation as for antagomiRs can be achieved by using a strong promoter for recombinant miRNA sponge expression [79].

Sanchez et al performed a stable miR-7 KD by using a miR-7 sponge construct and miR-7 depleted CHO-K1 population showed 40 % improved viable cell density as well as a 2-fold higher SEAP titer than controls during Fed-batch cultivation [80]. In a follow up study miR-7 sponge depletion in CHO DP12 cells even yielded 65 % increase in cell proliferation in parallel to 3-fold higher IgG titer [81]. Here, additional proteomic profiling revealed that miR-7 KD resulted in an overrepresentation of proteins involved in the Akt pathway and ribosome biogenesis, which might be the underlying molecular mechanisms for the observed phenotype.

Another promising miRNA KD candidate for CHO cell line engineering that negatively effects cell growth and proliferation is miR-378 [28]. After stable KD of this anti-proliferative miRNA by sponge decoy constructs, Costello et al. observed an increase in viable peak cell density of almost 59 %, nevertheless IgG titer or specific productivity were not increased significantly during batch cultivation. As potential target gene of miR-378 Ubiquitin carboxyl-terminal hydrolase 14 (Usp14) was identified by applying proteomic analysis [82, 83].

Taking advantage of anti-cancer research data, Kelly et al. investigated the anti-proliferative role of miR-34a, which is described to act as tumor-suppressive. Therefore, the authors reduced endogenous levels of this highly conserved miRNA by sponge constructs in order to boost cell growth and proliferation reaching a 2-fold higher titer with SEAP secreting CHO-K1 cells. However, miR-34a KD was accompanied with a minor negative impact on cell growth which is contradictory to previous observations from cancer cell lines [84].

In line with previous miRNA-sponge KD studies Pairawan et al. reduced endogenous levels of miR-15a and miR-16-1 which are known to control apoptosis and cell proliferation. During batch cultivation of IgG secreting CHO-K1 cells, a 3-fold increase in product titer was observed and the potential de-repression of Bcl-2 as molecular basis hypothesized [85].

Another major work focused on metabolic engineering of CHO cells by depletion of miR-23 based on its previously described involvement in the control of glutamate metabolism [86]. The authors intended to direct CHO-K1 SEAP secreting cells towards higher oxidative metabolism rates and miR-23 depleted cell lines displayed a 30 % higher mitochondrial activity and a 3-fold increased SEAP titer. The authors identified potential target genes, among them LETM1, an inner membrane protein of mitochondria, and IDH1 which is important for the tricarboxylic acid cycle [86]. This was the first study that altered global cell metabolism only by alteration of a single miRNA which pointed towards the potential of miRNAs in CHO production cell line engineering.

4.3 CRISPR/Cas9 Knockout of Unfavorable microRNAs Improve Growth and Productivity of CHO Cells

An alternative to KD of adverse acting miRNAs is the full knock out of the miRNA encoding region in the genome of CHO cells. For this strategy, several particularities have to be considered due to the non-coding nature of miRNAs. While the insertion of a single point mutation will generate impaired function due to frameshift mutations in protein coding genes, this is not necessarily applicable for miRNAs due to their non-coding nature. Even though point mutations in the loop region of pri- and pre-miRNAs can impair hairpin formation or recognition by proteins as Drosha and Dicer proteins and thus inhibit further processing of the miRNA, it is very difficult to generate cell lines that exhibit a total loss of the respective mature miRNA by introducing point mutations. More promising is therefore the total removal of the entire miRNA gene from the genome. In addition, miRNAs are often encoded in introns of protein encoding genes or in miRNA clusters. In these cases it has to be verified that splicing of the host gene transcript harboring the miRNA or miRNAs in close proximity are not affected in order to avoid false positive results by unintended gene editing [87].

Two publications exist that describe the KO of miRNAs with CRISPR/Cas9. The first study from Kellner et al. knocked out miR-27 a/b which is presumably associated with cell growth and recombinant productivity as it was identified to be downregulated during hypothermic cultivation conditions of CHO cells [11, 88]. Kellner et al. designed sgRNAs binding either the seed region or in close proximity to the seed region of miR-27. Of all investigated cells up to 88 % possessed insertions or deletions at the desired site. Most of the resulting single cell clones exhibited a downregulation of endogenous mature miR-27 levels rather than complete loss of mature miRNAs. Nevertheless, the authors observed an increase in viable cell count and viability at the end of batch and fed-batch cultivation in miR-27b reduced cell pools emphasizing the strategy of miRNA KOs [88].

In the second approach from Raab et al. miR-744 was targeted due to its detrimental effects on recombinant CHO cell productivity as elucidated by a large miRNA screen [32]. In contrast to Kellner et al. the authors deleted the entire miRNA encoding region from the genome of CHO cells by applying a combination of two flanking sgRNAs. Here, miR-744 expression was completely eliminated and clonal miR-744 KO cell lines were generated exhibiting a significantly increased antibody titer during batch cultivation in comparison to controls [89]. All KD and KO studies in CHO production cell lines described above are summarized in Table 3.

Table 3 Overview of all miRNA KD or KO studies conducted in CHO production cell lines

miRNA	Nature of depletion	Depletion approach	Phenotypic effect	Molecular mechanism	Publications
miR-7	Knock Down	antagomiR miR-sponge	Viability proliferation productivity	Genes involved in ribosome biogenesis & Akt pathway	Barron et al. (2011), Sanchez et al. (2014), and Coleman et al. (2019)
miR-30 family	Knock Down	antagomiR	Proliferation productivity	Regulation of Ubiquitin pathway	Fischer et al. (2014, 2015)
miR-483	Knock Down	antagomiR	productivity	Genes involved in cell survival & protein expression	Emmerling et al. (2015)
miR-92a	Knock Down	antagomiR	productivity	Cholesterol metabolism	Loh et al. (2017)
miR-466h	Knock Down	antagomiR shRNA vector	Anti-apoptotic	Delayed Caspase-3/7 activity	Druz et al. (2011, 2013)
miR-23	Knock Down	miR-sponge	Oxidative metabolism	LETM1 & IDH1 involved in mitochondrial activity	Kelly et al. (2015)
miR-378	Knock Down	miR-sponge	viability proliferation	Usp14 increases proliferation	Costello et al. (2018) and Coleman et al. (2018)
miR-34a	Knock Down	miR-sponge	Anti-proliferative	Genes involved in cell cycle progression (CCND1, CDK6)	Kelly et al. (2014)
miR-15a	Knock Down	miR-sponge	Apoptosis, proliferation	Maybe involvement of Bcl-2	Pairawan et al. (2019)
miR-16-1	Knock Down	miR-sponge	Apoptosis, proliferation	Maybe involvement of Bcl-2	Pairawan et al. (2019)
miR-744	Knock Out	CRISPR/Cas9	proliferation productivity		Raab et al. (2019)
miR-27a / b	Knock Out	CRISPR/Cas9	viability proliferation productivity		Kellner et al. (2018)

5 Conclusion: MicroRNAs Are Powerful Cell Line Engineering Tools

During the past decades CHO cells have emerged as predominant biopharmaceutical production hosts [1]. Although major progress has been made in order to increase productivity and product quality since their isolation in 1958, cell line development faces new challenging tasks like the increasing demand for biological therapeutics as biopharmaceuticals, biosimilars, and new complex antibody formats or superior product quality [3, 90]. While various traditional cell line engineering attempts focused on either overexpression, KD or KO of a single protein encoding gene, miRNAs offer the possibility to influence the expression of hundreds of mRNAs by just customizing the expression of a single miRNA acting as a global regulator of cellular transcription [4, 38, 59].

Although the majority of miRNAs are considered to act redundantly, numerous overexpression, KD and KO studies have proven the potential of miRNAs for the purpose of CHO production cell line engineering as exemplified above [91]. Nevertheless, comparably little is known about the detailed mechanisms, networks and target genes by which miRNAs exert their respective phenotypic function. Hence, more research needs to be conducted to predict miRNA targets more reliably despite their heterogeneous nature of target recognition. At present databases like miRTarBase, miRWalk2.0 or miRBase collect, organize and provide valuable information about miRNAs and their respective target genes either they are experimentally validated or only predicted by *in silico* algorithms [92–94]. Once more knowledge of miRNAs and their respective target mRNAs has been collected, pathway analysis based on transcriptomic and proteomic datasets might enable the detailed prediction of valid downstream effects of certain miRNA expression patterns.

Funding/Acknowledgement Statement Florian Klingler, Nikolas Zeh and Nadja Raab are funded by the German Federal Ministry of Education and Research, funding program Forschung an Fachhochschulen, contract numbers 13FH162PA6 (Nikolas Zeh, Nadja Raab) and 13FH635IA6 (Florian Klingler).

Conflict of Interest Statement There are no conflicts of interest.

References

1. Walsh G (2018) Biopharmaceutical benchmarks 2018. Nat Biotechnol 36:1136–1145
2. Kim JY, Kim Y-G, Lee GM (2012) CHO cells in biotechnology for production of recombinant proteins: current state and further potential. Appl Microbiol Biotechnol 93:917–930
3. Wurm FM (2004) Production of recombinant protein therapeutics in cultivated mammalian cells. Nat Biotechnol 22:1393–1398
4. Fischer S, Handrick R, Otte K (2015) The art of CHO cell engineering: a comprehensive retrospect and future perspectives. Biotechnol Adv 33:1878–1896

5. Fischer S, Otte K (2019) CHO cell engineering for improved process performance and product quality. In: Cell culture engineering. Wiley, pp 207–250
6. Ketting RF, Fischer SEJ, Bernstein E, Sijen T, Hannon GJ, Plasterk RHA (2001) Dicer functions in RNA interference and in synthesis of small RNA involved in developmental timing in C. elegans. Genes Dev 15:2654–2659
7. Provost P, Dishart D, Doucet J, Frendewey D, Samuelsson B, Rådmark O (2002) Ribonuclease activity and RNA binding of recombinant human Dicer. EMBO J 21:5864–5874
8. Hammond SM (2001) Argonaute2, a link between genetic and biochemical analyses of RNAi. Science 293:1146–1150
9. Hutvagner G (2005) Small RNA asymmetry in RNAi: function in RISC assembly and gene regulation. FEBS Lett 579:5850–5857
10. Müller D, Katinger H, Grillari J (2008) MicroRNAs as targets for engineering of CHO cell factories. Trends Biotechnol 26:359–365
11. Gammell P, Barron N, Kumar N, Clynes M (2007) Initial identification of low temperature and culture stage induction of miRNA expression in suspension CHO-K1 cells. J Biotechnol 130:213–218
12. Krichevsky AM, Gabriely G (2008) miR-21: a small multi-faceted RNA. J Cell Mol Med 13:39–53
13. Johnson KC, Jacob NM, Nissom PM, Hackl M, Lee LH, Yap M, Hu W (2011) Conserved microRNAs in Chinese hamster ovary cell lines. Biotechnol Bioeng 108:475–480
14. Hackl M, Jakobi T, Blom J, Doppmeier D, Brinkrolf K, Szczepanowski R, Bernhart SH, Siederdissen CH, JAH B, Wieser M, Kunert R, Jeffs S, Hofacker IL, Goesmann A, Pühler A, Borth N, Grillari J (2011) Next-generation sequencing of the Chinese hamster ovary microRNA transcriptome: Identification, annotation and profiling of microRNAs as targets for cellular engineering. J Biotechnol 153:62–75
15. Creighton CJ, Fountain MD, Yu Z, Nagaraja AK, Zhu H, Khan M, Olokpa E, Zariff A, Gunaratne PH, Matzuk MM, Anderson ML (2010) Molecular profiling uncovers a p53-associated role for microRNA-31 in inhibiting the proliferation of serous ovarian carcinomas and other cancers. Cancer Res 70:1906–1915
16. Dai R, Li J, Liu Y, Yan D, Chen S, Duan C, Liu X, He T, Li H (2010) miR-221/222 suppression protects against endoplasmic reticulum stress-induced apoptosis via p27(Kip1)- and MEK/ERK-mediated cell cycle regulation. Biol Chem 391:791–801
17. Grillari J, Hackl M, Grillari-Voglauer R (2010) miR-17-92 cluster: ups and downs in cancer and aging. Biogerontology 11:501–506
18. Hackl M, Brunner S, Fortschegger K, Schreiner C, Micutkova L, Mück C, Laschober GT, Lepperdinger G, Sampson N, Berger P, Herndler-Brandstetter D, Wieser M, Kühnel H, Strasser A, Rinnerthaler M, Breitenbach M, Mildner M, Eckhart L, Tschachler E, Trost A, Bauer JW, Papak C, Trajanoski Z, Scheideler M, Grillari-Voglauer R, Grubeck-Loebenstein B, Jansen-Dürr P, Grillari J (2010) miR-17, miR-19b, miR-20a, and miR-106a are down-regulated in human aging. Aging Cell 9:291–296
19. Ivanov SV, Goparaju CMV, Lopez P, Zavadil J, Toren-Haritan G, Rosenwald S, Hoshen M, Chajut A, Cohen D, Pass HI (2010) Pro-tumorigenic effects of miR-31 loss in mesothelioma. J Biol Chem 285:22809–22817
20. Stiefel F, Fischer S, Sczyrba A, Otte K, Hesse F (2016) miRNA profiling of high, low and non-producing CHO cells during biphasic fed-batch cultivation reveals process relevant targets for host cell engineering. J Biotechnol 225:31–43
21. Lin N, Davis A, Bahr S, Borgschulte T, Achtien K, Kayser K (2011) Profiling highly conserved microrna expression in recombinant IgG-producing and parental Chinese hamster ovary cells. Biotechnol Prog 27:1163–1171
22. Park J-K, Lee EJ, Esau C, Schmittgen TD (2009) Antisense inhibition of microRNA-21 or -221 arrests cell cycle, induces apoptosis, and sensitizes the effects of gemcitabine in pancreatic adenocarcinoma. Pancreas 38:e190–e199
23. Mazur X, Fussenegger M, Renner WA, Bailey JE (1998) Higher productivity of growth-arrested Chinese hamster ovary cells expressing the cyclin-dependent kinase inhibitor p27. Biotechnol Prog 14:705–713

24. Meents H, Enenkel B, Werner RG, Fussenegger M (2002) p27Kip1-mediated controlled proliferation technology increases constitutive sICAM production in CHO-DUKX adapted for growth in suspension and serum-free media. Biotechnol Bioeng 79:619–627
25. Mendell JT (2008) miRiad roles for the miR-17-92 cluster in development and disease. Cell 133:217–222
26. Hammond S, Swanberg JC, Polson SW, Lee KH (2012) Profiling conserved microRNA expression in recombinant CHO cell lines using Illumina sequencing. Biotechnol Bioeng 109:1371–1375
27. Maccani A, Hackl M, Leitner C, Steinfellner W, Graf AB, Tatto NE, Karbiener M, Scheideler M, Grillari J, Mattanovich D, Kunert R, Borth N, Grabherr R, Ernst W (2014) Identification of microRNAs specific for high producer CHO cell lines using steady-state cultivation. Appl Microbiol Biotechnol 98:7535–7548
28. Clarke C, Henry M, Doolan P, Kelly S, Aherne S, Sanchez N, Kelly P, Kinsella P, Breen L, Madden SF, Zhang L, Leonard M, Clynes M, Meleady P, Barron N (2012) Integrated miRNA, mRNA and protein expression analysis reveals the role of post-transcriptional regulation in controlling CHO cell growth rate. BMC Genomics 13:656
29. Xu X, Nagarajan H, Lewis NE, Pan S, Cai Z, Liu X, Chen W, Xie M, Wang W, Hammond S, Andersen MR, Neff N, Passarelli B, Koh W, Fan HC, Wang J, Gui Y, Lee KH, Betenbaugh MJ, Quake SR, Famili I, Palsson BO, Wang J (2011) The genomic sequence of the Chinese hamster ovary (CHO)-K1 cell line. Nat Biotechnol 29:735–741
30. Jadhav V, Hackl M, Hernandez Bort JA, Wieser M, Harreither E, Kunert R, Borth N, Grillari J (2012) A screening method to assess biological effects of microRNA overexpression in Chinese hamster ovary cells. Biotechnol Bioeng 109:1376–1385
31. Strotbek M, Florin L, Koenitzer J, Tolstrup A, Kaufmann H, Hausser A, Olayioye MA (2013) Stable microRNA expression enhances therapeutic antibody productivity of Chinese hamster ovary cells. Metab Eng 20:157–166
32. Fischer S, Buck T, Wagner A, Ehrhart C, Giancaterino J, Mang S, Schad M, Mathias S, Aschrafi A, Handrick R, Otte K (2014) A functional high-content miRNA screen identifies miR-30 family to boost recombinant protein production in CHO cells. Biotechnol J 9:1279–1292
33. Barron N, Kumar N, Sanchez N, Doolan P, Clarke C, Meleady P, O'Sullivan F, Clynes M (2011) Engineering CHO cell growth and recombinant protein productivity by overexpression of miR-7. J Biotechnol 151:204–211
34. Jadhav V, Hackl M, Druz A, Shridhar S, Chung CY, Heffner KM, Kreil DP, Betenbaugh M, Shiloach J, Barron N, Grillari J, Borth N (2013) CHO microRNA engineering is growing up: recent successes and future challenges. Biotechnol Adv 31:1501–1513
35. Schoellhorn M, Fischer S, Wagner A, Handrick R, Otte K (2017) miR-143 targets MAPK7 in CHO cells and induces a hyperproductive phenotype to enhance production of difficult-to-express proteins. Biotechnol Prog 33:1046–1058
36. Weis BL, Guth N, Fischer S, Wissing S, Fradin S, Holzmann K-H, Handrick R, Otte K (2018) Stable miRNA overexpression in human CAP cells: engineering alternative production systems for advanced manufacturing of biologics using miR-136 and miR-3074. Biotechnol Bioeng 115:2027–2038
37. Inwood S, Betenbaugh MJ, Shiloach J (2018) Methods for using small non-coding rnas to improve recombinant protein expression in mammalian cells. Genes (Basel) 9
38. Hackl M, Borth N, Grillari J (2012) miRNAs – pathway engineering of CHO cell factories that avoids translational burdening. Trends Biotechnol 30:405–406
39. Pieper LA, Strotbek M, Wenger T, Olayioye MA, Hausser A (2017) ATF6β-based fine-tuning of the unfolded protein response enhances therapeutic antibody productivity of Chinese hamster ovary cells. Biotechnol Bioeng 114:1310–1318
40. Okada N, Lin C-P, Ribeiro MC, Biton A, Lai G, He X, Bu P, Vogel H, Jablons DM, Keller AC, Wilkinson JE, He B, Speed TP, He L (2014) A positive feedback between p53 and miR-34 miRNAs mediates tumor suppression. Genes Dev 28:438–450
41. Mannironi C, Camon J, De Vito F, Biundo A, De Stefano ME, Persiconi I, Bozzoni I, Fragapane P, Mele A, Presutti C (2013) Acute stress alters amygdala microRNA miR-135a and miR-124 expression: inferences for corticosteroid dependent stress response. PLoS One 8:e73385

42. Seyhan AA (2016) A multiplexed miRNA and transgene expression platform for simultaneous repression and expression of protein coding sequences. Mol BioSyst 12:295–312
43. Greber D, Fussenegger M (2007) Multi-gene engineering: simultaneous expression and knockdown of six genes off a single platform. Biotechnol Bioeng 96:821–834
44. Xiao S, Chen Y-C, Betenbaugh MJ, Martin SE, Shiloach J (2015) MiRNA mimic screen for improved expression of functional neurotensin receptor from HEK 293 cells. Biotechnol Bioeng 112:1632–1643
45. Inwood S, Abaandou L, Betenbaugh M, Shiloach J (2020) Improved protein expression in HEK293 cells by over-expressing miR-22 and knocking-out its target gene, HIPK1. New Biotechnol 54:28–33
46. Inwood S, Buehler E, Betenbaugh M, Lal M, Shiloach J (2018) Identifying HIPK1 as target of miR-22-3p enhancing recombinant protein production from HEK 293 cell by using microarray and HTP siRNA screen. Biotechnol J 13:1700342
47. Fischer S, Marquart KF, Pieper LA, Fieder J, Gamer M, Gorr I, Schulz P, Bradl H (2017) miRNA engineering of CHO cells facilitates production of difficult-to-express proteins and increases success in cell line development. Biotechnol Bioeng 114:1495–1510
48. Jadhav V, Hackl M, Klanert G, Hernandez Bort JA, Kunert R, Grillari J, Borth N (2014) Stable overexpression of miR-17 enhances recombinant protein production of CHO cells. J Biotechnol 175:38–44
49. Loh WP, Loo B, Zhou L, Zhang P, Lee DY, Yang Y, Lam KP (2014) Overexpression of microRNAs enhances recombinant protein production in Chinese hamster ovary cells. Biotechnol J 9:1140–1151
50. Loh WP, Yang Y, Lam KP (2017) miR-92a enhances recombinant protein productivity in CHO cells by increasing intracellular cholesterol levels. Biotechnol J 12:1600488
51. Fischer S, Paul AJ, Wagner A, Mathias S, Geiss M, Schandock F, Domnowski M, Zimmermann J, Handrick R, Hesse F, Otte K (2015) miR-2861 as novel HDAC5 inhibitor in CHO cells enhances productivity while maintaining product quality. Biotechnol Bioeng 112:2142–2153
52. Meyer H-J, Reilly D, Martin SE, Wong AW (2017) Identification of a novel miRNA that increases transient protein expression in combination with valproic acid. Biotechnol Prog 33:1139–1145
53. Emmerling VV, Fischer S, Stiefel F, Holzmann K, Handrick R, Hesse F, Hörer M, Kochanek S, Otte K (2016) Temperature-sensitive miR-483 is a conserved regulator of recombinant protein and viral vector production in mammalian cells. Biotechnol Bioeng 113:830–841
54. Pieper LA, Strotbek M, Wenger T, Gamer M, Olayioye MA, Hausser A (2017) Secretory pathway optimization of CHO producer cells by co-engineering of the mitosRNA-1978 target genes CerS2 and Tbc1D20. Metab Eng 40:69–79
55. Bort JAH, Hackl M, Höflmayer H, Jadhav V, Harreither E, Kumar N, Ernst W, Grillari J, Borth N (2012) Dynamic mRNA and miRNA profiling of CHO-K1 suspension cell cultures. Biotechnol J 7:500–515
56. Klanert G, Jadhav V, Shanmukam V, Diendorfer A, Karbiener M, Scheideler M, Bort JH, Grillari J, Hackl M, Borth N (2016) A signature of 12 microRNAs is robustly associated with growth rate in a variety of CHO cell lines. J Biotechnol 235:150–161
57. He H, Jazdzewski K, Li W, Liyanarachchi S, Nagy R, Volinia S, Calin GA, Liu C-g, Franssila K, Suster S, Kloos RT, Croce CM, de la Chapelle A (2005) The role of microRNA genes in papillary thyroid carcinoma. Proc Natl Acad Sci 102:19075–19080
58. Wong QW-L, Ching AKK, Chan AWH, Choy K-W, To K-F, Lai PBS, Wong N (2010) MiR-222 overexpression confers cell migratory advantages in hepatocellular carcinoma through enhancing AKT signaling. Clin Cancer Res 16:867–875
59. Valdés-Bango Curell R, Barron N (2018) Exploring the potential application of short non-coding RNA-based genetic circuits in Chinese hamster ovary cells. Biotechnol J 13:1700220
60. Navarro FJ, Baulcombe DC (2019) MiRNA-mediated regulation of synthetic gene circuits in the green alga chlamydomonas reinhardtii. ACS Synth Biol 8:358–370
61. Malphettes L, Fussenegger M (2006) Improved transgene expression fine-tuning in mammalian cells using a novel transcription–translation network. J Biotechnol 124:732–746

62. Lillacci G, Benenson Y, Khammash M (2018) Synthetic control systems for high performance gene expression in mammalian cells. Nucleic Acids Res 46:9855–9863
63. Kim YG, Cha J, Chandrasegaran S (1996) Hybrid restriction enzymes: zinc finger fusions to Fok I cleavage domain. Proc Natl Acad Sci U S A 93:1156–1160
64. Christian M, Cermak T, Doyle EL, Schmidt C, Zhang F, Hummel A, Bogdanove AJ, Voytas DF (2010) Targeting DNA double-strand breaks with TAL effector nucleases. Genetics 186:757–761
65. Jinek M, Chylinski K, Fonfara I, Hauer M, Doudna JA, Charpentier E (2012) A programmable dual-RNA-guided DNA endonuclease in adaptive bacterial immunity. Science 337:816–821
66. Ran FA, Hsu PD, Wright J, Agarwala V, Scott DA, Zhang F (2013) Genome engineering using the CRISPR-Cas9 system. Nat Protoc 8:2281–2308
67. Hutvágner G, Simard MJ, Mello CC, Zamore PD (2004) Sequence-specific inhibition of small RNA function. PLoS Biol 2:E98
68. Meister G, Landthaler M, Dorsett Y, Tuschl T (2004) Sequence-specific inhibition of microRNA- and siRNA-induced RNA silencing. RNA 10:544–550
69. Lamond AI, Sproat BS (1993) Antisense oligonucleotides made of 2'-O-alkylRNA: their properties and applications in RNA biochemistry. FEBS Lett 325:123–127
70. Cummins LL, Owens SR, Risen LM, Lesnik EA, Freier SM, McGee D, Guinosso CJ, Cook PD (1995) Characterization of fully 2'-modified oligoribonucleotide hetero- and homoduplex hybridization and nuclease sensitivity. Nucleic Acids Res 23:2019–2024
71. Krützfeldt J, Poy MN, Stoffel M (2006) Strategies to determine the biological function of microRNAs. Nat Genet 38(Suppl):S14–S19
72. Krützfeldt J, Rajewsky N, Braich R, Rajeev KG, Tuschl T, Manoharan M, Stoffel M (2005) Silencing of microRNAs in vivo with "antagomirs". Nature 438:685–689
73. Lorenz C, Hadwiger P, John M, Vornlocher H-P, Unverzagt C (2004) Steroid and lipid conjugates of siRNAs to enhance cellular uptake and gene silencing in liver cells. Bioorg Med Chem Lett 14:4975–4977
74. Soutschek J, Akinc A, Bramlage B, Charisse K, Constien R, Donoghue M, Elbashir S, Geick A, Hadwiger P, Harborth J, John M, Kesavan V, Lavine G, Pandey RK, Racie T, Rajeev KG, Röhl I, Toudjarska I, Wang G, Wuschko S, Bumcrot D, Koteliansky V, Limmer S, Manoharan M, Vornlocher H-P (2004) Therapeutic silencing of an endogenous gene by systemic administration of modified siRNAs. Nature 432:173–178
75. Fischer S, Mathias S, Schaz S, Emmerling VV, Buck T, Kleemann M, Hackl M, Grillari J, Aschrafi A, Handrick R, Otte K (2015) Enhanced protein production by microRNA-30 family in CHO cells is mediated by the modulation of the ubiquitin pathway. J Biotechnol 212:32–43
76. Druz A, Chu C, Majors B, Santuary R, Betenbaugh M, Shiloach J (2011) A novel microRNA mmu-miR-466h affects apoptosis regulation in mammalian cells. Biotechnol Bioeng 108:1651–1661
77. Paddison PJ, Caudy AA, Bernstein E, Hannon GJ, Conklin DS (2002) Short hairpin RNAs (shRNAs) induce sequence-specific silencing in mammalian cells. Genes Dev 16:948–958
78. Druz A, Son Y, Betenbaugh M, Shiloach J (2013) Stable inhibition of mmu-miR-466h-5p improves apoptosis resistance and protein production in CHO cells. Metab Eng 16:87–94
79. Ebert MS, Neilson JR, Sharp PA (2007) MicroRNA sponges: competitive inhibitors of small RNAs in mammalian cells. Nat Methods 4:721–726
80. Sanchez N, Kelly P, Gallagher C, Lao NT, Clarke C, Clynes M, Barron N (2014) CHO cell culture longevity and recombinant protein yield are enhanced by depletion of miR-7 activity via sponge decoy vectors. Biotechnol J 9:396–404
81. Coleman O, Suda S, Meiller J, Henry M, Riedl M, Barron N, Clynes M, Meleady P (2019) Increased growth rate and productivity following stable depletion of miR-7 in a mAb producing CHO cell line causes an increase in proteins associated with the Akt pathway and ribosome biogenesis. J Proteome 195:23–32

82. Coleman O, Costello A, Henry M, Lao NT, Barron N, Clynes M, Meleady P (2018) A proteomic profiling dataset of recombinant Chinese hamster ovary cells showing enhanced cellular growth following miR-378 depletion. Data Br 21:2679–2688
83. Costello A, Coleman O, Lao NT, Henry M, Meleady P, Barron N, Clynes M (2018) Depletion of endogenous miRNA-378-3p increases peak cell density of CHO DP12 cells and is correlated with elevated levels of ubiquitin carboxyl-terminal hydrolase 14. J Biotechnol 288:30–40
84. Kelly PS, Gallagher C, Clynes M, Barron N (2015) Conserved microRNA function as a basis for Chinese hamster ovary cell engineering. Biotechnol Lett 37:787–798
85. Pairawan MS, Bolhassani A, Rahimpour A (2019) Enhanced transient expression of an anti-CD52 monoclonal antibody in CHO cells through utilization of miRNA sponge technology. Res Pharm Sci 14:335–342
86. Kelly PS, Breen L, Gallagher C, Kelly S, Henry M, Lao NT, Meleady P, O'Gorman D, Clynes M, Barron N (2015) Re-programming CHO cell metabolism using miR-23 tips the balance towards a highly productive phenotype. Biotechnol J 10:1029–1040
87. Aquino-Jarquin G (2017) Emerging role of CRISPR/Cas9 technology for microRNAs editing in cancer research. Cancer Res 77:6812–6817
88. Kellner K, Solanki A, Amann T, Lao N, Barron N (2018) Targeting miRNAs with CRISPR/Cas9 to improve recombinant protein production of CHO cells. Methods Mol Biol 1850:221–235
89. Raab N, Mathias S, Alt K, Handrick R, Fischer S, Schmieder V, Jadhav V, Borth N, Otte K (2019) CRISPR/Cas9-mediated knockout of microRNA-744 improves antibody titer of CHO production cell lines. Biotechnol J 1800477
90. Puck TT, Cieciura SJ, Robinson A (1958) Genetics of somatic mammalian cells. J Exp Med 108:945–956
91. Fischer S, Handrick R, Aschrafi A, Otte K (2015) Unveiling the principle of microRNA-mediated redundancy in cellular pathway regulation. RNA Biol 12:238–247
92. Dweep H, Gretz N (2015) miRWalk2.0: a comprehensive atlas of microRNA-target interactions. Nat Methods 12:697
93. Griffiths-Jones S (2010) miRBase: microRNA sequences and annotation. Curr Protoc Bioinformatics. Chapter 12, Unit 12 9:1–10
94. Chou C-H, Shrestha S, Yang C-D, Chang N-W, Lin Y-L, Liao K-W, Huang W-C, Sun T-H, Tu S-J, Lee W-H, Chiew M-Y, Tai C-S, Wei T-Y, Tsai T-R, Huang H-T, Wang C-Y, Wu H-Y, Ho S-Y, Chen P-R, Chuang C-H, Hsieh P-J, Wu Y-S, Chen W-L, Li M-J, Wu Y-C, Huang X-Y, Ng FL, Buddhakosai W, Huang P-C, Lan K-C, Huang C-Y, Weng S-L, Cheng Y-N, Liang C, Hsu W-L, Huang H-D (2018) miRTarBase update 2018: a resource for experimentally validated microRNA-target interactions. Nucleic Acids Res 46:D296–D302

An Omic's Data-Driven Approach Towards Engineering Mammalian Cell Factories and Bioprocesses for Biopharmaceutical Production

Mauro Torres, Veronica Ortuzar, Alan J. Dickson, and Hirra Hussain

Abbreviations

AAV	Adeno-associated viral
ACE	Artificial chromosome expression
AMPK	AMP-activated protein kinase
CFL1	Cofilin
CHO	Chinese hamster ovary
CLD	Cell line development
CQA	Critical quality attribute
CRISPR/Cas9	Clustered regularly interspaced short palindromic repeats/RNA guided Cas9 nuclease
CRISPRa	CRISPR-based gene activation
dCas9	Cas9 endonuclease dead/Dead Cas9
DIGE	Difference gel electrophoresis
DMSO	Dimethyl suphoxide
EPO	Erythropoietin
ER	Endoplasmic reticulum
Erp27	Endoplasmic reticulum protein 27 kDa
Erp57	Endoplasmic reticulum protein 57 kDa
FBA	Flux balance analysis
HAC	Human Artificial Chromosome
HCP	Host cell protein

Mauro Torres and Hirra Hussain were Joint lead authors

M. Torres · V. Ortuzar · A. J. Dickson (✉) · H. Hussain
Manchester Institute of Biotechnology, Department of Chemical Engineering & Analytical Sciences, Faculty of Science and Engineering, University of Manchester, Manchester, UK
e-mail: alan.dickson@manchester.ac.uk

R. Pörtner (ed.), *Cell Culture Engineering and Technology*, Cell Engineering 10,
https://doi.org/10.1007/978-3-030-79871-0_4

HEK293	Human embryonic kidney-293
Hs3st1	Heparan sulphate 3-O-sulfotransferase 1
HSC60	Heat shock protein 60 kDa
HSC70	Heat shock protein 70 kDa
HT	High titre
IFN-γ	interferon gamma
IgG	Immunoglobulin G
iPSC	Induced pleuripotent stem cells
LC-MS/MS	Liquid chromatography with tandem mass spectrometry
lncRNA	Long non-coding RNA
LT	Low titre
MAARGE	Multiplexable activation of artificially repressed genes
mAbs	Monoclonal antibodies
MDH	Malate dehydrogenase
MFA	Metabolic flux analysis
miRNA/miR	microRNA
MPC1/2	Mitochondrial pyruvate carrier 1 and 2
MS	Mass spectrometry
mTOR	Mechanistic target of rapamycin
NDST2	N-deacetylase/N-sulfotransferase
Neo^R	Aminoglycoside phosphotransferase, neomycin resistance protein
NSD	Nucleotide sugar donor
PAGE	Polyacrylamide gel electrophoresis
PC	Pyruvate carboxylase
PCA	Principal component analysis
RI	Random integration
RNA-seq	RNA sequencing
SEAP	Secreted alkaline phosphatase
siRNA	Small-interference RNA
SSI	Site-specific integration
ST6GAL	α-2,6-syalyltransferase
TALEN	Transcription activator-like effector nuclease
TCA	Tricarboxylic acid cycle
TFRE	Transcription factor regulatory elements
UTR	Untranslated region
VCP	Valosin-containing protein
ZFN	Zinc-finger nuclease

1 Introduction

Mammalian cells have successfully served as industrial platforms for manufacturing different types of biopharmaceuticals that are critical therapies for the treatment of complex chronic diseases (e.g., cancer, autoimmune disorders). These biopharma-

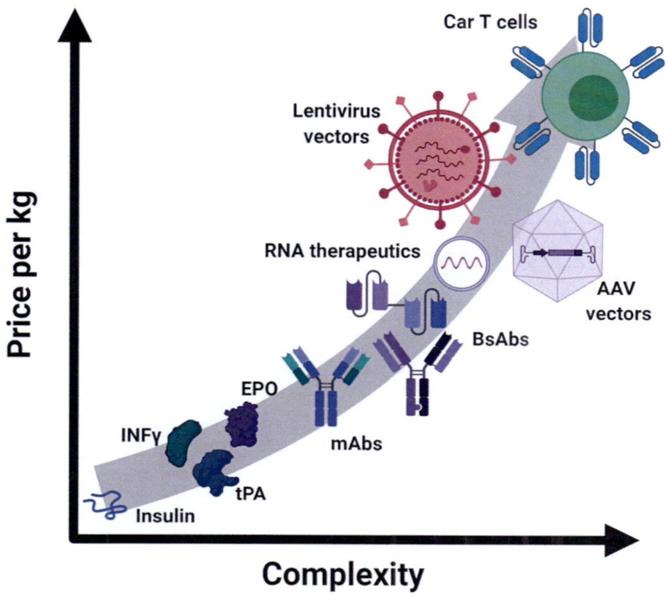

Fig. 1 Impact of the evolution of biopharmaceuticals on the manufacturability. This figure illustrates the exponential growth in the complexity (X-axis) of biopharmaceuticals, where complex molecular design, novel action mechanisms and manufacturing difficulties result in significantly increased product price (Y-axis)

ceuticals, including recombinant proteins (e.g., monoclonal antibodies) and viral particles (e.g., adeno-associated virus vectors, AAV), are largely produced using mammalian cell factories, with Chinese hamster ovary (CHO) and human embryonic kidney-293 (HEK293) cells as the predominant platforms. While the sector has managed to substantially improve cell densities, product yields and quality through trial and error ("brute force") approaches the capacity of these cell lines to turn recombinant genes into life-changing drugs is limited by intracellular processes (transcription, translation, processing, secretion). These intrinsic constraints are exacerbated by the development of a new generation of biopharmaceuticals, with novel designs and increased complexity for industrial manufacturing (e.g., multi-specific mAbs, complex fusion proteins, functional AAV vectors) (Fig. 1). The emergence of coronavirus disease 2019 (COVID-19) has highlighted the urgent need for production platforms that enable the rapid manufacturing of biotherapeutics on demand.

Even though we (as a community of scientists and industrialists) are aware of the cellular limitations, for productivity or product quality, we are less aware of the specific molecular events occurring within these cell factories that drive 'good' or 'bad' outcomes. Today, high-throughput omics technologies provide vast amounts of information of the cellular events, that increased our fundamental understanding

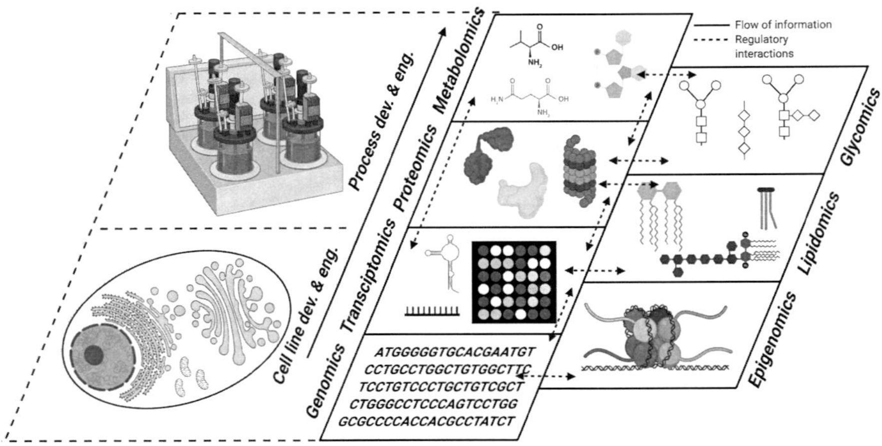

Fig. 2 Overview of the potential of omics technologies to increase the fundamental understanding of biological systems and to optimise cell factories and bioprocesses. This figure shows how omics technologies offers the potential to elucidate the intricate network of processes occurring within biological systems and to provide the rationale for cellular and process interventions that improve the product yields and quality

of these biological systems (enlighten the 'black box') and will change the paradigm by which cell factories can be engineered (Fig. 2). This Chapter focuses on the current state-of-the-art technologies that may be applied for designing engineering strategies for mammalian cell lines. These include the key lessons learnt from omics analysis on factors that corelate with productivity and product quality and how the combination of molecular and computational tools with 'omics data can rationalise intervention with mammalian cell factories and bioprocesses for enhanced production of biopharmaceuticals.

2 Molecular Approaches for Engineering Mammalian Cell Factories

2.1 Overexpression of Target Regulatory Genes

With the identification of gene targets for improving biotherapeutic production (Sect. 3), the overexpression of these genes is one of the most exploited approaches to improve mammalian cell lines. Similar to the method used to express recombinant proteins, the overexpression of gene targets is achieved through the delivery of an expression vector containing a codon-optimised version of a gene's cDNA under the control of a potent viral/cellular promoter and the presence of enhancer sequences. Incorporation of a selection marker (i.e., gene coding for antibiotic resistant or lacking metabolic enzyme) under the control of a week promoter allows the genomic integration of multiple copies of plasmid DNA and the generation of cell lines (pools) with heterogeneous expression of the recombinant gene (extensively reviewed by Gupta et al. [1]). The use of synthetic promoters to mediate gene expression in the host cell it has been also presented as an alternative engineering strategy [2]. While these approaches have been extensively used to overexpress several genes with beneficial consequences to the performance of mammalian cells, a single cell cloning process is needed to obtain cell lines with a desired phenotype and a homogenous expression of the specific target. To overcome these limitations and to target insertion for homogeneous expression, the use of semi-targeted transposase-based integration systems has been proposed [3, 4], allow transposition at a specific site of the genome or other expression systems. Several transposon systems have been designed for use in mammalian cells lines including *piggyBac*, Tol2 and *Sleeping Beauty* (Table 1).

An alternative to delivery of potential regulatory genes in standard DNA vectors is presented by the use of artificial chromosome expression (ACE), a technology that provides the potential to deliver a large genetic payload stable in the host cell without the need of genomic integration [10]. A series of studies have successfully used the ACE systems for expressing high levels of recombinant genes [11, 30, 31]. The capacity of ACE systems to incorporate multiple gene sequences (i.e., whole metabolic or signalling pathways) has opened the possibility to tailor-made cell factories, with desirable characteristics and phenotypes that allow for enhanced cellular performance [32]. While the generation of custom-built microbial factories have proven effective [33–35], this technology still needs to be evaluated in mammalian cells.

The emergence of CRISPR/Cas9 systems has provided a powerful and flexible tool for interventions of cellular genome. While often used as a genome-editing tool (Sect. 2.2), this system can be modified to genome incorporation and expression of a specific target genes. The generation of nuclease-null Cas9 (dCas9) combined with transcriptional activators (also known as CRISPR-based gene activation or CRISPRa) has been used to increase the endogenous expression of specific genes in mammalian cell hosts [36, 37]. Applications of CRISPRa in mammalian cell factories have focused on the upregulation of silenced glycosyltransferases [5] and UPR markers and anti-apoptotic genes [38]. Recently, Eisenhut et al. (2018) introduced a promising technology known as multiplexable activation of artificially repressed genes (or MAARGE), a sophisticated CRISPR/Cas9-based targeted integration system that enables the incorporation of multiple genes into the genome without laborious single cell cloning and screening process. However, additional efforts need to be undertaken to increase the expression of transgenes in order to make this novel mammalian expression system more robust.

An alternative method for expressing transgenes in mammalian cell hosts is via the direct transfection of *in vitro* transcribed mRNA, a technology that allows the gene target delivery without the undesired effects of plasmid-integration into the target genome. Despite the lower structural stability of mRNA compared to plasmid DNA [39], this delivery system brings forward several advantages, such

Table 1 Technologies for engineering mammalian cells lines

Approach	Method	Description	Reference(s)
Overexpression			
CRISPR/Cas9	CRISPRa (activation) CRISPRi (inhibition) MAARGE	The fusion of CRISPR/dCas9 with chromatin modification domains could provide epigenetic changes, or transcriptional activation and repression domains that can mediate transcriptional regulation.	[5–7]
		Multiple gene insertions from one plasmid in a single transfection.	[8]
Synthetic mammalian promoters	Promoter library design	Regulatory regions combine to generate a synthetic promoter that could enhance levels of gene expression.	[2, 9]
Artificial chromosome	Artificial Chromosome Expression (ACE) Human Artificial Chromosome (HAC)	Mammalian artificial chromosomes provide the delivery of large genetic payloads, that remain independent of the host genome, decreasing the risk of gene silencing positions.	[4, 10–12]
Transposon-based systems	*piggyBac* *Sleeping Beauty* Tol2	Regions flanked by direct repeats that allow the transposition of genes of interest. This technique allows a single copy of a recombinant DNA into one or multiple specific sites within the host genome.	[3, 13, 14]
RNAs	Synthetic circular RNAs	Circular RNA allows the ribosome to transcribe the protein message indefinitely, improve recombinant protein production by skipping the start process of the transcription. It may overload translational machinery.	[15]
Knock-out			
	Zinc-finger nucleases (ZFNs), transcription activator-like effector nucleases (TALENs), and CRISPR/Cas9	Genome editing capability to delete disadvantageous gene targets from the host genome.	[16–18]
Knockdown			
	Small interfering RNA (siRNA)/micro RNA (mRNA)	Operate within the RNA interference pathway to downregulate the expression of specific genes by degrading mRNA after transcription, preventing translation. siRNA is highly specific, and miRNA can target multiple sequences at one time.	[19–21]
	Long non-coding RNAs (lncRNA)	Regulates transcriptional machinery, sense protein-coding mRNAs without affecting the expression levels of the target mRNA	[22–25]
Directed evolution			
	Serum-free, nutrient feed, temperature adaptation, osmotic conditions Toxic by-product production	Reprogramming of cellular characteristics through the alteration of metabolic pathways that are translated into a specific phenotype, by non-invasive methods.	[26–29]

as considerably higher molar amounts of mRNA per transfection [40], no overload of the transcriptional machinery to avoid genetic and epigenetic controls and no requirements for nuclear translocation [41]. However, it may impose a significant load of the translational and post-translational machinery within cells. While this technology has been extensively used for the generation of induced pluripotent stem cells (iPSCs) [42], transient modification of cell phenotypes [43] and the development mRNA-vaccines [44], it has just recently been used in mammalian cell hosts for biopharmaceutical production [45–48]. Its large-scale implementation to enhance biopharmaceutical production by mammalian cells needs exploration [49].

2.2 Knock-Out of Gene Targets

In contrast to overexpressing genes, gene knock-out offers the possibility to delete disadvantageous targets from the host genome. The technologies for deleting specific genes have evolved from chemical- or physical-induced random mutagenesis to precise genome editing systems. The migration to targeted genome engineering has been promoted by the development of highly-specific technologies, such as involving the use of zinc-finger nucleases (ZFNs), transcription activator-like effector nucleases (TALENs) or CRISPR/Cas9 system [16–18]. These technologies have been used in mammalian cells to delete metabolic enzymes [50–53], glycosylation enzymes [54–57] and signalling molecules [58–60]. Amongst these techniques, CRISPR/Cas9 system has dominated investigations of the re-design of mammalian cell lines due to its rapid, cost-effective and easy-to-apply methodology, thus increasing the possibilities of alternatives engineering strategies (Sect. 2.1).

2.3 Knock-Down of Gene Targets

An alternative to gene knock-outs is gene silencing (or knock-down), an approach that decreases the amount of specific mRNA species without affecting the genome integrity. The most common technology used for gene silencing involves small-interference RNAs (siRNAs), small double-stranded RNA molecules (~20–24 bp length) that exhibit the complementary sequence of the target mRNA. The interaction between the small non-coding RNA with the specific mRNA target leads to degradation of the mRNA [19, 61]. Specific siRNAs have been used successfully to decrease the expression of metabolic enzymes or signalling proteins, resulting in enhanced culture performance of mammalian cells [20, 62–64]. However, due to high specificity, the application of siRNAs is limited to single targets [65].

Another technology based on RNA involves the use of microRNAs (miRNAs), small nucleotide molecules (~19–25 bp length) that bind to the 3′ untranslated region (3′UTR) of target gene transcripts by imperfect base-pairing interactions (only a section of the miRNA molecule binds to the target) that can inhibit translation of the target mRNA [65]. The lack of specificity in miRNAs enables their interaction with multiple gene targets (via common 3′UTRs). This generates the scenario on modulation of entire metabolic/signalling pathways, bringing forward the hypothesis of miRNAs as cellular tools to maintain homeostasis of, and integrate, multiple processes within cells [61]. Introduction of miRNAs into mammalian cells has led to improvements of cell growth, apoptosis and recombinant protein production [66–68]. However, given that the resulting interactions remain difficult to predict, the use of this technology may lead to undesired phenotypes beyond the original intention. Along the same line, several studies have proposed the applications of long non-coding RNAs (lncRNAs, ~200 bp) as regulatory tools to modulate the expression of multiple cellular events in mammalian cells. While lncRNAs have been correlated with growth and productivity in mammalian cells [22], there is a need for a broader understanding of the mechanisms that underpin the potential for lncRNA's to be used as cell engineering tools.

2.4 Directed Evolution of Cellular Phenotype

An alternative method to develop mammalian cell hosts is through 'directed evolution' methodologies. The aim of directed evolution is to reprogram cellular characteristics by altering genetic circuits and metabolic/signalling pathways underpinning a complex functional phenotype in a rapid and cost-effective manner. Applications of this methodology in mammalian cells include the adaptation to suspension growth under serum-free media [26, 69, 70], optimizing nutrient feed [71, 72], rapid proliferation [73], decreased production of metabolic waste-product (e.g. Lactate and ammonia) [27, 74, 75], hyperosmolarity [28], or cold-adapted cell lines [76, 77].

3 Lessons Learnt from 'Omics' Characterisation of CHO Cell Bioprocessing

3.1 Lesson I: Phenotype Instability Is an Inherent Feature of Mammalian Cells

Mammalian cells present an inherent genomic plasticity that may lead to unfavourable phenotypes and loss of the recombinant gene expression with negative effects on cellular productivities (known as production instability) [78–81]. As a tool to evaluate these phenotypic changes, extended cell line stability studies (2–3 months) are performed during the recombinant CHO cell line development (CLD) processes to ensure that the selected clone maintains overall product yields and quality along the manufacturing process [82]. This long and laborious procedure has promoted the identification of markers and molecular events (from genome and epigenome dynamics) correlated with production instability in mammalian cell factories. Initial genome drafts revealed that CHO cells undergo continuous changes in genome structure (e.g., chromosomal rearrangement, other karyotype variations) and sequence alterations (i.e., copy number variation) during routine cultivations [83, 84]. In particular, inherent genomic plasticity and multiple cell lineages with unique genomic landscapes have promoted researchers to continuously update of CHO-K1 and Chinese hamster genomes [85, 86], and sequence multiple industrially-relevant CHO cell variants, such as CHO-DG44, CHO-S and CHO-DXB11 [87–89]. Researchers have also taken cell line- and organelle-specific approaches to develop insights into the genetic diversity amongst CHO cell lines and their response to, and interaction, with the culture environment [90–95]. These data indicated that there is no single cause to explain production instability, but rather it arises as a consequence of a series of DNA sequence and genomic structure mutations because of continuous selection pressure (i.e., enriched medium for rapid cell growth, antibiotic/metabolic selection, different cultivation environment/scales). Omics studies have also evaluated epigenetic modifications (e.g., post-translational histone modifications, or DNA methylation) that have implications for DNA arrangement and gene accessibility. CHO cell lines are prone to large epigenetic changes during continuous cultivation (both in routine maintenance culture and in production batches) and exposure to different environmental conditions, affecting both the stability of recombinant gene expression, metabolism and cellular phenotype [73, 93, 96–99]. Specific changes in DNA methylation profile can switch on/off gene expression and the extent (and positioning) of methylation correlates with transcriptional activity [100], thus providing an explanation for the diverse production phenotypes of CHO cells. There is very significant interest in the application of genomic and epigenomic information to identify highly stable expression loci for the generation of new CHO cell hosts.

3.2 Lesson II: Post-translational Events Within Secretory Pathways Limit Cellular Productivity

Secretion of functional proteins with complete processing requires the coordinated action of multiple chaperones/enzymes within the secretory pathways, a long road with several compartments and regulatory checkpoints that have been proposed as limiting factors for the production of recombinant proteins [101, 102]. In this context, transcriptomic and proteomic studies have been powerful tools for development of fundamental understanding of the biology underpinning cellular productivity. A large number of studies have used different gene (e.g., microarrays, RNA-seq) and protein (e.g., 2D-DIGE, LC-MS/MS) profiling technologies to gain insights into mammalian cell factories during production of biopharmaceuticals.

Comparison between high and low producer cell lines (often at one single point during the exponential phase) has been used frequently to identify molecular components correlated with high productivities [103–107]. Other studies have focused on the effects of productivity enhancers, either molecules (e.g., sodium butyrate, DMSO) or environmental conditions (e.g., media/feeds, low temperature, hyperosmolarity), on the transcriptome/proteome profile [76, 108–111]. These studies have identified protein folding/secretion and cytoskeletal architecture as key biological functions correlated with high productive phenotypes across a diverse range of mammalian cell lines. Desirable qualities (in terms of cell growth, productivity and product quality) have been shown (via transcriptome and proteome profiling) to be correlated with the status of the secretory pathway, influencing the assembly, folding and processing of recombinant proteins. Therefore, many genes and proteins associated with post-translational events (from translocation to secretion) have provided the focus for potential targets for cell line engineering, with the caveat that these potential strategies remain cell- and protein- specific making them difficult to extrapolate to other cell factories.

Whilst transcriptome and proteome data have been used to provide mechanistic understanding of productivity of CHO cells, these profiling approaches have limitations for holistic explanation of the systems-level molecular events that set desirable cellular performance. This is particularly relevant for the study of the secretory pathways, orchestrated by specific transcriptional regulators and multiple enzyme and chaperones, whose expression varies dynamically according to the cell line and the surroundings [112, 113]. This limitation has been addressed with the development of computational tools and genome-scale models that are designed to allow the integration of multiple data sets of different nature (e.g., transcriptomic, proteomic, metabolomics) [114, 115]. Use of a multi-omics approaches aims to provide a more robust molecular description that takes into consideration the dynamic interplay between different levels of cellular integration. Recently, a genome-scale reconstruction of the secretory pathway highlighted the relevance of the post-translational events in the cellular productivity, delineating the metabolic costs and cellular machinery burden of each secreted protein in CHO, mouse and human cell lines [116]. The authors showed that highly secretory cells undergo a global adaptation that resulted in the decreased expression and secretion of energy-expensive host cell proteins, and provided a platform for simulating cellular interventions (knock-out/down) with the aim of enhancing performance of mammalian cells. In the context of rational design of mammalian cell factories, the combination of high-throughput ‘omics data with computational tools has the potential to revolutionise cell line development and opens possibilities for defining multiplex cell engineering strategies that target the secretory pathway (and essential ancillary reactions and cellular components), producing a systems-level shift towards a desired cellular phenotype.

3.3 *Lesson III: Efficient Metabolism Makes Mammalian Cell Factories More Effective*

Cellular metabolism encompasses the summation of all the biochemical reactions occurring within cells that support biological processes, by supplying biosynthetic building blocks (for biomass or protein production) and energy currency (in the form of ATP) as well as acting as regulatory elements in signalling pathways (e.g., mTOR, AMPK) [117]. While there is a generalised consensus about the relevance of cellular metabolism towards the performance of mammalian cell factories, we are still challenged by the metabolic definitions/signatures of a 'good' or 'optimal' bioprocess. Mass spectrometry (MS)-based metabolomics have become the tool of choice to analyse changes in concentration of (specific) metabolites both within cells and in the surrounding medium, thus providing substantial amount of information regarding the interaction of mammalian cell with their environment (culture medium) and characterisation of cell metabolism [118, 119]. Extracellular metabolite profiling monitors the main components of the culture medium (i.e., main carbon and energy sources, metabolic by-products, vitamins) that are critical for maintenance of growth and productivity [79, 118, 120–126]. The combination of this data with multivariate statistical analysis (e.g., principal component analysis [PCA] or partial least squares [PLS] variants) and/or stoichiometric metabolic modelling (e.g., metabolic flux analysis [MFA] or flux balance analysis [FBA]) has led to the identification of metabolic signatures, reflective of bioprocess status [127–134]. Particular interest has focused on the analysis of (specific) by-products of glucose metabolism (e.g., lactate, glycerol) and/or amino acid (e.g., ammonia, phenyl lactic acid, 2-hydroxybutyric acid, indole-3-carboxylate) metabolism that reveals catabolic imbalances that can impair the performance of mammalian cell cultures (extensively reviewed by Pereira et al. [135]). These valuable insights have promoted diverse process and cell engineering strategies focused on optimising cell metabolism using customised media/feeds or targeting the expression of metabolic enzymes.

Intracellular metabolite profiling offers insights of the physiological state on cells in culture via profiling metabolites indicative of energy metabolism, redox state, nucleotide synthesis and regulatory aspects of metabolic pathways [27, 70, 124, 136–138]. Additionally, the specific focus on lipids (*lipidomics*) has provided insights into regulatory control of cell growth, robustness and morphological status [139]. Some 'good' metabolic features (in terms of bioprocessing effectiveness) are emerging. For example, oxidative metabolic signatures (i.e., increased TCA cycle flux, favourable NADPH/NADP and GSH/GSSG ratios) underpin high-energy supply that leads to enhanced cellular specific productivity and prolonged culture lifespan. In contrast, a glycolytic metabolic signature sustains rapid growth of mammalian cells, but with an associated "cost" of high production of metabolic by-products (e.g., lactate) that may be detrimental to cell viability and product quality [124, 127, 128]. Extension to this type of knowledge will be crucial for identifying 'good' performance of clones during cell line development and scale-up processes.

However, there remains a gulf between defining metabolic signatures and clone selection or setting environmental conditions to support the desired equilibrium between high cell growth and productivity. We require further improvements in the metabolism monitoring technologies to increase our capacity to precisely assess and control metabolic processes and to design new cell lines with optimal metabolic signatures for biopharmaceuticals production.

3.4 Lesson IV: Product Quality Depends on an Intricate Network of Enzymes

Glycosylation (particularly N- and O-linked) is a hallmark for all secreted proteins and is a critical quality attribute (CQA) for biopharmaceuticals affecting their functioning and immunogenicity as therapeutics. The emergence of MS-based technologies has allowed the precise characterisation of glycan composition, indicating that CHO cells generate heterogeneous profiles of N- and O-linked glycosylation, with the precise profile being dependent on CHO cell variant [140], cell clone [141], culture medium/feed (e.g., glucose/glutamine levels, galactose/mannose supplementation) [142–144] and culture environment (e.g., pH, temperature, ammonia concentration) [143, 145–149]. Although glycan characterisation provides valuable information about the structure and quality of an expressed protein, the resultant heterogeneous glycosylation profile lacks the capacity to provide information about cellular pathways or culture conditions that lead to specific glycosylation profiles. Attachment and maturation of glycans are part of a complex biosynthesis/processing pathway involving a series of organelle-specific reactions that start within the endoplasmic reticulum (ER) and mature in the Golgi apparatus, via the action of multiple sequential and/or parallel processing enzyme pathways [150]. This complexity of potential reactions imposes a significant challenge for prediction of glycosylation profile or the extent of heterogeneity in the final product during bioprocessing. Therefore, global approaches will be necessary for a deeper understanding of the glycosylation machinery, an understanding that will need to integrate different layers of information (i.e., transcriptomics, proteomics, metabolomics, labelled-microscopy). Multi-omic approaches have been used to analyse the expression, activity and specificity of glycosyl transferases and hydrolases [151–153], to develop an understanding of the co-localisation enzymes and substrates (precursors) within compartments of the ER and the Golgi [154, 155] and to combine metabolic data into computational algorithms [146, 156]. In particular, the use of genome-scale and/or kinetic models in combination with process and cellular information have greatly contributed to a better understanding of the glycosylation machinery and set correlations and predictive systems that enable the association of inputs and outputs [116, 157–160]. With these resources, the intricacy of the glycoprotein processing network is being developed and, in turn, this is providing direction towards cellular/process engineering strategies to generate specific glycan structural profiles on the surface of recombinant proteins (an outcome that is particularly relevant for the manufacture of biosimilar and biobetter therapeutics) [161].

4 The Application of Data-Driven Approaches to Mammalian Cell Engineering

Omics technologies have been used to understand the biology of mammalian cells and to gain a better understanding of host cells and bioprocesses used to manufacture biopharmaceuticals. Earlier Sections have highlighted the types of engineering technologies and omics' studies, this Section focuses on examples where the lessons learnt from omics' studies have been translated into improved host cell systems and/or increased biopharmaceutical production through engineering approaches (summarised in Table 2).

4.1 Approach I: Increasing Stability of the Recombinant Gene Expression

The availability of detailed genome sequences and transcriptomic/epigenomic profiling, has enabled production instability of industrial mammalian cell lines to be addressed by use of site-specific integration (SSI), that targets recombinant gene incorporation to specific genomic loci with high expression, stability and desired epigenetic properties (also called hotspots or safe-harbours) [170]. Initial attempts to use SSI relied on the specific locus identified using genome and gene expression through screening of mammalian cell lines transfected with random integration (RI) protocols. These approaches have generated cell lines with high product yields using a single copy SSI, leading to comparable results to traditional random integration (RI) protocols [171, 172]. However, many potential genomic hot-spots still need to be experimentally validated in industrial settings [173]. Recent studies have searched for novel safe-harbours through systematic evaluation of the epigenetic signatures of mammalian cells, an approach that provides a clearer overview of gene transcription control within the context of the "living nucleus" and a potential map for identifying integration sites with maximum transgene production [93, 100]. Hilliard and Lee (2020) combined epigenomics and transcriptomics to analyse changes in the epigenome that occur during CHO cell line development and which can be related to different gene expression profiles in both host and recombinant CHO cell lines. The authors found that only 10% of the CHO genome contained transcriptionally permissive 3D chromatin structures with the enhanced genetic and epigenetic stabilities required for a desirable SSI [174]. These results provide a critical step towards further cellular interventions that increase the potential of

Table 2 Summary of studies that have applied Omics data in the engineering of mammalian cells

CHO Cell Line	Recombinant Protein(s)	Omics Study	Type of Engineering	Specific Targets	Phenotype	Reference(s)
CHO-GS cell system	IgG4	Metabolomics	Development of a nutrient feed and a cell culture feeding regime	Aspartate, asparagine, glutamate and pyruvate	*35% increase in cell biomass *2-fold increase in antibody titre	[70]
CHO	Recombinant IgG	Transcriptomics	Media supplementation	Increased lipid supplementation	*20% increase in protein titre	[162]
CHO	Recombinant antibody	Metabolomics	Media optimisation	Histidine	*increased cell productivity and protein quality	[71]
CHO^{BRI} (CHO DUXB11-derived)	a single-domain chimeric heavy chain camelid-derived antibody (EG2-hFc)	Proteomics	Nutrient supplementation of growth medium	Glucose, glutamine, proline, serine, cystine, asparagine, choline, and hypoxanthine	* $\sim$ 75% improvement to peak cell densities	[72]
CHO-K1	Monoclonal antibody	Transcriptomics and metabolomics	Development of a nutrient feed	Serine, threonine, tyrosine, lysine, histidine, glutamine, leucine, methionine, aspartate, cysteine, isoleucine, tryptophan	*increase in titre and cell productivity	[163]
CHO-K1	Monoclonal antibody	Transcriptomics and metabolomics	Development of a nutrient feed and media optimisation	Serine, threonine, tyrosine, lysine, leucine and valine	*increased specific cell productivity	[164]
Suspension CHO	Interferon-γ	Transcriptomics	Overexpression and knockdown	Fadd (upregulated in BC and FBC), Faim (downregulated in FBC), Alg-2 (upregulated in FBC), and requiem (upregulated in FBC)	*suppress apoptosis in culture. *prolonged culture viability *up to 2.5-fold increase in titre *improved quality - sialic acid content was maintained or enhanced	[165]
CHO-K1 CHO-DUKX	Monoclonal antibody	Transcriptomics and proteomics	Overexpression and knockdown	Valosin-containing protein (VCP)	*up to 2.1-fold increase in cell growth *knockdown of VCP decreases cell growth and viability	[20]

CHO-DG44	Recombinant monoclonal antibody (IgG) against rhesus D antigen	Metabolomics	Overexpression	Malate dehydrogenase II	*decrease in lactate secretion *increases in intracellular ATP and NADH *up to 1.9-fold increase in integral viable cell number *final mAb titer increased by 1.2-fold	[27]
Inducible CHO cell line	Erythropoietin	Proteomics	Overexpression	Heat shock protein 70 kDa and 60 kDa (HSC70, HSP60)	*10–15% enhanced cell concentration during serum-free adaptation and 15–33% reduction in adaptation time	[69]
CHO-K1	Human secreted alkaline phosphatase (SEAP)	Transcriptomics	Overexpression	microRNA-7 (miR-7)	*increased qP *decrease in cell growth	[166]
CHO-S	Heparin	Transcriptomics	Overexpression	Human N-deacetylase/N-sulfotransferase (NDST2) and mouse heparan sulfate 3-O-sulfotransferase 1 (Hs3st1)	*production of heparin in CHO cells	[167]
CHO-M (CHO-K1 derived)	Trastuzumab, bevacizumab, infliximab, interferon β protein and Etanercept (TNF receptor and fc-fusion)	Transcriptomics	Overexpression	Erp27, Erp57 and Foxa1	*increased cell density and viability *increased specific cell productivity *decreased amounts of reactive oxygen species (ROS)	[168]

(continued)

Table 2 (continued)

CHO Cell Line	Recombinant Protein(s)	Omics Study	Type of Engineering	Specific Targets	Phenotype	Reference(s)
CKO-K1	Human secreted alkaline phosphatase (SEAP)	Proteomics	siRNA knockdown	Cofilin	*increased cell productivity	[62, 63]
CS13–1.0 CHO cells	Chimeric IgG against S surface antigen of hepatitis B virus	Transcriptomics and translatome	siRNA knockdown	NeoR	*increased cell growth *increased antibody titre	[64]
CHO-S	n/a	Transcriptomics	Inhibition of miRNA	Mmu-miR- 466 h	*inhibition of the mmu- miR-466 h increased the expression levels of genes that resulted in increased cell viability and decreased Caspase-3/7 activation	[169]
CHO-DG44	IgG1	Transcriptomics	Stable expression of miRNAs	miR-557 and miR-1287	*increased the specific productivity of cells and resulted in a higher overall yield	[68]
CHO-S	Monoclonal antibody rituximab	Transcriptomics and proteomics	CRISPR knockout	Timp1, BGN, LGALS3BP, NID1.1, NID1.2, CTSD, Aga, Erp29, Gpr56, Tinagl1, Lgmn, Yeats2, Sparc and Lpl	*decreased host cell proteins (40–70%) *higher cell productivity and improved growth characteristics	[53]
CHO-K1	SEAP	Transcriptomics	Construction of synthetic promoters for biphasic production processes	Transcription factor regulatory elements (TFREs)	*increase in cell productivity of SEAP with synthetic promoters	[2]

SSI systems for generation of cell lines with high stability and expression of the transgenes.

A further application of omic technologies to increase recombinant gene expression utilises transcriptional information to design novel promoters. Johari et al., (2019) identified genes within the CHO genome that displayed high transcriptional activity under these different bioprocess environmental conditions. From these data, transcription factor regulatory elements (TFREs) were identified in the upstream regions of differentially-expressed genes and a specific subset of TFREs were functionally screened and were shown to support enhanced recombinant gene transcription in response to a switch to mild hypothermic growth conditions. Using such elements, the study generated novel synthetic promoters that were able to drive increased expression of recombinant genes in CHO cells, with an overall increase to cell productivity (up to 2.5-fold) [2]. This study exemplifies how omics enables re-design/tailored expression systems to develop improved manufacturing systems and processes.

4.2 Approach II: Enhancing Productivity by Targeting Secretory Pathways

Secretory pathways that convey recombinant proteins towards the extracellular environment consist of multiple intracellular compartments and are integrated by the action of several chaperones/enzymes that may limit post-translational events and overall production (Sect. 3.2). Transcriptomics and proteomics have provided great understanding of these events and suggest potential targets for enhancement of the overall process. For instance, Baik et al. (2011) investigated the intracellular proteome of recombinant CHO cells expressing Erythropoietin (EPO) in serum-supplemented and serum-free media. Proteomic profiling via 2D-PAGE and mass spectrometry analysis identified two chaperones, heat shock protein 70 kDa (HSC70) and 60 kDa (HSP60), as more highly expressed under serum-free conditions than in serum-containing medium, therefore directed them as potential cell engineering targets. Subsequent overexpression of HSC70 and HSP60, separately or together, led to an increased cell density (between 10% and 15%) and a decreased time for CHO cell adaptation to serum-free conditions [69]. Another study, using proteomic analysis of recombinant CHO cells, identified actin cytoskeleton regulator cofilin (CFL1) as a limiting factor for cell specific productivity of recombinant SEAP [62]. In a later study, the authors used siRNA to knockdown of CFL1 in CHO cells, resulting in an increase (80%) in recombinant protein specific productivity [63].

Comparative transcriptome analysis of recombinant CHO cells showing differential specific productivities has also suggested potential targets for cell engineering. For instance, transcriptomic analysis of CHO-K1 derived cell lines identified 32 potential target genes that were up-regulated in high producing clones – these candidates were involved in a variety of cell functions including signalling, protein folding, cytoskeleton organization, and cell survival [168]. Directed overexpression of two of the potential target genes in the ER (Erp27 and Erp57, which are chaperones that bind to unfolded proteins or are involved in di-sulphide bond formation, respectively) increased the cell density and culture viability. In addition, the production of a 'difficult-to-express' recombinant protein (interferonβ) was increased significantly, interpreted as a result of enhanced folding activity during processing and secretion [168]. An alternative strategy from the same study was overexpression of Foxa1 which was able to induce multiple metabolic changes to improve protein yields, decrease oxidative stress and improve cell growth.

At translational level, recombinant gene transcripts will compete with endogenous cellular transcripts at the level of the ribosome and this represents a potential molecular site for control of both recombinant gene expression and normal cellular regulatory events. Kallehauge et al. (2017) described a genome-wide study of protein translation (translatome) using ribosome profiling and an associated transcriptomic study (RNA-seq) for an antibody-producing CHO cell line. Whilst other studies have focused on global changes in translation, analysis of the translation of recombinant targets remains largely unexplored. This study showed that the recombinant mRNA sequestered up to 15% of actively translating ribosomes. Combined with transcriptomic analysis, the authors showed that the amount of transcript of the recombinant target influenced the cell-specific productivity. Using the associated datasets, the study examined the effects of limiting the expression of the Neo^R resistance marker, defining how much the load of an associated selection marker gene could have on recombinant gene expression. Knockdown of the Neo^R gene via siRNA increased cell growth and antibody production (18% increase in antibody yield). This work has generated an important paradigm surrounding the balance of translation for yields of the desired protein and sets an exemplar study to illustrate that shifting the transcriptional and translational capacity away from 'unnecessary' transcripts can increase the cellular ability to channel resources towards recombinant protein production [64]. Such studies combining genome-wide screening and multi-omics analysis provide a global view of the cell status and a powerful tool to identify how best to engineer the balance in cellular profiles to increase the capacity to produce biopharmaceuticals.

With the application of multi-omics analysis coupled with genome-scale modelling, researchers have shown the cellular burden that the secreted proteins impose on CHO cells and, specifically, to identify host cell proteins with increased metabolic costs [116]. This knowledge led to the design of a multiplex cell engineering strategy that created cell systems that are better producers and contain fewer process-related impurities such as host cell proteins. Kol et al., (2020) proposed eliminating host cell proteins would allow cellular resources to be channelled towards protein secretion, in particular recombinant protein production, whilst decreasing host-cell protein contamination in downstream processes. Their study a generated a series of 6, 11 and 14 protein knock-out clones via CRISPR/Cas9-mediated multiplex gene disruption, which resulted in between 40–70% decreased HCP content. Consequently, an improvement in antibody titre, quality and purity was observed [53]. Potentially this concept of modelling and gene editing can be a powerful approach to make better CHO cell factories with desirable consequences on the production and quality of biotherapeutics.

4.3 Approach III: Making Cell Metabolism More Efficient Through Metabolic Engineering

With the aim of improving the efficiency of cell metabolism to generate an enhanced manufacturing system, strategies have been suggested around channelling more metabolic intermediates into the mitochondria (for the TCA cycle) to enhance carbon utilisation and energy production. For instance, Chong et al., (2010) identified a potential bottleneck in the CHO cell TCA cycle from the accumulation of malate in culture medium, indicating a limitation in the conversion of malate to oxaloacetate (a reaction catalysed by malate dehydrogenase II [MDHII]). Overexpression of MDHII in CHO cells led to increased viable cell density, antibody production, higher amounts of intracellular ATP and NADH and decreased lactate production per cell [27]. Other approaches to increase TCA cycle activity have targeted the handling of pyruvate, a recognised metabolic bottleneck in CHO cells. Overexpression of pyruvate carboxylase (PC), which catalyse the conversion of pyruvate to oxaloacetate, significantly improved recombinant protein production and decreased lactate formation [175–177]. As an alternative, Bulté et al. [178] overexpressed mitochondrial pyruvate carriers (MPC1 and MPC2) in CHO cells, a strategy that resulted in an increased TCA cycle flux, decreased lactate production and increased r-protein production.

Metabolomics studies have identified several metabolic by-products that indicate loss of carbon from the inefficient catabolism of glucose and amino acids [135]. This knowledge has opened the possibilities of targeting specific enzymes to increase the efficiency of CHO cell metabolism. The most well-described exemplar by-products are lactate and ammonia, metabolites that present toxic effects during bioprocesses [179, 180]. The application of different process and metabolic engineering strategies have been applied successfully with resultant decreases in the production of these metabolites and improved culture performance of CHO cells [64, 74, 181–189]. Additionally, a series of intermediates or by-products of amino acid metabolism, have been identified as growth inhibitors of CHO cell growth [190]. The identification of these growth-inhibiting metabolites has promoted the development of a metabolic engineering strategy that completely eliminated the production of these compounds and enhanced cell growth and productivities in fed-batch cultures [191].

Analysis of culture limitations has led to the rational design of medium supplementation strategies or the design of nutrient feeds to overcome bottlenecks and boost cell culture performance [70, 72, 164]. Such strategies have proven to be effective and easily employed compared to genetic engineering approaches. Sellick et al. (2011) developed a nutrient feed based around four key amino acids (Table 2) which were observed to be depleted prior to the onset of the stationary phase. Use of the feed led to an increase in cell biomass and antibody titre [70]. Other studies have used multiple omics technologies to study cell culture processes, identified metabolic bottlenecks and designed feeds/supplements as a result. Blondeel et al., (2016) utilised both metabolomics and proteomic analysis to assess nutrient depletion and waste product accumulation following stable expression of a monoclonal antibody in CHO cell cultures. Subsequently, a nutrient feed was tailored to their specific process with 8 metabolites that were observed to be depleted in culture and use of this feed regime resulted in increased cell growth (~75% increase in peak cell density) [72].

Schaub et al. (2010) compared two fed-batch processes with an IgG-producing CHO cell line, one process was labelled as high titre (HT) and the other as low titre (LT). Transcriptomic analysis showed differences in gene expression between both processes over the time course of the fed-batch culture. In particular, gene expression of lipid metabolism pathways was upregulated in the HT process. In their study, the transcriptomic dataset led to design of a medium with increased lipid concentration, which, when added, resulted in a 20% increase in antibody titre [162]. Further, metabolomics studies using NMR measured and monitored intracellular and extracellular CHO cell metabolites and the data from that study directed the development of a proprietary growth medium which supported increased cell productivity and the protein quality [71]. The authors also showed a link between the depletion of histidine and decreased cell productivity.

Another study [163], applied a model to the metabolic data to identify limiting amino acids or those that significantly impact the recombinant target. The metabolic model was integrated and validated with transcriptomic analyses. Huang et al. (2020) described a genome-scale metabolic model to further understand their bioprocesses alongside transcriptomic analysis via RNA-sequencing. Using these models, strategies to optimise the culture medium were employed and verified experimentally. Such modelling and simulations aid in understanding processes and enable the development of new strategies to overcome limitations and minimise experimental testing of different combinations. In this case, the feed design strategies led to increased cell productivity [164]. Together these studies report an increase in cell growth and/or protein yields through medium deign and targeted nutrient feeds. Some overlap with the identified amino acids was seen between studies. However, no universal strategy was apparent potential due to different cell systems and recombinant proteins imposing different metabolic demands and consequent requirement for specific feeding regimes to achieve an efficient bioprocess.

4.4 Approach IV: Developing Specific Glycosylation Profiles Through Glycoengineering and Medium Design

Protein glycosylation status has proven to be a dynamic process during cell cultures, where the amounts of mannose and galactose species vary in a time-depending manner [157]. Sumit et al. (2019) undertook an integrative approach that employs multi-dimensional omics analyses (transcriptomic, metabolomics and glycomics) to analyse the glycosylation dynamics in recombinant CHO cells. The authors showed that changes to cellular metabolism (including central carbon metabolism and nucleotide sugar donor, NSD, biosynthesis) led to temporal bottlenecks in the addition of galactose and sialic acid. This knowledge enabled improvements in glycosylation heterogeneity by use of feeds with customised compositions of galactose, ManNAc and GlcNAc to bypass the impairment to biosynthetic pathways for NSDs [192]. For a refined control of glycosylation profiles, a synthetic biology approach has been used to redesign the glycosylation machinery. Chang et al. (2019) knocked out two glycosyltransferase genes and reintroduced synthetic glycosyltransferase genes under constitutive or inducible promoters. This allowing the production of antibodies with defined fucosylation (0–97%) and galactosylation (0–87%) contents [193]. These examples illustrate the possibilities for precise modification of N-glycosylation through process and genetic approaches that generate protein therapeutics with customised critical quality attributes.

CHO-based transcriptome analysis is a powerful approach for identification of the relationship between undesired product quality and the expression of specific metabolic/glycosylation enzymes. For instance, the lack of expression of α-2,6-syalyltransferase (ST6GAL) in CHO cells limited the production of recombinant protein with appropriate human glycosylation profile (sialic acid content) (Jenkins et al., 1996). Several groups have addressed this limitation by overexpressing ST6GAL in CHO cells that allowed these cell lines to generate the α-2,6-syalylated glycan residue [194–197]. Another example is associated with the lack or minimal expression of metabolic enzymes, N-deacetylase/N-sulfotransferase (NDST2) and heparan sulphate 3-O-sulfotransferase 1 (Hs3st1), critical for the synthesis and function of the anti-coagulant heparin [167]. This data led to engineered CHO cell systems overexpressing both enzymes, a strategy that resulted in the production of heparin with improved quality compared to previous efforts [167, 198].

4.5 Approach V: Improvement of Cell Growth Characteristics by Targeting Engineering of Apoptotic Pathways

Many studies have sought to improve cell growth and culture viability. One such study used a combination of transcriptomic and proteomic analysis of cell cultures with high and low growth rates to identify a panel of potential engineering candidates [20]. siRNA knockdown of these targets identified Valosin-containing

Protein (VCP) as having the biggest effect on cell culture growth and viability. Transient VCP overexpression resulted in increased cell growth and no impact on viability, further knockdown of this target had an adverse effect on culture growth validating the earlier observations that VCP was a key gene target to influence cell growth [20]. Stable overexpression of this target could potentially result in a better host cell phenotype with increased cell growth. Further, Wong et al. (2006a) reported that engineering of anti-apoptotic genes in CHO cell cultures delayed the onset of cell death and increased recombinant protein titres. Transcriptomic analysis of batch and fed-batch CHO cell cultures identified four differentially expressed genes (Fadd, Faim, Alg-2, and Requiem) [199]. In a later study, by the same group, the consequences of overexpression or knockdown of these genes were examined in a recombinant CHO cell line expressing human interferon gamma (IFN-γ) [165]. The data showed that targeting these anti-apoptotic genes conferred apoptosis resistance and enable prolonged cell cultures, improved cell viability and increased IFN-γ production and quality. However, the applicability of this strategy to other cell types and recombinant targets remains to be seen.

CHO-S cells grown in either fresh chemically-defined medium or nutrient-depleted were profiled by transcriptomic analysis [169]. In the depleted medium, cells showed increased caspase-3/7 activity, lowered culture viability and active apoptosis. Transcriptomic profiling via microarrays identified that 70 miRNAs were differentially expressed between medium types. In particular, mmu-miR-466 h, was identified as highly up-regulated in the depleted medium conditions and overexpression of mmu-miR-466 h decreased the amount of mRNA encoding several anti-apoptotic genes. Following analysis of the omics data set, the cells were transfected with an anti-miR-466 h which led to ~15% higher cell viability and decreased activation of caspase-3/7.

Another example was the identification of microRNA-7 (miR-7) as a potential target which promoted the development of a cell engineering strategy based of miR-7 overexpression that increased cell productivities [166]. However, using this strategy a decrease in cell growth was observed in response to miR7-overexpression suggesting that miR-7 expression may impact other processes such as protein translation or secretion in a temperature-dependent manner (Sect. 3.2). The use stable miRNA expression has gained attention to improve cell growth and productivity of CHO cell lines. Through a genome-wide miRNA screen, Strotbek et al., (2013) identified 9 miRNAs which correlated with an increase in IgG1 production. Expression of two of these identified miRNAs, miR-557 and miR-1287 increased viable cell density, protein titre and cell specific productivity [68]. Application of array analysis identified potentially unannotated miRNA sequences or those with unknown function. These example shows how omics can be used to identify the roles of miRNAs and how these can be engineered to overcome limitations on cell growth phenotypes and cell productivity.

5 Future Perspectives

'Omics datasets, coupled with engineering technologies presented here, are powerful tools for better understanding of cellular profiles and can be used to improve bioprocesses and increase biopharmaceutical production and quality. The use of omics has allowed researchers to gain large data sets that report on different CHO host cell backgrounds, distinct industrial bioprocesses, the impact of recombinant protein production on cell cultures and/or to better understand specific molecular events on a larger-scale at multiple cellular levels compared to previous technologies.

The methodologies used for omics analytics have improved vastly over recent years and the generation of data is no longer limiting. Currently, the 'bottleneck' mainly lies in the data interpretation using the databases and bioinformatic tools available and in the identification of the extent to which observations, amongst the very sizable amounts of data acquired, have functional significance or present viable targets for rewarding engineering. This step is often time-consuming, requiring follow-up studies to screen and validate potential markers and may well not bring the hoped-for rewards. The genomic databases used for bioinformatic analysis of CHO cells were initially limited and/or poorly annotated compared to other cell types e.g. Human cell lines. This restricted the certainty of interpretations. Technologies have advanced and CHO genome tools continue to be updated leading to a significant increase in our understanding of the predictability of engineering efforts.

Despite these limitations, many have sought to analyse and interpret omics data in CHO cells in response to different stimuli, but few studies have translated these observations into process development changes and/or increased biopharmaceutical production. To date the translation of omics analysis remains restricted and little is published in this area, compared to the significant number of papers that report omics data from CHO cells. Whilst there may be proprietary challenges in reporting the application of information to direct manufacturing outcome, it is also clear that we are moving to an era where the increased robustness of data and improved interpretation, along with technology to allow selective gene manipulation, is likely to generate improved translation of data to outcome.

The emergence of systems biology approaches to generate genome-wide models is proving to be a powerful resource to study cellular changes, identify potential limitations and guide engineering strategies to push expression hosts cells further to make 'super-producers". With the advancing technologies, the use of omics is becoming more accessible for researchers to use. Therefore, the field will benefit from bioinformatic tools that seamlessly compare and integrate multi-omic analyses such as genomic, transcriptomic, proteomic and metabolomic data. This will provide a holistic view of the cell and/or culture status and allow modelling/predictions of scenarios that would result in a more efficient process without the need for rigorous empirical screening and validation.

As we move towards increased innovation in biopharmaceutical production, as therapies become more complex with changing modes of manufacturing, this will challenge existing expression systems in different ways. The use of omics will prove to be a dominant force in order to characterise and understand these specific processes and develop methodologies with greater efficiency. With greater accessibility to such technologies and bioinformatic tools we anticipate further expansion in this area and greater application of this data into bioprocesses to create better and smarter platforms for biopharmaceutical production.

Acknowledgments The authors have been able to write this Chapter as a result of the understanding developed around this area through training and research support from a Becas Chile Scholarship and by funding from the Biotechnology and Biological Sciences Research Council (BBSRC; BB/R002096/1) and the Engineering and Physical Sciences Research Council (EPSRC; EP/T517252/1) of the UK.

References

1. Gupta K, Parasnis M, Jain R, Dandekar P (2019) Vector-related stratagems for enhanced monoclonal antibody production in mammalian cells. Biotechnol Adv 37:107415. https://doi.org/10.1016/j.biotechadv.2019.107415
2. Johari YB, Brown AJ, Alves CS, Zhou Y, Wright CM, Estes SD, Kshirsagar R, James DC (2019) CHO genome mining for synthetic promoter design. J Biotechnol 294:1–13. https://doi.org/10.1016/j.jbiotec.2019.01.015
3. Balasubramanian S, Rajendra Y, Baldi L, Hacker DL, Wurm FM (2016) Comparison of three transposons for the generation of highly productive recombinant CHO cell pools and cell lines. Biotechnol Bioeng 113:1234–1243. https://doi.org/10.1002/bit.25888
4. Schweickert PG, Cheng Z (2020) Application of genetic engineering in biotherapeutics development. J Pharm Innov 15:232–254. https://doi.org/10.1007/s12247-019-09411-6
5. Karottki KJ l C, Hefzi H, Xiong K, Shamie I, Hansen AH, Li S, Pedersen LE, Li S, Lee JS, Lee GM, Kildegaard HF, Lewis NE (2020) Awakening dormant glycosyltransferases in CHO cells with CRISPRa. Biotechnol Bioeng 117:593–598. https://doi.org/10.1002/bit.27199
6. Shen CC, Sung LY, Lin SY, Lin MW, Hu YC (2017) Enhancing protein production yield from Chinese hamster ovary cells by CRISPR interference. ACS Synth Biol 6:1509–1519. https://doi.org/10.1021/acssynbio.7b00020
7. Xiong K, Marquart KF, la Cour Karottki KJ, Li S, Shamie I, Lee JS, Gerling S, Yeo NC, Chavez A, Lee GM, Lewis NE, Kildegaard HF (2019) Reduced apoptosis in Chinese hamster ovary cells via optimized CRISPR interference. Biotechnol Bioeng 116:1813–1819. https://doi.org/10.1002/bit.26969
8. Eisenhut P, Klanert G, Weinguny M, Baier L, Jadhav V, Ivansson D, Borth N (2018) A CRISPR/Cas9 based engineering strategy for overexpression of multiple genes in Chinese hamster ovary cells. Metab Eng 48:72–81. https://doi.org/10.1016/j.ymben.2018.05.017
9. Brown AJ, Sweeney B, Mainwaring DO, James DC (2014) Synthetic promoters for CHO cell engineering. Biotechnol Bioeng 111:1638–1647. https://doi.org/10.1002/bit.25227
10. Greene A, Pascarelli K, Broccoli D, Perkins E (2019) Engineering synthetic chromosomes by sequential loading of multiple genomic payloads over 100 Kilobase pairs in size. Mol Ther – Methods Clin Dev 13:463–473. https://doi.org/10.1016/j.omtm.2019.04.006
11. Combs RG, Yu E, Roe S, Piatchek MB, Jones HL, Mott J, Kennard ML, Goosney DL, Monteith D (2011) Fed-batch bioreactor performance and cell line stability evaluation of the artificial chromosome expression technology expressing an IgG1 in chinese hamster ovary cells. Biotechnol Prog 27:201–208. https://doi.org/10.1002/btpr.505
12. Ohira T, Miyauchi K, Uno N, Shimizu N, Kazuki Y, Oshimura M, Kugoh H (2019) An efficient protein production system via gene amplification on a human artificial chromosome

and the chromosome transfer to CHO cells. Sci Rep 9:1–13. https://doi.org/10.1038/s41598-019-53116-2
13. Izsvák Z, Hackett PB, Cooper LJN, Ivics Z (2010) Translating sleeping beauty transposition into cellular therapies: victories and challenges. BioEssays 32:756–767
14. Rajendra Y, Balasubramanian S, Peery RB, Swartling JR, McCracken NA, Norris DL, Frye CC, Barnard GC (2017) Bioreactor scale up and protein product quality characterization of piggyBac transposon derived CHO pools. Biotechnol Prog 33:534–540. https://doi.org/10.1002/btpr.2447
15. Costello A, Lao NT, Barron N, Clynes M (2020) Reinventing the wheel: synthetic circular RNAs for mammalian cell engineering. Trends Biotechnol 38:217–230
16. Carroll D (2017) Genome editing: past, present, and future. Yale J Biol Med 90:653–659
17. Carroll D (2014) Genome engineering with targetable nucleases. Annu Rev Biochem 83:409–439
18. Komor AC, Badran AH, Liu DR (2017) CRISPR-based technologies for the manipulation of eukaryotic genomes. Cell 168:20–36
19. Vito D, Smales CM (2018) Engineering of the cellular translational machinery and non-coding RNAs to enhance CHO cell growth, recombinant product yields and quality. Curr Opin Chem Eng 22:199–208
20. Doolan P, Meleady P, Barron N, Henry M, Gallagher R, Gammell P, Melville M, Sinacore M, McCarthy K, Leonard M, Charlebois T, Clynes M (2010) Microarray and proteomics expression profiling identifies several candidates, including the Valosin-Containing Protein (VCP), involved in regulating high cellular growth rate in production CHO cell lines. Biotechnol Bioeng. https://doi.org/10.1002/bit.22670
21. Kwang Hong J, Lakshmanan M, Goudar C, Lee D-Y, Betenbaugh M, Titchener-Hooker N (2018) Towards next generation CHO cell line development and engineering by systems approaches. Curr Opin Chem Eng 22:1–10. https://doi.org/10.1016/j.coche.2018.08.002
22. Vito D, Eriksen JC, Skjødt C, Weilguny D, Rasmussen SK, Smales CM (2020) Defining lncRNAs correlated with CHO cell growth and IgG productivity by RNA-Seq. iScience:23. https://doi.org/10.1016/j.isci.2019.100785
23. Alessio E, Bonadio RS, Buson L, Chemello F, Cagnin S (2020) A single cell but many different transcripts: a journey into the world of long non-coding RNAs. Int J Mol Sci 21
24. Zucchelli S, Fasolo F, Russo R, Cimatti L, Patrucco L, Takahashi H, Jones MH, Santoro C, Sblattero D, Cotella D, Persichetti F, Carninci P, Gustincich S (2015) SINEUPs are modular antisense long non-coding RNAs that increase synthesis of target proteins in cells. Front Cell Neurosci 9:174. https://doi.org/10.3389/fncel.2015.00174
25. Zucchelli S, Patrucco L, Persichetti F, Gustincich S, Cotella D (2016) Engineering translation in mammalian cell factories to increase protein yield: the unexpected use of long non-coding SINEUP RNAs. Comput Struct Biotechnol J 14:404–410
26. Jukić S, Bubenik D, Pavlović N, Tušek AJ, Srček VG (2016) Adaptation of CHO cells in serum-free conditions for erythropoietin production: application of EVOP technique for process optimization. Biotechnol Appl Biochem 63:633–641. https://doi.org/10.1002/bab.1468
27. Chong WPK, Reddy SG, Yusufi FNK, Lee DY, Wong NSC, Heng CK, Yap MGS, Ho YS (2010) Metabolomics-driven approach for the improvement of Chinese hamster ovary cell growth: overexpression of malate dehydrogenase II. J Biotechnol 147:116–121. https://doi.org/10.1016/j.jbiotec.2010.03.018
28. Kamachi Y, Omasa T (2018) Development of hyper osmotic resistant CHO host cells for enhanced antibody production. J Biosci Bioeng 125:470–478. https://doi.org/10.1016/j.jbiosc.2017.11.002
29. Kishishita S, Katayama S, Kodaira K, Takagi Y, Matsuda H, Okamoto H, Takuma S, Hirashima C, Aoyagi H (2015) Optimization of chemically defined feed media for monoclonal antibody production in Chinese hamster ovary cells. J Biosci Bioeng 120:78–84. https://doi.org/10.1016/j.jbiosc.2014.11.022

30. Kennard ML, Goosney DL, Monteith D, Roe S, Fischer D, Mott J (2009) Auditioning of CHO host cell lines using the artificial chromosome expression (ACE) technology. Biotechnol Bioeng. https://doi.org/10.1002/bit.22407
31. Lindenbaum M, Perkins E, Csonka E, Fleming E, Garcia L, Greene A, Gung L, Hadlaczky G, Lee E, Leung J, MacDonald N, Maxwell A, Mills K, Monteith D, Perez CF, Shellard J, Stewart S, Stodola T, Vandenborre D, Vanderbyl S, Ledebur HC (2004) A mammalian artificial chromosome engineering system (ACE system) applicable to biopharmaceutical protein production, transgenesis and gene-based cell therapy. Nucleic Acids Res 32:e172–e172. https://doi.org/10.1093/nar/gnh169
32. Eisenstein M (2020) Building blocks for bespoke biology. Nature 578:633–635. https://doi.org/10.1038/d41586-020-00511-9
33. Luo Z, Hoffmann SA, Jiang S, Cai Y, Dai J (2020) Probing eukaryotic genome functions with synthetic chromosomes. Exp Cell Res 390:111936
34. Richardson SM, Mitchell LA, Stracquadanio G, Yang K, Dymond JS, DiCarlo JE, Lee D, Huang CLV, Chandrasegaran S, Cai Y, Boeke JD, Bader JS (2017) Design of a synthetic yeast genome. Science 355(80):1040–1044. https://doi.org/10.1126/science.aaf4557
35. Venetz JE, Del Medico L, Wölfle A, Schächle P, Bucher Y, Appert D, Tschan F, Flores-Tinoco CE, Van Kooten M, Guennoun R, Deutsch S, Christen M, Christen B (2019) Chemical synthesis rewriting of a bacterial genome to achieve design flexibility and biological functionality. Proc Natl Acad Sci U S A 116:8070–8079. https://doi.org/10.1073/pnas.1818259116
36. Forstnerič V, Oven I, Ogorevc J, Lainšček D, Praznik A, Lebar T, Jerala R, Horvat S (2019) CRISPRa-mediated FOXP3 gene upregulation in mammalian cells. Cell Biosci 9. https://doi.org/10.1186/s13578-019-0357-0
37. Wang G, Chow RD, Bai Z, Zhu L, Errami Y, Dai X, Dong MB, Ye L, Zhang X, Renauer PA, Park JJ, Shen L, Ye H, Fuchs CS, Chen S (2019) Multiplexed activation of endogenous genes by CRISPRa elicits potent antitumor immunity. Nat Immunol 20:1494–1505. https://doi.org/10.1038/s41590-019-0500-4
38. Martella A, Firth M, Taylor BJM, Göppert A, Cuomo EM, Roth RG, Dickson AJ, Fisher DI (2019) Systematic evaluation of CRISPRa and CRISPRi modalities enables development of a multiplexed, orthogonal gene activation and repression system. ACS Synth Biol 8:1998–2006. https://doi.org/10.1021/acssynbio.8b00527
39. Andreev DE, Terenin IM, Dmitriev SE, Shatsky IN (2016) Pros and cons of pDNA and mRNA transfection to study mRNA translation in mammalian cells. Gene 578:1–6
40. Kim TK, Eberwine JH (2010) Mammalian cell transfection: the present and the future. Anal Bioanal Chem 397:3173–3178. https://doi.org/10.1007/s00216-010-3821-6
41. Yamamoto A, Kormann M, Rosenecker J, Rudolph C (2009) Current prospects for mRNA gene delivery. Eur J Pharm Biopharm 71:484–489
42. Warren L, Manos PD, Ahfeldt T, Loh YH, Li H, Lau F, Ebina W, Mandal PK, Smith ZD, Meissner A, Daley GQ, Brack AS, Collins JJ, Cowan C, Schlaeger TM, Rossi DJ (2010) Highly efficient reprogramming to pluripotency and directed differentiation of human cells with synthetic modified mRNA. Cell Stem Cell 7:618–630. https://doi.org/10.1016/j.stem.2010.08.012
43. Caruso HG, Torikai H, Zhang L, Maiti S, Dai J, Do KA, Singh H, Huls H, Lee DA, Champlin RE, Heimberger AB, Cooper LJN (2016) Redirecting T-cell specificity to EGFR using mRNA to self-limit expression of chimeric antigen receptor. J Immunother 39:205–217. https://doi.org/10.1097/CJI.0000000000000126
44. Pardi N, Hogan MJ, Weissman D (2020) Recent advances in mRNA vaccine technology. Curr Opin Immunol 65:14–20
45. Coats MT, Bydlinski N, Maresch D, Diendorfer A, Klanert G, Borth N (2019) mRNA transfection into CHO-cells reveals production bottlenecks. Biotechnol J 1900198:1–7. https://doi.org/10.1002/biot.201900198
46. Costello A, Lao NT, Barron N, Clynes M (2019) Continuous translation of circularized mRNA improves recombinant protein titer. Metab Eng 52:284–292. https://doi.org/10.1016/j.ymben.2019.01.002

47. Costello A, Lao NT, Barron N, Clynes M (2019) Improved yield of rhEPO in CHO cells with synthetic 5′ UTR. Biotechnol Lett 41:231–239. https://doi.org/10.1007/s10529-018-2632-2
48. Arteta MY, Kjellman T, Bartesaghi S, Wallin S, Wu X, Kvist AJ, Dabkowska A, Székely N, Radulescu A, Bergenholtz J, Lindfors L (2018) Successful reprogramming of cellular protein production through mRNA delivered by functionalized lipid nanoparticles. Proc Natl Acad Sci U S A 115:E3351–E3360. https://doi.org/10.1073/pnas.1720542115
49. Van Hoecke L, Roose K (2019) How mRNA therapeutics are entering the monoclonal antibody field. J Transl Med 17:54
50. Sealover NR, Davis AM, Brooks JK, George HJ, Kayser KJ, Lin N (2013) Engineering Chinese Hamster Ovary (CHO) cells for producing recombinant proteins with simple glycoforms by zinc-finger nuclease (ZFN)-mediated gene knockout of mannosyl (alpha-1,3-)-glycoprotein beta-1, 2-N-acetylglucosaminyltransferase (Mgat1). J Biotechnol 167:24–32. https://doi.org/10.1016/j.jbiotec.2013.06.006
51. Chai YR, Cao XX, Ge MM, Mi CL, Guo X, Wang TY (2020) Knockout of cytidine monophosphate-N-acetylneuraminic acid hydroxylase in Chinese hamster ovary cells by CRISPR/Cas9-based gene-editing technology. Biochem Eng J 161:107663. https://doi.org/10.1016/j.bej.2020.107663
52. Fan L, Kadura I, Krebs LE, Hatfield CC, Shaw MM, Frye CC (2012) Improving the efficiency of CHO cell line generation using glutamine synthetase gene knockout cells. Biotechnol Bioeng 109:1007–1015. https://doi.org/10.1002/bit.24365
53. Kol S, Ley D, Wulff T, Decker M, Arnsdorf J, Schoffelen S, Hansen AH, Jensen TL, Gutierrez JM, Chiang AWT, Masson HO, Palsson BO, Voldborg BG, Pedersen LE, Kildegaard HF, Lee GM, Lewis NE (2020) Multiplex secretome engineering enhances recombinant protein production and purity. Nat Commun 11. https://doi.org/10.1038/s41467-020-15866-w
54. Yang Z, Wang S, Halim A, Schulz MA, Frodin M, Rahman SH, Vester-Christensen MB, Behrens C, Kristensen C, Vakhrushev SY, Bennett EP, Wandall HH, Clausen H (2015) Engineered CHO cells for production of diverse, homogeneous glycoproteins. Nat Biotechnol 33:842–844. https://doi.org/10.1038/nbt.3280
55. Byrne G, O'Rourke SM, Alexander DL, Yu B, Doran RC, Wright M, Chen Q, Azadi P, Berman PW (2018) CRISPR/Cas9 gene editing for the creation of an MGAT1-deficient CHO cell line to control HIV-1 vaccine glycosylation. PLoS Biol 16. https://doi.org/10.1371/journal.pbio.2005817
56. Schulz MA, Tian W, Mao Y, Van Coillie J, Sun L, Larsen JS, Chen YH, Kristensen C, Vakhrushev SY, Clausen H, Yang Z (2018) Glycoengineering design options for IgG1 in CHO cells using precise gene editing. Glycobiology 28:542–549. https://doi.org/10.1093/glycob/cwy022
57. Li SW, Wright M, Healey JF, Hutchinson JM, O'Rourke S, Mesa KA, Lollar P, Berman PW (2020) Gene editing in CHO cells to prevent proteolysis and enhance glycosylation: Production of HIV envelope proteins as vaccine immunogens. PLoS One:15. https://doi.org/10.1371/journal.pone.0233866
58. Cost GJ, Freyvert Y, Vafiadis A, Santiago Y, Miller JC, Rebar E, Collingwood TN, Snowden A, Gregory PD (2010) BAK and BAX deletion using zinc-finger nucleases yields apoptosis-resistant CHO cells. Biotechnol Bioeng 105:330–340. https://doi.org/10.1002/bit.22541
59. Ha TK, Hansen AH, Kildegaard HF, Lee GM (2020) Knockout of sialidase and pro-apoptotic genes in Chinese hamster ovary cells enables the production of recombinant human erythropoietin in fed-batch cultures. Metab Eng 57:182–192. https://doi.org/10.1016/j.ymben.2019.11.008
60. Henry MN, MacDonald MA, Orellana CA, Gray PP, Gillard M, Baker K, Nielsen LK, Marcellin E, Mahler S, Martínez VS (2020) Attenuating apoptosis in Chinese hamster ovary cells for improved biopharmaceutical production. Biotechnol Bioeng 117:1187–1203
61. Hong JK, Lakshmanan M, Goudar C, Lee DY (2018) Towards next generation CHO cell line development and engineering by systems approaches. Curr Opin Chem Eng 22:1–10
62. Hayduk EJ, Lee KH (2005) Cytochalasin D can improve heterologous protein productivity in adherent Chinese hamster ovary cells. Biotechnol Bioeng 90:354–364. https://doi.org/10.1002/bit.20438

63. Hammond S, Lee KH (2012) RNA interference of cofilin in Chinese hamster ovary cells improves recombinant protein productivity. Biotechnol Bioeng 109:528–535. https://doi.org/10.1002/bit.23322
64. Kallehauge TB, Li S, Pedersen LE, Ha TK, Ley D, Andersen MR, Kildegaard HF, Lee GM, Lewis NE (2017) Ribosome profiling-guided depletion of an mRNA increases cell growth rate and protein secretion. Sci Rep:7. https://doi.org/10.1038/srep40388
65. Valdés-Bango Curell R, Barron N (2018) Exploring the potential application of short non-coding RNA-based genetic circuits in Chinese hamster ovary cells. Biotechnol J 13:1–8. https://doi.org/10.1002/biot.201700220
66. Park J, Inwood S, Kruthiventi S, Jenkins J, Shiloach J, Betenbaugh M (2018) Progressing from transient to stable packaging cell lines for continuous production of lentiviral and gammaretroviral vectors. Curr Opin Chem Eng 22:128–137. https://doi.org/10.1016/j.coche.2018.09.007
67. Fischer S, Marquart KF, Pieper LA, Fieder J, Gamer M, Gorr I, Schulz P, Bradl H (2017) miRNA engineering of CHO cells facilitates production of difficult-to-express proteins and increases success in cell line development. Biotechnol Bioeng 114:1495–1510. https://doi.org/10.1002/bit.26280
68. Strotbek M, Florin L, Koenitzer J, Tolstrup A, Kaufmann H, Hausser A, Olayioye MA (2013) Stable microRNA expression enhances therapeutic antibody productivity of Chinese hamster ovary cells. Metab Eng 20:157–166. https://doi.org/10.1016/j.ymben.2013.10.005
69. Baik JY, Ha TK, Kim YH, Lee GM (2011) Proteomic understanding of intracellular responses of recombinant chinese hamster ovary cells adapted to grow in serum-free suspension culture. Biotechnol Prog. https://doi.org/10.1002/btpr.685
70. Sellick CA, Croxford AS, Maqsood AR, Stephens G, Westerhoff HV, Goodacre R, Dickson AJ (2011) Metabolite profiling of recombinant CHO cells: designing tailored feeding regimes that enhance recombinant antibody production. Biotechnol Bioeng 108:3025–3031. https://doi.org/10.1002/bit.23269
71. Bradley SA, Ouyang A, Purdie J, Smitka TA, Wang T, Kaerner A (2010) Fermentanomics: monitoring mammalian cell cultures with NMR spectroscopy. J Am Chem Soc. https://doi.org/10.1021/ja101962c
72. Blondeel EJM, Ho R, Schulze S, Sokolenko S, Guillemette SR, Slivac I, Durocher Y, Guillemette JG, McConkey BJ, Chang D, Aucoin MG (2016) An omics approach to rational feed: enhancing growth in CHO cultures with NMR metabolomics and 2D-DIGE proteomics. J Biotechnol 234:127–138. https://doi.org/10.1016/j.jbiotec.2016.07.027
73. Weinguny M, Eisenhut P, Klanert G, Virgolini N, Marx N, Jonsson A, Ivansson D, Lövgren A, Borth N (2020) Random epigenetic modulation of CHO cells by repeated knockdown of DNA methyltransferases increases population diversity and enables sorting of cells with higher production capacities. Biotechnol Bioeng. https://doi.org/10.1002/bit.27493
74. Freund NW, Croughan MS (2018) A simple method to reduce both lactic acid and ammonium production in industrial animal cell culture. Int J Mol Sci 19. https://doi.org/10.3390/ijms19020385
75. Buchsteiner M, Quek LE, Gray P, Nielsen LK (2018) Improving culture performance and antibody production in CHO cell culture processes by reducing the Warburg effect. Biotechnol Bioeng 115:2315–2327. https://doi.org/10.1002/bit.26724
76. Bedoya-López A, Estrada K, Sanchez-Flores A, Ramírez OT, Altamirano C, Segovia L, Miranda-Ríos J, Trujillo-Roldán MA, Valdez-Cruz NA (2016) Effect of temperature downshift on the transcriptomic responses of Chinese hamster ovary cells using recombinant human tissue plasminogen activator production culture. PLoS One 11:151529. https://doi.org/10.1371/journal.pone.0151529
77. Sunley K, Tharmalingam T, Butler M (2008) CHO cells adapted to hypothermic growth produce high yields of recombinant β-interferon. Biotechnol Prog 24:898–906. https://doi.org/10.1002/btpr.9
78. Kim M, O'Callaghan PM, Droms KA, James DC (2011) A mechanistic understanding of production instability in CHO cell lines expressing recombinant monoclonal antibodies. Biotechnol Bioeng 108:2434–2446. https://doi.org/10.1002/bit.23189

79. Chusainow J, Yang YS, Yeo JHM, Ton PC, Asvadi P, Wong NSC, Yap MGS (2009) A study of monoclonal antibody-producing CHO cell lines: what makes a stable high producer? Biotechnol Bioeng 102:1182–1196. https://doi.org/10.1002/bit.22158
80. Osterlehner A, Simmeth S, Göpfert U (2011) Promoter methylation and transgene copy numbers predict unstable protein production in recombinant chinese hamster ovary cell lines. Biotechnol Bioeng. https://doi.org/10.1002/bit.23216
81. Wippermann A, Noll T (2017) DNA methylation in CHO cells. J Biotechnol 258:206–210. https://doi.org/10.1016/j.jbiotec.2017.08.005
82. Wurm FM, Wurm MJ (2017) Cloning of CHO cells, productivity and genetic stability-a discussion. PRO. https://doi.org/10.3390/pr5020020
83. Xu X, Nagarajan H, Lewis NE, Pan S, Cai Z, Liu X, Chen W, Xie M, Wang W, Hammond S, Andersen MR, Neff N, Passarelli B, Koh W, Fan HC, Wang J, Gui Y, Lee KH, Betenbaugh MJ, Quake SR, Famili I, Palsson BO, Wang J (2011) The genomic sequence of the Chinese hamster ovary (CHO)-K1 cell line. Nat Biotechnol 29:735–741. https://doi.org/10.1038/nbt.1932
84. Baumann M, Vcelar S, Hernandez-Lopez I, Melcher M, Borth N (2016) Assessment of genomic rearrangements in Chinese Hamster Ovary (CHO) cells. New Biotechnol. https://doi.org/10.1016/j.nbt.2016.06.737
85. Rupp O, MacDonald ML, Li S, Dhiman H, Polson S, Griep S, Heffner K, Hernandez I, Brinkrolf K, Jadhav V, Samoudi M, Hao H, Kingham B, Goesmann A, Betenbaugh MJ, Lewis NE, Borth N, Lee KH (2018) A reference genome of the Chinese hamster based on a hybrid assembly strategy. Biotechnol Bioeng. https://doi.org/10.1002/bit.26722
86. Hilliard W, MacDonald ML, Lee KH (2020) Chromosome-scale scaffolds for the Chinese hamster reference genome assembly to facilitate the study of the CHO epigenome. Biotechnol Bioeng. https://doi.org/10.1002/bit.27432
87. Lewis NE, Liu X, Li Y, Nagarajan H, Yerganian G, O'Brien E, Bordbar A, Roth AM, Rosenbloom J, Bian C, Xie M, Chen W, Li N, Baycin-Hizal D, Latif H, Forster J, Betenbaugh MJ, Famili I, Xu X, Wang J, Palsson BO (2013) Genomic landscapes of Chinese hamster ovary cell lines as revealed by the Cricetulus griseus draft genome. Nat Biotechnol 31:759–765. https://doi.org/10.1038/nbt.2624
88. Kaas CS, Kristensen C, Betenbaugh MJ, Andersen MR (2015) Sequencing the CHO DXB11 genome reveals regional variations in genomic stability and haploidy. BMC Genomics. https://doi.org/10.1186/s12864-015-1391-x
89. Kimura S, Omasa T (2018) Genome sequence comparison between Chinese hamster ovary (CHO) DG44 cells and mouse using end sequences of CHO BAC clones based on BAC-FISH results. Cytotechnology. https://doi.org/10.1007/s10616-018-0233-5
90. Brinkrolf K, Rupp O, Laux H, Kollin F, Ernst W, Linke B, Kofler R, Romand S, Hesse F, Budach WE, Galosy S, Müller D, Noll T, Wienberg J, Jostock T, Leonard M, Grillari J, Tauch A, Goesmann A, Helk B, Mott JE, Pühler A, Borth N (2013) Chinese hamster genome sequenced from sorted chromosomes. Nat Biotechnol
91. Partridge MA, Davidson MM, Hei TK (2007) The complete nucleotide sequence of Chinese hamster (Cricetulus griseus) mitochondrial DNA. DNA Seq - J DNA Seq Mapp. https://doi.org/10.1080/10425170601101287
92. Kelly PS, Clarke C, Costello A, Monger C, Meiller J, Dhiman H, Borth N, Betenbaugh MJ, Clynes M, Barron N (2017) Ultra-deep next generation mitochondrial genome sequencing reveals widespread heteroplasmy in Chinese hamster ovary cells. Metab Eng. https://doi.org/10.1016/j.ymben.2017.02.001
93. Feichtinger J, Hernández I, Fischer C, Hanscho M, Auer N, Hackl M, Jadhav V, Baumann M, Krempl PM, Schmidl C, Farlik M, Schuster M, Merkel A, Sommer A, Heath S, Rico D, Bock C, Thallinger GG, Borth N (2016) Comprehensive genome and epigenome characterization of CHO cells in response to evolutionary pressures and over time. Biotechnol Bioeng 113:2241–2253. https://doi.org/10.1002/bit.25990
94. Wenger T, Melcher M, Hrdina A, Leisch F, Sagmeister R, Baumann M, Borth N, Betenbaugh M, Heffner K, Puklowski A, Jadhav V, Auer N, Vcelar S (2017) Karyotype variation of CHO

host cell lines over time in culture characterized by chromosome counting and chromosome painting. Biotechnol Bioeng 115:165–173. https://doi.org/10.1002/bit.26453
95. Bevan S, Schoenfelder S, Young RJ, Zhang L, Andrews S, Fraser P, O'Callaghan PM (2020) High-resolution three-dimensional chromatin profiling of the Chinese hamster ovary cell genome. Biotechnol Bioeng 51:bit.27607. https://doi.org/10.1002/bit.27607
96. Veith N, Ziehr H, MacLeod RAF, Reamon-Buettner SM (2016) Mechanisms underlying epigenetic and transcriptional heterogeneity in Chinese hamster ovary (CHO) cell lines. BMC Biotechnol. https://doi.org/10.1186/s12896-016-0238-0
97. Dhiman H, Gerstl MP, Ruckerbauer D, Hanscho M, Himmelbauer H, Clarke C, Barron N, Zanghellini J, Borth N (2019) Genetic and epigenetic variation across genes involved in energy metabolism and mitochondria of Chinese hamster ovary cell lines. Biotechnol J 14:1800681. https://doi.org/10.1002/biot.201800681
98. Wippermann A, Klausing S, Rupp O, Noll T, Hoffrogge R (2013) First CpG island microarray for genome-wide analyses of DNA methylation in Chinese hamster ovary cells: new insights into the epigenetic answer to butyrate treatment. BMC Proc. https://doi.org/10.1186/1753-6561-7-s6-o5
99. Wippermann A, Klausing S, Rupp O, Albaum SP, Büntemeyer H, Noll T, Hoffrogge R (2014) Establishment of a CpG island microarray for analyses of genome-wide DNA methylation in Chinese hamster ovary cells. Appl Microbiol Biotechnol. https://doi.org/10.1007/s00253-013-5282-2
100. Hernandez I, Dhiman H, Klanert G, Jadhav V, Auer N, Hanscho M, Baumann M, Esteve-Codina A, Dabad M, Gómez J, Alioto T, Merkel A, Raineri E, Heath S, Rico D, Borth N (2019) Epigenetic regulation of gene expression in Chinese hamster ovary cells in response to the changing environment of a batch culture. Biotechnol Bioeng 116:677–692. https://doi.org/10.1002/bit.26891
101. Mathias S, Fischer S, Handrick R, Fieder J, Schulz P, Bradl H, Gorr I, Gamer M, Otte K (2018) Visualisation of intracellular production bottlenecks in suspension-adapted CHO cells producing complex biopharmaceuticals using fluorescence microscopy. J Biotechnol 271:47–55. https://doi.org/10.1016/j.jbiotec.2018.02.009
102. Kaneyoshi K, Uchiyama K, Onitsuka M, Yamano N, Koga Y, Omasa T (2019) Analysis of intracellular IgG secretion in Chinese hamster ovary cells to improve IgG production. J Biosci Bioeng 127:107–113. https://doi.org/10.1016/j.jbiosc.2018.06.018
103. Charaniya S, Le H, Rangwala H, Mills K, Johnson K, Karypis G, Hu WS (2010) Mining manufacturing data for discovery of high productivity process characteristics. J Biotechnol 147:186–197. https://doi.org/10.1016/j.jbiotec.2010.04.005
104. Charaniya S, Karypis G, Hu WS (2009) Mining transcriptome data for function-trait relationship of hyper productivity of recombinant antibody. Biotechnol Bioeng 102:1654–1669. https://doi.org/10.1002/bit.22210
105. Clarke C, Doolan P, Barron N, Meleady P, O'Sullivan F, Gammell P, Melville M, Leonard M, Clynes M (2011) Predicting cell-specific productivity from CHO gene expression. J Biotechnol 151:159–165. https://doi.org/10.1016/j.jbiotec.2010.11.016
106. Hancock WS, Lyubarskaya Y, Madden H, Carlage T, Mhatre R, Zang L, Hincapie M (2009) Proteomic profiling of a high-producing Chinese hamster ovary cell culture. Anal Chem 81:7357–7362. https://doi.org/10.1021/ac900792z
107. Orellana CA, Marcellin E, Schulz BL, Nouwens AS, Gray PP, Nielsen LK (2015) High-antibody-producing chinese hamster ovary cells up-regulate intracellular protein transport and glutathione synthesis. J Proteome Res. https://doi.org/10.1021/pr501027c
108. Yee JC, de Leon GM, Philp RJ, Yap M, Hu W-S (2008) Genomic and proteomic exploration of CHO and hybridoma cells under sodium butyrate treatment. Biotechnol Bioeng 99:1186–1204. https://doi.org/10.1002/bit.21665
109. Yuk IH, Zhang JD, Ebeling M, Berrera M, Gomez N, Werz S, Meiringer C, Shao Z, Swanberg JC, Lee KH, Luo J, Szperalski B (2014) Effects of copper on CHO cells: insights from gene expression analyses. Biotechnol Prog 30:429–442. https://doi.org/10.1002/btpr.1868

110. Reinhart D, Damjanovic L, Castan A, Ernst W, Kunert R (2018) Differential gene expression of a feed-spiked super-producing CHO cell line. J Biotechnol 285:23–37. https://doi.org/10.1016/j.jbiotec.2018.08.013
111. Baek E, Lee JS, Lee GM (2018) Untangling the mechanism of 3-methyladenine in enhancing the specific productivity: transcriptome analysis of recombinant Chinese hamster ovary cells treated with 3-methyladenine. Biotechnol Bioeng 115:2243–2254. https://doi.org/10.1002/bit.26777
112. Torres M, Akhtar S, McKenzie EA, Dickson AJ (2020) Temperature down-shift modifies expression of UPR-/ERAD-related genes and enhances production of a chimeric fusion protein in CHO cells. Biotechnol J 2000081:2000081. https://doi.org/10.1002/biot.202000081
113. Prashad K, Mehra S (2015) Dynamics of unfolded protein response in recombinant CHO cells. Cytotechnology 67:237–254. https://doi.org/10.1007/s10616-013-9678-8
114. Kuo CC, Chiang AW, Shamie I, Samoudi M, Gutierrez JM, Lewis NE (2018) The emerging role of systems biology for engineering protein production in CHO cells. Curr Opin Biotechnol 51:64–69
115. Chen C, Le H, Goudar CT (2016) Integration of systems biology in cell line and process development for biopharmaceutical manufacturing. Biochem Eng J 107:11–17. https://doi.org/10.1016/j.bej.2015.11.013
116. Gutierrez JM, Feizi A, Li S, Kallehauge TB, Hefzi H, Grav LM, Ley D, Baycin Hizal D, Betenbaugh MJ, Voldborg B, Faustrup Kildegaard H, Min Lee G, Palsson BO, Nielsen J, Lewis NE (2020) Genome-scale reconstructions of the mammalian secretory pathway predict metabolic costs and limitations of protein secretion. Nat Commun 11:68. https://doi.org/10.1038/s41467-019-13867-y
117. Dickson AJ (2014) Enhancement of production of protein biopharmaceuticals by mammalian cell cultures: the metabolomics perspective. Curr Opin Biotechnol 30:73–79
118. Ho YS, Chong WPK, Lee D-Y, Chan ECY, Hiu AP, Thng SH (2012) LC-MS-based metabolic characterization of high monoclonal antibody-producing Chinese hamster ovary cells. Biotechnol Bioeng 109:3103–3111. https://doi.org/10.1002/bit.24580
119. Sellick CA, Hansen R, Stephens GM, Goodacre R, Dickson AJ (2011) Metabolite extraction from suspension-cultured mammalian cells for global metabolite profiling. Nat Protoc 6:1241–1249. https://doi.org/10.1038/nprot.2011.366
120. Ma N, Ellet JA, Okediadi C, Hermes P, McCormick E, Casnocha S (2009) A single nutrient feed supports both chemically defined NS0 and CHO fed-batch processes: improved productivity and lactate metabolism. Biotechnol Prog 25:1353–1363. https://doi.org/10.1002/btpr.238
121. Sengupta N, Rose ST, Morgan JA (2011) Metabolic flux analysis of CHO cell metabolism in the late non-growth phase. Biotechnol Bioeng 108:82–92. https://doi.org/10.1002/bit.22890
122. Dean J, Reddy P (2013) Metabolic analysis of antibody producing CHO cells in fed-batch production. Biotechnol Bioeng 110:1735–1747. https://doi.org/10.1002/bit.24826
123. Matuszczyk JC, Teleki A, Pfizenmaier J, Takors R (2015) Compartment-specific metabolomics for CHO reveals that ATP pools in mitochondria are much lower than in cytosol. Biotechnol J 10:1639–1650. https://doi.org/10.1002/biot.201500060
124. Sellick CA, Croxford AS, Maqsood AR, Stephens GM, Westerhoff HV, Goodacre R, Dickson AJ (2015) Metabolite profiling of CHO cells: molecular reflections of bioprocessing effectiveness. Biotechnol J 10:1434–1445. https://doi.org/10.1002/biot.201400664
125. Carinhas N, Duarte TM, Barreiro LC, Carrondo MJT, Alves PM, Teixeira AP (2013) Metabolic signatures of GS-CHO cell clones associated with butyrate treatment and culture phase transition. Biotechnol Bioeng 110:3244–3257. https://doi.org/10.1002/bit.24983
126. Teixeira AP, Carinhas N, Alves PM, Barreiro LC, Duarte TM, Carrondo MJT (2014) Metabolic responses of CHO cells to limitation of key amino acids. Biotechnol Bioeng 111:2095–2106. https://doi.org/10.1002/bit.25266
127. Templeton N, Smith KD, McAtee-Pereira AG, Dorai H, Betenbaugh MJ, Lang SE, Young JD (2017) Application of 13C flux analysis to identify high-productivity CHO metabolic phenotypes. Metab Eng 43:218–225. https://doi.org/10.1016/j.ymben.2017.01.008

128. Templeton N, Dean J, Reddy P, Young JD (2013) Peak antibody production is associated with increased oxidative metabolism in an industrially relevant fed-batch CHO cell culture. Biotechnol Bioeng 110:2013–2024. https://doi.org/10.1002/bit.24858
129. Templeton N, Xu S, Roush DJ, Chen H (2017) 13C metabolic flux analysis identifies limitations to increasing specific productivity in fed-batch and perfusion. Metab Eng 44:126–133. https://doi.org/10.1016/j.ymben.2017.09.010
130. Mulukutla BC, Gramer M, Hu WS (2012) On metabolic shift to lactate consumption in fed-batch culture of mammalian cells. Metab Eng 14:138–149. https://doi.org/10.1016/j.ymben.2011.12.006
131. Le T, Mulukutla BC, Yongky A, Hu W-S, Daoutidis P, Lee J (2015) Mechanism for multiplicity of steady states with distinct cell concentration in continuous culture of mammalian cells. Biotechnol Bioeng 112:1437–1445. https://doi.org/10.1002/bit.25566
132. Torres M, Berrios J, Rigual Y, Latorre Y, Vergara M, Dickson AJ, Altamirano C (2019) Metabolic flux analysis during galactose and lactate co-consumption reveals enhanced energy metabolism in continuous CHO cell cultures. Chem Eng Sci 205:201–211. https://doi.org/10.1016/j.ces.2019.04.049
133. Ahn WS, Antoniewicz MR (2011) Metabolic flux analysis of CHO cells at growth and non-growth phases using isotopic tracers and mass spectrometry. Metab Eng 13:598–609. https://doi.org/10.1016/j.ymben.2011.07.002
134. Lee D-Y, Yap MGS, Yusufi FNK, Selvarasu S, Lee YY, Ho YS, Chong WPK, Wong NSC (2012) Combined in silico modeling and metabolomics analysis to characterize fed-batch CHO cell culture. Biotechnol Bioeng 109:1415–1429. https://doi.org/10.1002/bit.24445
135. Pereira S, Kildegaard HF, Andersen MR (2018) Impact of CHO metabolism on cell growth and protein production: an overview of toxic and inhibiting metabolites and nutrients. Biotechnol J 13:1–13. https://doi.org/10.1002/biot.201700499
136. Chrysanthopoulos PK, Goudar CT, Klapa MI (2010) Metabolomics for high-resolution monitoring of the cellular physiological state in cell culture engineering. Metab Eng. https://doi.org/10.1016/j.ymben.2009.11.001
137. Dietmair S, Hodson MP, Quek LE, Timmins NE, Chrysanthopoulos P, Jacob SS, Gray P, Nielsen LK (2012) Metabolite profiling of CHO cells with different growth characteristics. Biotechnol Bioeng 109:1404–1414. https://doi.org/10.1002/bit.24496
138. Vodopivec M, Lah L, Narat M, Curk T (2019) Metabolomic profiling of CHO fed-batch growth phases at 10, 100, and 1,000 L. Biotechnol Bioeng 116:2720–2729. https://doi.org/10.1002/bit.27087
139. Budge JD, Knight TJ, Povey J, Roobol J, Brown IR, Singh G, Dean A, Turner S, Jaques CM, Young RJ, Racher AJ, Smales CM (2020) Engineering of Chinese hamster ovary cell lipid metabolism results in an expanded ER and enhanced recombinant biotherapeutic protein production. Metab Eng 57:203–216. https://doi.org/10.1016/j.ymben.2019.11.007
140. Reinhart D, Damjanovic L, Kaisermayer C, Sommeregger W, Gili A, Gasselhuber B, Castan A, Mayrhofer P, Grünwald-Gruber C, Kunert R (2019) Bioprocessing of recombinant CHO-K1, CHO-DG44, and CHO-S: CHO expression hosts favor either mAb production or biomass synthesis. Biotechnol J 14:1–11. https://doi.org/10.1002/biot.201700686
141. Van Berkel PHC, Gerritsen J, Perdok G, Valbjørn J, Vink T, Van De Winkel JGJ, Parren PWHI (2009) N-linked glycosylation is an important parameter for optimal selection of cell lines producing biopharmaceutical human IgG. Biotechnol Prog. https://doi.org/10.1002/btpr.92
142. van de Laar T, Rudd PM, O'Donoghue N, Davey GP, Burleigh SC, van Grunsven WM, Stroop CJ (2011) Synergizing metabolic flux analysis and nucleotide sugar metabolism to understand the control of glycosylation of recombinant protein in CHO cells. BMC Biotechnol 11:95. https://doi.org/10.1186/1472-6750-11-95
143. Fan Y, Jimenez Del Val I, Müller C, Wagtberg Sen J, Rasmussen SK, Kontoravdi C, Weilguny D, Andersen MR (2015) Amino acid and glucose metabolism in fed-batch CHO cell culture affects antibody production and glycosylation. Biotechnol Bioeng 112:521–535. https://doi.org/10.1002/bit.25450

144. Ehret J, Zimmermann M, Eichhorn T, Zimmer A (2019) Impact of cell culture media additives on IgG glycosylation produced in Chinese hamster ovary cells. Biotechnol Bioeng 116:816–830. https://doi.org/10.1002/bit.26904
145. Lin N, Mascarenhas J, Sealover NR, George HJ, Brooks J, Kayser KJ, Gau B, Yasa I, Azadi P, Archer-Hartmann S (2015) Chinese hamster ovary (CHO) host cell engineering to increase sialylation of recombinant therapeutic proteins by modulating sialyltransferase expression. Biotechnol Prog. https://doi.org/10.1002/btpr.2038
146. Villiger TK, Roulet A, Périlleux A, Stettler M, Broly H, Morbidelli M, Soos M (2016) Controlling the time evolution of mAb N-linked glycosylation, Part I: Microbioreactor experiments. Biotechnol Prog. https://doi.org/10.1002/btpr.2305
147. Gawlitzek M, Ryll T, Lofgren J, Sliwkowski MB (2000) Ammonium alters N-glycan structures of recombinant TNFR-IgG: degradative versus biosynthetic mechanisms. Biotechnol Bioeng. https://doi.org/10.1002/(SICI)1097-0290(20000620)68:6<637::AID-BIT6>3.0.CO;2-C
148. Ivarsson M, Noh H, Morbidelli M, Soos M (2015) Insights into pH-induced metabolic switch by flux balance analysis. Biotechnol Prog 31:347–357. https://doi.org/10.1002/btpr.2043
149. Sou SN, Sellick C, Lee K, Mason A, Kyriakopoulos S, Polizzi KM, Kontoravdi C (2015) How does mild hypothermia affect monoclonal antibody glycosylation? Biotechnol Bioeng 112:1165–1176. https://doi.org/10.1002/bit.25524
150. Fisher P, Thomas-Oates J, Wood AJ, Ungar D (2019) The N-glycosylation processing potential of the mammalian Golgi apparatus. Front Cell Dev Biol. https://doi.org/10.3389/fcell.2019.00157
151. Aikawa JI, Matsuo I, Ito Y (2012) In vitro mannose trimming property of human ER α-1,2 mannosidase i. Glycoconj J. https://doi.org/10.1007/s10719-011-9362-1
152. Ren J, Castellino FJ, Bretthauer RK (1997) Purification and properties of α-mannosidase II from Golgi-like membranes of baculovirus-infected Spodoptera frugiperda (IPLB-SF-21AE) cells. Biochem J. https://doi.org/10.1042/bj3240951
153. Yang Q, Wang LX (2016) Mammalian α-1, 6-fucosyltransferase (FUT8) is the sole enzyme responsible for the N-acetylglucosaminyltransferase I-independent core fucosylation of high-mannose N-glycans. J Biol Chem. https://doi.org/10.1074/jbc.M116.720789
154. Hassinen A, Rivinoja A, Kauppila A, Kellokumpu S (2010) Golgi N-glycosyltransferases form both homo- and heterodimeric enzyme complexes in live cells. J Biol Chem. https://doi.org/10.1074/jbc.M110.103184
155. Moremen KW, Tiemeyer M, Nairn AV (2012) Vertebrate protein glycosylation: diversity, synthesis and function. Nat Rev Mol Cell Biol
156. Sha S, Agarabi C, Brorson K, Lee DY, Yoon S (2016) N-glycosylation design and control of therapeutic monoclonal antibodies. Trends Biotechnol
157. Arigoni-Affolter I, Scibona E, Lin CW, Brühlmann D, Souquet J, Broly H, Aebi M (2019) Mechanistic reconstruction of glycoprotein secretion through monitoring of intracellular N-glycan processing. Sci Adv. https://doi.org/10.1126/sciadv.aax8930
158. Erklavec Zajec V, Novak U, Kastelic M, Japelj B, Lah L, Pohar A, Likozar B (2020) Dynamic multiscale metabolic network modeling of Chinese hamster ovary cell metabolism integrating N-linked glycosylation in industrial biopharmaceutical manufacturing. Biotechnol Bioeng. https://doi.org/10.1002/bit.27578
159. Stach CS, McCann MG, O'Brien CM, Le TS, Somia N, Chen X, Lee K, Fu HY, Daoutidis P, Zhao L, Hu WS, Smanski M (2019) Model-driven engineering of N-linked glycosylation in Chinese hamster ovary cells. ACS Synth Biol. https://doi.org/10.1021/acssynbio.9b00215
160. Jimenez del Val I, Nagy JM, Kontoravdi C (2011) A dynamic mathematical model for monoclonal antibody N-linked glycosylation and nucleotide sugar donor transport within a maturing Golgi apparatus. Biotechnol Prog. https://doi.org/10.1002/btpr.688
161. Duivelshof BL, Jiskoot W, Beck A, Veuthey JL, Guillarme D, D'Atri V (2019) Glycosylation of biosimilars: recent advances in analytical characterization and clinical implications. Anal Chim Acta. https://doi.org/10.1016/j.aca.2019.08.044

162. Schaub J, Clemens C, Schorn P, Hildebrandt T, Rust W, Mennerich D, Kaufmann H, Schulz TW (2010) CHO gene expression profiling in biopharmaceutical process analysis and design. Biotechnol Bioeng. https://doi.org/10.1002/bit.22549
163. Morris C, Polanco A, Yongky A, Xu J, Huang Z, Zhao J, McFarland KS, Park S, Warrack B, Reily M, Borys MC, Li Z, Yoon S (2020) Bigdata analytics identifies metabolic inhibitors and promoters for productivity improvement and optimization of monoclonal antibody (mAb) production process. Bioresour Bioprocess. https://doi.org/10.1186/s40643-020-00318-6
164. Huang Z, Xu J, Yongky A, Morris CS, Polanco AL, Reily M, Borys M, Li ZJ, Yoon S (2020) CHO cell productivity improvement by genome-scale modeling and pathway analysis: application to feed supplements. Biochem Eng J:107638. https://doi.org/10.1016/j.bej.2020.107638
165. Wong DCF, Wong KTK, Nissom PM, Heng CK, Yap MGS (2006) Targeting early apoptotic genes in batch and fed-batch CHO cell cultures. Biotechnol Bioeng. https://doi.org/10.1002/bit.20871
166. Barron N, Kumar N, Sanchez N, Doolan P, Clarke C, Meleady P, O'Sullivan F, Clynes M (2011) Engineering CHO cell growth and recombinant protein productivity by overexpression of miR-7. J Biotechnol. https://doi.org/10.1016/j.jbiotec.2010.12.005
167. Baik JY, Gasimli L, Yang B, Datta P, Zhang F, Glass CA, Esko JD, Linhardt RJ, Sharfstein ST (2012) Metabolic engineering of Chinese hamster ovary cells: towards a bioengineered heparin. Metab Eng. https://doi.org/10.1016/j.ymben.2012.01.008
168. Berger A, Le Fourn V, Masternak J, Regamey A, Bodenmann I, Girod PA, Mermod N (2020) Overexpression of transcription factor Foxa 1 and target genes remediate therapeutic protein production bottlenecks in Chinese hamster ovary cells. Biotechnol Bioeng 117:1101–1116. https://doi.org/10.1002/bit.27274
169. Druz A, Chu C, Majors B, Santuary R, Betenbaugh M, Shiloach J (2011) A novel microRNA mmu-miR-466h affects apoptosis regulation in mammalian cells. Biotechnol Bioeng. https://doi.org/10.1002/bit.23092
170. Zhou H, Liu Z-g, Sun Z, Huang Y, Yu W (2010) Generation of stable cell lines by site-specific integration of transgenes into engineered Chinese hamster ovary strains using an FLP-FRT system. J Biotechnol. https://doi.org/10.1016/j.jbiotec.2010.03.020
171. Inniss MC, Bandara K, Jusiak B, Lu TK, Weiss R, Wroblewska L, Zhang L (2017) A novel Bxb1 integrase RMCE system for high fidelity site-specific integration of mAb expression cassette in CHO cells. Biotechnol Bioeng 114:1837–1846. https://doi.org/10.1002/bit.26268
172. Zhang L, Inniss MC, Han S, Moffat M, Jones H, Zhang B, Cox WL, Rance JR, Young RJ (2015) Recombinase-mediated cassette exchange (RMCE) for monoclonal antibody expression in the commercially relevant CHOK1SV cell line. Biotechnol Prog. https://doi.org/10.1002/btpr.2175
173. Hamaker NK, Lee KH (2018) Site-specific integration ushers in a new era of precise CHO cell line engineering. Curr Opin Chem Eng 22:152–160. https://doi.org/10.1016/j.coche.2018.09.011
174. Hilliard W, Lee KH (2020) Systematic identification of Safe Harbor regions in the CHO genome through a comprehensive epigenome analysis. Biotechnol Bioeng. https://doi.org/10.1002/bit.27599
175. Vallée C, Durocher Y, Henry O (2014) Exploiting the metabolism of PYC expressing HEK293 cells in fed-batch cultures. J Biotechnol 169:63–70. https://doi.org/10.1016/j.jbiotec.2013.11.002
176. Irani N, Beccaria AJ, Wagner R (2002) Expression of recombinant cytoplasmic yeast pyruvate carboxylase for the improvement of the production of human erythropoietin by recombinant BHK-21 cells. J Biotechnol 93:269–282. https://doi.org/10.1016/S0168-1656(01)00409-6
177. Toussaint C, Henry O, Durocher Y (2016) Metabolic engineering of CHO cells to alter lactate metabolism during fed-batch cultures. J Biotechnol 217:122–131. https://doi.org/10.1016/j.jbiotec.2015.11.010
178. Bulté DB, Palomares LA, Parra CG, Martínez JA, Contreras MA, Noriega LG, Ramírez OT (2020) Overexpression of the mitochondrial pyruvate carrier reduces lactate production and

increases recombinant protein productivity in CHO cells. Biotechnol Bioeng 117:2633–2647. https://doi.org/10.1002/bit.27439
179. Torres M, Altamirano C, Dickson AJ (2018) Process and metabolic engineering perspectives of lactate production in mammalian cell cultures. Curr Opin Chem Eng 22:184–190. https://doi.org/10.1016/j.coche.2018.10.004
180. Ha TK, Lee GM (2015) Glutamine substitution: the role it can play to enhance therapeutic protein production. Pharm Bioprocess 3:249–261. https://doi.org/10.4155/pbp.15.6
181. Hefzi H, Lewis N (2017) Mammalian cells devoid of lactate dehydrogenase activity (WO 2017/192437 Al). World Intellect Prop. Organ
182. Schmitt J, Downey B, Beller J, Russell B, Quach A, Lyon D, Curran M, Mulukutla BC, Chu C (2019) Forecasting and control of lactate bifurcation in Chinese hamster ovary cell culture processes. Biotechnol Bioeng 116:2223–2235. https://doi.org/10.1002/bit.27015
183. Buchsteiner M, Quek L-E, Gray P, Nielsen LK (2018) Improving culture performance and antibody production in CHO cell culture processes by reducing the Warburg effect. Biotechnol Bioeng. https://doi.org/10.1002/bit.26724
184. Noh SM, Park JH, Lim MS, Kim JW, Lee GM (2017) Reduction of ammonia and lactate through the coupling of glutamine synthetase selection and downregulation of lactate dehydrogenase-A in CHO cells. Appl Microbiol Biotechnol 101:1035–1045. https://doi.org/10.1007/s00253-016-7876-y
185. Kim SH, Lee GM (2007) Down-regulation of lactate dehydrogenase-A by siRNAs for reduced lactic acid formation of Chinese hamster ovary cells producing thrombopoietin. Appl Microbiol Biotechnol 74:152–159. https://doi.org/10.1007/s00253-006-0654-5
186. Chen K, Liu Q, Xie L, Sharp PA, Wang DIC (2001) Engineering of a mammalian cell line for reduction of lactate formation and high monoclonal antibody production. Biotechnol Bioeng 72:55–61. https://doi.org/10.1002/1097-0290(20010105)72:1<55::AID-BIT8>3.0.CO;2-4
187. Karengera E, Robotham A, Kelly J, Durocher Y, De Crescenzo G, Henry O (2018) Concomitant reduction of lactate and ammonia accumulation in fed-batch cultures: impact on glycoprotein production and quality. Biotechnol Prog 34:494–504. https://doi.org/10.1002/btpr.2607
188. Gagnon M, Hiller G, Luan Y-T, Kittredge A, DeFelice J, Drapeau D (2011) High-End pH-controlled delivery of glucose effectively suppresses lactate accumulation in CHO Fed-batch cultures. Biotechnol Bioeng 108:1328–1337. https://doi.org/10.1002/bit.23072
189. Goldrick S, Lee K, Spencer C, Holmes W, Kuiper M, Turner R, Farid SS (2018) On-line control of glucose concentration in high-yielding mammalian cell cultures enabled through oxygen transfer rate measurements. Biotechnol J 13:1700607. https://doi.org/10.1002/biot.201700607
190. Mulukutla BC, Kale J, Kalomeris T, Jacobs M, Hiller GW (2017) Identification and control of novel growth inhibitors in fed-batch cultures of Chinese hamster ovary cells. Biotechnol Bioeng 114:1779–1790. https://doi.org/10.1002/bit.26313
191. Mulukutla BC, Geoffroy P, Mitchell J, Harrington C, Krishnan M, Kalomeris T, Morris C, Zhang L, Pegman P, Hiller GW (2019) Metabolic engineering of Chinese hamster ovary cells towards reduced biosynthesis and accumulation of novel growth inhibitors in fed-batch cultures. Metab Eng 54:54–68. https://doi.org/10.1016/j.ymben.2019.03.001
192. Sumit M, Dolatshahi S, Chu A-HA, Cote K, Scarcelli JJ, Marshall JK, Cornell RJ, Weiss R, Lauffenburger DA, Mulukutla BC, Figueroa B (2019) Dissecting N-glycosylation dynamics in Chinese hamster ovary cells fed-batch cultures using time course omics analyses. iScience 12:102–120. https://doi.org/10.1016/j.isci.2019.01.006
193. Chang MM, Gaidukov L, Jung G, Tseng WA, Scarcelli JJ, Cornell R, Marshall JK, Lyles JL, Sakorafas P, Chu A-HA, Cote K, Tzvetkova B, Dolatshahi S, Sumit M, Mulukutla BC, Lauffenburger DA, Figueroa B, Summers NM, Lu TK, Weiss R (2019) Small-molecule control of antibody N-glycosylation in engineered mammalian cells. Nat Chem Biol. https://doi.org/10.1038/s41589-019-0288-4
194. Marx N, Grünwald-Gruber C, Bydlinski N, Dhiman H, Ngoc Nguyen L, Klanert G, Borth N (2018) CRISPR-based targeted epigenetic editing enables gene expression modulation

of the silenced beta-galactoside alpha-2, 6-sialyltransferase 1 in CHO cells. Biotechnol J 13:1700217. https://doi.org/10.1002/biot.201700217
195. Bragonzi A, Distefano G, Buckberry LD, Acerbis G, Foglieni C, Lamotte D, Campi G, Marc A, Soria MR, Jenkins N, Monaco L (2000) A new Chinese hamster ovary cell line expressing α2,6-sialyltransferase used as universal host for the production of human-like sialylated recombinant glycoproteins. Biochim Biophys Acta-Gen Subj 1474:273–282. https://doi.org/10.1016/S0304-4165(00)00023-4
196. Fukuta K, Yokomatsu T, Abe R, Asanagi M, Makino T (2000) Genetic engineering of CHO cells producing human interferon-γ by transfection of sialyltransferases. Glycoconj J 17:895–904. https://doi.org/10.1023/A:1010977431061
197. Jassal R, Jenkins N, Charlwood J, Camilleri P, Jefferis R, Lund J (2001) Sialylation of human IgG-Fc carbohydrate by transfected rat α2,6-sialyltransferase. Biochem Biophys Res Commun 286:243–249. https://doi.org/10.1006/bbrc.2001.5382
198. Baik JY, Dahodwala H, Oduah E, Talman L, Gemmill TR, Gasimli L, Datta P, Yang B, Li G, Zhang F, Li L, Linhardt RJ, Campbell AM, Gorfien SF, Sharfstein ST (2015) Optimization of bioprocess conditions improves production of a CHO cell-derived, bioengineered heparin. Biotechnol J. https://doi.org/10.1002/biot.201400665
199. Wong DCF, Wong KTK, Lee YY, Morin PN, Heng CK, Yap MGS (2006) Transcriptional profiling of apoptotic pathways in batch and fed-batch CHO cell cultures. Biotechnol Bioeng. https://doi.org/10.1002/bit.20872

Redesigning Spent Media from Cell Culture Bioprocess to Feed New Bacterial Fermentations

Ciara Lynch, Lynda Jordan, and David J O' Connell

Abbreviations

BE Bioeconomy
CE Circular Economy
CBE Circular Bioeconomy
MAb Monoclonal Antibody
CHO Chinese Hamster Ovary

1 Introduction

A key goal of the emerging bioeconomy is the selective separation and extraction of valuable compounds from renewable materials, converting those resources into novel bio-based products and processes, with the potential to deliver market and industry-scalable sustainable resources. When we consider what is meant by a bioeconomy (BE) we can look to the European Commission who have been championing the development of this idea for the past decade. Their specific definition of the BE is *"the production of renewable biological resources and the conversion of these resources and waste streams into value added products, such as food, feed, bio-based products and bioenergy"* [1] Concurrently there has been a development of related terminology with regard to the Circular Economy (CE) with this being defined as *"minimizing the generation of waste and maintaining the value of products, materials and resources for as long as possible."* [2]. The

C. Lynch · L. Jordan · D. J. O'Connell (✉)
BiOrbic, Bioeconomy SFI Research Centre, O'Brien Centre for Science, University College Dublin, Dublin 4, Ireland
e-mail: david.oconnell@ucd.ie

R. Pörtner (ed.), *Cell Culture Engineering and Technology*, Cell Engineering 10,
https://doi.org/10.1007/978-3-030-79871-0_5

parallel descriptions of efforts to promote sustainability with industrial practice have then led to the adoption of the holistic term, the Circular Bioeconomy (CBE) [3]. The circular bioeconomy may then be thought of as a circular economy where non-renewable inputs to industrial systems are replaced by renewable biological resources with the European Commission defining this as the application of the CE concept to biological resources, products and materials [4–5]. It is also possible to look at this slightly differently and consider a CBE where non-renewable inputs are re-evaluated such that they are not replaced but redesigned such that they may become renewable to have a second life in delivering new products and materials.

The production of human monoclonal antibodies (MAb) for the treatment of chronic diseases has become the most substantial sector of the bioprocessing industry in the past decade, representing just over a quarter (27%) of all first-time biologic approvals from 2010 to 2014. This number rose to over half (53%) of all first-time approvals between 2015 and 2018. Production of these approved drug products increasingly relies on bioprocessing with mammalian cell lines. By 2018, 62 of 71 new biopharmaceutical active ingredients on the market were recombinant proteins, of which 52 (84%) are expressed in mammalian cell lines, with nine produced in microbial expression systems: Escherichia coli (five) or yeast (four) [6]. This has led to a significant intensification in activity of the bioprocessing industry globally with the consumption of millions of litres of cell culture media leading to the production of millions of tonnes of bioprocessing waste in a processing system, where once the drug product is extracted by affinity chromatography or related separation method, the liquid culture media is sent as waste to biokill containment. In 2018 over 16.5 million litres of active production capacity was estimated at more than 1,500 facilities worldwide including 6 million litres (37%) in the United States and Canada, 5.5 million litres in Western Europe (33%), and 4.7 million litres (25%) in the Asia–Pacific region, with a further 870,000 litres in China and 941,000 litres in India [7]. Approximately 10.2 million litres is based on the bioprocessing of mammalian cells primarily for MAb manufacture and nearly all using Chinese hamster ovary (CHO) host cells. At present the consignment of these millions of litres of cell culture media to waste immediately after the initial production of the drug product runs counter to the aims of the CBE, with a potentially valuable opportunity to investigate this waste stream, with its high water content and nutritional content being lost to containment as a hazardous material. The same systems biology approaches that have been employed to intensify the production of MAb products from CHO for example, may now be employed by researchers interested in developing the CBE to encompass the biopharmaceutical sector with a view to redesigning this waste as a valuable entity and converting this resource into a feed for new bio-based product manufacture.

2 Valorization of Waste with Bacteria

The successful application of systems biology approaches to the development of the CBE is exemplified by the engineering of the soil bacterium species *Pseudomonas putida*, which has been intensively developed through metabolic engineering strategies to take advantage of its robust nature to become a highly valuable tool for the conversion of waste streams to new value [8]. The manipulation of the genome of this organism through classical genetics approaches and more recently through the advent of genome editing technologies such as the CRISPR-Cas9 system has facilitated the suppression of undesirable characteristics and the amplification of properties that facilitate new biotechnology [9]. Genetic engineering of the metabolism of the Pseudomonas strain P. putida KT2440 is a particularly good example, resulting in the production of a wide range of target compounds of industrial value from native compounds such as polyhydroxyalkanoate for bioplastic manufacture (Fig. 1) to completely novel molecules [10]. The application of systems biology approaches in the past decade using new powerful analytic techniques for studying cellular metabolism, including accessing the central design elements of genomic/ transcriptomic/ proteomics as engineering instructions, have driven new efforts to modify the entire cell as well as cellular pathways used by the cell. This systems metabolic engineering opens up extraordinary possibilities in the use of bacteria for production of new products and also the use of waste streams as high value feed to generate these new products [11–12]. This synthetic biology knowhow driven by systems biology can support highly ambitious efforts such as the recycling of plastic waste, an enormous burden on the planet's ecosystem and an urgent problem for this and future generations [13]. Recycling of plastic and other non-conventional feedstocks arising from industrial production, including millions of litres of bioprocessing waste produced annually by the bioprocessing industry, is now not only of scientific interest but is demanded by the focus on developing the CBE.

3 Heterologous Expression of Recombinant Human Proteins Provides Alternative Waste Streams

In addition to the application of engineering strategies for recycling of environmental waste by bacterial species such as Pseudomonas, there is a long history of successful recombinant DNA engineering strategies that have led to the overproduction of human proteins for biotherapeutic application using the classical microbial 'cell factories' *Escherichia coli* and *Saccharomyces cerevisiae* [14]. Heterologous overexpression of recombinant human insulin for the treatment of diabetes was a major milestone in the use of DNA engineering. This process evolved from the initial production of each individual chain of the hormone protein in separate processes with subsequent chemical crosslinking to the production of the

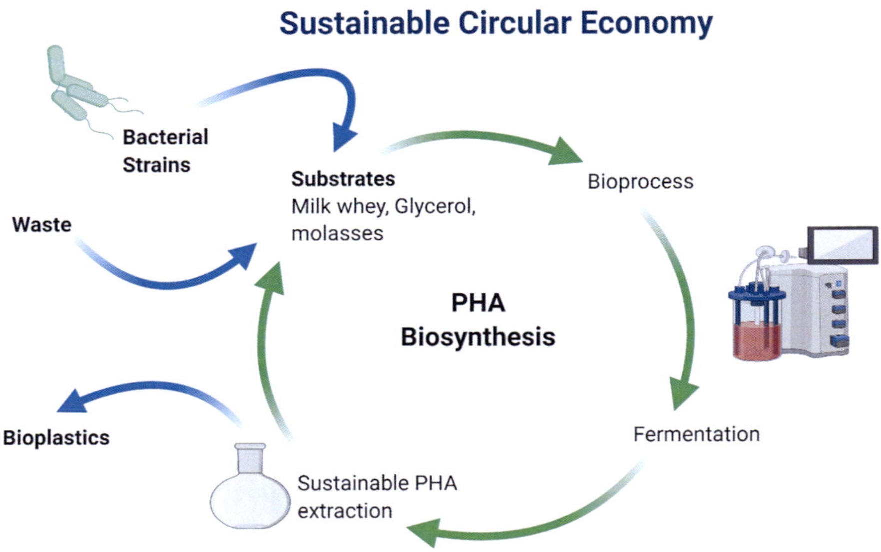

Fig. 1 Schematic diagram depicting the valorization of waste streams as feed for bacterial production of new products. Example shown is the production of the bioplastic polymer polyhydroxyalkanoate (PHA) using bacterial strains such as Pseudomonas species. Created with BioRender.com

authentic human proinsulin molecule [15–16]. Manipulating the microorganism to produce a human hormone that regulates a chronic human disease was the launchpad of the modern bioprocessing industry which has quickly gained on the traditional pharmaceutical sector in terms of value since the early 1980s. *Saccharomyces cerevi*siae, with eukaryotic cell architecture and efficient folding and secretory pathways, has also been further engineered for the successful preparation of some of the world's biggest selling replacement therapies with short acting insulin lispro and long acting insulin detemir [17]. However for complex macromolecules such as immunoglobulin G, the cell factory of choice is undoubtedly the CHO cell.

3.1 Expression of Human Proteins in CHO Cells

Chinese Hamster Ovary cells have been the principle system for the expression of human macromolecules such as immunoglobulins over the past 30 years, with 6 of the top 10 selling or blockbuster MAbs produced by these cells in regulated bioprocessing systems (Table 1). Key features of CHO cell bioprocessing include (i) the ability to grow in suspension culture to very high cell density (ii) production of high titres of product. Growth in serum-free, chemically defined media is an important enabler in determining reproducibility across different batches of cultures

Table 1 The top 10 biotherapeutic MAbs by value in the period 2014–2017 and their method of production

Rank	International Non-proprietary Name	Cumulative sales 2014–17 ($ billions)	Approved	Company	Type	Cell line
1	Adalimumab	62.6	2002	Abbvie	IgG1	CHO
2	Infliximab	35.6	1998	J&J	IgG1*	Sp2/0
3	Rituximab	29.1	1997	Roche	IgG1*	CHO
4	Trastuzumab	27.1	1998	Roche	IgG1	CHO
5	Bevacizumab	27.0	2004	Roche	IgG1	CHO
6	Ranibizumab	14.3	2006	Roche	Fab	E.coli
7	Denosumab	11.6	2010	Amgen	IgG2	CHO
8	Nivolumab	11.4	2014	BMS	IgG4	CHO
9	Eculizumab	10.7	2007	Alexion	IgG2/4	NS0
10	Golimumab	9.7	2009	Merck	IgG1	Sp2/0

*Indicates the molecule is chimeric

and possesses a much better safety profile than in media containing human or animal-derived proteins [18–19]. The intensification of upstream processing in the production of drug substance with this cell type began with mutations to suppress deleterious gene mediated events such as apoptosis [20]. Further mutational analysis helped to improve the titre and quality of antibody secreted by the cells through the silencing of selected genes. The incorporation of mutations in the dihydrofolate reductase (dhf r) gene led to the production of two of the most widely used CHO cell lines in the biopharmaceutical industry, CHODXB11 (also known as CHO-DUKX) and CHO-DG44 [21]. This event coincided with arguably the most significant step forward in the evolution of the systems biology-led approach to intensification of CHO cell bioprocessing, the sequencing of the CHO-K1 cell line [22]. This and the later advent of genomic engineering technologies has paved the way for maximising the productivity of this non-human cell line through systematic engineering of pathways and genes to drive production of human biotherapeutic proteins at multi-gram per litre scales resulting in multi-tonne yields of product annually [11, 23].

3.2 Expression of Human Proteins in HEK293 Cell Culture

The human cell line HEK293 has been used extensively in research and development laboratories and is very often used for functional interaction screening between human proteins, however it is also widely employed for the production of vaccines, anticancer drugs, and other clinically relevant drugs. Transient transfection of HEK293 cells is the most commonly used method for the commercial production of recombinant adeno-associated viruses (AAV) [24–25]. In addition advances in cell line development technology has allowed for increased productivity with human cell lines, leading to approved recombinant biotherapeutic products produced from HEK293 including agalsidase alfa for the treatment of Fabry disease and the first recombinant human coagulation factor VIII for the treatment of haemophilia [26–27]. Of particular interest during the ongoing COVID-19 pandemic is the production of new vaccines against this virus but also against other members of this virus family that cause SARS diseases. Research into the CoV spike receptor-binding domain in HEK293 cells has opened up the possibility of producing more effective immunogens that produce higher antibody titres and which can be produced in high yield bio process [28]. The versatility of this full human cell line model should see a significantly increasing volume of bioprocessing for human protein production in the coming decades.

3.3 Expression of Human Proteins in Non-mammalian Cell Culture

The Structural Genomics Consortium (SGC) is a group of laboratories in institutes and companies around the world that are focused on exploring less well-studied areas of the human genome. The baculovirus insect cell eukaryotic expression system was established in 2007 at the SGC as a measure to tackle challenging kinases, RNA-DNA processing proteins and integral membrane protein families [29]. The system has been used successfully with difficult to express proteins ranging from membrane bound proteins to complex enzymes that have proved intractable with other methods [30–31]. The success of this heterologous system is not confined to assisting with the resolution of crystal structures of proteins, but has been highly effective in the production of virus-like particles (VLPs) for the development of vaccines [19, 32].

The industrial scale of production of a growing array of proteins, from biotherapeutic drug molecules to VLPs, and structural genomics efforts using animal cell culture systems to eukaryotic insect cell expression systems, has led to a huge expansion of bioprocessing capability whether in small scale facilities for research and development or in the hundreds of thousands of litres capacity in bioprocessing facilities across the globe. This clearly opens up new opportunities into the research of how we should deal with this waste and identify new opportunities to use it as part of a circular bioeconomy.

4 Recycling of Waste Culture Media – Lessons from the Literature

A limited amount of research into reusing cell culture media has been explored for a variety of cell cultures, predominantly insect cell cultures, microalgae and yeast with a very small number of studies into reusing mammalian cell culture media. In general, the results have varied in terms of the utility of the spent media depending on the cell type used, however the general consensus from these studies suggests that (i) the use of partial recycled media is beneficial in secondary culture systems and (ii) inhibitory effects on cell growth using recycled media can also be mitigated by addition of supplements, or by using additional treatment steps on the media prior to use.

4.1 Recycling of Yeast Culture Media

Yeast fermentation is a major biotechnology system used to produce a wide range of products from biotherapeutic proteins to commodities at very large scales such as alcohols, oils, and amino acids. Reusing spent medium would be of distinct benefit with regards to reduction of industrial costs of media and reduced consumption of water. One study reported a wastewater reduction of 70% by medium recycle [33]. Most yeast species appeared to be largely unaffected for long periods of time in recycled media. For example, a *Saccharomyces cerevisiae* immobilised culture in a matrix, with recycled media supplemented with nutrients continuously pumped through the matrix, facilitated cell growth and alcohol production at equivalent levels for up to 40 days, followed by inhibition of the culture due to a build-up of ions and polysaccharides after this time point [34]. Media recycling in this system provides a highly interesting premise for evaluation of other bioprocesses.

Another yeast fermentation system *Apiotrichum curvatum* producing microbial lipids in a chemically defined medium, was recycled with centrifugation, supplemented with carbon and nitrogen sources, and sterilized before testing with ratios of spent medium to fresh medium of between 0.25, 0.5 or 0.75 recycled medium. The results showed that recycled medium addition did not negatively impact growth or lipid production until after the 3rd subsequent recycling step, where lipid production and cell dry weight dropped [35]. The effect of ions building up in the media was investigated using an ion exchange matrix, that successfully restored the lipid production and growth to a state similar to fresh media for a further two recycling steps. Upon testing a second media, whey permeate media, they found similar results, with a drop in productivity after the 2nd recycling step. These results appear to indicate that cellular metabolites in spent medium do not inhibit growth of yeast cultures, but that ion concentration plays a large factor. These studies provide very useful signposts for the redesign of spent media such as chemically defined media used in CHO cell bioprocess, in particular the monitoring of ion concentration levels in spent media.

4.2 Recycling of Microalgae Culture Media

Microalgae cultures are often used to produce feed for animals and is today a potential candidate for biofuel production [36]. So it is interesting to note that microalgae cultures, like yeast, appear to survive in low density cell cultures in a recycled medium. Furthermore, doing so could save up to 75% water and 62% on nutrients [37]. The area of using recycled media with regards to heterotrophic algae culture has been comprehensively reviewed [38]. One of the earliest studies with media recycling in microalgae cultures was completed using cultures of *Chorella pyrenoidosa* [39]. This study used low cell density cultures (< 3 g per litre of biomass), where the media was supplemented with urea, iron and calcium and fed back into the culture. The culture was maintained for over 72 days, with no build-up of auto-inhibitory molecules seen. However more recent studies show that cultures over 3 g per litre in density show a decrease in biomass obtained [40–41] while studies that stayed below this density were mostly uninhibited [37, 39, 42].

The study of low cell density cultures of *Chlorella vulgaris* used a media designed by the authors for cell culture reuse, substituting any free ions in the medium with ammonium instead, a chemical which microalgae assimilate for a nitrogen source [37]. Use of this medium appeared to negate much of the inhibitory effects of using a spent medium on the culture, even in high-productivity cultures, compared to their control medium. This minimal growth medium had a lower ion concentration after recycling than the control did. They estimated they could save 75% of water and 62% of nutrients in this manner. In a second study by the same team, they tested how many days they could continue the culture with no negative impact on biomass [42]. Interestingly, they actually had increases in biomass for the first 10 days of recycling, and then the levels returned to the original culture's biomass level. As this was a low cell density culture (~1.5 g per L) they were unaffected by the growth inhibition previous studies had found at high densities. They maintained the same growth and biomass production over 62 days of recycling, an impressive feat, using their specially designed media.

In summary, the culture density and associated quantities of free ions and secreted components in the spent media, were seen to impact on the benefit of reusing spent media and informs the investigative criteria for new study design with animal cell spent media recycling.

4.3 Recycling of Insect Cell Culture Media

One of the most active areas of research into the recycling of spent cell culture media is with insect cell cultures. It has long been known that using a small amount of spent media to supplement fresh media has beneficial effects on insect and mammalian cell cultures, attributed to the secretion of beneficial growth factors into the media by the culture cells. The extent to which spent media can be used to elicit the best growth response from these cultures is of particular relevance to a broader circular bioeconomy question. One of the earliest attempts at media recycling in insect cell culture immediately identified the presence of cytotoxic substances in the media which inhibited growth in most replicates studied [43]. They showed this in two separate insect cell lines, TN-368 and CP-169, but interestingly when the spent media were crossed over, the TN-368 spent media had no cytotoxic effect on the CP-169 cultures, leading to the hypothesis that the inhibitory substances may be cell line specific. Further assessment of various ratios of spent media to fresh media to test the growth of an insect cell line showed that the growth rate declined with increasing ratio of spent medium [44]. The spent media used in this study was from an eight-day culture and supplemented with 5% FBS. A 1:16 ratio of spent media to fresh media gave a growth rate similar to the fresh media alone, while 1:8 dropped by 8%, as did 1:5, and 1:1 gave a total reduction in growth rate of 67%. The effect of spent medium on protein production from insect cell expression systems was examined using two widely used cell lines, Sf-9 and High-Five, producing beta-galactosidase from E. coli [45]. Using the insect cell medium IPL-41, they investigated which supplements were optimal to add to the spent media by multiple combination experiments, whereby the optimal recipe was determined to contain yeastolate ultrafiltrate (8 g per L), glucose (2 g per L) and glutamine (8 mmol per L) plus 15–20% fresh medium. While their overall protein yield dropped, the biomass achieved was actually higher than the fresh media. The protein decrease could have been caused by a number of things, for example, it has been shown that ammonia accumulation occurs in spent media and has a negative impact on protein production by *Trichoplusia ni* cells [46]. The production of ion metabolites by cultures in spent media is again highlighted as an area of study by these findings.

Numerous studies have been completed attempting to determine not just the detrimental constituents of the spent media, but also the growth-promoting factors that may be present. Metalloproteinases were determined to be the only factor in spent media that was responsible for any and all growth-promoting effects in *Trichoplusia ni* cell cultures [47]. They tested this with conditioned media (spent media added), and in which they had inhibited metalloproteinases. None of the beneficial growth effects were seen in this culture, but when metalloproteinases were added, the benefits were restored. Since it has been shown that metalloproteinases are secreted into the media by insect cells and break down growth factor binding proteins, it was hypothesised by this group that metalloproteinases are necessary to release the growth factors in the spent media from their binding proteins [48–50]. Binding proteins of IGFs (insulin-like growth factors) have been found in the recycled mediums of the cell lines Sf-9 and High Five insect cells, which must therefore be broken down by secreted metalloproteinases in order to release the IGFs, which can then perform their growth-promoting function [49]. Cell cycle progression has also been found to be impacted by these autocrine secreted factors, further contributing to the beneficial effects of spent medium recycle [51].

Ex-Cell medium was examined for reuse possibilities using the High-Five insect cell line producing beta-galactosidase from *E. coli* [52]. Media was filtered through a 0.22 micron filter and supplementation with 0%, 25% and 50% spent media volumes were used to test effects on protein production. Production of beta-galactosidase increased in the 25% spent media test compared to the 0% control. Similar findings have been observed with a *Lepidopteran* Se301 cell line. Various percentages of spent media mixed with fresh media were tested, with measurement of culture growth rates and volume of baculovirus production, and it was found that these were only negatively affected when above a 60% spent media composition [53]. Most interestingly, there was actually a significant effect on baculovirus production at around 20% spent media composition, with a 90% increase in production compared to the fresh media control.

Taken together, it is obvious that using at least partial medium recycling has a beneficial effect on protein production in insect cell lines. It also does not appear to inhibit growth unless above a high ratio of spent media to fresh (>60%). These results were very instructive for this heterologous expression system, with conditioned media (recycled media) routinely added as a small percentage of the total culture volume as a result of these types of studies. These different cell systems also highlight that with further investigation into the composition of spent media there is significant potential to develop new methods to utilise the waste media streams from the bioprocessing of animal cell cultures.

4.4 Recycling of Mammalian Cell Culture Media

Mammalian cell culture is highly sensitive to extracellular metabolites and while recycling of cell culture media has been attempted, it has had very limited success. Early studies into recycling animal cell culture media witnessed a significant drop in cell growth from the first recycling attempt of 100% spent media [54]. In follow up studies, a partial medium recycle was proposed to mitigate some of the inhibitory effects on growth. Using a mouse hybridoma cell line and harvesting spent DMEM F-12 medium from cells producing an IgG like antibody, cells were grown in 80% spent medium for two subsequent recycling steps [55]. The first step did not see a significant drop in protein produced, at 29 mg/L compared to the control of 33 mg/L, but it did drop substantially for the second, at 18 mg/L. It was concluded that there is a greater negative effect of cytotoxic metabolites than positive effects from growth factors in the spent medium from mammalian cells at these high levels of spent media usage.

A later study proposed a smaller ratio of spent media of 63% to 37% fresh medium, which could possibly be used in subsequent recycling steps. Using mouse hybridoma cells over four subsequent recycling steps were tested with serum-free media with nutrient supplementation. Interestingly, these cells demonstrated a slight adaptation to the recycled medium, where the first and second had lower antibody production levels, which then recovered somewhat in the third and fourth. By the end of the recycling steps, 112 mg/L of protein yield was obtained, compared to just one round of fresh media giving 63 mg/L, highlighting a potentially useful benefit of media recycling in mammalian cells for the first time [56]. Spent media that had been run through a Protein G affinity column, was used in the culture of mouse hybridoma cells producing an IgG like protein. A percentage of the "recovered" media was employed, whereby spent media was concentrated 12-fold and then diluted 1:8 with fresh media. This approach was beneficial, with 60% recovered media giving a 33% increase in protein production compared to the fresh media [57]. The spent medium was proposed to contain multiple beneficial factors as well, though the mammalian cells appear more sensitive to high concentrations of spent medium than the insect cells. Some of these positive factors have already been identified, such as anti-apoptotic factors that some mammalian cell types secrete [58–59] or secreted growth factors such as platelet-derived growth factor (PDGF) [60].

For spent media from animal cell culture to prove valuable as a recycled waste stream to feed new bioprocesses, it will be necessary to carefully examine the chemical and protein composition of the media when recovered from the primary bioprocess. The identification of positive and negative growth factors and the requirements for supplementation will require careful empirical experimentation. The secondary bioprocess chosen to use this spent media will also require careful biochemical assessment to build a sustainable and valuable platform to harness the spent media waste to generate new value in the form of secondary bioprocessed products such as other recombinant proteins. One such organism that represents an excellent candidate to harness spent chemically defined media from mammalian cell bioprocess is *Escherichia coli*. It has been shown that the growth of *E. coli* is eventually inhibited by the production of acetic acid in culture and it has also been shown to produce much less acetic acid when growing in a chemically defined media, where carbon source is not in excess [60]. This type of fundamental understanding of the metabolism of one of the best studied 'microbial cell factories' along with the key signposts provided by previous studies in a range of other heterologous systems represents the blueprint for the applied study of animal cell culture waste for reuse in microbial fermentation as just one example (Fig. 2).

5 Analysis of Chemically Defined Spent Media

Utilising spent media as a potentially valuable waste stream requires thorough analysis of its composition. Many of the studies mentioned previously used multiple methods of cataloguing the various known nutrients within the recycled media which may have been depleted, or even the unknown cellular components left behind in such a media.

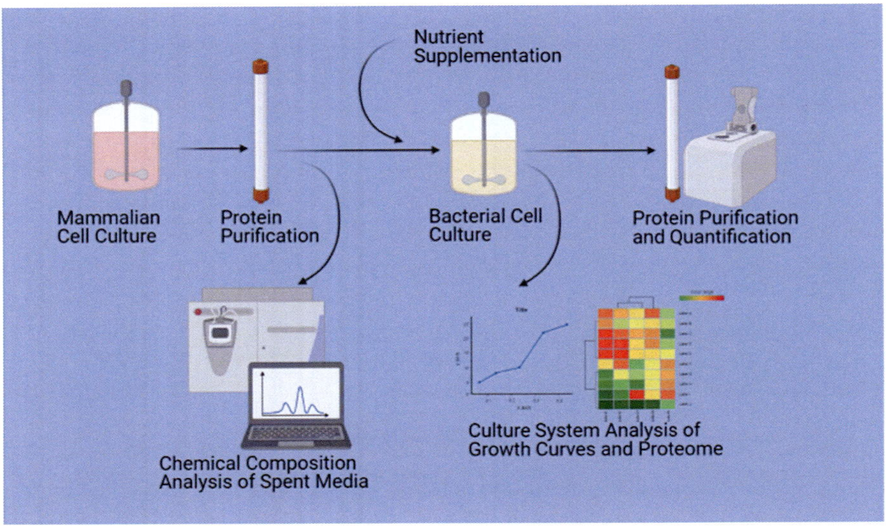

Fig. 2 Redesigning spent media from cell culture bioprocess to feed new bacterial fermentations. Harnessing a systems biology approach to analysis of spent cell culture media, fermentation conditions and protein production. Created with BioRender.com

5.1 Metabolomics

A benefit of using a chemically defined culture medium is that the concentration of nutrients within the media is known prior to consumption by cell cultures. It is therefore possible to analyse the remaining concentrations of these components to determine how much of each was consumed by the culture. This can be achieved through numerous analytical methods, principally Gas Chromatography (GC) and High-Performance Liquid Chromatography (HPLC). In the study of spent media effects on antibody production in mouse hybridoma cells, the free amino acids and glucose concentration in the spent media was analysed using HPLC, thereby finding the exact concentration of each which was required to be supplemented back into the media [56]. HPLC is a method best used to detect levels of a known component of the media, as you need to know its behaviour on a chromatographic peak, most usually achieved by running against known standards . While it can be used to detect the presence of impurities (unknown components), it cannot identify these components unless you know their properties in advance. Gas chromatography can be used in much the same way as HPLC to identify chemical components of the media that have known profiles [61]. When used in combination with mass spectrometry GC-MS can very sensitively identify metabolic patterns in culture media [62]. Glucose uptake can also be measured a number of ways, including using radiolabelled glucose analogues [63], or even using simple colorimetric assays [64]. The concentration of glucose, or other carbon sources, is known in defined mediums prior to use, so it is therefore feasible to find the concentration remaining in the media for supplementation purposes. Certain known proteins of known molecular weights can be tested using simple visual methods such as SDS-PAGE for an approximation of quantity based on band intensity, as was used to test for the depletion of transferrin in the spent media, a nutrient necessary for cultures producing an antibody like IgG [57].

5.2 *Proteomic Analysis of Unknown Components*

In many of the studies conducted into spent media usage, there were unknown growth-promoting or inhibiting factors within the media which impacted on further cell growth. As extracellular secretions are generally proteins, it is possible to analyse the media for the presence of these unknown proteins using mass spectrometry. Analysis of spent media from iPSC-derived neuronal cells to study the cell's secretome in Parkinson's models, using just 0.5 ml of spent media facilitated identification of over 500 proteins contained in the media [6]. Testing of a variety of preparation methods found that using media containing BSA hampered the identification of proteins by a 10-fold difference when compared with using B27-free media (contains no commercially added proteins). They also found that acetone precipitation was the most effective enrichment technique, yielding the most proteins, followed by TCA (trifluoracetic acid) precipitation. Mass spectrometry has been used to analyse secretomes for the presence of proteins in other disease models as well, such as metastatic cancers [65].

Proteomic analysis with peptide identification software, e.g., MaxQuant™, after mass spectrometry analysis is a critical component of these studies [66]. From this software, a file called "proteinGroups.txt" is generated with all of the information regarding identities and quantities of proteins in the data in tab delimited format. From here, the data can be analysed using an analysis software such as Perseus [67], or a bioinformatician may design their own code to gather relevant data from the file, such as the protein pathways significantly represented. Statistical analysis such as student's t-tests is then performed to check differences in protein concentration between samples to a level of significance (P-value <0.05). For this purpose, the UniProt ID mapping tool is invaluable, as lists of identifiers can be input to return most information about the proteins as an easily parsed tab delimited file, such as each protein's molecular weight, the pathway analysis, and whether it is a metabolite with catalytic or anabolic properties. All the information can be easily expressed in graph formats using Python's Matplotlib library, or R, with many good online resources for learning how, such as on GitHub, DataCamp, Rosalind, or StackOverflow among many others. Most of these courses are free and are a good use of free time during a pandemic, particularly if statistics is something you use often in your data, not just for proteomic or genomic data analysis.

5.3 Chemical Composition Analysis of Unknown Components

Some unknown components left behind in a spent media by cell cultures are not proteins, and therefore not detectable by mass spectrometry. These are generally carbohydrates, or lipids. There are certain analyses that can be run on these components to detect functional groups. For example, if testing for presence of short chain fatty acids such as acetic acid, gas chromatography or a carboxylic acid analyser can be used to estimate if any is present, and based on what functional groups it may contain, what the identity of that acid may be. Detecting unknown carbohydrates is more difficult and will need multiple tests to identify different functional groups present that can then help narrow down the possibilities, such as Benedict's test or an Iodine test.

While it is entirely possible to find the concentrations of known components in a spent medium and to identify unknown proteins, the difficulty lies in identifying unknown carbohydrates and lipids, and generally relies on functional group tests to find possible matches. Luckily, with the current advances in genome and proteome analysis, most cell lines secretome is very well studied and spent media components can therefore mostly be identified through mass spectrometry, SDS-PAGE and HPLC/GC.

6 Future Considerations for Developing CBE Research in the Bioprocessing Sector

Heterologous expression of recombinant human proteins in animal cell cultures at all scales, from investigative projects in research institutes to the production of drug substance by the biopharmaceutical industry, poses challenges and opportunities to create new value from the spent culture media waste stream that is being produced in ever increasing quantities globally. To participate in and help grow this opportunity these stakeholders will need to be encouraged to design new research strategies and new research funding schemes at one level, and will need to be incentivised to participate in the design of new policies with the aim of stimulating new practice by the global bioprocessing industry at another. There are sufficient precedents in the scientific literature that highlight the potential of turning waste from the bioprocess of cells into new value as part of a circular approach to changing the description of single use culture media to renewable media. The application of a systems biology led approach to the study of spent culture media and its components, and to the secondary culture systems that can utilise these waste streams as feed, has exciting potential to generate new value and place the bioprocessing field onto the Circular Bioeconomy map.

Funding Support The authors wish to thank Science Foundation Ireland and the EPSRC for joint funding of Ms Ciara Lynch through the Centre for Doctoral Training – Atoms to Products (18/EPSRC-CDT/3582) and the Science Foundation Ireland funding of Mrs Lynda Jordan through the BiOrbic Bioeconomy Research Centre (16/RC/3889).

Conflict of Interest Statement The authors have no conflict of interest to declare.

References

1. European Commission (2012) Innovating for sustainable growth: a bioeconomy for Europe. Brussels
2. European Commission (2015) Closing the loop – an EU action plan for the circular economy. European Commission, Brussels
3. Stegmann P, Londo M, Juninger M (2020) The circular bioeconomy: Its elements and role in European bioeconomy clusters. Resour Conserv Recycling X 6:100029
4. Temmes A, Peck P (2019) Do forest biorefineries fit with working principles of a circular bioeconomy? A case of Finnish and Swedish initiatives. Forest Policy Econ 110:1–12
5. European Commission (2017) Horizon 2020 Work Programme 2018–2020:9. Food Security, Sustainable Agriculture and Forestry, Marine and Maritime and Inland Water Research and the Bioeconomy, Brussels
6. Walsh G (2018) Biopharmaceutical benchmarks. Nat Biotechnol 36:1136–1145. https://doi.org/10.1038/nbt.4305
7. Rader RA, Langer ES (2018) Worldwide biopharmaceutical manufacturing capacity analysis: growth continues across the board. BioProcess International
8. Goff M, Nikodinovic-Runic J, O'Connor KE (2009) Characterization of temperature-sensitive and lipopolysaccharide overproducing transposon mutants of Pseudomonas putida CA-3 affected in PHA accumulation. FEMS Microbiol Lett 292(2):297–305. https://doi.org/10.1111/j.1574-6968.2009.01504.x
9. Aparicio T, de Lorenzo V, Martínez-García E (2018) CRISPR/Cas9-based counterselection boosts recombineering efficiency in Pseudomonas putida. Biotechnol J 13(5):1700161
10. Nikel PI, De Lorenzo V (2018) Pseudomonas putida as a functional chassis for industrial biocatalysis: from native biochemistry to trans-metabolism. Metab Eng 50:142–155
11. Lee SY, Kim HU (2015) Systems strategies for developing industrial microbial strains. Nat Biotechnol 33:1061–1072
12. Nielsen J (2017) Systems biology of metabolism. Annu Rev Biochem 86:245–275
13. Blank LM, Narancic T, Mampel J, Tiso T, O'Connor K (2020) Biotechnological upcycling of plastic waste and other non-conventional feedstocks in a circular economy. Curr Opin Biotechnol 62:212–219
14. Smanski M, Zhou H, Claesen J, Shen B, Fischbach MA, Voigt CA (2016) Synthetic biology to access and expand nature's chemical diversity. Nat Rev Microbiol 14:135–149
15. Johnson IS (1983) Human insulin from recombinant DNA technology. Science 219(4585):632–637
16. Chance RE, Frank BH (1993) Research, development, production, and safety of biosynthetic human insulin. Diabetes Care 16(Supplement 3):133–142
17. Ferrer-Miralles N, Domingo-Espín J, Corchero JL et al (2009) Microbial factories for recombinant pharmaceuticals. Microb Cell Factories 8:17
18. Kim JY, Kim YG, Lee GM (2012) CHO cells in biotechnology for production of recombinant proteins: current state and further potential. Appl Microbiol Biotechnol 93:917–930
19. Lai T, Yang Y, Ng SK (2013) Advances in mammalian cell line development technologies for recombinant protein production. Pharmaceuticals (Basel) 6:579–603

20. Wong DC et al (2006) Targeting early apoptotic genes in batch and fed-batch CHO cell cultures. Biotechnol Bioeng 95:350–361
21. Wurm FM, Hacker D (2011) First CHO genome. Nat Biotechnol 29:718–720
22. Xu X et al (2011) The genomic sequence of the Chinese hamster ovary (CHO)-K1 cell line. Nat Biotechnol 29:735–741
23. Kildegaard HF, Baycin-Hizal D, Lewis NE, Betenbaugh MJ (2013) The emerging CHO systems biology era: harnessing the omics revolution for biotechnology. Curr Opin Biotechnol 24(6):1102–1107. https://doi.org/10.1016/j.copbio.2013.02.007
24. Goehring A, Lee CH, Wang KH, Michel JC, Claxton DP, Baconguis I, Althoff T, Fischer S, Garcia KC, Gouaux E (2014) Screening and large-scale expression of membrane proteins in mammalian cells for structural studies. Nat Protoc 9(11):2574–2585. https://doi.org/10.1038/nprot.2014.173
25. Robert MA, Chahal PS, Audy A, Kamen A, Gilbert R, Gaillet B (2017) Manufacturing of recombinant adeno-associated viruses using mammalian expression platforms. Biotechnol J 12(3). https://doi.org/10.1002/biot.201600193
26. Beck M (2009) Agalsidase alfa for the treatment of Fabry disease: new data on clinical efficacy and safety. Expert Opin Biol Ther 9:255–261
27. Casademunt E, Martinelle K, Jernberg M et al (2012) The first recombinant human coagulation factor VIII of human origin: human cell line and manufacturing characteristics. Eur J Haematol 89:165–176
28. Dai L, Zheng T, Xu K, Han Y (2020) A universal design of betacoronavirus vaccines against COVID-19, MERS, and SARS. Cell 182(3):722–733
29. Mahajan P, Strain-Damerell C, Gileadi O, Burgess-Brown NA (2014) Medium-throughput production of recombinant human proteins: protein production in insect cells. Methods Mol Biol 1091:95–121. https://doi.org/10.1007/978-1-62703-691-7_6
30. Wang Q, Corey RA, Hedger G, Aryal P, Grieben M, Nasrallah C, Baronina A, Pike ACW, Shi J, Carpenter EP, Sansom MSP (2020) Lipid interactions of a ciliary membrane TRP channel: simulation and structural studies of polycystin-2. Structure 28(2):169–184.e5. https://doi.org/10.1016/j.str.2019.11.005
31. El Bakkouri M, Kouidmi I, Wernimont AK, Amani M (2019) Structures of the cGMP-dependent protein kinase in malaria parasites reveal a unique structural relay mechanism for activation. Proc Natl Acad Sci U S A 116(28):14164–14173
32. Vicente T, Roldão A, Peixoto C, Carrondo MJ, Alves PM (2011) Large-scale production and purification of VLP-based vaccines. J Invertebr Pathol 107(Suppl):S42–S48. https://doi.org/10.1016/j.jip.2011.05.004
33. Wang Y, Wei L, Jie B (2012) Repeated batch fermentation with water recycling and cell separation for microbial lipid production. Front Chem Sci Eng 6(4):453–460
34. Converti A, Perego P, Lodi A, Fiorito G, Del Borghi M, Ferraiolo G (1991) In-situ ethanol recovery and substrate recycling during continuous alcohol fermentation. Bioprocess Eng 7(1):3–10
35. Hsiao TY, Glatz CE, Glatz BA (1994) Broth recycle in a yeast fermentation. Biotechnol Bioeng 44(10):1228–1234
36. Abo BO, Odey EA, Bakayoko M, Kalakodio L (2019) Microalgae to biofuels production: a review on cultivation, application and renewable energy. Rev Environ Health 34(1):91–99
37. Hadj-Romdhane F, Jaouen P, Pruvost J, Grizeau D, Van Vooren G, Bourseau P (2012) Development and validation of a minimal growth medium for recycling chlorella vulgaris culture. Bioresour Technol 123:366–374
38. Lowrey J, Roberto E, Armenta BMS (2016) Nutrient and media recycling in heterotrophic microalgae cultures. Appl Microbiol Biotechnol 100(3):1061–1075
39. Leone DE (1963) growth of chlorella pyrenoidosa in recycled medium. Appl Microbiol 11:427–429
40. Lívanský K, Dedič K, Bínová J, Tichý V, Novotný P, Doucha J (1996) Influence of the nutrient solution recycling on the productivity of scenedesmus obliquus, utilization of nutrients and water in outdoor cultures. Algol Stud 81:105–113

41. Rodolfi L, Zittelli G, Barsanti L, Rosati G, Tredici MR (2003) Growth medium recycling in nannochloropsis sp. mass cultivation. Biomol Eng 20(4-6):243–248
42. Hadj-Romdhane F, Zheng X, Jaouen P, Pruvost J, Grizeau D, Croué JP, Bourseau P (2013) The culture of chlorella vulgaris in a recycled supernatant: effects on biomass production and medium quality. Bioresour Technol 132:285–292
43. Rechtoris C, McIntosh AH (1976) A cytotoxic substance in insect cell culture spent media. In Vitro 12(10):678–681
44. Kioukia N, Nienow AW, Emery AN, al-Rubeai M (1995) Physiological and environmental factors affecting the growth of insect cells and infection with baculovirus. J Biotechnol 38(3):243–251
45. Wu J, Ruan Q, Lam HYP (1998) Evaluation of spent medium recycle and nutrient feeding strategies for recombinant protein production in the insect cell–baculovirus process. J Biotechnol 66(2):109–116
46. Donaldson M, Wood HA, Kulakosky PC, Shuler ML (1999) Glycosylation of a recombinant protein in the Tn5B1-4 insect cell line: influence of ammonia, time of harvest, temperature, and dissolved oxygen. Biotechnol Bioeng 63(3):255–262
47. Eriksson U, Hassel J, Lüllau E, Häggström L (2005) Metalloproteinase activity is the sole factor responsible for the growth-promoting effect of conditioned medium in trichoplusia Ni insect cell cultures. J Biotechnol 119(1):76–86
48. Fowlkes JL, Winkler MK (2002) Exploring the interface between metallo-proteinase activity and growth factor and cytokine bioavailability. Cytokine Growth Factor Rev 13(3):277–287
49. Andersen AS, Per Hertz H, Schäffer L, Kristensen C (2000) A new secreted insect protein belonging to the immunoglobulin superfamily binds insulin and related peptides and inhibits their activities. J Biol Chem 275(22):16948–16953
50. Doverskog M, Tally M, Häggström L (1999) Constitutive secretion of an endogenous insulin-like peptide binding protein with high affinity for insulin in Spodoptera frugiperda (Sf9) cell cultures. Biochem Biophys Res Commun 265(3):674–679
51. Doverskog M, Bertram E, Ljunggren J, Ohman L, Sennerstam R, Häggström L (2000) Cell cycle progression in serum-free cultures of Sf9 insect cells: modulation by conditioned medium factors and implications for proliferation and productivity. Biotechnol Prog 16(5):837–846
52. Ikonomou L, Bastin G, Schneider Y, Agathos S (2004) Effect of partial medium replacement on cell growth and protein production for the High-Fivetrade Mark insect cell line. Cytotechnology 44(1-2):67–76
53. Beas-Catena A, Sánchez-Mirón A, Garía-Camacho F, Contreras-Gómez A, Molina-Grima E (2013) The effect of spent medium recycle on cell proliferation, metabolism and baculovirus production by the Lepidopteran Se301 cell line infected at very low MOI. J Microbiol Biotechnol 23(12):1747–1756
54. Mizrahi A, Avihoo A (1977) Growth medium utilization and its re-use for animal cell cultures. J Biol Stand 5(1):31–37
55. Kempken R, Büntemeyer H, Lehmann J (1991) The medium cycle bioreactor (MCB): monoclonal antibody production in a new economic production system. Cytotechnology 7(2):63–74
56. Riese U, Lütkemeyer D, Heidemann R, Büntemeyer H, Lehmann J (1994) Re-use of spent cell culture medium in pilot scale and rapid preparative purification with membrane chromatography. J Biotechnol 34(3):247–257
57. Ogawa A, Takada N, Terada S (2009) Effective antibody production by reusing culture medium previously used in antibody purification. Biosci Biotechnol Biochem 73(3):719–721
58. Melkonyan HS, Chang WC, Shapiro JP, Mahadevappa M, Fitzpatrick PA, Kiefer MC, Tomei LD, Umansky SR (1997) SARPs: a family of secreted apoptosis-related proteins. Proc Natl Acad Sci U S A 94(25):13636–13641
59. Polazzi E, Gianni T, Contestabile A (2001) Microglial cells protect cerebellar granule neurons from apoptosis: evidence for reciprocal signalling. Glia 36(3):271–280

60. Lauffenburger D, Cozens C (1989) Regulation of mammalian cell growth by autocrine growth factors: analysis of consequences for inoculum cell density effects. Biotechnol Bioeng 33(11):1365–1378
61. Brooks JB (1977) Detection of bacterial metabolites in spent culture media and body fluids by electron capture gas-liquid chromatography. Adv Chromatogr 15:1–31
62. Deshpande R, Yang TH, Heinzle E (2009) Towards a metabolic and isotopic steady state in CHO batch cultures for reliable isotope-based metabolic profiling. Biotechnol J 4:247–263
63. Yamamoto N et al (2011) Measurement of glucose uptake in cultured cells. Curr Protoc Pharmacol Chapter 12:Unit 12.14:1–22. https://doi.org/10.1002/0471141755.ph1214s55
64. Hulme CH, Westwood M, Myers JE, Heazell AE (2012) A high-throughput colorimetric-assay for monitoring glucose consumption by cultured trophoblast cells and placental tissue. Placenta 33(11):949–951. https://doi.org/10.1016/j.placenta.2012.08.001. Epub 2012 Aug 27. PMID: 22951137
65. Kwon OK, Jeon JM, Sung E, Na AY (2018) Comparative secretome profiling and mutant protein identification in metastatic prostate cancer cells by quantitative mass spectrometry-based proteomics. Cancer Genomics Proteomics 15(4):279–290. https://doi.org/10.21873/cgp.20086
66. Cox J, Mann M (2008) MaxQuant enables high peptide identification rates, individualized Ppb-Range mass accuracies and proteome-wide protein quantification. Nat Biotechnol 26(12):1367–1372
67. Tyanova S, Temu T, Sinitcyn P, Carlson A (2016) The Perseus computational platform for comprehensive analysis of proteomics data. Nat Methods 13(9):731–740

Monoclonal Antibody Glycoengineering for Biopharmaceutical Quality Assurance

Itzcóatl Gómez Aquino and Ioscani Jiménez del Val

1 Introduction

With a vast and continuously expanding market, biopharmaceutical companies are looking for ways to deliver ever-increasing quantities of therapeutic monoclonal antibodies (mAbs) to treat several life-threatening illnesses, including multiple types of cancer and autoimmune disorders (Table 1). Although advances in production cell line development, high-throughput bioprocess optimisation, and the uptake of non-traditional bioprocessing technologies (e.g., continuous integrated processes) have all contributed to meeting manufacturing capacity targets, ensuring mAb quality in terms of homogeneity, safety, and therapeutic efficacy remains a key challenge.

91 of the 98 currently approved mAb products (to the date of writing) are based on the gamma immunoglobulin (IgG) isotype, the best-selling of which are presented in Table 1. IgG mAbs are extremely complex molecules that consist of four disulphide-crosslinked peptide chains and two consensus asparagine (N)-linked glycosylation sites on their constant fragment (Fc) (Fig. 1). Although multiple features of these complex molecules may contribute to heterogeneity (e.g., amino acid sequence, 3D structure, glycosylation C-terminal lysine, amidation, etc.), glycosylation is, by far, the main source of mAb variability and principal driver for efficacy modulation [1]. Indeed, recent studies have found substantial lot-to-lot glycosylation variability in different commercial mAb preparations [2, 3]. Glycosylation heterogeneity has been closely linked with even modest excursions in a breadth of cell culture operating conditions [4–7], which in many cases, cannot be prevented or curbed during manufacturing campaigns.

I. G. Aquino · I. J. del Val (✉)
School of Chemical & Bioprocess Engineering, University College Dublin, Dublin, Ireland
e-mail: ioscani.jimenezdelval@ucd.ie

R. Pörtner (ed.), *Cell Culture Engineering and Technology*, Cell Engineering 10,
https://doi.org/10.1007/978-3-030-79871-0_6

Table 1 Top-20 selling mAb-based therapeutics of 2019[a]

Generic name	Target & Format	Indication	Production Host	Approval year	2019 Sales (US$M)
Adalimumab	Anti-TNFa human IgG1 mAb	Immune modulation	CHO	2002	19,169
Pembrolizumab	Anti-PD-1 humanized IgG4 mAb	Oncology	CHO	2014	11,121
Nivolumab	Anti-PD-1 human IgG4 mAb	Oncology	CHO	2014	7989
Aflibercept	VEGFr kinase inhibitor; human VEGF A&B receptors:Human IgG1 fc fusion	Ophthalmology & Oncology	CHO	2011	7650
Bevacizumab	Anti-VEGFr humanized IgG1 mAb	Oncology	CHO	2004	7638
Etanercept	TNFa inhibitor; human TNF receptor:Human IgG1 fc fusion	Immune modulation	CHO	1998	7218
Rituximab	Anti-CD20 chimeric IgG1 mAb	Immune modulation & oncology	CHO	1997	6540
Trastuzumab	Anti-HER2 humanized IgG1 mAb	Oncology	CHO	1998	6522
Ustekinumab	Anti-IL-12 & IL-23 human IgG1 mAb	Immune modulation	CHO	2009	6361
Infliximab	Anti-TNFa chimeric IgG1 mAb	Immune modulation	Sp2/0	1998	5030
Denosumab	Anti-RANKL human IgG2 mAb	Oncology	CHO	2010	4607
Eculizumab	Anti-complement factor C5 humanized IgG2/4 mAb	Immune modulation	GS-NS0	2007	3946
Pertuzumab	Anti-HER2 humanized IgG1 mAb	Oncology	CHO	2012	3804
Secukinumab	Anti-IL-17A human IgG1 mAb	Immune modulation	CHO	2015	3551
Golimumab	Anti-TNFa human IgG1 mAb	Immune modulation	CHO	2009	3406
Omalizumab	Anti-IgE humanized IgG1 mAb	Immune modulation	CHO	2003	3190
Daratumumab	Anti-CD38 human IgG1 mAb	Oncology	CHO	2015	2998
Abatacept	CD80 & CD86 inhibitor; CTLA-4:Human IgG1 fc fusion	Immune modulation	CHO	2005	2977
Dupilumab	Anti-IL-4Rα human IgG4 mAb	Immune modulation	CHO	2017	2322
Atezolizumab	Anti-PD-L1 humanized IgG1 mAb	Oncology	CHO	2016	2025
				Total	118,064

[a]The values were obtained from the publicly available 2019 end-of-year Investor Reports from the respective pharmaceutical companies

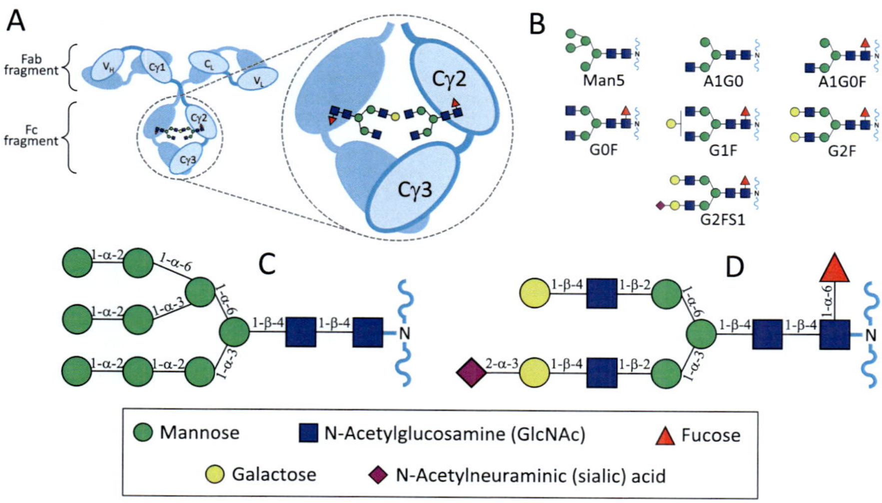

Fig. 1 IgG mAb structure, consensus glycosylation sites, and common mAb Fc N-linked glycans [17]

The basic structure of an IgG mAb molecule, including the antigen binding (Fab) and crystallisable (Fc) fragments and the two consensus Fc N-linked glycosylation sites present on its Fcγ domains are shown (**a**). The most common Fc N-glycans of mAbs [2, 9, 10], including shorthand notation, are presented (**b**). The structures and glycosidic bond configurations of Man9 (**c**) and a fully processed bi-antennary N-glycan (**d**) are also shown. The symbols outlined in the box denote the monosaccharide constituents of the glycans.

Concerns surrounding the glycosylation heterogeneity of mAbs are compounded by the well-documented impact protein-bound carbohydrates (glycans) have on the safety, serum half-life, and therapeutic mechanisms of mAbs [8]. In conjunction with its inherent variability, these facets have made glycosylation a widely acknowledged determinant of mAb product quality. It is therefore evident, from a regulatory and manufacturing perspective, that controlling glycosylation is of paramount importance to manufacture homogenous, safe, and highly efficacious mAbs.

To address the above mAb quality concerns, substantial academic and industrial research has centred around developing strategies to control the glycosylation of mAbs. These glycoengineering strategies can be split into two categories, based on the level at which they operate. *Cellular glycoengineering* strategies are those whereby production cells are genetically modified to control mAb glycosylation. *Metabolic glycoengineering* strategies involve the supplementation of cell culture media with small molecules to manipulate the metabolic pathways which underly glycosylation.

2 mAb N-Linked Glycosylation

The N-linked Fc glycans of mAbs are usually biantennary structures with an array of possible monosaccharide extensions, such as core fucose (Fuc), galactose (Gal), and N-acetylneuraminic acid (Neu5Ac) (Fig. 1b) [2, 9, 10]. Less processed glycans, such as those devoid of N-acetylglucosamine (GlcNAc) on their α-1,6 arm or the five mannose (Man5) structure (Fig. 1b, A1G0 or A1G0F), have also been observed on the Fc of commercial mAbs [9].

Although there is a fraction of polyclonal human IgG that is N-glycosylated in their Fab fragment [11], cetuximab is the only marketed mAb with Fab glycosylation [12]. Because Fab glycans are less likely to encounter the steric constraints of the Fc, they can be further processed to include GlcNAc bisection and increased sialylation [13, 14], lactosamine (β −1,2 GlcNAc-Gal) repeats, tandem α-1,3 galactose moieties (αGal), and up to four antennae [15]. These features can exert changes in the functional activity of IgGs, including in antigen binding, serum half-life, stability, and aggregation [16].

The relative abundance and ensuing variability of mAb-bound N-glycans arises from the intracellular process of glycosylation depicted in Fig. 2 [17]. It begins in the endoplasmic reticulum, when oligosaccharyltransferase (OST, EC 2.4.99.18) co-translationally adds the $Glc_3Man_9GlcNAc_2$ precursor oligosaccharide to available Asn-X-Ser/Thr amino acid sequons of the nascent polypeptide chain [18]. The three glucose (Glc) residues on the precursor oligosaccharide serve as markers for the calnexin-calreticulin cycle and thus aid in adequate protein folding [19]. Once correctly folded, the glycoprotein is exported, by means of vesicles, to the Golgi apparatus, where the glycan is sequentially modified by a number of enzyme-catalysed reactions.

First, α-mannosidase I (αManI, EC 3.2.1.24) trims off mannose residues from the Man9 glycans (Fig. 1b) entering Golgi to yield the Man5 structure (Fig. 1c). α-1,3 N-acetylglucosaminyltransferase (MGAT1, EC 2.4.1.101) then adds GlcNAc to the α-1,3 glycan arm. The resulting glycan can then be modified by α-mannosidase II (αManII, EC 3.2.1.114), which removes the two remaining mannose residues of the α-1,6 arm. From this point on, parallel reactions may occur, where GlcNAc is added to the α-1,6 arm by N-acetylglucosaminyltransferase II (MGAT2, EC 2.4.1.101) or core fucose is added by α-1,6 fucosyltransferase (α6FucT, Fut8, EC 2.4.1.68), as shown in Step 4b of Fig. 2. β-1,4 galactosyltransferase (β4GalT, EC 2.4.1.38) can now add Gal residues to the GlcNAcs on either arm of the glycan to produce mono or bi-galactosylated species. Finally, galactosylated glycans can be capped with Neu5Ac residues bound through α-2,6 (EC 2.4.99.1) or α-2,3 (EC 2.4.99.4) glycosidic linkages.

All monosaccharide addition reactions require nucleotide sugar donors (NSDs) as co-substrates. The four NSDs required for mAb N-linked glycosylation are uridine diphosphate N-acetylglucosamine (UDP-GlcNAc), guanidine diphosphate fucose (GDP-Fuc), uridine diphosphate galactose (UDP-Gal), and cytidine monophosphate N-acetylneuraminic acid (CMP-Neu5Ac). NSD are produced

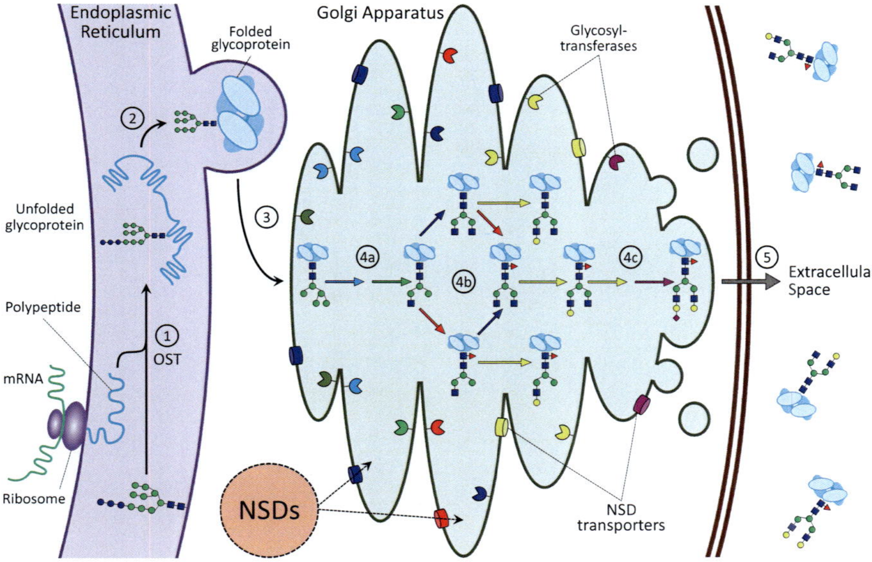

Fig. 2 The N-linked glycosylation process of mAbs in mammalian cells [18, 26]

The process begins in the Endoplasmic Reticulum, where oligosaccharyltransferase (OST) co-translationally transfers the $Glc_3Man_9GlcNAc_2$ precursor oligosaccharide to the mAb polypeptide backbone (Step 1). The glycan aids in protein folding (Step 2) [19] after which the mAb is exported to the Golgi apparatus (Step 3) for further processing. While the mAbs traverse Golgi, glycosylation enzymes sequentially transform their glycans from immature (high mannose) to complex biantennary structures (Steps 4a to 4c). Once the mAbs reach the trans-Golgi network, they are sorted and secreted to the extracellular environment (Step 5). Nucleotide Sugar Donors (NSDs), which are required co-substrates for monosaccharide addition reactions are synthesised in the cytosol/nucleus of the cell and transported into the secretory compartments through bespoke solute carriers (NSD transporters).

from common nutrients, such as glucose, galactose, and glucosamine via the biosynthetic pathway depicted in Fig. 3 [20, 21] and are transported via bespoke solute carriers (NSD transporters) into Golgi, where they are consumed in the glycosylation reactions. Each NSD has its own transporter: NGT (SLC35A3) for UDP-GlcNAc, GFT (SLC35C1) for GDP-Fuc, UGT (SLC35A2) for UDP-Gal, and CST (SLC35A1) for CMP-sialic acids [22].

In addition to the availability of cellular machinery and NSD co-substrates, N-glycan processing also depends on glycoprotein residence time within the secretory compartments [23, 24], as well as the activity and localisation of glycoenzymes and NSD transporters along Golgi. These aspects are inherently dynamic as they are heavily influenced by cell culture conditions, such as temperature, pH, ammonia accumulation, among others. Detailed reviews on the impact bioprocess conditions have on protein glycosylation are available elsewhere [25–27] and are, thus, beyond the scope of this chapter.

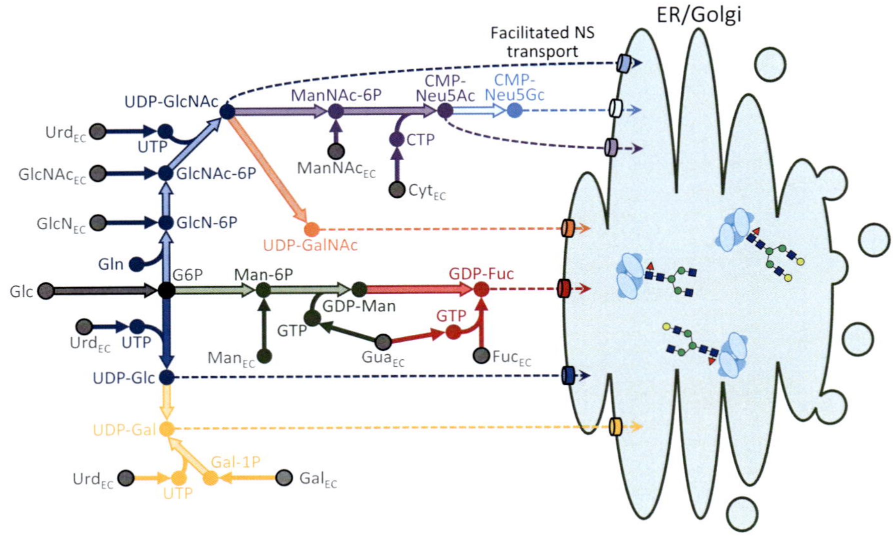

Fig. 3 Nucleotide sugar metabolism [20]
The most common NSD co-substrates for mAb N-glycan biosynthesis are uridine diphosphate GlcNAc (UDP-GlcNAc), guanosine diphosphate fucose (GDP-Fuc), uridine diphosphate galactose (UDP-Gal), and cytidine monophosphate Neu5Ac (CMP-Neu5Ac). NSDs are synthesised in the cytosol/nucleus from common metabolites, such as glucose (Glc), glutamine (Gln), glucosamine (GlcN) and nucleoside triphosphates (CTP, GTP, and UTP) [20, 21]. Precursors of the nucleoside (Cyt, Gua, and Urd) and saccharide (Glc, GlcN, GlcNAc, Man, Fuc, and Gal) components of NSDs can be fed during cell culture (denoted by the subscript EC) and incorporated into the NSD metabolic pathway. NSDs are transported into the lumen of ER or Golgi compartments by means of bespoke solute carriers: NGT (SLC35A3) for UDP-GlcNAc, GFT (SLC35C1) for GDP-Fuc, UGT (SLC35A2) for UDP-Gal, and CST (SLC35A1) for CMP-sialic acids [22].

Overall, protein glycosylation is a complex, non-template driven process that combines bespoke cellular machinery with metabolism to yield a breadth of possible glycan structures that vary in response to environmental stimuli. Indeed, protein glycosylation is thought to have evolved, at least in part, to provide proteins with an intrinsic degree of variability and functional adaptability [28]. Controlling protein glycosylation goes against these evolutionary purposes and, thus, poses many challenges in adequately engineering the machinery and metabolism of cells to produce optimal and consistent glycosylation profiles in therapeutic proteins.

3 N-Linked Glycans as Quality Attributes of mAbs

Different glycosylation motifs are known to heavily influence the safety, pharmacokinetics, and pharmacodynamics of therapeutic mAbs, properties which, in conjunction, are essential determinants of biopharmaceutical product quality.

Table 2 presents a summary of the impact glycosylation has on the key quality features of mAbs. Safety concerns are most commonly associated with the presence of non-human monosaccharides/glycosidic linkages. For example, α-Gal, a murine glycan motif, has caused fatal anaphylaxis in certain patient subpopulations undergoing cancer treatment with cetuximab [29].

Serum half-life (pharmacokinetics), on the other hand, is linked with the extent of glycan processing. Increased Neu5Ac content (higher glycan processing) has been widely reported to extend the half-life of therapeutic proteins in patient's sera [30]. In contrast, less processed glycans, such as the Man5 structure, result in reduced mAb serum half-life, possibly through mannose binding lectin-mediated clearance [31, 32].

Fc glycans play a crucial role in the therapeutic mechanisms of both cancer and anti-inflammatory mAbs. Absence of core fucosylation has been observed to enhance antibody-dependent cellular cytotoxicity (ADCC) up to 50-fold in *in vitro* assays [33], while increased galactosylation positively impacts both ADCC [34] and complement-dependent cytotoxicity (CDC) [35]. Increased sialylation, on the other hand, has been linked to the anti-inflammatory activity of intravenous gamma globulin (IVIG) therapies [36, 37].

The influence Fc N-glycans have on the therapeutic mechanisms of mAbs is associated with their role in defining the fragment's 3D structure [38]. The Fc reversibly binds to its receptors (FcγRs) on the surface of the patient's immune cells to trigger both ADCC and anti-inflammatory activity [39, 40], while it also binds to the C1q component of the complement system to elicit CDC [41]. All three therapeutic mechanisms rely on the affinity with which these Fc-mediated interactions occur. This affinity, in turn, is determined by the Fc's 3D structure and, therefore, by the relative abundance and monosaccharide composition of the N-glycans present therein.

4 Cellular mAb Glycoengineering

Due to the direct impact core fucosylation, galactosylation, and sialylation have on the safety, serum half-life, therapeutic efficacy, and glycosylation variability of commercial mAb products [42, 3], several cell engineering strategies have been developed to control these glycan motifs. Based on technology that was developed nearly two decades ago [43], there are currently five commercially-available mAbs, mogamulizumab [44], obinutuzumab [45], benralizumab [46], inebilizumab [47], and belantamab [48], where fucosylation has been genetically eliminated from the production cell lines. Based on the clinical advantage of these glycoengineered products, an ever-expanding knowledgebase on the interplay between glycosylation and therapeutic activity, and the advent of novel cell engineering approaches, over 20 additional glycoengineered mAbs are currently undergoing clinical evaluation [49]. This section will provide an in-depth review of the genetic engineering strategies that have been developed to modulate the glycosylation of mAbs.

Table 2 Glycosylation as a critical quality attribute of mAbs

Glycan moiety	Safety	Clearance (PK)	Efficacy (PD) as determined from mAb binding affinity for Fcγ receptors or the C1q protein of the complement system								
			ADCC			ADCP		CDC			Immune Modulation (IVIG)
Absent					[215, 248-252]					[248, 253]	
High mannose			[95, 98]		[251, 254, 255]					[95, 98, 256]	
Fucose				[257]	[33, 258-262]	[263]	[263-265]		[258, 259]	[50, 258, 266]	
Bisected			[70, 267, 268]			[265]		[212]			
Galactose		[269]	[41, 214, 254, 270]	[33, 212, 250]		[263]		[35, 212, 250, 271, 272]			[273]
αGal epitope	[274]										
Sialylation (Neu5Ac)				[214, 250, 275]	[36, 83, 85, 276, 277]					[278]	[279, 280]
Sialylation (Neu5Gc)	[84, 281]										

Negative effects of each glycan moiety on the safety, PK, and PD of mAbs are indicated in red, while positive effects are shown in green. Neutral effects are shaded in grey. Absence of colouring indicates interactions that have not yet been reported

4.1 Core Fucose Cellular Glycoengineering

In seminal work by Shinkawa and collaborators [33], absence of core fucosylation was reported to increase ADCC activity up to 50-fold in *in vitro* assays. Based on this key finding, the first fucose-engineered cell lines were developed by performing sequential homologous recombination with the Cre-LoxP system to disrupt both alleles of FUT8 (the gene for α6FucT) in CHO-DG44 cells. The strategy yielded a completely defucosylated anti-CD20 IgG1 mAb that achieved a > 100-fold increase in ADCC activity, when compared with Rituxan® [50]. Notably, the first approved glycoengineered mAb was created using the above FUT8 knockout strategy [44]. In subsequent studies, silencing of FUT8 with small interfering RNA (siRNA) resulted in a 20% reduction of α6FucT expression and a concomitant increase (between 10% and 60%) in defucosylated mAb glycoforms [51]. Short hairpin RNA (shRNA) has also been used to reduce fucosylation by targeting FUT8 [52].

Fucose cellular engineering has also been performed by disrupting the genes associated with GDP-Fuc biosynthesis. Heterologous expression of prokaryotic GDP-6-deoxy-D-lyxo-4-hexulose reductase, an enzyme that produces GDP-rhamnose from GDP-Fuc precursors, in CHO cells yielded 98% mAb defucosylation [53, 54], while stable transfection of GDP-6deoxy-D-talose synthetase reduced fucosylation to 20% by diverting the pathway to GDP-talose [55]. In another approach, the siRNA-mediated knockdown of GDP-mannose 4,6-dehydratase was unable to fully eliminate mAb fucosylation; however, when coupled with α6FucT silencing, mAb afucosylation increased from 3% to 98% [56].

More recently, gene editing tools, such as zinc finger nucleases (ZNFs), transcription activator-like effector nucleases (TALENs), and CRISPR-Cas9, have been used to knockout the genes for α6FucT [57–62] and the GDP-Fuc transporter (GFT) [63, 64]. CRISPR-Cas9 has been particularly successful in streamlining the cellular engineering workflow, achieving FUT8 indel frequencies of up to 99.7% [59] and enabling the simultaneous knockout of multiple glycogenes [65, 66]. A Cas9 alternative, Cpf1, provides similar efficiencies in reducing FUT8 activity by targeting the gene's endogenous promoter [67].

Another cellular glycoengineering strategy that has been used to abrogate core fucosylation is the ectopic expression of β1,4-N-acetylglucosaminyltransferase III (MGAT3) in mAb-producing cell lines where it is not constitutively expressed [43]. MGAT3 adds bisecting GlcNAc (bGlcNAc) to the core N-glycan (Fig. 4d) and disrupts fucosylation by competing with α6FucT activity: once bGlcNAc is added to the glycan, it can no longer be fucosylated [68]. MGAT3 was expressed in CHO cells that produced an anti-neuroblastoma mAb and, depending on its expression level, increasing bGlcNAc content was observed in the produced glycans [43]. Ectopic expression of MGAT3 in an anti-CD20 mAb-producing CHO cell line achieved a bGlcNAc content of between 48% and 71% and a concomitant 20-fold increase in ADCC activity [69]. The impact of MGAT3 expression on core fucosylation was later optimised by modifying the enzyme's transmembrane domain to ensure it localises before α6FucT along the Golgi apparatus [70]. Ectopic MGAT3 expression has led to the commercialisation of the second glycoengineered mAb, Gazyva®, which is currently prescribed as a first-line treatment of chronic lymphocytic leukaemia [71].

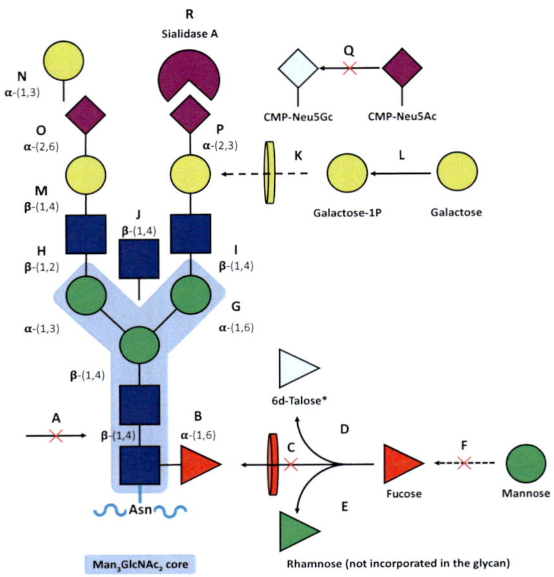

A. endo-β-N-acetylglucosaminidase (endoT)
- Co-expression [102]

B. Fucosyltransferase VIII (Fut8)
- Disruption by homologous recombination [50]
- Knock-down: siRNA [51, 56], shRNA [52]
- Knock-out: ZFN [57, 60], TALENs [61], and CRISPR-Cas9 [62, 65]/ CRISPR-CPf1 [67].
- Endogenous promoter disruption [67]
- Targeted integration with controlled expression [247]

C. Fucose transporter (SLC35C1)
- Knock-down siRNA [56]
- Knock-out CRISPR/Cas9 [63], ZFNs [64], and TALENs

D. Talose synthetase (GTS)
- Ectopic expression [55]

E. GDP-6-deoxy-D-lyxo-4-hexulose reductase (RMD)
- Co-expression [53], ectopic expression [54], and transient expression [134]

F. Mannose dehydratase (GMD)
- Knock-down: siRNA [56]

G. Mannosidase (Man2A1/2A2)
- Ectopic co-expression [66]
- Knock-out: CRISPR/Cas9 [87]

H. N-acetylglucosaminyltransferase I (Mgat I)
- Knock-out: ZFN [96]

I. N-acetylglucosaminyltransferase II (Mgat II)
- Ectopic expression [74]
- Knock-out: CRISPR/Cas9 [87]

J. N-acetylglucosaminyltransferase III (Mgat III)
- Co-expression [43], ectopic co-expression [66], and transient expression [70].

K. Galactose transporter (Slc35D1)
- Ectopic expression [74]

L. Galactokinase (GalK1)
- Ectopic expression [74]

M. Galactosyltransferase (β4galT I-IV)
- Transient expression [134], ectopic expression [72-74] targeted integration with controlled expression [247]
- Knock-out ZFN, TALENs, and CRISPR-Cas9 [61, 75, 76]

N. Galactosyltransferase (Ggta I)
- Knock-out: ZFN [78]

O. Sialyltransferase I (ST6Gal I)
- Ectopic expression [73, 86], targeted integration [87], gene activation [90], and overexpression [92, 93]

P. Sialyltransferase IV/VI (ST3Gal IV/VI)
- Ectopic expression [72] and targeted integration [87]
- CRISPR-Cas9 knock-out: [86, 87]

Q. CMP-Neu5Ac hydroxylase (CMAH)
- Knock-out: CRISPR/Cas9 [78]
- Knock-down: asRNA [88]

R. Sialidase A (Neu4)
- Overexpression [85]

Fig. 4 Overview of cellular glycoengineering strategies On right hand side, we list the glycosylation enzymes that have been engineered to date. The cell engineering strategy (disruption, overexpression, co-expression, knock-down and knock-out) and gene editing tool used (siRNA: small interfering RNA; shRNA: short hairpin RNA; CRISPR-Cas9: Clustered regularly interspaced short palindromic repeats-Cas9 nuclease complex; ZFN: Zinc Finger Nucleases; TALENs: Transcription Activator-Like Effector Nucleases) are also listed. The left-hand side presents the glycan motifs that have been engineered using the listed strategies. The identity of individual monosaccharides and glycosidic bond configurations is also shown. The symbols for each monosaccharide are the same as in Figure 1 The core N-glycan structure ($Man_3GlcNAc_2$) is highlighted in the shaded box.

4.2 Galactosylation Engineering

Cellular glycoengineering strategies have targeted galactosylation to enhance the therapeutic mechanisms and eliminate undesired immunogenic responses of mAbs. In order to achieve enhanced ADCC [34], CDC [35], and immune modulation by providing additional sialylation sites [37, 36], increased galactosylation has been achieved by ectopic expression of human β4GalTI [72–74]; conversely, knocking-out this gene leads to poor galactosylation levels [75, 76]. The presence of αGal motifs, which commonly cause adverse immunogenic responses in patients, has been eliminated in CHO cells by ZFN-mediated biallelic inactivation of Ggta1 [77, 78], which is the gene that codes for α3GalT.

Galactosylation of mAb Fc glycans can also be altered by targeting genes that are peripheral to the glycosylation process. Enhanced mAb galactosylation has been achieved by overexpression of phosphate synthase II, aspartate transcarbamoylase, and dihydrooratase, which are enzymes involved in *de novo* pyrimidine biosynthesis. Overexpression of these enzymes resulted in higher UTP biosynthesis which, in turn, increased the intracellular pool of UDP-Gal by between 200% to 350% [79]. In a similar approach, overexpression of uridine diphosphatase was observed to improve the galactosylation of a CTLA4-IgG fusion protein [80]. A recent model-driven cell engineering study achieved significant reduction of G0 glycoforms with over 65% bi-galactosylation by co-expressing genes involved with precursor production, NSD biosynthesis and its transport, as well as β4GalTI [74].

Reducing the accumulation of metabolic by-products that impact the activity and localisation of β4GalT may also enhance mAb Fc galactosylation. Reduced lactate accumulation via the down-regulation of lactate dehydrogenase in a GS-CHO cell line producing rituximab resulted in a 20% increase in G1F and G2F glycoforms [81]. Reduction of ammonia accumulation, e.g. by the simultaneous overexpression of carbamoyl phosphate synthetase I and ornithine transcarbamoylase, has also been found to enhance mAb Fc galactosylation [82].

4.3 Sialylation

Terminal Fc glycan sialylation has been reported to reduce the ADCC activity of mAbs by up to 20-fold [83] and is, therefore, an undesirable motif in oncology products. In addition, the most commonly used cell lines for mAb manufacturing produce non-human sialylation: CHO cells add Neu5Ac with the non-human α-2,3 linkage, whereas murine cells may add N-glycolylneuraminic acid (Neu5Gc), a xenoantigenic sialic acid [84]. For the above reasons, glycoengineering strategies have been developed to eliminate sialylation in oncological mAbs by ectopic expression of sialidases [85], knocking out α3SiaT with CRISPR/Cas9 [86, 87] or by siRNA silencing of the Neu5Gc biosynthetic gene, CMP-N-acetylneuraminic acid hydroxylase [88].

In contrast, the presence of α-2,6 linked Neu5Ac in Fc glycans has been reported to be beneficial for the anti-inflammatory activity of intravenous IgG therapies [36, 89]. To this end, enhancing α6Neu5Ac sialylation has been achieved by ectopic expression or gene activation of α6SiaT in CHO cells [90–92, 86]. On its own, overexpression of α6SiaT achieved up to 18% of mono- and di-sialylated mAb glycoforms and, when combined with the knockout of α3SiaT, up to 77% of α-2,6 sialylated Fc glycans was achieved [93].

Increasing the pool of available ManNAc by overexpressing UDP-N-acetylglucosamine 2-epiemerase/N-acetylmannosamine kinase in conjunction with α6SiaT substantially increases the amount of α-2,6 sialylated glycans [94]. Furthermore, mAb Fc glycoforms with over 85% α-2,6 sialylation were obtained with simultaneous overexpression of α6SiaT and β4GalT in CHO [73]. It is worth noting that β4GalT was crucial in achieving such a high level, indicating that the absence of galactose residues may be a limiting factor for sialylation.

4.4 Man5

mAbs with Man5 Fc glycans have been reported to present enhanced cytotoxic activity [95]. Due to where this glycan motif is produced along the glycosylation reaction network (Fig. 2), cellular engineering strategies to enrich the proportion of this glycoform have relied on knocking out the MGAT1 glycosyltransferase. Up to 90% Man5 mAb Fc glycoforms have been produced through ZFN-mediated knockout of MGAT1 in CHO cells [96]. It must be noted that this methodology has been developed to produce mAbs that are not intended for therapeutic use. This is likely due to the potentially negative effects of Man5 glycans in therapeutic products, which range from increased serum clearance rates [97, 32, 98] to potentially exacerbated immune responses [99].

4.5 Aglycosylation and Glycan Truncation

Due to the variability that may arise from glycosylation, complete removal of mAb Fc glycans has also been explored. To assess the impact of Fc aglycosylation on the effector functions of mAbs, a library of IgG heavy chain mutations lacking the consensus N-glycosylation sequon has been developed. This work found that two point mutations in the heavy chain yielded Fc's with a three-fold higher binding affinity for FcγRIIa than glycosylated mAbs, thus demonstrating that aglycosylated mAbs may present similar or even improved cytotoxic therapeutic properties [100]. Atezolizumab, an Fc-engineered monoclonal antibody devoid of glycans, has been FDA approved as a second-line therapy for advanced bladder cancer, but its therapeutic mechanism does not rely on Fc-mediated effector functions [101].

GlycoDelete, an elegant strategy to produce mAbs with homogenous glycosylation, has been developed by Meuris and collaborators [102]. GlycoDelete generates a homogenous pool of GlcNAc-Gal-Neu5Ac trisaccharide glycans by knocking out MGAT1 and ectopic expression of endo-β-N-acetylglucosaminidase (EndoT) fused to the transmembrane domain of α6SiaT. mAbs produced with GlycoDelete cells present reduced serum clearance, similar CD20 cell binding, reduced FcγR binding affinity, and lower ADCC activity when compared to the non-glycoengineered mAb. Due to the possibility of impaired effector function activity, the GlycoDelete strategy may be more suitable to produce antigen-neutralizing mAbs and not those which rely on FcγR-mediated cytotoxic effector functions.

5 Metabolic Glycoengineering

The metabolism of mAb production cell lines can be manipulated to steer glycan profiles towards a desired distribution. The majority of metabolic glycoengineering strategies involve feeding metabolic precursors of NSD biosynthesis during cell culture. A simplified NSD biosynthetic network is presented in Fig. 3, where it can be seen that all NSDs involved in glycosylation are synthesised from the metabolites (e.g., glucose, glutamine, and glucosamine) which drive central carbon metabolism, cell growth, and mAb biosynthesis.

More specifically, metabolic glycoengineering involves supplementing cell culture with monosaccharides, nucleosides, and/or glycosylation enzyme co-factors (e.g., Mn^{2+} for MGAT1, MGAT2, and β4GalT) to increase the production of specific glycan motifs. Culture supplementation with cytidine, guanosine, uridine, mannose, glucosamine, N-acetylmannosamine, and galactose in concentrations ranging from 0.2–40 mM has been performed to control the site occupancy [103, 104], branching [105, 106], galactosylation [107], and sialylation [108–110] of therapeutic glycoproteins, such as IL-Mu6, tPA, IFN-γ, and erythropoietin.

The key challenge of metabolic glycoengineering is that the fed monosaccharides [111] and nucleosides [112] may arrest cell growth and, in some cases, reduce volumetric productivity [113].

Metabolic pathways beyond NSD biosynthesis indirectly influence mAb glycosylation by generating by-products that influence glycosylation machinery activity. For example, CHO cells consume excess glutamine towards TCA intermediates, via glutamate anaplerosis [114]. In this context, glutamine consumption produces a substantial amount of ammonia which, in turn, reduces mAb galactosylation and sialylation by inhibiting the catalytic activity [115] and, potentially, Golgi localisation [116] of the β4GalT and αSiaT enzymes.

This section reviews the metabolic glycoengineering strategies which have been deployed to modulate the fucosylation, galactosylation, sialylation, and Man5 glycosylation of mAbs. These mAb metabolic glycoengineering strategies are summarised in Fig. 5.

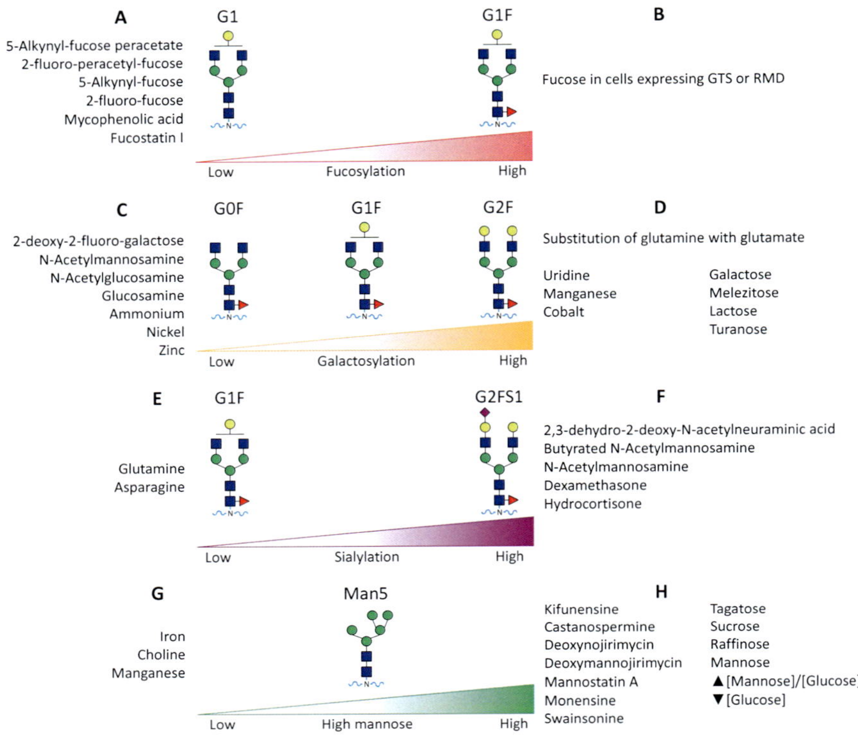

A. Lower fucosylation
- 5-Alkynyl-fucose peracetate [118]
- 2-fluoro-peracetyl-fucose [118]
- 5-Alkynyl-fucose [117, 118]
- 2-fluoro-fucose [117, 118]
- Mycophenolic acid [119]
- Fucostatin I [120]

B. Increase fucosylation
- Fucose in cells expressing GTS or RMD [55]

C. Lower galactosylation
- 2-deoxy-2-fluoro-galactose [134]
- N-Acetylmannosamine [121]
- N-Acetylglucosamine [122, 133]
- Glucosamine [133]
- Ammonia [124]
- Nickel [126]
- Zinc [125]

D. Increase galactosylation
- Substitution of Gln with Glu [131]
- Uridine [124]
- Manganese [124]
- Galactose [121, 122]
- UMG [113, 123]
- Melezitose [128]Lactose [128]
- Turanose [128]
- Cobalt [126, 127]

E. Lower sialylation
- Glutamine & asparagine due to ammonia production [146, 147]

F. Increase sialylation
- 2,3-dehydro-2-deoxy-N-acetylneuraminic acid [137]
- Butyrated N-Acetylmannosamine [135]
- N-Acetylmannosamine [108, 110]
- Dexamethasone [139]
- Hydrocortisone [140]

G. Lower mannosylation
- Iron [151]
- Choline [152]
- Manganese [149]

H. Increase mannosylation
- Kifunensine [98, 156]
- Castanospermine [136]
- Deoxynojirimycin [136]
- Deoxymannojirimycin [136]
- Mannostatin A [136]
- Monensine [136, 159]
- Swainsonine [158]
- Tagatose [136, 155]
- Sucrose [136, 155]
- Raffinose [128, 136]
- Mannose [122, 154]
- ▲[Mannose]/[Glucose] [153]
- ▼[Glucose] [161]
- ▼[Glutamine] [130]

Fig. 5 Overview of metabolic glycoengineering strategies The compounds, including amino acids, carbohydrates, nucleotide bases, co-factors, and steroids, that have been reported to modulate glycosylation are shown. The arrows indicate increases (▲) or decreases (▼). Components in brackets represent concentrations. The glycan notation is the same as in Fig. 1.

5.1 Fucosylation

mAb α-1,6 fucosylation can be curbed by feeding cultured cells with fluorinated and/or acetylated fucose analogues (e.g., 2-fluorofucose, 5-alkynylfucose and acetylated derivatives) to inhibit α-1,6 fucosyltransferase activity (Fut8) [117]. Although highly successful in abrogating mAb fucosylation to enhance ADCC activity, recent work has found that a small fraction of these analogues can be transferred to mAb glycans [118].

Reduced fucosylation has also been achieved by supplementing cell culture with mycophenolic acid [119], an inhibitor of GTP biosynthesis, and fucostatin I [120], which is a potent inhibitor of GDP-Mannose dehydratase (an enzyme along the GDP-Man to GDP-Fuc biosynthetic pathway, Fig. 3). Conversely, desired levels of mAb fucosylation can be obtained by feeding fucose to cells deficient in the *de novo* GDP-Fucose biosynthetic pathway [55].

5.2 Galactosylation

Feeding of uridine, manganese, and galactose (UMG) separately or, more effectively, as a cocktail is a proven method to increase mAb β-1,4 galactosylation. A 6% increase in β-1,4 galactosylation was achieved when 10 mM galactose was fed to GS-NS0 cells producing a humanised IgG mAb [121]. In a more recent study, Kildegaard et al. [122] obtained a 25% increase in mAb galactosylation when CHO cells were fed 20 mM galactose. This group also reported a 4% reduction in mAb galactosylation when the cells were fed 20 mM GlcNAc. Importantly, these feeding strategies led to statistically significant changes to cell growth and/or mAb productivity.

The most successful and widely used glycoengineering strategy to modulate galactosylation is UMG feeding, where uridine (U) and galactose (G) aim to overcome limitations in UDP-Gal biosynthesis, and Mn^{2+} (M) is added to enhance β4GalT activity (Mn2+ is the enzyme's co-factor). With UMG feeding, a shift in mAb galactosylation, from 3% to 23% in G1F, was achieved, independently of bioreactor scale [123]. Subsequent work used a DoE statistical modelling framework to predict the UMG doses required to achieve targeted mAb galactosylation content [113]. In another example, the individual effects of $MnCl_2$, galactose, and NH_4Cl feeding on the galactosylation of an IgG mAb produced by CHO-K1 cells were studied [124]. This work found that, for this cell line, feeding $MnCl_2$ and galactose were the key drivers for mAb galactosylation.

The effect of metal ion feeding on mAb galactosylation has also been reported. Zn^{2+} concentrations above 100 μM [125] and Ni^{2+} concentrations above 500 μM [126] have been reported to reduce mAb galactosylation by inhibiting β4GalT activity. Co^{2+} concentrations of up to 50 μM [126, 127] mildly increase mAb galactosylation without affecting cell growth or product yield. Feeding of additional, non-traditional carbohydrates (e.g., melezitose, lactose, and turanose) are also reported to slightly increase mAb galactosylation with negligible reductions in cell growth and product yield [128].

Metabolic galactosylation engineering strategies beyond UMG feeding have also been explored. Interestingly, severe glucose starvation has been reported to yield increased mAb galactosylation [23]. Although somewhat unintuitive, these authors report that increased mAb galactosylation occurred due to increased intracellular UDP-Gal availability caused by reductions in cell growth and mAb cell-specific productivity. This conclusion is supported by stoichiometric calculations which indicate that the predominant sink of UDP-Gal is cellular N- and O-galactosylation [129]. This inverse relationship between productivity and galactosylation has also been reported in work where a multifactorial statistical framework was used to optimise the yield of a bevacizumab biosimilar [130]. Here, mAb productivity was maximised by feeding aspartate, glutamate, arginine, and glycine but also yielded a shift towards lower levels of product galactosylation.

In another metabolic glycoengineering strategy, glutamine was substituted by glutamate in cell culture media to yield a 67% reduction in NH_4^+ and a concomitant 10% increase in galactosylated (G1F and G2F) mAb glycoforms. Reduced NH_4^+ also increased mAb titre 1.7-fold [131]. The buffering capacity of other amino acids, such as threonine, proline, and glycine, has also been leveraged to counteract the detrimental effect NH_4^+ has on glycosylation [132].

Although increasing galactosylation (due to its positive influence on PK and PD) has been the most common aim, metabolic glycoengineering strategies have also been developed to abrogate galactosylation in order to reduce product variability. With this goal, glucosamine and N-acetylglucosamine (GlcNAc) have been fed to cell culture to yield mAbs with a-galactosylated glycans, mainly G0F glycans [133]. Interestingly, GlcNAc feeding had no significant effect on cell growth and productivity. Analogously to fucosylation inhibitors, substantial reductions in mAb galactosylation have been obtained by supplementing culture media with 0.5 mM of 2-deoxy-2-fluoro-d-galactose (2FG), a fluorinated galactose analogue which inhibits β4GalT activity [134]. A ~ 20% reduction in product titre was observed with 2FG feeding.

5.3 Sialylation

As outlined previously, low levels of Fc sialylation are typically observed due to limited space between the heavy chains which form the crystallisable fragment of γ-globulin mAbs. However, CMP-Neu5Ac precursor feeding strategies have been deployed to increase the sialylation of surface-exposed glycans within other therapeutic glycoproteins and mAb Fab glycans. Specifically, cytidine along with 20 mM N-acetylmannosamine (ManNAc) [108, 110] and butyrated ManNAc [135] have been fed to increase the sialylation of IFN-γ (by 15%) and EPO (by 32%). Similarly, Ehret et al. [136] used ManNAc feeding to fine-tune mAb Fab sialylation.

Sialidase inhibitors are also used to control therapeutic protein sialylation. During culture, cells release sialidase enzymes which are known to remove Neu5Ac residues from product glycans. Supplementing cell culture with sialidase inhibitors, such as 2,3-dehydro-2-deoxy-N-acetylneuraminic acid [137] and N-acetyl-2,3-dehydro-2-Deoxyneuraminic acid [136], has resulted in increased product sialylation.

Although the mechanisms are not yet fully understood, small lipophilic hormones, including steroids, are known to influence glycoprotein productivity and glycosylation [138]. Supplementation of culture media with dexamethasone enhances the sialylation of a CTLA4-Ig fusion protein by eliciting increased β4GalT and α3SiaT expression in CHO cells [139]. Similarly, culture supplementation with hydrocortisone is reported to increase product sialylation in a dose and time-dependent manner [140].

Sodium butyrate (NaBu) is a well-studied enhancer of recombinant protein productivity in mammalian cells. NaBu is known to enhance gene accessibility and, thereby, increase mRNA transcription and overall protein synthesis [141]. Although many studies report an inverse relationship between protein productivity and sialylation [142–144], other work has found that total sialylation remains constant [145] after NaBu treatment, albeit with a lower Neu5Gc/Neu5Ac ratio.

As with galactosylation, anaplerotic consumption of ammoniogenic amino acids (e.g., glutamine and asparagine) can indirectly impact product sialylation by shifting intracellular pH and, ultimately, inhibit the expression [146], activity [147], and possibly, the Golgi localisation [116] of β4GalT and αSiaT. Similarly to fucosylation and galactosylation, peracetylated and fluorinated Neu5Ac analogues have been used to reduce glycoprotein sialylation [148].

5.4 *Man5*

Due to its deleterious effect on serum half-life, efforts to reduce the production of mAbs with Man5 glycans has centred around enhancing the activity of MGAT1, the enzyme which catalyses the only reaction where Man5 is consumed during glycan processing in Golgi. Mn^{2+}, which is the co-factor for MGAT1, has been fed to cell culture to enhance the enzyme's activity and reduce Man5 production [149]. However, when combined with the use of galactose and fructose as glucose alternatives during cell culture, Mn^{2+} has resulted in increased Man5 levels [150]. Culture supplementation with Fe^{2+} decreases Man5 production [151] and feeding the lipid precursor, choline, has also been used to reduce the production of Man5 glycoforms [152].

Because mAbs bearing Man5 glycans may yield enhanced ADCC activity by reducing fucosylation [95, 97], strategies to enrich production of this glycoform have been developed. Huang et al. [153] increased Man5 production by controlling the mannose:glucose ratio in cell culture media. Feeding mannose also increases the fraction of Man5 glycoforms [154]. Culture media supplementation with other atypical carbon sources (e.g., raffinose, tagatose, and sucrose) has also been used to increase production of Man5 mAbs [128, 155].

Kifunesine, an alkaloid known to inhibit αManI activity has been used to produce oligomannose [98, 156] mAb glycoforms. Although kifunesine is the most effective, other glycosidase and mannosidase inhibitors (e.g., swainsonine, castanospermine, deoxynojirimycin, deoxymannojirimycin, monensine, and mannostatin A) have also been used to inhibit production of complex glycans [136, 157–159].

5.5 *Aglycosylation (Site Occupancy)*

Because mAb Fc aglycosylation in absence of protein backbone engineering is known to severely hinder effector functions [100], metabolic strategies to reduce or eliminate mAb Fc glycosylation have been limited to investigating the metabolic drivers of N-glycan site occupancy. Predictably, glucose availability plays a pivotal role in N-glycan site occupancy: 24 hours of glucose starvation have been reported to compromise mAb Fc glycan site occupancy (52% aglycosylation), via altered biosynthesis of the $Glc_3Man_9GlcNAc_2$ dolichol-linked precursor oligosaccharide [160]. Similar results, which mechanistically link mAb Fc aglycosylation with glucose starvation through reductions in intracellular nucleotide sugar pools, have also been reported [161]. Feeding culture with tunicamycin, a potent inhibitor of OST (EC 2.4.99.18) has also been used to abrogate mAb Fc N-glycan occupancy [162].

6 Glycoengineering of Alternative Production Systems

Historically, mammalian cells – in particular, CHO cells – have dominated as the expression platform of choice for mAb production. 18 of the 20 best-selling mAbs (Table 1) and over 60% of all commercialised mAbs are produced through CHO cell culture, with the remainder being manufactured using mouse myeloma-derived cells (e.g., NS0 and Sp2/0) [163].

CHO cells are the dominant production host for mAbs due to their ability to produce 'human-like' post-translational modifications (PTMs) and their ability to grow at high densities in chemically defined (animal component-free) culture media [164, 165]. A wealth of knowledge in producing biopharmaceuticals with CHO cells has accumulated over the past 30 years (since the approval of therapeutic tPA in 1986) and has resulted in a proven safety and manufacturability track record with regulatory agencies.

Immortalised human cell lines have the advantage of native PTM machinery and are capable of growing in serum free, suspension culture [166]. However, a key concern associated with human cell lines is their susceptibility of being contaminated by human pathogenic viruses, an issue that is minimised with CHO cells [167]. Despite this challenge, some therapeutic glycoproteins are manufactured using human cell lines, such as HEK283 (e.g., recombinant human Factor VIII for the treatment of haemophilia) and HT-1080 (e.g., recombinant replacement enzymes) [166]. Human cells are required to manufacture these products, likely due to particular molecular features or PTMs (e.g., sulphation). In addition, otilimab, an immune modulation mAb produced with Per.C6 cells (derived from immortalised human retina cells), is currently undergoing Phase III clinical trials [168].

Bacterial, fungal, plant, and insect cells are proven platforms for therapeutic protein production. However, therapeutic proteins manufactured with these cells have been restricted to simpler molecules because wild type versions of these cell types lack the cellular machinery required to produce human-like PTMs. More recently, the ongoing expansion of novel synthetic and systems biology tools has enabled the use of engineered bacterial, fungal, plant, and insect cells as exciting alternative options for mAb production.

6.1 Bacterial Cell Glycoengineering

Non-engineered *E. coli* has been restricted to the production of non-glycosylated antibody fragments (e.g., certolizumab-pegol and ranibizumab) [169] or nanobodies (caplacizumab) [170]. Although the N-glycosylation pathway of *C. jejuni* [171] and, more recently, eukaryotic glycosyltransferases [172] have been engineered into *E. coli*, the glycan profiles produced with these glycoengineered cells is still far from being 'human-like'.

6.2 Fungal Cell Glycoengineering

Because hypermannosylation has been identified as a significant constraint for therapeutic glycoprotein production in wild type yeast cells [173], *P. pastoris*, *S. cerevisiae*, *H. polymorpha*, and *Y. lipolytica* have all been extensively glycoengineered to reduce this xenoantigenic glycan motif by knocking out genes for α-1,6 mannosyltransferase [174, 175]. Beyond reducing hypermannosylation, yeast cells have been glycoengineered to produce human-like glycosylation through ectopic expression of branching [176], elongation [177], and mannose-trimming enzymes [175], which in many cases have been modified to include Golgi localisation components to ensure that the glycosylation reactions occur in the appropriate sequence [178, 179].

Although the above glycoengineering efforts successfully yield human-like glycosylation, residual amounts of hypermannosylated glycans are still produced [180, 181]. Hamilton et al. [182] obtained fully human complex glycans on erythropoietin produced by *P. pastoris* but their glycoengineering strategy required multiple endogenous enzyme knockouts and the ectopic expression of 14 heterologous genes.

6.3 *Plant Cell Glycoengineering*

Plant-based expression systems have been tested to produce antibodies since the late 80s [183], but native plant glycosylation contains xenoantigenic monosaccharide residues, such as β1,2-xylose and α1,3-fucose [184] and Lewis A epitopes [185]. These non-human sugar residues can be easily abrogated by knocking-out/down the corresponding glycosyltransferase genes [186–188].

Further humanisation of plant glycoprofiles includes overexpressing galactosyltransferase enzymes [189] and the Neu5Ac biosynthetic pathway [190]. Zmapp, an anti-Ebola mAb cocktail produced by transgenic tobacco plants [191], was granted 'Fast-Track' FDA status in 2015 [192]; however, it was deemed less effective than Regeneron's CHO-derived REGN-EB3 [193], which was recently FDA approved for the treatment of this life-threatening viral infection.

6.4 *Insect Cell Glycoengineering*

The baculovirus-insect cell system may also offer a cost-effective alternative for mAb production, with lepidopteran cells (e.g., *Sf*9, *Sf*21, and *Trichoplusia ni*) or living silkworm larvae being the most investigated production hosts [194]. Native insect cell glycosylation consists of paucimannose structures [195], which arise from low expression levels of MGAT2 [196] and other glycan elongation enzymes [197, 198]. In addition, wild type insect cells produce xenoantigenic α-1,3 fucosylation and express glycosidases, which break down human-like glycosylation motifs such as β1,2-GlcNAc [199].

The two strategies to glycoengineer insect cell lines involve either transfecting the missing glycogenes into the cells [200–202] or co-transfecting these genes, via baculovirus infection, alongside the protein of interest [203–206]. Although these strategies have yielded >80% complex mAb glycans with high levels of galactosylation (50%), a significant amount of paucimannose glycans remain [206]. Fucosylation (both α1,3- and α1,6-linked) has been abrogated in insect cell lines through the ectopic expression of GDP-6-deoxy-D-lyxo-4-hexulose reductase (RMD), which eliminates GDP-Fuc biosynthesis [207].

7 *In Vitro* Glycoengineering

In vitro glycoengineering, which involves the *ex vivo* chemoenzymatic modification of (glyco)protein intermediates, has recently garnered interest because it yields very homogenous product glycans from highly controllable processes. mAb *in vitro* glycoengineering centres around enzymatic modifications because the branching, stereo-, and regiochemistry of N-glycans makes purely chemical modifications exceptionally challenging [208]. Depending on the (glyco)protein produced upstream, glycans can be trimmed and then elongated or fully synthesised while bound to the peptide backbone [209], but they can also be obtained from a different source (commonly egg-yolk glycopeptide [210, 211]) and transferred, as a unit, onto the mAb backbone.

Traditional *in vitro* glycoengineering has targeted galactosylation by incubating the mAb with purified galactosyltransferase, the galactose donor (UDP-Gal), and the enzyme co-factor ($MnCl_2$) [212–214]. *In vitro* methods to enhance sialylation – either maximising α-2,3 Neu5Ac content or swapping it to human-like (α-2,6 Neu5Ac) – use sialidases (to remove α-2,3 Neu5Ac), sialyltransferases, and CMP-Neu5Ac [134, 214, 215].

In contrast, *in vitro* modulation of mAb fucosylation is more problematic, as native fucosidases and fucosyltransferases are unable to modify complex glycans, most likely due to steric hindrance effects [216]. Fucose addition has recently been made possible with an AlfC α-fucosidase (E274A) mutant, which can fucosylate intact N-glycoproteins [216]. *In vitro* fucosidic bond cleavage of mature mAb glycoforms (which would yield enhanced ADCC activity) is still an obstacle, as engineered fucosidases are yet to achieve desirable levels of activity [217].

As can be observed, the components required for glycan extension via *in vitro* glycoengineering are those involved in the biological reaction: (*i*) an activated nucleotide sugar donor (NSD), (*ii*) a glycosyltransferase, and (*iii*) the carbohydrate acceptor. Importantly, mammalian glycosyltransferases are, themselves, glycoproteins whose activity depends on their carbohydrates [218]. Therefore, the glycosyltransferases used for *in vitro* glycoengineering are either isolated from animal sources (which is undesirable in a cGMP environment) or are produced recombinantly using a mammalian expression platform, which may add to manufacturing costs.

In addition, the NSDs (e.g., UDP-Gal, CMP-Neu5Ac, and GDP-Fuc) are required to be at reasonably high purity for *in vitro* glycoengineering purposes, which could also increase manufacturing costs. Furthermore, the use of additional processing steps would necessitate in-process controls to ensure that the product does not suffer undesired chemical modifications (e.g., oxidation and deamidation) or degradation due to prolonged incubation times [219].

Alternative *in vitro* glycoengineering approaches consist of chemoenzymatic methods where IgG-optimised endoglycosidases are used to cleave the native N-glycan [220, 221] and subsequently act as ligases to attach an oxazoline-derivatised carbohydrate obtained from a sialoglycoprotein [222]. However, because the primary function of these enzymes is hydrolytic, their transglycosylation activity is limited and achieves low yields [221]. To overcome this challenge, the hydrolytic activity of endoglycosidases has been abrogated to yield so-called glycosynthases [223, 224] which can selectively and efficiently transglycosylate Fc and Fab domains in mAbs [225–227]. As long as the antibody contains the innermost GlcNAc (with or without fucose) bound to the peptide backbone, glycosynthases can be used for the chemoenzymatic remodelling of mAbs produced in any platform, including insect [228] or yeast cells [229].

Importantly for mAbs whose therapeutic mechanisms rely on ADCC, the carbohydrate produced after endoglycosidase treatment may have exposed fucose, which can readily be removed using fucosidases [230, 217]. The endoglycosidase/glycosynthase technology has been integrated with cellular glycoengineering, where *N. benthamiana* was engineered to produce an endoglycosidase to perform *in vivo* deglycosylation of rituximab which later served as the transglycosylation acceptor of A2G2 glycans [231]. Finally, such chemoenzymatic strategies can be used to introduce unnatural glycan structures [232] or for the production of glycosite-specific antibody-drug conjugates [233, 234].

Enzymatic and chemoenzymatic techniques have been used extensively to test the biological activity and impact of a particular glycoform on mAb conformation [41, 215, 235, 228, 38], but their use at the industrial scale is yet to be demonstrated. However, a proof of concept for the manufacture of one kilogram of G2-glycan remodelled mAb was reported by Warnock et al. in 2005 [236]. Recent work, which paves the way towards deployment of *in vitro* glycoengineering at larger scales, includes solid-state glycan remodelling on mAbs captured on Protein A resin. This strategy enables glycotransferase reutilisation and eliminates the need for additional purification steps [237]. In the same context, endoglycosydases and glycosynthases have been immobilised on a matrix to aid the chemoenzymatic process [238].

8 mAb Glycoengineering: Future Perspectives

The importance of glycosylation for the overall quality of therapeutic mAbs has led to the development of a multitude of strategies to steer this post translational modification towards profiles that ensure safety and enhanced therapeutic properties. Although five glycoengineered mAbs are currently marketed, next generation technologies are being developed to tackle the glycosylation quality assurance challenges of next generation mAb processes/products.

Three key drivers for ongoing glycoengineering efforts are: (*i*) the emergence of biosimilar mAbs, which are required to match the glycosylation profiles of the originator products, (*ii*) bioprocess intensification strategies (e.g., perfusion bioreactors) where conventional glycosylation control strategies may be difficult to deploy, and (*iii*) the potential of personalising mAb therapies based on patient and product glycosylation profiles. A distinct challenge associated with these three drivers is that the target glycosylation profiles are unlikely to be absolute (i.e., fully afucosylated, galactosylated, or sialylated). Therefore, next generation glycoengineering strategies should enable the fine tuning of mAb glycosylation profiles across a broad spectrum and should also be dynamically responsive to address excursions during bioprocessing.

As with any control system, schemes for real-time tuning of mAb glycosylation require at least three components: a sensor, a control signal, and an actuator. In this context, the glycoengineering strategies reviewed in the preceding sections of this chapter can be viewed as the actuators, whereas computational models of glycosylation can be leveraged as 'soft sensors' and simultaneously be used to define the magnitude and duration of the control stimuli to maintain product glycosylation at desired setpoints. Indeed, substantial work has gone into developing sophisticated computational models of therapeutic protein glycosylation, where mechanistic, flux-based, and statistical frameworks have been developed to link bioprocess conditions with product glycosylation with the ultimate aim of delivering robust glycosylation control strategies [239].

Given the vast progress in developing glycoengineering strategies and computational models of glycosylation, a seamless integration of both elements will directly address the quality assurance challenges associated with next generation therapeutic mAb products. Indeed, recent publications have already explored such synergies, where a kinetic model of glycosylation has been used to define dynamic metabolic glycoengineering strategies (UMG feeding regimes) that maximise β4-galactosylation of mAbs while minimising negative effects on product yield [240]. In another example, statistical modelling has been used to develop and operate a bioprocess with the aim of achieving mAb glycosylation biosimilarity [241].

More recently, the Glycan Residues Balance Analysis (GReBA) model [242] was used to control mAb glycosylation by adjusting the feed composition in a perfusion bioreactor [243], a mechanistic model was used to predict the effect of amino acid and copper supplementation on the glycosylation and yield of a mAb product [244], and a hybrid computational model, where mechanistic elements have been coupled with a machine learning algorithm, has been used to predict the influence of cell culture conditions and cellular glycoengineering events on product glycosylation [245].

A gap that remains to be bridged is real-time, model-based regulation of cellular glycoengineering. Real-time tuning of glycogene expression based on model-based soft sensors is most promising given the implicit flexibility and robustness, as well as the breadth of glycoforms that could be achieved. Glycosylation models have indeed been used to propose non-tuneable cellular glycoengineering strategies to minimise deviations from target mAb glycan profiles. For example, the level and extent of cellular glycoengineering required to achieve therapeutic protein biosimilarity have been defined using a flux-based glycosylation model [246]. Similarly, a mechanistic glycosylation model was used to identify the level of MGAT2 overexpression required to eliminate the production of deleterious Man5 mAb glycoforms [24]. The next step forward would be to synergise computational models with dynamically tuneable cell engineering strategies similar to those recently developed for controlling the α6-fucosylation and β4-galactosylation of therapeutic mAbs [247]. Such an integration may extend the use of tuneable cellular glycoengineering strategies from drug discovery to manufacturing and enable near-perfect control of therapeutic mAb glycosylation.

9 Concluding Remarks

The rapidly expanding cellular and metabolic glycoengineering toolbox is primed to deliver unprecedented levels of quality control for current and next generation mAb products. The path towards fully realising robust control of mAb glycosylation resides in exploiting novel synthetic biology platforms (e.g., CRISPR-based tools and inducible gene circuits [247]), combining cellular and metabolic glycoengineering, and leveraging the breadth of available computational models of glycosylation. The development of new glycoengineering tools and synergising them across the different phases of the glycosylation process will greatly enhance mAb product quality assurance and may lead to novel products where glycans can be tuned to achieve optimal therapeutic outcomes in individual patients (glycopersonalised medicine).

Disclosure of Interests All authors declare they have no conflict of interest.

Ethical Approval This chapter does not contain any studies with human participants or animals performed by any of the authors.

References

1. Higel F, Seidl A, Sörgel F, Friess W (2016) N-glycosylation heterogeneity and the influence on structure, function and pharmacokinetics of monoclonal antibodies and Fc fusion proteins. Eur J Pharm Biopharm 100:94–100. https://doi.org/10.1016/j.ejpb.2016.01.005
2. Planinc A, Dejaegher B, Vander Heyden Y, Viaene J, Van Praet S, Rappez F, Van Antwerpen P, Delporte C (2017) Batch-to-batch N-glycosylation study of infliximab, trastuzumab and bevacizumab, and stability study of bevacizumab. Eur J Hosp Pharm-S P 24(5):286–292. https://doi.org/10.1136/ejhpharm-2016-001022
3. Raju TS, Jordan RE (2012) Galactosylation variations in marketed therapeutic antibodies. MAbs 4(3):385–391. https://doi.org/10.4161/mabs.19868
4. Kim S-M, Chang K-H, Oh DJ (2018) Effect of environmental parameters on glycosylation of recombinant immunoglobulin G produced from recombinant CHO cells. Biotechnol Bioprocess Eng 23(4):456–464. https://doi.org/10.1007/s12257-018-0109-8
5. Jiang R, Chen H, Xu S (2018) pH excursions impact CHO cell culture performance and antibody N-linked glycosylation. Bioprocess Biosyst Eng 41(12):1731–1741. https://doi.org/10.1007/s00449-018-1996-y
6. Brunner M, Fricke J, Kroll P, Herwig C (2017) Investigation of the interactions of critical scale-up parameters (pH, pO(2) and pCO(2)) on CHO batch performance and critical quality attributes. Bioprocess Biosyst Eng 40(2):251–263. https://doi.org/10.1007/s00449-016-1693-7
7. Ivarsson M, Villiger TK, Morbidelli M, Soos M (2014) Evaluating the impact of cell culture process parameters on monoclonal antibody N-glycosylation. J Biotechnol 188:88–96. https://doi.org/10.1016/j.jbiotec.2014.08.026
8. Jefferis R (2009) Glycosylation as a strategy to improve antibody-based therapeutics. Nat Rev Drug Discov 8(3):226–234. https://doi.org/10.1038/nrd2804

9. Stadlmann J, Pabst M, Kolarich D, Kunert R, Altmann F (2008) Analysis of immunoglobulin glycosylation by LC-ESI-MS of glycopeptides and oligosaccharides. Proteomics 8(14):2858–2871. https://doi.org/10.1002/pmic.200700968
10. Nakano M, Higo D, Arai E, Nakagawa T, Kakehi K, Taniguchi N, Kondo A (2009) Capillary electrophoresis-electrospray ionization mass spectrometry for rapid and sensitive N-glycan analysis of glycoproteins as 9-fluorenylmethyl derivatives. Glycobiology 19(2):135–143. https://doi.org/10.1093/glycob/cwn115
11. Holland M, Yagi H, Takahashi N, Kato K, Savage CO, Goodall DM, Jefferis R (2006) Differential glycosylation of polyclonal IgG, IgG-Fc and IgG-Fab isolated from the sera of patients with ANCA-associated systemic vasculitis. Biochim Biophys Acta 1760(4):669–677. https://doi.org/10.1016/j.bbagen.2005.11.021
12. Magdelaine-Beuzelin C, Kaas Q, Wehbi V, Ohresser M, Jefferis R, Lefranc MP, Watier H (2007) Structure-function relationships of the variable domains of monoclonal antibodies approved for cancer treatment. Crit Rev Oncol Hematol 64(3):210–225. https://doi.org/10.1016/j.critrevonc.2007.04.011
13. Hafkenscheid L, Bondt A, Scherer HU, Huizinga TW, Wuhrer M, Toes RE, Rombouts Y (2017) Structural analysis of variable domain glycosylation of anti-Citrullinated protein antibodies in rheumatoid arthritis reveals the presence of highly Sialylated Glycans. Mol Cell Proteomics 16(2):278–287. https://doi.org/10.1074/mcp.M116.062919
14. Bondt A, Rombouts Y, Selman MH, Hensbergen PJ, Reiding KR, Hazes JM, Dolhain RJ, Wuhrer M (2014) Immunoglobulin G (IgG) Fab glycosylation analysis using a new mass spectrometric high-throughput profiling method reveals pregnancy-associated changes. Mol Cell Proteomics 13(11):3029–3039. https://doi.org/10.1074/mcp.M114.039537
15. Janin-Bussat MC, Tonini L, Huillet C, Colas O, Klinguer-Hamour C, Corvaia N, Beck A (2013) Cetuximab Fab and Fc N-glycan fast characterization using IdeS digestion and liquid chromatography coupled to electrospray ionization mass spectrometry. Methods Mol Biol 988:93–113. https://doi.org/10.1007/978-1-62703-327-5_7
16. van de Bovenkamp FS, Hafkenscheid L, Rispens T, Rombouts Y (2016) The emerging importance of IgG fab glycosylation in immunity. J Immunol 196(4):1435–1441. https://doi.org/10.4049/jimmunol.1502136
17. Stanley P, Taniguchi N, Aebi M (2015) N-Glycans. In: Varki A, Cummings RD, Esko JD et al (eds) Essentials of Glycobiology, Cold Spring Harbor (NY), pp 99–111. https://doi.org/10.1101/glycobiology.3e.009
18. Kornfeld R, Kornfeld S (1985) Assembly of asparagine-linked oligosaccharides. Annu Rev Biochem 54:631–664. https://doi.org/10.1146/annurev.biochem.54.1.631
19. Ellgaard L, Helenius A (2003) Quality control in the endoplasmic reticulum. Nat Rev Mol Cell Biol 4(3):181–191. https://doi.org/10.1038/nrm1052
20. Murrell MF, Yarema KJ, Levchenko A (2004) The systems biology of glycosylation. Chembiochem 5(10):1334–1347. https://doi.org/10.1002/cbic.200400143
21. Burleigh SC, van de Laar T, Stroop CJM, van Grunsven WMJ, O'Donoghue N, Rudd PM, Davey GP (2011) Synergizing metabolic flux analysis and nucleotide sugar metabolism to understand the control of glycosylation of recombinant protein in CHO cells. BMC Biotechnol 11. https://doi.org/10.1186/1472-6750-11-95
22. Hadley B, Maggioni A, Ashikov A, Day CJ, Haselhorst T, Tiralongo J (2014) Structure and function of nucleotide sugar transporters: current progress. Comput Struct Biotechnol J 10(16):23–32. https://doi.org/10.1016/j.csbj.2014.05.003
23. Fan Y, Jimenez Del Val I, Muller C, Lund AM, Sen JW, Rasmussen SK, Kontoravdi C, Baycin-Hizal D, Betenbaugh MJ, Weilguny D, Andersen MR (2015) A multi-pronged investigation into the effect of glucose starvation and culture duration on fed-batch CHO cell culture. Biotechnol Bioeng 112(10):2172–2184. https://doi.org/10.1002/bit.25620
24. del Val IJ, Fan Y, Weilguny D (2016) Dynamics of immature mAb glycoform secretion during CHO cell culture: an integrated modelling framework. Biotechnol J 11(5):610–623. https://doi.org/10.1002/biot.201400663

25. Hossler P, Khattak SF, Li ZJ (2009) Optimal and consistent protein glycosylation in mammalian cell culture. Glycobiology 19(9):936–949
26. del Val IJ, Kontoravdi C, Nagy JM (2010) Towards the implementation of quality by design to the production of therapeutic monoclonal antibodies with desired glycosylation patterns. Biotechnol Progr 26(6):1505–1527. https://doi.org/10.1002/Btpr.470
27. Jedrzejewski PM, Del Val IJ, Polizzi KM, Kontoravdi C (2013) Applying quality by design to glycoprotein therapeutics: experimental and computational efforts of process control. Pharm BioProc 1(1):51–69
28. Springer SA, Gagneux P (2013) Glycan evolution in response to collaboration, conflict, and constraint. J Biol Chem 288(10):6904–6911. https://doi.org/10.1074/jbc.R112.424523
29. Steinke JW, Platts-Mills TA, Commins SP (2015) The alpha-gal story: lessons learned from connecting the dots. J Allergy Clin Immunol 135(3):589–596. https://doi.org/10.1016/j.jaci.2014.12.1947
30. Varki A (2008) Sialic acids in human health and disease. Trends Mol Med 14(8):351–360. https://doi.org/10.1016/j.molmed.2008.06.002
31. Li Y, Monine M, Huang Y, Swann P, Nestorov I, Lyubarskaya Y (2016) Quantitation and pharmacokinetic modeling of therapeutic antibody quality attributes in human studies. MAbs 8(6):1079–1087. https://doi.org/10.1080/19420862.2016.1186322
32. Goetze AM, Liu YD, Zhang Z, Shah B, Lee E, Bondarenko PV, Flynn GC (2011) High-mannose glycans on the Fc region of therapeutic IgG antibodies increase serum clearance in humans. Glycobiology 21(7):949–959. https://doi.org/10.1093/glycob/cwr027
33. Shinkawa T, Nakamura K, Yamane N, Shoji-Hosaka E, Kanda Y, Sakurada M, Uchida K, Anazawa H, Satoh M, Yamasaki M, Hanai N, Shitara K (2003) The absence of fucose but not the presence of galactose or bisecting N-acetylglucosamine of human IgG1 complex-type oligosaccharides shows the critical role of enhancing antibody-dependent cellular cytotoxicity. J Biol Chem 278(5):3466–3473. https://doi.org/10.1074/jbc.M210665200
34. Thomann M, Reckermann K, Reusch D, Prasser J, Tejada ML (2016) Fc-galactosylation modulates antibody-dependent cellular cytotoxicity of therapeutic antibodies. Mol Immunol 73:69–75. https://doi.org/10.1016/j.molimm.2016.03.002
35. Peschke B, Keller CW, Weber P, Quast I, Lunemann JD (2017) Fc-Galactosylation of human immunoglobulin gamma isotypes improves C1q binding and enhances complement-dependent cytotoxicity. Front Immunol 8:646. https://doi.org/10.3389/fimmu.2017.00646
36. Kaneko Y, Nimmerjahn F, Ravetch JV (2006) Anti-inflammatory activity of immunoglobulin G resulting from Fc sialylation. Science 313(5787):670–673
37. Washburn N, Schwab I, Ortiz D, Bhatnagar N, Lansing JC, Medeiros A, Tyler S, Mekala D, Cochran E, Sarvaiya H, Garofalo K, Meccariello R, Meador JW 3rd, Rutitzky L, Schultes BC, Ling L, Avery W, Nimmerjahn F, Manning AM, Kaundinya GV, Bosques CJ (2015) Controlled tetra-Fc sialylation of IVIg results in a drug candidate with consistent enhanced anti-inflammatory activity. Proc Natl Acad Sci U S A 112(11):E1297–E1306. https://doi.org/10.1073/pnas.1422481112
38. Krapp S, Mimura Y, Jefferis R, Huber R, Sondermann P (2003) Structural analysis of human IgG-Fc glycoforms reveals a correlation between glycosylation and structural integrity. J Mol Biol 325(5):979–989
39. Nimmerjahn F, Ravetch JV (2010) Antibody-mediated modulation of immune responses. Immunol Rev 236:265–275. https://doi.org/10.1111/j.1600-065X.2010.00910.x
40. Nimmerjahn F, Ravetch JV (2008) Fcgamma receptors as regulators of immune responses. Nat Rev Immunol 8(1):34–47. https://doi.org/10.1038/nri2206
41. Dekkers G, Treffers L, Plomp R, Bentlage AEH, de Boer M, Koeleman CAM, Lissenberg-Thunnissen SN, Visser R, Brouwer M, Mok JY, Matlung H, van den Berg TK, van Esch WJE, Kuijpers TW, Wouters D, Rispens T, Wuhrer M, Vidarsson G (2017) Decoding the human immunoglobulin G-glycan repertoire reveals a Spectrum of Fc-receptor- and complement-mediated-effector activities. Front Immunol 8:877. https://doi.org/10.3389/fimmu.2017.00877

42. Schiestl M, Stangler T, Torella C, Cepeljnik T, Toll H, Grau R (2011) Acceptable changes in quality attributes of glycosylated biopharmaceuticals. Nat Biotechnol 29(4):310–312. https://doi.org/10.1038/nbt.1839
43. Umaña P, Jean-Mairet J, Moudry R, Amstutz H, Bailey JE (1999) Engineered glycoforms of an antineuroblastoma IgG1 with optimized antibody-dependent cellular cytotoxic activity. Nat Biotechnol 17(2):176–180
44. Beck A, Reichert JM (2012) Marketing approval of mogamulizumab: a triumph for glyco-engineering. MAbs 4(4):419–425. https://doi.org/10.4161/mabs.20996
45. Cameron F, McCormack PL (2014) Obinutuzumab: first global approval. Drugs 74(1):147–154. https://doi.org/10.1007/s40265-013-0167-3
46. Kolbeck R, Kozhich A, Koike M, Peng L, Andersson CK, Damschroder MM, Reed JL, Woods R, Dall'acqua WW, Stephens GL, Erjefalt JS, Bjermer L, Humbles AA, Gossage D, Wu H, Kiener PA, Spitalny GL, Mackay CR, Molfino NA, Coyle AJ (2010) MEDI-563, a humanized anti-IL-5 receptor alpha mAb with enhanced antibody-dependent cell-mediated cytotoxicity function. J Allergy Clin Immunol 125(6):1344–1353. https://doi.org/10.1016/j.jaci.2010.04.004
47. Herbst R, Wang Y, Gallagher S, Mittereder N, Kuta E, Damschroder M, Woods R, Rowe DC, Cheng L, Cook K, Evans K, Sims GP, Pfarr DS, Bowen MA, Dall'Acqua W, Shlomchik M, Tedder TF, Kiener P, Jallal B, Wu H, Coyle AJ (2010) B-cell depletion in vitro and in vivo with an afucosylated anti-CD19 antibody. J Pharmacol Exp Ther 335(1):213–222. https://doi.org/10.1124/jpet.110.168062
48. Dolgin E (2020) Anti-BCMA therapy endorsed, despite eye toxicity. Cancer Discov. https://doi.org/10.1158/2159-8290.Cd-nb2020-074
49. Chiang AW, Li S, Spahn PN, Richelle A, Kuo CC, Samoudi M, Lewis NE (2016) Modulating carbohydrate-protein interactions through glycoengineering of monoclonal antibodies to impact cancer physiology. Curr Opin Struct Biol 40:104–111. https://doi.org/10.1016/j.sbi.2016.08.008
50. Yamane-Ohnuki N, Kinoshita S, Inoue-Urakubo M, Kusunoki M, Iida S, Nakano R, Wakitani M, Niwa R, Sakurada M, Uchida K, Shitara K, Satoh M (2004) Establishment of FUT8 knockout Chinese hamster ovary cells: an ideal host cell line for producing completely defucosylated antibodies with enhanced antibody-dependent cellular cytotoxicity. Biotechnol Bioeng 87(5):614–622. https://doi.org/10.1002/bit.20151
51. Mori K, Kuni-Kamochi R, Yamane-Ohnuki N, Wakitani M, Yamano K, Imai H, Kanda Y, Niwa R, Iida S, Uchida K, Shitara K, Satoh M (2004) Engineering Chinese hamster ovary cells to maximize effector function of produced antibodies using FUT8 siRNA. Biotechnol Bioeng 88(7):901–908. https://doi.org/10.1002/bit.20326
52. Beuger V, Kunkele KP, Koll H, Gartner A, Bahner M, Burtscher H, Klein C (2009) Short-hairpin-RNA-mediated silencing of fucosyltransferase 8 in Chinese-hamster ovary cells for the production of antibodies with enhanced antibody immune effector function. Biotechnol Appl Biochem 53(Pt 1):31–37. https://doi.org/10.1042/BA20080220
53. von Horsten HH, Ogorek C, Blanchard V, Demmler C, Giese C, Winkler K, Kaup M, Berger M, Jordan I, Sandig V (2010) Production of non-fucosylated antibodies by co-expression of heterologous GDP-6-deoxy-D-lyxo-4-hexulose reductase. Glycobiology 20(12):1607–1618. https://doi.org/10.1093/glycob/cwq109
54. Roy G, Martin T, Barnes A, Wang J, Jimenez RB, Rice M, Li L, Feng H, Zhang S, Chaerkady R, Wu H, Marelli M, Hatton D, Zhu J, Bowen MA (2018) A novel bicistronic gene design couples stable cell line selection with a fucose switch in a designer CHO host to produce native and afucosylated glycoform antibodies. MAbs 10(3):416–430. https://doi.org/10.1080/19420862.2018.1433975
55. Kelly RM, Kowle RL, Lian Z, Strifler BA, Witcher DR, Parekh BS, Wang T, Frye CC (2018) Modulation of IgG1 immunoeffector function by glycoengineering of the GDP-fucose biosynthesis pathway. Biotechnol Bioeng 115(3):705–718. https://doi.org/10.1002/bit.26496
56. Imai-Nishiya H, Mori K, Inoue M, Wakitani M, Iida S, Shitara K, Satoh M (2007) Double knockdown of alpha1,6-fucosyltransferase (FUT8) and GDP-mannose 4,6-dehydratase

(GMD) in antibody-producing cells: a new strategy for generating fully non-fucosylated therapeutic antibodies with enhanced ADCC. BMC Biotechnol 7:84. https://doi.org/10.1186/1472-6750-7-84
57. Malphettes L, Freyvert Y, Chang J, Liu PQ, Chan E, Miller JC, Zhou Z, Nguyen T, Tsai C, Snowden AW, Collingwood TN, Gregory PD, Cost GJ (2010) Highly efficient deletion of FUT8 in CHO cell lines using zinc-finger nucleases yields cells that produce completely nonfucosylated antibodies. Biotechnol Bioeng 106(5):774–783. https://doi.org/10.1002/bit.22751
58. Cristea S, Freyvert Y, Santiago Y, Holmes MC, Urnov FD, Gregory PD, Cost GJ (2013) In vivo cleavage of transgene donors promotes nuclease-mediated targeted integration. Biotechnol Bioeng 110(3):871–880. https://doi.org/10.1002/bit.24733
59. Ronda C, Pedersen LE, Hansen HG, Kallehauge TB, Betenbaugh MJ, Nielsen AT, Kildegaard HF (2014) Accelerating genome editing in CHO cells using CRISPR Cas9 and CRISPy, a web-based target finding tool. Biotechnol Bioeng 111(8):1604–1616. https://doi.org/10.1002/bit.25233
60. Liu PQ, Chan EM, Cost GJ, Zhang L, Wang J, Miller JC, Guschin DY, Reik A, Holmes MC, Mott JE, Collingwood TN, Gregory PD (2010) Generation of a triple-gene knockout mammalian cell line using engineered zinc-finger nucleases. Biotechnol Bioeng 106(1):97–105. https://doi.org/10.1002/bit.22654
61. Yang Z, Wang S, Halim A, Schulz MA, Frodin M, Rahman SH, Vester-Christensen MB, Behrens C, Kristensen C, Vakhrushev SY, Bennett EP, Wandall HH, Clausen H (2015) Engineered CHO cells for production of diverse, homogeneous glycoproteins. Nat Biotechnol 33(8):842–844. https://doi.org/10.1038/nbt.3280
62. Sun T, Li C, Han L, Jiang H, Xie Y, Zhang B, Qian X, Lu H, Zhu J (2015) Functional knockout of FUT8 in Chinese hamster ovary cells using CRISPR/Cas9 to produce a defucosylated antibody. Eng Life Sci 15(6):660–666. https://doi.org/10.1002/elsc.201400218
63. Chan KF, Shahreel W, Wan C, Teo G, Hayati N, Tay SJ, Tong WH, Yang Y, Rudd PM, Zhang P, Song Z (2016) Inactivation of GDP-fucose transporter gene (Slc35c1) in CHO cells by ZFNs, TALENs and CRISPR-Cas9 for production of fucose-free antibodies. Biotechnol J 11(3):399–414. https://doi.org/10.1002/biot.201500331
64. Zhang P, Haryadi R, Chan KF, Teo G, Goh J, Pereira NA, Feng H, Song Z (2012) Identification of functional elements of the GDP-fucose transporter SLC35C1 using a novel Chinese hamster ovary mutant. Glycobiology 22(7):897–911. https://doi.org/10.1093/glycob/cws064
65. Grav LM, Lee JS, Gerling S, Kallehauge TB, Hansen AH, Kol S, Lee GM, Pedersen LE, Kildegaard HF (2015) One-step generation of triple knockout CHO cell lines using CRISPR/Cas9 and fluorescent enrichment. Biotechnol J 10(9):1446–1456. https://doi.org/10.1002/biot.201500027
66. Popp O, Moser S, Zielonka J, Ruger P, Hansen S, Plottner O (2018) Development of a pre-glycoengineered CHO-K1 host cell line for the expression of antibodies with enhanced Fc mediated effector function. MAbs 10(2):290–303. https://doi.org/10.1080/19420862.2017.1405203
67. Schmieder V, Bydlinski N, Strasser R, Baumann M, Kildegaard HF, Jadhav V, Borth N (2018) Enhanced genome editing tools for multi-gene deletion Knock-out approaches using paired CRISPR sgRNAs in CHO cells. Biotechnol J 13(3):e1700211. https://doi.org/10.1002/biot.201700211
68. Fukuta K, Abe R, Yokomatsu T, Omae F, Asanagi M, Makino T (2000) Control of bisecting GlcNAc addition to N-linked sugar chains. J Biol Chem 275(31):23456–23461. https://doi.org/10.1074/jbc.M002693200
69. Davies J, Jiang L, Pan L, LaBarre MJ, Anderson D, Reff M (2001) Expression of GnTIII in a recombinant anti-CD20 CHO production cell line: expression of antibodies with altered Glycoforms leads to an increase in ADCC through higher affinity for Fc RIII. Biotechnol Bioeng 74(4):288–294

70. Ferrara C, Brunker P, Suter T, Moser S, Puntener U, Umana P (2006) Modulation of therapeutic antibody effector functions by glycosylation engineering: influence of Golgi enzyme localization domain and co-expression of heterologous beta1, 4-N-acetylglucosaminyltransferase III and Golgi alpha-mannosidase II. Biotechnol Bioeng 93(5):851–861. https://doi.org/10.1002/bit.20777
71. Goede V, Fischer K, Engelke A, Schlag R, Lepretre S, Montero LF, Montillo M, Fegan C, Asikanius E, Humphrey K, Fingerle-Rowson G, Hallek M (2015) Obinutuzumab as frontline treatment of chronic lymphocytic leukemia: updated results of the CLL11 study. Leukemia 29(7):1602–1604. https://doi.org/10.1038/leu.2015.14
72. Weikert S, Papac D, Briggs J, Cowfer D, Tom S, Gawlitzek M, Lofgren J, Mehta S, Chisholm V, Modi N, Eppler S, Carroll K, Chamow S, Peers D, Berman P, Krummen L (1999) Engineering Chinese hamster ovary cells to maximize sialic acid content of recombinant glycoproteins. Nat Biotechnol 17(1116–21)
73. Raymond C, Robotham A, Spearman M, Butler M, Kelly J, Durocher Y (2015) Production of alpha2,6-sialylated IgG1 in CHO cells. MAbs 7(3):571–583. https://doi.org/10.1080/19420862.2015.1029215
74. Stach CS, McCann MG, O'Brien CM, Le TS, Somia N, Chen X, Lee K, Fu HY, Daoutidis P, Zhao L, Hu WS, Smanski M (2019) Model-driven engineering of N-linked glycosylation in Chinese hamster ovary cells. ACS Synth Biol 8(11):2524–2535. https://doi.org/10.1021/acssynbio.9b00215
75. Amann T, Hansen AH, Kol S, Lee GM, Andersen MR, Kildegaard HF (2018) CRISPR/Cas9-multiplexed editing of Chinese hamster ovary B4Gal-T1, 2, 3, and 4 tailors N-glycan profiles of therapeutics and secreted host cell proteins. Biotechnol J 13(10):e1800111. https://doi.org/10.1002/biot.201800111
76. Bydlinski N, Maresch D, Schmieder V, Klanert G, Strasser R, Borth N (2018) The contributions of individual galactosyltransferases to protein specific N-glycan processing in Chinese hamster ovary cells. J Biotechnol 282:101–110. https://doi.org/10.1016/j.jbiotec.2018.07.015
77. Lin N, George HJ, Mascarenhas J, Collingwood TN, Kayser KJ, Achtien K (2013) Cells deficient in cmp-n-acetylneuraminic acid hydroxylase and/or glycoprotein alpha-1,3-galactosyltransferase. Google Patents
78. Lin NAN, George Henry J, Mascarenhas J, Collingwood Trevor N, Kayser Kevin J, Achtien K (2018) Cells deficient in Cmp-n-acetylneuraminic acid hydroxylase and/or glycoprotein Alpha-1,3-galactosyltransferase. EP Patent EP 2726604 B1, 2018/04/04
79. Ryll T (2001) Improved galactosylation of recombinant glycoproteins. Google Patents,
80. Collins BE, Guo T, Thiruneelakantapillai L, Millea K, Bulik DA (2012) Production of glycoproteins. Google Patents,
81. Noh SM, Park JH, Lim MS, Kim JW, Lee GM (2017) Reduction of ammonia and lactate through the coupling of glutamine synthetase selection and downregulation of lactate dehydrogenase-A in CHO cells. Appl Microbiol Biotechnol 101(3):1035–1045. https://doi.org/10.1007/s00253-016-7876-y
82. Park H-S, Kim I-H, Kim I-Y, Kim K-H, Kim H-J (2000) Expression of carbamoyl phosphate synthetase I and ornithine transcarbamoylase genes in Chinese hamster ovary dhfr-cells decreases accumulation of ammonium ion in culture media. J Biotechnol 81(2–3):129–140. https://doi.org/10.1016/S0168-1656(00)00282-0
83. Scallon BJ, Tam SH, McCarthy SG, Cai AN, Raju TS (2007) Higher levels of sialylated Fc glycans in immunoglobulin G molecules can adversely impact functionality. Mol Immunol 44(7):1524–1534. https://doi.org/10.1016/j.molimm.2006.09.005
84. Ghaderi D, Taylor RE, Padler-Karavani V, Diaz S, Varki A (2010) Implications of the presence of N-glycolylneuraminic acid in recombinant therapeutic glycoproteins. Nat Biotechnol 28(8):863–867. https://doi.org/10.1038/nbt.1651
85. Naso MF, Tam SH, Scallon BJ, Raju TS (2010) Engineering host cell lines to reduce terminal sialylation of secreted antibodies. MAbs 2(5):519–527. https://doi.org/10.4161/mabs.2.5.13078

86. Chung CY, Wang Q, Yang S, Yin B, Zhang H, Betenbaugh M (2017) Integrated genome and protein editing swaps alpha-2,6 sialylation for alpha-2,3 sialic acid on recombinant antibodies from CHO. Biotechnol J 12(2). https://doi.org/10.1002/biot.201600502
87. Schulz MA, Tian W, Mao Y, Van Coillie J, Sun L, Larsen JS, Chen YH, Kristensen C, Vakhrushev SY, Clausen H, Yang Z (2018) Glycoengineering design options for IgG1 in CHO cells using precise gene editing. Glycobiology 28(7):542–549. https://doi.org/10.1093/glycob/cwy022
88. Chenu S, Grégoire A, Malykh Y, Visvikis A, Monaco L, Shaw L, Schauer R, Marc A, Goergen J-L (2003) Reduction of CMP-N-acetylneuraminic acid hydroxylase activity in engineered Chinese hamster ovary cells using an antisense-RNA strategy. Biochim Biophys Acta Gen Subj 1622(2):133–144. https://doi.org/10.1016/s0304-4165(03)00137-5
89. Jefferis R (2006) A sugar switch for anti-inflammatory antibodies. Nat Biotech 24(10):1230–1231
90. Marx N, Grunwald-Gruber C, Bydlinski N, Dhiman H, Ngoc Nguyen L, Klanert G, Borth N (2018) CRISPR-based targeted epigenetic editing enables gene expression modulation of the silenced Beta-Galactoside Alpha-2,6-Sialyltransferase 1 in CHO cells. Biotechnol J 13(10):e1700217. https://doi.org/10.1002/biot.201700217
91. KJlC K, Hefzi H, Xiong K, Shamie I, Hansen AH, Li S, Pedersen LE, Li S, Lee JS, Lee GM, Kildegaard HF, Lewis NE (2020) Awakening dormant glycosyltransferases in CHO cells with CRISPRa. Biotechnol Bioeng 117(2):593–598. https://doi.org/10.1002/bit.27199
92. Onitsuka M, Kim WD, Ozaki H, Kawaguchi A, Honda K, Kajiura H, Fujiyama K, Asano R, Kumagai I, Ohtake H, Omasa T (2012) Enhancement of sialylation on humanized IgG-like bispecific antibody by overexpression of alpha2,6-sialyltransferase derived from Chinese hamster ovary cells. Appl Microbiol Biotechnol 94(1):69–80. https://doi.org/10.1007/s00253-011-3814-1
93. Lin N, Mascarenhas J, Sealover NR, George HJ, Brooks J, Kayser KJ, Gau B, Yasa I, Azadi P, Archer-Hartmann S (2015) Chinese hamster ovary (CHO) host cell engineering to increase sialylation of recombinant therapeutic proteins by modulating sialyltransferase expression. Biotechnol Prog 31(2):334–346. https://doi.org/10.1002/btpr.2038
94. Nguyen TS, Misaki R, Ohashi T, Fujiyama K (2020) Enhancement of sialylation in rIgG in glyco-engineered Chinese hamster ovary cells. Cytotechnology 72(3):343–355. https://doi.org/10.1007/s10616-020-00381-z
95. Zhou Q, Shankara S, Roy A, Qiu H, Estes S, McVie-Wylie A, Culm-Merdek K, Park A, Pan C, Edmunds T (2008) Development of a simple and rapid method for producing non-fucosylated oligomannose containing antibodies with increased effector function. Biotechnol Bioeng 99(3):652–665. https://doi.org/10.1002/bit.21598
96. Sealover NR, Davis AM, Brooks JK, George HJ, Kayser KJ, Lin N (2013) Engineering Chinese hamster ovary (CHO) cells for producing recombinant proteins with simple glycoforms by zinc-finger nuclease (ZFN)-mediated gene knockout of mannosyl (alpha-1,3-)-glycoprotein beta-1,2-N-acetylglucosaminyltransferase (Mgat1). J Biotechnol 167(1):24–32. https://doi.org/10.1016/j.jbiotec.2013.06.006
97. Kanda Y, Yamada T, Mori K, Okazaki A, Inoue M, Kitajima-Miyama K, Kuni-Kamochi R, Nakano R, Yano K, Kakita S, Shitara K, Satoh M (2007) Comparison of biological activity among nonfucosylated therapeutic IgG1 antibodies with three different N-linked Fc oligosaccharides: the high-mannose, hybrid, and complex types. Glycobiology 17(1):104–118. https://doi.org/10.1093/glycob/cwl057
98. Yu M, Brown D, Reed C, Chung S, Lutman J, Stefanich E, Wong A, Stephan J-P, Bayer R (2012) Production, characterization, and pharmacokinetic properties of antibodies with N-linked mannose-5 glycans. mAbs 4(4):475–487. https://doi.org/10.4161/mabs.20737
99. Malhotra R, Wormald MR, Rudd PM, Fischer PB, Dwek RA, Sim RB (1995) Glycosylation changes of IgG associated with rheumatoid arthritis can activate complement via the mannose-binding protein. Nat Med 1(3):237–243. https://doi.org/10.1038/nm0395-237

100. Sazinsky SL, Ott RG, Silver NW, Tidor B, Ravetch JV, Wittrup KD (2008) Aglycosylated immunoglobulin G(1) variants productively engage activating Fc receptors. Proc Natl Acad Sci U S A 105(51):20167–20172. https://doi.org/10.1073/pnas.0809257105
101. Inman BA, Longo TA, Ramalingam S, Harrison MR (2017) Atezolizumab: A PD-L1-blocking antibody for bladder Cancer. Clin Cancer Res 23(8):1886–1890. https://doi.org/10.1158/1078-0432.Ccr-16-1417
102. Meuris L, Santens F, Elson G, Festjens N, Boone M, Dos Santos A, Devos S, Rousseau F, Plets E, Houthuys E, Malinge P, Magistrelli G, Cons L, Chatel L, Devreese B, Callewaert N (2014) GlycoDelete engineering of mammalian cells simplifies N-glycosylation of recombinant proteins. Nat Biotechnol 32(5):485–489. https://doi.org/10.1038/nbt.2885
103. Gawlitzek M, Estacio M, Fürch T, Kiss R (2009) Identification of cell culture conditions to control N-glycosylation site-occupancy of recombinant glycoproteins expressed in CHO cells. Biotechnol Bioeng 103(6):1164–1175. https://doi.org/10.1002/bit.22348
104. Gawlitzek M, Conradt HS, Wagner R (1995) Effect of different cell culture conditions on the polypeptide integrity and N-glycosylation of a recombinant model glycoprotein. Biotechnol Bioeng 46(6):536–544. https://doi.org/10.1002/bit.260460606
105. Gawlitzek M, Valley U, Wagner R (1998) Ammonium ion and glucosamine dependent increases of oligosaccharide complexity in recombinant glycoproteins secreted from cultivated BHK-21 cells. Biotechnol Bioeng 57(5):518–528. https://doi.org/10.1002/(SICI)1097-0290(19980305)57:5<518::AID-BIT3>3.0.CO;2-J
106. Yang M, Butler M (2002) Effects of Ammonia and glucosamine on the heterogeneity of erythropoietin Glycoforms. Biotechnol Prog 18(1):129–138. https://doi.org/10.1021/bp0101334
107. Liu J, Wang J, Fan L, Chen X, Hu D, Deng X, Poon HF, Wang H, Liu X, Tan WS (2015) Galactose supplementation enhance sialylation of recombinant Fc-fusion protein in CHO cell: an insight into the role of galactosylation in sialylation. World J Microbiol Biotechnol 31(7):1147–1156. https://doi.org/10.1007/s11274-015-1864-8
108. Gu X, Wang DIC (1998) Improvement of interferon-γ sialylation in Chinese hamster ovary cell culture by feeding of N-acetylmannosamine. Biotechnol Bioeng 58(6):642–648. https://doi.org/10.1002/(SICI)1097-0290(19980620)58:6<642::AID-BIT10>3.0.CO;2-9
109. Baker KN, Rendall MH, Hills AE, Hoare M, Freedman RB, James DC (2001) Metabolic control of recombinant protein N-glycan processing in NS0 and CHO cells. Biotechnol Bioeng 73(3):188–202. https://doi.org/10.1002/bit.1051
110. Wong NSC, Wati L, Nissom PM, Feng HT, Lee MM, Yap MGS (2010) An investigation of intracellular glycosylation activities in CHO cells: effects of nucleotide sugar precursor feeding. Biotechnol Bioeng 107(2):321–336. https://doi.org/10.1002/bit.22812
111. Altamirano C, Cairo JJ, Godia F (2001) Decoupling cell growth and product formation in Chinese hamster ovary cells through metabolic control. Biotechnol Bioeng 76(4):351–360. https://doi.org/10.1002/bit.10096
112. Carvalhal AV, Santos SS, Calado J, Haury M, Carrondo MJT (2003) Cell growth arrest by nucleotides, nucleosides and bases as a tool for improved production of recombinant proteins. Biotechnol Prog 19(1):69–83. https://doi.org/10.1021/bp0255917
113. Grainger RK, James DC (2013) CHO cell line specific prediction and control of recombinant monoclonal antibody N-glycosylation. Biotechnol Bioeng 110(11):2970–2983. https://doi.org/10.1002/bit.24959
114. Wahrheit J, Nicolae A, Heinzle E (2014) Dynamics of growth and metabolism controlled by glutamine availability in Chinese hamster ovary cells. Appl Microbiol Biotechnol 98(4):1771–1783. https://doi.org/10.1007/s00253-013-5452-2
115. Aghamohseni H, Ohadi K, Spearman M, Krahn N, Moo-Young M, Scharer JM, Butler M, Budman HM (2014) Effects of nutrient levels and average culture pH on the glycosylation pattern of camelid-humanized monoclonal antibody. J Biotechnol 186:98–109. https://doi.org/10.1016/j.jbiotec.2014.05.024
116. Rivinoja A, Hassinen A, Kokkonen N, Kauppila A, Kellokumpu S (2009) Elevated Golgi pH impairs terminal N-glycosylation by inducing mislocalization of Golgi glycosyltransferases. J Cell Physiol 220(1):144–154. https://doi.org/10.1002/jcp.21744

117. Okeley NM, Alley SC, Anderson ME, Boursalian TE, Burke PJ, Emmerton KM, Jeffrey SC, Klussman K, Law C-L, Sussman D, Toki BE, Westendorf L, Zeng W, Zhang X, Benjamin DR, Senter PD (2013) Development of orally active inhibitors of protein and cellular fucosylation. Proc Natl Acad Sci 110(14):5404–5409. https://doi.org/10.1073/pnas.1222263110
118. Zimmermann M, Ehret J, Kolmar H, Zimmer A (2019) Impact of acetylated and non-acetylated Fucose analogues on IgG glycosylation. Antibodies (Basel) 8(1):9. https://doi.org/10.3390/antib8010009
119. Zhang A, Tsang VL, Markely LR, Kurt L, Huang YM, Prajapati S, Kshirsagar R (2016) Identifying the differences in mechanisms of mycophenolic acid controlling fucose content of glycoproteins expressed in different CHO cell lines. Biotechnol Bioeng 113(11):2367–2376. https://doi.org/10.1002/bit.25995
120. Allen JG, Mujacic M, Frohn MJ, Pickrell AJ, Kodama P, Bagal D, San Miguel T, Sickmier EA, Osgood S, Swietlow A, Li V, Jordan JB, Kim KW, Rousseau AC, Kim YJ, Caille S, Achmatowicz M, Thiel O, Fotsch CH, Reddy P, McCarter JD (2016) Facile modulation of antibody Fucosylation with small molecule Fucostatin inhibitors and Cocrystal structure with GDP-mannose 4,6-dehydratase. ACS Chem Biol 11(10):2734–2743. https://doi.org/10.1021/acschembio.6b00460
121. Hills AE, Patel A, Boyd P, James DC (2001) Metabolic control of recombinant monoclonal antibody N-glycosylation in GS-NS0 cells. Biotechnol Bioeng 75(2):239–251. https://doi.org/10.1002/bit.10022
122. Kildegaard HF, Fan Y, Sen JW, Larsen B, Andersen MR (2016) Glycoprofiling effects of media additives on IgG produced by CHO cells in fed-batch bioreactors. Biotechnol Bioeng 113(2):359–366. https://doi.org/10.1002/bit.25715
123. Gramer MJ, Eckblad JJ, Donahue R, Brown J, Shultz C, Vickerman K, Priem P, van den Bremer ET, Gerritsen J, van Berkel PH (2011) Modulation of antibody galactosylation through feeding of uridine, manganese chloride, and galactose. Biotechnol Bioeng 108(7):1591–1602. https://doi.org/10.1002/bit.23075
124. St. Amand MM, Radhakrishnan D, Robinson AS, Ogunnaike BA (2014) Identification of manipulated variables for a glycosylation control strategy. Biotechnol Bioeng 111(10):1957–1970. https://doi.org/10.1002/bit.25251
125. Prabhu A, Gadre R, Gadgil M (2018) Zinc supplementation decreases galactosylation of recombinant IgG in CHO cells. Appl Microbiol Biotechnol 102(14):5989–5999. https://doi.org/10.1007/s00253-018-9064-8
126. Prabhu A, Gadgil M (2019) Nickel and cobalt affect galactosylation of recombinant IgG expressed in CHO cells. Biometals 32(1):11–19. https://doi.org/10.1007/s10534-018-0152-0
127. Hossler P, Racicot C (2015) Targeted shifting of protein glycosylation profiles in mammalian cell culture through media supplementation of cobalt. J Glycobiol:03. https://doi.org/10.4172/2168-958X.1000108
128. Hossler P, Racicot C, Chumsae C, McDermott S, Cochran K (2017) Cell culture media supplementation of infrequently used sugars for the targeted shifting of protein glycosylation profiles. Biotechnol Prog 33(2):511–522. https://doi.org/10.1002/btpr.2429
129. del Val IJ, Polizzi KM, Kontoravdi C (2016) A theoretical estimate for nucleotide sugar demand towards Chinese hamster ovary cellular glycosylation. Sci Rep 6:28547. https://doi.org/10.1038/srep28547
130. Torkashvand F, Vaziri B, Maleknia S, Heydari A, Vossoughi M, Davami F, Mahboudi F (2015) Designed amino acid feed in improvement of production and quality targets of a therapeutic monoclonal antibody. PLoS One 10(10):e0140597. https://doi.org/10.1371/journal.pone.0140597
131. Hong JK, Cho SM, Yoon SK (2010) Substitution of glutamine by glutamate enhances production and galactosylation of recombinant IgG in Chinese hamster ovary cells. Appl Microbiol Biotechnol 88(4):869–876. https://doi.org/10.1007/s00253-010-2790-1
132. Chen P, Harcum SW (2005) Effects of amino acid additions on ammonium stressed CHO cells. J Biotechnol 117(3):277–286. https://doi.org/10.1016/j.jbiotec.2005.02.003

133. Blondeel EJ, Braasch K, McGill T, Chang D, Engel C, Spearman M, Butler M, Aucoin MG (2015) Tuning a MAb glycan profile in cell culture: supplementing N-acetylglucosamine to favour G0 glycans without compromising productivity and cell growth. J Biotechnol 214:105–112. https://doi.org/10.1016/j.jbiotec.2015.09.014
134. Dekkers G, Plomp R, Koeleman CA, Visser R, von Horsten HH, Sandig V, Rispens T, Wuhrer M, Vidarsson G (2016) Multi-level glyco-engineering techniques to generate IgG with defined Fc-glycans. Sci Rep 6:36964. https://doi.org/10.1038/srep36964
135. Yin B, Wang Q, Chung CY, Bhattacharya R, Ren X, Tang J, Yarema KJ, Betenbaugh MJ (2017) A novel sugar analog enhances sialic acid production and biotherapeutic sialylation in CHO cells. Biotechnol Bioeng 114(8):1899–1902. https://doi.org/10.1002/bit.26291
136. Ehret J, Zimmermann M, Eichhorn T, Zimmer A (2019) Impact of cell culture media additives on IgG glycosylation produced in Chinese hamster ovary cells. Biotechnol Bioeng 116(4):816–830. https://doi.org/10.1002/bit.26904
137. Gramer MJ, Goochee CF, Chock VY, Brousseau DT, Sliwkowski MB (1995) Removal of sialic acid from a glycoprotein in CHO cell culture supernatant by action of an extracellular CHO cell sialidase. Biotechnology (N Y) 13(7):692–698. https://doi.org/10.1038/nbt0795-692
138. Medveďová L, Farkaš R (2004) Hormonal control of protein glycosylation: role of steroids and related lipophilic ligands. Endocr Regul 38(2):65–79
139. Jing Y, Qian Y, Li ZJ (2010) Sialylation enhancement of CTLA4-Ig fusion protein in Chinese hamster ovary cells by dexamethasone. Biotechnol Bioeng 107(3):488–496. https://doi.org/10.1002/bit.22827
140. Rouiller Y, Perilleux A, Marsaut M, Stettler M, Vesin MN, Broly H (2012) Effect of hydrocortisone on the production and glycosylation of an Fc-fusion protein in CHO cell cultures. Biotechnol Prog 28(3):803–813. https://doi.org/10.1002/btpr.1530
141. Jiang Z, Sharfstein ST (2008) Sodium butyrate stimulates monoclonal antibody over-expression in CHO cells by improving gene accessibility. Biotechnol Bioeng 100(1):189–194. https://doi.org/10.1002/bit.21726
142. Crowell CK, Qin Q, Grampp GE, Radcliffe RA, Rogers GN, Scheinman RI (2008) Sodium butyrate alters erythropoietin glycosylation via multiple mechanisms. Biotechnol Bioeng 99(1):201–213. https://doi.org/10.1002/bit.21539
143. Rodriguez J, Spearman M, Huzel N, Butler M (2005) Enhanced production of monomeric interferon-β by CHO cells through the control of culture conditions. Biotechnol Prog 21(1):22–30. https://doi.org/10.1021/bp049807b
144. Sung YH, Song YJ, Lim SW, Chung JY, Lee GM (2004) Effect of sodium butyrate on the production, heterogeneity and biological activity of human thrombopoietin by recombinant Chinese hamster ovary cells. J Biotechnol 112(3):323–335. https://doi.org/10.1016/j.jbiotec.2004.05.003
145. Chen F, Kou T, Fan L, Zhou Y, Ye Z, Zhao L, Tan W-S (2011) The combined effect of sodium butyrate and low culture temperature on the production, sialylation, and biological activity of an antibody produced in CHO cells. Biotechnol Bioprocess Eng 16(6):1157–1165. https://doi.org/10.1007/s12257-011-0069-8
146. Chen P, Harcum SW (2006) Effects of elevated ammonium on glycosylation gene expression in CHO cells. Metab Eng 8(2):123–132. https://doi.org/10.1016/j.ymben.2005.10.002
147. Gawlitzek M, Ryll T, Lofgren J, Sliwkowski MB (2000) Ammonium alters N-glycan structures of recombinant TNFR-IgG: degradative versus biosynthetic mechanisms. Biotechnol Bioeng 68(6):637–646. https://doi.org/10.1002/(sici)1097-0290(20000620)68:6<637::Aid-bit6>3.0.Co;2-c
148. Rillahan CD, Antonopoulos A, Lefort CT, Sonon R, Azadi P, Ley K, Dell A, Haslam SM, Paulson JC (2012) Global metabolic inhibitors of sialyl- and fucosyltransferases remodel the glycome. Nat Chem Biol 8(7):661–668. https://doi.org/10.1038/nchembio.999
149. Pacis E, Yu M, Autsen J, Bayer R, Li F (2011) Effects of cell culture conditions on antibody N-linked glycosylation—what affects high mannose 5 glycoform. Biotechnol Bioeng 108(10):2348–2358. https://doi.org/10.1002/bit.23200

150. Surve T, Gadgil M (2015) Manganese increases high mannose glycoform on monoclonal antibody expressed in CHO when glucose is absent or limiting: implications for use of alternate sugars. Biotechnol Prog 31(2):460–467. https://doi.org/10.1002/btpr.2029
151. Wentz Alane E, Hemmavanh DON, Matuck Joseph G (2017) Use of metal ions for modulation of protein glycosylation profiles of recombinant proteins. US Patent US 9598667 B2, 2017/03/21
152. Kuwae S, Miyakawa I, Doi T (2018) Development of a chemically defined platform fed-batch culture media for monoclonal antibody-producing CHO cell lines with optimized choline content. Cytotechnology 70(3):939–948. https://doi.org/10.1007/s10616-017-0185-1
153. Huang CJ, Lin H, Yang JX (2015) A robust method for increasing fc glycan high mannose level of recombinant antibodies. Biotechnol Bioeng 112(6):1200–1209. https://doi.org/10.1002/bit.25534
154. Slade PG, Caspary RG, Nargund S, Huang CJ (2016) Mannose metabolism in recombinant CHO cells and its effect on IgG glycosylation. Biotechnol Bioeng 113(7):1468–1480. https://doi.org/10.1002/bit.25924
155. Hossler P, McDermott S, Racicot C, Chumsae C, Raharimampionona H, Zhou Y, Ouellette D, Matuck J, Correia I, Fann J, Li J (2014) Cell culture media supplementation of uncommonly used sugars sucrose and tagatose for the targeted shifting of protein glycosylation profiles of recombinant protein therapeutics. Biotechnol Prog 30(6):1419–1431. https://doi.org/10.1002/btpr.1968
156. van Berkel PH, Gerritsen J, van Voskuilen E, Perdok G, Vink T, van de Winkel JG, Parren PW (2010) Rapid production of recombinant human IgG with improved ADCC effector function in a transient expression system. Biotechnol Bioeng 105(2):350–357. https://doi.org/10.1002/bit.22535
157. Elbein AD, Pan YT, Solf R, Vosbeck K (1983) Effect of swainsonine, an inhibitor of glycoprotein processing, on cultured mammalian cells. J Cell Physiol 115(3):265–275. https://doi.org/10.1002/jcp.1041150309
158. Krahn N, Spearman M, Meier M, Dorion-Thibaudeau J, McDougall M, Patel TR, De Crescenzo G, Durocher Y, Stetefeld J, Butler M (2017) Inhibition of glycosylation on a camelid antibody uniquely affects its FcγRI binding activity. Eur J Pharm Sci 96:428–439. https://doi.org/10.1016/j.ejps.2016.09.040
159. Pande S, Rahardjo A, Livingston B, Mujacic M (2015) Monensin, a small molecule ionophore, can be used to increase high mannose levels on monoclonal antibodies generated by Chinese hamster ovary production cell-lines. Biotechnol Bioeng 112(7):1383–1394. https://doi.org/10.1002/bit.25551
160. Liu B, Spearman M, Doering J, Lattova E, Perreault H, Butler M (2014) The availability of glucose to CHO cells affects the intracellular lipid-linked oligosaccharide distribution, site occupancy and the N-glycosylation profile of a monoclonal antibody. J Biotechnol 170:17–27. https://doi.org/10.1016/j.jbiotec.2013.11.007
161. Villacres C, Tayi VS, Lattova E, Perreault H, Butler M (2015) Low glucose depletes glycan precursors, reduces site occupancy and galactosylation of a monoclonal antibody in CHO cell culture. Biotechnol J 10(7):1051–1066. https://doi.org/10.1002/biot.201400662
162. Walker MR, Lund J, Thompson KM, Jefferis R (1989) Aglycosylation of human IgG1 and IgG3 monoclonal antibodies can eliminate recognition by human cells expressing Fc gamma RI and/or Fc gamma RII receptors. Biochem J 259(2):347–353. https://doi.org/10.1042/bj2590347
163. Grilo AL, Mantalaris A (2019) The increasingly human and profitable monoclonal antibody market. Trends Biotechnol 37(1):9–16. https://doi.org/10.1016/j.tibtech.2018.05.014
164. Levinson Arthur D, Pennica D, Kohr William J, Vehar Gordon A, Goeddel David V, Yelverton Elizabeth M, Simonsen Christian C (2001) Human Tpa production using vectors coding for Dhfr protein. US Patent US 6261837 B1, 2001/07/17
165. Wurm FCHO Quasispecies—implications for manufacturing processes. Vol 1. MDPI. https://doi.org/10.3390/pr1030296

166. Dumont J, Euwart D, Mei B, Estes S, Kshirsagar R (2016) Human cell lines for biopharmaceutical manufacturing: history, status, and future perspectives. Crit Rev Biotechnol 36(6):1110–1122. https://doi.org/10.3109/07388551.2015.1084266
167. Barone PW, Wiebe ME, Leung JC, Hussein ITM, Keumurian FJ, Bouressa J, Brussel A, Chen D, Chong M, Dehghani H, Gerentes L, Gilbert J, Gold D, Kiss R, Kreil TR, Labatut R, Li Y, Müllberg J, Mallet L, Menzel C, Moody M, Monpoeho S, Murphy M, Plavsic M, Roth NJ, Roush D, Ruffing M, Schicho R, Snyder R, Stark D, Zhang C, Wolfrum J, Sinskey AJ, Springs SL (2020) Viral contamination in biologic manufacture and implications for emerging therapies. Nat Biotechnol 38(5):563–572. https://doi.org/10.1038/s41587-020-0507-2
168. Crotti C, Agape E, Becciolini A, Biggioggero M, Favalli EG (2019) Targeting granulocyte-monocyte Colony-stimulating factor signaling in rheumatoid arthritis: future prospects. Drugs 79(16):1741–1755. https://doi.org/10.1007/s40265-019-01192-z
169. Walsh G (2018) Biopharmaceutical benchmarks 2018. Nat Biotechnol 36(12):1136–1145. https://doi.org/10.1038/nbt.4305
170. Sargentini-Maier ML, De Decker P, Tersteeg C, Canvin J, Callewaert F, De Winter H (2019) Clinical pharmacology of caplacizumab for the treatment of patients with acquired thrombotic thrombocytopenic purpura. Expert Rev Clin Pharmacol 12(6):537–545. https://doi.org/10.1080/17512433.2019.1607293
171. Wacker M, Linton D, Hitchen PG, Nita-Lazar M, Haslam SM, North SJ, Panico M, Morris HR, Dell A, Wren BW, Aebi M (2002) N-linked glycosylation in campylobacter jejuni and its functional transfer into E. coli. Science 298(5599):1790–1793
172. Valderrama-Rincon JD, Fisher AC, Merritt JH, Fan Y-Y, Reading CA, Chhiba K, Heiss C, Azadi P, Aebi M, DeLisa MP (2012) An engineered eukaryotic protein glycosylation pathway in Escherichia coli. Nat Chem Biol 8(5):434–436. https://doi.org/10.1038/nchembio.921
173. Hamilton SR, Gerngross TU (2007) Glycosylation engineering in yeast: the advent of fully humanized yeast. Curr Opin Biotechnol 18(5):387–392. https://doi.org/10.1016/j.copbio.2007.09.001
174. Jacobs PP, Geysens S, Vervecken W, Contreras R, Callewaert N (2009) Engineering complex-type N-glycosylation in Pichia pastoris using GlycoSwitch technology. Nat Protoc 4(1):58–70. https://doi.org/10.1038/nprot.2008.213
175. De Pourcq K, Vervecken W, Dewerte I, Valevska A, Van Hecke A, Callewaert N (2012) Engineering the yeast Yarrowia lipolytica for the production of therapeutic proteins homogeneously glycosylated with Man8GlcNAc2 and Man5GlcNAc2. Microb Cell Factories 11(1):53. https://doi.org/10.1186/1475-2859-11-53
176. Wang H, Song HL, Wang Q, Qiu BS (2013) Expression of glycoproteins bearing complex human-like glycans with galactose terminal in Hansenula polymorpha. World J Microbiol Biotechnol 29(3):447–458. https://doi.org/10.1007/s11274-012-1197-9
177. Hamilton SR, Bobrowicz P, Bobrowicz B, Davidson RC, Li H, Mitchell T, Nett JH, Rausch S, Stadheim TA, Wischnewski H, Wildt S, Gerngross TU (2003) Production of complex human glycoproteins in yeast. Science 301(5637):1244–1246. https://doi.org/10.1126/science.1088166
178. Choi B-K, Bobrowicz P, Davidson RC, Hamilton SR, Kung DH, Li H, Miele RG, Nett JH, Wildt S, Gerngross TU (2003) Use of combinatorial genetic libraries to humanize N-linked glycosylation in the yeast Pichia pastoris. Proc Natl Acad Sci 100(9):5022–5027. https://doi.org/10.1073/pnas.0931263100
179. Nasab FP, Aebi M, Bernhard G, Frey AD (2012) A combined system to engineer glycosylation efficiency and glycan structure in Saccharomyces cerevisiae. Appl Environ Microbiol:AEM:02817–02812. https://doi.org/10.1128/aem.02817-12
180. Ye J, Ly J, Watts K, Hsu A, Walker A, McLaughlin K, Berdichevsky M, Prinz B, Sean Kersey D, d'Anjou M, Pollard D, Potgieter T (2011) Optimization of a glycoengineered Pichia pastoris cultivation process for commercial antibody production. Biotechnol Prog 27(6):1744–1750. https://doi.org/10.1002/btpr.695
181. Potgieter TI, Cukan M, Drummond JE, Houston-Cummings NR, Jiang Y, Li F, Lynaugh H, Mallem M, McKelvey TW, Mitchell T, Nylen A, Rittenhour A, Stadheim TA, Zha D,

d'Anjou M (2009) Production of monoclonal antibodies by glycoengineered Pichia pastoris. J Biotechnol 139(4):318–325. https://doi.org/10.1016/j.jbiotec.2008.12.015
182. Hamilton SR, Davidson RC, Sethuraman N, Nett JH, Jiang Y, Rios S, Bobrowicz P, Stadheim TA, Li H, Choi BK, Hopkins D, Wischnewski H, Roser J, Mitchell T, Strawbridge RR, Hoopes J, Wildt S, Gerngross TU (2006) Humanization of yeast to produce complex terminally sialylated glycoproteins. Science 313(5792):1441–1443. https://doi.org/10.1126/science.1130256
183. Hiatt A, Caffferkey R, Bowdish K (1989) Production of antibodies in transgenic plants. Nature 342(6245):76–78. https://doi.org/10.1038/342076a0
184. Bardor M, Faveeuw C, Fitchette A-C, Gilbert D, Galas L, Trottein F, Faye L, Lerouge P (2003) Immunoreactivity in mammals of two typical plant glyco-epitopes, core α(1,3)-fucose and core xylose. Glycobiology 13(6):427–434. https://doi.org/10.1093/glycob/cwg024
185. Fitchette-Lainé A-C, Gomord V, Cabanes M, Michalski J-C, Saint Macary M, Foucher B, Cavelier B, Hawes C, Lerouge P, Faye L (1997) N-glycans harboring the Lewis a epitope are expressed at the surface of plant cells. Plant J 12(6):1411–1417. https://doi.org/10.1046/j.1365-313x.1997.12061411.x
186. Schähs M, Strasser R, Stadlmann J, Kunert R, Rademacher T, Steinkellner H (2007) Production of a monoclonal antibody in plants with a humanized N-glycosylation pattern. Plant Biotechnol J 5(5):657–663. https://doi.org/10.1111/j.1467-7652.2007.00273.x
187. Strasser R, Altmann F, Mach L, Glössl J, Steinkellner H (2004) Generation of Arabidopsis thaliana plants with complex N-glycans lacking β1,2-linked xylose and core α1,3-linked fucose. FEBS Lett 561(1–3):132–136. https://doi.org/10.1016/s0014-5793(04)00150-4
188. Cox KM, Sterling JD, Regan JT, Gasdaska JR, Frantz KK, Peele CG, Black A, Passmore D, Moldovan-Loomis C, Srinivasan M, Cuison S, Cardarelli PM, Dickey LF (2006) Glycan optimization of a human monoclonal antibody in the aquatic plant Lemna minor. Nat Biotechnol 24(12):1591–1597. https://doi.org/10.1038/nbt1260
189. Strasser R, Castilho A, Stadlmann J, Kunert R, Quendler H, Gattinger P, Jez J, Rademacher T, Altmann F, Mach L, Steinkellner H (2009) Improved virus neutralization by plant-produced anti-HIV antibodies with a homogeneous beta1,4-galactosylated N-glycan profile. J Biol Chem 284(31):20479–20485. https://doi.org/10.1074/jbc.M109.014126
190. Castilho A, Pabst M, Leonard R, Veit C, Altmann F, Mach L, Glössl J, Strasser R, Steinkellner H (2008) Construction of a functional CMP-sialic acid biosynthesis pathway in Arabidopsis. Plant Physiol 147(1):331–339. https://doi.org/10.1104/pp.108.117572
191. Qiu X, Wong G, Audet J, Bello A, Fernando L, Alimonti JB, Fausther-Bovendo H, Wei H, Aviles J, Hiatt E, Johnson A, Morton J, Swope K, Bohorov O, Bohorova N, Goodman C, Kim D, Pauly MH, Velasco J, Pettitt J, Olinger GG, Whaley K, Xu B, Strong JE, Zeitlin L, Kobinger GP (2014) Reversion of advanced Ebola virus disease in nonhuman primates with ZMapp. Nature 514(7520):47–53. https://doi.org/10.1038/nature13777
192. Largent EA (2016) EBOLA and FDA: reviewing the response to the 2014 outbreak, to find lessons for the future. J Law Biosci 3(3):489–537. https://doi.org/10.1093/jlb/lsw046
193. New antibodies best ZMapp in Ebola trial (2019) Nature. Biotechnology 37(10):1105–1105. https://doi.org/10.1038/s41587-019-0284-y
194. Cérutti M, Golay J (2012) Lepidopteran cells, an alternative for the production of recombinant antibodies? mAbs 4(3):294–309. https://doi.org/10.4161/mabs.19942
195. Shi X, Jarvis DL (2007) Protein N-glycosylation in the baculovirus-insect cell system. Curr Drug Targets 8(10):1116–1125. https://doi.org/10.2174/138945007782151360
196. Altmann F, Kornfeld G, Dalik T, Staudacher E, Glössl J (1993) Processing of asparagine-linked oligosaccharides in insect cells. N-acetylglucosaminyltransferase I and II activities in cultured lepidopteran cells. Glycobiology 3(6):619–625. https://doi.org/10.1093/glycob/3.6.619
197. Abdul-Rahman B, Ailor E, Jarvis D, Betenbaugh M, Lee YC (2002) Beta-(1 -> 4)-galactosyltransferase activity in native and engineered insect cells measured with time-resolved europium fluorescence. Carbohydr Res 337(21–23):2181–2186. https://doi.org/10.1016/s0008-6215(02)00260-4

198. Hollister JR, Jarvis DL (2001) Engineering lepidopteran insect cells for sialoglycoprotein production by genetic transformation with mammalian beta 1,4-galactosyltransferase and alpha 2,6-sialyltransferase genes. Glycobiology 11(1):1–9. https://doi.org/10.1093/glycob/11.1.1
199. Altmann F, Schwihla H, Staudacher E, Glössl J, März L (1995) Insect cells contain an unusual, membrane-bound beta-N-acetylglucosaminidase probably involved in the processing of protein N-glycans. J Biol Chem 270(29):17344–17349. https://doi.org/10.1074/jbc.270.29.17344
200. Hollister JR, Shaper JH, Jarvis DL (1998) Stable expression of mammalian β1,4-galactosyltransferase extends the N-glycosylation pathway in insect cells. Glycobiology 8(5):473–480. https://doi.org/10.1093/glycob/8.5.473
201. Hollister J, Grabenhorst E, Nimtz M, Conradt H, Jarvis DL (2002) Engineering the protein N-glycosylation pathway in insect cells for production of biantennary, complex N-glycans. Biochemistry 41(50):15093–15104. https://doi.org/10.1021/bi026455d
202. Aumiller JJ, Mabashi-Asazuma H, Hillar A, Shi X, Jarvis DL (2012) A new glycoengineered insect cell line with an inducibly mammalianized protein N-glycosylation pathway. Glycobiology 22(3):417–428. https://doi.org/10.1093/glycob/cwr160
203. Wagner R, Liedtke S, Kretzschmar E, Geyer H, Geyer R, Klenk H-D (1996) Elongation of the N-glycans of fowl plague virus hemagglutinin expressed in Spodoptera frugiperda (Sf9) cells by coexpression of human β1,2-N-acetylglucosaminyltransferase I. Glycobiology 6(2):165–175. https://doi.org/10.1093/glycob/6.2.165
204. Jarvis DL, Howe D, Aumiller JJ (2001) Novel baculovirus expression vectors that provide sialylation of recombinant glycoproteins in lepidopteran insect cells. J Virol 75(13):6223–6227. https://doi.org/10.1128/jvi.75.13.6223-6227.2001
205. Hang GD, Chen CJ, Lin CY, Chen HC, Chen H (2003) Improvement of glycosylation in insect cells with mammalian glycosyltransferases. J Biotechnol 102(1):61–71. https://doi.org/10.1016/s0168-1656(02)00364-4
206. Palmberger D, Wilson IBH, Berger I, Grabherr R, Rendic D (2012) SweetBac: a new approach for the production of Mammalianised glycoproteins in insect cells. PLoS One 7(4):e34226. https://doi.org/10.1371/journal.pone.0034226
207. Mabashi-Asazuma H, Kuo C-W, Khoo K-H, Jarvis DL (2013) A novel baculovirus vector for the production of nonfucosylated recombinant glycoproteins in insect cells. Glycobiology 24(3):325–340. https://doi.org/10.1093/glycob/cwt161
208. Seeberger PH, Overkleeft HS (2015) Chemical synthesis of Glycans and Glycoconjugates. In: Varki A, Cummings RD, Esko JD et al (eds) Essentials of Glycobiology. Cold Spring Harbor Laboratory Press, New York, pp 681–679. https://doi.org/10.1101/glycobiology.3e.053
209. Fairbanks AJ (2018) Synthetic and semi-synthetic approaches to unprotected N-glycan oxazolines. Beilstein J Org Chem 14:416–429. https://doi.org/10.3762/bjoc.14.30
210. Liu L, Prudden AR, Bosman GP, Boons G-J (2017) Improved isolation and characterization procedure of sialylglycopeptide from egg yolk powder. Carbohydr Res 452:122–128. https://doi.org/10.1016/j.carres.2017.10.001
211. Alagesan K, Kolarich D (2019) Improved strategy for large scale isolation of sialylglycopeptide (SGP) from egg yolk powder. MethodsX 6:773–778. https://doi.org/10.1016/j.mex.2019.04.007
212. Hodoniczky J, Zheng YZ, James DC (2005) Control of recombinant monoclonal antibody effector functions by Fc N-glycan remodeling in vitro. Biotechnol Prog 21(6):1644–1652. https://doi.org/10.1021/bp050228w
213. Raju TS, Briggs JB, Chamow SM, Winkler ME, Jones AJS (2001) Glycoengineering of therapeutic glycoproteins: in vitro Galactosylation and sialylation of glycoproteins with terminal N-Acetylglucosamine and galactose residues. Biochemistry 40(30):8868–8876. https://doi.org/10.1021/bi010475i
214. Thomann M, Schlothauer T, Dashivets T, Malik S, Avenal C, Bulau P, Rüger P, Reusch D (2015) In vitro glycoengineering of IgG1 and its effect on Fc receptor binding and ADCC activity. PLoS One 10(8):e0134949. https://doi.org/10.1371/journal.pone.0134949

215. Mimura Y, Church S, Ghirlando R, Ashton PR, Dong S, Goodall M, Lund J, Jefferis R (2000) The influence of glycosylation on the thermal stability and effector function expression of human IgG1-Fc: properties of a series of truncated glycoforms. Mol Immunol 37(12–13):697–706. https://doi.org/10.1016/s0161-5890(00)00105-x
216. Li C, Zhu S, Ma C, Wang L-X (2017) Designer α1,6-Fucosidase mutants enable direct Core Fucosylation of intact N-Glycopeptides and N-glycoproteins. J Am Chem Soc 139(42):15074–15087. https://doi.org/10.1021/jacs.7b07906
217. Tsai T-I, Li S-T, Liu C-P, Chen KY, Shivatare SS, Lin C-W, Liao S-F, Lin C-W, Hsu T-L, Wu Y-T, Tsai M-H, Lai M-Y, Lin N-H, Wu C-Y, Wong C-H (2017) An effective bacterial Fucosidase for glycoprotein remodeling. ACS Chem Biol 12(1):63–72. https://doi.org/10.1021/acschembio.6b00821
218. Mikolajczyk K, Kaczmarek R, Czerwinski M (2020) How glycosylation affects glycosylation: the role of N-glycans in glycosyltransferase activity. Glycobiology. https://doi.org/10.1093/glycob/cwaa041
219. Zheng S, Qiu D, Adams M, Li J, Mantri RV, Gandhi R (2017) Investigating the degradation behaviors of a therapeutic monoclonal antibody associated with pH and buffer species. AAPS PharmSciTech 18(1):42–48. https://doi.org/10.1208/s12249-015-0403-0
220. Collin M, Olsén A (2001) EndoS, a novel secreted protein from Streptococcus pyogenes with endoglycosidase activity on human IgG. EMBO J 20(12):3046–3055. https://doi.org/10.1093/emboj/20.12.3046
221. Goodfellow JJ, Baruah K, Yamamoto K, Bonomelli C, Krishna B, Harvey DJ, Crispin M, Scanlan CN, Davis BG (2012) An endoglycosidase with alternative glycan specificity allows broadened glycoprotein Remodelling. J Am Chem Soc 134(19):8030–8033. https://doi.org/10.1021/ja301334b
222. Sun B, Bao W, Tian X, Li M, Liu H, Dong J, Huang W (2014) A simplified procedure for gram-scale production of sialylglycopeptide (SGP) from egg yolks and subsequent semi-synthesis of Man3GlcNAc oxazoline. Carbohydr Res 396:62–69. https://doi.org/10.1016/j.carres.2014.07.013
223. Fan SQ, Huang W, Wang LX (2012) Remarkable transglycosylation activity of glycosynthase mutants of endo-D, an endo-β-N-acetylglucosaminidase from Streptococcus pneumoniae. J Biol Chem 287(14):11272–11281. https://doi.org/10.1074/jbc.M112.340497
224. Huang W, Giddens J, Fan S-Q, Toonstra C, Wang L-X (2012) Chemoenzymatic Glycoengineering of intact IgG antibodies for gain of functions. J Am Chem Soc 134(29):12308–12318. https://doi.org/10.1021/ja3051266
225. Giddens JP, Lomino JV, DiLillo DJ, Ravetch JV, Wang L-X (2018) Site-selective chemoenzymatic glycoengineering of Fab and Fc glycans of a therapeutic antibody. Proc Natl Acad Sci 115(47):12023–12027. https://doi.org/10.1073/pnas.1812833115
226. Li T, Tong X, Yang Q, Giddens JP, Wang L-X (2016) Glycosynthase mutants of endoglycosidase S2 show potent Transglycosylation activity and remarkably relaxed substrate specificity for antibody glycosylation remodeling. J Biol Chem 291(32):16508–16518. https://doi.org/10.1074/jbc.M116.738765
227. Giddens JP, Lomino JV, Amin MN, Wang LX (2016) Endo-F3 Glycosynthase mutants enable Chemoenzymatic synthesis of Core-fucosylated Triantennary complex type Glycopeptides and glycoproteins. J Biol Chem 291(17):9356–9370. https://doi.org/10.1074/jbc.M116.721597
228. Kurogochi M, Mori M, Osumi K, Tojino M, S-i S, Takashima S, Hirose Y, Tsukimura W, Mizuno M, Amano J, Matsuda A, Tomita M, Takayanagi A, Shoda S-I, Shirai T (2015) Glycoengineered monoclonal antibodies with homogeneous glycan (M3, G0, G2, and A2) using a Chemoenzymatic approach have different affinities for FcγRIIIa and variable antibody-dependent cellular cytotoxicity activities. PLoS One 10(7):e0132848. https://doi.org/10.1371/journal.pone.0132848
229. Liu CP, Tsai TI, Cheng T, Shivatare VS, Wu CY, Wu CY, Wong CH (2018) Glycoengineering of antibody (Herceptin) through yeast expression and in vitro enzymatic glycosylation. Proc Natl Acad Sci U S A 115(4):720–725. https://doi.org/10.1073/pnas.1718172115

230. Li C, Li T, Wang L-X (2018) Chemoenzymatic Defucosylation of therapeutic antibodies for enhanced effector functions using bacterial α-Fucosidases. Methods in molecular biology (Clifton, NJ) 1827:367-380. https://doi.org/10.1007/978-1-4939-8648-4_19
231. Bennett LD, Yang Q, Berquist BR, Giddens JP, Ren Z, Kommineni V, Murray RP, White EL, Holtz BR, Wang LX, Marcel S (2018) Implementation of glycan remodeling to plant-made therapeutic antibodies. Int J Mol Sci 19(2). https://doi.org/10.3390/ijms19020421
232. Shivatare SS, Huang L-Y, Zeng Y-F, Liao J-Y, You T-H, Wang S-Y, Cheng T, Chiu C-W, Chao P, Chen L-T, Tsai T-I, Huang C-C, Wu C-Y, Lin N-H, Wong C-H (2018) Development of glycosynthases with broad glycan specificity for the efficient glyco-remodeling of antibodies. Chem Commun 54(48):6161–6164. https://doi.org/10.1039/C8CC03384F
233. Tang F, Wang L-X, Huang W (2017) Chemoenzymatic synthesis of glycoengineered IgG antibodies and glycosite-specific antibody–drug conjugates. Nat Protoc 12(8):1702–1721. https://doi.org/10.1038/nprot.2017.058
234. Thompson P, Ezeadi E, Hutchinson I, Fleming R, Bezabeh B, Lin J, Mao S, Chen C, Masterson L, Zhong H, Toader D, Howard P, Wu H, Gao C, Dimasi N (2016) Straightforward Glycoengineering approach to site-specific antibody-Pyrrolobenzodiazepine conjugates. ACS Med Chem Lett 7(11):1005–1008. https://doi.org/10.1021/acsmedchemlett.6b00278
235. Li T, DiLillo DJ, Bournazos S, Giddens JP, Ravetch JV, Wang L-X (2017) Modulating IgG effector function by Fc glycan engineering. Proc Natl Acad Sci 114(13):3485–3490. https://doi.org/10.1073/pnas.1702173114
236. Warnock D, Bai X, Autote K, Gonzales J, Kinealy K, Yan B, Qian J, Stevenson T, Zopf D, Bayer RJ (2005) In vitro galactosylation of human IgG at 1 kg scale using recombinant galactosyltransferase. Biotechnol Bioeng 92(7):831–842. https://doi.org/10.1002/bit.20658
237. Tayi VS, Butler M (2018) Solid-phase enzymatic remodeling produces high yields of single Glycoform antibodies. Biotechnol J 13(4):e1700381. https://doi.org/10.1002/biot.201700381
238. Li T, Li C, Quan DN, Bentley WE, Wang LX (2018) Site-specific immobilization of endoglycosidases for streamlined chemoenzymatic glycan remodeling of antibodies. Carbohydr Res 458-459:77–84. https://doi.org/10.1016/j.carres.2018.02.007
239. Kontoravdi C, Jimenez del Val I (2018) Computational tools for predicting and controlling the glycosylation of biopharmaceuticals. Curr Opin Chem Eng 22:89–97. https://doi.org/10.1016/j.coche.2018.08.007
240. Kotidis P, Jedrzejewski P, Sou SN, Sellick C, Polizzi K, Del Val IJ, Kontoravdi C (2019) Model-based optimization of antibody galactosylation in CHO cell culture. Biotechnol Bioeng 116(7):1612–1626. https://doi.org/10.1002/bit.26960
241. Sokolov M, Morbidelli M, Butte A, Souquet J, Broly H (2018) Sequential multivariate cell culture modeling at multiple scales supports systematic shaping of a monoclonal antibody toward a quality target. Biotechnol J 13(4):e1700461. https://doi.org/10.1002/biot.201700461
242. Zhang L, Wang M, Castan A, Stevenson J, Chatzissavidou N, Hjalmarsson H, Vilaplana F, Chotteau V (2020) Glycan residues balance analysis - GReBA: a novel model for the N-linked glycosylation of IgG produced by CHO cells. Metab Eng 57:118–128. https://doi.org/10.1016/j.ymben.2019.08.016
243. Zhang L, Schwarz H, Wang M, Castan A, Hjalmarsson H, Chotteau V (2020) Control of IgG glycosylation in CHO cell perfusion cultures by GReBA mathematical model supported by a novel targeted feed. TAFE Metab Eng. https://doi.org/10.1016/j.ymben.2020.11.004
244. Luo Y, Lovelett RJ, Price JV, Radhakrishnan D, Barnthouse K, Hu P, Schaefer E, Cunningham J, Lee KH, Shivappa RB, Ogunnaike BA (2020) Modeling the effect of amino acids and copper on monoclonal antibody productivity and glycosylation: a modular approach. Biotechnol J:e2000261. https://doi.org/10.1002/biot.202000261

245. Kotidis P, Kontoravdi C (2020) Harnessing the potential of artificial neural networks for predicting protein glycosylation. Metab Eng Commun 10:e00131. https://doi.org/10.1016/j.mec.2020.e00131
246. Spahn PN, Hansen AH, Kol S, Voldborg BG, Lewis NE (2017) Predictive glycoengineering of biosimilars using a Markov chain glycosylation model. Biotechnol J 12(2). https://doi.org/10.1002/biot.201600489
247. Chang MM, Gaidukov L, Jung G, Tseng WA, Scarcelli JJ, Cornell R, Marshall JK, Lyles JL, Sakorafas P, Chu AA, Cote K, Tzvetkova B, Dolatshahi S, Sumit M, Mulukutla BC, Lauffenburger DA, Figueroa B Jr, Summers NM, Lu TK, Weiss R (2019) Small-molecule control of antibody N-glycosylation in engineered mammalian cells. Nat Chem Biol 15(7):730–736. https://doi.org/10.1038/s41589-019-0288-4
248. Nose M, Wigzell H (1983) Biological significance of carbohydrate chains on monoclonal antibodies. Proc Natl Acad Sci U S A 80(21):6632–6636. https://doi.org/10.1073/pnas.80.21.6632
249. Leatherbarrow RJ, Rademacher TW, Dwek RA, Woof JM, Clark A, Burton DR, Richardson N, Feinstein A (1985) Effector functions of a monoclonal aglycosylated mouse IgG2a: binding and activation of complement component C1 and interaction with human monocyte Fc receptor. Mol Immunol 22(4):407–415. https://doi.org/10.1016/0161-5890(85)90125-7
250. Boyd P, Lines A, Patel AK (1995) The effect of the removal of sialic acid, galactose and total carbohydrate on the functional activity of Campath-1H. Mol Immunol 32(17–18):1311–1318
251. Yamaguchi Y, Nishimura M, Nagano M, Yagi H, Sasakawa H, Uchida K, Shitara K, Kato K (2006) Glycoform-dependent conformational alteration of the fc region of human immunoglobulin G1 as revealed by NMR spectroscopy. Biochim Biophys Acta 1760(4):693–700. https://doi.org/10.1016/j.bbagen.2005.10.002
252. Tao MH, Morrison SL (1989) Studies of aglycosylated chimeric mouse-human IgG. Role of carbohydrate in the structure and effector functions mediated by the human IgG constant region. J Immunol 143(8):2595–2601
253. Lund J, Takahashi N, Pound JD, Goodall M, Jefferis R (1996) Multiple interactions of IgG with its core oligosaccharide can modulate recognition by complement and human Fc gamma receptor I and influence the synthesis of its oligosaccharide chains. J Immunol 157(11):4963–4969
254. Dashivets T, Thomann M, Rueger P, Knaupp A, Buchner J, Schlothauer T (2015) Multi-angle effector function analysis of human monoclonal IgG Glycovariants. PLoS One 10(12):e0143520. https://doi.org/10.1371/journal.pone.0143520
255. Kanda Y, Yamada T, Mori K, Okazaki A, Inoue M, Kitajima-Miyama K, Kuni-Kamochi R, Nakano R, Yano K, Kakita S, Shitara K, Satoh M (2006) Comparison of biological activity among nonfucosylated therapeutic IgG1 antibodies with three different N-linked Fc oligosaccharides: the high-mannose, hybrid, and complex types. Glycobiology 17(1):104–118. https://doi.org/10.1093/glycob/cwl057
256. Wright A, Morrison SL (1998) Effect of C2-associated carbohydrate structure on Ig effector function: studies with chimeric mouse-human IgG1 antibodies in glycosylation mutants of Chinese hamster ovary cells. J Immunol 160(7):3393–3402
257. Peipp M, Lammerts van Bueren JJ, Schneider-Merck T, Bleeker WWK, Dechant M, Beyer T, Repp R, van Berkel PHC, Vink T, van de Winkel JGJ, Parren PWHI, Valerius T (2008) Antibody fucosylation differentially impacts cytotoxicity mediated by NK and PMN effector cells. Blood 112(6):2390–2399. https://doi.org/10.1182/blood-2008-03-144600
258. Shields RL, Lai J, Keck R, O'Connell LY, Hong K, Meng YG, Weikert SH, Presta LG (2002) Lack of fucose on human IgG1 N-linked oligosaccharide improves binding to human Fcgamma RIII and antibody-dependent cellular toxicity. J Biol Chem 277(30):26733–26740. https://doi.org/10.1074/jbc.M202069200
259. Niwa R, Natsume A, Uehara A, Wakitani M, Iida S, Uchida K, Satoh M, Shitara K (2005) IgG subclass-independent improvement of antibody-dependent cellular cytotoxicity by fucose removal from Asn297-linked oligosaccharides. J Immunol Methods 306(1–2):151–160. https://doi.org/10.1016/j.jim.2005.08.009

260. Suzuki E, Niwa R, Saji S, Muta M, Hirose M, Iida S, Shiotsu Y, Satoh M, Shitara K, Kondo M, Toi M (2007) A nonfucosylated anti-HER2 antibody augments antibody-dependent cellular cytotoxicity in breast cancer patients. Clin Cancer Res 13(6):1875–1882. https://doi.org/10.1158/1078-0432.Ccr-06-1335
261. Masuda K, Kubota T, Kaneko E, Iida S, Wakitani M, Kobayashi-Natsume Y, Kubota A, Shitara K, Nakamura K (2007) Enhanced binding affinity for FcgammaRIIIa of fucose-negative antibody is sufficient to induce maximal antibody-dependent cellular cytotoxicity. Mol Immunol 44(12):3122–3131. https://doi.org/10.1016/j.molimm.2007.02.005
262. Okazaki A, Shoji-Hosaka E, Nakamura K, Wakitani M, Uchida K, Kakita S, Tsumoto K, Kumagai I, Shitara K (2004) Fucose depletion from human IgG1 oligosaccharide enhances binding enthalpy and association rate between IgG1 and FcgammaRIIIa. J Mol Biol 336(5):1239–1249. https://doi.org/10.1016/j.jmb.2004.01.007
263. Kuhns S, Shu J, Xiang C, Guzman RD, Zhang Q, Bretzlaff W, Miscalichi N, Kalenian K, Joubert M (2020) Differential influence on antibody dependent cellular phagocytosis by different glycoforms on therapeutic monoclonal antibodies. J Biotechnol 317:5–15. https://doi.org/10.1016/j.jbiotec.2020.04.017
264. Herter S, Birk MC, Klein C, Gerdes C, Umana P, Bacac M (2014) Glycoengineering of therapeutic antibodies enhances monocyte/macrophage-mediated phagocytosis and cytotoxicity. J Immunol 192(5):2252–2260. https://doi.org/10.4049/jimmunol.1301249
265. Golay J, Da Roit F, Bologna L, Ferrara C, Leusen JH, Rambaldi A, Klein C, Introna M (2013) Glycoengineered CD20 antibody obinutuzumab activates neutrophils and mediates phagocytosis through CD16B more efficiently than rituximab. Blood 122(20):3482–3491. https://doi.org/10.1182/blood-2013-05-504043
266. Chung S, Quarmby V, Gao X, Ying Y, Lin L, Reed C, Fong C, Lau W, Qiu ZJ, Shen A, Vanderlaan M, Song A (2012) Quantitative evaluation of fucose reducing effects in a humanized antibody on Fcγ receptor binding and antibody-dependent cell-mediated cytotoxicity activities. mAbs 4(3):326–340. https://doi.org/10.4161/mabs.19941
267. Davies J, Jiang L, Pan L-Z, LaBarre MJ, Anderson D, Reff M (2001) Expression of GnTIII in a recombinant anti-CD20 CHO production cell line: expression of antibodies with altered glycoforms leads to an increase in ADCC through higher affinity for FCγRIII. Biotechnol Bioeng 74(4):288–294. https://doi.org/10.1002/bit.1119
268. Umaña P, Jean-Mairet J, Moudry R, Amstutz H, Bailey JE (1999) Engineered glycoforms of an antineuroblastoma IgG1 with optimized antibody-dependent cellular cytotoxic activity. Nat Biotechnol 17(2):176–180. https://doi.org/10.1038/6179
269. Huang L, Biolsi S, Bales KR, Kuchibhotla U (2006) Impact of variable domain glycosylation on antibody clearance: an LC/MS characterization. Anal Biochem 349(2):197–207. https://doi.org/10.1016/j.ab.2005.11.012
270. Subedi GP, Barb AW (2016) The immunoglobulin G1 N-glycan composition affects binding to each low affinity Fc γ receptor. MAbs 8(8):1512–1524. https://doi.org/10.1080/19420862.2016.1218586
271. Tsuchiya N, Endo T, Matsuta K, Yoshinoya S, Aikawa T, Kosuge E, Takeuchi F, Miyamoto T, Kobata A (1989) Effects of galactose depletion from oligosaccharide chains on immunological activities of human IgG. J Rheumatol 16(3):285–290
272. Nimmerjahn F, Anthony RM, Ravetch JV (2007) Agalactosylated IgG antibodies depend on cellular Fc receptors for in vivo activity. Proc Natl Acad Sci U S A 104(20):8433–8437. https://doi.org/10.1073/pnas.0702936104
273. Karsten CM, Pandey MK, Figge J, Kilchenstein R, Taylor PR, Rosas M, McDonald JU, Orr SJ, Berger M, Petzold D, Blanchard V, Winkler A, Hess C, Reid DM, Majoul IV, Strait RT, Harris NL, Köhl G, Wex E, Ludwig R, Zillikens D, Nimmerjahn F, Finkelman FD, Brown GD, Ehlers M, Köhl J (2012) Anti-inflammatory activity of IgG1 mediated by Fc galactosylation and association of FcγRIIB and dectin-1. Nat Med 18(9):1401–1406. https://doi.org/10.1038/nm.2862

274. Chung CH, Mirakhur B, Chan E, Le QT, Berlin J, Morse M, Murphy BA, Satinover SM, Hosen J, Mauro D, Slebos RJ, Zhou Q, Gold D, Hatley T, Hicklin DJ, Platts-Mills TA (2008) Cetuximab-induced anaphylaxis and IgE specific for galactose-alpha-1,3-galactose. N Engl J Med 358(11):1109–1117. https://doi.org/10.1056/NEJMoa074943
275. Jassal R, Jenkins N, Charlwood J, Camilleri P, Jefferis R, Lund J (2001) Sialylation of human IgG-Fc carbohydrate by transfected rat alpha2,6-sialyltransferase. Biochem Biophys Res Commun 286(2):243–249. https://doi.org/10.1006/bbrc.2001.5382
276. Kumpel BM, Rademacher TW, Rook GA, Williams PJ, Wilson IB (1994) Galactosylation of human IgG monoclonal anti-D produced by EBV-transformed B-lymphoblastoid cell lines is dependent on culture method and affects Fc receptor-mediated functional activity. Hum Antibodies Hybridomas 5(3–4):143–151
277. Mimura Y, Sondermann P, Ghirlando R, Lund J, Young SP, Goodall M, Jefferis R (2001) Role of oligosaccharide residues of IgG1-Fc in Fc gamma RIIb binding. J Biol Chem 276(49):45539–45547. https://doi.org/10.1074/jbc.M107478200
278. Quast I, Keller CW, Maurer MA, Giddens JP, Tackenberg B, Wang L-X, Münz C, Nimmerjahn F, Dalakas MC, Lünemann JD (2015) Sialylation of IgG Fc domain impairs complement-dependent cytotoxicity. J Clin Invest 125(11):4160–4170. https://doi.org/10.1172/JCI82695
279. Anthony RM, Nimmerjahn F, Ashline DJ, Reinhold VN, Paulson JC, Ravetch JV (2008) Recapitulation of IVIG anti-inflammatory activity with a recombinant IgG Fc. Science 320(5874):373–376. https://doi.org/10.1126/science.1154315
280. Samuelsson A, Towers TL, Ravetch JV (2001) Anti-inflammatory activity of IVIG mediated through the inhibitory Fc receptor. Science 291(5503):484–486. https://doi.org/10.1126/science.291.5503.484
281. Zhu A, Hurst R (2002) Anti-N-glycolylneuraminic acid antibodies identified in healthy human serum. Xenotransplantation 9(6):376–381. https://doi.org/10.1034/j.1399-3089.2002.02138.x

Next-Generation Cell Engineering Platform for Improving Recombinant Protein Production in Mammalian Cells

Sung Wook Shin, Minji Kyeong, and Jae Seong Lee

Abbreviations

ADCC	antibody-dependent cell-mediated cytotoxicity
Alb-EPO	albumin-erythropoietin
ALT1	alanine transferase 1
BAK	BCL2-antagonist/killer
BAX	BCL2-associated X protein
BCAA	branched-chain amino acid
BCAT1	branched chain aminotransferase 1
BCL-2	B-cell lymphoma 2
BCL-XL	B-cell lymphoma XL
BECN1	Beclin-1
BMP-4	bone morphogenetic protein-4
Cas9	CRISPR-associated protein 9
CERT	ceramide transfer protein
CHO	Chinese hamster ovary
CMAH	cytidine monophospho-N-acetylneuraminic acid hydroxylase
CRISPR	Clustered Regularly Interspaced Short Palindromic Repeats
CRISPRa	CRISPR activation
CRISPRi	CRISPR interference
CYLD	cylindromatosis
dCas	dead Cas

Sung Wook Shin and Minji Kyeong have contributed equally to this work.

S. W. Shin · M. Kyeong · J. S. Lee (✉)
Department of Molecular Science and Technology, Ajou University, Suwon, Republic of Korea
e-mail: jaeseonglee@ajou.ac.kr

R. Pörtner (ed.), *Cell Culture Engineering and Technology*, Cell Engineering 10,
https://doi.org/10.1007/978-3-030-79871-0_7

DHFR	dihydrofolate reductase
DSBs	DNA double-strand breaks
dsRNAs	double-stranded RNAs
DTE	difficult to express
ER	endoplasmic reticulum
FAIM	Fas apoptotic inhibitory molecule
FUT8	α-1,6-fucosyltransferase
GEMs	genome-scale models
GlcNAC	N-acetylglucosamine
GnTIII	β1,4-N-acetylglucosaminyltransferase III
GOI	gene of interest
HCPs	host cell proteins
HR	homologous recombination
HSP27	heat-shock protein 27
HSP70	heat-shock protein 70
IFN-γ	interferon-γ
IGF1	insulin-like growth factor 1
Indel	insertion/deletion
IVCC	integral of viable cell concentration
KRAB	Krüppel-associated box
mAb	monoclonal antibody
MDH II	malate dehydrogenase II
Mgat1	N-acetylglucosaminyltransferases I
Mgat4	N-acetylglucosamine transferase IV
miRNAs	microRNAs
MPC	mitochondrial pyruvate carrier
mRNA	messenger RNA
Neu5Gc	N-glycolylneuraminic acid
NHEJ	non-homologous end-joining
PDI	protein disulfide isomerase
PTMs	post-translational modifications
PYC	pyruvate carboxylase
rCHO	recombinant CHO
RISC	RNA-induced silencing complex
RMCE	recombinase-mediated cassette exchange
RNAi	RNA interference
ROS	reactive oxygen species
SAM	synergistic activation mediator
SCD1	stearoyl-CoA desaturase 1
SEAP	secreted alkaline phosphatase
SGE	stable gene expression
shRNA	small hairpin RNA
siRNA	small interfering RNA
SNAP-23	synaptosome-associated protein of 23 kDa
SNARE	soluble N-ethylmaleimide-sensitive factor receptor

SREBF1	sterol regulatory element-binding factor 1
ST6GAL1	ST6 β-galactoside -2,6-sialyltransferase 1
TALENs	transcription activator-like effector nucleases
TAUT	taurine transporter
TFs	transcription factors
TGE	transient gene expression
TI	targeted integration
TPO	Thrombopoietin
TSS	transcription start site
UPR	unfolded protein response
VAMP8	vesicle-associated membrane protein 8
VCD	viable cell density
VKD	Vitamin K-dependent
XBP1s	spliced form of X-box binding protein 1
YY1	Yin Yang1
ZFNs	zinc-finger nucleases

1 Introduction

Mammalian cells, including Chinese hamster ovary (CHO) cells, are successfully employed as a manufacturing platform to produce high-value biopharmaceuticals. Owing to their unique properties, including metabolic plasticity, ease of growth in a large-scale suspension culture, and ability to synthesize complex human-like proteins, mammalian cell lines are preferred for the production of recombinant therapeutic proteins [1, 2]. Since the first use of mammalian cells for biopharmaceutical manufacturing, vast improvements have been made in host cell line engineering, expression vector engineering, transgene amplification systems, clone selection/screening technologies, and process development, including cell culture media and optimization of culture environmental parameters over the last two to three decades [1, 3, 4]. Such technical advances result in high volumetric productivity and product titers of over 5–10 g/L in monoclonal antibody (mAb) production in the biopharmaceutical industry [4]. However, the productivity of most therapeutic proteins remains much lower than that of mAbs, and there remains a need to develop more efficient mammalian cell expression systems. Moreover, a number of highly potent multi-specific Ab formats, including bispecific antibodies and fusion proteins, have been developed as next-generation Ab products and have become more popular in the biopharmaceutical industry [5, 6]. Despite the superior clinical benefits of non-native format proteins, they present various manufacturing issues, including product instability, low expression levels, and complex purification processes, making them difficult to express (DTE) for current expression systems [5, 7]. Therefore, producing DTE proteins in a cost-effective manner will likely require substantial cell engineering.

Historical trends in biotherapeutic production reveal that bioprocess development followed by cell engineering has mainly contributed to continuous improvements in volumetric productivity [8]. The primary focus in cell engineering is to enhance per-cell yields by modulating metabolism, secretory pathways, and cell death pathways. At present, the focus has shifted from higher productivity to desired product quality control, such as *N*-glycosylation, protein charge, or amino acid sequence, while maintaining the long-term stability of cell lines at cell line development and large-scale production stages [9]. To date, most advancements in cell engineering have been achieved by engineering a single or limited number of targets that are selected based on previous studies. Engineering target genes with known functions are introduced transiently or stably through random transgene integration, depending on the purpose of the engineering and cell engineering methodologies. Recent advances in high-throughput omics data (genomic, transcriptomic, proteomic, metabolomic, lipidomic, glycomic, and epigenomic) profiling [9, 10] and the advent of genome editing tools [11, 12] provide novel opportunities for next-generation cell engineering aimed at unraveling the molecular basis of efficient recombinant protein production and systematic cell engineering in a rational manner.

In the following sections, we summarize the current cell engineering workflow while describing major steps and considerations (Fig. 1). We then detail the relevant state-of-the-art cell engineering approaches as well as the classification of engineering targets and their identification. Furthermore, we review how the cell engineering platform has been applied in CHO cells to improve the production of biopharmaceuticals. Finally, we provide practical considerations for cell engineering including the expression level, inducibility, origin, and expression method of engineering target genes. As a comprehensive summary of cell engineering efforts has been published recently [13, 14], we devote special attention to general guidelines for cell engineering and next-generation cell engineering platforms within the mammalian systems/synthetic biology framework.

2 Cell Engineering Workflow: Overview

Mammalian cell engineering aimed at optimizing the production of given proteins requires a holistic understanding of complex mammalian cell physiology, from high transcription/translation of the production target gene to high rates of protein secretion with tailor-made post-translational modifications (PTMs). A recent meta-review of cell factory engineering has provided generic engineering strategies for protein expression in eukaryotic host cells [15]. Other than approaches to optimize the expression of transgenes through expression vector engineering, some of which include promoter engineering, gene fusion, UTR engineering, codon optimization, and secretion signal engineering, cell engineering strategies can be classified into the following two categories: optimizing product titers and quality [16]. Two cell culture variables—the time integral of viable cell concentration (IVCC) and specific productivity (q_p; pg/cell/day or pcd)—contribute to the final therapeutic protein

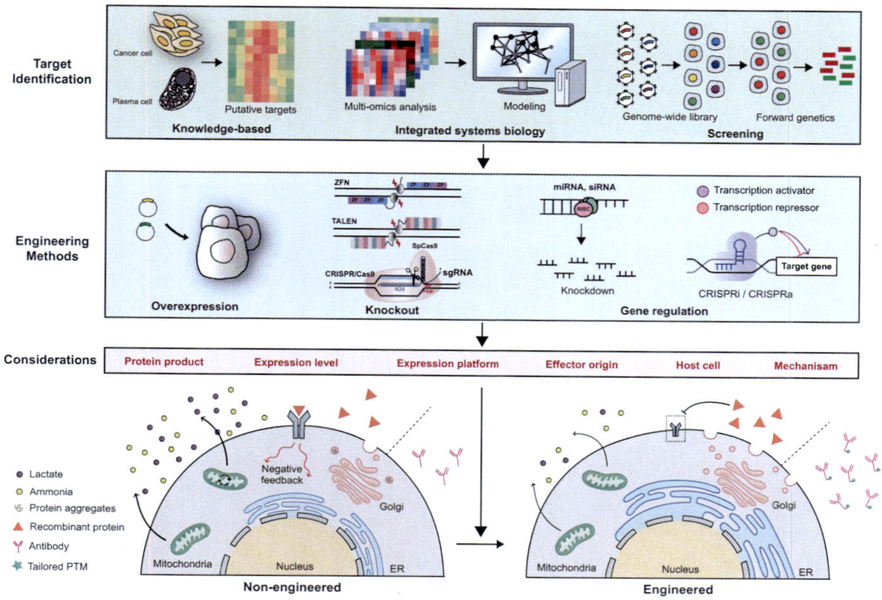

Fig. 1 Schematic workflow and strategies for mammalian cell engineering. The development of superior host cell lines can be achieved by modulating cellular pathways. Identification of engineering target gene(s) is based on three major approaches; knowledge-based, systems biology, and screening. Modulation of effector gene expression is achieved by applying appropriate engineering methods such as overexpression, knockout, knockdown, and CRISPR-mediated gene regulation. Considering context-specific engineering criteria is desirable. The phenotypic gain- or loss- of function engineering allows for the establishment of next-generation mammalian cell factories with regard to culture performance (viable cell density and culture duration), metabolism (toxic by-products), protein synthesis/secretory capacity (specific productivity, q_p), and product quality (post translational modification (PTM))

titer [16]. Improving IVCC can be achieved through two main approaches: (i) prolonging culture longevity and (ii) accelerating cell proliferation by increasing the specific growth rate (μ) and the maximum viable cell concentration. Improving q_p means that the production capacity of cells expands, leading to an increased production rate. For the optimization of protein quality, the modulation of PTM, particularly glycosylation, is required to obtain and maintain desired product quality profiles. Quality control strategies depend on the properties and applications of target proteins. Within the context of target proteins and production platforms, certain cell engineering strategies can be applied to various rate-limiting steps to overcome bottlenecks. Engineering methods and specific considerations largely rely on target cellular mechanisms and pathways, more specifically, engineering target gene candidates, which will be discussed in detail below.

3 Engineering Target Discovery: Knowledge-Based, Modeling, and Screening

For ideal mammalian cell lines for protein production, the identification and selection of proper cell engineering targets is crucial. Modifications in the expression level of target genes are expected to lead to desired phenotypic traits, such as robust growth, high viability and titer, and defined PTMs.

Basic and translational research regarding the biological functions of genes and proteins has primarily provided putative targets. For example, continuous cell lines showed metabolic flux rewiring from aerobic glycolysis for proliferation to mitochondrial oxidative phosphorylation during the stationary phase [17]. To reduce the accumulation of toxic by-products derived from glycolysis, several studies have targeted metabolic pathway genes associated with increased glycolysis that are conserved across different cell and tissue types and are frequently observed in cancer and transformed cell lines [18, 19]. Cancer cells also show increased expression of anti-apoptotic machinery to circumvent cell death [20], which has been successfully implemented for improved viability in industrial mammalian cell lines. Moreover, professional antibody-secretory plasma cells possess dramatically expanded metabolic and secretory machinery [21, 22]. Cellular mechanisms underlying B-cell differentiation into plasma cells provide a fascinating insight into engineering cellular capacity for increased protein production [23]. However, engineering knowledge-based targets has remained predominantly empirical.

Recent advances in systems biology approaches have provided insights into molecular mechanisms and phenotypic differences across host cell lines and high/low producers [8, 9]. Multiomics data have been used to identify multiple engineering targets associated with desired attributes. Depending on the type of individual omics data profiling, data analysis has provided signatures of molecules; genes, proteins, and metabolites, shown in recombinant cell lines with desired traits. Furthermore, computational modeling approaches have enabled more informative analysis of omics data by linking the genotype to the phenotype on a mechanistic basis [9, 24]. With the development of genome-scale models (GEMs) of industrial mammalian cells, including CHO, detailed secretory pathways coupled to metabolism and interaction maps of major cellular processes have been developed [25–28]. Such models allow a holistic understanding of the cellular pathways and simulation of cell growth and protein production. CHO metabolic/secretory GEMs quantitatively estimate the metabolic costs of producing therapeutic proteins and successfully predict growth rates and engineering effects in recombinant CHO (rCHO) cell lines [25, 26]. By integrating multi-omics data with GEMs, it is possible to identify more reliable engineering targets. For example, an integrative analysis of the transcriptome and CHO metabolic GEMs revealed several metabolic features observed in the antibody-producing rCHO cell line, leading to the prediction of the potential overexpression targets [9, 24].

In addition to knowledge-based modeling, forward genetic screens have been used not only to interrogate gene function but also to identify novel targets [29].

The introduction of genome-wide mutagenesis directs desired phenotypic changes under certain phenotypic selections of interest, followed by identification of the responsible genes. Traditional forward screens have relied on random mutagenesis and a laborious process to identify causality [29]. This process has enabled the establishment of useful mammalian host cell lines, such as dihydrofolate reductase (DHFR)-deficient CHO host cell lines [30], that allow efficient gene amplification, leading to higher productivity. The advent of RNA interference (RNAi) has improved the forward screens by gene knockdown rather than knockout of gene function [31]. RNAi screening has identified non-intuitive engineering targets that affect productivity in CHO cell lines [32]. Despite its useful features, the inefficient knockdown of the target gene and substantial off-target effects have led to false-positive results and poor reproducibility [33–35]. The development of Clustered Regularly Interspaced Short Palindromic Repeats (CRISPR)-based genome editing technologies has expanded the toolbox for large-scale forward screens [36, 37]. In several side-by-side comparison studies of the two methods, CRISPR screening demonstrated increased sensitivity, consistent phenotypes, and fewer off-targets compared to RNAi screening [34, 35]. Combining the results of two screening methods further improves the identification of reliable candidate hits, showing the complementary features of each technology [35]. In addition to the knockout-based loss-of-function screen, variations of CRISPR approaches, such as CRISPR interference (CRISPRi) and activation (CRISPRa), have expanded applications that allow the regulation of endogenous transcription, thus, making genome-wide gain-of-function screens feasible [33, 38]. Recently, genome-wide CRISPR knockout screening platforms have been successfully applied to CHO cell lines [39, 40]. The first CRISPR knockout screen in CHO targeted metabolic pathways under an industrially relevant selection pressure, glutamine deprivation, yielding both predictable and novel targets [39]. Rather than classical lentiviral CRISPR screening methods, the latest genome-wide virus-free CRISPR screening platform offers easy access for researchers with reduced noise for the identification of targets in protein production [40].

4 Cell Engineering Approaches: Engineering Methods

To overcome naturally occurring cellular limitations, numerous genetic engineering approaches have been attempted to alter cellular processes and mitigate bottlenecks in mammalian cells [13]. In this section, we summarize host cell engineering technologies adopted to improve general or specific attributes of CHO cell factories, including the introduction of beneficial genes, knockout or knockdown of deleterious endogenous genes, and CRISPR-mediated gene expression regulation.

4.1 Effector Gene Overexpression

Once putative genes that appeared to be advantageous for the production of recombinant proteins have been identified, one engineering strategy is the ectopic overexpression of beneficial genes, either transiently or stably. Both expression platforms offer better performance to mammalian cell lines depending on the engineering contexts, such as host cells, therapeutic proteins, and expression levels of target genes. CHO codon-optimized complementary DNA, which consists only of a protein coding sequence, can be cloned into a mammalian expression vector [41–43], and is commonly introduced into mammalian cells via either lipofection or electroporation. Upon transfection, transient gene expression (TGE) can lead to improved therapeutic protein production. However, engineering effects can vary depending on transfection method/efficiency, and a limited time period for estimating the phenotype makes stable gene expression (SGE) more common practice in mammalian cells. In case of SGE, transfected pool cells undergo antibiotic selection for approximately one to three weeks in selection media, leading to the generation of a polyclonal pool of cells where random integration of the expression vector into their chromosome has occurred. From heterogeneous mixed pool cells showing phenotypic differences due to varying levels of gene of interest (GOI), homogenous clonal cell lines are generated by various single-cell cloning methods, such as limiting dilution. Candidate clones with desired stable and robust engineered phenotypes, such as high IVCC and q_p as well as tailored PTMs, are finally screened/selected. Furthermore, expression levels can be controlled by inducible gene expression using inducible promoters within individual clones, which can exhibit *bona fide* engineering effects without clone-to-clone variation.

4.2 Gene Knockout

In addition to overexpression, the knockout of deleterious endogenous genes is a useful method to improve the general and specific features of host cells. Chemical or UV irradiation-induced random mutagenesis has traditionally been applied for nonspecific DNA knockouts. Historically, CHO-K1 host cells turn into DHFR-negative phenotypic cell lines referred to as CHO-DXB11 by random mutagenesis [44]. Since 1987, genetically engineered CHO-DXB11 cell lines have been widely used for the development of high-producing stable cell lines via the DHFR/MTX gene amplification system. In contrast, introducing target-specific loss-of-function mutations was first performed via the homologous recombination (HR) pathway. However, extremely low naturally occurring HR events limit its application in mammalian cell engineering [45–48]. Along with publicly available genome sequences of Chinese hamster and CHO cell lines since 2011 [49–52], the advancement of highly efficient genome editing technologies has provided a robust and precise engineering platform for the sequence-specific modification of the CHO genome. Such genome editing tools include zinc-finger nucleases (ZFNs), transcription activator-like effector nucleases (TALENs), and the most state-of-the-art CRISPR/CRISPR-associated protein 9 (Cas9) system [53, 54]. These nucleases introduce site-specific DNA double-strand breaks (DSBs), which trigger two main endogenous DNA repair pathways; homology-directed repair and non-homologous end-joining (NHEJ) [55]. In the case of the NHEJ pathway, discontinuous DNA ends are directly ligated, which frequently leaves scars, i.e. nucleotide insertion/deletion (indels) near DSB regions, leading to site-specific loss-of-function mutations [11]. ZFN- and CRISPR/Cas9-mediated α-1,6-fucosyltransferase (*FUT8*) gene knockout in CHO host cell lines successfully leads to the production of nonfucosylated IgG, which strengthens antibody-dependent cellular cytotoxicity (ADCC) [56, 57].

Among genome editing technologies, CRISPR/Cas9, which is derived from the bacterial immune system, is dominantly implemented in the CHO community over ZFNs and TALENs due to its design simplicity, high targeting efficiency, cost-effectiveness, and even multiplexing [12, 58]. Therefore, generating an improved CHO host cell line via the CRISPR-Cas9 and NHEJ pathways is a straightforward and novel methodology that supports rational loss-of-function engineering.

4.3 Gene Knockdown via RNA Interference

Instead of the knockout of endogenous genes, RNAi technology provides post-transcriptional gene silencing (i.e., knockdown of gene expression) and has been extensively exploited to disrupt gene expression in CHO cells [59].

Since its first discovery in *Caenorhabditis elegans* [60], the RNAi gene silencing strategy frequently utilizes small double-stranded RNAs (dsRNAs) that are either chemically synthesized small interfering RNAs (siRNAs) or plasmid-mediated small hairpin RNA (shRNA). Exogenous dsRNAs delivered into the cytoplasm are cleaved by the RNase III-like enzyme Dicer. The resulting single-stranded 20–25 bp RNA molecules loaded onto Argonaute-2 completely bind to target messenger RNA (mRNA) in conjunction with the RNA-induced silencing complex (RISC), which causes target mRNA degradation (single mRNA target) [61]. siRNA-mediated gene silencing is extensively exploited in the CHO field to improve cellular behavior by controlling various pathways involved in cell death, metabolism, secretion, glycosylation, etc. [62–67].

On the other hand, eukaryotic cells express another class of small non-coding RNAs, referred to as microRNAs (miRNAs) [68]. miRNAs are short endogenous regulatory RNAs (~22 bp) that are highly conserved among species and play an important role in epigenetic cellular regulation [69–71]. Long hairpin-structured primary miRNAs are initially transcribed in the nucleus and cleaved by the nuclear RNase III enzyme Drosha, generating a short precursor miRNA [72, 73]. Following transport into the cytoplasm, further processing of miRNA creates miRISC by a mechanism similar to that described in siRNA, leading to mRNA translational repression [74]. Due to the imperfect complementary binding property of miRNA, a single miRNA can theoretically target several hundred potential mRNAs (3′ UTR region of target mRNA (2–7 bp)), which expands its applicability in terms of mammalian cell engineering, especially for DTE production [75–79]. Given that recent transcriptome evidence implies that general host cell engineering is more likely to be fulfilled by controlling many global gene expressions, miRNA technology has recently emerged as an appealing tool for general CHO host cell engineering [80–82]. Unlike knockout approaches, such non-coding RNAs can control mRNA translation efficiency in mammalian cells, thus, the effect of the downregulation of essential genes on cellular phenotypes can be studied [83–85].

4.4 Endogenous Gene Expression Regulation via the CRISPR-dCas Toolbox

In addition to RNAi technology, more recent CRISPR-associated catalytically dead Cas (dCas) techniques; CRISPRi or CRISPRa, allow the interpretation of endogenous gene function, genome-wide screens, synthetic genetic circuits, and modulation of cellular pathways [33, 38, 86, 87]. Along with the complementary use of dCas12a (Type V), nuclease activity-null dCas9 (Type II) is currently of particular interest, as it retains only RNA-guided DNA-binding characteristics without introducing genomic DSBs [88, 89]. These site-specific DNA-binding proteins offer a robust platform for the recruitment of various effector proteins involved in epigenetic modifications, genome imaging, transcription regulation, etc. [90, 91].

Although endogenous gene repression is feasible in bacterial cells by binding dCas9 alone to promoter regions, transcriptionally repressive domain-fused dCas9 was developed for efficient CRISPRi applications in mammalian cells. The Krüppel-associated box (KRAB) domain of human Kox1 exhibits substantial transcriptional repressor activity, thus, dCas9-KRAB is widely used for targeting the transcription start site (TSS) of endogenous genes [91–95]. Recently, another class of CRISPR, Cas13d (Type VI), was exploited for gene knockdown (RNA targeting) and implicated the potential of CRISPR-Cas13d for CHO cell engineering [96].

On the other hand, CRISPRa facilitates gain-of-function engineering through transcriptional activator domains, including herpes simplex viral protein 16 (VP16), four or 10 copies of VP16 (VP64 or VP160), and nuclear factor-κB transactivation domain (p65) [93, 97–99].

Synergistic properties of CRISPRi and CRISPRa with a cluster of guide RNAs [98, 100, 101], and multi-copy effector recruitment platforms, particularly in CRISPRa, such as the SunTag system, dCas9-VPR (VP64-p65-Rta), and synergistic activation mediator (SAM) systems, have greatly widened their applications in various cell types [102–106]. Furthermore, for better applications, the careful design of guide RNAs for targeting TSS and evading nucleosome occupancy is important for high reproducibility and efficiency [105, 107, 108].

In summary, various novel engineering technologies have enabled loss- or gain-of-function engineering and the selection of appropriate strategies will guide the generation of superior CHO cell factories [109].

5 Target Classification: General vs. Specific Attributes

Appropriate selection of cell engineering targets is essential to modify the properties of cells in a desired way. Numerous cell engineering approaches have aimed to modulate a variety of cellular processes or signaling pathways. The engineering targets can be classified into two types according to the engineering purpose; general and product-specific attributes. The general attributes are primarily aimed at enhancing product titers by improving IVCC and q_p. These include fast cell growth, resistance to cell death, minimized by-product formation, and improved target protein synthesis and secretory capacity, which are useful for enhancing the general performance of cells regardless of target proteins. The product-specific attributes are highly related to target proteins, primarily involved in product quality control. Tailor-made glycosylation requires modifications in specific PTMs, and such engineering targets may not be applicable to different types of products. Rather than the product quality of target proteins, product-specific attributes need to be considered to determine whether target proteins negatively affect cell performance through the activation/inhibition of specific signaling pathways. In this section, we delineate the applications of cell engineering approaches in mammalian cells with a number of examples in CHO cells (Table 1).

5.1 Cell Proliferation and Death

The increase in protein production using mammalian cells is directly related to the number of viable cells and culture longevity of the producing cells. Therefore, the modulation of cell proliferation and cell death can be the most intuitive engineering approach, and result in the improvement of the general attributes of mammalian cells. Cell proliferation increases the number of cells and is defined by the balance between cell division and death. Usually, cell proliferation regulation arrests a specific cell cycle phase or accelerates cell division rates to avoid long lag phase. Cultivation under mild hypothermia conditions, modification of the expression levels of the proteins related to cell cycle regulation (e.g., cyclin-dependent kinases, SIPL1, and cofilin), or chemical treatment can be used to arrest the cell cycle [110–114]. The G1 phase of the cell cycle is one of the ideal phases for increasing recombinant protein production because ribosomes or translation-related genes are generally produced during this period [115, 116]. On the other hand, several studies have reported that the S phase is the most reasonable step for increasing the production of proteins, such as interferon-γ (IFN-γ) [117]. Therefore, these studies

Table 1 Mammalian cell engineering strategies for enhanced recombinant protein production

Cellular Pathway	Engineering Method	Target gene	Engineering Phenotype	Recombinant Protein Product	References
Cell proliferation and death	[a]Overexpression	Human cdkl3	Fast transition from the G0/G1 phases to the S phase	–	[111]
		Human cox15	Increased maximum viable cell numbers	–	[111]
		Human SIPL1	Improved cell proliferation and migration by PTEN inactivation that leads AKT activation	–	[113]
		Human BCL-2 and CHO Beclin-1	Extended culture longevity and higher viable cell density (VCD)	–	[128]
		Human BCL-2	Reduced apoptosis	mAb	[218]
		CHO HSP27 and HSP70	Extended culture time and 2.5-fold improvement in IFNγ titer	IFNγ	[135]
		CHO HSP27	2-fold higher VCD and mAb titer	mAb	[136]
		CHO FAIM	80% increased VCD and 2.5-fold enhanced IFNγ titer with higher apoptosis resistance	IFNγ	[137]
		CHO Atg9A	Reduced autophagy reduction with co-expression upstream autophagy regulators	–	[142]
		Rat Cyclophilin B	Reduced cell death by attenuating ER stress	–	[167]
	Knockout	BAX/BAK	Increased apoptosis resistance and five-fold increase in IgG titer	mAb	[125]
		BAX, BAK, and FUT8	Prolonged culture longevity by attenuating apoptosis	mAb	[58]
		BAX, BAK, and sialidase genes	Improved heavily sialylated glycoproteins	Recombinant human erythropoietin	[126]

		CYLD	50% improved product titer	mAb	[138]
		*CASP8AP2	2.5-fold enhanced the specific protein productivity	SEAP	[134]
		Caspase 3 and 7	Enhanced cell viability and 55% increase in thrombopoietin (TPO) titer	Human TPO	[66]
		IGF1 receptors	10-fold increase in protein production	Human IGF1	[143]
		BMP receptors	Enhanced expression level of BMP-4 up to 2.4-fold	BMP-4	[144]
	Knockdown	Cofilin	Improved specific productivity	SEAP tPA	[112]
		BAK/BAX	Enhanced apoptosis resistance during fed-batch culture	mAb	[96]
Cell metabolism	[a]Overexpression	Mouse MPC1 and MPC2	Reduced lactate production and increased titer	SEAP mAb	[147]
		Yeast PYC2	Improved mAb titer up to 5% and galactosylation up to 2.5-folds	mAb	[148]
			Increased mAb expression level up to 70% with improved product quality		[149]
		CHO ALT1	Higher IgG yield in shorter cultivation time	mAb	[154]
		CHO TAUT	Improved cell viability and IgG titer	mAb	[155]
		CHO MDH II	Enhanced IVCC with increased intracellular ATP and NADH		[156]
	Knockout	HCPs	Improved protein purity	–	[146]
		BCAT1	Enhanced cell growth and productivities in fed-batch cultures	–	[151]
		BCAT1 and BCAT2	Improved cell growth	–	[153]
Secretory capacity	[a]Overexpression	Human BiP, ATF6C, XBP1, and cyclophilin B	Increased expression of difficult-to-express (DTE) mAbs and reduced cell growth	mAb	[162]

(continued)

Table 1 (continued)

Cellular Pathway	Engineering Method	Target gene	Engineering Phenotype	Recombinant Protein Product	References
		Human SNAP-23 and VAMP8	Improved productivity	SEAP	[172]
		Human CERT	Increased specific productivity	Human serum albumin mAb	[170]
		*human Munc18b	Increased secretion	SEAP	[171]
		Human signaling receptor protein 14	Improved production of DTE proteins	mAb	[205]
		Human SLY1, Munc18c, and XBP1	Increased mAb production	mAb	[173]
		Mouse SCD1 and SREBF1	Enhanced protein yields	mAb	[174]
		CHO YY1	Increased antibody titer up to six-fold	SEAP VEGF165 mAb	[201]
		Human XBP1	Higher endoplasmic reticulum content and increased protein production	$VEGF_{121}$ SEAP	[163]
				mAb	[164]
			Increased specific protein production	TPO mAb	[165]
		CHO ATF4	Increased specific mAb production	mAb	[166]
	Knockout	PIGBOS	Modulation of cellular sensitivity toward ER stress	–	[169]
		CerS2 and Tbc1D20	Improved mAb productivity	mAb	[63]
PTM	[a]Overexpression	Human ST6GAL1	Human-like N-glycans	rhA1AT rhC1INH	[186]
		Rat GnTIII and human ManII	Nonfucosylated mAbs with complex N-glycans	mAb	[229]
		Rat GnTIII	mAb expression with increased bisecting glycan chains that lead high ADCC	mAb	[181]

		Human Mgat1 and Mgat4	Extension of antennary structures with optimal expression ratio of Mgat1:Mgat4	Alb-EPO	[183]
		Human soluble form of PACE	Increased the secretion capability with correct PTM	BMP-7	[197]
	Knockout	CMAH	80% lower hydroxylase activity	–	[187]
			Minimized immune response in human	rEPO	[188]
		MGAT1	Production of mAb with Man5	mAb	[195]
		FUT8	Enhanced ADCC with nonfucosylated mAbs	mAb	[56]
					[192]
					[193]
	Knockdown	FUT8		mAb	[190]
					[96]

[a]Effector origin is indicated in overexpression engineering
[b]Human cell line was used

suggest that the appropriate phase for the increase in protein production due to cell cycle arrest differs depending on the cell line or type of protein produced.

In addition, cell division rates can be improved, leading to an increase in cell proliferation, through the mitigation of translation bottlenecks that occur when proteins are overexpressed. mRNAs of rapidly expressed proteins during cell proliferation tend to be encoded by rare codons rather than common codons. Ribosome occupancy profiling and proteomics measurements have revealed that mRNA with rare codons is translated before mRNA with common codons during cell proliferation. An approximately 30% improvement in protein production is achieved when common codons are replaced with rare codons using this property [118].

Cell death is defined as the termination of all cellular functions, including growth, proliferation, and metabolism. Cell death is divided into unprogrammed cell death, traditionally referred to as necrosis, and programmed cell death, including apoptosis, autophagy, pyroptosis, and necroptosis, which is a new form of necrosis. Unprogrammed cell death is an accidental cell death caused by unexpected external cell injury, while programmed cell death is regulated cell death by defined molecular mechanisms that are driven by cellular stress resulting from nutrient depletion and metabolic by-product accumulation [119–121]. The cells undergoing the cell death process have problems with tardy protein synthesis rates and low protein quality. Therefore, delaying cell death by increasing the number of viable cells and culture longevity can increase protein production efficiency. Apoptosis is the most common form of cell death and is composed of two main pathways; the extrinsic (death receptor) pathway and the intrinsic (mitochondrial) pathway [122, 123]. The intrinsic pathway is activated by intracellular stress, such as metabolic or hypoxic stress, and the extrinsic pathway is induced by external stimuli, such as the binding between ligands and receptors that activates the caspase pathway [122]. The most well-known central regulator is the BCL-2 family, which makes up critical intracellular checkpoints within apoptosis [124]. The BCL-2 family consists of pro-apoptotic proteins, such as BCL2-associated X protein (BAX) and BCL2-antagonist/killer (BAK), and anti-apoptotic proteins, such as B-cell lymphoma 2 (BCL-2) and B-cell lymphoma XL (BCL-XL). Knockdown/knockout of pro-apoptotic proteins or overexpression of anti-apoptotic proteins is a common way to improve resistance to apoptosis [58–128]. In addition, caspases, which are a family of endoproteases that play an essential role in cell death, can be engineered to inhibit apoptosis [129, 130]. Caspases that control apoptosis consist of initiator caspases (caspase-2, −8, −9, and −10) and executioner caspases (caspase-3, −6, and −7). The initiator caspases comprise one or more functional domains and are activated by the intrinsic or extrinsic pathway [130, 131]. After activation of the initiator caspases, they cleave the executioner caspases that exist as procaspase dimers. Activated executioner caspases are responsible for the degradation of diverse proteins that occur during apoptosis. [129, 132, 133]. Therefore, the knockout of caspases, which are responsible for the final process of apoptosis, can increase cell longevity by delaying apoptosis [66, 134]. Other types of anti-apoptotic proteins, including apoptosis and caspase activation inhibitor, Fas apoptotic inhibitory molecule (FAIM), heat-shock protein 27 (HSP27), heat-shock protein 70 (HSP70) and deubiquitinase cylindromatosis (CYLD), have been used as engineering targets for the delayed onset of apoptotic cell death [135–138]. New types of apoptotic regulators have been discovered, such as FAM83G, which may be promising targets for more effective prevention of cell death [139].

As well as apoptosis, autophagy contributes to cell death in mammalian cells. Under stressful cell culture conditions, such as nutrient deprivation and chemical treatments, autophagy is triggered together with apoptosis [140]. Autophagy is a global catabolic process involving the degradation and recycling of cellular constituents. Although the role of autophagy in CHO cell cultures is not yet fully understood, it is considered to be mainly related to cell survival as an adaptive response to stress conditions [140, 141]. In contrast to various anti-apoptosis engineering strategies, targeting autophagy in CHO cell culture is limited. Overexpression of core autophagy-related genes, such as Beclin-1 (BECN1) and ATG9A has been attempted to modulate autophagy in CHO cells [128, 142]. Interestingly, co-overexpression of BCL-2 and BECN1 results in the simultaneous targeting of both apoptosis and autophagy, and delays cell death more efficiently than the overexpression of BCL-2 alone [128]. Given the potential of autophagy engineering for the efficient protection of mammalian cells, the management of autophagy is a promising strategy to manage cell death in mammalian cell cultures.

Rather than targeting genes directly involved in cell proliferation and cell death pathways, indirect regulation of cell growth and recombinant protein production can be achieved by controlling signaling pathways induced by target recombinant proteins. The production of insulin-like growth factor 1 (IGF1) and recombinant bone morphogenetic protein-4 (BMP-4) activates intracellular signaling pathways mediated by the IGF1 and BMP receptors, respectively, which results in growth inhibition and low product titers [143, 144]. To overcome the negative effects of product-driven signaling pathways, two studies knocked out the relevant receptors, thus, significantly increased the expression levels of IGF1 and rhBMP-4 [143, 144]. These studies emphasize the importance of protein-specific cell engineering in addition to engineering general targets in specific signaling pathways.

5.2 *Cell Metabolism*

Cellular metabolism is one of the most studied processes for high-yield production while aiming to reduce by-product formation and improve metabolic activity in mammalian cells. This process is the sum of all chemical changes that provide energy and necessary components to essential processes, including the synthesis of new molecules and the breakdown and removal of other molecules within the cell. Mammalian cells, including CHO cells, have low glucose catabolism, resulting in the accumulation of toxic by-products, such as ammonia and lactic acid, during the culture process, causing cell growth problems [145]. These problems can be alleviated by regulating either the expression level of specific transporter or process-related impurities such as host cell proteins (HCPs) [146, 147]. For example, overexpression of the subunits of the mitochondrial pyruvate carrier (MPC), which introduces pyruvate into mitochondria, reduces the lactate yield by up to 50%, and increases protein expression by up to 40% compared to the control. In addition, enhanced cell growth rates and maximum viable cell density are observed [147]. These results can also be produced by increasing the pyruvate flux to the TCA cycle through the overexpression of codon-optimized yeast cytosolic pyruvate carboxylase (PYC2) [148, 149]. Reducing the amount of lactic acid or ammonia prolongs the cell growth phase, but cell division eventually halts due to the accumulation of other by-products. This phenomenon suggests that other growth inhibitory factors are present in later stages [150, 151]. In this case, the knockdown of aminotransferases, such as branched chain aminotransferase 1 (BCAT1) is another method of preventing the accumulation of additional by-products [151, 152]. BCATs are enzymes that initiate the branched-chain amino acid (BCAA: phenylalanine, tyrosine, valine, leucine, and isoleucine) catabolic pathway. Since ammonia or lactic acid is manufactured as an intermediate or by-product from the catabolism of BCAAs, excessive catabolism of BCAAs can interfere with cell growth [151]. Therefore, the knockdown of BCATs prevents the accumulation of additional by-products and improves cell growth [151–153]. On the other hand, protein yield can be increased by overexpressing aminotransferase. Pyruvate and glutamate are formed without by-product accumulation and provide a higher mAb yield in a shorter culture period when alanine transferase 1 (ALT1) and taurine transporter (TAUT) are co-expressed in CHO cells [154]. As TAUT uptakes alanine and taurine efficiently, co-overexpression of ALT1, the enzyme that produces pyruvate and glutamate by catalyzing reversible transamination between alanine and 2-oxoglutarate, can successfully utilize by-product alanine with minimized lactate production [154, 155]. Furthermore, cells overexpressing malate dehydrogenase II (MDH II) can produce more ATP through the TCA cycle using excess malate present in the culture medium [156]. Due to the properties of carbon metabolism (TCA cycle, pentose phosphate pathway, and glycolysis), the methods described above can affect almost all cellular pathways, including cell death, glycosylation, and secretory pathways. Therefore, when engineering metabolic target genes, the interaction between each process must be considered [157]. Finally, high expression of recombinant proteins accompanies increased energy production and improved secretory capacity, which can lead to an increase in reactive oxygen species (ROS). In this case, the expression of antioxidant enzymes or transcription factors (TFs) involved in oxidative stress can be targeted to mitigate the negative effects of ROS on cell growth and protein production [158]. Recently, it has been reported that overexpression of Foxa1, a pioneering TF, protects cells from oxidative stress and increases the production of both easy- and difficult-to-express proteins [159].

5.3 Secretory Capacity

More than 25% of proteins synthesized in mammalian cells are secreted through the secretory pathway, which consists of the endoplasmic reticulum (ER), Golgi apparatus, and endomembrane system [26]. Therefore, increasing the secretion efficiency of cells is an appropriate method of increasing the yield of recombinant proteins. For high secretory capacity, cell engineering has aimed to reduce the stress generated by protein synthesis or control the expression of proteins constituting the secretory pathway to alleviate bottlenecks in the secretion process. Among the organelles involved in the secretory pathway, the ER is the most crucial for protein production as it is directly involved in protein synthesis, folding, and transport. When the ectopic expression of recombinant proteins occurs, excessive protein synthesis acts as ER stress, triggering the unfolded protein response (UPR) pathway. The UPR signaling pathway is directly involved in solving problems caused by redundant protein synthesis, including protein aggregation or misfolding. Therefore, this pathway helps increase protein production efficiency and maintain protein quality [160]. In addition, since most folding problems occur when DTE proteins are synthesized, the overexpression of proteins or TFs involved in the UPR pathway can increase DTE protein production. The mammalian UPR pathway is composed of IRE1α, PERK, and ATF6 signaling cascades. IRE1α, PERK, and ATF6, also referred to as ER stress sensors, exist in an inactive state under non-ER stress conditions through physical interactions with BiP [160, 161]. Under ER stress conditions, the ER chaperone BiP binds to unfolded and misfolded proteins, thereby releasing PERK and IRE1α, which are subsequently homodimerized, and ATF6, which is translocated to the Golgi apparatus [160]. The released ER stress sensors activate the downstream TFs; the spliced form of X-box binding protein 1 (XBP1s), eukaryotic translation initiation factor 2 subunit α, and ATF6c (ATF6p50). Since the sensor proteins and activated TFs play critical roles in the UPR, they are over-expressed to relieve ER stress and improve secretory capacity [162–166]. Examples of engineering UPR-related factors are as follows: The expression of BiP, ATF6c, or XBP1s increases the q_p of DTE mAb, and as one of the UPR transactivators, the expression of cyclophilin B increases IVCC by regulating protein folding and attenuating ER stress [162–167]. In addition to the UPR pathway markers, similar effects can be achieved with the expression of the mitochondrial microprotein (e.g., 54-amino acid microprotein PIGBOS) or mitochondrial-associated ER membrane protein that regulates ER stress [168, 169].

Apart from relieving ER stress, various genes involved in regulating the fusion of secretory vesicles to the plasma membrane can be targeted to improve secretory capacity. For instance, the heterologous expression of ceramide transfer protein (CERT), which is associated with protein kinase D-dependent protein transport from the Golgi to the plasma membrane, elevates specific productivities of mAbs [170]. Transgenic expression of Munc18b, one of the key proteins involved in the fusion of vesicles from the Golgi to the plasma membrane, increases secreted alkaline phosphatase (SEAP) production [171]. In addition, the extracellular soluble N-ethylmaleimide-sensitive factor receptor (SNARE) protein is another key protein for the vesicle fusion to the target membrane. Ectopic and stable expression of some SNARE proteins, such as the synaptosome-associated protein of 23 kDa (SNAP-23) and vesicle-associated membrane protein 8 (VAMP8), successfully enhance SEAP production [172, 173].

In addition, some genes in CHO cells encode proteins that are transcribed and translated but are not involved in cell growth or proliferation. Expressing these proteins can cause bottlenecks in secretory pathways. Therefore, knockdown of this gene using shRNA or miRNA increases the protein yield without negatively affecting cell survival [63]. Finally, there is a positive correlation between ER size and secretory capacity. For instance, Budge et al. showed that the expression of sterol regulatory element-binding factor 1 (SREBF1) and stearoyl-CoA desaturase 1 (SCD1), which are involved in lipid biosynthesis, improves lipid metabolism in CHO, thereby expanding ER size and increasing protein yield by 1.5 ~ nine fold [174].

5.4 *Post-Translational Modification*

When producing a therapeutic protein, it is essential to maintain the quality of the protein and increase the protein expression level. Although CHO is the most commonly used host cell due to its human-like PTMs, some types of human glycosylation, such as α-2,6-sialylation and α-1,3/4-fucosylation, do not occur in CHO [175, 176]. Moreover, CHO produces antigenic glycans, such as N-glycolylneuraminic acid (Neu5Gc) and galactose-α1,3-galactose, which are not present in humans [175, 177–179]. However, in small quantities, they can trigger immune responses and reduce efficacy in humans. In this respect, modulating PTMs, especially N-glycosylation, can be an excellent method of maintaining product quality. To produce fully humanized N-glycans, regulation of the expression level of various enzymes associated with PTM is utilized [50, 176, 180]. Some examples of these enzymes are described below.

β1,4-N-acetylglucosaminyltransferase III (GnTIII) catalyzes the addition of an N-acetylglucosamine (GlcNAC) residue to a bisecting position of N-linked oligosaccharide chains, being one of the representative enzymes in glycosylation pathways. Overexpression of GnTIII leads to the expression of antibodies with enhanced bisecting glycan chains, resulting in an approximately 10- to 20-fold lower antibody dosage with high ADCC [181].

Moreover, sialylated glycoproteins have a longer half-life than desialylated glycoproteins [182, 183]. In contrast to human cells, CHO does not express α-2,6-sialyltransferase and only expresses α-2,3-sialyltransferase. Consequently, CHO is limited to making glycoproteins precisely the same as humans [14, 184]. Therefore, expressing deficient sialyltransferase can enhance the bioactivity of therapeutic proteins. For instance, overexpression of ST6 β-galactoside α-2,6-sialyltransferase 1 (ST6GAL1) could lead to the production of highly sialylated IgG1 up to 70% [185].

In addition, when combining overexpression of ST6GAL1 with CRISPR/Cas9-mediated disruption of 10 genes involved in the glycosylation pathway in CHO, the engineered cells produce rhA1AT and rhC1INH with homogeneous N-linked glycans, which are similar to the structures of native human plasma A1AT and C1INH [186].

As in the above results, product quality can be improved through effector gene expression. By applying the opposite rationale, tailored glycosylation patterns can be obtained by gene knockout/knockdown. Knockout of Cmah encoding cytidine monophospho-N-acetylneuraminic acid hydroxylase (CMAH) inhibits the conversion of Neu5Ac to Neu5Gc in CHO cells, resulting in the production of fewer recombinant proteins containing antigenic Neu5Gc residues [187, 188]. Furthermore, knockout/knockdown of *FUT8*, which is responsible for attaching core fucose to N-glycans, allows the production of non-fucosylated mAbs with enhanced ADCC, even at lower antibody doses as described above [56, 189–194].

Even if engineering is carried out with the same target gene, it is necessary to selectively apply an appropriate engineering approach according to the product type. For example, vaccines should be easily detected by receptors on antigen-presenting cells in the body, while therapeutics should not trigger an immune response. Therefore, N-acetylglucosaminyltransferases I (Mgat1) is an example of a target whose engineering method should be different depending on the product type. In albumin-erythropoietin (Alb-EPO)-expressing cell lines, the co-expression of N-acetylglucosamine transferase IV (Mgat4) and Mgat1 catalyze the production of highly sialylated Alb-EPO with an enhanced half-life, avoiding immune responses in vivo [183]. On the other hand, Mgat1 knockout cells manufacture simple structure glycoproteins with terminal mannose residues. Glycoproteins with a terminal mannose residue are rapidly detected by the mannose receptor, and this property helps improve the efficacy of vaccines [195].

In addition to improving the protein quality, the modulation of PTMs can be utilized to improve product yields. Insufficient unique PTMs, such as the cleavage of propeptides of recombinant proteins and γ-carboxylation, pose a challenge to achieving high productivity of some types of recombinant proteins. Vitamin K-dependent (VKD) clotting factors undergo a unique PTM known as γ-glutamyl carboxylation. To achieve high productivities of VKD proteins while maintaining high specific activity, the cellular γ-carboxylation capacity has been improved by extensive cell engineering [196]. Additionally, for the efficient production of propeptide-containing recombinant proteins, another unique PTM, cleavage of propeptides, is required to debottleneck in mammalian cells. The ectopic over-expression of the cleaving enzyme PACE, a ubiquitously expressed mammalian proprotein convertase, increased the secretion capability and percentage of correctly processed recombinant human BMP-7 in CHO cells [197], which supports the importance of product-specific cell engineering targeting PTMs.

6 Considerations

Many engineering studies have altered cellular pathways and demonstrated that improved characteristics of recombinant mammalian cells can be accomplished by regulating engineering targets [13]. However, despite reported positive effects, the modulation of effector gene(s) often results in inconsistent results (no or negative effects). These contradictory results may be partly explained by several reasons and engineers should be mindful of the fact that cellular engineering outcomes are context dependent [198]. In the following section, we describe practical considerations when it comes to modifying the target gene(s) to provide general guidelines for mammalian cell engineering.

6.1 Expression Level

First, the outcome of gain-of-function engineering of the host cell line is determined not only by the effector gene expression but also its expression level [198–200] . For instance, the overexpression of Yin Yang1 (YY1) TF, a polycomb group family protein, led to the maximized product titer of different biomolecules at the optimal YY1 expression level [201]. Furthermore, UPR-related TFs, including XBP1s and ATF6c, boosted the volumetric productivity of Fc-fusion proteins only at lower expression levels [202]. Such TF engineering simultaneously modifies the expression level of several downstream genes, which is regarded as a general effector with less product specificity. The general effectors seem to be more attractive targets for the global reprogramming of cellular pathways than the co-expression of limited number of effector genes.

To increase secretory capacity, the overexpression of protein disulfide isomerase (PDI), an ER-resident molecular chaperone, is an interesting engineering approach. However, the modulation of a single effector protein often results in inconsistent effects. One possible explanation for these results is the non-optimal expression level of PDI. As the engineering effects of PDIs are highly product-specific depending on whether/where post-transcriptional bottlenecks are present, elaborate engineering efforts involving the systematic comparison of expression levels are required to obtain true-positive results by single (or preferentially multiple) effector gene(s) [203–207].

6.2 Expression Platform

Effector protein expression platforms are another criterion. The precise control of expression levels can be performed by dose-dependent plasmid transfection in TGE [206] or by different promoters (in terms of strength and inducibility) in

TGE/SGE [200, 208]. TGE of effector proteins generates polyclonal cell pools, which represent the average engineering phenotype. Although TGE is useful for engineering target screening, potent engineering targets can be masked in TGE engineering. Therefore, engineering effects should be validated using SGE in clonal cell lines.

Compared to constitutive expression, inducible mammalian gene expression systems, including the tetracycline system and cumate gene-switches, is an ideal expression platform for the production of toxic products as well as for discriminating growth and production phases of the cell culture [209–212]. The main advantages of inducible systems are temporal and reversible transgene expression by external or intrinsic inducers within guaranteed clonality [213]. These characteristics allow dose-dependent as well as dynamic expression control, which is more likely to reduce false-positive results.

Recently, apart from traditional random integration, there has been growing emphasis on the application of CRISPR/Cas9- and recombinase-mediated cassette exchange (RMCE)-mediated site-specific integration of transgenes at predetermined genomic loci. This system possesses several advantages in terms of predictable and reproducible expression control [214–217]. The stable expression of engineering target genes via the site-specific integration is limited, except for dynamic expression control of BCL-2 by CRISPR/Cas9-mediated gene tagging method at the endogenous P21 locus [218]. Dual-RMCE and CRISPR/Cas9-mediated double targeted integration (TI) is feasible in CHO cells [219, 220], thus, TI-based expression may represent advanced cell engineering platforms that enable the elaborate expression level control of multiple targets in CHO cell engineering [207].

6.3 *Origin of Effector Genes*

Expressed effectors interact with host biomolecules (e.g., TFs for cofactors/DNA and noncoding RNAs for mRNA) within a given host environment. Therefore, engineering effects may differ with the origin of effector genes, depending on the molecular mechanism. Both human- and CHO-derived PDIs and XBP1s have been successfully utilized to improve q_p of mAbs in CHO cells [162, 164, 165, 204, 221–223], although human orthologs are more often reported. However, the overexpression of human YY1 TF abolishes engineering effects, whereas CHO-derived YY1 exclusively improves the product titer of different biomolecules in CHO cell lines [201]. This result implies the significance of host-specific interactions of effector proteins, thus, considering the gene origin is often desirable depending on the function of engineering targets. General effector genes, such as TFs and miRNAs are more likely to interact with host biomolecules, thus, to fit natural interactions, the autologous effector origin (CHO) would be a better choice than orthologous (human/mouse) ones in CHO cell engineering [13].

Host environments may differ even in CHO *quasispecies* [224], referring to diverse CHO host cell lines from the perspective of genotypic and phenotypic differences, such as specific growth rate, ER size, and mitochondrial mass, probably due to copy-number variations, karyotypes, and single nucleotide polymorphisms [49, 198, 225–227]. Although the direct relationship has not yet been systematically explored, it is noteworthy that engineering impacts on different host cell may differ depending on the origin of the genes.

7 Conclusions

Since the great success of the mammalian cell-based expression of recombinant therapeutic proteins, mostly mAbs, the biopharmaceutical industry is facing the next challenges that should be resolved to continue the current success story. The recent emergence of various DTE proteins and remarkably accelerated approvals of biopharmaceuticals, including biosimilar products, while accompanying the trend of continuously increasing the use of mammalian over nonmammalian expression systems [2] pose both challenges and opportunities for cost-effective biopharmaceutical development and production. Furthermore, more important than the manufacturing capacity and cost of production is to reduce the timeline from product development to clinic and marker. As the era of personalized medicine approaches, a platform is needed that rapidly produces small amounts of various therapeutic proteins. In the context of outbreaks of pandemic infectious diseases that have become increasingly common in recent decades, a robust and advanced production platform will be directly translated to life-saving therapies and maximal patient benefits [228]. Next-generation cell engineering technologies, such as precise genome editing and gene expression regulation technologies, could provide appropriate solutions by substantially improving the current industrial mammalian cell lines in a more productive and efficient direction. Integrating emerging trends in mammalian systems and synthetic biology will further facilitate the development of next-generation mammalian cell factories, promising the successful supply of biopharmaceuticals in the upcoming era.

Acknowledgments This work was supported by the NRF funded by the Korean government (2021R1A2C4002733 and 2019R1A6A1A11051471). The authors declare no competing financial interests. Neither ethical approval nor informed consent was required for this study.

References

1. Ha TK, Lee JS, Lee GM (2019) Platform Technology for Therapeutic Protein Production. Cell Culture Engineering: Recombinant Protein Production:1–22
2. Walsh G (2018) Biopharmaceutical benchmarks 2018. Nat Biotechnol 36(12):1136–1145

3. Lai T, Yang Y, Ng SK (2013) Advances in mammalian cell line development technologies for recombinant protein production. Pharmaceuticals 6(5):579–603
4. Li F, Vijayasankaran N, Shen AY, Kiss R, Amanullah A (2010) Cell culture processes for monoclonal antibody production. MAbs 2(5):466–479. https://doi.org/10.4161/mabs.2.5.12720
5. Wang Q, Chen Y, Park J, Liu X, Hu Y, Wang T, McFarland K, Betenbaugh MJ (2019) Design and production of bispecific antibodies. Antibodies 8(3):43
6. Spiess C, Zhai Q, Carter PJ (2015) Alternative molecular formats and therapeutic applications for bispecific antibodies. Mol Immunol 67(2):95–106
7. Mathias S, Fischer S, Handrick R, Fieder J, Schulz P, Bradl H, Gorr I, Gamer M, Otte K (2018) Visualisation of intracellular production bottlenecks in suspension-adapted CHO cells producing complex biopharmaceuticals using fluorescence microscopy. J Biotechnol 271:47–55
8. Kuo C-C, Chiang AW, Shamie I, Samoudi M, Gutierrez JM, Lewis NE (2018) The emerging role of systems biology for engineering protein production in CHO cells. Curr Opin Biotechnol 51:64–69
9. Hong JK, Lakshmanan M, Goudar C, Lee D-Y (2018) Towards next generation CHO cell line development and engineering by systems approaches. Curr Opin Chem Eng 22:1–10
10. Kildegaard HF, Baycin-Hizal D, Lewis NE, Betenbaugh MJ (2013) The emerging CHO systems biology era: harnessing the 'omics revolution for biotechnology. Curr Opin Biotechnol 24(6):1102–1107
11. Kim H, Kim J-S (2014) A guide to genome engineering with programmable nucleases. Nat Rev Genet 15(5):321–334
12. Lee JS, Grav LM, Lewis NE, Faustrup Kildegaard H (2015) CRISPR/Cas9-mediated genome engineering of CHO cell factories: application and perspectives. Biotechnol J 10(7):979–994
13. Fischer S, Handrick R, Otte K (2015) The art of CHO cell engineering: a comprehensive retrospect and future perspectives. Biotechnol Adv 33(8):1878–1896
14. Fischer S, Otte K (2019) CHO cell engineering for improved process performance and product quality. Cell Culture Engineering: Recombinant Protein Production:207–250
15. Davy AM, Kildegaard HF, Andersen MR (2017) Cell factory engineering. Cell Syst 4(3):262–275
16. Baek E, Kim CL, Park JH, Lee GM (2015) Cell engineering for therapeutic protein production. In: Animal cell culture. Springer, pp. 565–590
17. Young JD (2013) Metabolic flux rewiring in mammalian cell cultures. Curr Opin Biotechnol 24(6):1108–1115
18. Mulukutla BC, Khan S, Lange A, Hu W-S (2010) Glucose metabolism in mammalian cell culture: new insights for tweaking vintage pathways. Trends Biotechnol 28(9):476–484
19. Dang CV, Semenza GL (1999) Oncogenic alterations of metabolism. Trends Biochem Sci 24(2):68–72
20. Labi V, Erlacher M (2015) How cell death shapes cancer. Cell Death Dis 6(3):e1675–e1675
21. Shaffer AL, Shapiro-Shelef M, Iwakoshi NN, Lee A-H, Qian S-B, Zhao H, Yu X, Yang L, Tan BK, Rosenwald A, Hurt EM, Petroulakis E, Sonenberg N, Yewdell JW, Calame K, Glimcher LH, Staudt LM (2004) XBP1, downstream of Blimp-1, expands the secretory apparatus and other organelles, and increases protein synthesis in plasma cell differentiation. Immunity 21(1):81–93. https://doi.org/10.1016/j.immuni.2004.06.010
22. Lam WY, Bhattacharya D (2018) Metabolic links between plasma cell survival, secretion, and stress. Trends Immunol 39(1):19–27
23. Dinnis DM, James DC (2005) Engineering mammalian cell factories for improved recombinant monoclonal antibody production: lessons from nature? Biotechnol Bioeng 91(2):180–189
24. Yusufi FNK, Lakshmanan M, Ho YS, Loo BLW, Ariyaratne P, Yang Y, Ng SK, Tan TRM, Yeo HC, Lim HL, Ng SW, Hiu AP, Chow CP, Wan C, Chen S, Teo G, Song G, Chin JX, Ruan X, Sung KWK, Hu W-S, Yap MGS, Bardor M, Nagarajan N, Lee D-Y (2017) Mammalian systems biotechnology reveals global cellular adaptations in a recombinant CHO cell line. Cell Syst 4(5):530–542. https://doi.org/10.1016/j.cels.2017.04.009

25. Hefzi H, Ang KS, Hanscho M, Bordbar A, Ruckerbauer D, Lakshmanan M, Orellana CA, Baycin-Hizal D, Huang Y, Ley D (2016) A consensus genome-scale reconstruction of Chinese hamster ovary cell metabolism. Cell Syst 3(5):434–443
26. Gutierrez JM, Feizi A, Li S, Kallehauge TB, Hefzi H, Grav LM, Ley D, Hizal DB, Betenbaugh MJ, Voldborg B (2020) Genome-scale reconstructions of the mammalian secretory pathway predict metabolic costs and limitations of protein secretion. Nat Commun 11(1):1–10
27. Quek L-E, Dietmair S, Hanscho M, Martínez VS, Borth N, Nielsen LK (2014) Reducing Recon 2 for steady-state flux analysis of HEK cell culture. J Biotechnol 184:172–178
28. Lund AM, Kaas CS, Brandl J, Pedersen LE, Kildegaard HF, Kristensen C, Andersen MR (2017) Network reconstruction of the mouse secretory pathway applied on CHO cell transcriptome data. BMC Syst Biol 11(1):37
29. Grimm S (2004) The art and design of genetic screens: mammalian culture cells. Nat Rev Genet 5(3):179–189
30. Urlaub G, Käs E, Carothers AM, Chasin LA (1983) Deletion of the diploid dihydrofolate reductase locus from cultured mammalian cells. Cell 33(2):405–412
31. Boutros M, Ahringer J (2008) The art and design of genetic screens: RNA interference. Nat Rev Genet 9(7):554–566
32. Klanert G, Fernandez DJ, Weinguny M, Eisenhut P, Bühler E, Melcher M, Titus SA, Diendorfer AB, Gludovacz E, Jadhav V (2019) A cross-species whole genome siRNA screen in suspension-cultured Chinese hamster ovary cells identifies novel engineering targets. Sci Rep 9(1):1–11
33. Schuster A, Erasimus H, Fritah S, Nazarov PV, van Dyck E, Niclou SP, Golebiewska A (2019) RNAi/CRISPR screens: from a pool to a valid hit. Trends Biotechnol 37(1):38–55
34. Evers B, Jastrzebski K, Heijmans JP, Grernrum W, Beijersbergen RL, Bernards R (2016) CRISPR knockout screening outperforms shRNA and CRISPRi in identifying essential genes. Nat Biotechnol 34(6):631–633
35. Morgens DW, Deans RM, Li A, Bassik MC (2016) Systematic comparison of CRISPR/Cas9 and RNAi screens for essential genes. Nat Biotechnol 34(6):634–636
36. Shalem O, Sanjana NE, Hartenian E, Shi X, Scott DA, Mikkelsen TS, Heckl D, Ebert BL, Root DE, Doench JG (2014) Genome-scale CRISPR-Cas9 knockout screening in human cells. Science 343(6166):84–87
37. Wang T, Wei JJ, Sabatini DM, Lander ES (2014) Genetic screens in human cells using the CRISPR-Cas9 system. Science 343(6166):80–84
38. Gilbert LA, Horlbeck MA, Adamson B, Villalta JE, Chen Y, Whitehead EH, Guimaraes C, Panning B, Ploegh HL, Bassik MC (2014) Genome-scale CRISPR-mediated control of gene repression and activation. Cell 159(3):647–661
39. Karottki KJC, Hefzi H, Li S, Pedersen LE, Spahn PN, Joshi C, Ruckerbauer D, Bort JAH, Thomas A, Lee JS, Borth N, Lee GM, Kildegaard HF, Lewis NE (2021) A metabolic CRISPR-Cas9 screen in Chinese hamster ovary cells identifies glutamine-sensitive genes. Metab Eng 66:114–122
40. Xiong K, Karottki KJC, Hefzi H, Li S, Grav LM, Li S, Spahn P, Lee JS, Ventina I, Lee GM, Lewis NE, Pedersen LE (2021) An optimized genome-wide, virus-free CRISPR screen for mammalian cells. Cell Rep Methods 1(4):100062
41. Ou K-C, Wang C-Y, Liu K-T, Chen Y-L, Chen Y-C, Lai M-D, Yen M-C (2014) Optimization protein productivity of human interleukin-2 through codon usage, gene copy number and intracellular tRNA concentration in CHO cells. Biochem Biophys Res Commun 454(2):347–352
42. Chung BK-S, Yusufi FN, Yang Y, Lee D-Y (2013) Enhanced expression of codon optimized interferon gamma in CHO cells. J Biotechnol 167(3):326–333
43. Magistrelli G, Poitevin Y, Schlosser F, Pontini G, Malinge P, Josserand S, Corbier M, Fischer N (2017) Optimizing assembly and production of native bispecific antibodies by codon de-optimization. In: MAbs. vol 2. Taylor & Francis, pp 231–239
44. Urlaub G, Chasin LA (1980) Isolation of Chinese hamster cell mutants deficient in dihydrofolate reductase activity. Proc Natl Acad Sci 77(7):4216–4220

45. Krämer O, Klausing S, Noll T (2010) Methods in mammalian cell line engineering: from random mutagenesis to sequence-specific approaches. Appl Microbiol Biotechnol 88(2):425–436
46. Durai S, Mani M, Kandavelou K, Wu J, Porteus MH, Chandrasegaran S (2005) Zinc finger nucleases: custom-designed molecular scissors for genome engineering of plant and mammalian cells. Nucleic Acids Res 33(18):5978–5990
47. Kandavelou K, Ramalingam S, London V, Mani M, Wu J, Alexeev V, Civin CI, Chandrasegaran S (2009) Targeted manipulation of mammalian genomes using designed zinc finger nucleases. Biochem Biophys Res Commun 388(1):56–61
48. Vasquez KM, Marburger K, Intody Z, Wilson JH (2001) Manipulating the mammalian genome by homologous recombination. Proc Natl Acad Sci 98(15):8403–8410
49. Lewis NE, Liu X, Li Y, Nagarajan H, Yerganian G, O'brien E, Bordbar A, Roth AM, Rosenbloom J, Bian C (2013) Genomic landscapes of Chinese hamster ovary cell lines as revealed by the Cricetulus griseus draft genome. Nat Biotechnol 31(8):759–765
50. Xu X, Nagarajan H, Lewis NE, Pan S, Cai Z, Liu X, Chen W, Xie M, Wang W, Hammond S (2011) The genomic sequence of the Chinese hamster ovary (CHO)-K1 cell line. Nat Biotechnol 29(8):735
51. Brinkrolf K, Rupp O, Laux H, Kollin F, Ernst W, Linke B, Kofler R, Romand S, Hesse F, Budach WE (2013) Chinese hamster genome sequenced from sorted chromosomes. Nat Biotechnol 31(8):694–695
52. Kremkow BG, Baik JY, MacDonald ML, Lee KH (2015) CHOgenome. Org 2.0: genome resources and website updates. Biotechnol J 10(7):931–938
53. Gaj T, Gersbach CA, Barbas CF III (2013) ZFN, TALEN, and CRISPR/Cas-based methods for genome engineering. Trends Biotechnol 31(7):397–405
54. Gupta SK, Shukla P (2017) Gene editing for cell engineering: trends and applications. Crit Rev Biotechnol 37(5):672–684
55. Lieber MR (2010) The mechanism of double-strand DNA break repair by the nonhomologous DNA end-joining pathway. Annu Rev Biochem 79:181–211
56. Malphettes L, Freyvert Y, Chang J, Liu PQ, Chan E, Miller JC, Zhou Z, Nguyen T, Tsai C, Snowden AW (2010) Highly efficient deletion of FUT8 in CHO cell lines using zinc-finger nucleases yields cells that produce completely nonfucosylated antibodies. Biotechnol Bioeng 106(5):774–783
57. Zong H, Han L, Ding K, Wang J, Sun T, Zhang X, Cagliero C, Jiang H, Xie Y, Xu J (2017) Producing defucosylated antibodies with enhanced in vitro antibody-dependent cellular cytotoxicity via FUT8 knockout CHO-S cells. Eng Life Sci 17(7):801–808
58. Grav LM, Lee JS, Gerling S, Kallehauge TB, Hansen AH, Kol S, Lee GM, Pedersen LE, Kildegaard HF (2015) One-step generation of triple knockout CHO cell lines using CRISPR/Cas9 and fluorescent enrichment. Biotechnol J 10(9):1446–1456
59. Wu S-C (2009) RNA interference technology to improve recombinant protein production in Chinese hamster ovary cells. Biotechnol Adv 27(4):417–422
60. Fire A, Xu S, Montgomery MK, Kostas SA, Driver SE, Mello CC (1998) Potent and specific genetic interference by double-stranded RNA in Caenorhabditis elegans. Nature 391(6669):806–811
61. Bagasra O, Prilliman KR (2004) RNA interference: the molecular immune system. J Mol Histol 35(6):545–553
62. Kallehauge TB, Li S, Pedersen LE, Ha TK, Ley D, Andersen MR, Kildegaard HF, Lee GM, Lewis NE (2017) Ribosome profiling-guided depletion of an mRNA increases cell growth rate and protein secretion. Sci Rep 7(1):1–12
63. Pieper LA, Strotbek M, Wenger T, Gamer M, Olayioye MA, Hausser A (2017) Secretory pathway optimization of CHO producer cells by co-engineering of the mitosRNA-1978 target genes CerS2 and Tbc1D20. Metab Eng 40:69–79
64. Zhou M, Crawford Y, Ng D, Tung J, Pynn AF, Meier A, Yuk IH, Vijayasankaran N, Leach K, Joly J (2011) Decreasing lactate level and increasing antibody production in Chinese hamster ovary cells (CHO) by reducing the expression of lactate dehydrogenase and pyruvate dehydrogenase kinases. J Biotechnol 153(1–2):27–34

65. Kim SH, Lee GM (2007) Down-regulation of lactate dehydrogenase-A by siRNAs for reduced lactic acid formation of Chinese hamster ovary cells producing thrombopoietin. Appl Microbiol Biotechnol 74(1):152–159
66. Sung YH, Lee JS, Park SH, Koo J, Lee GM (2007) Influence of co-down-regulation of caspase-3 and caspase-7 by siRNAs on sodium butyrate-induced apoptotic cell death of Chinese hamster ovary cells producing thrombopoietin. Metab Eng 9(5–6):452–464
67. Zhang M, Koskie K, Ross JS, Kayser KJ, Caple MV (2010) Enhancing glycoprotein sialylation by targeted gene silencing in mammalian cells. Biotechnol Bioeng 105(6):1094–1105
68. Pillai RS, Bhattacharyya SN, Filipowicz W (2007) Repression of protein synthesis by miRNAs: how many mechanisms? Trends Cell Biol 17(3):118–126
69. Bartel DP (2018) Metazoan micrornas. Cell 173(1):20–51
70. Ibáñez-Ventoso C, Vora M, Driscoll M (2008) Sequence relationships among C. elegans, D. melanogaster and human microRNAs highlight the extensive conservation of microRNAs in biology. PLoS One 3(7):e2818
71. Pasquinelli AE (2012) MicroRNAs and their targets: recognition, regulation and an emerging reciprocal relationship. Nat Rev Genet 13(4):271–282
72. Denli AM, Tops BB, Plasterk RH, Ketting RF, Hannon GJ (2004) Processing of primary microRNAs by the microprocessor complex. Nature 432(7014):231–235
73. Gregory RI, K-p Y, Amuthan G, Chendrimada T, Doratotaj B, Cooch N, Shiekhattar R (2004) The microprocessor complex mediates the genesis of microRNAs. Nature 432(7014):235–240
74. Ha M, Kim VN (2014) Regulation of microRNA biogenesis. Nat Rev Mol Cell Biol 15(8):509–524
75. Barron N, Sanchez N, Kelly P, Clynes M (2011) MicroRNAs: tiny targets for engineering CHO cell phenotypes? Biotechnol Lett 33(1):11–21
76. Hackl M, Borth N, Grillari J (2012) miRNAs–pathway engineering of CHO cell factories that avoids translational burdening. Trends Biotechnol 30(8):405–406
77. Fischer S, Marquart KF, Pieper LA, Fieder J, Gamer M, Gorr I, Schulz P, Bradl H (2017) miRNA engineering of CHO cells facilitates production of difficult-to-express proteins and increases success in cell line development. Biotechnol Bioeng 114(7):1495–1510
78. Klanert G, Jadhav V, Shanmukam V, Diendorfer A, Karbiener M, Scheideler M, Bort JH, Grillari J, Hackl M, Borth N (2016) A signature of 12 microRNAs is robustly associated with growth rate in a variety of CHO cell lines. J Biotechnol 235:150–161
79. Schoellhorn M, Fischer S, Wagner A, Handrick R, Otte K (2017) miR-143 targets MAPK7 in CHO cells and induces a hyperproductive phenotype to enhance production of difficult-to-express proteins. Biotechnol Prog 33(4):1046–1058
80. Valdés-Bango Curell R, Barron N (2018) Exploring the potential application of short non-coding RNA-based genetic circuits in Chinese hamster ovary cells. Biotechnol J 13(10):1700220
81. Fischer S, Handrick R, Aschrafi A, Otte K (2015) Unveiling the principle of microRNA-mediated redundancy in cellular pathway regulation. RNA Biol 12(3):238–247
82. Müller D, Katinger H, Grillari J (2008) MicroRNAs as targets for engineering of CHO cell factories. Trends Biotechnol 26(7):359–365
83. Vito D, Smales CM (2018) The long non-coding RNA transcriptome landscape in CHO cells under batch and fed-batch conditions. Biotechnol J 13(10):1800122
84. Vito D, Smales CM (2018) Engineering of the cellular translational machinery and non-coding RNAs to enhance CHO cell growth, recombinant product yields and quality. Curr Opin Chem Eng 22:199–208
85. Stiefel F, Fischer S, Hackl M, Handrick R, Hesse F, Borth N, Otte K, Grillari J (2015) Noncoding RNAs, post-transcriptional RNA operons and Chinese hamster ovary cells. Pharm Bioprocess 3(3):227–247
86. Dominguez AA, Lim WA, Qi LS (2016) Beyond editing: repurposing CRISPR–Cas9 for precision genome regulation and interrogation. Nat Rev Mol Cell Biol 17(1):5

87. Nakamura M, Srinivasan P, Chavez M, Carter MA, Dominguez AA, La Russa M, Lau MB, Abbott TR, Xu X, Zhao D (2019) Anti-CRISPR-mediated control of gene editing and synthetic circuits in eukaryotic cells. Nat Commun 10(1):1–11
88. Jinek M, Chylinski K, Fonfara I, Hauer M, Doudna JA, Charpentier E (2012) A programmable dual-RNA–guided DNA endonuclease in adaptive bacterial immunity. Science 337(6096):816–821
89. Gasiunas G, Barrangou R, Horvath P, Siksnys V (2012) Cas9–crRNA ribonucleoprotein complex mediates specific DNA cleavage for adaptive immunity in bacteria. Proc Natl Acad Sci 109(39):E2579–E2586
90. Xu X, Qi LS (2019) A CRISPR–dCas toolbox for genetic engineering and synthetic biology. J Mol Biol 431(1):34–47
91. Pickar-Oliver A, Gersbach CA (2019) The next generation of CRISPR–Cas technologies and applications. Nat Rev Mol Cell Biol 20(8):490–507
92. Qi LS, Larson MH, Gilbert LA, Doudna JA, Weissman JS, Arkin AP, Lim WA (2013) Repurposing CRISPR as an RNA-guided platform for sequence-specific control of gene expression. Cell 152(5):1173–1183
93. Gilbert LA, Larson MH, Morsut L, Liu Z, Brar GA, Torres SE, Stern-Ginossar N, Brandman O, Whitehead EH, Doudna JA (2013) CRISPR-mediated modular RNA-guided regulation of transcription in eukaryotes. Cell 154(2):442–451
94. Gao X, Tsang JC, Gaba F, Wu D, Lu L, Liu P (2014) Comparison of TALE designer transcription factors and the CRISPR/dCas9 in regulation of gene expression by targeting enhancers. Nucleic Acids Res 42(20):e155–e155
95. Shen C-C, Sung L-Y, Lin S-Y, Lin M-W, Hu Y-C (2017) Enhancing protein production yield from chinese hamster ovary cells by CRISPR interference. ACS Synth Biol 6(8):1509–1519
96. Shen C-C, Lin M-W, Nguyen BKT, Chang C-W, Shih J-R, Nguyen MTT, Chang Y-H, Hu Y-C (2020) CRISPR-Cas13d for gene knockdown and engineering of CHO cells. ACS Synthetic Biology
97. Perez-Pinera P, Kocak DD, Vockley CM, Adler AF, Kabadi AM, Polstein LR, Thakore PI, Glass KA, Ousterout DG, Leong KW (2013) RNA-guided gene activation by CRISPR-Cas9–based transcription factors. Nat Methods 10(10):973–976
98. Maeder ML, Linder SJ, Cascio VM, Fu Y, Ho QH, Joung JK (2013) CRISPR RNA–guided activation of endogenous human genes. Nat Methods 10(10):977–979
99. Farzadfard F, Perli SD, Lu TK (2013) Tunable and multifunctional eukaryotic transcription factors based on CRISPR/Cas. ACS Synth Biol 2(10):604–613
100. Kleinjan DA, Wardrope C, Sou SN, Rosser SJ (2017) Drug-tunable multidimensional synthetic gene control using inducible degron-tagged dCas9 effectors. Nat Commun 8(1):1–9
101. Cheng AW, Wang H, Yang H, Shi L, Katz Y, Theunissen TW, Rangarajan S, Shivalila CS, Dadon DB, Jaenisch R (2013) Multiplexed activation of endogenous genes by CRISPR-on, an RNA-guided transcriptional activator system. Cell Res 23(10):1163–1171
102. Tanenbaum ME, Gilbert LA, Qi LS, Weissman JS, Vale RD (2014) A protein-tagging system for signal amplification in gene expression and fluorescence imaging. Cell 159(3):635–646
103. Chavez A, Scheiman J, Vora S, Pruitt BW, Tuttle M, Iyer EP, Lin S, Kiani S, Guzman CD, Wiegand DJ (2015) Highly efficient Cas9-mediated transcriptional programming. Nat Methods 12(4):326–328
104. Konermann S, Brigham MD, Trevino AE, Joung J, Abudayyeh OO, Barcena C, Hsu PD, Habib N, Gootenberg JS, Nishimasu H (2015) Genome-scale transcriptional activation by an engineered CRISPR-Cas9 complex. Nature 517(7536):583–588
105. Chavez A, Tuttle M, Pruitt BW, Ewen-Campen B, Chari R, Ter-Ovanesyan D, Haque SJ, Cecchi RJ, Kowal EJ, Buchthal J (2016) Comparison of Cas9 activators in multiple species. Nat Methods 13(7):563–567
106. Sajwan S, Mannervik M (2019) Gene activation by dCas9-CBP and the SAM system differ in target preference. Sci Rep 9(1):1–11
107. Horlbeck MA, Gilbert LA, Villalta JE, Adamson B, Pak RA, Chen Y, Fields AP, Park CY, Corn JE, Kampmann M (2016) Compact and highly active next-generation libraries for CRISPR-mediated gene repression and activation. elife 5:e19760

108. Horlbeck MA, Witkowsky LB, Guglielmi B, Replogle JM, Gilbert LA, Villalta JE, Torigoe SE, Tjian R, Weissman JS (2016) Nucleosomes impede Cas9 access to DNA in vivo and in vitro. Elife 5:e12677
109. Boettcher M, McManus MT (2015) Choosing the right tool for the job: RNAi, TALEN, or CRISPR. Mol Cell 58(4):575–585
110. Kido M, Idogaki H, Nishikawa K, Omasa T (2020) Low-concentration staurosporine improves recombinant antibody productivity in Chinese hamster ovary cells without inducing cell death. J Biosci Bioeng
111. Jaluria P, Betenbaugh MJ, Konstantopoulos K, Shiloach J (2007) Enhancement of cell proliferation in various mammalian cell lines by gene insertion of a cyclin-dependent kinase homolog. BMC Biotechnol 7:71–71
112. Hammond S, Lee KH (2012) RNA interference of cofilin in Chinese hamster ovary cells improves recombinant protein productivity. Biotechnol Bioeng 109(2):528–535
113. De Melo J, Wu V, He L, Yan J, Tang D (2014) SIPL1 enhances the proliferation, attachment, and migration of CHO cells by inhibiting PTEN function. Int J Mol Med 34(3):835–841. https://doi.org/10.3892/ijmm.2014.1840
114. Lee Y-J, Keng PC (2005) Studying the effects of actin cytoskeletal destabilization on cell cycle by cofilin overexpression. Mol Biotechnol 31(1):1–10
115. Kumar N, Gammell P, Clynes M (2007) Proliferation control strategies to improve productivity and survival during CHO based production culture. Cytotechnology 53(1–3):33–46
116. Moore A, Mercer J, Dutina G, Donahue CJ, Bauer KD, Mather JP, Etcheverry T, Ryll T (1997) Effects of temperature shift on cell cycle, apoptosis and nucleotide pools in CHO cell batch cultues. Cytotechnology 23(1–3):47–54
117. Fox SR, Tan HK, Tan MC, Wong SNC, Yap MG, Wang DI (2005) A detailed understanding of the enhanced hypothermic productivity of interferon-γ by Chinese-hamster ovary cells. Biotechnol Appl Biochem 41(3):255–264
118. Guimaraes JC, Mittal N, Gnann A, Jedlinski D, Riba A, Buczak K, Schmidt A, Zavolan M (2020) A rare codon-based translational program of cell proliferation. Genome Biol 21(1):1–20
119. Wu W, Liu P, Li J (2012) Necroptosis: an emerging form of programmed cell death. Crit Rev Oncol Hematol 82(3):249–258
120. D'Arcy MS (2019) Cell death: a review of the major forms of apoptosis, necrosis and autophagy. Cell Biol Int 43(6):582–592
121. Tang D, Kang R, Berghe TV, Vandenabeele P, Kroemer G (2019) The molecular machinery of regulated cell death. Cell Res 29(5):347–364
122. Grilo AL, Mantalaris A (2019) Apoptosis: a mammalian cell bioprocessing perspective. Biotechnol Adv 37(3):459–475
123. Elmore S (2007) Apoptosis: a review of programmed cell death. Toxicol Pathol 35(4):495–516
124. Krampe B, Al-Rubeai M (2010) Cell death in mammalian cell culture: molecular mechanisms and cell line engineering strategies. Cytotechnology 62(3):175–188
125. Cost GJ, Freyvert Y, Vafiadis A, Santiago Y, Miller JC, Rebar E, Collingwood TN, Snowden A, Gregory PD (2010) BAK and BAX deletion using zinc-finger nucleases yields apoptosis-resistant CHO cells. Biotechnol Bioeng 105(2):330–340
126. Ha TK, Hansen AH, Kildegaard HF, Lee GM (2020) Knockout of sialidase and pro-apoptotic genes in Chinese hamster ovary cells enables the production of recombinant human erythropoietin in fed-batch cultures. Metab Eng 57:182–192
127. Henry MN, MacDonald MA, Orellana CA, Gray PP, Gillard M, Baker K, Nielsen LK, Marcellin E, Mahler S, Martínez VS (2020) Attenuating apoptosis in Chinese hamster ovary cells for improved biopharmaceutical production. Biotechnol Bioeng 117(4):1187–1203
128. Lee JS, Ha TK, Park JH, Lee GM (2013) Anti-cell death engineering of CHO cells: co-overexpression of Bcl-2 for apoptosis inhibition, Beclin-1 for autophagy induction. Biotechnol Bioeng 110(8):2195–2207

129. McIlwain DR, Berger T, Mak TW (2013) Caspase functions in cell death and disease. Cold Spring Harb Perspect Biol 5(4):a008656
130. Shalini S, Dorstyn L, Dawar S, Kumar S (2015) Old, new and emerging functions of caspases. Cell Death Diff 22(4):526–539
131. Riedl SJ, Shi Y (2004) Molecular mechanisms of caspase regulation during apoptosis. Nat Rev Mol Cell Biol 5(11):897–907
132. Walsh JG, Cullen SP, Sheridan C, Lüthi AU, Gerner C, Martin SJ (2008) Executioner caspase-3 and caspase-7 are functionally distinct proteases. Proc Natl Acad Sci 105(35):12815–12819
133. Slee EA, Adrain C, Martin SJ (2001) Executioner caspase-3,-6, and-7 perform distinct, non-redundant roles during the demolition phase of apoptosis. J Biol Chem 276(10):7320–7326
134. Abaandou L, Sharma AK, Shiloach J (2020) Knockout of the caspase 8 associated protein 2 gene improves recombinant protein expression in HEK293 cells through up-regulation of the cyclin-dependent kinase inhibitor 2A gene. Biotechnol Bioeng
135. Lee YY, Wong KT, Tan J, Toh PC, Mao Y, Brusic V, Yap MG (2009) Overexpression of heat shock proteins (HSPs) in CHO cells for extended culture viability and improved recombinant protein production. J Biotechnol 143(1):34–43
136. Tan JG, Lee YY, Wang T, Yap MG, Tan TW, Ng SK (2015) Heat shock protein 27 overexpression in CHO cells modulates apoptosis pathways and delays activation of caspases to improve recombinant monoclonal antibody titre in fed-batch bioreactors. Biotechnol J 10(5):790–800
137. Wong DCF, Wong KTK, Nissom PM, Heng CK, Yap MGS (2006) Targeting early apoptotic genes in batch and fed-batch CHO cell cultures. Biotechnol Bioeng 95(3):350–361
138. Lu Y, Zhou Q, Han Q, Wu P, Zhang L, Zhu L, Weaver DT, Xu C, Zhang B (2018) Inactivation of deubiquitinase CYLD enhances therapeutic antibody production in Chinese hamster ovary cells. Appl Microbiol Biotechnol 102(14):6081–6093
139. Okada J, Sunaga N, Yamada E, Saito T, Ozawa A, Nakajima Y, Okada K, Pessin JE, Okada S, Yamada M (2020) FAM83G is a novel inducer of apoptosis. Molecules 25(12):2810
140. Lee JS, Lee GM (2012) Rapamycin treatment inhibits CHO cell death in a serum-free suspension culture by autophagy induction. Biotechnol Bioeng 109(12):3093–3102
141. Kim YJ, Baek E, Lee JS, Lee GM (2013) Autophagy and its implication in Chinese hamster ovary cell culture. Biotechnol Lett 35(11):1753–1763
142. Lee JS, Lee GM (2012) Estimation of autophagy pathway genes for autophagy induction: overexpression of Atg9A does not induce autophagy in recombinant Chinese hamster ovary cells. Biochem Eng J 68:221–226
143. Romand S, Jostock T, Fornaro M, Schmidt J, Ritter A, Wilms B, Laux H (2016) Improving expression of recombinant human IGF-1 using IGF-1R knockout CHO cell lines. Biotechnol Bioeng 113(5):1094–1101
144. Kim CL, Lee GM (2019) Improving recombinant bone morphogenetic protein-4 (BMP-4) production by autoregulatory feedback loop removal using BMP receptor-knockout CHO cell lines. Metab Eng 52:57–67
145. Pereira S, Kildegaard HF, Andersen MR (2018) Impact of CHO metabolism on cell growth and protein production: an overview of toxic and inhibiting metabolites and nutrients. Biotechnol J 13(3):1700499
146. Kol S, Ley D, Wulff T, Decker M, Arnsdorf J, Schoffelen S, Hansen AH, Jensen TL, Gutierrez JM, Chiang AW (2020) Multiplex secretome engineering enhances recombinant protein production and purity. Nat Commun 11(1):1–10
147. Bulté DB, Palomares LA, Parra CG, Martínez JA, Contreras MA, Noriega LG, Ramírez OT (2020) Overexpression of the mitochondrial pyruvate carrier reduces lactate production and increases recombinant protein productivity in CHO cells. Biotechnol Bioeng 117(9):2633–2647
148. Gupta SK, Sharma A, Kushwaha H, Shukla P (2017) Over-expression of a codon optimized yeast cytosolic pyruvate carboxylase (PYC2) in CHO cells for an augmented lactate metabolism. Front Pharmacol 8:463

149. Gupta SK, Srivastava SK, Sharma A, Nalage VH, Salvi D, Kushwaha H, Chitnis NB, Shukla P (2017) Metabolic engineering of CHO cells for the development of a robust protein production platform. PLoS One 12(8):e0181455
150. Mulukutla BC, Kale J, Kalomeris T, Jacobs M, Hiller GW (2017) Identification and control of novel growth inhibitors in fed-batch cultures of Chinese hamster ovary cells. Biotechnol Bioeng 114(8):1779–1790
151. Mulukutla BC, Mitchell J, Geoffroy P, Harrington C, Krishnan M, Kalomeris T, Morris C, Zhang L, Pegman P, Hiller GW (2019) Metabolic engineering of Chinese hamster ovary cells towards reduced biosynthesis and accumulation of novel growth inhibitors in fed-batch cultures. Metab Eng 54:54–68
152. Papathanassiu AE, Ko J-H, Imprialou M, Bagnati M, Srivastava PK, Vu HA, Cucchi D, McAdoo SP, Ananieva EA, Mauro C (2017) BCAT1 controls metabolic reprogramming in activated human macrophages and is associated with inflammatory diseases. Nat Commun 8:16040
153. Pereira S, Ley D, Schubert M, Grav LM, Kildegaard HF, Andersen MR (2019) BCAT1 and BCAT2 disruption in CHO cells has cell line-dependent effects. J Biotechnol 306:24–31
154. Tabuchi H, Sugiyama T (2013) Cooverexpression of alanine aminotransferase 1 in Chinese hamster ovary cells overexpressing taurine transporter further stimulates metabolism and enhances product yield. Biotechnol Bioeng 110(8):2208–2215
155. Tabuchi H, Sugiyama T, Tanaka S, Tainaka S (2010) Overexpression of taurine transporter in Chinese hamster ovary cells can enhance cell viability and product yield, while promoting glutamine consumption. Biotechnol Bioeng 107(6):998–1003
156. Chong WP, Reddy SG, Yusufi FN, Lee D-Y, Wong NS, Heng CK, Yap MG, Ho YS (2010) Metabolomics-driven approach for the improvement of Chinese hamster ovary cell growth: overexpression of malate dehydrogenase II. J Biotechnol 147(2):116–121
157. Richelle A, Lewis NE (2017) Improvements in protein production in mammalian cells from targeted metabolic engineering. Curr Opin Syst Biol 6:1–6
158. Chevallier V, Andersen MR, Malphettes L (2020) Oxidative stress-alleviating strategies to improve recombinant protein production in CHO cells. Biotechnol Bioeng 117(4):1172–1186
159. Berger A, Le Fourn V, Masternak J, Regamey A, Bodenmann I, Girod PA, Mermod N (2020) Overexpression of transcription factor Foxa1 and target genes remediate therapeutic protein production bottlenecks in Chinese hamster ovary cells. Biotechnol Bioeng 117(4):1101–1116
160. Hetz C, Zhang K, Kaufman RJ (2020) Mechanisms, regulation and functions of the unfolded protein response. Nat Rev Mol Cell Biol:1–18
161. Bertolotti A, Zhang Y, Hendershot LM, Harding HP, Ron D (2000) Dynamic interaction of BiP and ER stress transducers in the unfolded-protein response. Nat Cell Biol 2(6):326–332
162. Pybus LP, Dean G, West NR, Smith A, Daramola O, Field R, Wilkinson SJ, James DC (2014) Model-directed engineering of "difficult-to-express" monoclonal antibody production by Chinese hamster ovary cells. Biotechnol Bioeng 111(2):372–385
163. Tigges M, Fussenegger M (2006) Xbp1-based engineering of secretory capacity enhances the productivity of Chinese hamster ovary cells. Metab Eng 8(3):264–272
164. Becker E, Florin L, Pfizenmaier K, Kaufmann H (2008) An XBP-1 dependent bottle-neck in production of IgG subtype antibodies in chemically defined serum-free Chinese hamster ovary (CHO) fed-batch processes. J Biotechnol 135(2):217–223
165. Gulis G, Simi KCR, de Toledo RR, Maranhao AQ, Brigido MM (2014) Optimization of heterologous protein production in Chinese hamster ovary cells under overexpression of spliced form of human X-box binding protein. BMC Biotechnol 14(1):26
166. Haredy AM, Nishizawa A, Honda K, Ohya T, Ohtake H, Omasa T (2013) Improved antibody production in Chinese hamster ovary cells by ATF4 overexpression. Cytotechnology 65(6):993–1002
167. Kim J, Choi TG, Ding Y, Kim Y, Ha KS, Lee KH, Kang I, Ha J, Kaufman RJ, Lee J (2008) Overexpressed cyclophilin B suppresses apoptosis associated with ROS and Ca2+ homeostasis after ER stress. J Cell Sci 121(21):3636–3648

168. Tsai S-Y, Hayashi T, Mori T, Su T-P (2009) Sigma-1 receptor chaperones and diseases. Central Nervous Syst Agents Med Chem (Formerly Current Medicinal Chemistry-Central Nervous System Agents) 9(3):184–189
169. Chu Q, Martinez TF, Novak SW, Donaldson CJ, Tan D, Vaughan JM, Chang T, Diedrich JK, Andrade L, Kim A (2019) Regulation of the ER stress response by a mitochondrial microprotein. Nat Commun 10(1):1–13
170. Florin L, Pegel A, Becker E, Hausser A, Olayioye MA, Kaufmann H (2009) Heterologous expression of the lipid transfer protein CERT increases therapeutic protein productivity of mammalian cells. J Biotechnol 141(1–2):84–90
171. Peng R-W, Guetg C, Tigges M, Fussenegger M (2010) The vesicle-trafficking protein munc18b increases the secretory capacity of mammalian cells. Metab Eng 12(1):18–25
172. Peng RW, Abellan E, Fussenegger M (2011) Differential effect of exocytic SNAREs on the production of recombinant proteins in mammalian cells. Biotechnol Bioeng 108(3):611–620
173. Peng RW, Fussenegger M (2009) Molecular engineering of exocytic vesicle traffic enhances the productivity of Chinese hamster ovary cells. Biotechnol Bioeng 102(4):1170–1181
174. Budge JD, Knight TJ, Povey J, Roobol J, Brown IR, Singh G, Dean A, Turner S, Jaques CM, Young RJ (2020) Engineering of Chinese hamster ovary cell lipid metabolism results in an expanded ER and enhanced recombinant biotherapeutic protein production. Metab Eng 57:203–216
175. Lalonde M-E, Durocher Y (2017) Therapeutic glycoprotein production in mammalian cells. J Biotechnol 251:128–140
176. Amann T, Schmieder V, Faustrup Kildegaard H, Borth N, Andersen MR (2019) Genetic engineering approaches to improve posttranslational modification of biopharmaceuticals in different production platforms. Biotechnol Bioeng 116(10):2778–2796
177. Ghaderi D, Taylor RE, Padler-Karavani V, Diaz S, Varki A (2010) Implications of the presence of N-glycolylneuraminic acid in recombinant therapeutic glycoproteins. Nat Biotechnol 28(8):863–867
178. Bosques CJ, Collins BE, Meador JW, Sarvaiya H, Murphy JL, DelloRusso G, Bulik DA, Hsu I-H, Washburn N, Sipsey SF (2010) Chinese hamster ovary cells can produce galactose-α-1, 3-galactose antigens on proteins. Nat Biotechnol 28(11):1153–1156
179. Butler M, Spearman M (2014) The choice of mammalian cell host and possibilities for glycosylation engineering. Curr Opin Biotechnol 30:107–112
180. Alves CS, Prajapati S (2015) Optimizing Chinese hamster ovary cell line development via targeted control of N-glycosylation. Pharm Bioproces 3(7):443–461
181. Davies J, Jiang L, Pan LZ, LaBarre MJ, Anderson D, Reff M (2001) Expression of GnTIII in a recombinant anti-CD20 CHO production cell line: expression of antibodies with altered glycoforms leads to an increase in ADCC through higher affinity for FCγRIII. Biotechnol Bioeng 74(4):288–294
182. Ghaderi D, Zhang M, Hurtado-Ziola N, Varki A (2012) Production platforms for biotherapeutic glycoproteins. Occurrence, impact, and challenges of non-human sialylation. Biotechnol Genet Eng Rev 28(1):147–176
183. Cha H-M, Lim J-H, Yeon J-H, Hwang J-M, Kim D-I (2017) Co-overexpression of Mgat1 and Mgat4 in CHO cells for production of highly sialylated albumin-erythropoietin. Enzym Microb Technol 103:53–58
184. Butler M (2005) Animal cell cultures: recent achievements and perspectives in the production of biopharmaceuticals. Appl Microbiol Biotechnol 68(3):283–291
185. Raymond C, Robotham A, Kelly J, Lattová E, Perreault H, Durocher Y (2012) Production of highly sialylated monoclonal antibodies, vol 397. InTech Rijeka, Croatia
186. Amann T, Hansen AH, Kol S, Hansen HG, Arnsdorf J, Nallapareddy S, Voldborg B, Lee GM, Andersen MR, Kildegaard HF (2019) Glyco-engineered CHO cell lines producing alpha-1-antitrypsin and C1 esterase inhibitor with fully humanized N-glycosylation profiles. Metab Eng 52:143–152
187. Chenu S, Grégoire A, Malykh Y, Visvikis A, Monaco L, Shaw L, Schauer R, Marc A, Goergen J-L (2003) Reduction of CMP-N-acetylneuraminic acid hydroxylase activity in engineered

Chinese hamster ovary cells using an antisense-RNA strategy. Biochim Biophys Acta (BBA)-General Sub 1622(2):133–144
188. Chai Y-R, Caob X-X, Ge M-M, Mi C-L, Guo X, Wang T-Y (2020) Knockout of cytidine monophosphate-N-acetylneuraminic acid hydroxylase in Chinese hamster ovary cells by CRISPR/Cas9-based gene-editing technology. Biochem Eng J:107663
189. Yamane-Ohnuki N, Satoh M (2009) Production of therapeutic antibodies with controlled fucosylation. In: MAbs. vol 3. Taylor & Francis, pp 230–236
190. Mori K, Kuni-Kamochi R, Yamane-Ohnuki N, Wakitani M, Yamano K, Imai H, Kanda Y, Niwa R, Iida S, Uchida K (2004) Engineering Chinese hamster ovary cells to maximize effector function of produced antibodies using FUT8 siRNA. Biotechnol Bioeng 88(7):901–908
191. Tejwani V, Andersen MR, Nam JH, Sharfstein ST (2018) Glycoengineering in CHO cells: advances in systems biology. Biotechnol J 13(3):1700234
192. Yamane-Ohnuki N, Kinoshita S, Inoue-Urakubo M, Kusunoki M, Iida S, Nakano R, Wakitani M, Niwa R, Sakurada M, Uchida K (2004) Establishment of FUT8 knockout Chinese hamster ovary cells: an ideal host cell line for producing completely defucosylated antibodies with enhanced antibody-dependent cellular cytotoxicity. Biotechnol Bioeng 87(5):614–622
193. Cristea S, Freyvert Y, Santiago Y, Holmes MC, Urnov FD, Gregory PD, Cost GJ (2013) In vivo cleavage of transgene donors promotes nuclease-mediated targeted integration. Biotechnol Bioeng 110(3):871–880
194. Imai-Nishiya H, Mori K, Inoue M, Wakitani M, Iida S, Shitara K, Satoh M (2007) Double knockdown of α1, 6-fucosyltransferase (FUT8) and GDP-mannose 4, 6-dehydratase (GMD) in antibody-producing cells: a new strategy for generating fully non-fucosylated therapeutic antibodies with enhanced ADCC. BMC Biotechnol 7(1):84
195. Sealover NR, Davis AM, Brooks JK, George HJ, Kayser KJ, Lin N (2013) Engineering Chinese hamster ovary (CHO) cells for producing recombinant proteins with simple glycoforms by zinc-finger nuclease (ZFN)—mediated gene knockout of mannosyl (alpha-1, 3-)-glycoprotein beta-1, 2-N-acetylglucosaminyltransferase (Mgat1). J Biotechnol 167(1):24–32
196. Kumar SR (2015) Industrial production of clotting factors: challenges of expression, and choice of host cells. Biotechnol J 10(7):995–1004
197. Sathyamurthy M, Lee JS, Park JH, Kim YJ, Jeong JY, Jang JW, Lee GM (2013) Overexpression of PACEsol improves BMP-7 processing in recombinant CHO cells. J Biotechnol 164(2):336–339
198. Hansen HG, Pristovšek N, Kildegaard HF, Lee GM (2017) Improving the secretory capacity of Chinese hamster ovary cells by ectopic expression of effector genes: lessons learned and future directions. Biotechnol Adv 35(1):64–76
199. Xiao S, Shiloach J, Betenbaugh MJ (2014) Engineering cells to improve protein expression. Curr Opin Struct Biol 26:32–38
200. Brown AJ, James DC (2016) Precision control of recombinant gene transcription for CHO cell synthetic biology. Biotechnol Adv 34(5):492–503
201. Tastanova A, Schulz A, Folcher M, Tolstrup A, Puklowski A, Kaufmann H, Fussenegger M (2016) Overexpression of YY1 increases the protein production in mammalian cells. J Biotechnol 219:72–85
202. Johari YB, Estes SD, Alves CS, Sinacore MS, James DC (2015) Integrated cell and process engineering for improved transient production of a difficult-to-express fusion protein by CHO cells. Biotechnol Bioeng 112(12):2527–2542
203. Mohan C, Lee GM (2010) Effect of inducible co-overexpression of protein disulfide isomerase and endoplasmic reticulum oxidoreductase on the specific antibody productivity of recombinant Chinese hamster ovary cells. Biotechnol Bioeng 107(2):337–346
204. Cain K, Peters S, Hailu H, Sweeney B, Stephens P, Heads J, Sarkar K, Ventom A, Page C, Dickson A (2013) A CHO cell line engineered to express XBP1 and ERO1-Lα has increased levels of transient protein expression. Biotechnol Prog 29(3):697–706
205. Le Fourn V, Girod P-A, Buceta M, Regamey A, Mermod N (2014) CHO cell engineering to prevent polypeptide aggregation and improve therapeutic protein secretion. Metab Eng 21:91–102

206. Cartwright JF, Arnall CL, Patel YD, Barber NO, Lovelady CS, Rosignoli G, Harris CL, Dunn S, Field RP, Dean G (2020) A platform for context-specific genetic engineering of recombinant protein production by CHO cells. J Biotechnol
207. Eisenhut P, Klanert G, Weinguny M, Baier L, Jadhav V, Ivansson D, Borth N (2018) A CRISPR/Cas9 based engineering strategy for overexpression of multiple genes in Chinese hamster ovary cells. Metab Eng 48:72–81
208. Qin JY, Zhang L, Clift KL, Hulur I, Xiang AP, Ren B-Z, Lahn BT (2010) Systematic comparison of constitutive promoters and the doxycycline-inducible promoter. PLoS One 5(5):e10611
209. Lam C, Santell L, Wilson B, Yim M, Louie S, Tang D, Shaw D, Chan P, Lazarus RA, Snedecor B (2017) Taming hyperactive h DN ase I: stable inducible expression of a hyperactive salt- and actin-resistant variant of human deoxyribonuclease I in CHO cells. Biotechnol Prog 33(2):523–533
210. Misaghi S, Chang J, Snedecor B (2014) It's time to regulate: coping with product-induced nongenetic clonal instability in CHO cell lines via regulated protein expression. Biotechnol Prog 30(6):1432–1440
211. Sheikholeslami Z, Jolicoeur M, Henry O (2013) The impact of the timing of induction on the metabolism and productivity of CHO cells in culture. Biochem Eng J 79:162–171
212. Mellahi K, Brochu D, Gilbert M, Perrier M, Ansorge S, Durocher Y, Henry O (2019) Assessment of fed-batch cultivation strategies for an inducible CHO cell line. J Biotechnol 298:45–56
213. Kallunki T, Barisic M, Jäättelä M, Liu B (2019) How to choose the right inducible gene expression system for mammalian studies? Cell 8(8):796
214. Hamaker NK, Lee KH (2018) Site-specific integration ushers in a new era of precise CHO cell line engineering. Curr Opin Chem Eng 22:152
215. Lee JS, Kallehauge TB, Pedersen LE, Kildegaard HF (2015) Site-specific integration in CHO cells mediated by CRISPR/Cas9 and homology-directed DNA repair pathway. Sci Rep 5:8572
216. Pristovsek N, Nallapareddy S, Grav LM, Hefzi H, Lewis NE, Rugbjerg P, Hansen HG, Lee GM, Andersen MR, Kildegaard HF (2019) Systematic evaluation of site-specific recombinant gene expression for programmable mammalian cell engineering. ACS Synth Biol 8(4):758–774
217. Shin SW, Lee JS (2020) CHO cell line development and engineering via site-specific integration: challenges and opportunities. Biotechnol Bioprocess Eng:1–13
218. Lee Y, Kwak JM, Lee JS (2020) Endogenous p21-dependent transgene control for CHO cell engineering. ACS Synth Biol 9(7):1572–1580
219. Sergeeva D, Lee GM, Nielsen LK, Grav LM (2020) Multicopy targeted integration for accelerated development of high-producing Chinese hamster ovary cells. ACS Synth Biol 9(9):2546–2561
220. Shin SW, Lee JS (2020) Optimized CRISPR/Cas9 strategy for homology-directed multiple targeted integration of transgenes in CHO cells. Biotechnol Bioeng 117(6):1895–1903
221. Mohan C, Park SH, Chung JY, Lee GM (2007) Effect of doxycycline-regulated protein disulfide isomerase expression on the specific productivity of recombinant CHO cells: Thrombopoietin and antibody. Biotechnol Bioeng 98(3):611–615
222. Hansen HG, Nilsson CN, Lund AM, Kol S, Grav LM, Lundqvist M, Rockberg J, Lee GM, Andersen MR, Kildegaard HF (2015) Versatile microscale screening platform for improving recombinant protein productivity in Chinese hamster ovary cells. Sci Rep 5(1):1–12
223. Ku SC, Ng DT, Yap MG, Chao SH (2008) Effects of overexpression of X-box binding protein 1 on recombinant protein production in Chinese hamster ovary and NS0 myeloma cells. Biotechnol Bioeng 99(1):155–164
224. Wurm FM (2013) CHO quasispecies—implications for manufacturing processes. PRO 1(3):296–311
225. Kaas CS, Kristensen C, Betenbaugh MJ, Andersen MR (2015) Sequencing the CHO DXB11 genome reveals regional variations in genomic stability and haploidy. BMC Genomics 16(1):160

226. Hu Z, Guo D, Yip SS, Zhan D, Misaghi S, Joly JC, Snedecor BR, Shen AY (2013) Chinese hamster ovary K1 host cell enables stable cell line development for antibody molecules which are difficult to express in DUXB11-derived dihydrofolate reductase deficient host cell. Biotechnol Prog 29(4):980–985
227. Golabgir A, Gutierrez JM, Hefzi H, Li S, Palsson BO, Herwig C, Lewis NE (2016) Quantitative feature extraction from the Chinese hamster ovary bioprocess bibliome using a novel meta-analysis workflow. Biotechnol Adv 34(5):621–633
228. Kelley B (2020) Developing therapeutic monoclonal antibodies at pandemic pace. Nat Biotechnol 38(5):540–545
229. Ferrara C, Brünker P, Suter T, Moser S, Püntener U, Umaña P (2006) Modulation of therapeutic antibody effector functions by glycosylation engineering: influence of Golgi enzyme localization domain and co-expression of heterologous β1, 4-N-acetylglucosaminyltransferase III and Golgi α-mannosidase II. Biotechnol Bioeng 93(5):851–861

Part III
Development of Cell-based Technologies and Therapeutics

Manufacturing Human Pluripotent Stem Cells and Differentiated Progenitors

Svetlan Vassilev and Steve Kah Weng Oh

Abbreviations

ACF	animal component free
bFGF	basic fibroblast growth factor
CAR	chimeric antigen receptor
CDM	chemically defined media
CFD	computational fluid dynamics
cGMP	current good manufacturing practice
CM	cardiomyocyte
CM	conditioned media
CVD	cardiovascular disease
DO	dissolved oxygen
EB	embryoid body
ECM	extracellular matrix
FDA	United States Food and Drug Administration

S.V. wrote the manuscript and generated tables and figures. S.O. conceived the name and focus of the chapter, provided papers for review, and critically revised the manuscript.

S. Vassilev
Stem Cell Group, Bioprocessing Technology Institute, A*STAR (Agency for Science, Technology, and Research), Singapore, Singapore

Department of Microbiology and Immunology, National University of Singapore, Singapore, Singapore
e-mail: tp-svetvass@bti.a-star.edu.sg

S. K. W. Oh (✉)
Stem Cell Group, Bioprocessing Technology Institute, A*STAR (Agency for Science, Technology, and Research), Singapore, Singapore
e-mail: steve_oh@bti.a-star.edu.sg

R. Pörtner (ed.), *Cell Culture Engineering and Technology*, Cell Engineering 10,
https://doi.org/10.1007/978-3-030-79871-0_8

GAiT	global alliance for iPSC therapies
HARV	high aspect rotating vessel
hESC	human embryonic stem cell
hiPSC	human induced pluripotent stem cell
hPSC	human pluripotent stem cell
HSC	hematopoietic stem cell
ICH	International Council for Harmonisation of Technical Requirements for Pharmaceuticals for Human Use
ISCBI	international stem cell banking initiative
KOSR	knockout serum replacement
MC	microcarrier
MCB	master cell bank
MEF	mouse embryonic fibroblast
MEF-CM	mouse embryonic fibroblast-conditioned media
MK	megakaryocyte
NEAA	non-essential amino acids
NPC	neural progenitor cell
PI3K	phosphoinositide-3-kinase
QbD	quality by design
RCCS	rotating cell culture system
RWB	rotating wall bioreactor
STLV	slow turning lateral vessel
STR	stirred tank reactor
TGF-β	transforming growth factor beta
VWB	vertical wheel bioreactor
WCB	working cell bank
XFM	xeno free media

1 Introduction

1.1 Human Pluripotent Stem Cells

Human pluripotent stem cells (hPSCs) represent the gateway to the twenty-first century's most promising therapeutic development. They offer the promise of curative treatments that can restore functionality to damaged organs, fight cancers and target diseases like never before. Additionally, they can be utilized as a model of human disease in vitro, augmenting our understanding of the underlying causes of pathology and our drug screening and development capabilities [1].

Pluripotent stem cells have two key abilities that promote their use as therapeutic and research agents – they can replenish themselves and expand continuously, and they can differentiate into all individual cell types in the body. Two major groups of cells have been described in great detail over the last few decades – human embryonic stem cells (hESCs) and induced pluripotent stem cells (hiPSCs)

[2]. Mouse ESCs were first cultured in vitro by Evans and Kaufman in 1981 [3]. Later, the first hESCs were isolated in 1998 by J. Thomson [4]. However, challenges pertaining to the ethical considerations surrounding their isolation from human embryos presented a barrier to their large-scale development and use. In 2007, the groundbreaking work of Yamanaka and colleagues brought forth an alternative – using a combination of just four transcription factors, they were able to induce somatic cells to return to a state of pluripotency, with a strikingly similar genetic profile to hESCs [5]. These induced pluripotent stem cells offer an enticing alternative to hESCs through their removal of ethical concerns, their low reported immunogenicity and expansion and differentiation potential [6].

Human ESCs and iPSCs harbour enormous potential to be used in the field of cell therapy. Their differentiation capacity offers the ability to produce all cells of the adult human body in vitro, from cardiac cells, through hepatocytes to neuronal progenitors and all cells in between [7]. However, as the field matures, major overarching challenges must be addressed to ensure the safety, efficacy, and acceptable overall costs of manufacture [8]. These can be broadly categorized as those pertaining to the cells themselves (e.g. robust and reproducible differentiation processes, maintenance of pluripotent state), and the difficulties of manufacturing them at industrially relevant scales at acceptable cost without compromising their long-term quality [9].

1.2 Challenges in Large Scale Manufacturing of iPSCs

One of the major difficulties of hPSC clinical translation is the extremely large dose requirements to produce the desired therapeutic effect. Present-day therapies regularly require upwards of 10^8 cells, or 100 million [10]. Some cell therapies, like blood transfusions, consist of an even larger number of cells – one unit of blood contains 2×10^{12} red blood cells (RBCs) [11]. In order to deliver such a large quantity of cells, current lab-scale production processes in the millilitres must be scaled up to litres and tens of litres.

Production of clinical grade hPSCs and differentiated cells in high doses requires large amounts of media containing expensive recombinant proteins [12]. Current processes for hPSC expansion often require frequent media changes over multiple passages in order to obtain the necessary cell yield, which contributes greatly to the high cost of manufacturing. Additionally, any manufacturing system for a product intended for human clinical use must adhere strictly to current Good Manufacturing Practices (cGMP), in order to minimize the immunogenic risk of introducing foreign agents/contaminants into the final therapeutic [13]. Media formulations and cytokine and growth factor cocktails must therefore strive to remove the use of xenogeneic materials and look towards the use of expensive recombinant human proteins and/or small molecules [14]. Challenges concerning hPSC and final product storage, cell line sourcing, and in process monitoring and control must also be addressed if hPSCs are to meet the demands of the regenerative medicine market [15].

This chapter aims to provide a thorough overview of the manufacturing challenges surrounding hPSC production at clinically relevant scales. We present a detailed summary of the current strategies employed in hPSC scale-up and consider the advances that have been made in media formulations and culture conditions on the road to robust and cGMP compliant processes. Additionally, we evaluate the major limitations of current platforms and discuss the obstacles that are yet to be solved on the quest of delivering the therapeutic promise of stem cells.

2 Strategies for hPSC Manufacturing

2.1 2D Cultures

Human pluripotent stem cell culture and expansion has traditionally been performed in 2D monolayer systems, where cells are grown on tissue culture plates and flasks [16]. To facilitate cell adhesion, vessels are coated with feeder cells (such as inactivated fibroblasts) [4] or, more recently, defined matrices (such as recombinant proteins, Matrigel and others) [17]. Two-dimensional culture is well-suited to laboratory scale studies but faces significant challenges in meeting the demands of industrial scale manufacturing [18]. Methods have been proposed for increasing the expansion capacity of 2D culture by simply adding additional culture plates in parallel as needed (termed "scale-out" methods) [19]. Several systems have been developed with this approach in mind, such as the CellSTACK culture chambers or Cell Factories [20].

Unfortunately, scale-out approaches generally do not solve the inherent issues of 2D culture. Having multiple plates/flasks in parallel (i.e. individual "lots") significantly increases the labour required to maintain and passage the hPSCs, as each vessel must be checked individually [21]. Additionally, a considerable amount of physical space is required, as temperature, humidity and gas control is vital to cell expansion. Scale-out approaches inherently possess less capacity for automation of process parameters such as dissolved oxygen (DO), glucose and cytokine concentration and pH [9], although some studies have been published describing varying degrees of automation [22, 23]. Furthermore, 2D culture is often performed statically, which, left unchecked, can lead to the formation of gradients along the culture vessel, exposing cells to varying quantities of nutrients, oxygen, and inhibitory factors [24]. The adherent nature of the 2D culture also necessitates the use of enzymatic or chemical dissociation, imposing an additional burden on cell harvesting [25].

The difficulty in process parameter control presents a particular issue with regard to cGMP manufacturing. Having multiple vessels in parallel with little automation can lead to so-called lot-to-lot variability, wherein cells in different plates are exposed to unique conditions [26]. Lot-to-lot variability leads to a heterogeneous and difficult to accurately define final product and obstructs its validation from agencies such as the FDA [15].

2.2 3D Cultures

Compared to the more established 2D systems, three-dimensional suspension culture strategies offer many enticing benefits for industrial manufacture of hPSCs. The use of bioreactor vessels allows a more controllable scale-up as opposed to the scale-out approach discussed above. This combats lot-to-lot variability and leads to a more homogeneous final product [27]. Additionally, dynamic suspension conditions are easier to employ, via impellers or other means (depending on the type of vessel used), which reduces undesirable gradient formation and improves the circulation of nutrients, growth factors, and oxygen [28]. This in turn allows the culture to reach much higher cell density than 2D systems can usually achieve, improving the yield of cells per volume of media used and reducing the overall cost of the process and resulting therapy [29]. Additionally, bioreactor-based process scale-up provides opportunities for the use of in-process monitoring equipment, leading to better control of the culture conditions, higher yields, and improved quality of cells. However, one notable advantage of 2D flasks over 3D systems in the ability to visually inspect cells directly using microscopy [30].

Multiple studies have suggested that 3D culture systems may also provide a more "native" cell signalling microenvironment within the bioreactor [31–33]. Two-dimensional monolayer culture plates restrict the cell-cell interactions of hPSCs, while 3D manufacturing may promote an environment that more closely resembles conditions during embryonic development [34]. There is emerging evidence that 3D suspensions enhance the effectiveness of spatiotemporal signals that drive the proliferation and differentiation of the hPSCs. For example, a study by Branco and colleagues in 2019 suggested that hiPSC-derived cardiomyocytes mature earlier and quicker in a 3D suspension culture when compared to a tissue culture plate monolayer system, possibly because the 3D platform better recapitulated the normal cardiac tissue differentiation during embryogenesis [31].

3 Approaches for 3D Culture

3.1 Free Aggregate Suspension

One of the most common ways of culturing hPSCs in 3D systems is through the spontaneous formation of free-floating cellular aggregates in suspension. Expansion of hPSCs as aggregates eliminates the need for extracellular matrix (ECM) coatings, as cells instead adhere to each other, forming spheroids [35]. This is normally accomplished by hPSC dissociation into single cells and culturing them in the presence of ROCK inhibitor, which serves to prevent anoikis (dissociation-triggered apoptosis) [36].

Three-dimensional aggregate cultures have successfully been utilized for the expansion of hPSCs to close to industrial titres [37]. Multiple lab-scale studies have demonstrated the feasibility of achieving clinically-relevant expansion, notably by Abbasalizadeh and colleagues, who showed a robust 1-month expansion process that could in principle produce 2×10^9 hiPSCs using only four 200 mL spinner flasks [38]. These studies have subsequently been effectively translated to the pilot scale in bioreactors, as performed by Kwok et al. in 2017, who utilized a single-use Mobius bioreactor with a working volume of 1 L to produce a total of 2×10^9 hiPSCs in a single run over 7 days. The cells produced maintained their expression of key pluripotency markers SOX2 and OCT4 and retained their ability to spontaneously differentiate into the three germ layers [39]. Table 1 provides a summary of hPSCs cultured as 3D aggregates.

Notable advantages of 3D aggregate cultures include a relatively lower overall cost compared to other methods, due to the elimination of attachment substrates and the relatively simple downstream harvesting of expanded cells [18]. However, care must be taken to ensure even aggregate size distribution and to prevent the formation for large aggregates, as larger clusters of cells inhibit mass transfer to the core of the aggregate, potentially leading to spontaneous differentiation and necrosis [54]. Studies have reported that aggregate size should not exceed ~300–350 μm to prevent decreased cell viability and slow expansion rates [55]. Bioreactor control parameters should be optimized accordingly – depending on the type of system chosen for expansion, variations in impeller type, geometry and average energy dissipation rate can be adjusted to maintain aggregate size. Recent work has also suggested that short-term exposure to retinoic acid can support hiPSC expansion in 3D aggregate cultures by potentially activating the PI3K signalling cascade and maintaining expression of pluripotency-related transcription factors (TFs) [56].

3.2 Microcarrier Cultures

Microcarrier suspension cultures involve expanding hPSCs as adherent cells on small beads, typically 100–400 μm in diameter [57]. Compared to 2D plate culture, microcarrier (MC) platforms provide a substantial increase in the surface area available for cell adhesion (and thus expansion) while retaining the advantages of 3D suspension systems, including enhanced scalability and process parameter control [58]. Depending on the product used, 1 g of microcarrier beads could represent a similar overall surface area as fifteen 75 cm^2 tissue culture flasks [59]. The large surface area provided allows for very high cell density culture, but careful consideration must be given to the cell-to-surface ratio used in bioreactors, as this will dictate the maximal achievable density [18]. Importantly, initial seeding densities and seeding conditions should be optimized to promote MC-hPSC inter-

Table 1 Summary of studies in which free aggregate suspension cultures were utilized, sorted by their year of publication

Cell type	Vessel	Working volume	Media	Seeding density	Max cell number	Fold expansion	References
hiPSC	Spinner Flask Single-use VWB	100 mL	mTeSr1	2×10^4 cells/mL	6.4×10^5 cells/mL	32	[40]
hiPSC	Mobius® 3-L Single-use Bioreactor	1 L	mTeSR1	2×10^5 cells/mL	2×10^6 cells/mL	10	[39]
hiPSC	Spinner Flask	100 mL	mTeSR1	2×10^4–8×10^4 cells/mL	2.6×10^5–6×10^5 cells/mL	7.6–13	[41]
hESC hiPSC	DasGip cellferm-pro bioreactor	200 mL	Cellartis DEF-CS Xeno-Free Culture Medium	2.5×10^5–5×10^5 cells/mL	4.7×10^6 cells/mL	19	[42]
hiPSC	BioLevitator™ Bioreactor	50 mL	mTeSR1 or E8	7.5×10^4 cells/mL	3.7×10^5 cells/mL	4.9	[43]
hiPSC	CELLSPIN system	100 mL	E8	4×10^5–5×10^5 cells/mL	1.5×10^6 cells/mL	3.7	[44]
hiPSC	Spinner Flask	100 mL	mTeSR1	3×10^5 cells/mL	1.1×10^6 cells/mL	3.6	[45]
hiPSC	Spinner Flask	100 mL	DMEM/F12 + GlutaMAX + bFGF	2×10^5–1×10^6 cells/mL	0.6×10^6–1.8×10^6 cells/mL	1.8–4	[38]
hiPSC	Cellferm® pro Parallel Bioreactor System	100 mL	mTeSR1	4–5 $\times 10^5$ cells/mL	2.4×10^6 cells/mL	5.5	[46]
hESC	Spinner Flask	60 mL 500 mL	StemPro hESC SFM	2.5×10^5 cells/mL	1×10^6 cells/mL	4	[47]
hESC hiPSC	Spinner Flask	100 mL	mTeSR1	1×10^6 cells/mL	3×10^6 cells/mL	3	[48]
hESChiPSC	Spinner Flask	25 mL	DMEM/F12 + KOSR + NEAA + GlutaMAX + bFGF	1.06×10^5 cells/mL	1.9×10^6 cells/mL	17.7	[49]

(continued)

Table 1 (continued)

Cell type	Vessel	Working volume	Media	Seeding density	Max cell number	Fold expansion	References
hESChiPSC	Erlenmeyer Flask	25 mL	mTeSR1	3.3×10^4 cells/mL	1.3×10^5 cells/mL	4	[50]
hESC	Erlenmeyer Flask	25 mL	DMEM/F12 + KOSR + NEAA + L-glutamine + bFGF + IL6RIL6 chimera	1.3×10^6 cells/mL	3.25×10^7 cells/mL	25	[51]
hESC	Spinner Flask	50 mL	mTeSR1	1×10^6 cells/mL	2.4×10^6 cells/mL	2–3	[28]
hESC	NDS Stirred bioreactor	100 mL	mTeSR1	1.8×10^4 cells/mL	0.45×10^6 cells/mL	25	[37]
hESC	STLV	55 mL	DMEM-KO + 20%FBS + Glutamine + NEAA	5×10^5 cells/mL	NR	N/A	[52]
hESC	HARV STLV	50 mL55 mL	DMEM-KO + 20%FBS + Glutamine + NEAA	$0.1–1.2 \times 10^6$ cells/mL	NR	N/A	[53]

hiPSC human induced pluripotent stem cell, *hESC* human embryonic stem cell, *VWB* vertical wheel bioreactor, *bFGF* basic fibroblast growth factor, *KOSR* knockout serum replacement, *NEAA* non-essential amino acids, *STLV* slow turning lateral vessel, *HARV* high-aspect rotating vessel action and adhesion [60]. For example, Storm et al. demonstrated that intermittent stirring during the first 24 h of MC culture resulted in better overall efficiency when compared to continuous stirring [61].

Microcarrier materials must be carefully optimized for a successful industrial process and key consideration must be given to both the core material of the bead, as well as the coating used to support cell adhesion (if any). MCs have been produced from a variety of materials, such as cellulose, polystyrene, gelatine, and others [62–64] with comparable density to the media; the use of biodegradable materials could in theory allow the MCs to be used as a delivery system in vivo by transporting the differentiated cells to targeted tissues and organs [65]. hPSC growth can also be affected by the surface charge of the beads. Our group has previously reported successful expansion of hPSCs on recombinant human laminin-coated polystyrene microcarriers, achieving a 15-fold expansion over a 7-day culture while maintaining expression of pluripotent markers and eventually differentiating into functional erythroblasts [66]. MC systems have also successfully been used to expand hPSCs to clinically relevant titres (over 2×10^6 cells/mL) in a 3 L single-use BioBLU reactor, with an average 93-fold expansion across 10 runs [67]. Table 2 summarizes the progress made by key studies using various microcarriers and vessels.

However, microcarrier systems are not devoid of challenges in hPSC culture. MC systems can come with unexpected costs associated with the production of the microcarriers themselves, and recombinant coatings carry an additional investment and can suffer from inconsistency issues [18, 60]. Additionally, unlike free-floating 3D aggregates, MC-cultured hPSCs must be chemically or enzymatically dissociated from the beads upon completion of the process, which can have an effect on cell viability [80]. Pluripotent cells can form multiple layers of cells on a single bead, which adds an additional challenge in recovering as many cells as possible due to the possibility of undesirable spontaneous differentiation and genetic alterations [81, 82]. Further downstream operations must be employed to efficiently separate the MCs from the cells (e.g. additional filtration steps), prompting the development of xeno-free dissolvable microcarriers to circumvent the issue [83]. Alternatively, other groups have focused their efforts on reusing microcarriers following efficient capture and sterilization, which can in theory alleviate some of the production costs [84]. MC-hPSC cultures are also highly affected by the hydrodynamic conditions within the bioreactor – careful choice and maintenance of the agitation rate can lessen the shear stress experienced between the beads and minimize collisions [55].

3.3 *Cell Encapsulation*

3.3.1 Microencapsulation

Microencapsulation technology can be considered a subset of microcarrier systems and involves the capture of hPSCs within the beads themselves. Capsules can be manufactured using a variety of polymeric materials such as agarose, polystyrene,

Table 2 Overview of suspension culture studies performed using microcarriers, ordered by year of publication

Cell type	Vessel	Working volume	Media	MC type	Seeding density	Max cell number	Fold expansion	References
hiPSC	BioBLU Single-use Bioreactor	3 L	L7™ TFO2	SoloHill plastic + L7™ hPSC Matrix	$2–4 \times 10^4$ cells/mL	$>2 \times 10^6$ cells/mL	93	[67]
hiPSC	Spinner Flask	15 mL	mTeSR1	PAS-M, PAS-HydroM	5×10^6 cells/mL	3.3×10^8 cells/mL	65.7	[68]
hESChiPSC	Spinner Flask	25 mL	mTeSR1	SoloHill Plastic/Plastic Plus + LN521/LN111/PLL	2×10^5 cells/mL	1.85×10^6 cells/mL	9	[69]
hESC	Spinner Flask Cellferm-pro Bioreactor	60 mL 180 mL	mTeSR1	Synthemax II Hydrogel/Polystyrene	1×10^5 cells/mL	2×10^6 cells/mL	20	[70]
hESChiPSC	Spinner Flask	50 mL	TeSR2	SoloHill Polystyrene + Collagen	8×10^4 cells/mL	1.6×10^6 cells/mL	20	[71]
hESChiPSC	Spinner Flask	25 mL	mTeSR1	Polystyrene + PLL/VN/LN	2×10^5 cells/mL	$\sim 3 \times 10^6$ cells/mL	15	[63]
hiPSC	Spinner Flask	15 mL	mTeSR1	Synthemax II	$8.3–1.7 \times 10^5$ cells/mL	NR	N/A	[72]
hESC	Spinner Flask	25 mL	MEF-CM	DE53 + Laminin/Matrigel	4×10^5 cells/mL	3.4×10^6 cells/mL	8.5	[73]
hESC	BIOSTAT® STR	300 mL	DMEM-KO + KOSR + NEAA + GlutaMAX + bFGF	Cytodex III	4.5×10^5 cells/mL	2.2×10^6 cells/mL	4.9	[74]
hESC	Spinner Flask	25 mL	mTeSR1	DE53 + Matrigel	6×10^5 cells/mL	2.5×10^6 cells/mL	4.2	[75]

hESC	Spinner Flask	30 mL	DMEM-KO + KOSR + bFGF	CultiSpher-S/SoloHill Collagen/SoloHill FACT	6×10^4 cells/mL	NR	N/A	[61]
hESC	Spinner Flask	25 mL	CM	Cellulose MCs + Matrigel	6×10^5 cells/mL	3.5×10^6 cells/mL	5.8	[76]
hiPSC	Spinner Flask	NR	MF-CM	NR	$2.5–10 \times 10^4$ cells/mL	$3–4 \times 10^5$ cells/mL	~7.5	[77]
hESC	Spinner Flask	66 mL	DMEM/F-12 + KOSR + Glutamine + NEAA + bFGF	Cytodex III	$2–2.5 \times 10^5$ cells/mL	1.5×10^6 cells/mL	7.5	[78]
hESC	Spinner Flask	80 mL	DMEM-KO + KOSR + NEAA + bFGF	Hillex II	6.25×10^4 cells/mL	1.4×10^5 cells/mL	2.2	[79]

MEF-CM mouse embryonic fibroblast-conditioned media, *CM* conditioned media, *NR* not reported, *N/A* not applicable cellulose, and others [85]. A key advantage of cell encapsulation is the diminished effect of shear stress in suspension culture [86]. However, many of the difficulties with using traditional surface-adherent MCs are still present, such as the high manufacturing cost of the spheres. Additionally, surface area limitations and difficulty isolating cells from the microcapsules are also of concern [18].

3.3.2 Hydrogels

Hydrogel encapsulation has been shown to provide a protected and efficient environment for hPSC culture [87]. Several studies have explored the possibility of expanding hPSCs in 3D hydrogels, notably by Lei and Shaffer in 2013, who showed that human induced pluripotent cells can effectively be expanded and differentiated into neural progenitors with only 50 mL thermoresponsive hydrogel yielding 10^9 cells [88]. However, issues with hydrogel and medium compatibility and hydrogel sphere size were reported and remain poorly studied.

4 Process Engineering for 3D Systems

As three-dimensional culture systems offer the most promise of expanding hPSCs to clinically relevant numbers, many potential vessels and platforms have been studied to determine the optimal configurations for a competitive industrial process. In order to ensure robustness and scalability, the choice of culture vessel is critical; an economically viable process should (1) produce a large number of hPSCs while maintaining their pluripotency and undifferentiated state, (2) allow for ease of lineage differentiation within the vessel, (3) provide good options for in-process monitoring and control, and (4) be scalable to the litres and tens of litres scale.

4.1 Choice of Bioreactor

The inception of an hPSC manufacturing process begins with a choice of a suitable culture vessel. Several bioreactor configurations have been extensively studied for their ability to facilitate large-scale expansion of hPSCs.

4.1.1 Spinner Flasks

Spinner flasks have been a staple in hPSC culture and have seen extensive use for lab-scale studies of hPSC expansion. The most notable advantage of spinner cultures over traditional 2D monolayers is the spinner's ability to sustain a homogeneous environment within the vessel through agitation. Spinners are also widely available, relatively cheap, and straightforward to set up and run [89]. They are relatively basic, but well-understood and generally well-tolerated by cells – Wang and colleagues successfully expanded hiPSCs to over 1.5×10^6 cells/mL in a CELLSPIN spinner flask system for over 10 passages while maintaining pluripotency markers and an undifferentiated karyotype [44]. However, spinner flasks suffer disadvantages when it comes to scaling up, most notably a limited capacity for online monitoring of process parameters such as DO and pH. As such, they are best suited to laboratory and small-scale process optimization studies, or as an intermittent vessel in a seed train of a large-scale industrial bioprocess [60].

4.1.2 Stirred Tank Bioreactors

Stirred tank bioreactors (STRs) are the most prevalent reactors in the biomanufacturing industry. They have seen widespread as the culture platform of choice for both bacterial and subsequently mammalian industrial-scale culture [90]. Given their dominance in biomanufacturing, STRs have been studied, characterised, and developed extensively; both more traditional stainless steel and more recent single-use systems are widely available in scales ranging from millilitres to the tens of thousands of litres [91]. Scale-up can be achieved by straightforward means, e.g. on the basis of constant power dissipation per unit volume or constant tip speed [92]. As the name suggests, STRs are outfitted with an impeller in order to ensure effective mixing and to keep cells in suspension, with various designs bringing their own advantages and shortcomings [93]. These have been thoroughly described elsewhere and will not be discussed further in this review.

A major advantage of stirred tank reactors is the ease of implementing process monitoring and control tools. Tanks can be outfitted with probes for online monitoring of culture conditions (DO, pH, temperature, pCO_2) and analysers for nutrient, ammonia, and lactate level measurement [9, 94]. Additionally, ports allow for efficient non-destructive and non-interrupting sampling of the culture [9]. Precise and efficient monitoring of hPSC cultures is vital to optimize conditions to effectively preserve pluripotency and minimize undesirable differentiation. Of particular importance in scaling up STR hPSC processes are the hydrodynamic impact on cells within the suspension and the oxygen mass transfer effects. Traditionally, STRs used for the manufacturing of biologics have generally sacrificed some cell quality for increased product titres; however, shear stress can severely impact hPSC viability and pluripotency [44, 75]. Studies performed in spinner flasks have attempted to elucidate optimal agitation speeds for different culture methods – for example, for a 50 mL spinner microcarrier hiPSC process, Badenes et al. determined an optimal rate of 30–70 RPM [95]. Small-scale culture vessels typically rely exclusively on oxygen delivered via the reactor headspace; however, increasing the dimensions of the vessel can eventually outstrip the ability of the headspace to provide sufficient dissolved oxygen, leading to a requirement for sparging or increasing the agitation rate, which can generate a feedback loop into shear stress generation [96]. A remarkable recent study performed by Shafa and colleagues in

2019 demonstrates the use of Computational Fluid Dynamics (CFD) to aid in the scaling up of hPSC culture. Using modelling and simulation of the hydrodynamic conditions within the bioreactor, the study demonstrated scale up of an iPSC-to-cardiomyocyte differentiation system from a small-scale 125 mL spinner flask to a 3 L BioBLU STR with comparable results [97].

4.1.3 Others

Rotating Wall

Rotating wall bioreactors (RWBs, or Rotating Cell Culture Systems, RCCS) have been developed as an alternative to STRs in an attempt to address issues with shear stress within the reactor. Two main systems have seen use – high aspect rotating vessels (HARV) and slow turning lateral vessels (STLV) – with both comprising a cylindrical reactor, slowly rotating along its horizontal axis with no headspace provided [53]. Studies performed with hESCs by Gerecht-Nir et al. and Côme et al. demonstrated some of the advantages of the system, mainly the low hydrodynamic shear stress imposed by the slow rotation and acceptable mixing, resulting in threefold expansion of hESCs by the former study [52, 53]. However, notable issues with the system have also been reported, including the formation of air bubbles (which can damage cells) and high technical and engineering complexity [98]. This presents a challenge to the system's scalability and thus RWBs have generally not been adopted for hPSC culture.

Vertical Wheel

Vertical wheel bioreactors (VWB) provide mixing using a large vertical impeller that generates both radial and axial flows and a concave U-shaped vessel. A relatively novel system, it has recently been shown as an effective and scalable platform for hPSC culture, providing excellent homogenisation of culture contents with low shear stress at lower power input required [99]. VWBs have been used with both aggregate and microcarrier cultures, with Nogueira et al. achieving 1.2×10^6 cells/mL maximum density and Rodrigues et al. attaining similar densities in a xeno-free microcarrier-based process in an 80 mL/300 mL working volume PBS MINI VWB, respectively [100, 101]. Vertical wheel reactors have been developed with working volumes of up to 50 L and can be used as single-use systems as well.

4.2 Operation Modes

The choice of feeding strategy has a profound impact on the scale-up and expansion capability of hPSCs and their physiological state within the vessel. Several key operation modes have previously been utilized for hPSC culture in 3D systems, namely batch, fed-batch, repeated-batch, and perfusion.

4.2.1 Batch

The simplest approach, a batch process, involves supplying the required nutrients and growth factors in bulk at the beginning of the run, with no further alterations to the culture media throughout [9]. Running bioreactors in batch mode offers a simple and straightforward approach with low risk of contamination. However, hPSC cultures are virtually incompatible with batch operations. Stem cells require near daily supplementation with growth factors to maintain an undifferentiated state and cannot easily tolerate high nutrient concentrations [102]. Additionally, accumulation of lactate and other waste and/or toxic components over time reduces cell viability significantly [84]. Studies have attempted to perform hPSC culture in batch mode at low inoculation densities of 2×10^4 cells/mL but have found that the volumetric yield of cells per mL of media used is essentially futile for the design of a clinically relevant bioprocess [103].

4.2.2 Fed-Batch

Fed-batch cultivation involves addition of a constant stream of one or more components (e.g. glucose, amino acids, etc) throughout the culture, resulting in an overall increase in the volume within the vessel over the duration of the process. Fed-batch operation has seen widespread use in the biomanufacturing industry, as it allows cells to reach higher densities by delaying nutrient exhaustion and allowing cells to remain in exponential growth [104]. However, to date, no studies on hPSC cultivation have used this method due to the sensitivity of stem cells to rapidly accumulating inhibitory metabolites and the limited stability of essential cytokines and growth factors [9].

4.2.3 Repeated-Batch

Repeated-batch has established itself as one of the main operation modes of choice for hPSC cultivation. Beginning with an initial batch phase, the repeated-batch approach consists of multiple feeding cycles (usually every 24 h) with partial or complete media change over the culture duration [105]. This provides the cells with a renewed supply of essential nutrients and factors and allows them to maintain their pluripotency and expansion. Repeated-batch cultivation has been successfully and extensively employed for both microcarrier and aggregate suspensions with cell yields reaching over 2×10^6 cells per mL [74, 95, 106].

However, media exchanges by their nature result in drastic and sudden shifts in the culture conditions, in terms of pH, nutrient, metabolite, and growth factor concentrations. This results in an interruption to the cultivation process which some have reported results in a linear growth curve, as opposed to the desired exponential growth necessary for industrial manufacture [9].

4.2.4 Perfusion

Perfusion systems are the most complex mode of operation for cell cultivation. During perfusion (or continuous) culture, cells are retained within the vessel while the media is continuously exchanged, providing a constant stream of fresh nutrients and growth factors while simultaneously removing waste metabolites such as lactate [107, 108]. This, in theory, ensures the cells remain in as homogeneous environment as possible during the culture period. Unlike repeated-batch processes, perfusion does not require the interruption of the culture, and can theoretically support the highest cell densities of all operation modes described thus far [109]. A study by Kropp and colleagues performed a comparison between repeated-batch and perfusion systems using two hiPSC lines expanded in 300 mL single-use BioBLU bioreactors. They demonstrated that the perfusion system resulted in a 47% higher final cell density (2.85×10^6 cells/mL) compared to using repeated-batch (1.94×10^6 cells/mL) [105].

In practice, however, such continuous systems carry significant drawbacks that must be overcome if they are to become industrially viable. Continuous media exchanges carry the highest risk of process contamination (with batches being the lowest), as media is constantly circulating in and out of the vessel [109]. Furthermore, it is a very mechanically complex and media-intensive process, requiring vast amounts of expensive formulations to remain productive [110]. This complexity significantly increases the cost of maintaining such a process and is currently one of the main reasons it is considered inefficient for large-scale operations. Additionally, cell retention using filters or gravimetric control presents some challenges due to the delicate nature of hPSCs [74, 105] (Fig. 1).

4.3 Towards cGMP Compliant Manufacturing

Since their initial discovery and isolation, hPSCs have been promoted as the next generation of regenerative medicine, both as a source of derived cells and as a therapeutic product themselves. In order to fulfil this promise and make their way into clinical use, hPSCs must be produced with strict accordance to the highest cGMP standards. This requirement exists in order to eliminate, to the best of our ability, immunogenic incidents, and introduction of foreign materials (such as animal viruses or prions) when stem cells are administered into human patients, and to ensure the highest possible quality of the final cell material [117–119]. Careful

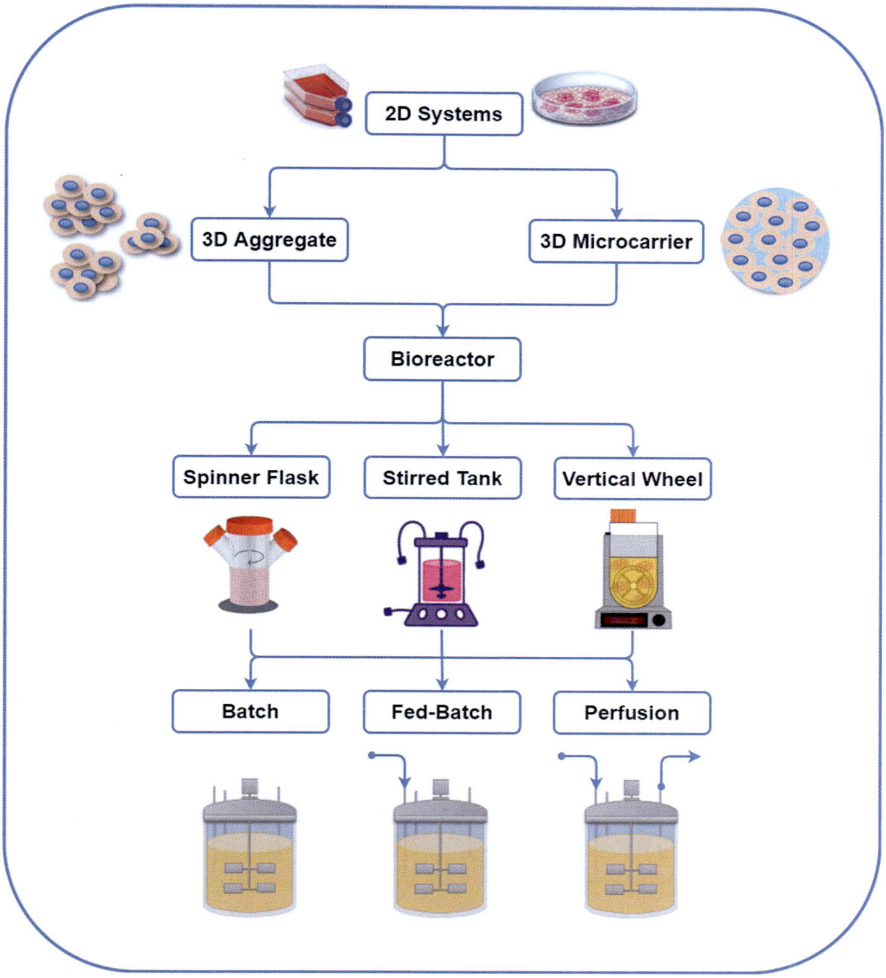

Fig. 1 Approaches to the scalable manufacturing of hPSCs. Initial systems for pluripotent cell manufacture were developed for 2D culture plates and flasks. Subsequently, scale-up was achieved through the use of free-floating hPSC aggregates or cells cultured adherently on microcarriers coated with ECM proteins. Initially, suspension cultures were demonstrated in spinner flasks in laboratory scale experiments; however, studies have since demonstrated the scalability of hPSC manufacture in stirred tanks and novel systems such as vertical wheel bioreactors. Adapted from [100, 111–116]

consideration must be given to the entirety of a manufacturing process, from the raw materials, through cell reprogramming strategies, to the design of the production facility and beyond.

4.3.1 Xenogeneic, Xeno- and Animal Component-Free

In practical terms, production processes that rely on materials derived from any species other than humans can be considered xenogeneic [84]. Xenogeneic processes introduce components that are foreign to the human immune system and can have deleterious effects if inadvertently introduced into a human patient. Due to the adherent nature of hPSCs, initial attempts to expand them in 2D static culture involved the use of murine embryonic fibroblasts (MEF) as feeder layers on culture dishes using a basal media such as DMEM and serum replacement, amino acids, nitrogen supplement and other growth factors; MEF-conditioned media has also seen frequent use [4, 82]. However, over the last few decades, hPSC culture has seen a steady shift from undefined, xenogeneic conditions, to more or fully defined xeno-free systems, using improvements to culture media formulations, extracellular matrix proteins and reprogramming strategies.

Media Optimizations

Stem cell culture media has seen progressive development in recent years, with studies suggesting that employing defined and xeno-free compositions could have a positive impact on differentiation potential. For example, a more recent study from 2019 demonstrated that the use of xeno-derived serum and animal components in the hPSC culturing media resulted in altered cell morphology and gene expression, and reduced their overall expansion potential [120]. The use of undefined animal or human serum in media can also produce unintended variability between batches [121]. Therefore, research has been focused on elucidating the essential growth factors and cytokines necessary for the activation of signalling pathways involved in maintaining pluripotency, such as the transforming growth factor-β (TFG-β) or basic fibroblast growth factor (bFGF) pathways [122]. In particular, it has been demonstrated that bFGF is essential for hESC self-renewal [82, 123].

The development of chemically defined media (CDM) has propelled hPSC culture towards cGMP compliance; examples of a CDM that has seen widespread use in the research and development field is STEMCELL's mTeSR1, created for use as a standardized formulation for feeder-free hPSC maintenance [124]. Multiple studies have effectively utilized it for hPSC expansion [39, 40, 68, 69]. However, it contains a bovine-derived albumin source, making it unsuitable for clinical use. Further advances in media formulations have led to the design of fully defined xeno-free (XFM), such as Gibco's StemPro SFM series and STEMCELL's TeSR2, which contain no components derived from non-human sources [18]. Using TeSR2, Fan et al. demonstrated a 20-fold expansion of hiPSCs in a xeno-free microcarrier suspension system [71]. Finally, animal component-free media (ACF) has recently seen widespread adoption due to its fully defined, fully animal-(and human) free formulation, making it suitable for processes destined for clinical applications. Notable examples of ACF include Gibco's Essential 8, STEMCELL's TeSR-E8 and Lonza's L7hPSC [121]. Recent studies with hiPSCs have demonstrated the

Table 3 List of popular media formulations by grade

Grade	Developer	Brand
CDM	STEMCELL	mTeSR1
XFM	STEMCELL	TeSR2
		mTeSR3D
	Gibco	StemPro SFM
		StemPro SFM Xenofree
ACF	STEMCELL	TeSR-E8
	Gibco	Essential 8
	Lonza	L7 hPSC

Adapted from [18]

successful use of these formulations, for example by Hamad and colleagues, who produced ~1 × 10^6/mL functional cardiomyocytes using E8 media in a DASGIP parallel bioreactor system [125]. An additional benefit of these formulations is the fact that they are only composed (as the names suggest) of the minimal factors necessary for stem cell maintenance, leading to a dramatic reduction in cost, up to 50% compared to TeSR2 [82] (Table 3).

ECM Optimizations

Extracellular matrix proteins allow hPSCs to attach and adhere to their surroundings. The first reported xeno-free hESC culture process was developed by Richards et al., who substituted MEF feeders in favor of human epithelial feeders [126]. However, feeder cells are inherently inefficient for large scale hPSC cultures due to their variability, safety concerns and scalability issues [15]. Feeder-free systems have since been developed to circumvent this using ECM coatings made of recombinant mouse and human peptides or entirely synthetic polymers [27, 127].

As previously stated, 3D aggregate suspension cultures circumvent the need for ECM optimizations. However, in microcarrier-based cultures beads have been developed with ECM protein coatings, most commonly laminin [128], collagen [129], fibronectin [130], vitronectin [131] or a combination therein. Early studies were performed similar to 2D cultures, using Matrigel-coated MCs [127]. Matrigel is, however, animal source-derived, and studies have been performed to elucidate the components responsible for supporting hPSC adhesion and stemness. Notably, various laminin isoforms have been shown to bind integrin $\alpha6\beta1$, mediating cell adhesion [128]; similar studies have recognised vitronectin and e-cadherin as important players in adhesion and self-renewal [131, 132]. Consequently, MCs coated with recombinant human isoforms of these proteins have been developed as a xeno-free alternative. Recent advances in the production of xeno-free materials has also prompted the adoption of synthetic polymers, such as the Synthemax II microcarriers (see Table 4). For instance, Silva et al. utilized Synthemax II MCs in a Cellferm-pro bioreactor with a working volume of 180 mL to expand hESCs 20-fold over 10 days [70].

Table 4 Common microcarriers used in hPSC manufacturing

Brand	Material	Surface	Developer
Synthemax II	Polystyrene	Synthemax II coating	Corning Inc
Cytodex 1	Dextran	DEAE	GE Healthcare
Cytodex 3	Dextran	Collagen	
Hillex II	Polystyrene	Cationic	SoloHill Engineering Inc
Plastic	Polystyrene	Neutral	
Plastic Plus	Polystyrene	Cationic	
DE-52	Cellulose	DEAE	Whatman
DE-53	Cellulose	DEAE	

Adapted from [60, 84]

4.3.2 Quality Control

Unlike established biologic drugs, where purified proteins secreted by cells during culture are the eventual therapeutic, the cells themselves represent the final product in hPSC processes. Rigorous testing with appropriate robustness and reproducibility is essential for the quality control of stem cell therapies [133]. Pluripotent cells respond quickly to unfavourable conditions in their microenvironment and must be actively monitored using both offline and online methods to ensure their stemness and differentiation potential remain unaffected, and these methods must be able to be integrated with 3D suspension culture for the purposes of cGMP-compliant scale-up [134].

Offline Monitoring

Offline techniques aim to assess critical hPSC parameters such as cell density and viability, aggregate characteristics, and pluripotency marker expression. Trypan Blue exclusion microscopy (along with other dye exclusion assays) are a staple technique used frequently to assess cell viability and concentration [135]. Flow cytometry and immunohistochemistry are regularly employed to quantify expression of various pluripotency markers, such as NANOG, SOX2, SSEA4, OCT4, TRA-1-60 and TRA-1-81 [39, 136]. Interestingly, a 2017 study by D'Antonio and colleagues suggested the use of fluorescence cell barcoding (FCB) coupled with flow cytometry as a high-throughput means of assessing pluripotency [137]. Assessing the metabolic state of hPSCs is another important quality control measure – high performance liquid chromatography (HPLC) and mass spectrometry (MS) have been employed to reveal key metabolic markers connected with pluripotency [18, 138]; real-time or quantitative PCR, coupled with ELISA screens are frequently used to detect expression of stemness-associated genes [27, 139, 140]. Additionally, hPSCs differentiation potential is commonly tested using embryoid body (EB) formation assays to determine whether the cells are able to generate cells from all 3 germ layers [141].

Online Monitoring

However, in order to progress towards industrial scale processes with multiple simultaneous batches, online monitoring approaches must be developed and optimized [94]. Offline assays, while accurate, should be complemented with online ones for successful industrial-scale bioprocesses. Nonautomated sample collection and preparation is slow and labour-intensive, and entails additional costs due to the requirement of specialized kit [18]. More importantly, they are usually destructive and impose a significant delay between sampling and availability of results, reducing their usefulness at large scale [142]. Ideally, technologies can be implemented that allow for real-time assessment of culture and cell conditions. Such sensors have been developed and are in use for the measurement of pH, oxygen, and glucose concentration, while spectroscopy has been employed for online metabolite monitoring [94, 143]. Three-dimensional fluorescence microscopy has been suggested for assessment of hPSC aggregates in culture, while label-free ion MS and Raman spectroscopy are potential candidates for non-invasive examination of pluripotency and other key factors [144, 145]. For instance, in 2016, Brauchle et al. described the use of Raman microspectroscopy to characterize hPSCs and identify differentiating cardiomyocytes [146].

Cryopreservation and Stability

Preservation of pluripotent stem cells is an essential requirement for ensuring the widespread availability of cells for clinical applications. Expanded cells can be preserved under cGMP as either master or working cell banks (MCB/WCB) and later thawed, expanded, and differentiated as needed [27, 82]. In-depth understanding of the effects of cryopreservation on hPSCs is vital for the creation of well-characterized starting materials for expansion and the reduction of process variability downstream. A recent study by Shafa et al. examined the long-term effects of cryopreservation on hPSC genomic stability, expansion and differentiation potential, and telomerase activity. Thawed cells displayed normal karyotype and were able to proliferate in both two- and three-dimensional cultures after 5 years of cryopreservation [147].

GMP Banks of hPSCs

One crucial avenue of ensuring the consistent supply of high-quality stem cells on a global scale is the generation of cGMP-compliant human stem cell banks [148]. Multiple studies have successfully explored strategies for creating hPSC banks [47, 149], but in order to promote a global hPSC trade network, a standardized framework must be implemented. To this end, projects such as the International Stem Cell Banking Initiative (ISCBI) have taken it upon themselves to deliver a harmonized set of principles to guide manufacturers in the preparation of hPSC banks [150]. Over the last decade, the ISCBI has championed the creation of a unified global system of standards detailing the best practices in banking of research- and clinical-grade hPSCs using state-of-the-art scientific methods [151, 152]. Guidelines encompass the complete journey of hPSC lines, from their ethical acquisition/generation, through tests assessing gene expression profile, cell identity, pluripotency, and differentiation capacity, to their distribution and shipment globally. Other initiatives have also been founded with similar intent, like the Global Alliance for iPSC therapies (GAiT), which serves as a resource for scientists and organisations working on iPSC-based therapies [153].

Quality by Design

As hPSC therapies draw closer to the general population, the biopharmaceutical industry and the regulatory agencies overseeing it have begun promoting a defined Quality by Design (QbD) approach to stem cell biomanufacturing [154, 155]. Based on the International Council for Harmonisation of Technical Requirements for Pharmaceuticals for Human Use (ICH) guidelines regarding the safe and responsible production of therapeutics [156], the QbD strategy promotes the use of risk assessment and robust and reliable analytical techniques to ensure consistency in product management. This system encourages project engineers and scientists to view and design manufacturing processes from an end-to-end standpoint as opposed to a chain of individual unit operations. Designing hPSC therapies with the QbD framework will result in closed, automatized systems with the ability to respond rapidly to and mitigate variations, thus ensuring a smooth and lean biomanufacturing process [155].

5 Advances in Production of iPSC-Derived Progenitors

One of the most promising aspects of hPSC manufacturing is the potential for the large-scale derivation of clinically useful cell types. Differentiation of embryonic and induced stem cells into functional cell types is itself a challenging and lengthy process. Since their isolation, an ever-increasing number of studies have demonstrated hPSC commitment towards a wide array of functional cells from all three germ layers (ecto-, endo- and mesoderm), including lineages of the brain, blood, liver, and many others (Fig. 2).

5.1 Cardiomyocytes

Perhaps one of the most therapeutically relevant cell types for modern medicine, hPSC-derived cardiomyocytes (CMs) have the potential to revolutionize the way

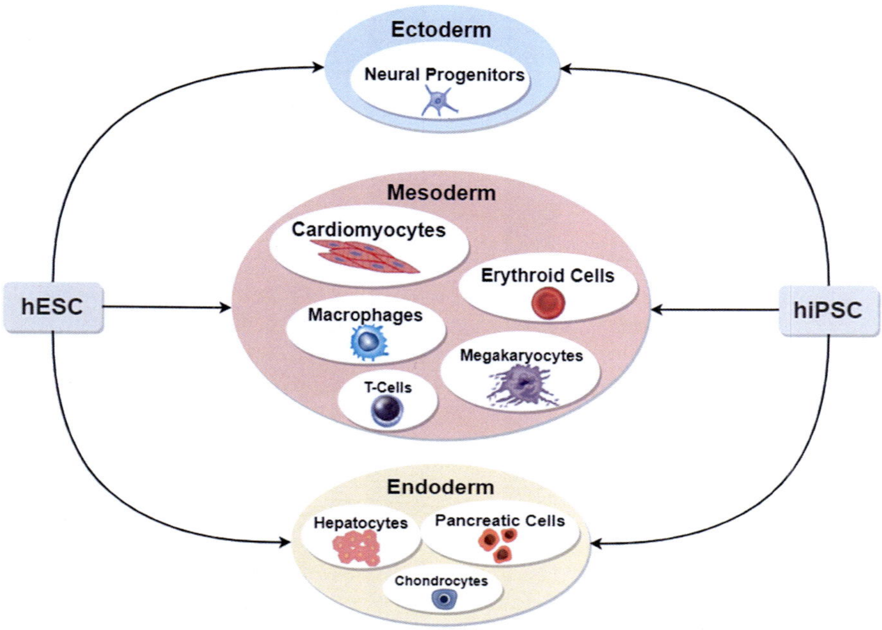

Fig. 2 Graphical illustration of cells and lineages derived from human embryonic stem cells (hESCs) and human induced pluripotent stem cells (hiPSCs). Notable cell types that have been obtained from directed differentiation of hPSCs in 3D culture include cardiomyocytes, erythrocytes, megakaryocytes, macrophages, neural progenitor cells, pancreatic cells, and hepatic cells. Adapted from [157–165]

cardiovascular diseases (CVDs) are treated. CVDs are widely regarded as the world's leading cause of death, surpassing even cancer [166]. Cardiac muscle tissue is under constant dynamic stress and is unable to proliferate [87]. Their loss during myocardial infarctions cannot be replenished by the body and CM engraftment is usually the only way to restore functionality to the damaged tissues [167]. However, donor shortages around the world have prompted researchers to turn towards pluripotent stem cells as a potentially limitless starting material for the production of functional cardiomyocytes.

Generation of CMs from hPSCs has received much attention due to the vast market availability for such therapies. Due to the number of potential recipients, many 3D scale-up strategies for hPSC-CM derivation have been explored, including both aggregate and microcarrier suspension systems [31, 81, 83, 95, 97, 100, 168–170]. Early reports in 2008 by Niebruegge et al. demonstrated aggregate hESC differentiation into mesoderm and eventually CMs in 125 mL DasGip cellferm-pro bioreactors with a yield of 6.3×10^5 cells/mL [171]. More recent studies using the same bioreactor system resulted in a titre of ~1×10^6 cells/mL using fully defined xeno-free E8 media [125]. Microcarrier cultures have also shown considerable scale-up promise, as reported by our group in 2020, using a 300 mL Biostat B-DCU to differentiate hiPSCs into 7.1 × 10^6 CMs per mL on Geltrex-coated Cytodex I microcarriers [172].

5.2 *Hematopoietic Cells*

Human pluripotent stem cell-derived hematopoietic stem cells (HSCs) and their lineages are an enticing direction for stem cell therapies. HSCs have a very high potential therapeutic value, given that they can be utilized to reconstitute a patient's hematopoietic system as a direct transplant, or to give rise to differentiated blood cells, for instance erythrocytes for blood transfusion, mature immune cells for immunotherapy and many others [173]. Much progress has been achieved in elucidating the pathways required for the generation of engraftable hPSC-HSCs, for instance by forced overexpression of key transcription factors (such as RUNX1, SPI1, FOSB and GFI1) or by attempting to recapitulate embryonic development and the hematopoietic niche within the bone marrow ex vivo [174–176]. Unfortunately, such attempts have demonstrated varying success with the generation of long-term repopulating HSCs and are difficult to scale up.

5.2.1 Erythroid Cells

Development of several key HSC-lineage therapies has seen extensive progress. Among these, the generation of hPSC-derived universal red blood cells has seen a large number of publications as the field has matured [177–181]. Lu et al. published a method for the differentiation of hESCs into functional erythrocytes in 6-well plates, producing an impressive 10^{10}–10^{11} cells per well [182]. However, scale-up cannot be achieved in plate culture, forcing researchers to focus on the development of 3D systems. Using a microcarrier embryoid body (EB) approach, our group has developed a scalable and fully defined xeno-free approach for the generation of red blood cells [66, 183]. This method has recently been demonstrated in fully agitated conditions, generating an average of ~7 × 10^8 hemoglobinized erythroid cells over 27 days of culture in 125 mL spinner flasks [184].

5.2.2 Macrophages

Another possible route for HSC-derived cells is the manufacture of macrophages. Macrophages play important roles in many diseases, such as hereditary alveolar proteinosis [185]. As such, they represent a potential source of therapeutically relevant cells. In 2008, Karlsson published the first robust and streamlined protocol for the generation of macrophages from hESCs without the need of coculture [186]. Further studies have led to gradual improvements in the protocol, with multiple groups reporting reproducible continuous generation of mature cells in small scale culture plates [187–189]. However, progress with 3D bioreactors has been slow. Recently, Ackermann and colleagues demonstrated the first scale-up of macrophage generation in 250 mL DASbox Mini bioreactors with online monitoring of pH, DO, temperature and biomass, resulting in continuous production of 1–3 $\times$ 10^7 macrophages per week over >5 weeks in culture [190].

5.2.3 Megakaryocytes

Megakaryocytes (MKs) are highly therapeutically relevant cells, mainly for the generation of platelets for transfusion use [191]. Traditionally, platelets have been isolated from donor blood, which suffers from limited donor numbers, contamination, and a limited shelf life [192]. Derivation of platelet-producing MKs in bioreactors would alleviate many of these issues and provide a consistent supply of platelets to patients with thrombocytopaenia. Multiple studies have demonstrated the ability to produce MKs and platelets on a small scale, for example in 3D collagen scaffolds or microfluidic bioreactor-on-a-chip devices [193, 194]. Importantly, in 2018 Eicke and colleagues described a 3D laminin 521-coated microcarrier based method of producing MK cells in 50 mL spinner flasks, generating a total of ~1.9 $\times$ 10^8 functional cells [195]. More research is needed, however, to produce clinically relevant numbers of platelets, as these studies only reported the production of up to 30–70 platelets per MK cell.

5.2.4 T Cells

The field of immunotherapy has gained significant traction over the last few years as CAR-T cell therapies such as Yescarta® and Kymriah® have secured regulatory approval and have entered clinical use after showing remarkable efficacy in patient trials [196]. However, both therapies carry staggering costs, notably from their difficult manufacturing process that relies on leukapheresis of peripheral blood to isolate autologous T cells, delivering a chimeric antigen receptor (CAR) to the T cells, and reinjecting them into the patient. HPSCs could potentially help alleviate this problem and provide a more straightforward and cheaper alternative. Monolayer systems have been described for hPSC to T cell specification [197], but significant progress in T cell suspension cultures is thus far lacking. Agitated bioreactor culture of T cells has been demonstrated by Costariol et al., but this method was carried out with T cells from healthy donors [198]. New advances in the field of HSC manufacturing will hopefully allow T-cell derivation in bioreactors and their eventual use as off-the-shelf products [173, 199, 200].

5.3 Neural Cells

In vitro generation of human neural cells can potentially be of great clinical importance for ex vivo disease modelling, drug discovery and potentially even treatment of stroke patients using tissue grafts of functional neural cells [201]. In particular, hPSCs-derived neural progenitor cells (NPCs) have received much attention because of their ability to mature into neurons. Generation of NPCs from hESCs has previously been described in lab studies [88, 202]. However, much remains to be addressed in order to enhance production into the pilot and industrial scales. Several studies have previously demonstrated the cultivation of hPSC-NPCs in spinner flasks using both aggregation and microcarrier approaches, with cell yields between 7×10^5–8×10^6 cells/mL [203–205]. Single-use vertical-wheel reactors have also been reported for NPC generation [101]. In another step forward for hPSC-NPC scale-up, Koenig et al. conducted a study using a 125 mL DASbox mini bioreactor to obtain 2×10^8 neural cells in a 3D aggregation suspension system [206].

5.4 Others

Although not the object of as much attention, significant progress has been made for the scalable production of multiple other cell types with applications in regenerative medicine. For instance, the in vitro generation of pancreatic and islet-like cells could have a marked effect on the treatment strategies for patients with Type I diabetes. To that end, several studies have shown the efficient differentiation of pancreatic cells in scalable suspension culture [207, 208]. Mihara and colleagues were able to produce up to 1.6×10^8 pancreatic progenitor cells (PPCs) in a 100 mL STR using the aggregation method [209], while Yabe et al. showed the hiPSC differentiation into islet-like cells in 30 mL spinner flasks to generate 5–6×10^7 total cells per vessel [210]. Other studies have focused on expanding cells of the definitive endoderm [129, 211], the hepatic lineage [212–214] and chondrocytes [215].

6 Future Prospects and Conclusion

Pluripotent stem cell culture has seen significant improvements over the last few decades. As regenerative medicine has slowly transitioned from fiction to possibility, traditional two-dimensional hPSC culture systems have proven themselves largely unfit for the task of expanding and maintaining the vast quantities of cells necessary for therapeutic purposes. Issues with space- and cost-efficiency, labour-intensive culture and batch-to-batch variability have precluded their scale-up into industrial manufacturing of pluripotent cells [9].

Research and development across the pharmaceutical industry has instead shifted towards the promise of scalable three-dimensional manufacturing technologies in the form of suspension bioreactors. Extensive research have been put into adapting bioreactor platforms currently used for mammalian cell culture for hPSC cultivation instead, but elucidating the exact conditions and mechanisms by which high-quality and robust expansion can be achieved remains an unsolved task [60]. The main focus of current work is in hPSC cultivation performed either as free-floating aggregates or as adherent cells on microcarrier beads. Both have shown great promise for hPSC expansion and differentiation into specific lineages of interest, but significant challenges will need to be addressed for therapies to emerge on the market [18].

In order for the field of hPSC manufacturing in suspension culture to fully mature, issues pertaining to process variability must be solved. Bioreactor and agitator configurations, dimensions and operating parameters must be critically evaluated – traditional stirred tanks have been shown to negatively impact delicate hPSCs, while newer variants such as the vertical-wheel bioreactor could potentially provide excellent mixing in a low-shear environment [55, 101]. New technologies for online monitoring of culture conditions are also slowly being developed, which will allow better and quicker understanding of changing process parameters and adequate and timely responses to avoid batch failures. This will be of vital importance going forward, as clinical-grade hPSC cultivation moves towards clinical production and stringent adherence to cGMP standards becomes a requirement [18]. Media and extracellular matrix optimizations have already allowed for xeno- and animal-free culture systems to be employed at lab and pilot scale, but more work is needed to reduce the cost of growth factors and ECM proteins for a commercially viable bioprocess [82].

However, despite major obstacles on the road ahead, there is ample cause for a positive outlook on the stem cell biomanufacturing field. Researchers have successfully demonstrated expansion of highly enriched populations of therapeutically relevant cell types, such as cardiomyocytes, erythrocytes, pancreatic cells, and more. Scale-up issues are gradually being overcome, with many studies now being performed in multiple litre-bioreactors. Xeno-free systems have become an industry standard and are capable of maintaining hPSCs in their undifferentiated state for prolonged culture times, and new understanding of the underlying molecular and genetic effects of culture conditions have allowed gradual optimizations for more robust and reproducible systems. Ensuring the safety and therapeutic efficacy of these complicated systems is an enormous challenge, but the studies presented herein encourages the viewpoint that slowly but surely stem cell therapies will achieve their promise of revolutionizing human medicine.

Acknowledgements We thank the Agency for Science Technology and Research (A∗STAR) as well as the SINGA scholarship for providing the funding for this work.

References

1. Bragança J, Lopes JA, Mendes-Silva L, Santos JMA (2019) Induced pluripotent stem cells, a giant leap for mankind therapeutic applications. World J Stem Cells 11:421–430. https://doi.org/10.4252/wjsc.v11.i7.421
2. Romito A, Cobellis G (2016) Pluripotent stem cells: current understanding and future directions. Stem Cells Int 2016:1–20. https://doi.org/10.1155/2016/9451492
3. Evans MJ, Kaufman MH (1981) Establishment in culture of pluripotential cells from mouse embryos. Nature 292:154–156. https://doi.org/10.1038/292154a0
4. Thomson JA (1998) Embryonic stem cell lines derived from human blastocysts. Science 282:1145–1147. https://doi.org/10.1126/science.282.5391.1145
5. Takahashi K, Tanabe K, Ohnuki M et al (2007) Induction of pluripotent stem cells from adult human fibroblasts by defined factors. Cell 131:861–872. https://doi.org/10.1016/j.cell.2007.11.019
6. Chhabra A (2017) Inherent immunogenicity or lack thereof of pluripotent stem cells: implications for cell replacement therapy. Front Immunol 8:993. https://doi.org/10.3389/fimmu.2017.00993
7. Ilic D, Devito L, Miere C, Codognotto S (2015) Human embryonic and induced pluripotent stem cells in clinical trials: table 1. Br Med Bull 116:ldv045. https://doi.org/10.1093/bmb/ldv045
8. Baghbaderani BA, Tian X, Neo BH et al (2015) cGMP-manufactured human induced pluripotent stem cells are available for pre-clinical and clinical applications. Stem Cell Rep 5:647–659. https://doi.org/10.1016/j.stemcr.2015.08.015
9. Kropp C, Massai D, Zweigerdt R (2017) Progress and challenges in large-scale expansion of human pluripotent stem cells. Process Biochem 59:244–254. https://doi.org/10.1016/j.procbio.2016.09.032
10. Fan Y, Wu J, Ashok P et al (2015) Production of human pluripotent stem cell therapeutics under defined xeno-free conditions: progress and challenges. Stem Cell Rev Rep 11:96–109. https://doi.org/10.1007/s12015-014-9544-x
11. Lee E, Sivalingam J, Lim ZR et al (2018) Review: in vitro generation of red blood cells for transfusion medicine: progress, prospects and challenges. Biotechnol Adv 36:2118–2128. https://doi.org/10.1016/j.biotechadv.2018.09.006
12. Rodrigues CAV, Fernandes TG, Diogo MM et al (2011) Stem cell cultivation in bioreactors. Biotechnol Adv 29:815–829. https://doi.org/10.1016/j.biotechadv.2011.06.009
13. Trounson A, DeWitt ND (2016) Pluripotent stem cells progressing to the clinic. Nat Rev Mol Cell Biol 17:194–200. https://doi.org/10.1038/nrm.2016.10
14. Mannello F, Tonti GA (2007) Concise review: no breakthroughs for human mesenchymal and embryonic stem cell culture: conditioned medium, feeder layer, or feeder-free; medium with fetal calf serum, human serum, or enriched plasma; serum-free, serum replacement nonconditioned medium, or ad hoc formula? All that glitters is not gold! Stem Cells 25:1603–1609. https://doi.org/10.1634/stemcells.2007-0127
15. Dashtban M, Panchalingam KM, Shafa M, Ahmadian Baghbaderani B (2020) Addressing manufacturing challenges for commercialization of iPSC-based therapies. Springer US, New York
16. Villa-Diaz LG, Ross AM, Lahann J, Krebsbach PH (2013) Concise review: the evolution of human pluripotent stem cell culture: from feeder cells to synthetic coatings. Stem Cells 31:1–7. https://doi.org/10.1002/stem.1260
17. Crook JM, Peura TT, Kravets L et al (2007) The generation of six clinical-grade human embryonic stem cell lines. Cell Stem Cell 1:490–494. https://doi.org/10.1016/j.stem.2007.10.004
18. Polanco A, Kuang B, Yoon S (2020) Bioprocess technologies that preserve the quality of iPSCs. Trends Biotechnol 38:1128. S0167779920300664. https://doi.org/10.1016/j.tibtech.2020.03.006

19. Eaker S, Armant M, Brandwein H et al (2013) Concise review: guidance in developing commercializable autologous/patient-specific cell therapy manufacturing. Stem Cells Transl Med 2:871–883. https://doi.org/10.5966/sctm.2013-0050
20. Tirughana R, Metz MZ, Li Z et al (2018) GMP production and scale-up of adherent neural stem cells with a quantum cell expansion system. Mol Ther – Methods Clin Dev 10:48–56. https://doi.org/10.1016/j.omtm.2018.05.006
21. McKee C, Chaudhry GR (2017) Advances and challenges in stem cell culture. Colloids Surf B: Biointerfaces 159:62–77. https://doi.org/10.1016/j.colsurfb.2017.07.051
22. Terstegge S, Laufenberg I, Pochert J et al (2007) Automated maintenance of embryonic stem cell cultures. Biotechnol Bioeng 96:195–201. https://doi.org/10.1002/bit.21061
23. Thomas RJ, Anderson D, Chandra A et al (2009) Automated, scalable culture of human embryonic stem cells in feeder-free conditions. Biotechnol Bioeng 102:1636–1644. https://doi.org/10.1002/bit.22187
24. Want AJ, Nienow AW, Hewitt CJ, Coopman K (2012) Large-scale expansion and exploitation of pluripotent stem cells for regenerative medicine purposes: beyond the T flask. Regen Med 7:71–84. https://doi.org/10.2217/rme.11.101
25. Liu W, Deng C, Godoy-Parejo C et al (2019) Developments in cell culture systems for human pluripotent stem cells. World J Stem Cells 11:968–981. https://doi.org/10.4252/wjsc.v11.i11.968
26. Veraitch FS, Scott R, Wong J-W et al (2008) The impact of manual processing on the expansion and directed differentiation of embryonic stem cells. Biotechnol Bioeng 99:1216–1229. https://doi.org/10.1002/bit.21673
27. Abbasalizadeh S, Baharvand H (2013) Technological progress and challenges towards cGMP manufacturing of human pluripotent stem cells based therapeutic products for allogeneic and autologous cell therapies. Biotechnol Adv 31:1600–1623. https://doi.org/10.1016/j.biotechadv.2013.08.009
28. Singh H, Mok P, Balakrishnan T et al (2010) Up-scaling single cell-inoculated suspension culture of human embryonic stem cells. Stem Cell Res 4:165–179. https://doi.org/10.1016/j.scr.2010.03.001
29. Zweigerdt R (2009) Large scale production of stem cells and their derivatives. In: Martin U (ed) Engineering of stem cells. Springer Berlin Heidelberg, Berlin, Heidelberg, pp 201–235
30. Jensen C, Teng Y (2020) Is it time to start transitioning from 2D to 3D cell culture? Front Mol Biosci 7:33. https://doi.org/10.3389/fmolb.2020.00033
31. Branco MA, Cotovio JP, Rodrigues CAV et al (2019) Transcriptomic analysis of 3D cardiac differentiation of human induced pluripotent stem cells reveals faster cardiomyocyte maturation compared to 2D culture. Sci Rep 9:9229. https://doi.org/10.1038/s41598-019-45047-9
32. Hodge AJ, Zhong J, Lipke EA (2016) Enhanced stem cell-derived cardiomyocyte differentiation in suspension culture by delivery of nitric oxide using *S* -nitrosocysteine: no enhances cardiomyocyte differentiation. Biotechnol Bioeng 113:882–894. https://doi.org/10.1002/bit.25849
33. Li Y (2014) Dendritic cells derived from pluripotent stem cells: potential of large scale production. World J Stem Cells 6:1. https://doi.org/10.4252/wjsc.v6.i1.1
34. Guo N-N, Liu L-P, Zheng Y-W, Li Y-M (2020) Inducing human induced pluripotent stem cell differentiation through embryoid bodies: a practical and stable approach. World J Stem Cells 12:25–34. https://doi.org/10.4252/wjsc.v12.i1.25
35. Chen VC, Ye J, Shukla P et al (2015) Development of a scalable suspension culture for cardiac differentiation from human pluripotent stem cells. Stem Cell Res 15:365–375. https://doi.org/10.1016/j.scr.2015.08.002
36. Watanabe K, Ueno M, Kamiya D et al (2007) A ROCK inhibitor permits survival of dissociated human embryonic stem cells. Nat Biotechnol 25:681–686. https://doi.org/10.1038/nbt1310
37. Krawetz R, Taiani JT, Liu S et al (2010) Large-scale expansion of pluripotent human embryonic stem cells in stirred-suspension bioreactors. Tissue Eng Part C Methods 16:573–582. https://doi.org/10.1089/ten.tec.2009.0228

38. Abbasalizadeh S, Larijani MR, Samadian A, Baharvand H (2012) Bioprocess development for mass production of size-controlled human pluripotent stem cell aggregates in stirred suspension bioreactor. Tissue Eng Part C Methods 18:831–851. https://doi.org/10.1089/ten.tec.2012.0161
39. Kwok CK, Ueda Y, Kadari A et al (2018) Scalable stirred suspension culture for the generation of billions of human induced pluripotent stem cells using single-use bioreactors. J Tissue Eng Regen Med 12:e1076. https://doi.org/10.1002/term.2435
40. Borys BS, So T, Colter J et al (2020) Optimized serial expansion of human induced pluripotent stem cells using low-density inoculation to generate clinically relevant quantities in vertical-wheel bioreactors. Stem Cells Transl Med 9:1036. sctm.19-0406. https://doi.org/10.1002/sctm.19-0406
41. Meng G, Liu S, Poon A, Rancourt DE (2017) Optimizing human induced pluripotent stem cell expansion in stirred-suspension culture. Stem Cells Dev 26:1804–1817. https://doi.org/10.1089/scd.2017.0090
42. Abecasis B, Aguiar T, Arnault É et al (2017) Expansion of 3D human induced pluripotent stem cell aggregates in bioreactors: bioprocess intensification and scaling-up approaches. J Biotechnol 246:81–93. https://doi.org/10.1016/j.jbiotec.2017.01.004
43. Elanzew A, Sommer A, Pusch-Klein A et al (2015) A reproducible and versatile system for the dynamic expansion of human pluripotent stem cells in suspension. Biotechnol J 10:1589–1599. https://doi.org/10.1002/biot.201400757
44. Wang Y, Chou B-K, Dowey S et al (2013) Scalable expansion of human induced pluripotent stem cells in the defined xeno-free E8 medium under adherent and suspension culture conditions. Stem Cell Res 11:1103–1116. https://doi.org/10.1016/j.scr.2013.07.011
45. Haraguchi Y, Matsuura K, Shimizu T et al (2015) Simple suspension culture system of human iPS cells maintaining their pluripotency for cardiac cell sheet engineering: expansion/cardiac differentiation of hiPSCs by suspension culture. J Tissue Eng Regen Med 9:1363–1375. https://doi.org/10.1002/term.1761
46. Olmer R, Lange A, Selzer S et al (2012) Suspension culture of human pluripotent stem cells in controlled, stirred bioreactors. Tissue Eng Part C Methods 18:772–784. https://doi.org/10.1089/ten.tec.2011.0717
47. Chen VC, Couture SM, Ye J et al (2012) Scalable GMP compliant suspension culture system for human ES cells. Stem Cell Res 8:388–402. https://doi.org/10.1016/j.scr.2012.02.001
48. Zweigerdt R, Olmer R, Singh H et al (2011) Scalable expansion of human pluripotent stem cells in suspension culture. Nat Protoc 6:689–700. https://doi.org/10.1038/nprot.2011.318
49. Amit M, Laevsky I, Miropolsky Y et al (2011) Dynamic suspension culture for scalable expansion of undifferentiated human pluripotent stem cells. Nat Protoc 6:572–579. https://doi.org/10.1038/nprot.2011.325
50. Olmer R, Haase A, Merkert S et al (2010) Long term expansion of undifferentiated human iPS and ES cells in suspension culture using a defined medium. Stem Cell Res 5:51–64. https://doi.org/10.1016/j.scr.2010.03.005
51. Amit M, Chebath J, Margulets V et al (2010) Suspension culture of undifferentiated human embryonic and induced pluripotent stem cells. Stem Cell Rev Rep 6:248–259. https://doi.org/10.1007/s12015-010-9149-y
52. Côme J, Nissan X, Aubry L et al (2008) Improvement of culture conditions of human embryoid bodies using a controlled perfused and dialyzed bioreactor system. Tissue Eng Part C Methods 14:289–298. https://doi.org/10.1089/ten.tec.2008.0029
53. Gerecht-Nir S, Cohen S, Itskovitz-Eldor J (2004) Bioreactor cultivation enhances the efficiency of human embryoid body (hEB) formation and differentiation. Biotechnol Bioeng 86:493–502. https://doi.org/10.1002/bit.20045
54. Van Winkle AP, Gates ID, Kallos MS (2012) Mass transfer limitations in embryoid bodies during human embryonic stem cell differentiation. Cells Tissues Organs 196:34–47. https://doi.org/10.1159/000330691

55. Kinney MA, Sargent CY, McDevitt TC (2011) The multiparametric effects of hydrodynamic environments on stem cell culture. Tissue Eng Part B Rev 17:249–262. https://doi.org/10.1089/ten.teb.2011.0040
56. De Angelis MT, Parrotta EI, Santamaria G, Cuda G (2018) Short-term retinoic acid treatment sustains pluripotency and suppresses differentiation of human induced pluripotent stem cells. Cell Death Dis 9:6. https://doi.org/10.1038/s41419-017-0028-1
57. Koh B, Sulaiman N, Fauzi MB et al (2020) Three dimensional microcarrier system in mesenchymal stem cell culture: a systematic review. Cell Biosci 10:75. https://doi.org/10.1186/s13578-020-00438-8
58. Tavassoli H, Alhosseini SN, Tay A et al (2018) Large-scale production of stem cells utilizing microcarriers: a biomaterials engineering perspective from academic research to commercialized products. Biomaterials 181:333–346. https://doi.org/10.1016/j.biomaterials.2018.07.016
59. Malda J, Frondoza CG (2006) Microcarriers in the engineering of cartilage and bone. Trends Biotechnol 24:299–304. https://doi.org/10.1016/j.tibtech.2006.04.009
60. Badenes SM, Fernandes TG, Rodrigues CAV et al (2016) Microcarrier-based platforms for in vitro expansion and differentiation of human pluripotent stem cells in bioreactor culture systems. J Biotechnol 234:71–82. https://doi.org/10.1016/j.jbiotec.2016.07.023
61. Storm MP, Orchard CB, Bone HK et al (2010) Three-dimensional culture systems for the expansion of pluripotent embryonic stem cells. Biotechnol Bioeng 107:683–695. https://doi.org/10.1002/bit.22850
62. Frondoza C, Sohrabi A, Hungerford D (1996) Human chondrocytes proliferate and produce matrix components in microcarrier suspension culture. Biomaterials 17:879–888. https://doi.org/10.1016/0142-9612(96)83283-2
63. Lam AT-L, Li J, Chen AK-L et al (2014) Cationic surface charge combined with either vitronectin or laminin dictates the evolution of human embryonic stem cells/microcarrier aggregates and cell growth in agitated cultures. Stem Cells Dev 23:1688–1703. https://doi.org/10.1089/scd.2013.0645
64. Malda J, Kreijveld E, Temenoff JS et al (2003) Expansion of human nasal chondrocytes on macroporous microcarriers enhances redifferentiation. Biomaterials 24:5153–5161. https://doi.org/10.1016/S0142-9612(03)00428-9
65. Zhang Z, Eyster TW, Ma PX (2016) Nanostructured injectable cell microcarriers for tissue regeneration. Nanomedicine 11:1611–1628. https://doi.org/10.2217/nnm-2016-0083
66. Sivalingam J, Lam AT-L, Chen HY et al (2016) Superior red blood cell generation from human pluripotent stem cells through a novel microcarrier-based embryoid body platform. Tissue Eng Part C Methods 22:765–780. https://doi.org/10.1089/ten.tec.2015.0579
67. Pandey PR, Tomney A, Woon MT et al (2019) End-to-end platform for human pluripotent stem cell manufacturing. Int J Mol Sci 21:89. https://doi.org/10.3390/ijms21010089
68. Badenes SM, Fernandes TG, Miranda CC et al (2017) Long-term expansion of human induced pluripotent stem cells in a microcarrier-based dynamic system: long-term expansion of hiPS cells on microcarriers. J Chem Technol Biotechnol 92:492–503. https://doi.org/10.1002/jctb.5074
69. Lam AT-L, Li J, Chen AK-L et al (2015) Improved human pluripotent stem cell attachment and spreading on xeno-free Laminin-521-coated microcarriers results in efficient growth in agitated cultures. BioResearch Open Access 4:242–257. https://doi.org/10.1089/biores.2015.0010
70. Silva MM, Rodrigues AF, Correia C et al (2015) Robust expansion of human pluripotent stem cells: integration of bioprocess design with transcriptomic and metabolomic characterization: hPSC expansion bioprocess and "omics" tools. Stem Cells Transl Med 4:731–742. https://doi.org/10.5966/sctm.2014-0270
71. Fan Y, Hsiung M, Cheng C, Tzanakakis ES (2013) Facile engineering of xeno-free microcarriers for the scalable cultivation of human pluripotent stem cells in stirred suspension. Tissue Eng Part A 20:588. 131128071850006. https://doi.org/10.1089/ten.tea.2013.0219

72. Badenes SM, Fernandes TG, Rodrigues CAV et al (2014) Scalable expansion of human-induced pluripotent stem cells in xeno-free microcarriers. In: Turksen K (ed) Stem cells and good manufacturing practices. Springer New York, New York, pp 23–29
73. Chen AK-L, Chen X, Choo ABH et al (2011) Critical microcarrier properties affecting the expansion of undifferentiated human embryonic stem cells. Stem Cell Res 7:97–111. https://doi.org/10.1016/j.scr.2011.04.007
74. Serra M, Brito C, Sousa MFQ et al (2010) Improving expansion of pluripotent human embryonic stem cells in perfused bioreactors through oxygen control. J Biotechnol 148:208–215. https://doi.org/10.1016/j.jbiotec.2010.06.015
75. Leung HW, Chen A, Choo ABH et al (2011) Agitation can induce differentiation of human pluripotent stem cells in microcarrier cultures. Tissue Eng Part C Methods 17:165–172. https://doi.org/10.1089/ten.tec.2010.0320
76. Oh SKW, Chen AK, Mok Y et al (2009) Long-term microcarrier suspension cultures of human embryonic stem cells. Stem Cell Res 2:219–230. https://doi.org/10.1016/j.scr.2009.02.005
77. Kehoe DE, Jing D, Lock LT, Tzanakakis ES (2010) Scalable stirred-suspension bioreactor culture of human pluripotent stem cells. Tissue Eng Part A 16:405–421. https://doi.org/10.1089/ten.tea.2009.0454
78. Fernandes AM, Marinho PAN, Sartore RC et al (2009) Successful scale-up of human embryonic stem cell production in a stirred microcarrier culture system. Braz J Med Biol Res 42:515–522. https://doi.org/10.1590/S0100-879X2009000600007
79. Phillips BW, Horne R, Lay TS et al (2008) Attachment and growth of human embryonic stem cells on microcarriers. J Biotechnol 138:24–32. https://doi.org/10.1016/j.jbiotec.2008.07.1997
80. Yang HS, Jeon O, Bhang SH et al (2010) Suspension culture of mammalian cells using thermosensitive microcarrier that allows cell detachment without proteolytic enzyme treatment. Cell Transplant 19:1123–1132. https://doi.org/10.3727/096368910X516664
81. Lecina M, Ting S, Choo A et al (2010) Scalable platform for human embryonic stem cell differentiation to cardiomyocytes in suspended microcarrier cultures. Tissue Eng Part C Methods 16:1609–1619. https://doi.org/10.1089/ten.tec.2010.0104
82. Wang Y, Cheng L, Gerecht S (2014) Efficient and scalable expansion of human pluripotent stem cells under clinically compliant settings: a view in 2013. Ann Biomed Eng 42:1357–1372. https://doi.org/10.1007/s10439-013-0921-4
83. Rodrigues AL, Rodrigues CAV, Gomes AR et al (2019) Dissolvable microcarriers allow scalable expansion and harvesting of human induced pluripotent stem cells under xeno-free conditions. Biotechnol J 14:1800461. https://doi.org/10.1002/biot.201800461
84. Badenes SM, Fernandes-Platzgummer A, Rodrigues CAV et al (2016) Microcarrier culture systems for stem cell manufacturing. In: Stem cell manufacturing. Elsevier, Amsterdam, pp 77–104
85. Gasperini L, Mano JF, Reis RL (2014) Natural polymers for the microencapsulation of cells. J R Soc Interface 11:20140817. https://doi.org/10.1098/rsif.2014.0817
86. Hernández RM, Orive G, Murua A, Pedraz JL (2010) Microcapsules and microcarriers for in situ cell delivery☆. Adv Drug Deliv Rev 62:711–730. https://doi.org/10.1016/j.addr.2010.02.004
87. Le MNT, Hasegawa K (2019) Expansion culture of human pluripotent stem cells and production of cardiomyocytes. Bioengineering 6:48. https://doi.org/10.3390/bioengineering6020048
88. Lei Y, Schaffer DV (2013) A fully defined and scalable 3D culture system for human pluripotent stem cell expansion and differentiation. Proc Natl Acad Sci 110:E5039–E5048. https://doi.org/10.1073/pnas.1309408110
89. Gupta P, Ismadi M-Z, Verma PJ et al (2016) Optimization of agitation speed in spinner flask for microcarrier structural integrity and expansion of induced pluripotent stem cells. Cytotechnology 68:45–59. https://doi.org/10.1007/s10616-014-9750-z
90. Chu L, Robinson DK (2001) Industrial choices for protein production by large-scale cell culture. Curr Opin Biotechnol 12:180–187. https://doi.org/10.1016/S0958-1669(00)00197-X

91. Schirmer C, Nussbaumer T, Schöb R et al (2018) Development, engineering and biological characterization of stirred tank bioreactors. In: Yeh M-K, Chen Y-C (eds) Biopharmaceuticals. InTech, London
92. Doran PM (2013) Reactor engineering. In: Bioprocess engineering principles. Elsevier, London, pp 761–852
93. Serra M, Brito C, Correia C, Alves PM (2012) Process engineering of human pluripotent stem cells for clinical application. Trends Biotechnol 30:350–359. https://doi.org/10.1016/j.tibtech.2012.03.003
94. dos Santos FF, Andrade PZ, da Silva CL, Cabral JMS (2013) Bioreactor design for clinical-grade expansion of stem cells. Biotechnol J 8:644–654. https://doi.org/10.1002/biot.201200373
95. Badenes SM, Fernandes TG, Cordeiro CSM et al (2016) Defined essential 8^{TM} medium and vitronectin efficiently support scalable xeno-free expansion of human induced pluripotent stem cells in stirred microcarrier culture systems. PLoS One 11:e0151264. https://doi.org/10.1371/journal.pone.0151264
96. Papoutsakis E (1991) Fluid-mechanical damage of animal cells in bioreactors. Trends Biotechnol 9:427–437. https://doi.org/10.1016/0167-7799(91)90145-8
97. Shafa M, Panchalingam KM, Walsh T et al (2019) Computational fluid dynamics modeling, a novel, and effective approach for developing scalable cell therapy manufacturing processes. Biotechnol Bioeng 116:3228–3241. https://doi.org/10.1002/bit.27159
98. Phelan MA, Gianforcaro AL, Gerstenhaber JA, Lelkes PI (2019) An air bubble-isolating rotating wall vessel bioreactor for improved spheroid/organoid formation. Tissue Eng Part C Methods 25:479–488. https://doi.org/10.1089/ten.tec.2019.0088
99. Croughan MS, Giroux D, Fang D, Lee B (2016) Novel single-use bioreactors for scale-up of anchorage-dependent cell manufacturing for cell therapies. In: Stem cell manufacturing. Elsevier, Boston, pp 105–139
100. Nogueira DES, Rodrigues CAV, Carvalho MS et al (2019) Strategies for the expansion of human induced pluripotent stem cells as aggregates in single-use Vertical-Wheel^{TM} bioreactors. J Biol Eng 13:74. https://doi.org/10.1186/s13036-019-0204-1
101. Rodrigues CA, Silva TP, Nogueira DE et al (2018) Scalable culture of human induced pluripotent cells on microcarriers under xeno-free conditions using single-use vertical-wheel^{TM} bioreactors: expansion of human induced pluripotent stem cells in vertical-wheel bioreactors. J Chem Technol Biotechnol 93:3597–3606. https://doi.org/10.1002/jctb.5738
102. Marinho PA, Chailangkarn T, Muotri AR (2015) Systematic optimization of human pluripotent stem cells media using design of experiments. Sci Rep 5:9834. https://doi.org/10.1038/srep09834
103. Hunt MM, Meng G, Rancourt DE et al (2014) Factorial experimental design for the culture of human embryonic stem cells as aggregates in stirred suspension bioreactors reveals the potential for interaction effects between bioprocess parameters. Tissue Eng Part C Methods 20:76–89. https://doi.org/10.1089/ten.tec.2013.0040
104. Yang WC (2019) Mammalian fed-batch cell culture for biopharmaceuticals. In: Lee GM, Faustrup Kildegaard H, Lee SY et al (eds) Cell culture engineering, 1st edn. Wiley, Weinheim, pp 313–345
105. Kropp C, Kempf H, Halloin C et al (2016) Impact of feeding strategies on the scalable expansion of human pluripotent stem cells in single-use stirred tank bioreactors: hPSC expansion in perfused single-use bioreactors. Stem Cells Transl Med 5:1289–1301. https://doi.org/10.5966/sctm.2015-0253
106. Bardy J, Chen AK, Lim YM et al (2013) Microcarrier suspension cultures for high-density expansion and differentiation of human pluripotent stem cells to neural progenitor cells. Tissue Eng Part C Methods 19:166–180. https://doi.org/10.1089/ten.tec.2012.0146
107. Castilho LR, Medronho RA (2002) Cell retention devices for suspended-cell perfusion cultures. In: Schügerl K, Zeng A-P, Aunins JG et al (eds) Tools and applications of biochemical engineering science. Springer Berlin Heidelberg, Berlin, Heidelberg, pp 129–169

108. Fong WJ, Tan HL, Choo A, Oh SKW (2005) Perfusion cultures of human embryonic stem cells. Bioprocess Biosyst Eng 27:381–387. https://doi.org/10.1007/s00449-005-0421-5
109. Castilho LR (2014) Continuous animal cell perfusion processes: the first step toward integrated continuous biomanufacturing. In: Subramanian G (ed) Continuous processing in pharmaceutical manufacturing. Wiley-VCH Verlag GmbH & Co. KGaA, Weinheim, pp 115–154
110. Yeo D, Kiparissides A, Cha JM et al (2013) Improving embryonic stem cell expansion through the combination of perfusion and bioprocess model design. PLoS One 8:e81728. https://doi.org/10.1371/journal.pone.0081728
111. Alhaque S, Themis M, Rashidi H (2018) Three-dimensional cell culture: from evolution to revolution. Philos Trans R Soc B Biol Sci 373:20170216. https://doi.org/10.1098/rstb.2017.0216
112. Chan SW, Rizwan M, Yim EKF (2020) Emerging methods for enhancing pluripotent stem cell expansion. Front Cell Dev Biol 8:70. https://doi.org/10.3389/fcell.2020.00070
113. García-Rendón A, Garibay-Escobar A, Guzmán R, Tejeda-Mansir A (2018) Plasmid-DNA lipid nanovaccines. In: Lipid nanocarriers for drug targeting. Elsevier, Amsterdam, pp 231–267
114. Merck (2020) Cell therapy manufacturing expansion vessels | Sigma-Aldrich. https://www.sigmaaldrich.com/technical-documents/articles/biology/cell-therapy-expansion-vessels.html. Accessed 26 Aug 2020
115. NIH (2020) Appendix C: human embryonic stem cells and human embryonic germ cells | stemcells.nih.gov. https://stemcells.nih.gov/info/2001report/appendixC.htm. Accessed 26 Aug 2020
116. ThermoFisher (2020) Thermo scientific nunc EasYFlask cell culture flasks – dishes, plates and flasks. https://www.fishersci.com/shop/products/nunc-cell-culture-treated-easyflasks/p-4524963. Accessed 26 Aug 2020
117. Andrews PW, Cavagnaro J, Deans R et al (2014) Harmonizing standards for producing clinical-grade therapies from pluripotent stem cells. Nat Biotechnol 32:724–726. https://doi.org/10.1038/nbt.2973
118. Migliaccio G, Pintus C (2012) Role of the EU framework in regulation of stem cell-based products. In: Weyand B, Dominici M, Hass R et al (eds) Mesenchymal stem cells – basics and clinical application II. Springer Berlin Heidelberg, Berlin, Heidelberg, pp 287–299
119. Yu G, Kamano Y, Wang F et al (2015) Feeder cell sources and feeder-free methods for human iPS cell culture. In: Sasaki K, Suzuki O, Takahashi N (eds) Interface oral health science 2014. Springer Japan, Tokyo, pp 145–159
120. McGrath M, Tam E, Sladkova M et al (2019) GMP-compatible and xeno-free cultivation of mesenchymal progenitors derived from human-induced pluripotent stem cells. Stem Cell Res Ther 10:11. https://doi.org/10.1186/s13287-018-1119-3
121. Chen G, Gulbranson DR, Hou Z et al (2011) Chemically defined conditions for human iPSC derivation and culture. Nat Methods 8:424–429. https://doi.org/10.1038/nmeth.1593
122. Pera MF, Tam PPL (2010) Extrinsic regulation of pluripotent stem cells. Nature 465:713–720. https://doi.org/10.1038/nature09228
123. Manning BD, Cantley LC (2007) AKT/PKB signaling: navigating downstream. Cell 129:1261–1274. https://doi.org/10.1016/j.cell.2007.06.009
124. The International Stem Cell Initiative Consortium, Akopian V, Andrews PW et al (2010) Comparison of defined culture systems for feeder cell free propagation of human embryonic stem cells. In Vitro Cell Dev Biol Anim 46:247–258. https://doi.org/10.1007/s11626-010-9297-z
125. Hamad S, Derichsweiler D, Papadopoulos S et al (2019) Generation of human induced pluripotent stem cell-derived cardiomyocytes in 2D monolayer and scalable 3D suspension bioreactor cultures with reduced batch-to-batch variations. Theranostics 9:7222–7238. https://doi.org/10.7150/thno.32058

126. Richards M, Fong C-Y, Chan W-K et al (2002) Human feeders support prolonged undifferentiated growth of human inner cell masses and embryonic stem cells. Nat Biotechnol 20:933–936. https://doi.org/10.1038/nbt726
127. Xu C, Inokuma MS, Denham J et al (2001) Feeder-free growth of undifferentiated human embryonic stem cells. Nat Biotechnol 19:971–974. https://doi.org/10.1038/nbt1001-971
128. Miyazaki T, Futaki S, Hasegawa K et al (2008) Recombinant human laminin isoforms can support the undifferentiated growth of human embryonic stem cells. Biochem Biophys Res Commun 375:27–32. https://doi.org/10.1016/j.bbrc.2008.07.111
129. Lock LT, Tzanakakis ES (2009) Expansion and differentiation of human embryonic stem cells to endoderm progeny in a microcarrier stirred-suspension culture. Tissue Eng Part A 15:2051–2063. https://doi.org/10.1089/ten.tea.2008.0455
130. Melkoumian Z, Weber JL, Weber DM et al (2010) Synthetic peptide-acrylate surfaces for long-term self-renewal and cardiomyocyte differentiation of human embryonic stem cells. Nat Biotechnol 28:606–610. https://doi.org/10.1038/nbt.1629
131. Braam SR, Zeinstra L, Litjens S et al (2008) Recombinant Vitronectin is a functionally defined substrate that supports human embryonic stem cell self-renewal via αVβ5 integrin. Stem Cells 26:2257–2265. https://doi.org/10.1634/stemcells.2008-0291
132. Nagaoka M, Si-Tayeb K, Akaike T, Duncan SA (2010) Culture of human pluripotent stem cells using completely defined conditions on a recombinant E-cadherin substratum. BMC Dev Biol 10:60. https://doi.org/10.1186/1471-213X-10-60
133. Stacey GN, Crook JM, Hei D, Ludwig T (2013) Banking human induced pluripotent stem cells: lessons learned from embryonic stem cells? Cell Stem Cell 13:385–388. https://doi.org/10.1016/j.stem.2013.09.007
134. Guilak F, Cohen DM, Estes BT et al (2009) Control of stem cell fate by physical interactions with the extracellular matrix. Cell Stem Cell 5:17–26. https://doi.org/10.1016/j.stem.2009.06.016
135. Ho MSH, Fryga A, Laslett AL (2011) Flow cytometric analysis of human pluripotent stem cells. In: Schwartz PH, Wesselschmidt RL (eds) Human pluripotent stem cells. Humana Press, Totowa, pp 221–230
136. Fong CY, Peh GSL, Gauthaman K, Bongso A (2009) Separation of SSEA-4 and TRA-1–60 labelled undifferentiated human embryonic stem cells from a heterogeneous cell population using magnetic-activated cell sorting (MACS) and fluorescence-activated cell sorting (FACS). Stem Cell Rev Rep 5:72–80. https://doi.org/10.1007/s12015-009-9054-4
137. D'Antonio M, Woodruff G, Nathanson JL et al (2017) High-throughput and cost-effective characterization of induced pluripotent stem cells. Stem Cell Rep 8:1101–1111. https://doi.org/10.1016/j.stemcr.2017.03.011
138. Vaňhara P, Kučera L, Prokeš L et al (2018) Intact cell mass spectrometry as a quality control tool for revealing minute phenotypic changes of cultured human embryonic stem cells: mass spectrometry quality control of cell cultures. Stem Cells Transl Med 7:109–114. https://doi.org/10.1002/sctm.17-0107
139. Kim HJ, Kim U-J, Vunjak-Novakovic G et al (2005) Influence of macroporous protein scaffolds on bone tissue engineering from bone marrow stem cells. Biomaterials 26:4442–4452. https://doi.org/10.1016/j.biomaterials.2004.11.013
140. Quinn KP, Bellas E, Fourligas N et al (2012) Characterization of metabolic changes associated with the functional development of 3D engineered tissues by non-invasive, dynamic measurement of individual cell redox ratios. Biomaterials 33:5341–5348. https://doi.org/10.1016/j.biomaterials.2012.04.024
141. Kurosawa H (2007) Methods for inducing embryoid body formation: in vitro differentiation system of embryonic stem cells. J Biosci Bioeng 103:389–398. https://doi.org/10.1263/jbb.103.389
142. Suhito IR, Angeline N, Choo S-S et al (2018) Nanobiosensing platforms for real-time and non-invasive monitoring of stem cell pluripotency and differentiation. Sensors 18:2755. https://doi.org/10.3390/s18092755

143. Esmonde-White KA, Cuellar M, Uerpmann C et al (2017) Raman spectroscopy as a process analytical technology for pharmaceutical manufacturing and bioprocessing. Anal Bioanal Chem 409:637–649. https://doi.org/10.1007/s00216-016-9824-1
144. Ilin Y, Kraft ML (2015) Secondary ion mass spectrometry and Raman spectroscopy for tissue engineering applications. Curr Opin Biotechnol 31:108–116. https://doi.org/10.1016/j.copbio.2014.10.011
145. von Erlach TC, Hedegaard MAB, Stevens MM (2015) High resolution Raman spectroscopy mapping of stem cell micropatterns. Analyst 140:1798–1803. https://doi.org/10.1039/C4AN02346C
146. Brauchle E, Schenke-Layland K (2013) Raman spectroscopy in biomedicine – non-invasive in vitro analysis of cells and extracellular matrix components in tissues. Biotechnol J 8:288–297. https://doi.org/10.1002/biot.201200163
147. Shafa M, Walsh T, Panchalingam KM et al (2019) Long-term stability and differentiation potential of cryopreserved cGMP-compliant human induced pluripotent stem cells. Int J Mol Sci 21:108. https://doi.org/10.3390/ijms21010108
148. Crook JM, Stacey GN (2014) Setting quality standards for stem cell banking, research and translation: the international stem cell banking initiative. In: Ilic D (ed) Stem cell banking. Springer New York, New York, pp 3–9
149. Alvarez-Palomo B, Vives J, Casaroli-Marano RP et al (2019) Adapting cord blood collection and banking standard operating procedures for HLA-homozygous induced pluripotent stem cells production and banking for clinical application. J Clin Med 8:476. https://doi.org/10.3390/jcm8040476
150. The International Stem Cell Banking Initiative (2009) Consensus guidance for banking and supply of human embryonic stem cell lines for research purposes. Stem Cell Rev Rep 5:301–314. https://doi.org/10.1007/s12015-009-9085-x
151. Crook JM, Hei D, Stacey G (2010) The international stem cell banking initiative (ISCBI): raising standards to bank on. In Vitro Cell Dev Biol Anim 46:169–172. https://doi.org/10.1007/s11626-010-9301-7
152. Kim J-H, Kurtz A, Yuan B-Z et al (2017) Report of the international stem cell banking initiative workshop activity: current hurdles and progress in seed-stock banking of human pluripotent stem cells: international stem cell banking initiative. Stem Cells Transl Med 6:1956–1962. https://doi.org/10.1002/sctm.17-0144
153. Kim J-H, Alderton A, Crook JM et al (2019) A report from a workshop of the international stem cell banking initiative, held in collaboration of global alliance for iPSC therapies and the Harvard Stem Cell Institute, Boston, 2017: international stem cell banking initiative. Stem Cells 37:1130–1135. https://doi.org/10.1002/stem.3003
154. Lipsitz YY, Timmins NE, Zandstra PW (2016) Quality cell therapy manufacturing by design. Nat Biotechnol 34:393–400. https://doi.org/10.1038/nbt.3525
155. Rao G, Moreira A, Brorson K (2009) Disposable bioprocessing: the future has arrived. Biotechnol Bioeng 102:348–356. https://doi.org/10.1002/bit.22192
156. ICH (2009) ICH harmonised tripartite guideline pharamaceutical development Q8(R2)
157. Casarosa S, Bozzi Y, Conti L (2014) Neural stem cells: ready for therapeutic applications? Mol Cell Ther 2:31. https://doi.org/10.1186/2052-8426-2-31
158. Davis RP, van den Berg CW, Casini S et al (2011) Pluripotent stem cell models of cardiac disease and their implication for drug discovery and development. Trends Mol Med 17:475–484. https://doi.org/10.1016/j.molmed.2011.05.001
159. Ishikawa T, Kobayashi M, Yanagi S et al (2015) Human induced hepatic lineage-oriented stem cells: autonomous specification of human iPS cells toward hepatocyte-like cells without any exogenous differentiation factors. PLoS One 10:e0123193. https://doi.org/10.1371/journal.pone.0123193
160. Lonza (2020) Tuning macrophages for immunotherapy. In: Lonza. https://www7.lonza.com/products-services/bioresearch/primary-cells/hematopoietic-cells/hematopoietic-knowledge-center/macrophages.aspx.html. Accessed 6 Sep 2020

161. McArthur K, Chappaz S, Kile BT (2018) Apoptosis in megakaryocytes and platelets: the life and death of a lineage. Blood 131:605–610. https://doi.org/10.1182/blood-2017-11-742684
162. National Cancer Institute (2020) PDQ cancer information summaries. In: PDQ cancer information summary. https://www.ncbi.nlm.nih.gov/books/NBK82221/. Accessed 6 Sep 2020
163. National Cancer Institute (2020) National Cancer Institute Erythrocyte. In: National Cancer Institute. https://www.cancer.gov/publications/dictionaries/cancer-terms/def/erythrocyte. Accessed 6 Sep 2020
164. Takahashi Y, Takebe T, Taniguchi H (2016) Engineering pancreatic tissues from stem cells towards therapy. Regen Ther 3:15–23. https://doi.org/10.1016/j.reth.2016.01.002
165. Toh WS, Foldager CB, Pei M, Hui JHP (2014) Advances in mesenchymal stem cell-based strategies for cartilage repair and regeneration. Stem Cell Rev Rep 10:686–696. https://doi.org/10.1007/s12015-014-9526-z
166. Dimmeler S (2011) Cardiovascular disease review series. EMBO Mol Med 3:697–697. https://doi.org/10.1002/emmm.201100182
167. Madigan M, Atoui R (2018) Therapeutic use of stem cells for myocardial infarction. Bioengineering 5:28. https://doi.org/10.3390/bioengineering5020028
168. Jing D, Parikh A, Tzanakakis ES (2010) Cardiac cell generation from encapsulated embryonic stem cells in static and scalable culture systems. Cell Transplant 19:1397–1412. https://doi.org/10.3727/096368910X513955
169. Lam AT-L, Chen AK-L, Li J et al (2014) Conjoint propagation and differentiation of human embryonic stem cells to cardiomyocytes in a defined microcarrier spinner culture. Stem Cell Res Ther 5:110. https://doi.org/10.1186/scrt498
170. Ting S, Chen A, Reuveny S, Oh S (2014) An intermittent rocking platform for integrated expansion and differentiation of human pluripotent stem cells to cardiomyocytes in suspended microcarrier cultures. Stem Cell Res 13:202–213. https://doi.org/10.1016/j.scr.2014.06.002
171. Niebruegge S, Bauwens CL, Peerani R et al (2009) Generation of human embryonic stem cell-derived mesoderm and cardiac cells using size-specified aggregates in an oxygen-controlled bioreactor. Biotechnol Bioeng 102:493–507. https://doi.org/10.1002/bit.22065
172. Laco F, Lam AT-L, Woo T-L et al (2020) Selection of human induced pluripotent stem cells lines optimization of cardiomyocytes differentiation in an integrated suspension microcarrier bioreactor. Stem Cell Res Ther 11:118. https://doi.org/10.1186/s13287-020-01618-6
173. Haake K, Ackermann M, Lachmann N (2019) Concise review: towards the clinical translation of induced pluripotent stem cell-derived blood cells— *ready for take-off*. Stem Cells Transl Med 8:332–339. https://doi.org/10.1002/sctm.18-0134
174. Lu S-J, Kelley T, Feng Q et al (2013) 3D microcarrier system for efficient differentiation of human induced pluripotent stem cells into hematopoietic cells without feeders and serum. Regen Med 8:413–424. https://doi.org/10.2217/rme.13.36
175. Sandler VM, Lis R, Liu Y et al (2014) Reprogramming human endothelial cells to haematopoietic cells requires vascular induction. Nature 511:312–318. https://doi.org/10.1038/nature13547
176. Wilkinson AC, Igarashi KJ, Nakauchi H (2020) Haematopoietic stem cell self-renewal in vivo and ex vivo. Nat Rev Genet 21:541–554. https://doi.org/10.1038/s41576-020-0241-0
177. Christaki E-E, Politou M, Antonelou M et al (2019) Ex vivo generation of transfusable red blood cells from various stem cell sources: a concise revisit of where we are now. Transfus Apher Sci 58:108–112. https://doi.org/10.1016/j.transci.2018.12.015
178. Dias J, Gumenyuk M, Kang H et al (2011) Generation of red blood cells from human induced pluripotent stem cells. Stem Cells Dev 20:1639–1647. https://doi.org/10.1089/scd.2011.0078
179. Garate Z, Quintana-Bustamante O, Crane AM et al (2015) Generation of a high number of healthy erythroid cells from gene-edited pyruvate kinase deficiency patient-specific induced pluripotent stem cells. Stem Cell Rep 5:1053–1066. https://doi.org/10.1016/j.stemcr.2015.10.002

180. Hirose S, Takayama N, Nakamura S et al (2013) Immortalization of erythroblasts by c-MYC and BCL-XL enables large-scale erythrocyte production from human pluripotent stem cells. Stem Cell Rep 1:499–508. https://doi.org/10.1016/j.stemcr.2013.10.010
181. Olivier EN, Marenah L, McCahill A et al (2016) High-efficiency serum-free feeder-free erythroid differentiation of human pluripotent stem cells using small molecules: SF/FF erythroid differentiation of hPSCs. Stem Cells Transl Med 5:1394–1405. https://doi.org/10.5966/sctm.2015-0371
182. Lu S-J, Feng Q, Park JS et al (2008) Biologic properties and enucleation of red blood cells from human embryonic stem cells. Blood 112:4475–4484. https://doi.org/10.1182/blood-2008-05-157198
183. Sivalingam J, Chen HY, Yang B-X et al (2018) Improved erythroid differentiation of multiple human pluripotent stem cell lines in microcarrier culture by modulation of Wnt/β-catenin signaling. Haematologica 103:e279–e283. https://doi.org/10.3324/haematol.2017.180919
184. Sivalingam J, Yu S, Lim ZR et al (2020) A scalable suspension platform for generating high density cultures of universal red blood cells from human induced pluripotent stem cells. Stem Cell Rep 16:182
185. Happle C, Lachmann N, Ackermann M et al (2018) Pulmonary transplantation of human induced pluripotent stem cell–derived macrophages ameliorates pulmonary alveolar proteinosis. Am J Respir Crit Care Med 198:350–360. https://doi.org/10.1164/rccm.201708-1562OC
186. Karlsson KR, Cowley S, Martinez FO et al (2008) Homogeneous monocytes and macrophages from human embryonic stem cells following coculture-free differentiation in M-CSF and IL-3. Exp Hematol 36:1167–1175. https://doi.org/10.1016/j.exphem.2008.04.009
187. Gutbier S, Wanke F, Dahm N et al (2020) Large-scale production of human iPSC-derived macrophages for drug screening. Int J Mol Sci 21:4808. https://doi.org/10.3390/ijms21134808
188. Nenasheva T, Gerasimova T, Serdyuk Y et al (2020) Macrophages derived from human induced pluripotent stem cells are low-activated "naïve-like" cells capable of restricting mycobacteria growth. Front Immunol 11:1016. https://doi.org/10.3389/fimmu.2020.01016
189. Shi J, Xue C, Liu W, Zhang H (2018) Differentiation of human-induced pluripotent stem cells to macrophages for disease modeling and functional genomics. Curr Protoc Stem Cell Biol 48:e74. https://doi.org/10.1002/cpsc.74
190. Ackermann M, Kempf H, Hetzel M et al (2018) Bioreactor-based mass production of human iPSC-derived macrophages enables immunotherapies against bacterial airway infections. Nat Commun 9:5088. https://doi.org/10.1038/s41467-018-07570-7
191. Baigger A, Blasczyk R, Figueiredo C (2017) Towards the manufacture of megakaryocytes and platelets for clinical application. Transfus Med Hemother 44:165–173. https://doi.org/10.1159/000477261
192. Ohto H, Nollet KE (2011) Overview on platelet preservation: better controls over storage lesion. Transfus Apher Sci 44:321–325. https://doi.org/10.1016/j.transci.2011.03.008
193. Shepherd JH, Howard D, Waller AK et al (2018) Structurally graduated collagen scaffolds applied to the ex vivo generation of platelets from human pluripotent stem cell-derived megakaryocytes: enhancing production and purity. Biomaterials 182:135–144. https://doi.org/10.1016/j.biomaterials.2018.08.019
194. Thon JN, Dykstra BJ, Beaulieu LM (2017) Platelet bioreactor: accelerated evolution of design and manufacture. Platelets 28:472–477. https://doi.org/10.1080/09537104.2016.1265922
195. Eicke D, Baigger A, Schulze K et al (2018) Large-scale production of megakaryocytes in microcarrier-supported stirred suspension bioreactors. Sci Rep 8:10146. https://doi.org/10.1038/s41598-018-28459-x
196. Dai X, Mei Y, Nie J, Bai Z (2019) Scaling up the manufacturing process of adoptive T cell immunotherapy. Biotechnol J 14:1800239. https://doi.org/10.1002/biot.201800239
197. Montel-Hagen A, Crooks GM (2019) From pluripotent stem cells to T cells. Exp Hematol 71:24–31. https://doi.org/10.1016/j.exphem.2018.12.001

198. Costariol E, Rotondi M, Amini A et al (2019) Establishing the scalable manufacture of primary human T-cells in an automated stirred-tank bioreactor. Biotechnol Bioeng 116:2488–2502. https://doi.org/10.1002/bit.27088
199. Ferreira AF, Calin GA, Picanço-Castro V et al (2018) Hematopoietic stem cells from induced pluripotent stem cells – considering the role of microRNA as a cell differentiation regulator. J Cell Sci 131:jcs203018. https://doi.org/10.1242/jcs.203018
200. Li F (2017) Seeking the "mother of blood" from human pluripotent stem cells: are we there yet? Genes Dis 4:123–124. https://doi.org/10.1016/j.gendis.2017.07.005
201. Zhao X, Moore DL (2018) Neural stem cells: developmental mechanisms and disease modeling. Cell Tissue Res 371:1–6. https://doi.org/10.1007/s00441-017-2738-1
202. Qiu L, Lim YM, Chen AK et al (2016) Microcarrier-expanded neural progenitor cells can survive, differentiate, and innervate host neurons better when transplanted as aggregates. Cell Transplant 25:1343–1357. https://doi.org/10.3727/096368915X690378
203. Miranda CC, Fernandes TG, Diogo MM, Cabral JMS (2016) Scaling up a chemically-defined aggregate-based suspension culture system for neural commitment of human pluripotent stem cells. Biotechnol J 11:1628–1638. https://doi.org/10.1002/biot.201600446
204. Rigamonti A, Repetti GG, Sun C et al (2016) Large-scale production of mature neurons from human pluripotent stem cells in a three-dimensional suspension culture system. Stem Cell Rep 6:993–1008. https://doi.org/10.1016/j.stemcr.2016.05.010
205. Yan Y, Song L, Madinya J et al (2018) Derivation of cortical spheroids from human induced pluripotent stem cells in a suspension bioreactor. Tissue Eng Part A 24:418–431. https://doi.org/10.1089/ten.tea.2016.0400
206. Koenig L, Ramme A, Faust D et al (2018) Production of human induced pluripotent stem cellderived cortical neurospheres in the DASbox® mini bioreactor system
207. Pagliuca FW, Millman JR, Gürtler M et al (2014) Generation of functional human pancreatic β cells in vitro. Cell 159:428–439. https://doi.org/10.1016/j.cell.2014.09.040
208. Rezania A, Bruin JE, Arora P et al (2014) Reversal of diabetes with insulin-producing cells derived in vitro from human pluripotent stem cells. Nat Biotechnol 32:1121–1133. https://doi.org/10.1038/nbt.3033
209. Mihara Y, Matsuura K, Sakamoto Y et al (2017) Production of pancreatic progenitor cells from human induced pluripotent stem cells using a three-dimensional suspension bioreactor system: pancreatic progenitor cells produced by suspension bioreactor system. J Tissue Eng Regen Med 11:3193–3201. https://doi.org/10.1002/term.2228
210. Yabe SG, Fukuda S, Nishida J et al (2019) Induction of functional islet-like cells from human iPS cells by suspension culture. Regen Ther 10:69–76. https://doi.org/10.1016/j.reth.2018.11.003
211. Sahabian A, Sgodda M, Naujok O et al (2019) Chemically-defined, xeno-free, scalable production of hPSC-derived definitive endoderm aggregates with multi-lineage differentiation potential. Cell 8:1571. https://doi.org/10.3390/cells8121571
212. Meier F, Freyer N, Brzeszczynska J et al (2017) Hepatic differentiation of human iPSCs in different 3D models: a comparative study. Int J Mol Med 40:1759. https://doi.org/10.3892/ijmm.2017.3190
213. Park Y, Chen Y, Ordovas L, Verfaillie CM (2014) Hepatic differentiation of human embryonic stem cells on microcarriers. J Biotechnol 174:39–48. https://doi.org/10.1016/j.jbiotec.2014.01.025
214. Vosough M, Omidinia E, Kadivar M et al (2013) Generation of functional hepatocyte-like cells from human pluripotent stem cells in a scalable suspension culture. Stem Cells Dev 22:2693–2705. https://doi.org/10.1089/scd.2013.0088
215. Lach MS, Wroblewska J, Kulcenty K et al (2019) Chondrogenic differentiation of pluripotent stem cells under controllable serum-free conditions. Int J Mol Sci 20:2711. https://doi.org/10.3390/ijms20112711

Biomanufacturing of Mesenchymal Stromal Cells for Therapeutic Applications

Ross A. Marklein, Morgan Mantay, Cheryl Gomillion, and James N. Warnock

Abbreviations

AT-MSC	Adipose tissue derived mesenchymal stromal cell
bFGF	Basic fibroblast growth factor
BLA	Biologics License Application
BM-MSC	Bone marrow derived mesenchymal stromal cell
CMC	Chemistry, Manufacturing and Controls
CPP	Critical process parameters
CQA	Critical quality attribute
DMEM	Dulbecco's Modified Eagle Medium
EGF	Epidermal growth factor
EV	Extracellular vesicle
FBS	Fetal Bovine Serum
FDA	Food and Drug Administration
FGF-2	Fibroblast growth factor 2
HGF	Hepatocyte growth factor
HGF	Human growth factor
HIF-1	Hypoxia-inducible factor 1-alpha
HLA	Human leukocyte antigen
hMSC	Human mesenchymal stromal cell
IGF	Insulin-like growth factor
iMSC	Induced mesenchymal stromal cell
IND	Investigational New Drug
iPSC	Induced pluripotent stem cell

R. A. Marklein · M. Mantay · C. Gomillion · J. N. Warnock (✉)
School of Chemical, Materials and Biomedical Engineering, University of Georgia, Athens, GA, USA
e-mail: james.warnock@uga.edu

R. Pörtner (ed.), *Cell Culture Engineering and Technology*, Cell Engineering 10,
https://doi.org/10.1007/978-3-030-79871-0_9

ISCT	International Society for Cell and Gene Therapy
MSC	Mesenchymal stromal cell
NSF	National Science Foundation
PBR	Packed bed reactor
PDGF	Platelet-derived growth factor
PLGA	Poly (lactic-co-glycolic acid)
QbD	Quality-by-design
TGF-β	Transforming growth factor beta
UC-MSC	Umbilic cord derived mesenchymal stromal cell
VEGF	Vascular endothelial growth factor

1 Introduction

Mesenchymal stromal cells (MSCs) are one of the most widely investigated cellular therapies. A wealth of preclinical and clinical studies demonstrate their ability to both directly (through differentiation into multiple cell-types) and indirectly (through secretion of trophic and immunomodulatory factors) treat a wide array of degenerative diseases [1–7]. MSCs are currently being utilized in hundreds of clinical trials worldwide for applications such as tissue engineering replacements of damaged bone and cartilage [8], modulation of immune diseases [7], and delivery of targeted cancer therapeutics [9]. MSCs are well tolerated in both autologous and allogeneic settings and their long-standing safety profile has enabled rapid translation of preclinical results into promising therapeutic candidates in the clinic. Furthermore, MSCs can be derived from different tissue sources (bone marrow, adipose, umbilical cord, etc.) and readily expanded *in vitro* to achieve clinically-relevant cell numbers necessary to treat the large number of patients affected by diseases with no currently available cures.

Although the MSC acronym was initially termed "Mesenchymal Stem Cell [10]," the definition and interpretation of how MSCs exert their therapeutic effect has shifted over the past decade towards an emphasis on their secretion of paracrine, pro-regenerative factors [4, 11, 12] instead of their stem cell-like ability to differentiate into multiple lineages and replace diseased or damaged tissues, such as bone or cartilage. While investigations into their tissue engineering potential have continued and made substantial progress, a position statement by the International Society for Cell and Gene Therapy (ISCT) has indicated that 'mesenchymal stromal cell' is the preferred nomenclature and more appropriate to encompass the broad spectrum of MSC functions [13]. Therapeutic applications for MSCs can be broadly characterized as (1) Direct: whereby MSCs differentiate into osteoblasts, chondrocytes, or adipocytes and produce extracellular matrix, or (2) Indirect: MSCs induce trophic, pro-regenerative effects on damaged tissues or modulate the immune system without integrating into host tissues. For direct tissue engineering applications, MSCs are expanded to large numbers and directly implanted into the injury site, or combined with a biomaterial scaffold to better promote tissue repair and integration. This would be the case for degenerated cartilage (due to osteoarthritis or traumatic injury) or large bone defects/injuries [8, 14, 15]. For indirect regenerative medicine applications, MSCs or MSC-derived secreted factors are administered either systemically or at the site of injury to facilitate/promote host cell repair processes, such as stem cell differentiation [16, 17] or angiogenesis [18]. Alternatively, MSCs or MSC-derived secreted factors may modulate immune cells such as T cells, macrophages, or microglia [19–22]. These mechanisms are believed to occur in a 'hit-and-run' manner with the therapeutic effect being observed long after the MSCs have been cleared and are no longer present in the body. This behavior is especially promising for degenerative disease with an inflammatory component such as osteoarthritis [8, 23], diabetes, and neuroinflammatory diseases [24], such as Parkinson's Disease, Alzheimer's Disease, and Multiple Sclerosis.

While an abundance of literature and clinical data motivates the widespread use of MSC therapeutics, there are few approved MSC therapies worldwide. The lack of successful clinical translation can be attributed to the following challenges with MSC manufacturing:

- MSC Functional Heterogeneity: Defined as differences in MSC therapeutic function between MSCs derived from different donors, tissues, and manufacturing conditions
- Lack of Critical Quality Attributes (CQAs): CQAs are properties or characteristics of a cell therapy product that should fall within specified limits to ensure adequate safety and efficacy [25].
- MSC Manufacturing Scale-Up: Translation from traditional 2D flasks/hyperflasks to larger-scale formats such as 3D bioreactors and biomaterial systems.
- Standardization: To ensure consistent, reproducible manufacturing, standard reference materials and standardized assays to assess MSC quality must be developed [3, 26].

This chapter will focus on MSC manufacturing considerations and how efforts are being made to address these unique, interrelated challenges and ultimately accelerate translation of MSC therapies towards safe and effective licensed products that are readily available for patients.

2 Overview of MSC Manufacturing

A generalized overview of the critical steps and components of MSC manufacturing are highlighted in Fig. 1. Broadly, MSC manufacturing can be divided into three stages: (1) Selection and isolation of MSCs from a specific donor/tissue source, (2) Expansion of MSCs, and (3) Final formulation of MSC product and therapeutic administration. Due to the lack of CQAs and standardized approaches for manufacturing MSCs, significant heterogeneity can be introduced at all stages of manufacturing and thus the final product administered to a patient can vary tremendously in terms of safety and efficacy.

Donor and tissue source for MSCs can significantly impact MSC quality in terms of safety and efficacy. Although undifferentiated MSCs can be well tolerated in an allogeneic setting, the use of autologous MSCs for bone and cartilage repair is

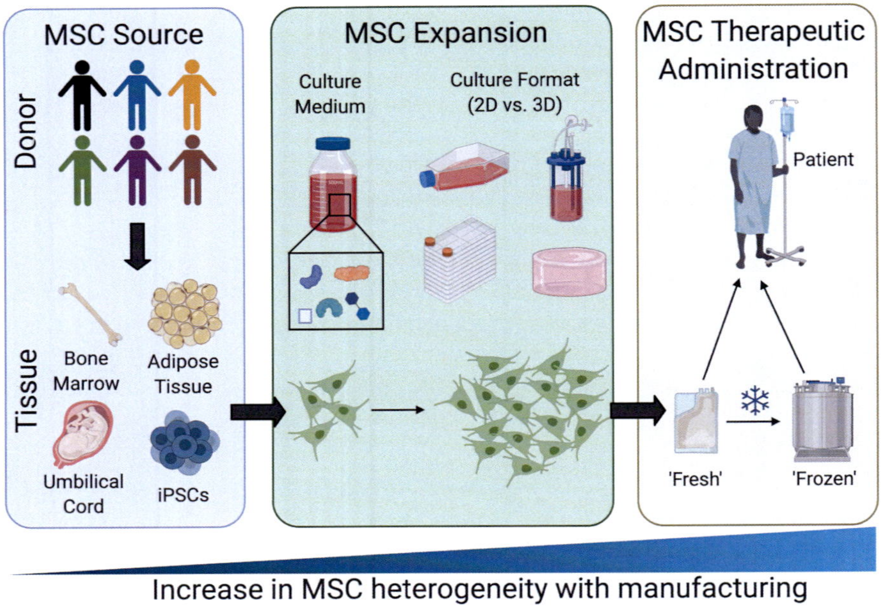

Fig. 1 Overview of MSC manufacturing

preferred as the process of MSC differentiation into osteoblasts and chondrocytes induces changes in HLA expression (e.g. HLA-DR [27]) and lymphocyte activation [28] that can potentially result in an immune response in an allogeneic transplantation setting [29–31]. Although there is evidence that allogeneic, differentiated MSCs do not elicit a significant immune response *in vitro* [32], HLA-matching is often not performed for clinical trials and it is unclear whether allogeneic differentiated MSC therapies do not elicit an undesired immune response [29]. Autologous MSC manufacturing can present major challenges compared to allogeneic MSC manufacturing as the considerable patient-patient variability and inability to treat multiple patients with the same product can result in increased production costs and product heterogeneity [33]. The ability to manufacture large quantities of MSCs in an allogeneic setting thus represents a more attractive strategy as it would enable an 'off-the-shelf' therapy and thus a more consistent MSC product than a unique autologous MSC product manufactured for each patient. However, significant donor-donor variability can be observed in terms of therapeutically-relevant functions such as production of immunomodulatory factors IDO [34] and PD-L1 [35], suppression of activated T cells [20], and treatment of critical limb ischemia (i.e. angiogenesis [36]). Therefore, it is essential to develop improved methods for screening donors from which high quality MSCs can be derived.

MSCs derived from different tissue sources have also been shown to vary in their functional capacity. For example, bone marrow derived MSCs (BM-MSCs) were found to better promote wound closure [37] and endothelial tube formation [38] *in vitro* than umbilical cord derived MSCs (UC-MSCs); however, in another study

[39] UC-MSCs were found to have lower doubling times (higher proliferation rates) and greater production of immunomodulatory factors than BM-MSCs. Furthermore, exosomes produced by BM-MSCs, UC-MSCs, and adipose tissue derived MSCs (AT-MSCs) possess distinct proteomic signatures with notable gene ontological differences observed in terms of Notch and integrin signaling (high in BM-MSC exosomes), glycosaminoglycan and bone development (high in AT-MSCs), and laminin-binding and vasculature development (high in UC-MSCs) [40]. Although these tissue-dependent differences have been observed, there are no comprehensive, definitive studies demonstrating MSCs derived from one specific tissue type's preference or 'superiority' over other MSC tissue sources and thus future work is warranted to better understand how donor and tissue variability impact final MSC product quality. Induced pluripotent stem cell (iPSC) derived MSCs (iMSCs) are also emerging as a promising alternative to tissue-derived MSCs as iPSCs represent a potentially limitless source of isogenic MSCs [41, 42]. However, the non-standardized approaches for differentiating iMSCs from iPSCs adds another layer of manufacturing heterogeneity that must be addressed and efforts are being made to better understand MSC identity and origin as this technology evolves [43, 44].

To manufacture clinically-relevant numbers of MSCs (10^{10}–10^{13} cells depending on application) [45], *ex vivo* expansion is necessary once MSCs have been obtained from a given donor/tissue source. This process of expansion typically involves multiple passages where MSCs are allowed to grow until achieving high confluency (~70–90%) and then subsequently harvested (using reagents such as trypsin, collagenase, or TrypLE) prior to replating (or inoculating in the case of a bioreactor) at lower densities ('subculture') to enable further expansion and mitigate contact inhibition. Although the generalized process of passaging is commonly employed, there are no standardized protocols employed throughout the expansion process in terms of reagents (i.e., culture medium), critical process parameters (CPPs, such as seeding density, culture time, pH, etc.), and culture format/vessels (2D vs. 3D, surface modifications, etc.) Differences in cell dissociation reagents can impact MSC surface marker expression (CD73, CD105, CD140a) [46] while changes in MSC seeding density impact production of extracellular vesicles (EVs) [47] and can also affect MSC metabolism (e.g. favoring oxidative phosphorylation versus glycolysis) [48]. Furthermore, MSC manufacturers often have different cell banking strategies whereby master cell banks, working cell banks, and final MSC products are cryopreserved at different stages of manufacturing to enable better quality control and manufacture of consistent product lots [49]. Depending on the clinical application, significant cell expansion may be required to treat large number of patients and therefore MSCs will be subjected to extended culture in terms of additional passages and cryopreservation steps. MSCs exhibit decreased function (T cell suppression [20], osteogenesis [50], adipogenesis [51], and chondrogenesis [52]) and increased senescence [53] at later passages and it is therefore critical to develop optimized, standardized culture platforms that enable rapid MSC expansion while also maintaining the desired therapeutic function across multiple passages.

Once expanded, the final MSC product can be formulated and administered in many ways depending on the desired therapeutic application. In the case of tissue engineering applications, MSCs can be locally injected into the injury site or combined with a biomaterial (e.g. hyaluronic acid hydrogel or porous PLGA scaffold) and immediately implanted or cultured *ex vivo* to promote differentiation and matrix deposition prior to implantation [54–57]. For regenerative medicine applications where MSCs are hypothesized to exert their therapeutic effect through secreted factors, the route of administration can include intravenous (most common), intranasal, intraperitoneal, intrathecal, and intra-articular [58–62]. Although the majority of MSCs injected intravenously are rapidly cleared (on the order of days) and predominantly end up trapped in the lungs, their therapeutic effect can be observed not only in the lungs [2], but in other remote tissue locations, for example in the case of neuroinflammation [24], graft-versus-host disease [63], and post-myocardial infarction [60]. Finally, the effects of cryopreservation on MSC function have been demonstrated in a number of studies with decreased immunomodulatory function (both *in vitro* [64] and *in vivo* [59]) and post-implantation survival [58] observed for cryopreserved MSCs used immediately post-thaw versus non-cryopreserved MSCs or cryopreserved MSCs that are culture rescued post-thaw (Figs. 2 and 3). As cryopreservation enables 'off-the-shelf' potential through long-term storage and greater manufacturing scale than producing 'fresh,' non-cryopreserved MSC products, there is great interest in developing better approaches for understanding how cryopreservation negatively impacts MSC quality and whether new strategies can be employed to mitigate its effects. Ultimately, decentralized (or 'redistributed') manufacturing [65, 66] of cryopreserved MSC products represents the most promising approach as requiring additional manufacturing steps to be performed at individual clinical sites by different operators can introduce additional heterogeneity that may contribute to inconsistent clinical outcomes.

3 Manufacturing Considerations

3.1 Culture Medium Considerations

Culture medium for expanding MSCs (and other cell therapies) represents one of the most significant costs in product development. MSCs are adherent cells that grow robustly in a number of cell culture formats and environments; however, this has introduced significant heterogeneity in terms of manufacturing as countless culture media formulations have been used. Because the ISCT MSC criteria established in 2006 [67] do no effectively discriminate between MSCs with 'good' or 'bad' therapeutic potential, using these criteria (i.e. plastic adherence, tri-lineage differentiation, and expression of generic fibroblast-like surface markers) to evaluate and develop effective culture media has been a major challenge facing MSC clinical translation.

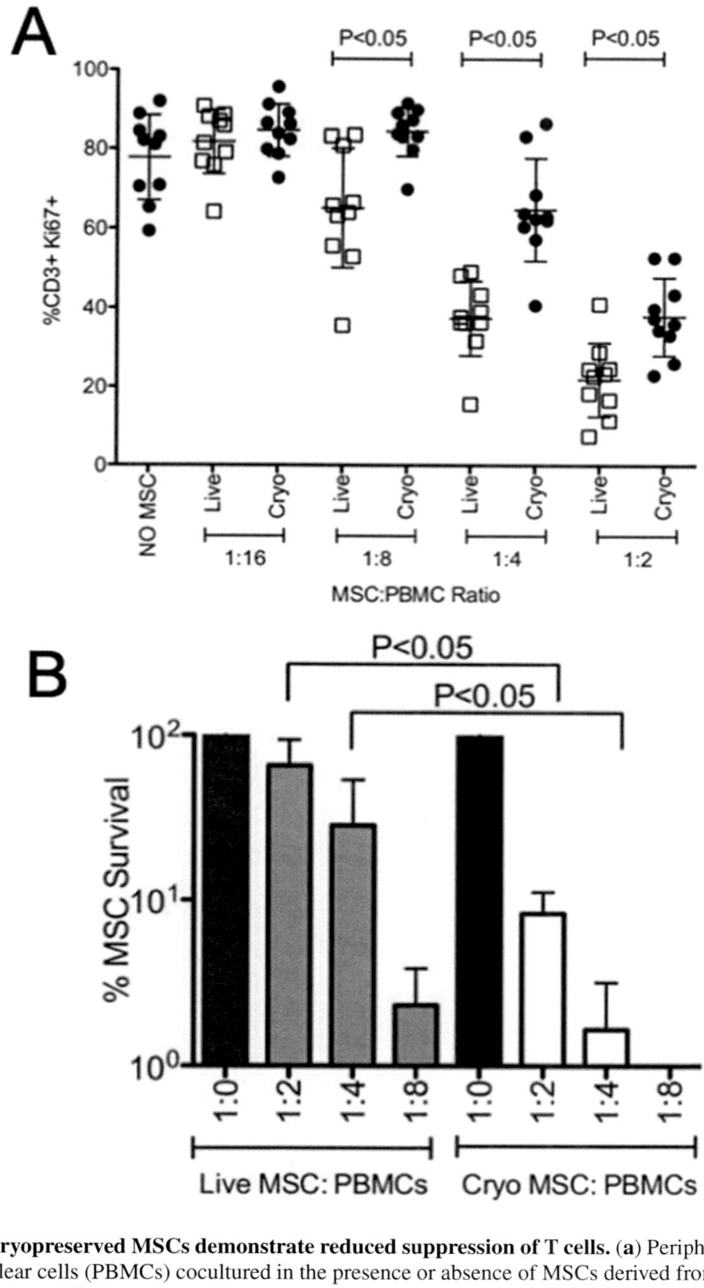

Fig. 2 Cryopreserved MSCs demonstrate reduced suppression of T cells. (**a**) Peripheral blood mononuclear cells (PBMCs) cocultured in the presence or absence of MSCs derived from actively growin culture (Live) or thawed from cryopreservation (Cryo) were stimulated with SEB. Four days post, T cell proliferation was measured by Ki67 intracellular staining. Cumulative % of T cell proliferation (CD3+/Ki67+) in the presence of variable MSC and PBMC ratio is shown. Lower %CD3+/Ki67+ values indicate great suppression of T cells by MSCs. (**b**) Frozen-thawed MSCs are susceptible to lysis by activated T cells. CFSE labeled Live and Cryo MSCs were cocultured with SEB activated PBMCs at indicated ratios. Plate bound MSCs from the coculture were trypsinized and event counts were recorded in flow cytometry with the normalization of counting beads. Live and Cryo MSC count in the absence of PBMCs were used for normalization and calculation of % survival of Live and Cryo MSCs in the presence of PBMCs. (Taken from Chinnadurai et al. [64], rights permitted)

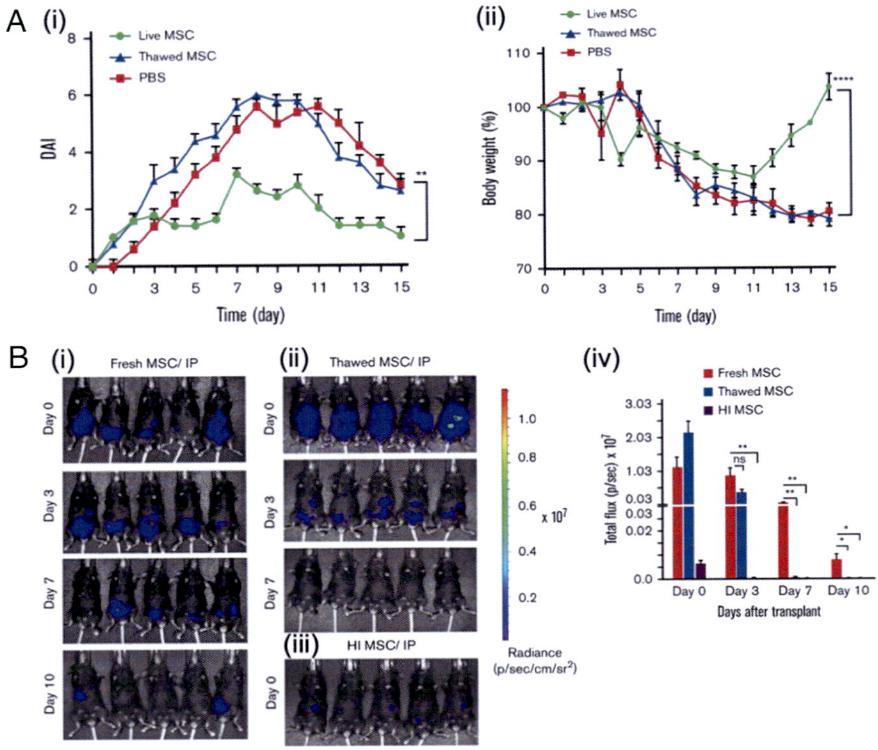

Fig. 3 Cryopreserved MSCs have reduced in vivo therapeutic potential. (**a**) Live and thawed (previously cryoprserved) MSCs were injected in C57BL/6 mice in DSS-induced colitis mouse model and monitored for both disease activity index (DAI, (i)) and body weight (ii). (**b**) Dwell time of intraperitoneal delivered fresh (live), thawed (cryopreserved), and heat –inactivated (HI) MSCs assessed based on bioluminsecene measured as total flux (photons/sec). Longitudinal in vivo images of representative mice shown at selected time points for fresh (i), thawed (ii), and HI (iii) MSCs. BL was measured from all in vivo ROIs at each time point as total flux. (Taken from Giri and Galipeau [59], rights permitted)

MSC culture media can be broadly categorized into undefined and chemically-defined media, with undefined being further subdivided into xenogenic and xeno-fee media. Early MSC research primarily utilized xenogenic, undefined medium consisting of a commercial base medium (e.g. alpha-MEM or DMEM), antibiotics, amino acid supplements (e.g. L-glutamine), and animal-derived serum (fetal bovine serum (FBS) being the most common) [68]. While this formulation is still widely used in academic research today due to historical use and relative low cost (compared to xeno-free and chemically-defined media), there are concerns in terms of safety and product consistency. Specifically, the use of animal-derived products requires additional testing to ensure absence of adventitious agents, immunogenic factors and other contaminants [69]. Furthermore, the unknown composition of FBS from different lots and vendors has been shown to produce MSCs with varied proliferation and therapeutically-relevant functions [70]. In an effort to address safety concerns associated with the use of xenogenic media, xeno-free alternatives to FBS such as human serum or human platelet lysate [71]. While these human-derived media supplements can produce MSCs with similar (or even improved) function compared to MSCs manufactured using FBS-containing media, there is also a poor understanding of the exact composition and thus significant lot-lot differences have been observed [72, 73]. To address the safety concerns of xenogenic media, as well as the consistency issues of xeno-free, undefined media, efforts have been made to create chemically-defined media with the exact amount of each component known. MSCs manufactured using chemically-defined medium have been shown to perform similar to MSCs cultured in undefined medium (in terms of proliferation and function) [70, 74], but the cost for chemically-defined media is significantly higher. Furthermore, it is still unknown what the exact media component requirements are for MSC manufacturing in terms of growth factors, metabolites, lipids, etc., and it is likely that unique defined media may be required for different therapeutic applications to effectively 'select' for MSCs with desired properties.

To better understand the impact of media formulations on MSC behavior, researchers have investigated the effects of specific cell signaling pathways and regulators of MSC growth and metabolism. Growth factors including basic fibroblast growth factor (bFGF), epidermal growth factor (EGF), insulin-like growth factor (IGF), platelet-derived growth factor (PDGF), and hepatocyte growth factor (HGF) have all been implicated in maintenance of MSC function across extended passages [75–77]. EGF and bFGF are readily incorporated into defined MSC media and have well-known roles in regulation of ERK/Akt signaling, and addition of both of these growth factors has been demonstrated to increase MSC proliferation rate, as well as mitigate the onset of senescence that typically occurs at later passages [75, 78–80]. Wnt signaling also plays a key role in maintenance of MSC function during manufacturing and Wnt agonism (through addition of small molecules or growth factors like Wnt3a) and downstream effects (increased beta-catenin) have been associated with greater MSC proliferation [81]. However, differences in the role or extent of Wnt signaling can be associated with MSCs derived from different sources [82] and therefore it is important to understand how these unique tissue-dependent differences in signaling can guide selection of MSCs with desired properties. Members of the transforming growth factor beta (TGF-β) family also play key roles in MSC growth and function through canonical SMAD signaling and non-canonical pathways (e.g. ERK, Akt, p38, and Rac1). A number of studies have suggested that inhibition of TGF-β signaling during MSC expansion can lead to increased clonogenicity, proliferation, secretion of immunomodulatory and angiogenic factors, as well as decreased senescence and apoptosis [80, 83–86]. These effects have been associated with epigenetic changes (i.e. chromatin accessibility [86, 87]), as well as metabolic changes associated with a decrease in mitochondrial reactive oxygen species (ROS) [83]. It is worth noting that while inhibition of TGF-β signaling during expansion is associated with improved MSC proliferation and maintenance of 'stemness,' TGF-β signaling is required for MSC chondrogenesis [88, 89].

The ability to manufacture large numbers of MSCs is critically dependent on our understanding of metabolism and how it relates to differences in selection of culture medium. Extensive *ex vivo* expansion of MSCs can be associated with a characteristic 'metabolic shift' from glycolysis at early passages to oxidative phosphorylation at later passages [90]. Control of MSC metabolism can occur through addition (or inhibition) of exogenous growth factors (such TGF-β, FGF-2, EGF), addition of specific metabolites or small molecules, or culturing at different levels of O_2 (i.e. normoxia vs. hypoxia) [75, 77, 90–93]. Specific examples are shown in Fig. 4. Differences in MSC metabolism (i.e. metabolic heterogeneity) can be attributed not only to extended culture in nutrient rich, high oxygen environments, but also to inherent differences in MSCs derived from different tissues and donors [94]. Additionally, MSCs derived from diseased, atherosclerotic patients possessed significant mitochondrial dysfunction, produced higher levels of ROS, and had diminished immunomodulatory function (in terms of T cell suppression) than MSCs derived from healthy patients [95]. During culture expansion there are well-documented increases in ROS levels [90], which leads to reduced autophagy and mitophagy [91], and contributes to the onset of senescence. Efforts have been made to identify regulators of MSC metabolism specifically with a focus on pathways with known roles in ROS production, mitochondrial function, and oxygen utilization. For example, addition of ROS scavengers (such as N-acetyl-L-cysteine) reduced ROS levels in MSCs from diseased patients and resulted in decreased doubling times (i.e. higher proliferation rates), decreased secretion of pro-inflammatory cytokines, and improved immunomodulatory function [95]. The addition of small molecule inhibitors of Akt/mTOR (e.g. rapamycin and LY294002) signaling, which has been shown to positively correlate with ROS levels, can result in a delayed onset of senescence and long-term maintenance of MSC functions such as osteogenesis and secretion of immunomodulatory factors [93, 96]. Culturing MSCs in hypoxia (typically <5% O_2) also has a pronounced effect on MSC metabolism through sustained activation and stabilization of hypoxia-inducible factor 1-alpha (HIF-1) [97]. Hypoxic culture can result in decreased production of ROS, secretion of pro-regenerative growth factors (e.g. HGF, bFGF, VEGF, IGF-1), and improved survival post-transplantation [97]. Efforts are also being made to develop hypoxia-mimetic conditions through introduction of small molecule drugs such as desferrioxamine and HIF-prolyl hydroxylase inhibitors [90, 98]. As MSC manufacturers continue to increase the scale and complexity of culture systems to meet clinical need, a greater understanding of the metabolic demands of MSCs is needed and will drive further development and refinement of culture media in the future.

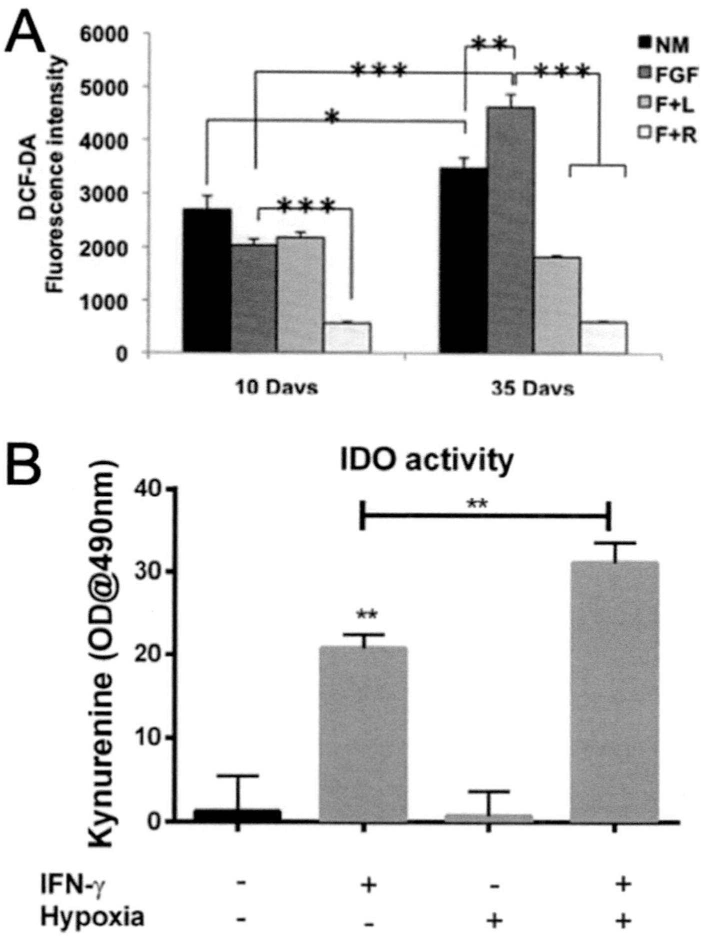

Fig. 4 Control of MSC metabolism during manufacturing. (**a**) MSCs treated with normal medium (NM), fibroblast growth factor-2 (FGF-2, F), FGF-2 in the presence of PI3K/Akt inhibitor LY294002 (10 μM) (F + L), FGF-2 in the presence of mTOR inhibitor rapamycin (10 nM) (F + R) for up to 35 days. Total reactive oxygen species (ROS) was determined by staining with H2DCFDA (DCF-DA) following flow cytometry analysis. (Taken from Gharibi et al. [96], rights permitted). (**b**) Indoleamine 2,3-dioxygenase (IDO) activity of MSCs cultured with or without IFN-γ (40 ng/mL) in either normoxic or hypoxic (2% O2) conditions for 24 h. (Taken from Liu et al. [93], rights permitted)

3.2 Bioreactors for Manufacturing MSCs and Scaleup Considerations

As mentioned previously, a single dose of MSCs will require 10^{10}–10^{13} cells for autologous or allogeneic therapies in clinical settings depending on the specific

Table 1 Ranges of hMSCs used for clinical applications

Therapeutic target	Range of cells	References
Acute respiratory distress syndrome (ARD)	3×10^6 cells/kg	[94]
Amyotrophic lateral sclerosis (ALS)	11–12 $\times 10^5$ cells/patient 2×10^5 cells/kg	[106, 107]
Diabetes (type 1)	2.1–3.6 × 106 cells/kg	[108]
Multiple sclerosis	1–2 $\times 10^6$ cells/kg	[109, 110]
Idiopathic Parkinson's disease (iPD)	1×10^6, 3×10^6, 6×10^6, or 10×10^6 cells/kg	[39]
Spinal cord injury	8×10^6, 40×10^6, or 50×10^6 cells/patient *depending on location and time point of clinical trial*	[111]

application [45, 99] and other variables, such as patient body weight. Table 1 summarizes reported numbers of MSCs used for selected clinical applications, and affirms the significant magnitude of cells that may be required. A typical bone marrow biopsy may yield a very low retrieval percentage (~0.01%) of hMSCs acquired from a single-donor bone marrow sample, [100–102]. Alternatively, adipose tissue has been shown to be more accessible and yield a greater percentage of MSCs, with the percent range of stem cells from biopsies ranging from 1% to 10% [103–105].

These yields are still several orders of magnitude less than the number of cells required for *in vivo* therapy, thus, *ex vivo* expansion of cells is essential for large-scale manufacturing of MSCs for clinical application. This biomanufacturing of cells may be pursued via scale-up or scale-out approaches, where cell expansion is achieved by using significantly larger individual culture vessels (i.e., increasing batch size), or by multiplying the number of individual culture vessels used, respectively. In practice, scale-up approaches have been preferred for production of "off-the-shelf" allogeneic therapies, while scale-out approaches preferable for autologous therapies [112]. However, whether a scale-up or scale-out approach is taken, employment of appropriate culture conditions is essential for attaining sufficient quantities of MSCs for clinically-relevant dosages while simultaneously ensuring cell phenotype, potency, and therapeutic efficacy.

In addition to the biochemical and physiochemical parameters essential for successful growth of MSCs, proper physical processing and culturing considerations are equally important. MSCs are anchorage-dependent cells, requiring a suitable culture surface to support their adhesion and facilitate subsequent processes of cell proliferation and differentiation. Traditional cell culturing approaches are based on the use of static, two-dimensional (2D) planar vessels, consisting of Petri dishes, multi-well plates, T-flasks, and roller bottles to obtain monolayer cultures. These vessel surfaces are commonly made of plastics, such as polystyrene, that have been treated for tissue culture via chemical or physical processes (e.g., surface functionalization, plasma treatment, etc.) to increase substrate hydrophilicity to support cell adhesion [113–116]. The expansion potential of MSCs is constrained by the amount of surface area available in the particular vessel used, such that large-scale production of these adherent cells will require large surface area to support sufficient cell proliferation.

For 2D culture, increased surface area can be accomplished via multilayer T-flasks, such as the Corning® High Yield PERformance Flask (HYPERFlask®) cell culture vessel with 1720 cm^2 available growth area, Corning® CellSTACK® Cell Culture Chambers with up to 25,440 cm^2 cell growth area available in their largest 40-stack size, or Thermo Scientific™ Nunc™ Cell Factory™ systems with up to 25,280 cm^2 growth area in its 40-tray option, among others [117–119]. In most cases, large-scale manufacturing of cells cultured in 2D will require numerous vessels to achieve clinical-level doses. This need will also necessitate sufficient physical space and infrastructure for vessel incubators and cleanrooms to minimize susceptibility to contamination and to meet ISO qualifications, requiring extensive financial investment for facilities [45, 99]. Further, if automated systems are not in place, significant manual labor and time will be required for processing, with all of these factors together driving the resulting cost of these therapies and deterring potential accessibility for patients [45, 120–122]. In addition to these physical constraints, long-term culturing of cells in 2D presents other challenges related to cell quality and function, where prolonged expansion can result in senescence with increasing passages [120, 123]. More specifically, in the case of MSCs undergoing prolonged expansion in 2D, additional concerns related to maintenance of their phenotype, potency, and functionality exist when these cells are cultured *ex vivo* due to their sensitivity. Particularly, findings from previous studies have demonstrated decrease in the proliferation potential of cells, changes in hMSC phenotype, or a decrease in their differentiation or immunomodulatory potential over population doublings, which can result in changes to their therapeutic efficacy when implanted *in vivo* [75, 124–130]. MSC heterogeneity is again of concern with 2D expansion since individual flasks constitute its own batch, which may contribute to batch-to-batch variability [123]. Evaluation and characterization of these cells in 2D culture is also challenging, since microscopic evaluation is commonly used for assessing monolayer cultures, and observation of MSCs may not be possible with large vessels using a typical microscope.

To circumvent these limitations of conventional 2D monolayer flasks for MSC expansion, alternative culturing strategies have been explored, largely consisting of dynamic bioreactor systems for three-dimensional (3D) microcarrier-based cultures. Such systems have been investigated to support cell yields up to trillions of cells, while reducing potential costs for manufacturing sufficient clinical doses through shortened culture times [99, 123]. The specific type of bioreactor selected will depend on whether or not the process is a batch or continuous production, or a static or dynamic culture system [131]. A batch production is defined as a reactor that is filled or charged with reagents that will be left inside the vessel as the reaction is carried out with no inflow or outflow until the reaction is complete; whereas a continuous production refers to a vessel that has a continuous inflow of reagents and a continuous outflow of product, as it is assumed to be appropriately mixed while operating at steady-state [132].

Bioreactors, have become an integral part of the scale-up production process by allowing for potentially automated culturing platforms where culture parameters, such as culture pH, temperature, and oxygen supply, can be controlled, and robust well-defined populations of cells can be obtained [133–136]. MSCs have been shown to respond actively to their environment, and the dynamic environments of bioreactors that mimic *in vivo* conditions can result in the production of bioactive factors in the MSC secretome and impact the expansion, fate, and differentiation potential of MSCs, where enhanced differentiation to a certain lineage has been observed depending on the culture environment [137–140]. Expansion of MSCs under hypoxic conditions, with oxygen levels ranging from 2–5% within an incubator environment, has been shown to yield improved growth rates (Fig. 5) as a result of reduced oxygen consumption and production of reactive oxygen species. In addition, hypoxic conditions can result in improved maintenance of differentiation potential and decreased apoptosis, thus demonstrating the importance of system control for these cells during manufacturing [141–148].

Various types of dynamic bioreactors exist (Fig. 6), including 3D microcarrier-based stirred tank (or spinner flask systems if little or no monitoring system incorporated), Wave (or rocking) bioreactors, and Vertical-Wheel™ bioreactors that are mechanically-driven. In addition, other hydraulically-driven dynamic systems include hollow fiber and packed bed bioreactors. Stirred tank systems make for an attractive alternative to traditional static culture systems because they are simple, easy to scale-up, provide excellent gas exchange to cell cultures and monitoring of critical culture parameters is straightforward [136]. The use of stirred bioreactors can support shorter culture times, resulting in reduced manufacturing costs over time. These closed systems and the availability of commercially-available disposable vessels, such as single-use plastic spinner flasks, can support greater culture efficacy by reducing possible contamination risks [123, 134, 149].

Scalability of stirred tank systems has also been sufficient to yield large batches of MSCs ranging from 5 L to 50 L (or larger, up to 500 L), dependent on bioreactor size and microcarrier density [150–155], of which a number of studies have been previously reviewed [134, 135, 156]. Most importantly, the properties of MSCs cultured under 3D dynamic conditions can be altered and controlled for specific applications by altering fluid properties through varying impeller type/size, rotational speed, agitation rates, etc. [113, 134, 140, 157–159]. Stirred tank and spinner flask cultures can support cell aggregates in suspension, cells encapsulated in microspheres, and cells seeded onto coated microcarriers [160–164]. MSC aggregates can form when the cells adhere to one another when there is no suitable surface present to support cell adhesion [165], and aggregate formation has been attempted for MSC expansion [166, 167]. However, it is difficult to control the expansion scale-up of MSC aggregate culture because it is almost impossible to control the aggregate size. In addition, aggregate culture can also damage the final cell product, as the increase in aggregate size and density can result in decreased

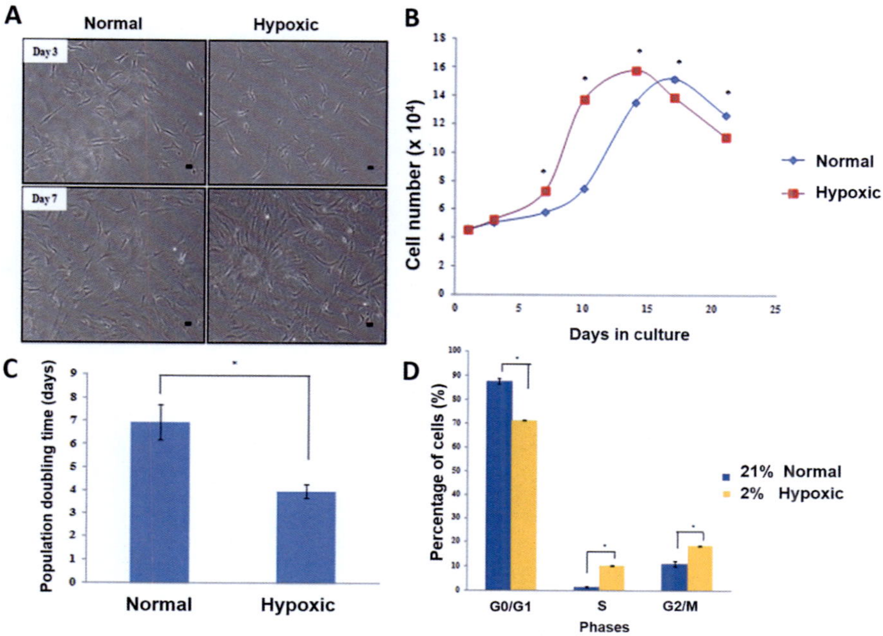

Fig. 5 MSCs cultured under hypoxic conditions exhibit improved outcomes relative to normal atmospheric oxygen conditions. (**a**) Human adipose-derived stromal cells (ASCs) exhibited typical fibroblast-like morphology for both culture conditions. (**b**) Growth curves for cells showed an increased cell proliferation rate *under in situ* normoxia. (**c**) Population doubling time was significantly lower under in situ normoxia than at atmospheric O2 concentration. (**d**) Cell cycle analysis showed the percentage of ASCs in S-phase was higher under *in situ* normoxia as compared to atmospheric O2 concentration. (Adapted from Choi et al. [147] rights permitted under Creative Commons license)

mass transport of nutrients to the center, limiting cell proliferation and potentially damaging the cells [168, 169].

Another dynamic platform that has become more widely used for large-scale manufacturing is the Wave bioreactor system, which consists of inflatable flexible plastic bags as culture vessels that are placed on a rocking platform [170–172]. While impeller motion can negatively impact cells in stirred vessels if not optimized, these concerns are alleviated with the agitation provided via the rocking motion of Wave bioreactors, making this a suitable culturing method for shear-sensitive cells. These bioreactors are reported to provide good nutrient distribution, higher oxygen transfer than spinner flasks and perform comparable to stirred tank bioreactors (up to 100 L in volume) [170, 171, 173, 174]. The disposable plastic bags are appealing for addressing sterility concerns via a closed system and bags are available up to 500 L, which can further help reduce manufacturing costs [173]. Original application of these bioreactors was commonly used for suspension cultures, [136, 170, 175] however, recent studies have demonstrated successful expansion of anchorage-

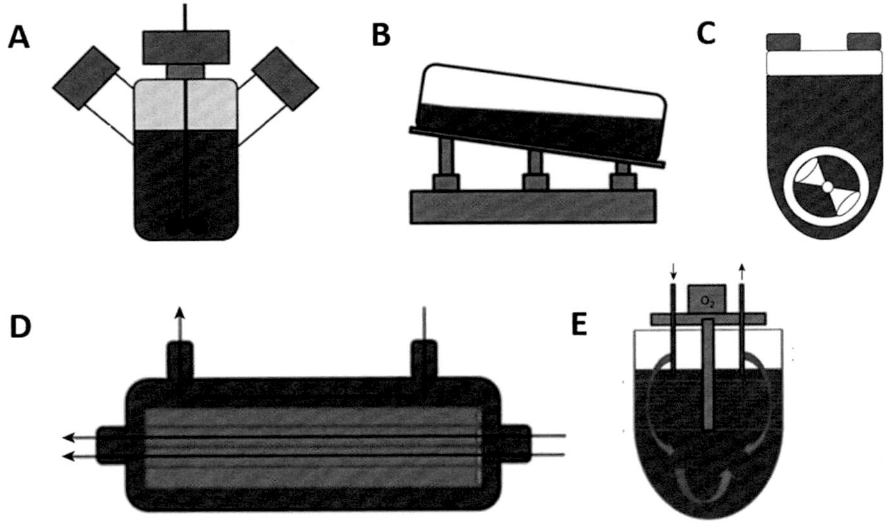

Fig. 6 Various types of bioreactors exist for MSC culture and expansion. Schematic representation of specific vessels, including (**a**) stirred tank/spinner flasks, (**b**) wave bioreactors, (**c**) Vertical-Wheel bioreactors, (**d**) hollow fiber bioreactors, and (**e**) packed/fixed bed bioreactors. (Adapted from Colao et al. [149] with permission)

dependent cells using microcarrier-supported cultures in Wave bioreactors [172]. Studies evaluating MSC expansion using Wave bioreactors have been limited to an investigation of placental hMSC expansion using macroporous gelatin microcarriers, where feasibility of the Wave bioreactor approach for MSC expansion was demonstrated [176]. Additional studies for optimization of this approach to ensure MSC efficacy and function are required.

With a focus on increasing the efficiency of culture mixing, dynamic culturing has also been demonstrated using the more recently developed Vertical Wheel bioreactor design. These systems consist of a vessel with a U-shaped bottom and a vertically-rotating impeller/wheel that rotates around a horizonal axis [177]. In comparison to the typical stirred bioreactors, the vertical mixing allows lower agitation rates, and subsequently, lower effects due to shear stress, which would could be advantageous for MSC culture since shear stress may impact cell viability and phenotype [123]. In addition, higher mass transfer rates are observed, as well as more homogenous suspensions. This particular system also supports favorable scalability conditions via single-use large-volume vessels, ranging from 60 mL up to 500 L [113]. Studies comparing use of Vertical wheel bioreactors to static planar cultures showed successful culture of MSCs from umbilical cord and adipose tissue using the Vertical Wheel system, which was significantly less expensive than static cultures [178]. Additional MSC culture optimization studies have also explored fed-batch culturing in xeno-free media using the Vertical Wheel system, with hMSCs successfully maintaining proliferation, phenotypic and functional characteristics, providing a protocol for future scale-up approaches [179].

In addition to the mechanically-driven stirred, Wave, and Vertical Wheel bioreactor systems, hydraulically-driven perfusion-based bioreactors, such as hollow fiber and fixed/packed bed bioreactors have been applied for MSC expansion [127]. Perfusion bioreactors feature a vessel where the media is directed to flow through a construct that is either porous or 3D, which can help overcome the expansion bottlenecks seen in monolayer cell culture methods by increasing oxygen and nutrient transport in the system [180, 181]. Studies have shown that cell metabolism is crucial to stem cells retaining pluripotency, or multipotency depending on the source, and perfusion bioreactors can provide a constant flow of fresh media that removes inhibitory metabolic byproducts, resulting in a stable metabolic system for high-density expansion [182]. Perfusion-based systems have been utilized with specific cell phenotypes for amplification via the use of external cues for cell conditioning. For example, a novel bioreactor system for the purpose of conditioning chondrocyte cells used in autologous chondrocyte implants was developed, and the flow of the media was used to provide hydrostatic pressure to the cells for potentially increased protein production and gene expression [183], where the estimated pressure loads on cartilage of 0.1–0.2 MPa were achievable in the system [184].

Hollow fiber bioreactors are perfusion-like bioreactors that present an opportunity to produce a high density of cells before MSCs cross the threshold for expansion before losing key stem cell characteristics such as potency [185]. Hollow fiber bioreactors are tubular vessels containing one or a bundle of polymer fibers that culture medium is perfused through [186]. The hollow fibers can be used for different transports, such as supplying fresh nutrients and oxygen or removing spent medium and cell products [186–188]. The set-up of hollow fiber bioreactors creates a closed environment comparable with *in vivo* conditions and a large surface area for anchorage-dependent stem cells, which satisfies one of the key necessities of dynamic culture [189]. Hollow fiber bioreactors have been used for MSC expansion, cell product formation, and vector production, with the intent of finding a culturing platform to integrate at the industry level of scale-up [190–193]. Studies have demonstrated use of hollow fiber bioreactors for improving metabolic function of MSC-derived hepatocytes, where a perfused 3D culturing environment mimicking the natural gradient of the liver found in the body, led to upregulation of certain genes and development of a more heterogeneous cell population [194, 195].

Packed bed bioreactors (PBRs) are cylindrical vessels that contain a layer or packing of material, such as microcarriers, that support cell adhesion and facilitate the circulation of media through the "packed bed" to provide the necessary nutrients, oxygen delivery and removal of metabolic byproducts [132, 196]. PBRs offer a controlled environment for stem cell expansion in a small volume, which is cost-effective in terms of materials and medium [196]. This vessel type is also beneficial for anchorage-dependent, or immobilized, stem cells because it offers a low shear stress environment [137]. It is important to consider how the vessel's culture system can affect the cells as shear stress has been studied to observe its effects such as reduction or alteration of the stem-like properties and differentiation lineage and the overall quality of the final stem cell population after expansion [197, 198]. However, the most significant limitation of PBRs for MSCs is the maximum media flow rate [199], where studies have shown that this mechanical stress can be beneficial if adequately regulated, such as in the case of osteogenic differentiation, where shear stress was shown to upregulate MSCs predetermined lineage preference to bone if the shear stress reached 0.015 Pa [200].

Specifically for MSCs, a scalable packed bed style bioreactor was developed by Osiecki, et al. for the isolation and subsequent expansion of MSCs, which featured a plasma-treated and fused bed of polystyrene beads encased in a shell that was gas permeable to enhance gas exchange [199]. The MSCs were expanded in the system, but had a lower growth rate than the control flasks and an uneven distribution of the cells due to the seeding methods; however, the cells did retain their osteogenic potential [199]. This serves as an example of the benefits of a packed bed style bioreactor as a closed system with potential for automation, which can overcome nutrient gradients with a high enough flow rate. Unfortunately, cell harvesting can be very difficult as the system has to be flushed with a trypsin-EDTA solution to cleave the cells from the culturing surface. This can be insufficient, thus resulting in lower cell yields for this type of bioreactor system and compromising its economic feasibility for large-scale manufacturing.

3.3 Selection of Microcarriers

By far, of the existing dynamic systems used for cell expansion, 3D microcarrier-based culturing approaches are most commonly used for cell expansion and preferred for clinical and commercial-scale MSC production. Thus, it is essential to understand key aspects regarding their use. Microcarrier-based culture relies on a high surface-area-to-volume ratio to produce a high density of stem cells in the final product. Thus, the smaller the microcarrier, then the higher the potential surface area present, which increases the potential growth area [201]. This characteristic makes microcarriers ideal for process scale-up as higher densities can be achieved in a reduced medium volume, which can lower the cost of expansion [202]. Microcarriers can be prepared from biomaterials, including ceramics, and synthetic or natural polymers depending on the needed characteristics, length of culture and time and application. Initially, synthetic microcarriers made from glass or polymers, such as polystyrene, and polylactic acid, allowed for sufficient mechanical strength and consistency in the microcarrier properties from batch-to-batch. However, these types of microcarriers maintained limited cell attachment proficiency due to the lack of sites for anchorage-dependent cells to adhere to [203, 204]. This limitation has been overcome by preparing microcarriers either by coating with extracellular matrix proteins (e.g., collagen, fibronectin, etc.) or a hydrogel mixture containing appropriate ligands, or by chemically enhancing the surface to promote cell adhesion for cell expansion purposes [205, 206]. Natural biomaterials for microcarriers are widely available and inexpensive. Natural materials such as gelatin, collagen, cellulose, pectin, alginate, and chitosan also have excellent biocompatibility as they do not induce an adverse biological response from the stem cells, and have been used for microcarrier-based applications with MSCs [203, 204, 207–209]. These characteristics are specifically applicable to commercially-available microcarriers, such as Cultispher, Cytophor, Cytodex, Synthemax, Solohill, and other formulations currently on the market [99, 127, 210–213].

Microcarriers are also advantageous because properties such as their surface chemistry, topography, porosity, geometric shape, matrix coating, and mechanical characteristics (Fig. 7) can be augmented to control cell attachment for better expansion and to potentially enhance the phenotype expression of MSCs [208, 214–216]. In particular, MSCs are sensitive to mechanical forces and stress, such that they are able to sense and respond to substrate stiffness through various cell adhesion signaling pathways. This sensitivity is highly relevant for MSC culture where alterations in phenotype and multilineage potential are observed. Changes in mechanical stiffness have also been associated with fluctuations in secretion of trophic factors present in the MSC secretome [123, 153, 154, 217]. For these reasons, when selecting scale-up approaches, careful consideration must be given to materials selected and the specific type of bioreactor that will be used for MSC expansion.

The use of microcarrier-based approaches and bioreactors for successful MSC expansion is dependent upon several important parameters, including cell seeding (inoculation) density, microcarrier concentration, length of time for culture, agitation mode and speed, in-line monitoring or sampling of metabolites, and harvesting strategy [174, 218]. Without standardized prescriptive processes for MSC expansion, these scale-up and scale-out approaches contribute further to MSC heterogeneity and subsequent challenges regarding their clinical use. For example, the different types of materials and culturing approaches used for microcarrier-based systems, contribute to increased heterogeneity since different types of materials may support varied attachment efficiency, proliferation rates, and subsequent cell potency. Furthermore, cell separation and detachment can be complicated for certain materials, negatively affecting cell harvest yields and therapeutic efficacy. Concerns related to reproducibility due to heterogeneity of culturing methods and MSC response are also linked to economic cost for manufacturing cells for clinical applications. Extensive cost and economic analyses have resulted in what is described as a technology S-Curve for cell therapy manufacture, which relates R&D effort and investment required for a company to performance (i.e. yield) of cells attained in a lot [45]. Specific considerations related to cell expansion, such as expansion platform and raw materials (Table 2), are contributors to the cost of manufacturing clinical cell therapies [33, 99]. Based on estimates and projections, microcarrier-based expansion would be most cost effective for projected yields, with reported estimates of $700/dose for microcarriers versus $15,000/dose for multi-layer vessels to obtain a dose of 10^9 cells. For future improved MSC therapeutic manufacturing, it will therefore, be essential to identify optimized

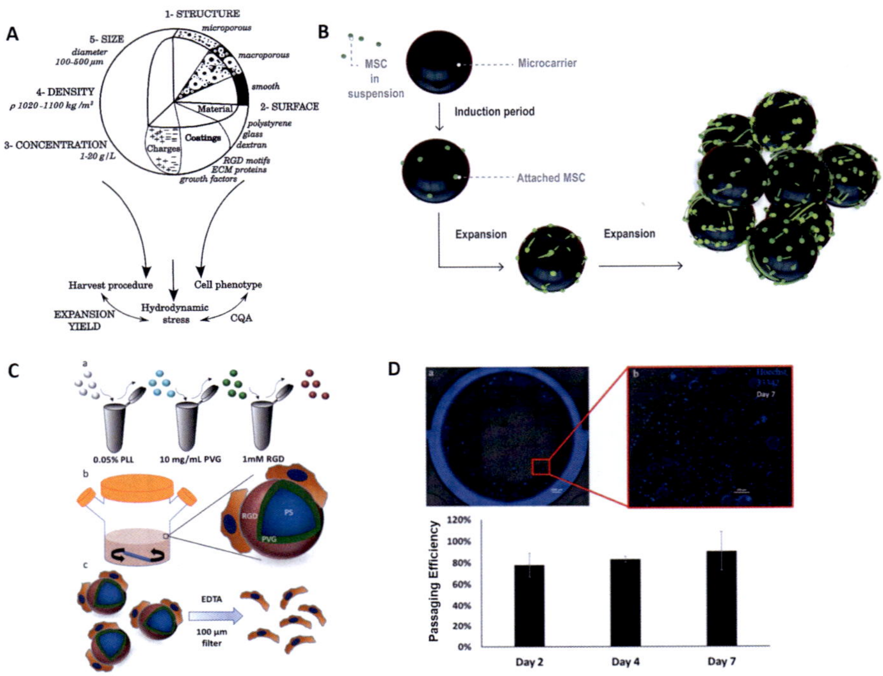

Fig. 7 Microcarrier-based cultures support MSC expansion. (**a**) Five main characteristics associated with microcarriers impact hMSC expansion, including microcarrier type, surface properties, concentration, density, and size. (Taken from Martin et al. [218] with permission.) (**b**) Visual representation of MSC expansion on microcarriers over time within a bioreactor. MSCs initially attach at low coverage and spread over the induction period. The cells then enter a growth phase and expand to cover a large proportion of the microcarrier surface area. (Adapted from Cherian et al. [123]. Rights permitted under Creative Commons license). (**c**) Polystyrene microcarriers are functionalized using a sequential anchoring process of PLL adsorption, followed by PVG copolymer anchoring, and RGD functionalization. hMSC are then cultured in xeno-free media in a stirred flask bioreactor, and then passaged using EDTA, with cells separated from the microcarriers using a cell strainer. (**d**) EDTA passaging effective over culture period; fluorescent micrographs of hMSC grown on PVG + RGD microcarriers cultured xeno-free media and passaged using EDTA. Quantification of passaging efficiency (percentage of cells removed from microcarriers by EDTA) was defined as the percentage of total nuclei that were detached from microcarriers after EDTA exposure. (**c** and **d** reprinted with permission from (Krutty et al.). Copyright (2021) American Chemical Society)

culture approaches based on material advances for enhanced microcarrier substrates and technological advances for more efficient bioreactor systems.

As culturing methods trend towards the use of serum-free and xeno-free media formulations to increase clinical application, the need for materials that can support cell functions has grown. Commercially-available T-flasks and multi-layer flasks prepared with coatings of adhesive proteins, such as Synthemax® by Corning and CELLstart™ by ThermoFisher Scientific are commonly used for 2D planar

Table 2 Considerations for cell expansion and material costs in cell manufacturing

Topic	Key cost of goods considerations
Cell expansion	
Operational performance and lot size	Use estimates of expansion yields or harvest densities and downstream yields to determine if number of expansion units per lot is practical for each technology choice
Quality control and regulatory compliance	Automated or manual processing and open or closed processing are key decisions. Currently, most processes involve manual handling. Automated and closed processing may reduce costs and improve cell quality
Scalability	Multi-layer vessels can reach production limitations at higher cell production numbers. Determine by using the S-curve method [45] if the desired lot sizes over a product life cycle can be met by planar technologies or whether a switch to bioreactor technologies is required
Material costs	
Safety	Start with materials that are deemed by regulatory authorities in key target markets to be safe ancillary materials
Reproducibility	Consistently use the same materials, from the same sources, to reduce the risk of variance resulting from changes in materials used in the cells produced
Physiological requirements	When possible, replace complex, animal-sourced materials from media formulations with well-defined components
Supply chain	Create and test strategies for "bioequivalence" as part of the chemistry, manufacturing, and controls for those reagents at risk of supply disruption

Adapted from Lipsitz et al. [33] with permission

culture under serum-free conditions [113]. For 3D microcarrier substrates, similar approaches have been applied through the use of microcarriers coated with adhesive proteins [216, 219–223]. Specifically, for MSCs, however, recent studies have evaluated coatings to not only support cell adhesion, but coatings that can affect specific properties of MSCs, such as modulating the MSC secretome *in vitro*. For example, a layer-by-layer approach was used to form heparin/collagen surface coatings for culturing hMSCs. The coatings were shown to support hMSC proliferation in low-serum conditions and enhance production of cytokines when stimulated with interferon gamma, indicating modulation of immunosuppressive potency [224]. Synthetic chemically-defined microcarriers, formed by coating glass or polystyrene microcarriers with layered adhesive peptides, have also been explored for supporting MSCs in xeno-free conditions and allowing harvesting without the use of enzymes (Fig. 7c and d). Cells were able to adhere and proliferate on the microcarriers, and differentiation potential was maintained post-harvest [225, 226].

Other biomaterial advances towards optimizing MSC culture have focused on downstream processing, where improved harvesting strategies could support better cell yields, quality and purity of cultures, and subsequently increase accessibility and therapeutic outcomes. In particular, thermally responsive microcarriers that allow non-invasive detachment of cultured cells through temperature reduction without proteolytic enzyme treatment have been explored for increasing cell harvest yields [227]. Gelatin-based dissolvable microcarriers formed via droplet microfluidics have shown potential for support of MSC culture comparable to commercially-available microcarriers, with harvesting improved through direct dissolution using a Pronase enzyme solution [228]. Meanwhile, dispersible and dissolvable porous gelatin-based microcarrier tablets have been investigated for large-scale manufacturing of MSCs. The 3D TableTrix is a tablet form made of tightly packed microcarriers, which disperse into individual microcarriers when contacting an aqueous solution. Cells harvested using a digestion solution were shown to retain MSC phenotype and differentiation potential [229].

4 MSC Quality Control

Due to their functional heterogeneity, quality control represents one of the most significant challenges hindering translation of MSC therapies. MSC products are regulated as biologics by the U.S. Food and Drug Administration (FDA) and require completion of clinical trials, conducted under FDA-cleared investigational new drug (IND) applications, to demonstrate their safety and efficacy for a given indication. The chemistry, manufacturing, and controls (CMC) section of an IND provides specific details on the donor/tissue source of MSCs, protocols describing manufacturing steps from source material to final product formulation, as well as descriptions of all assays and release criteria included as part of the sponsor's quality control (21 CFR 312.23(a)(7)(iv)(b)). Quality control consists of analytical methods that measure product identity, purity, potency, and safety along with acceptable limits and release criteria. Although MSC safety and tolerability is well-established (in both autologous and allogeneic settings), identification of CQAs predictive of their therapeutic function (i.e. efficacy) in both preclinical and clinical studies has been challenging due to their complex, multifaceted mechanisms of action.

Although the majority of clinical trials investigating MSCs focus on their paracrine actions (i.e. immunomodulation and secretion of angiogenic and pro-regenerative factors), MSCs are still predominantly characterized using the ISCT criteria established in 2006 (Table 3) [67]. However, there are efforts to revise these criteria particularly with respect to the hypothesized MSC mechanisms of action as the surface markers and adherence criteria can be generally ascribed to all fibroblast-like cells.

Furthermore, qualitatively tri-lineage differentiation is likely irrelevant in the context of T cell suppression [34] or macrophage efferocytosis [230], which are becoming increasingly recognized as primary MSC mechanisms of action.

Table 3 Summary of criteria to identify MSC

(1) Adherence to plastic in standard culture conditions		
(2) Phenotype	Positive (> 95%+)	Negative (<2%+)
	CD105	CD45
	CD73	CD34
	CD90	CD14 or CD11b
		CD79a or CD19
		HLA-DR
(3) *In vitro* differentiation: Osteoblasts, adipocytes, chondroblasts (demonstrated by staining of *in vitro* cell culture)		

Adapted from Dominici et al. [67]

As increasing evidence mounts that the ISCT criteria do not effectively predict therapeutically-relevant MSC functions, there have been attempts to identify novel CQAs (Fig. 8). In terms of surface markers, MSCs expressing higher levels of CD146 showed higher production of the immunomodulatory factor indoleamine 2,3-dioxygenase (IDO), as well as improved *in vitro* (T cell modulation) and *in vivo* (decreased synovitis) function [231]. CD264 expression in MSCs was shown to be indicative of cellular aging as MSC lots with higher CD264 expression were senescent, had longer doubling times, and exhibited significantly reduced osteogenesis and adipogenesis [232]. Surface marker and intracellular expression of immunomodulatory factors such as PD-L1 (CD274) and IDO have also shown utility in predicting MSC immunosuppressive capacity, which agrees well with their known role in regulation of T cell activation. As MSC secreted factors are believed to play a significant role in their mechanism of action, researchers have also explored their potential as CQAs. For example, MSC production of Prostaglandin E2 (PGE2), a potent immunomodulatory metabolite, was associated with improved outcomes in a murine model of traumatic brain injury [233], while MSC secretion of TSG-6 was positively correlated with improved outcomes inflammation in a corneal injury [234]. Within the nucleus, the transcription factor GATA6 is associated with cellular aging as increased GATA6 with increasing passage was observed and associated with decreased proliferation and tri-lineage differentiation potential [235]. Recently, high content imaging based approaches have emerged to characterize MSC heterogeneity and identify novel CQAs. Using functionally-relevant stimuli, researchers were able to identify morphological features of MSCs that predicted osteogenesis [50], chondrogenesis [52], and immunosuppression [20] from multiple donors and at early and late passages. Other imaging based approaches were able to predict MSC proliferation [236] and senescence [237], while more advanced live imaging approaches are being developed to monitor MSC clonality, migration, and the emergence of heterogeneity [238, 239].

While many of these approaches have demonstrated utility in predicting MSC quality (in terms of proliferative potential and multifunctionality), there still exists unmet needs in terms of in-process monitoring and standardization. Many of the approaches highlighted above are either endpoint, destructive assessments or they

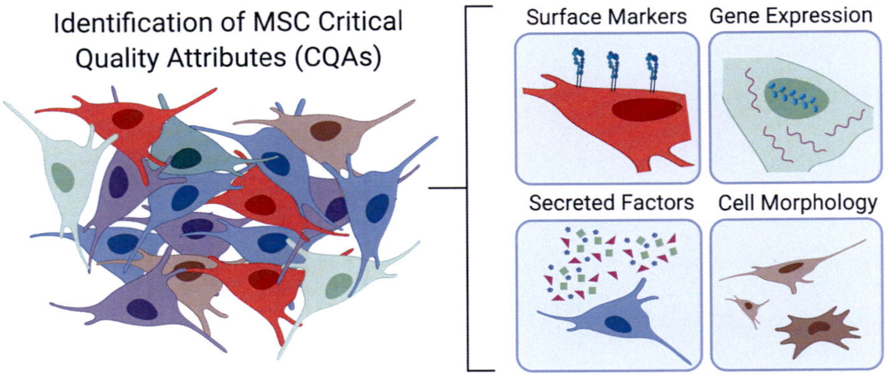

Fig. 8 Identification of MSC CQAs. MSCs can be characterized using various assays to assess expression of cell surface markers, gene expression, secretion of proteins/metabolites, and cell morphology

cannot readily be performed on the final product in a timely manner prior to patient administration. In larger scale bioreactor formats, assessment of dissolved oxygen, lactate, and ammonium enables in-process, non-destructive monitoring of MSC energy consumption [240]; however, these measures provide a more general assessment of MSC growth (and overall cell yield) and likely not sufficient to predict therapeutic function (i.e. potency). In terms of standardization, there are no well-established CQAs in the MSC field, as well as no reference standard MSC materials to compare between (and within) labs. Demonstrating product comparability [241] is challenging when sponsors make significant manufacturing stages during product development and having standardized functional (potency) assays, as well as reference materials, would not only improve quality control for sponsors of MSC trials, but also aid FDA regulators responsible for reviewing CMC sections of INDs and BLAs. iPS-derived MSCs represent a promising source of reference MSCs for standardization as an isogenic iPS cell-line can proliferate indefinitely and could be subsequently differentiated into MSCs in a controlled manner and distributed to different MSC manufacturers [41].

5 Conclusions

Mesenchymal stromal cells hold tremendous potential to treat a vast array of diseases and injuries. However, the commercial viability of these therapies will be contingent on our ability to manufacture cells in sufficient quantities with proven and consistent quality and efficacy. The sections in this chapter highlight the state-of-the-art technology currently available and the challenges associated with each of them, respectively. Many of these challenges are complex and unique to cell therapies; for most biopharmaceuticals, such as monoclonal antibodies, recombinant proteins or viral vaccines, the cell acts as a factory to produce the product. In contrast, for cell therapies the cell is the product, and as a living entity its properties and function can change with every manipulation, requiring a whole new paradigm for large-scale manufacturing and quality control. Additionally, very little standardization, automation, and in-line quality control exists across the field for cell collection, cell characterization and processing, cell identity markers, potency assays, and storage solutions. For most MSC-based therapies, little is known about their critical quality attributes (CQAs) that render them safe and effective for specific disease indications. Furthermore, little is known about which manufacturing or process variables can yield a consistent and reproducible product with appropriate CQAs. Without the ability to implement Quality-by-Design (QbD) manufacturing processes, large-scale, reproducible, low-cost production of high-quality and safe MSCs cannot be achieved, lessening the potential impact of therapies.

Overcoming these challenges will require a multi-disciplinary approach incorporating cell-culture and cell-biology, bioprocess engineering, electrical and mechanical engineering focused on sensors and automation in bioreactors, manufacturing and supply chain logistics, regulatory knowledge, and data sciences, AI, and predictive analytics. It is also essential to have national and international private-public collaborations between industry and academia, as well as close involvement of federal regulatory and standardization agencies. Several such consortia have recently been established in the U.S., which include the National Institute for Innovation in Manufacturing of Biopharmaceuticals (NIIMBL), the Advanced Regenerative Manufacturing Institute's (ARMI) BioFabUSA, and the NSF Engineering Research Centers (ERCs) for Cell Manufacturing Technologies (CMaT) and Cellular Metamaterials (Cell-Met). These networks, and many others like them, will be critical for conducting the essential research necessary to bring the enormous potential of MSC therapies to patients around the world.

References

1. Martin I, Galipeau J, Kessler C, Blanc KL, Dazzi F (2019) Challenges for mesenchymal stromal cell therapies. Sci Transl Med 11:eaat2189. https://doi.org/10.1126/scitranslmed.aat2189
2. Johnson CL, Soeder Y, Dahlke MH (2017) Concise review: mesenchymal stromal cell-based approaches for the treatment of acute respiratory distress and sepsis syndromes. Stem Cells Transl Med 6:1141. https://doi.org/10.1002/sctm.16-0415
3. Samsonraj RM, Raghunath M, Nurcombe V, Hui JH, van Wijnen AJ, Cool SM (2017) Concise review: multifaceted characterization of human mesenchymal stem cells for use in regenerative medicine. Stem Cells Transl Med 6(12):2173–2185. https://doi.org/10.1002/sctm.17-0129
4. Zhao T, Sun F, Liu J, Ding T, She J, Mao F, Xu W, Qian H, Yan Y (2019) Emerging role of mesenchymal stem cell-derived exosomes in regenerative medicine. Curr Stem Cell Res Ther 14(6):482–494. https://doi.org/10.2174/1574888x14666190228103230

5. Varderidou-Minasian S, Lorenowicz MJ (2020) Mesenchymal stromal/stem cell-derived extracellular vesicles in tissue repair: challenges and opportunities. Theranostics 10(13):5979–5997. https://doi.org/10.7150/thno.40122
6. Mendicino M, Bailey AM, Wonnacott K, Puri RK, Bauer SR (2014) MSC-based product characterization for clinical trials: an FDA perspective. Cell Stem Cell 14(2):141–145. https://doi.org/10.1016/j.stem.2014.01.013
7. Levy O, Kuai R, Siren EMJ, Bhere D, Milton Y, Nissar N, Biasio MD, Heinelt M, Reeve B, Abdi R, Alturki M, Fallatah M, Almalik A, Alhasan AH, Shah K, Karp JM (2020) Shattering barriers toward clinically meaningful MSC therapies. Sci Adv 6:1–19
8. Confalonieri D, Schwab A, Walles H, Ehlicke F (2018) Advanced therapy medicinal products: a guide for bone marrow-derived MSC application in bone and cartilage tissue engineering. Tissue Eng Part B Rev 24(2):155–169. https://doi.org/10.1089/ten.teb.2017.0305
9. Xie X, Wu H, Li M, Chen X, Xu X, Ni W, Lu C, Ni R, Bao B, Xiao M (2019) Progress in the application of exosomes as therapeutic vectors in tumor-targeted therapy. Cytotherapy 21(5):509–524. https://doi.org/10.1016/j.jcyt.2019.01.001
10. Caplan AI (1991) Mesenchymal stem cells. J Orthop Res 9:641. https://doi.org/10.1002/jor.1100090504
11. Phinney DG (2012) Functional heterogeneity of mesenchymal stem cells: implications for cell therapy. J Cell Biochem 113:2806–2812. https://doi.org/10.1002/jcb.24166
12. Phinney DG, Galipeau J (2019) Manufacturing mesenchymal stromal cells for clinical applications: a survey of good manufacturing practices at U.S. academic centers. Cytotherapy 21(7):782–792. https://doi.org/10.1016/j.jcyt.2019.04.003
13. Viswanathan S, Shi Y, Galipeau J, Krampera M, Leblanc K, Martin I, Nolta J, Phinney DG, Sensebe L (2019) Mesenchymal stem versus stromal cells: International Society for Cell & Gene Therapy (ISCT®) mesenchymal stromal cell committee position statement on nomenclature. Cytotherapy 21(10):1019–1024. https://doi.org/10.1016/j.jcyt.2019.08.002
14. Li YY, Lam KL, Chen AD, Zhang W, Chan BP (2019) Collagen microencapsulation recapitulates mesenchymal condensation and potentiates chondrogenesis of human mesenchymal stem cells – a matrix-driven in vitro model of early skeletogenesis. Biomaterials 213(January):119210–119210. https://doi.org/10.1016/j.biomaterials.2019.05.021
15. Raisin S, Belamie E, Morille M (2016) Non-viral gene activated matrices for mesenchymal stem cells based tissue engineering of bone and cartilage. Biomaterials 104:223–237. https://doi.org/10.1016/j.biomaterials.2016.07.017
16. He J, Zhang N, Zhu Y, Jin R, Wu F (2021) MSC spheroids-loaded collagen hydrogels simultaneously promote neuronal differentiation and suppress inflammatory reaction through PI3K-Akt signaling pathway. Biomaterials 265(October 2020):120448–120448. https://doi.org/10.1016/j.biomaterials.2020.120448
17. Jarmalavičiute A, Tunaitis V, Pivoraite U, Venalis A, Pivoriunas A (2015) Exosomes from dental pulp stem cells rescue human dopaminergic neurons from 6-hydroxy-dopamine-induced apoptosis. Cytotherapy 17(7):932–939. https://doi.org/10.1016/j.jcyt.2014.07.013
18. Pinto DS, Ahsan T, Serra J, Fernandes-Platzgummer A, Cabral JMS, da Silva CL (2020) Modulation of the in vitro angiogenic potential of human mesenchymal stromal cells from different tissue sources. J Cell Physiol 235(10):7224–7238. https://doi.org/10.1002/jcp.29622
19. Bracho-Sanchez E, Hassanzadeh A, Brusko MA, Wallet MA, Keselowsky BG (2019) Dendritic cells treated with exogenous indoleamine 2,3-dioxygenase maintain an immature phenotype and suppress antigen-specific T cell proliferation. J Immunol Regen Med 5(September 2018):100015–100015. https://doi.org/10.1016/j.regen.2019.100015
20. Marklein RA, Klinker MW, Drake KA, Polikowsky HG, Lessey-Morillon EC, Bauer SR (2019) Morphological profiling using machine learning reveals emergent subpopulations of interferon-γ–stimulated mesenchymal stromal cells that predict immunosuppression. Cytotherapy 21(1):17–31. https://doi.org/10.1016/j.jcyt.2018.10.008
21. Nicotra T, Desnos A, Halimi J, Antonot H, Reppel L, Belmas T, Freton A, Stranieri F, Mebarki M, Larghero J, Cras A, Faivre L (2020) Mesenchymal stem/stromal cell quality control: validation of mixed lymphocyte reaction assay using flow cytometry according to ICH Q2(R1). Stem Cell Res Ther 11(1):1–10. https://doi.org/10.1186/s13287-020-01947-6

22. Noh MY, Lim SM, Oh K-W, Cho K-A, Park J, Kim K-S, Lee S-J, Kwon M-S, Kim SH (2016) Mesenchymal stem cells modulate the functional properties of microglia via TGF-β secretion. Stem Cells Transl Med 5:1538. https://doi.org/10.5966/sctm.2015-0217
23. Contentin R, Demoor M, Concari M, Desancé M, Audigié F, Branly T, Galéra P (2020) Comparison of the chondrogenic potential of mesenchymal stem cells derived from bone marrow and umbilical cord blood intended for cartilage tissue engineering. Stem Cell Rev Rep 16(1):126–143. https://doi.org/10.1007/s12015-019-09914-2
24. Galieva LR, James V, Mukhamedshina YO, Rizvanov AA (2019) Therapeutic potential of extracellular vesicles for the treatment of nerve disorders. Front Neurosci 13(March):1–9. https://doi.org/10.3389/fnins.2019.00163
25. Lipsitz YY, Timmins NE, Zandstra PW (2016) Quality cell therapy manufacturing by design. Nat Biotechnol 34(4):393–400. https://doi.org/10.1038/nbt.3525
26. Galipeau J, Krampera M, Barrett J, Dazzi F, Deans RJ, DeBruijn J, Dominici M, Fibbe WE, Gee AP, Gimble JM, Hematti P, Koh MBC, LeBlanc K, Martin I, McNiece IK, Mendicino M, Oh S, Ortiz L, Phinney DG, Planat V, Shi Y, Stroncek DF, Viswanathan S, Weiss DJ, Sensebe L (2015) International society for cellular therapy perspective on immune functional assays for mesenchymal stromal cells as potency release criterion for advanced phase clinical trials. Cytotherapy 18:151. https://doi.org/10.1016/j.jcyt.2015.11.008
27. Technau A, Froelich K, Hagen R, Kleinsasser N (2011) Adipose tissue-derived stem cells show both immunogenic and immunosuppressive properties after chondrogenic differentiation. Cytotherapy 13(3):310–317. https://doi.org/10.3109/14653249.2010.504769
28. Mukonoweshuro B, Brown CJF, Fisher J, Ingham E (2014) Immunogenicity of undifferentiated and differentiated allogeneic mouse mesenchymal stem cells. J Tissue Eng 5(X):204173141453425. https://doi.org/10.1177/2041731414534255
29. Berglund AK, Fortier LA, Antczak DF, Schnabel LV (2017) Immunoprivileged no more: measuring the immunogenicity of allogeneic adult mesenchymal stem cells. Stem Cell Res Ther 8(1):1–7. https://doi.org/10.1186/s13287-017-0742-8
30. Grau-vorster M, Laitinen A, Nystedt J, Vives J (2019) HLA-DR expression in clinical-grade bone marrow-derived multipotent mesenchymal stromal cells: a two-site study. Stem Cell Res Ther 9:1–8
31. Grau-Vorster M, Rodr´iGuez L, Torrents-Zapata S, Vivas D, Codinach M, Blanco M, Oliver-Vila I, Garc´iA-LÓPez JOAN, Vives J (2019) Levels of IL-17F and IL-33 correlate with HLA-DR activation in clinical-grade human bone marrow–derived multipotent mesenchymal stromal cell expansion cultures. Cytotherapy 21(1):32–40. https://doi.org/10.1016/j.jcyt.2018.09.009
32. Kiernan CH, KleinJan A, Peeters M, Wolvius EB, Farrell E, Brama PAJ (2018) Allogeneic chondrogenically differentiated human bone marrow stromal cells do not induce dendritic cell maturation. J Tissue Eng Regen Med 12(6):1530–1540. https://doi.org/10.1002/term.2682
33. Lipsitz YY, Milligan WD, Fitzpatrick I, Stalmeijer E, Farid SS, Tan KY, Smith D, Perry R, Carmen J, Chen A, Mooney C, Fink J (2017) A roadmap for cost-of-goods planning to guide economic production of cell therapy products. Cytotherapy 19(12):1383–1391. https://doi.org/10.1016/j.jcyt.2017.06.009
34. François M, Romieu-Mourez R, Li M, Galipeau J (2012) Human MSC suppression correlates with cytokine induction of indoleamine 2,3-dioxygenase and bystander M2 macrophage differentiation. Mol Ther 20(1):187–195. https://doi.org/10.1038/mt.2011.189
35. Guan Q, Li Y, Shpiruk T, Bhagwat S, Wall DA (2018) Inducible indoleamine 2,3-dioxygenase 1 and programmed death ligand 1 expression as the potency marker for mesenchymal stromal cells. Cytotherapy 20:639. https://doi.org/10.1016/j.jcyt.2018.02.003
36. Kang I, Lee BC, Choi SW, Lee JY, Kim JJ, Kim BE, Kim DH, Lee SE, Shin N, Seo Y, Kim HS, Kim DI, Kang KS (2018) Donor-dependent variation of human umbilical cord blood mesenchymal stem cells in response to hypoxic preconditioning and amelioration of limb ischemia. Exp Mol Med 50(4):1. https://doi.org/10.1038/s12276-017-0014-9
37. Burja B, Barlič A, Erman A, Mrak-Poljšak K, Tomšič M, Sodin-Semrl S, Lakota K (2020) Human mesenchymal stromal cells from different tissues exhibit unique responses to

different inflammatory stimuli. Curr Res Transl Med 68(4):217–224. https://doi.org/10.1016/j.retram.2020.05.006
38. Du WJ, Chi Y, Yang ZX, Li ZJ, Cui JJ, Song BQ, Li X, Yang SG, Han ZB, Han ZC (2016) Heterogeneity of proangiogenic features in mesenchymal stem cells derived from bone marrow, adipose tissue, umbilical cord, and placenta. Stem Cell Res Ther 7(1):1–11. https://doi.org/10.1186/s13287-016-0418-9
39. Mennan C, Garcia J, Roberts S, Hulme C, Wright K (2019) A comprehensive characterisation of large-scale expanded human bone marrow and umbilical cord mesenchymal stem cells. Stem Cell Res Ther 10(1):1–15. https://doi.org/10.1186/s13287-019-1202-4
40. Wang Z-g, He Z-y, Liang S, Yang Q, Cheng P, A-m C (2020) Comprehensive proteomic analysis of exosomes derived from human bone marrow, adipose tissue, and umbilical cord mesenchymal stem cells. Stem Cell Res Ther 11(1):1–11. https://doi.org/10.1186/s13287-020-02032-8
41. Prockop DJ (2017) The exciting prospects of new therapies with mesenchymal stromal cells. Cytotherapy 19:1. Elsevier B.V. https://doi.org/10.1016/j.jcyt.2016.09.008
42. Spitzhorn LS, Megges M, Wruck W, Rahman MS, Otte J, Degistirici Ö, Meisel R, Sorg RV, Oreffo ROC, Adjaye J (2019) Human iPSC-derived MSCs (iMSCs) from aged individuals acquire a rejuvenation signature. Stem Cell Res Ther 10(1):1–18. https://doi.org/10.1186/s13287-019-1209-x
43. Diederichs S, Tuan RS (2014) Functional comparison of human-induced pluripotent stem cell-derived mesenchymal cells and bone marrow-derived mesenchymal stromal cells from the same donor. Stem Cells Dev 23(14):1594–1610. https://doi.org/10.1089/scd.2013.0477
44. Slukvin II, Kumar A (2018) The mesenchymoangioblast, mesodermal precursor for mesenchymal and endothelial cells. Cell Mol Life Sci 75(19):3507–3520. https://doi.org/10.1007/s00018-018-2871-3
45. Simaria AS, Hassan S, Varadaraju H, Rowley J, Warren K, Vanek P, Farid SS (2014) Allogeneic cell therapy bioprocess economics and optimization: single-use cell expansion technologies. Biotechnol Bioeng 111(1):69–83. https://doi.org/10.1002/bit.25008
46. Tsuji K, Ojima M, Otabe K, Horie M, Koga H, Sekiya I, Muneta T (2017) Effects of different cell-detaching methods on the viability and cell surface antigen expression of synovial mesenchymal stem cells. Cell Transplant 26(6):1089–1102. https://doi.org/10.3727/096368917X694831
47. Patel DB, Gray KM, Santharam Y, Lamichhane TN, Stroka KM, Jay SM (2017) Impact of cell culture parameters on production and vascularization bioactivity of mesenchymal stem cell-derived extracellular vesicles. Bioeng Transl Med 2(2):170–179. https://doi.org/10.1002/btm2.10065
48. Liu Y, Muñoz N, Bunnell BA, Logan TM, Ma T (2015) Density-dependent metabolic heterogeneity in human mesenchymal stem cells. Stem Cells 33:3368. https://doi.org/10.1002/stem.2097
49. Oja S, Kaartinen T, Ahti M, Korhonen M, Laitinen A, Nystedt J (2019) The utilization of freezing steps in mesenchymal stromal cell (MSC) manufacturing: potential impact on quality and cell functionality attributes. Front Immunol 10(July):1–14. https://doi.org/10.3389/fimmu.2019.01627
50. Marklein RA, Lo Surdo JL, Bellayr IH, Godil SA, Puri RK, Bauer SR (2016) High content imaging of early morphological signatures predicts long term mineralization capacity of human mesenchymal stem cells upon osteogenic induction. Stem Cells 34(4):935–947. https://doi.org/10.1002/stem.2322
51. Lo Surdo JL, Millis BA, Bauer SR (2013) Automated microscopy as a quantitative method to measure differences in adipogenic differentiation in preparations of human mesenchymal stromal cells. Cytotherapy 15(12):1527–1540. https://doi.org/10.1016/j.jcyt.2013.04.010

52. Lam J, Bellayr IH, Marklein RA, Bauer SR, Puri RK, Sung KE (2018) Functional profiling of chondrogenically induced multipotent stromal cell aggregates reveals transcriptomic and emergent morphological phenotypes predictive of differentiation capacity. Stem Cells Transl Med 7(9):664–675. https://doi.org/10.1002/sctm.18-0065
53. Yin L, Wu Y, Yang Z, Tee CA, Denslin V, Lai Z, Lim CT, Lee EH, Han J (2018) Microfluidic label-free selection of mesenchymal stem cell subpopulation during culture expansion extends the chondrogenic potential: in vitro. Lab Chip 18(6):878–889. https://doi.org/10.1039/c7lc01005b
54. Deng Y, Lei G, Lin Z, Yang Y, Lin H, Tuan RS (2019) Engineering hyaline cartilage from mesenchymal stem cells with low hypertrophy potential via modulation of culture conditions and Wnt/β-catenin pathway. Biomaterials 192(November 2018):569–578. https://doi.org/10.1016/j.biomaterials.2018.11.036
55. Guo T, Lim CG, Noshin M, Ringel JP, Fisher JP (2018) 3D printing bioactive PLGA scaffolds using DMSO as a removable solvent. Bioprinting 10(November):1–8. https://doi.org/10.1016/j.bprint.2018.e00038
56. Harada N, Watanabe Y, Sato K, Abe S, Yamanaka K, Sakai Y, Kaneko T, Matsushita T (2014) Bone regeneration in a massive rat femur defect through endochondral ossification achieved with chondrogenically differentiated MSCs in a degradable scaffold. Biomaterials 35(27):7800–7810. https://doi.org/10.1016/j.biomaterials.2014.05.052
57. Kim M, Erickson IE, Huang AH, Garrity ST, Mauck RL, Steinberg DR (2018) Donor variation and optimization of human mesenchymal stem cell chondrogenesis in hyaluronic acid. Tissue Eng Part A 24(21–22):1693–1703. https://doi.org/10.1089/ten.tea.2017.0520
58. Braid LR, Wood CA, Wiese DM, Ford BN (2018) Intramuscular administration potentiates extended dwell time of mesenchymal stromal cells compared to other routes. Cytotherapy 20(2):232–244. https://doi.org/10.1016/j.jcyt.2017.09.013
59. Giri J, Galipeau J (2020) Mesenchymal stromal cell therapeutic potency is dependent upon viability, route of delivery, and immune match. Blood Adv 4(9):1987–1997. https://doi.org/10.1182/bloodadvances.2020001711
60. Kanelidis AJ, Premer C, Lopez J, Balkan W, Hare JM (2017) Route of delivery modulates the efficacy of mesenchymal stem cell therapy for myocardial infarction: a meta-analysis of preclinical studies and clinical trials. Circ Res 120(7):1139–1150. https://doi.org/10.1161/CIRCRESAHA.116.309819
61. Losurdo M, Pedrazzoli M, D'Agostino C, Elia CA, Massenzio F, Lonati E, Mauri M, Rizzi L, Molteni L, Bresciani E, Dander E, D'Amico G, Bulbarelli A, Torsello A, Matteoli M, Buffelli M, Coco S (2020) Intranasal delivery of mesenchymal stem cell-derived extracellular vesicles exerts immunomodulatory and neuroprotective effects in a 3xTg model of Alzheimer's disease. Stem Cells Transl Med 9(September 2019):1–17. https://doi.org/10.1002/sctm.19-0327
62. Oh KW, Noh MY, Kwon MS, Kim HY, Oh S, Park J, Kim HJ, Ki CS, Kim SH (2018) Repeated intrathecal mesenchymal stem cells for amyotrophic lateral sclerosis. Ann Neurol 84(3):361–373. https://doi.org/10.1002/ana.25302
63. Lim J, Heo J, Ju H, Shin JW, Kim YH, Lee S, Yu HY, Ryu CM, Yun HD, Song S, Hong KS, Chung HM, Kim HR, Roe JS, Choi K, Kim IG, Jeong EM, Shin DM (2020) Glutathione dynamics determine the therapeutic efficacy of mesenchymal stem cells for graft-versus-host disease via CREB1-NRF2 pathway. Sci Adv 6(16):1–18. https://doi.org/10.1126/sciadv.aba1334
64. Chinnadurai R, Copland IB, Garcia MA, Petersen CT, Lewis CN, Waller EK, Kirk AD, Galipeau J (2016) Cryopreserved mesenchymal stromal cells are susceptible to T-cell mediated apoptosis which is partly rescued by IFNγ licensing. Stem Cells 34(9):2429–2442. https://doi.org/10.1002/stem.2415
65. Harrison RP, Rafiq QA, Medcalf N (2018) Centralised versus decentralised manufacturing and the delivery of healthcare products: a United Kingdom exemplar. Cytotherapy 20(6):873–890. https://doi.org/10.1016/j.jcyt.2018.05.003

66. Harrison RP, Ruck S, Rafiq QA, Medcalf N (2018) Decentralised manufacturing of cell and gene therapy products: learning from other healthcare sectors. Biotechnol Adv 36(2):345–357. https://doi.org/10.1016/j.biotechadv.2017.12.013
67. Dominici M, Le Blanc K, Mueller I, Slaper-Cortenbach I, Marini F, Krause D, Deans R, Keating A, Prockop D, Horwitz E (2006) Minimal criteria for defining multipotent mesenchymal stromal cells. The International Society for cellular therapy position statement. Cytotherapy 8(4):315–317. https://doi.org/10.1080/14653240600855905
68. Cantor JR (2019) The rise of physiologic media. Trends Cell Biol 29(11):854–861. https://doi.org/10.1016/j.tcb.2019.08.009
69. Karnieli O, Friedner OM, Allickson JG, Zhang N, Jung S, Fiorentini D, Abraham E, Eaker SS, Yong TK, Chan A, Griffiths S, When Ak OS, Karnieli O (2017) A consensus introduction to serum replacements and serum-free media for cellular therapies. Cytotherapy 19(2):155–169. https://doi.org/10.1016/j.jcyt.2016.11.011
70. Barilani M, Lavazza C, Boldrin V, Ragni E, Parazzi V, Crosti M, Montelatici E, Giordano R, Lazzari L (2016) A chemically defined medium-based strategy to efficiently generate clinically relevant cord blood mesenchymal stromal colonies. Cell Transplant 25(8):1501–1514. https://doi.org/10.3727/096368916X690827
71. Dessels C, Potgieter M, Pepper MS (2016) Making the switch: alternatives to fetal bovine serum for adipose-derived stromal cell expansion. Front Cell Dev Biol 4(October):1–10. https://doi.org/10.3389/fcell.2016.00115
72. Allen AB, Butts EB, Copland IB, Stevens HY, Guldberg RE (2017) Human platelet lysate supplementation of mesenchymal stromal cell delivery: issues of xenogenicity and species variability. J Tissue Eng Regen Med 11(10):2876–2884. https://doi.org/10.1002/term.2191
73. Ren J, Ward D, Chen S, Tran K, Jin P, Sabatino M, Robey PG, Stroncek DF (2018) Comparison of human bone marrow stromal cells cultured in human platelet growth factors and fetal bovine serum. J Transl Med 16(1):1–15. https://doi.org/10.1186/s12967-018-1400-3
74. Xu J, Lian W, Chen J, Li W, Li L, Huang Z (2020) Chemical-defined medium supporting the expansion of human mesenchymal stem cells. Stem Cell Res Ther 11(1):1–11. https://doi.org/10.1186/s13287-020-01641-7
75. Gharibi B, Hughes FJ (2012) Effects of medium supplements on proliferation, differentiation potential, and in vitro expansion of mesenchymal stem cells. Stem Cells Transl Med 1(11):771–782. https://doi.org/10.5966/sctm.2010-0031
76. Hoch AI, Leach JK (2014) Concise review: optimizing expansion of bone marrow mesenchymal stem/stromal cells for clinical applications. Stem Cells Transl Med 4:412. https://doi.org/10.5966/sctm.2013-0196
77. Rodrigues M, Griffith LG, Wells A (2010) Growth factor regulation of proliferation and survival of multipotential stromal cells. Stem Cell Res Ther 1(4):1–12. https://doi.org/10.1186/scrt32
78. Rodrigues M, Blair H, Stockdale L, Griffith L, Wells A (2013) Surface tethered epidermal growth factor protects proliferating and differentiating multipotential stromal cells from FasL-induced apoptosis. Stem Cells 31(1):104–116. https://doi.org/10.1002/stem.1215
79. Tamama K, Fan VH, Griffith LG, Blair HC, Wells A (2006) Epidermal growth factor as a candidate for ex vivo expansion of bone marrow-derived mesenchymal stem cells. Stem Cells 24(3):686–695. https://doi.org/10.1634/stemcells.2005-0176
80. Zhang C, Guo H, Yang C, Chen Q, Huang J, Liu L, Zhang Y, Jin S, Song A, Yang P (2019) The biological behavior optimization of human periodontal ligament stem cells via preconditioning by the combined application of fibroblast growth factor-2 and A83-01 in in vitro culture expansion. J Transl Med 17(1):1–11. https://doi.org/10.1186/s12967-019-1799-1
81. Hoffman MD, Benoit DSW (2015) Agonism of Wnt-β-catenin signalling promotes mesenchymal stem cell (MSC) expansion. J Tissue Eng Regen Med 9:E13. https://doi.org/10.1002/term.1736

82. Paciejewska MM, Maijenburg MW, Gilissen C, Kleijer M, Vermeul K, Weijer K, Veltman JA, Von Lindern M, Van Der Schoot CE, Voermans C (2016) Different balance of Wnt signaling in adult and fetal bone marrow-derived mesenchymal stromal cells. Stem Cells Dev 25(12):934–947. https://doi.org/10.1089/scd.2015.0263
83. Carrillo-Gálvez AB, Gálvez-Peisl S, González-Correa JE, de Haro-Carrillo M, Ayllón V, Carmona-Sáez P, Ramos-Mejía V, Galindo-Moreno P, Cara FE, Granados-Principal S, Muñoz P, Martin F, Anderson P (2020) GARP is a key molecule for mesenchymal stromal cell responses to TGF-β and fundamental to control mitochondrial ROS levels. Stem Cells Transl Med 9(5):636–650. https://doi.org/10.1002/sctm.19-0372
84. Dubon MJ, Yu J, Choi S, Park KS (2018) Transforming growth factor β induces bone marrow mesenchymal stem cell migration via noncanonical signals and N-cadherin. J Cell Physiol 233(1):201–213. https://doi.org/10.1002/jcp.25863
85. Gurung S, Werkmeister JA, Gargett CE (2015) Inhibition of transforming growth factor-β receptor signaling promotes culture expansion of undifferentiated human endometrial mesenchymal stem/stromal cells. Sci Rep 5(September):27–31. https://doi.org/10.1038/srep15042
86. Lucciola R, Vrljicak P, Gurung S, Filby C, Darzi S, Muter J, Ott S, Brosens JJ, Gargett CE (2020) Impact of sustained transforming growth factor-β receptor inhibition on chromatin accessibility and gene expression in cultured human endometrial MSC. Front Cell Dev Biol 8(September):1–15. https://doi.org/10.3389/fcell.2020.567610
87. Pasumarthy KK, Doni Jayavelu N, Kilpinen L, Andrus C, Battle SL, Korhonen M, Lehenkari P, Lund R, Laitinen S, Hawkins RD (2017) Methylome analysis of human bone marrow MSCs reveals extensive age- and culture-induced changes at distal regulatory elements. Stem Cell Rep 9(3):999–1015. https://doi.org/10.1016/j.stemcr.2017.07.018
88. Li J, Pei M (2012) Cell senescence: a challenge in cartilage engineering and regeneration. Tissue Eng Part B Rev 18(4):270–287. https://doi.org/10.1089/ten.teb.2011.0583
89. Sathy BN, Daly A, Gonzalez-Fernandez T, Olvera D, Cunniffe G, McCarthy HO, Dunne N, Jeon O, Alsberg E, Donahue TLH, Kelly DJ (2019) Hypoxia mimicking hydrogels to regulate the fate of transplanted stem cells. Acta Biomater 88:314–324. https://doi.org/10.1016/j.actbio.2019.02.042
90. Yuan X, Logan TM, Ma T (2019) Metabolism in human mesenchymal stromal cells: a missing link between HMSC biomanufacturing and therapy? Front Immunol 10(May):1–11. https://doi.org/10.3389/fimmu.2019.00977
91. Liu Y, Muñoz N, Tsai AC, Logan TM, Ma T (2017) Metabolic reconfiguration supports reacquisition of primitive phenotype in human mesenchymal stem cell aggregates. Stem Cells 35:398. https://doi.org/10.1002/stem.2510
92. Gharibi B, Ghuman MS, Hughes FJ (2012) Akt- and Erk-mediated regulation of proliferation and differentiation during PDGFRβ-induced MSC self-renewal. J Cell Mol Med 16(11):2789–2801. https://doi.org/10.1111/j.1582-4934.2012.01602.x
93. Liu Y, Yuan X, Muñoz N, Logan TM, Ma T (2019) Commitment to aerobic glycolysis sustains immunosuppression of human mesenchymal stem cells. Stem Cells Transl Med 8(1):93–106. https://doi.org/10.1002/sctm.18-0070
94. Lee SJ, Yi TG, Ahn SH, Lim DK, Kim S, Lee HJ, Cho YK, Lim JY, Sung JH, Yun JH, Lim J, Song SU, Kwon SW (2018) Comparative study on metabolite level in tissue-specific human mesenchymal stem cells by an ultra-performance liquid chromatography quadrupole time of flight mass spectrometry. Anal Chim Acta 1024:112. https://doi.org/10.1016/j.aca.2018.04.018
95. Mancini OK, Lora M, Cuillerier A, Shum-Tim D, Hamdy R, Burelle Y, Servant MJ, Stochaj U, Colmegna I (2018) Mitochondrial oxidative stress reduces the immunopotency of mesenchymal stromal cells in adults with coronary artery disease. Circ Res 122(2):255–266. https://doi.org/10.1161/CIRCRESAHA.117.311400
96. Gharibi B, Farzadi S, Ghuman M, Hughes FJ (2014) Inhibition of Akt/mTOR attenuates age-related changes in mesenchymal stem cells. Stem Cells 32:2256. https://doi.org/10.1002/stem.1709

97. Liu Y, Ma T (2015) Metabolic regulation of mesenchymal stem cell in expansion and therapeutic application. Biotechnol Prog 31(2):468–481. https://doi.org/10.1002/btpr.2034
98. Fujisawa K, Takami T, Okada S, Hara K, Matsumoto T, Yamamoto N, Yamasaki T, Sakaida I (2018) Analysis of metabolomic changes in mesenchymal stem cells on treatment with desferrioxamine as a hypoxia mimetic compared with hypoxic conditions. Stem Cells 36:1226. https://doi.org/10.1002/stem.2826
99. Pereira Chilima TD, Moncaubeig F, Farid SS (2018) Impact of allogeneic stem cell manufacturing decisions on cost of goods, process robustness and reimbursement. Biochem Eng J 137:132–151. https://doi.org/10.1016/j.bej.2018.04.017
100. Vidal MA, Kilroy GE, Johnson JR, Lopez MJ, Moore RM, Gimble JM (2006) Cell growth characteristics and differentiation frequency of adherent equine bone marrow–derived mesenchymal stromal cells: adipogenic and osteogenic capacity. Vet Surg 35(7):601–610
101. Chamberlain G, Fox J, Ashton B, Middleton J (2007) Concise review: mesenchymal stem cells: their phenotype, differentiation capacity, immunological features, and potential for homing. Stem Cells 25(11):2739–2749
102. Pittenger MF, Mackay AM, Beck SC, Jaiswal RK, Douglas R, Mosca JD, Moorman MA, Simonetti DW, Craig S, Marshak DR (1999) Multilineage potential of adult human mesenchymal stem cells. Science 284(5411):143. https://doi.org/10.1126/science.284.5411.143
103. Bunnell BA, Flaat M, Gagliardi C, Patel B, Ripoll C (2008) Adipose-derived stem cells: isolation, expansion and differentiation. Methods 45(2):115–120
104. Baer PC, Geiger H (2012) Adipose-derived mesenchymal stromal/stem cells: tissue localization, characterization, and heterogeneity. Stem Cells Int 2012:812693
105. Zuk PA, Zhu M, Mizuno H, Huang J, Futrell JW, Katz AJ, Benhaim P, Lorenz HP, Hedrick MH (2001) Multilineage cells from human adipose tissue: implications for cell-based therapies. Tissue Eng 7(2):211–228. https://doi.org/10.1089/107632701300062859
106. Mazzini L, Ferrero I, Luparello V, Rustichelli D, Gunetti M, Mareschi K, Testa L, Stecco A, Tarletti R, Miglioretti M (2010) Mesenchymal stem cell transplantation in amyotrophic lateral sclerosis: a phase I clinical trial. Exp Neurol 223(1):229–237
107. Nabavi SM, Arab L, Jarooghi N, Bolurieh T, Abbasi F, Mardpour S, Azimyian V, Moeininia F, Maroufizadeh S, Sanjari L (2018) Safety, feasibility of intravenous and intrathecal injection of autologous bone marrow derived mesenchymal stromal cells in patients with amyotrophic lateral sclerosis: an open label phase I clinical trial. Cell J 20(4):592
108. Carlsson P-O, Schwarcz E, Korsgren O, Le Blanc K (2015) Preserved β-cell function in type 1 diabetes by mesenchymal stromal cells. Diabetes 64(2):587–592
109. Connick P, Kolappan M, Crawley C, Webber DJ, Patani R, Michell AW, Du M-Q, Luan S-L, Altmann DR, Thompson AJ (2012) Autologous mesenchymal stem cells for the treatment of secondary progressive multiple sclerosis: an open-label phase 2a proof-of-concept study. Lancet Neurol 11(2):150–156
110. Connick P, Kolappan M, Patani R, Scott MA, Crawley C, He X-L, Richardson K, Barber K, Webber DJ, Wheeler-Kingshott CA (2011) The mesenchymal stem cells in multiple sclerosis (MSCIMS) trial protocol and baseline cohort characteristics: an open-label pre-test: post-test study with blinded outcome assessments. Trials 12(1):62
111. Jarocha D, Milczarek O, Wedrychowicz A, Kwiatkowski S, Majka M (2015) Continuous improvement after multiple mesenchymal stem cell transplantations in a patient with complete spinal cord injury. Cell Transplant 24(4):661–672
112. Roh KH, Nerem RM, Roy K (2016) Biomanufacturing of therapeutic cells: state of the art, current challenges, and future perspectives. Annu Rev Chem Biomol Eng 7:455–478. https://doi.org/10.1146/annurev-chembioeng-080615-033559
113. de Almeida Fuzeta M, de Matos Branco AD, Fernandes-Platzgummer A, da Silva CL, Cabral JMS (2020) Addressing the manufacturing challenges of cell-based therapies. Adv Biochem Eng Biotechnol 171:225–278. https://doi.org/10.1007/10_2019_118
114. Larsson A, Dérand H (2002) Stability of polycarbonate and polystyrene surfaces after hydrophilization with high intensity oxygen RF plasma. J Colloid Interface Sci 246(1):214–221

115. Jokinen V, Suvanto P, Franssila S (2012) Oxygen and nitrogen plasma hydrophilization and hydrophobic recovery of polymers. Biomicrofluidics 6(1):016501
116. Lerman MJ, Lembong J, Muramoto S, Gillen G, Fisher JP (2018) The evolution of polystyrene as a cell culture material. Tissue Eng B Rev 24(5):359–372
117. Sigma-Aldrich Co. (2016) Fundamental Techniques in cell culture laboratory handbook. Available via https://www.sigmaaldrich.com/deepweb/assets/sigmaaldrich/marketing/global/documents/425/663/fundamental-techniques-in-cell-culture.pdf. Accessed September 3, 2021
118. Fajardo-Orduña GR, Mayani H, Castro-Manrreza ME, Flores-Figueroa E, Flores-Guzmán P, Arriaga-Pizano L, Piña-Sánchez P, Hernández-Estévez E, Castell-Rodríguez AE, Chávez-Rueda AK, Legorreta-Haquet MV, Santiago-Osorio E, Montesinos JJ (2016) Bone marrow mesenchymal stromal cells from clinical scale culture: in vitro evaluation of their differentiation, hematopoietic support, and immunosuppressive capacities. Stem Cells Dev 25(17):1299–1310. https://doi.org/10.1089/scd.2016.0071
119. ThermoSCientific (2017) Culture of Human mesenchymal stem cells in nunc cell factory systems. Available via https://assets.thermofisher.com/TFS-Assets/BPD/Application-Notes/culture-human-mesenchymal-stem-app-note.pdf. Accessed September 3, 2021.
120. Abbasalizadeh S, Baharvand H (2013) Technological progress and challenges towards cGMP manufacturing of human pluripotent stem cells based therapeutic products for allogeneic and autologous cell therapies. Biotechnol Adv 31(8):1600–1623
121. Chen AK-L, Reuveny S, Oh SKW (2013) Application of human mesenchymal and pluripotent stem cell microcarrier cultures in cellular therapy: achievements and future direction. Biotechnol Adv 31(7):1032–1046
122. Serra M, Brito C, Correia C, Alves PM (2012) Process engineering of human pluripotent stem cells for clinical application. Trends Biotechnol 30(6):350–359
123. Cherian DS, Bhuvan T, Meagher L, Heng TSP (2020) Biological considerations in scaling up therapeutic cell manufacturing. Front Pharmacol 11:654. https://doi.org/10.3389/fphar.2020.00654
124. Hoch A, Leach JK (2014) Concise review: optimizing expansion of bone marrow mesenchymal stem/stromal cells for clinical applications. Stem Cells Transl Med 3(5):643–652
125. Turinetto V, Vitale E, Giachino C (2016) Senescence in human mesenchymal stem cells: functional changes and implications in stem cell-based therapy. Int J Mol Sci 17(7):1164
126. Bara JJ, Richards RG, Alini M, Stoddart MJ (2014) Concise review: bone marrow-derived mesenchymal stem cells change phenotype following in vitro culture: implications for basic research and the clinic. Stem Cells 32(7):1713–1723. https://doi.org/10.1002/stem.1649
127. Jossen V, van den Bos C, Eibl R, Eibl D (2018) Manufacturing human mesenchymal stem cells at clinical scale: process and regulatory challenges. Appl Microbiol Biotechnol 102(9):3981–3994. https://doi.org/10.1007/s00253-018-8912-x
128. Wagner W, Horn P, Castoldi M, Diehlmann A, Bork S, Saffrich R, Benes V, Blake J, Pfister S, Eckstein V, Ho AD (2008) Replicative senescence of mesenchymal stem cells: a continuous and organized process. PLoS One 3(5):e2213–e2213. https://doi.org/10.1371/journal.pone.0002213
129. Banfi A, Muraglia A, Dozin B, Mastrogiacomo M, Cancedda R, Quarto R (2000) Proliferation kinetics and differentiation potential of ex vivo expanded human bone marrow stromal cells: implications for their use in cell therapy. Exp Hematol 28(6):707–715. https://doi.org/10.1016/s0301-472x(00)00160-0
130. von Bahr L, Sundberg B, Lönnies L, Sander B, Karbach H, Hägglund H, Ljungman P, Gustafsson B, Karlsson H, Le Blanc K, Ringdén O (2012) Long-term complications, immunologic effects, and role of passage for outcome in mesenchymal stromal cell therapy. Biol Blood Marrow Transplant: J Am Soc Blood Marrow Transplant 18(4):557–564. https://doi.org/10.1016/j.bbmt.2011.07.023
131. Konstantinov KB, Cooney CL (2015) White paper on continuous bioprocessing May 20–21 2014 continuous manufacturing symposium. J Pharm Sci 104(3):813–820
132. Fogler HS (1999) Elements of chemical reaction engineering. Pearson, Boston

133. Kirouac DC, Zandstra PW (2008) The systematic production of cells for cell therapies. Cell Stem Cell 3(4):369–381
134. Silva Couto P, Rotondi MC, Bersenev A, Hewitt CJ, Nienow AW, Verter F, Rafiq QA (2020) Expansion of human mesenchymal stem/stromal cells (hMSCs) in bioreactors using microcarriers: lessons learnt and what the future holds. Biotechnol Adv 45:107636. https://doi.org/10.1016/j.biotechadv.2020.107636
135. Hassan MNFB, Yazid MD, Yunus MHM, Chowdhury SR, Lokanathan Y, Idrus RBH, Ng AMH, Law JX (2020) Large-scale expansion of human mesenchymal stem cells. Stem Cells Int 2020:9529465. https://doi.org/10.1155/2020/9529465
136. Warnock JN, Al-Rubeai M (2006) Bioreactor systems for the production of biopharmaceuticals from animal cells. Biotechnol Appl Biochem 45(Pt 1):1–12. https://doi.org/10.1042/ba20050233
137. Yeatts AB, Choquette DT, Fisher JP (2013) Bioreactors to influence stem cell fate: augmentation of mesenchymal stem cell signaling pathways via dynamic culture systems. Biochim Biophys Acta (BBA)-Gen Subj 1830(2):2470–2480
138. Teixeira FG, Panchalingam KM, Assunção-Silva R, Serra SC, Mendes-Pinheiro B, Patrício P, Jung S, Anjo SI, Manadas B, Pinto L (2016) Modulation of the mesenchymal stem cell secretome using computer-controlled bioreactors: impact on neuronal cell proliferation, survival and differentiation. Sci Rep 6:27791
139. Brindley D, Moorthy K, Lee J-H, Mason C, Kim H-W, Wall I (2011) Bioprocess forces and their impact on cell behavior: implications for bone regeneration therapy. J Tissue Eng 2011:620247
140. Bunpetch V, Wu H, Zhang S, Ouyang H (2017) From "bench to bedside": current advancement on large-scale production of mesenchymal stem cells. Stem Cells Dev 26(22):1662–1673. https://doi.org/10.1089/scd.2017.0104
141. Bonab MM, Alimoghaddam K, Talebian F, Ghaffari SH, Ghavamzadeh A, Nikbin B (2006) Aging of mesenchymal stem cell in vitro. BMC Cell Biol 7:14. https://doi.org/10.1186/1471-2121-7-14
142. Krinner A, Zscharnack M, Bader A, Drasdo D, Galle J (2009) Impact of oxygen environment on mesenchymal stem cell expansion and chondrogenic differentiation. Cell Prolif 42(4):471–484. https://doi.org/10.1111/j.1365-2184.2009.00621.x
143. Fotia C, Massa A, Boriani F, Baldini N, Granchi D (2015) Hypoxia enhances proliferation and stemness of human adipose-derived mesenchymal stem cells. Cytotechnology 67(6):1073–1084. https://doi.org/10.1007/s10616-014-9731-2
144. Choi JR, Yong KW, Wan Safwani WKZ (2017) Effect of hypoxia on human adipose-derived mesenchymal stem cells and its potential clinical applications. Cell Mol Life Sci 74(14):2587–2600. https://doi.org/10.1007/s00018-017-2484-2
145. Feng Y, Zhu M, Dangelmajer S, Lee YM, Wijesekera O, Castellanos CX, Denduluri A, Chaichana KL, Li Q, Zhang H, Levchenko A, Guerrero-Cazares H, Quiñones-Hinojosa A (2014) Hypoxia-cultured human adipose-derived mesenchymal stem cells are non-oncogenic and have enhanced viability, motility, and tropism to brain cancer. Cell Death Dis 5(12):e1567–e1567. https://doi.org/10.1038/cddis.2014.521
146. Dos Santos F, Andrade PZ, Boura JS, Abecasis MM, da Silva CL, Cabral JM (2010) Ex vivo expansion of human mesenchymal stem cells: a more effective cell proliferation kinetics and metabolism under hypoxia. J Cell Physiol 223(1):27–35. https://doi.org/10.1002/jcp.21987
147. Choi JR, Pingguan-Murphy B, Wan Abas WAB, Yong KW, Poon CT, Noor Azmi MA, Omar SZ, Chua KH, Xu F, Wan Safwani WKZ (2015) In situ normoxia enhances survival and proliferation rate of human adipose tissue-derived stromal cells without increasing the risk of tumourigenesis. PLoS One 10(1):e0115034. https://doi.org/10.1371/journal.pone.0115034
148. Oliveira PH, Boura JS, Abecasis MM, Gimble JM, da Silva CL, Cabral JM (2012) Impact of hypoxia and long-term cultivation on the genomic stability and mitochondrial performance of ex vivo expanded human stem/stromal cells. Stem Cell Res 9(3):225–236. https://doi.org/10.1016/j.scr.2012.07.001

149. Colao IL, Corteling R, Bracewell D, Wall I (2018) Manufacturing exosomes: a promising therapeutic platform. Trends Mol Med 24(3):242–256. https://doi.org/10.1016/j.molmed.2018.01.006
150. Schirmaier C, Jossen V, Kaiser SC, Jüngerkes F, Brill S, Safavi-Nab A, Siehoff A, van den Bos C, Eibl D, Eibl R (2014) Scale-up of adipose tissue-derived mesenchymal stem cell production in stirred single-use bioreactors under low-serum conditions. Eng Life Sci 14(3):292–303. https://doi.org/10.1002/elsc.201300134
151. Lawson T, Kehoe DE, Schnitzler AC, Rapiejko PJ, Der KA, Philbrick K, Punreddy S, Rigby S, Smith R, Feng Q, Murrell JR, Rook MS (2017) Process development for expansion of human mesenchymal stromal cells in a 50L single-use stirred tank bioreactor. Biochem Eng J 120:49–62. https://doi.org/10.1016/j.bej.2016.11.020
152. Rafiq QA, Brosnan KM, Coopman K, Nienow AW, Hewitt CJ (2013) Culture of human mesenchymal stem cells on microcarriers in a 5 l stirred-tank bioreactor. Biotechnol Lett 35(8):1233–1245. https://doi.org/10.1007/s10529-013-1211-9
153. Robb KP, Fitzgerald JC, Barry F, Viswanathan S (2019) Mesenchymal stromal cell therapy: progress in manufacturing and assessments of potency. Cytotherapy 21(3):289–306. https://doi.org/10.1016/j.jcyt.2018.10.014
154. Ma T, Tsai A-C, Liu Y (2016) Biomanufacturing of human mesenchymal stem cells in cell therapy: influence of microenvironment on scalable expansion in bioreactors. Biochem Eng J 108:44–50. https://doi.org/10.1016/j.bej.2015.07.014
155. Tsai AC, Jeske R, Chen X, Yuan X, Li Y (2020) Influence of microenvironment on mesenchymal stem cell therapeutic potency: from planar culture to microcarriers. Front Bioeng Biotechnol 8:640. https://doi.org/10.3389/fbioe.2020.00640
156. Pigeau GM, Csaszar E, Dulgar-Tulloch A (2018) Commercial scale manufacturing of allogeneic cell therapy. Front Med 5:233. https://doi.org/10.3389/fmed.2018.00233
157. Riboldi SA, Bertoldi S, Mantero S (2017) In vitro dynamic culture of cell-biomaterial constructs. In: Characterization of polymeric biomaterials. Elsevier, Duxford, pp 339–363
158. Bunpetch V, Zhang Z-Y, Zhang X, Han S, Zongyou P, Wu H, Hong-Wei O (2019) Strategies for MSC expansion and MSC-based microtissue for bone regeneration. Biomaterials 196:67–79. https://doi.org/10.1016/j.biomaterials.2017.11.023
159. Hupfeld J, Gorr IH, Schwald C, Beaucamp N, Wiechmann K, Kuentzer K, Huss R, Rieger B, Neubauer M, Wegmeyer H (2014) Modulation of mesenchymal stromal cell characteristics by microcarrier culture in bioreactors. Biotechnol Bioeng 111(11):2290–2302. https://doi.org/10.1002/bit.25281
160. Borys BS, Roberts EL, Le A, Kallos MS (2018) Scale-up of embryonic stem cell aggregate stirred suspension bioreactor culture enabled by computational fluid dynamics modeling. Biochem Eng J 133:157–167
161. Hookway TA, Butts JC, Lee E, Tang H, McDevitt TC (2016) Aggregate formation and suspension culture of human pluripotent stem cells and differentiated progeny. Methods 101:11–20
162. Kehoe DE, Jing D, Lock LT, Tzanakakis ES (2009) Scalable stirred-suspension bioreactor culture of human pluripotent stem cells. Tissue Eng A 16(2):405–421
163. Lam AT-L, Chen AK-L, Ting SQ-P, Reuveny S, Oh SK-W (2016) Integrated processes for expansion and differentiation of human pluripotent stem cells in suspended microcarriers cultures. Biochem Biophys Res Commun 473(3):764–768
164. Oh SK, Chen AK, Mok Y, Chen X, Lim U-M, Chin A, Choo AB, Reuveny S (2009) Long-term microcarrier suspension cultures of human embryonic stem cells. Stem Cell Res 2(3):219–230
165. Ashok P, Fan Y, Rostami MR, Tzanakakis ES (2015) Aggregate and microcarrier cultures of human pluripotent stem cells in stirred-suspension systems. In: Bioreactors in stem cell biology. Springer, New York, pp 35–52
166. Allen LM, Matyas J, Ungrin M, Hart DA, Sen A (2019) Serum-free culture of human mesenchymal stem cell aggregates in suspension bioreactors for tissue engineering applications. Stem Cells Int 2019:4607461. https://doi.org/10.1155/2019/4607461

167. Bijonowski BM, Yuan X, Jeske R, Li Y, Grant SC (2020) Cyclical aggregation extends in vitro expansion potential of human mesenchymal stem cells. Sci Rep 10(1):20448. https://doi.org/10.1038/s41598-020-77288-4
168. McKee C, Chaudhry GR (2017) Advances and challenges in stem cell culture. Colloids Surf B: Biointerfaces 159:62–77
169. Merten O-W (2015) Advances in cell culture: anchorage dependence. Philos Trans R Soc B: Biol Sci 370(1661):1
170. Weber W, Weber E, Geisse S, Memmert K (2002) Optimisation of protein expression and establishment of the wave bioreactor for Baculovirus/insect cell culture. Cytotechnology 38(1–3):77–85. https://doi.org/10.1023/a:1021102015070
171. Godara P, McFarland CD, Nordon RE (2008) Design of bioreactors for mesenchymal stem cell tissue engineering. J Chem Technol Biotechnol: Int Res Process Environ Clean Technol 83(4):408–420
172. Kumar A, Starly B (2015) Large scale industrialized cell expansion: producing the critical raw material for biofabrication processes. Biofabrication 7(4):044103. https://doi.org/10.1088/1758-5090/7/4/044103
173. Mizukami A, Swiech K (2018) Mesenchymal stromal cells: from discovery to manufacturing and commercialization. Stem Cells Int 2018:4083921. https://doi.org/10.1155/2018/4083921
174. Schnitzler AC, Verma A, Kehoe DE, Jing D, Murrell JR, Der KA, Aysola M, Rapiejko PJ, Punreddy S, Rook MS (2016) Bioprocessing of human mesenchymal stem/stromal cells for therapeutic use: current technologies and challenges. Biochem Eng J 108:3–13. https://doi.org/10.1016/j.bej.2015.08.014
175. Eibl R, Werner S, Eibl D (2009) Bag bioreactor based on wave-induced motion: characteristics and applications. Adv Biochem Eng Biotechnol 115:55–87. https://doi.org/10.1007/10_2008_15
176. Timmins NE, Kiel M, Günther M, Heazlewood C, Doran MR, Brooke G, Atkinson K (2012) Closed system isolation and scalable expansion of human placental mesenchymal stem cells. Biotechnol Bioeng 109(7):1817–1826. https://doi.org/10.1002/bit.24425
177. Croughan MS, Giroux D, Fang D, Lee B (2016) Chapter 5 – novel single-use bioreactors for scale-up of anchorage-dependent cell manufacturing for cell therapies. In: Cabral JMS, Lobato de Silva C, Chase LG, Margarida Diogo M (eds) Stem cell manufacturing. Elsevier, Boston, pp. 105–139. doi:https://doi.org/10.1016/B978-0-444-63265-4.00005-4
178. de Sousa PD, Bandeiras C, de Almeida FM, Rodrigues CAV, Jung S, Hashimura Y, Tseng RJ, Milligan W, Lee B, Ferreira FC, Lobato da Silva C, Cabral JMS (2019) Scalable manufacturing of human mesenchymal stromal cells in the vertical-wheel bioreactor system: an experimental and economic approach. Biotechnol J 14(8):e1800716. https://doi.org/10.1002/biot.201800716
179. Lembong J, Kirian R, Takacs JD, Olsen TR, Lock LT, Rowley JA, Ahsan T (2020) Bioreactor parameters for microcarrier-based human MSC expansion under xeno-free conditions in a vertical-wheel system. Bioengineering (Basel) 7(3):73. https://doi.org/10.3390/bioengineering7030073
180. Hoch AI, Duhr R, Di Maggio N, Mehrkens A, Jakob M, Wendt D (2017) Expansion of bone marrow mesenchymal stromal cells in perfused 3D ceramic scaffolds enhances in vivo bone formation. Biotechnol J 12(12):1700071
181. Guyot Y, Luyten F, Schrooten J, Papantoniou I, Geris L (2015) A three-dimensional computational fluid dynamics model of shear stress distribution during neotissue growth in a perfusion bioreactor. Biotechnol Bioeng 112(12):2591–2600
182. Yeo D, Kiparissides A, Cha JM, Aguilar-Gallardo C, Polak JM, Tsiridis E, Pistikopoulos EN, Mantalaris A (2013) Improving embryonic stem cell expansion through the combination of perfusion and bioprocess model design. PLoS One 8(12):e81728
183. Elder BD, Athanasiou KA (2009) Hydrostatic pressure in articular cartilage tissue engineering: from chondrocytes to tissue regeneration. Tissue Eng B Rev 15(1):43–53

184. Schroeder C, Hoelzer A, Zhu G, Woiczinski M, Betz OB, Graf H, Mayer-Wagner S, Mueller PE (2016) A closed loop perfusion bioreactor for dynamic hydrostatic pressure loading and cartilage tissue engineering. J Mech Med Biol 16(03):1650025
185. Lambrechts T, Papantoniou I, Rice B, Schrooten J, Luyten FP, Aerts J-M (2016) Large-scale progenitor cell expansion for multiple donors in a monitored hollow fibre bioreactor. Cytotherapy 18(9):1219–1233
186. Chapman LA, Whiteley JP, Byrne HM, Waters SL, Shipley RJ (2017) Mathematical modelling of cell layer growth in a hollow fibre bioreactor. J Theor Biol 418:36–56
187. Ahmed HMM, Salerno S, Piscioneri A, Khakpour S, Giorno L, De Bartolo L (2017) Human liver microtissue spheroids in hollow fiber membrane bioreactor. Colloids Surf B: Biointerfaces 160:272–280
188. Knöspel F, Freyer N, Stecklum M, Gerlach JC, Zeilinger K (2016) Periodic harvesting of embryonic stem cells from a hollow-fiber membrane based four-compartment bioreactor. Biotechnol Prog 32(1):141–151
189. Naghib SD, Di Maio FP, De Bartolo L, Curcio E, Di Renzo A (2018) Automation and control system for fluid dynamic stability in hollow-fiber membrane bioreactor for cell culture. J Chem Technol Biotechnol 93(3):710–719
190. Sheu J, Beltzer J, Fury B, Wilczek K, Tobin S, Falconer D, Nolta J, Bauer G (2015) Large-scale production of lentiviral vector in a closed system hollow fiber bioreactor. Mol Ther Methods Clin Dev 2:15020
191. Pihl AF, Offersgaard AF, Mathiesen CK, Prentoe J, Fahnøe U, Krarup H, Bukh J, Gottwein JM (2018) High density Huh7. 5 cell hollow fiber bioreactor culture for high-yield production of hepatitis C virus and studies of antivirals. Sci Rep 8(1):17505
192. Yan IK, Shukla N, Borrelli DA, Patel T (2018) Use of a hollow fiber bioreactor to collect extracellular vesicles from cells in culture. In: Extracellular RNA. Springer, New York, pp 35–41
193. Jones M, Nankervis B, Frank N, Vang B, DiLorenzo T, Hill D, Coeshott C (2018) CD146 expression, as a surrogate biomarker for human mesenchymal stromal cell multilineage differentiation, is preserved during cell expansion in an automated hollow-fiber membrane bioreactor. Pharm Bioprocess 6(3):93–105
194. Cipriano M, Freyer N, Knöspel F, Oliveira NG, Barcia R, Cruz PE, Cruz H, Castro M, Santos JM, Zeilinger K (2017) Self-assembled 3D spheroids and hollow-fibre bioreactors improve MSC-derived hepatocyte-like cell maturation in vitro. Arch Toxicol 91(4):1815–1832
195. Godoy P, Hewitt NJ, Albrecht U, Andersen ME, Ansari N, Bhattacharya S, Bode JG, Bolleyn J, Borner C, Boettger J (2013) Recent advances in 2D and 3D in vitro systems using primary hepatocytes, alternative hepatocyte sources and non-parenchymal liver cells and their use in investigating mechanisms of hepatotoxicity, cell signaling and ADME. Arch Toxicol 87(8):1315–1530
196. Spier MR, Vandenberghe L, Medeiros ABP, Soccol CR (2011) Application of different types of bioreactors in bioprocesses. In: Bioreactors: design, properties and applications, pp 53–87
197. Guilak F, Cohen DM, Estes BT, Gimble JM, Liedtke W, Chen CS (2009) Control of stem cell fate by physical interactions with the extracellular matrix. Cell Stem Cell 5(1):17–26
198. Knippenberg M, Helder MN, Zandieh Doulabi B, Semeins CM, Wuisman PI, Klein-Nulend J (2005) Adipose tissue-derived mesenchymal stem cells acquire bone cell-like responsiveness to fluid shear stress on osteogenic stimulation. Tissue Eng 11(11–12):1780–1788
199. Osiecki MJ, McElwain SD, Lott WB (2018) Modelling mesenchymal stromal cell growth in a packed bed bioreactor with a gas permeable wall. PLoS One 13(8):e0202079
200. Li D, Tang T, Lu J, Dai K (2009) Effects of flow shear stress and mass transport on the construction of a large-scale tissue-engineered bone in a perfusion bioreactor. Tissue Eng A 15(10):2773–2783
201. Eibes G, dos Santos F, Andrade PZ, Boura JS, Abecasis MM, da Silva CL, Cabral JM (2010) Maximizing the ex vivo expansion of human mesenchymal stem cells using a microcarrier-based stirred culture system. J Biotechnol 146(4):194–197

202. Badenes S, Fernandes-Platzgummer A, Rodrigues C, Diogo M, da Silva C, Cabral J (2016) Microcarrier culture systems for stem cell manufacturing. In: Stem cell manufacturing. Elsevier, Amsterdam, Netherlands, pp 77–104
203. Li B, Wang X, Wang Y, Gou W, Yuan X, Peng J, Guo Q, Lu S (2015) Past, present, and future of microcarrier-based tissue engineering. J Orthop Transl 3(2):51–57
204. Hossain KMZ, Patel U, Ahmed I (2015) Development of microspheres for biomedical applications: a review. Progress Biomater 4(1):1–19
205. Hervy M, Weber JL, Pecheul M, Dolley-Sonneville P, Henry D, Zhou Y, Melkoumian Z (2014) Long term expansion of bone marrow-derived hMSCs on novel synthetic microcarriers in xeno-free, defined conditions. PLoS One 9(3):e92120
206. Dias AD, Elicson JM, Murphy WL (2017) Microcarriers with synthetic hydrogel surfaces for stem cell expansion. Adv Healthc Mater 6(16):1700072
207. Atala A, Lanza R, Mikos T, Nerem R (2008) Principles of regenerative medicine. Academic, Amsterdam
208. Tavassoli H, Alhosseini SN, Tay A, Chan PP, Oh SKW, Warkiani ME (2018) Large-scale production of stem cells utilizing microcarriers: a biomaterials engineering perspective from academic research to commercialized products. Biomaterials 181:333
209. de Soure AM, Fernandes-Platzgummer A, da Silva CL, Cabral JMS (2016) Scalable microcarrier-based manufacturing of mesenchymal stem/stromal cells. J Biotechnol 236:88–109. https://doi.org/10.1016/j.jbiotec.2016.08.007
210. Koh B, Sulaiman N, Fauzi MB, Law JX, Ng MH, Idrus RBH, Yazid MD (2020) Three dimensional microcarrier system in mesenchymal stem cell culture: a systematic review. Cell Biosci 10(1):75. https://doi.org/10.1186/s13578-020-00438-8
211. Rafiq QA, Coopman K, CJJCOiCE H (2013) Scale-up of human mesenchymal stem cell culture: current technologies and future challenges. Curr Opin Chem Eng 2(1):8–16
212. Tavassoli H, Alhosseini SN, Tay A, Chan PPY, Weng Oh SK, Warkiani ME (2018) Large-scale production of stem cells utilizing microcarriers: a biomaterials engineering perspective from academic research to commercialized products. Biomaterials 181:333–346. https://doi.org/10.1016/j.biomaterials.2018.07.016
213. Rafiq QA, Coopman K, Nienow AW, Hewitt CJ (2016) Systematic microcarrier screening and agitated culture conditions improves human mesenchymal stem cell yield in bioreactors. Biotechnol J 11(4):473–486. https://doi.org/10.1002/biot.201400862
214. Malda J, Frondoza CG (2006) Microcarriers in the engineering of cartilage and bone. Trends Biotechnol 24(7):299–304
215. Yu C, Kornmuller A, Brown C, Hoare T, Flynn LE (2017) Decellularized adipose tissue microcarriers as a dynamic culture platform for human adipose-derived stem/stromal cell expansion. Biomaterials 120:66–80
216. Heathman TR, Glyn VA, Picken A, Rafiq QA, Coopman K, Nienow AW, Kara B, Hewitt CJ (2015) Expansion, harvest and cryopreservation of human mesenchymal stem cells in a serum-free microcarrier process. Biotechnol Bioeng 112(8):1696–1707
217. Ogle ME, Doron G, Levy MJ, Temenoff JS (2020) Hydrogel culture surface stiffness modulates mesenchymal stromal cell secretome and alters senescence. Tissue Eng A 26(23–24):1259–1271. https://doi.org/10.1089/ten.tea.2020.0030
218. Martin C, Olmos É, Collignon M-L, De Isla N, Blanchard F, Chevalot I, Marc A, Guedon E (2017) Revisiting MSC expansion from critical quality attributes to critical culture process parameters. Process Biochem 59:231–243. https://doi.org/10.1016/j.procbio.2016.04.017
219. Heathman TR, Nienow AW, McCall MJ, Coopman K, Kara B, Hewitt CJ (2015) The translation of cell-based therapies: clinical landscape and manufacturing challenges. Regen Med 10(1):49–64. https://doi.org/10.2217/rme.14.73
220. Leber J, Barekzai J, Blumenstock M, Pospisil B, Salzig D, Czermak P (2017) Microcarrier choice and bead-to-bead transfer for human mesenchymal stem cells in serum-containing and chemically defined media. Process Biochem 59:255–265. https://doi.org/10.1016/j.procbio.2017.03.017
221. Carmelo JG, Fernandes-Platzgummer A, Diogo MM, da Silva CL, Cabral JM (2015) A xeno-free microcarrier-based stirred culture system for the scalable expansion of human

mesenchymal stem/stromal cells isolated from bone marrow and adipose tissue. Biotechnol J 10(8):1235–1247. https://doi.org/10.1002/biot.201400586

222. Tozetti PA, Caruso SR, Mizukami A, Fernandes TR, da Silva FB, Traina F, Covas DT, Orellana MD, Swiech K (2017) Expansion strategies for human mesenchymal stromal cells culture under xeno-free conditions. Biotechnol Prog 33(5):1358–1367. https://doi.org/10.1002/btpr.2494
223. Dos Santos F, Campbell A, Fernandes-Platzgummer A, Andrade PZ, Gimble JM, Wen Y, Boucher S, Vemuri MC, da Silva CL, Cabral JM (2014) A xenogeneic-free bioreactor system for the clinical-scale expansion of human mesenchymal stem/stromal cells. Biotechnol Bioeng 111(6):1116–1127. https://doi.org/10.1002/bit.25187
224. Cifuentes SJ, Priyadarshani P, Castilla-Casadiego DA, Mortensen LJ, Almodóvar J, Domenech M (2020) Heparin/collagen surface coatings modulate the growth, secretome, and morphology of human mesenchymal stromal cell response to interferon-gamma. J Biomed Mater Res A 109:951. https://doi.org/10.1002/jbm.a.37085
225. Krutty JD, Koesser K, Schwartz S, Yun J, Murphy WL, Gopalan P (2021) Xeno-free bioreactor culture of human mesenchymal stromal cells on chemically defined microcarriers. ACS Biomater Sci Eng 7(2):617–625. https://doi.org/10.1021/acsbiomaterials.0c00663
226. Krutty JD, Dias AD, Yun J, Murphy WL, Gopalan P (2019) Synthetic, chemically defined polymer-coated microcarriers for the expansion of human mesenchymal stem cells. Macromol Biosci 19(2):e1800299. https://doi.org/10.1002/mabi.201800299
227. Tamura A, Kobayashi J, Yamato M, Okano T (2012) Thermally responsive microcarriers with optimal poly(N-isopropylacrylamide) grafted density for facilitating cell adhesion/detachment in suspension culture. Acta Biomater 8(11):3904–3913. https://doi.org/10.1016/j.actbio.2012.07.006
228. Ng EX, Wang M, Neo SH, Tee CA, Chen C-H, Van Vliet KJ (2020) Dissolvable gelatin-based microcarriers generated through droplet microfluidics for expansion and culture of mesenchymal stromal cells. Biotechnol J 16(3):2000048. https://doi.org/10.1002/biot.202000048
229. Yan X, Zhang K, Yang Y, Deng D, Lyu C, Xu H, Liu W, Du Y (2020) Dispersible and dissolvable porous microcarrier tablets enable efficient large-scale human mesenchymal stem cell expansion. Tissue Eng Part C Methods 26(5):263–275. https://doi.org/10.1089/ten.tec.2020.0039
230. Min H, Xu L, Parrott R, Overall CC, Lillich M, Rabjohns EM, Rampersad RR, Tarrant TK, Meadows N, Fernandez-Castaneda A, Gaultier A, Kurtzberg J, Filiano AJ (2020) Mesenchymal stromal cells reprogram monocytes and macrophages with processing bodies. Stem Cells (Dayton, Ohio) 39(May):1–14. https://doi.org/10.1002/stem.3292
231. Bowles AC, Kouroupis D, Willman MA, Perucca Orfei C, Agarwal A, Correa D (2020) Signature quality attributes of CD146+ mesenchymal stem/stromal cells correlate with high therapeutic and secretory potency. Stem Cells 38(8):1034–1049. https://doi.org/10.1002/stem.3196
232. Madsen SD, Russell KC, Tucker HA, Glowacki J, Bunnell BA, O'Connor KC (2017) Decoy TRAIL receptor CD264: a cell surface marker of cellular aging for human bone marrow-derived mesenchymal stem cells. Stem Cell Res Ther 8(1):1–14. https://doi.org/10.1186/s13287-017-0649-4
233. Kota DJ, Prabhakara KS, Toledano-Furman N, Bhattarai D, Chen Q, DiCarlo B, Smith P, Triolo F, Wenzel PL, Cox CS, Olson SD (2017) Prostaglandin E2 indicates therapeutic efficacy of mesenchymal stem cells in experimental traumatic brain injury. Stem Cells 35(5):1416–1430. https://doi.org/10.1002/stem.2603
234. Lee RH, Yu JM, Foskett AM, Peltier G, Reneau JC, Bazhanov N, Oh JY, Prockop DJ (2014) TSG-6 as a biomarker to predict efficacy of human mesenchymal stem/progenitor cells (hMSCs) in modulating sterile inflammation in vivo. Proc Natl Acad Sci U S A 111(47):16766–16771. https://doi.org/10.1073/pnas.1416121111
235. Jiao H, Walczak BE, Lee MS, Lemieux ME, Li WJ (2020) GATA6 regulates aging of human mesenchymal stem/stromal cells. Stem Cells 39(March):1–16. https://doi.org/10.1002/stem.3297

236. Kowal JM, Schmal H, Halekoh U, Hjelmborg JB, Kassem M (2020) Single-cell high-content imaging parameters predict functional phenotype of cultured human bone marrow stromal stem cells. Stem Cells Transl Med 9:189. https://doi.org/10.1002/sctm.19-0171
237. Ang J, Lee YA, Raghothaman D, Jayaraman P, Teo KL, Khan FJ, Reuveny S, Chang YT, Kang NY, Oh S (2019) Rapid detection of senescent mesenchymal stromal cells by a fluorescent probe. Biotechnol J 14(10):1–10. https://doi.org/10.1002/biot.201800691
238. Bertolo A, Gemperli A, Gruber M, Gantenbein B, Baur M, Pötzel T, Stoyanov J (2015) In vitro cell motility as a potential mesenchymal stem cell marker for multipotency. Stem Cells Transl Med 4:84. https://doi.org/10.5966/sctm.2014-0156
239. Rennerfeldt DA, Raminhos JS, Leff SM, Manning P, Van Vliet KJ (2019) Emergent heterogeneity in putative mesenchymal stem cell colonies: single-cell time lapsed analysis. PLoS One 14:e0213452. https://doi.org/10.1371/journal.pone.0213452
240. Nienow AW, Coopman K, Heathman TRJ, Rafiq QA, Hewitt CJ (2016) Bioreactor engineering fundamentals for stem cell manufacturing. In: Stem cell manufacturing, pp 43–75. https://doi.org/10.1016/B978-0-444-63265-4.00003-0
241. Food US, Drug Administration/Center for Biologics E, Research (2005) Q5E comparability of biotechnological/biological products subject to changes in their manufacturing process. In: ICH quality guidelines: an implementation guide. Wiley, Hoboken. https://doi.org/10.1089/blr.2005.24.627

Process Design for Human Mesenchymal Stem Cell Products in Stirred-Tank Bioreactors

Jan Barekzai, Florian Petry, Peter Czermak, and Denise Salzig

Abbreviations

ad	Adipose tissue
APIs	Active pharmaceutical ingredients
ATMPs	Advanced therapy medicinal products
bm	Bone marrow
CBMP	Cell-based medicinal products
CD	Cluster of differentiation
CPP	Critical process parameter
CQA	Critical quality attribute
DO	Dissolved oxygen
ECM	Extracellular matrix
EMA	European medicines agency
EVs	Extracellular vesicles
FBS	Fetal bovine serum
FDA	Food and drug administration
GMP	Good manufacturing practice
hESC	Human embryonic stem cell
IL	Interleukin
INF	Interferon
ISCT	International society of cell therapy
MSCs	Mesenchymal stem cells
PAT	Process analytical technology
PX	Passage number X

J. Barekzai · F. Petry · P. Czermak · D. Salzig (✉)
Institute of Bioprocess Engineering and Pharmaceutical Technology IBPT), University of Applied Sciences Mittelhessen (THM), Gießen, Germany
e-mail: denise.salzig@lse.thm.de

R. Pörtner (ed.), *Cell Culture Engineering and Technology*, Cell Engineering 10,
https://doi.org/10.1007/978-3-030-79871-0_10

QbD	Quality by design
QTPP	Quality target product profile
R&D	Research and development
sCTMP	Somatic cell therapy medicinal products
STR	Stirred-tank bioreactor
TNF	tumor necrosis factor
uc	Umbilical cord

Nomenclature

N_c	Critical agitation rate
d_S	Diameter of the stirrer
d_T	Diameter of the tank
h_S	Height of the stirrer
k_La	Volumetric mass transfer coefficient
T	Temperature
p	Pressure

1 Introduction

Human mesenchymal stem cells (MSCs) for cell therapy are classed as advanced therapy medicinal products (ATMPs), which are defined as medicines for human use that are based on genes, cells, or tissue engineering, excluding vaccines. The global ATMP sector is growing rapidly, with over 900 companies worldwide, 1060 clinical trials involving ATMPs, and 14 products already approved for the market [1]. In a molecular context, ATMPs are highly complex products, even more so than biologicals such as antibodies or insulin. The active pharmaceutical ingredients (APIs) of ATMPs are complex entities such as viable cells and/or infectious viruses, which require elaborate and costly characterization. The regulation of ATMPs is not globally harmonized. The two major regulatory authorities, the US Food and Drug Administration (FDA) and the European Medicines Agency (EMA), differ slightly in terms of ATMP subclasses. The FDA classification is relatively broad and covers two major groups: gene therapy products and cellular therapy products. However, the EMA differentiates between gene therapy medicinal products and cell-based medicinal products (CBMPs), and further divides CBMPs into somatic cell therapy medicinal products (sCTMPs) and tissue engineered products. The fourth class in the EMA classification is the combined ATMPs, featuring mixtures of other product types [2]. The largest class is the sCTMPs, representing 39% of all ATMPs (www.grandviewresearch.com data from 2019) and 18% of all sCTMPs contain MSCs as the API [3].

The therapeutic application of MSCs requires an average of 416 million cells per dose [4], far exceeding the number of cells that can be isolated by tissue aspiration. All MSC manufacturing processes must therefore include an *in vitro* expansion step. One option for MSC expansion is the use of static culture vessels such as T-flasks, which are common consumables in many research and development (R&D) laboratories. However, assuming that 13.9 million MSCs can be produced per T-175 flask (175 cm^2 growth surface), 30 such flasks would be needed per dose per patient. A phase I clinical trial with 20 patients would therefore require 600 flasks, a phase II trial with 200 patients would require 6000, and a phase III trial with 2000 patients would require 60,000. Cultivation of the latter would require 450 standard CO_2 incubators (160 L) and 130 trained staff. This simple calculation shows the limitations of the so-called scale-out approach. Bioreactors are therefore preferable for the scale-up of MSC cultivation for clinical trials. For comparison, the MSCs required for a phase III trial involving 2000 patients can be produced using microcarriers in one stirred-tank bioreactor with a working volume of 1050 L. This is not only more economical but also allows the precise control the MSC microenvironment, which is necessary to ensure the functionality of the final MSC product.

2 MSC-Based Products Are Non-typical Stem Cell Products

MSCs have been studied for several decades, but a precise definition has been surprisingly challenging. In 2006, the International Society of Cell Therapy (ISCT) defined minimal criteria that must be met before cells can be defined as MSCs. Such cells must (i) show plastic adherence; (ii) express the cluster of differentiation (CD) surface markers CD73, CD90 and CD105, but not CD11b, CD14, CD19, CD34, CD45 or HLA-DR; and (iii) be able to differentiate into cartilage, bone, and fat cells *in vitro* [5]. To define MSCs as “stem cells” is misleading because MSCs *in vivo* show non-typical stem cell behavior. Stem cells are capable of both self-renewal and differentiation *in vivo*, whereas MSCs are only capable of self-renewal and do not differentiate *in vivo*. Instead, MSCs stimulate local stem cells to differentiate and to regenerate the destroyed or dysfunctional tissue. Therefore, the therapeutic benefit of MSCs reflects the properties of their secretome.

The MSC secretome comprises a pool of cytokines, chemokines, growth factors and extracellular vesicles (EVs) carrying proteins, lipids, and various RNAs, and differs widely among MSC isolates and subpopulations. MSCs can modulate immune cells, reduce inflammation, apoptosis, or fibrosis, and improve angiogenesis [6]. These modes of action are clinically relevant, as seen when surveying the clinical trials involving MSCs. There are currently 374 phase I, 314 phase II and 45 phase III trials with MSCs as the API (www.clinicaltrials.gov, search term *mesenchymal stem cells*, 2021). A quarter of these trials are in the field of immunology, using the immunomodulatory properties of MSCs to treat conditions such as Crohn’s disease, graft-vs-host disease, or immunodeficiency. Another significant proportion of the trials exploit the anti-inflammatory effect of MSCs to treat rheumatic diseases such as osteoarthritis, rheumatoid arthritis, and ankylosing spondylitis. Their anti-apoptotic potential is being used in clinical trials targeting stroke and cardiac defects. As well as inhibiting cardiomyocyte apoptosis, MSCs are useful in cardiac therapy because they secrete growth factors such as VEGF and improve angiogenesis [7]. Since 2016, MSCs have also been defined as medicinal signaling cells, which properly reflects their therapeutic activity [8].

For the clinical use of MSCs, manufacturing (including *in vitro* expansion) must follow FDA/EMA guidance and must produce a sufficient quantity of viable cells. However, the most important goal is to ensure the MSCs are therapeutically functional, due to their role as the API. The definition of therapeutic functionality differs for each therapeutic approach, and the *in vitro* MSC expansion step must therefore be adapted for each MSC product. It is important to understand if and how the various process parameters affect the MSC product, allowing the critical process parameters to be tightly controlled, thus ensuring reproducibility, standardization, and economic efficiency.

3 MSC Functionality Is Determined by the Microenvironment

The successful manufacturing of functional MSCs is primarily dependent on the microenvironment *in vitro*. MSCs are found in various human tissues. They were initially isolated from bone marrow (bm-MSCs) based on their plastic adherence, but today they are usually isolated from adipose tissue (ad-MSCs) or umbilical cord blood (uc-MSCs), which are more accessible [9]. MSCs are also found in various other adult, fetal and perinatal tissues [10]. Regardless of their origin, isolated MSCs are heterogeneous and polyclonal cells, but even monoclonal MSCs become heterogeneous during *in vitro* expansion [11]. MSCs have different growth rates depending on their source, but even MSCs from the same source tissue but different donors show different growth performance [12]. For example, ad-MSCs proliferate more quickly than bm-MSCs [13], and juvenile uc-MSCs proliferate more quickly than adult MSCs [14]. Furthermore, potency is often dependent on origin. The immunomodulatory activity of bm-MSCs exceeds that of other MSCs [13], whereas ad-MSCs show stronger immunosuppressive effects than bm-MSCs [15], Wharton-jelly MSCs inhibit mitogen-induced T-cell responses to a greater extent [16], and uc-MSCs show the highest angiogenic capacity *in vitro* [17]. These differences reflect the microenvironment of the cells *in vivo*, which defines the functionality and properties of MSCs *in vitro*.

MSCs are influenced by several factors *in vivo*, including other MSCs, other cells (e.g., neighboring cells, immune and cancer cells and their EVs), the inflammatory regulators in the environment, the components, stiffness, elasticity, and topography of the surrounding extracellular matrix (ECM), nutrients (e.g., glucose, lipids, oxygen, and trace elements), waste products, and soluble factors such as chemokines, cytokines, and hormones (Fig. 1). These factors clearly differ between bone marrow, adipose tissue and the umbilical cord. MSCs are surrounded by other cells, with which they communicate via surface receptors, soluble factors,

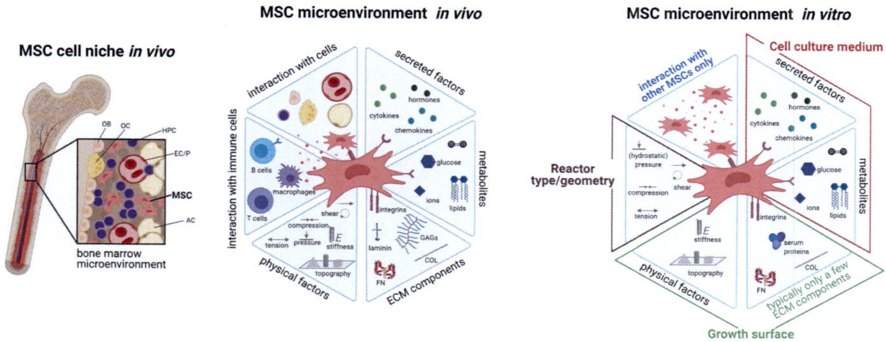

Fig. 1 The properties of MSCs are defined by their niche and microenvironment. Left panel: MSCs *in vivo* are affected by their tissue of origin (in this case, bone marrow) and the corresponding microenvironment. MSCs are influenced by surrounding cells such as osteoblasts (OB), osteoclasts (OC), hematopoietic stem cells (HPC), endothelial cells/pericytes (EC/P), adipocytes (AC) and immune cells such as T cells, B cells and macrophages. Middle panel: As well as interactions with cells, including other MSCs, the properties of MSCs are influenced by physical factors such as shear and pressure, extracellular matrix (ECM) components such as fibronectin (FN), glycosaminoglycans (GAGs) and collagen (COL), metabolites such as glucose, ions and lipids, and secreted factors such as hormones, chemokines and cytokines. Right panel: The *in vivo* microenvironment can be only partly imitated *in vitro*. The only cellular interactions are with other MSCs. Most physical factors (shear and pressure) are attributed to the reactor design, whereas the ECM components are restricted and mainly provided by the growth surface. The cell culture medium provides certain metabolites and secreted factors, and its composition is very flexible

and EVs. Accordingly, MSCs are often described as donor cells (providing EVs to other cells types) but they also act as recipients and their behavior is thereby modulated by their neighbors. Each cell produces a secretome that forms a microenvironment affecting surrounding cells, including MSCs [18]. For example, neural cells (and their EVs) facilitate MSC neuronal induction [19], whereas endothelial cells and their EVs influence MSC proliferation, migration, and the secretion of soluble factors such as matrix metalloproteases (MMP-1 and MMP-3), chemokine ligand 2 (CCL-2), and interleukin (IL)-6 [20]. Immune cells such as monocytes also communicate with MSCs. Lipopolysaccharide-activated monocytes secrete soluble factors and EVs that modulate the MSC phenotype [21]. MSCs also react to inflammation or cancer. An acute inflammatory environment induces the immunosuppressive effect of MSCs, whereas a chronic inflammatory environment causes pro-inflammatory behavior [22]. EVs from cancer cells stimulate MSCs to produce and secrete inflammatory cytokines such as IL-6, IL-8, and monocyte chemoattractant protein (MCP)-1 [23]. Cancer stem cells can also induce epigenetic changes in recipient cells. MSCs are attracted to the tumor environment and change their phenotype, becoming pro-tumorigenic. This correlates with the overexpression of genes involved in cell migration, ECM remodeling, angiogenesis and tumor growth [24].

As stated above, MSCs can remodel the ECM but the ECM also exerts a reciprocal influence. In their natural niche, MSCs form cell–cell connections via cadherins and connexins, and also interact with ECM components. MSCs are influenced by the biochemical constituents of the ECM, but also by its stiffness and topography [25]. The ECM surrounding ad-MSCs and bm-MSCs induces changes in MSC quantity, morphology, and function [26]. MSC proliferation is enhanced when the origin of the MSC matches the tissue origin of the cultured ECM [27]. Each tissue has a certain concentration of ECM components but also a certain stiffness and elasticity. MSCs usually originate from very soft tissues such as marrow, or soft tissues such as fat. MSCs adhere only weakly in these tissues, which is necessary to maintain their self-renewing capability. Such MSCs are characterized by only low levels of integrin-mediated signaling through focal adhesion kinase (FAK), retaining levels of extracellular signal-regulated kinase (ERK1/2) to support growth but not differentiation [28].

Given the above, it is clear that the interactions between MSCs and their microenvironment are very complex. Expanding MSCs beyond their natural niche induces massive changes in their microenvironment, which means that every step during *in vitro* cultivation has a non-neutral effect on MSC biological properties. An *in vitro* expansion process must therefore mimic the cellular niche to ensure that MSCs remain functional and therapeutically active. However, several aspects of the natural niche cannot be replicated *in vitro* (Fig. 1). Interactions with other cell types and the immune system are absent *in vitro*, so the only intercellular interactions are with other MSCs. Interactions involving cell–cell connections, cytoskeletal elements, the ECM, and overall tissue topography can have profound effects on multipotent MSCs. Harvesting MSCs from a bone marrow niche with its condensed cell-rich environment and culturing them *in vitro* removes the cell–cell cadherin and connexin connections and replaces them with cell–substrate and cell–matrix interactions as the cells produce more ECM [29].

A certain set of physical and ECM interactions can be replaced *in vitro* by the growth surface/matrix. Several growth surfaces are available, including the planar (often polystyrene) surfaces found in T-flasks, curved growth surfaces such as microcarriers, and 3D matrices such as porous microcarriers, hydrogels, and spheroids. The planar polystyrene surface does not imitate the natural niche very well because it is stiff, with an elasticity module ($E = 1$–10 GPa) much higher than that of the natural MSC matrix (bm-MSC $E < 0.3$ kPa; ad-MSC $E = 2$–6 kPa) [30]. MSCs grown on stiff, planar surfaces show limited expansion potential and lower differentiation capacity with increased passage number. Moreover, the stiffness of the growth surface/matrix can alter transcription [31] and thus modulate MSC behavior and differentiation [32]. This suggests that MSCs retain an environmental memory, meaning that transcription is still altered even if MSCs are transferred from T-flaks to a softer substrate later [33]. The whole MSC manufacturing process must therefore be considered in this context because the routine passaging of MSCs on planar polystyrene surfaces may inadvertently and permanently alter their phenotype and functionality.

On planar surfaces, the ECM and other proteins build up as a continuous layer which establishes an apical-basal polarity and a restricted adhesion to the *x*–*y* plane. The adhesion is often very strong and spaced over large distances (~5 μm), which promotes unwanted differentiation because MSCs undergoing osteogenesis require stronger adhesion [28]. In contrast, curved surfaces transduce cytoskeletal changes that influence MSC migration and differentiation [34]. The cell–cell interactions are less strong, with fewer adhesion points. The 3D materials are often softer than flasks, and the ECM forms discrete fibrils. MSCs can establish 3D networks without polarity, and spreading is sterically hindered. With topography and elasticity/stiffness, a certain set of physical factors are defined by the growth surface/matrix. Other physical factors, such as tension, pressure, compression and especially shear, are provided by the bioreactor system. Static culture vessels apply few if any of these factors, whereas bioreactors provide a more nurturing environment where the strength of the factors can be controlled by the reactor geometry, conditions, and equipment setup.

Soluble components such as cytokines, chemokines, hormones, nutrients, trace elements, ions, and lipids are provided *in vitro* by the cell culture medium. This requires a detailed understanding what MSCs really need and how the components influence MSC properties. In the absence of such knowledge, fetal bovine serum (FBS) was formerly used as an obligate component of MSC culture medium. Although its exact composition is undefined, FBS offers various adhesion factors, cytokines, EVs, hormones, and protease inhibitors that support cell growth [35]. As our understanding of cell requirements has improved, and given more recent ethical and regulatory concerns, FBS is no longer allowed for the manufacture of clinical MSC products and serum-free or chemically defined media are preferred.

The plasticity of MSCs (their ability to change due to the conditions in their microenvironment) can be exploited. If the effect of environmental parameters is known, the bioprocess can be designed to trigger the production of MSCs with a certain therapeutic function. Specific cultivation conditions should therefore be defined to prevent differentiation into unwanted cell linages and the total loss of therapeutic potential.

4 Bioreactor Systems Can Create the Appropriate Microenvironment for MSC Expansion *In Vitro*

The transition from laboratory-scale experiments to industrial biomanufacturing processes is hampered by the intricacy of MSCs and their interactions with the microenvironment. To ensure a consistent and standardized manufacturing process, the development is grounded in the quality-by-design (QbD) principle, which provides a rational framework and combines scientific knowledge from biological and engineering perspectives. This requires a clear definition of the quality target product profile (QTPP). The product attributes are known as critical quality attributes (CQAs) and define the product in terms of physical, chemical, and biological properties. Therefore, the identity, purity, and potency of each MSC product is tightly controlled but varies for individual MSC products and their therapeutic indications, which cannot be transferred from one product to another. In short, the identity of MSCs is determined by morphological and phenotypic analysis, meaning the presence or absence of specific surface markers as described above. The potency of MSCs is generally dependent on therapeutic indications. Several potency assays are available but only specific assays are applicable for individual MSC products. For example, the assessment of a differentiation potency assay is only appropriate for MSCs that develop their therapeutic mechanism based on tissue formation. Sterility and purity, meaning the absence of contaminants such as unwanted cell types, particles, or pathogens, must be proven to ensure safety and efficacy [6]. Every process parameter that influences the CQAs of an MSC product is described as a critical process parameter (CPP). Ideally, CPPs are controlled throughout the manufacturing process, but some CPPs are difficult to access depending on the bioreactor design.

Static cultures in T-flasks or hyperflasks are not only difficult to scale up, they also lack process control. This led to the commercial development of large-scale planar bioreactor systems with an integrated stirrer and pH and dissolved oxygen (DO) control, providing surface areas of up to 12.24 m^2, equivalent to ~700 T-175 flasks (Table 1). Although these systems are suitable for MSC expansion, drawbacks include limited monitoring of cell growth, dissimilarity with *in vivo* conditions, and labor-intensive and time-consuming operations. Nevertheless, these systems can be used for the preparation of MSCs for phase I and II clinical studies involving only a small number of patients.

Table 1 Large-scale single-use bioreactors potentially suitable for the manufacture of MSC-sEVs. All bioreactor systems monitor pH, temperature, gas flow and stirring speed except the NANT XL, which is not aerated or stirred. Suspension bioreactors were prepared with 15 g/L microcarriers giving a specific surface area of 360 cm^2/g

Bioreactor system	Process mode	Monitored parameters	Growth surface [m^2]	Working volume [L]	Supplier
Planar bioreactors					
NANT XL	Batch	CO_2, confluency	0.318	–	VivaBioCell
Xpansion200	Batch	pH, DO, confluency	12.24	21.9	Pall
Hollow-Fiber bioreactors					
Quantum	Perfusion	T, DO	2.1	0.189	Terumo
Fixed bed bioreactors					
BioBLU 5p	Perfusion	T, pH, DO	18	3.75	Eppendorf
iCellis	Perfusion	T, pH, DO, biomass, p	500	25	Pall
Scale-X nitro	Perfusion	pH, T, DO, p	600	–	Univercells
Suspension bioreactors					
Mobius 2000 L	Batch	T, pH, DO, biomass	1080	2000	Merck
Xcellerex XDR 2000	Batch	T, pH, DO, biomass	1080	2000	Cytiva
Allegro STR 2000 L	Batch	T, pH, DO, biomass	1080	2000	Pall

In contrast to planar bioreactor systems, MSCs in hollow-fiber and fixed-bed or packed-bed bioreactors create a 3D microenvironment [36]. Fixed-bed and packed-bed bioreactors provide a large surface area for cell growth over a bed of macro carriers, which the cells use as a substrate. Metabolites are provided by the constant supply of fresh medium, and waste products are removed continually. In hollow-fiber bioreactors, MSCs grow in the interstitial spaces of a cartridge of hollow fibers that mimic blood capillaries and thus simultaneously deliver metabolites while removing waste products [37]. Both bioreactor systems can prevent the inhibition of cell growth by the buildup of toxic metabolites, and the process can therefore be extended for several months, increasing MSC yields and economic efficiency [38]. Both bioreactor systems achieve a high yield of cells per unit volume because the cells grow very densely, close to physiological conditions, helping to maintain their CQAs. However, these bioreactor systems must still overcome challenges hindering large-scale manufacturing, including (i) heterogeneous cell distribution; (ii) reduced metabolite availability and waste product removal due to high cell densities and insufficient diffusive mass transfer; (iii) lack of direct cell growth monitoring, relying instead on metabolism-derived approximations such as mass balance of oxygen levels; and most importantly (iv) low harvesting efficiency [39, 40]. Given the high cell densities, the enzymatic contact surface is restricted and long incubation times are required for detachment, which reduces cell viability [41, 42].

The importance of harvesting for MSC manufacturing has led to the introduction of suspension bioreactors such as stirred-tank reactors (STRs) for the large-scale expansion of MSCs. For example, in a 50-L STR with a working volume of 35 L, a 50-fold expansion was achieved with a final yield of 2.6×10^{10} cells [43]. For industrial-scale manufacturing, several disposable STRs are commercially available (Table 1). MSCs are anchorage-dependent cells, so the growth surface is generally increased by the use of mostly spherical microcarriers with cell-specific properties to encourage attachment, proliferation and harvesting. However, microcarriers that ensure proper attachment and proliferation are not necessarily suitable for biomanufacturing processes when the cell is the API. For example, MSCs attach strongly to Cytodex I microcarriers but the harvesting efficiency is only ~20% [44]. Therefore, a well-designed microcarrier screening process should include attachment, proliferation, harvesting kinetics and MSC functionality. Once suitable microcarriers are identified, the exponential growth phase can be extended by bead-to-bead transfer without enzymatic treatment, ensuring high cell yields, surface-to-volume ratios and economic efficiency over a range of scales [45]. Furthermore, STRs do not suffer from the disadvantage of heterogeneous cell distribution as seen in hollow-fiber and fixed-bed/packed-bed bioreactors. Convective mass transport prevails instead of diffusive processes, ensuring the sufficient availability of nutrients and oxygen. The homogeneous cell distribution in STRs also allows representative sampling if necessary. Most importantly, STRs are compatible with process analytical technology (PAT) to guarantee process control [46]. In addition to online controlled parameters such as pH, temperature and DO, and the offline measurement of glucose levels, cell growth, viability and size can be monitored online by impedance spectroscopy [47]. This online technology ensures process transparency and control during MSC biomanufacturing. STRs are therefore the most suitable bioreactors for the manufacture of MSCs as products because of their process flexibility, economy and tight control of CPPs, allowing them to meet CQAs with low batch-to-batch variations. We therefore focus below on CPPs for MSC expansion in STRs.

5 CPPs for MSC Expansion in STRs

5.1 Cell-Related Parameters for MSCs

Regardless of the bioreactor system, the medium, growth surface and other cell-related parameters have a profound impact on the success of MSC expansion. As stated above, cell–cell interactions *in vitro* are restricted to MSCs because no other cells are present. The relevant CPPs include the MSC source, age and density. The source and donor of the MSCs should be fixed, because important MSC properties such as doubling time are strongly dependent on this parameter. For example, under the same cultivation conditions and medium, uc-MSCs had a significantly shorter doubling time (4 days) than adult MSC (7 days) [48].

MSC age is also important because aging (population doubling *in vitro*) causes MSCs to increase in volume [49], proliferate more slowly, begin to lose the expression of MSC markers, and become more fibroblast-like in morphology [50]. MSCs reach senescence *in vitro* after a source-dependent number of doublings, for example ~50 in the case of uc-MSCs [51] and ~ 70 in the case of ad-MSCs [52]. The MSC expansion process should therefore be started with a distinct population doubling and/or stopped before the population doubling limit is reached. This limitation can be overcome using an immortalized MSC line if the line displays the desired therapeutic functions.

For MSC expansion, the initial cell density and final cell density must be standardized in order to reach the same number of population doublings during one passage [53]. The final cell density is restricted by the growth surface area and the efficiency of harvesting, but is typically in the range 5×10^4–1×10^5 cells/cm^2. The initial cell density varies from 100 to 10,000 cells/cm^2. MSCs derived from initial high-density cultures feature a larger number of flat cells and the proliferation rate is lower. However, the initial density should not be too low, because cultures initially plated at a density of 10–100 cells/cm^2 do not expand effectively [54]. The initial cell density must be chosen carefully because it also affects cell age, given that cells with lower initial densities require additional rounds of doubling to reach the final cell density. In an expansion process requiring several passages, these cumulative age differences lead to different cell populations even though the passage number remains the same. Furthermore, even monoclonal MSCs become heterogeneous during expansion. If the initial seeding density is too low, the risk increases that certain MSC subgroups may overgrow the general MSC population. The fastest growing MSC subgroup is not necessarily the one with the best therapeutic potential. Accordingly, cellular dynamics during the MSC expansion process must be monitored carefully. A fast growth rate is not sufficient alone and the therapeutic efficacy of the expanded MSCs must be considered as well.

MSCs *in vivo* are surrounded by several cells, so replicating this effect *in vitro* by providing the corresponding EVs may be beneficial. EVs from differentiated cells, immune cells and cancer cells can all modulate the properties of MSCs [18]. Other interactions, with living or inactivated bacteria, can increase the absolute number of MSCs, improve their immunomodulatory properties, and promote the expression of anti-inflammatory factors [55]. The easiest parameters to control *in vitro* are the physicochemical factors, which are mainly related to the culture medium and the growth surface.

5.2 Physicochemical Parameters for MSCs

MSCs respond to physical parameters such as hydrostatic pressure, tensile stress, compression, vibration, and ultrasound by modifying their transcriptional profiles (mechanotranscription). Many of these factors can promote MSC differentiation [56], but it is unclear whether they can also influence the fate of undifferentiated MSCs, or affect their proliferation or functionality. Given that mechanical stimuli are part of the natural MSC niche, such factors are likely to play a key role in the biological and structural responses of MSCs.

More is known about the impact of chemical/biochemical factors on MSCs. The availability of oxygen in the natural cell niche is low (2–7% pO_2) [57], whereas many bioreactors strive to achieve atmospheric oxygen conditions (21% pO_2). High oxygen levels promote the generation of reactive oxygen species (ROS) which damage MSCs and induce apoptosis [58]. Many studies have therefore highlighted the need to cultivate MSCs under hypoxic conditions from isolation until transplantation [59–65]. Hypoxia (typically 2–5% pO_2) is known to increase bm-MSC density, inhibit senescence and maintain the undifferentiated state [66–69]. Even the composition of MSC-derived EVs changes during hypoxia, reflecting the upregulation of hypoxia inducible factor 1 α (HIF-1α) and miR-126, improving the therapeutic efficacy of bone fracture healing [70].

After oxygen, the second most important requirement for MSCs is glucose. Although a low glucose concentration (5.5 mM) is maintained *in vivo* [71], many cell culture media contain high levels of glucose (22 mM). The effects of high glucose levels have been reported, with conflicting claims, but there is evidence for a limited impact on MSC proliferation and function [72–74]. Low glucose levels (5.5 mM) slightly increased the frequency of apoptosis in ad-MSCs [75] but weakly promoted the proliferation of bm-MSC [76]. High glucose levels may be a pathological trigger for MSCs, creating disease-specific microenvironments in conditions such as diabetes. Other physicochemical factors such as pH and osmolarity are also associated with diseases, and these factors must be kept within physiological ranges to ensure the health of MSCs cultivated *in vitro*. Even weak acidity (pH 6.8) and hyperosmolarity (485 mOsm) can inhibit the proliferation of ad-MSCs [75] and bm-MSCs [77], and promote necrosis. Trace elements and metal ions are essential for MSCs, but some metal ions promote differentiation (e.g., Mg^{2+} promotes osteogenesis and Li^+ promotes myogenesis [78]).

Cytokines are potent regulators of MSC behavior *in vivo* and *in vitro*. The priming of MSCs by cytokines *in vitro* has been described in detail. Interferon (IFN)-γ and tumor necrosis factor (TNF)-α are the most prominent inducers of immunosuppressive MSC behavior, promoting survival and proliferation [79], but interleukins such as IL-1α, IL-1β and IL-2 also induce an immunosuppressive phenotype [22]. Stromal cell-derived factor 1 (CXCL12/SCDF-1) is a chemotactic for MSCs, promoting survival, proliferation, and paracrine functions [80]. The microenvironment *in vivo* combines several cytokines and each cytokine has a concentration-dependent effect [78]. Therefore, a design-of-experiments (DoE) approach may be useful to evaluate the impact of cytokines on the therapeutic function of MSCs, allowing the identification of concentration-dependent effects and also interactions between two or more growth factors.

5.3 Microcarriers Provide the Growth Surface for MSCs in a STR

Although part of the physicochemical parameters, we discuss the MSC growth surface/matrix separately because it is essential for MSC expansion. MSCs are strictly anchorage dependent and will undergo a form of programmed cell death known as anoikis if a substrate is unavailable. In a STR, the growth surface is often provided in form of microcarriers, which are small beads (100–300 μm in diameter) with a similar density to the medium, allowing homogenous distribution in the bioreactor by stirring. Microcarriers are considered as a form of 3D cultivation, but the cells nevertheless grow as a monolayer on the curved surfaces, so the term pseudo-3D is more appropriate. Microcarriers can be classified as porous or non-porous. Porous microcarriers mimic 3D cell–cell interactions more accurately than their non-porous counterparts, but the surface of the latter can be modified (e.g., by coating with ECM molecules) to enhance cell attachment, or by physical treatment to change the surface charge and wettability [81].

Most microcarriers recommended for human MSC expansion are commercial non-porous beads with a polystyrene core and various coatings or surface treatments. These are very stiff and the coatings, if present, are generally not thick enough to enable full control over the surface mechanical stiffness sensed by the cells. A coating must be 10–20 μm thick to mask the stiffness of the underlying substrate [82]. The influence of microcarrier stiffness on MSC properties has not been evaluated in detail, perhaps because the curvature effect on mechanical stress makes the results difficult to interpret. MSCs may therefore be less sensitive to the stiffness of microcarriers than planar surfaces [83]. The interaction between MSCs and microcarriers is responsible for cell attachment, proliferation, and detachment.

MSCs from different sources, and even MSCs from the same source but different donors, have different surface-attachment requirements and properties [45]. This explains the broad range of microcarrier types, and the selection of appropriate carriers requires prior knowledge or attachment experiments.

Although porous microcarriers may imitate *in vivo* conditions more accurately, non-porous microcarriers allow more efficient cell harvesting. All microcarriers facilitate cell attachment and proliferation, but it remains challenging to harvest cells efficiently without damaging them [41]. The proteolytic enzymes used for cell passaging and tissue digestion may damage the ECM and thus affect the corresponding signaling pathways, ultimately affecting MSC behavior [84].

Microcarriers offer a simple and efficient way to expand MSCs and produce clinically relevant numbers of cells with the required characteristics [42]. Commercial microcarriers do not provide all the benefits of the natural MSC niche but can generate vigorous MSCs with potent therapeutic functionality. Several investigations have tailored microcarriers for MSC expansion, aiming to mimic the natural niche more precisely, for example by adjusting material stiffness, coating the surface with more natural ECM structures, and using dissolvable microcarriers to improve the efficiency of harvesting.

5.4 *Equipment-Related Parameters for MSCs*

Agitation in STRs is usually achieved by placing the impeller near the bottom of the vessel, generating the driving force for a convective flow regime that homogenizes the culture microenvironment, disperses gas and nutrients, ensures sufficient mixing, and reduces laminar boundary layers. These conditions are important for MSC bioprocessing, but agitation also generates shear forces and other forces that can have a profound effect on MSC growth or functionality. The influence of shear forces on cell proliferation and functionality has been described in 2D models (flow chambers), but with some contradictory results. Whereas some studies reported a positive effect on proliferation, others reported a decline [85]. This shows that every MSC product and manufacturing process must be characterized individually and no overall correlation between MSC products and the CPP "agitation" is valid. All STRs should limit the force experienced by cells to reduce the likelihood of cell damage and maintain CQAs such as functionality. The forces acting on cells growing on microcarriers are associated with hydrodynamic stress as well as cell–carrier and carrier–carrier collisions [86]. The resistance of MSCs to dynamic forces in a STR can be estimated using Kolmogorov's turbulence theory, which explains that stress acting on MSCs and microcarriers is caused by eddies similar in size to the cells/microcarriers and the distance between microcarriers [87]. These eddies do not cause cell damage if they exceed a critical size ($\geq 60\%$ of the cell or microcarrier diameter). Kolmogorov's theory is valid for a turbulent regime, but most MSC expansion processes in STRs are found within the transitional range, making this approximation inaccurate [46, 88].

Although cell stress must be limited, the power input must be sufficient to achieve a homogeneous microcarrier suspension. An agitation rate that is just sufficient to suspend the microcarriers means they do not remain on the bottom surface of the vessel for more than one second. A further increase can achieve the critical agitation rate (Nc) where microcarriers are homogeneously distributed. Nc can be approximated based on the microcarrier concentration, STR geometry and a stirrer-dependent constant [89]. The ability to achieve homogeneity while maintaining low cell stress is strongly influenced by the stirrer type.

Different stirrer types can be categorized by their fluid pumping characteristics. Radial impellers such as the Rushton turbine generally have high power inputs but low suspension capabilities. Axial pumping stirrers such as marine propellers or impellers are more suitable for MSCs grown on microcarriers. They facilitate bottom-to-top fluid movement and hence the Nc is low, which minimizes cell stress. However, many subtypes of stirrers have been developed by combining axial and radial fluid characteristics, such as the three-segmented pitched-blade stirrer. Fluid movement within the bioreactor is also affected by the interplay with the bioreactor setup. The stirrer diameter to tank diameter ratio (d_S/d_T) and the stirrer height to diameter ratio (h_S/d_S) are important parameters. The d_S/d_T ratio should be at least 0.4 to guarantee sufficient mixing characteristics, especially in large-scale processes, whereas h_S/d_S should be high to ensure homogeneous power dissipation in the bioreactor [90]. Nc can be reduced further using a fully baffled system [91] because baffles covert tangential flow to axial/radial flow, thus increasing suspension capabilities and homogeneity.

MSC biomanufacturing processes must ensure a sufficient oxygen and nutrient supply. Aeration systems can be divided broadly into headspace, membrane and bubble aeration setups. Headspace and membrane aeration systems are sufficient for small-scale experiments, but bubble aeration by means of a sparger is necessary at larger scales [92]. With the help of a sparger, a higher oxygen transfer rate (k_La) is achieved by increasing the interfacial area between the gas and liquid phases [93].

Bubble aeration can influence MSC growth because the high local velocities caused by rising and bursting bubbles generate shear stress and are responsible for foaming and cell entrapment. The sparger must therefore be chosen carefully. Macrospargers produce large bubbles with small interfacial areas and high local velocities, while microspargers produce smaller bubbles with a homogeneous size distribution and a large interfacial area, thus increasing the k_La [93]. However, excess oxygen induces oxidative stress by generating ROS, which disrupt biochemical processes [94]. The rational selection of aeration systems can be achieved by characterizing the oxygen demand of the cells. Primary MSCs consume oxygen at the rate of 90–100 fmol/(cell·h) whereas an immortalized cell line has a much higher demand of 300 fmol/(cell·h) [36, 95]. This highlights the importance of process design, in which STRs are customized and adjusted to specific MSC needs to ensure that CQAs are maintained.

6 Production of MSC-Derived EVs

The therapeutic effect of MSCs is mainly conferred by the secretome, particularly EVs, resulting in growing interest in the use of EVs as cell-free therapeutics. EVs were originally considered as waste products, but their therapeutic potential has been confirmed. The therapeutic application of EVs overcomes the drawbacks of manufacturing viable cells and the complexity of transfusion processes. EVs are more robust than cells, and more stable during storage and transport, thus maintaining their therapeutic efficacy [96, 97].

EVs are divided into three broad categories differing in size and therapeutic potential. Exosomes (30–100 nm) and microvesicles (50–1000 nm) are the most suitable as therapeutics, whereas apoptotic bodies have limited applicability [98]. Exosomes are derived from the budding endosomal membrane and are matured as intraluminal vesicles within the lumen of multivesicular endosomes (MVEs). The MVEs are transported within the endosomal system, and fuse with the cell surface for EVs release. Microvesicles are formed by the outward budding and fission of the plasma membrane and the release of EVs into the extracellular space. Both exosomes and microvesicles are positive for CD9 and CD81, whereas CD37, CD63, CD53 and CD151 are only found on exosomes [99]. The nomenclature and classification cannot be based on size and functionality alone because there are major differences in biogenesis, but the separation of EVs based on biogenesis is unrealistic. Therapeutically active vesicles in the size range 40–200 nm are therefore described as small EVs (MSC-sEVs), as recommended by MISEV2018 [100]. MSC-sEVs contain proteins, lipids, and various RNA molecules that can elicit responses from recipient cells. The positive effect of MSC-sEVs has clearly been shown over short and long distances. MSC-sEVs inhibit inflammation, apoptosis, and fibrosis, but enhance angiogenesis and tissue regeneration [98].

Like EVs in general, MSC-sEVs are communication vehicles that influence the state and functionality of neighboring recipient cells. The cargo of MSC-sEVs has been investigated to determine the bioactive molecules responsible for their therapeutic functionality, and has been classified based on molecular and cellular functions such as transcription factors, chemokines, cytokines, growth factors and miRNAs. However, the results of different genomic, proteomic, metabolomic and glycomic studies have differed considerably. This reflects the physiological diversity of MSCs, which respond to triggers in their environment (such as inflammation or hypoxia) by adjusting their metabolism and secreting MSC-sEVs representing the physiological state of the donor cell. The cargo is therefore highly sensitive to stimuli in the microenvironment and thus to *in vitro* process parameters. For the comparison of MSC-sEVs, it is therefore necessary to consider process parameters as well as the intrinsic nature of the donor cell [101]. The medium composition is also important, because the production of MSC-sEVs can be boosted by reducing the concentration of FBS and oxygen levels or increasing pro-inflammatory factors and shear rates [102].

The sensitivity of MSCs (and MSC-sEV composition) to the microenvironment is yet not fully understood. These first approaches to process design by triggering MSC-sEV production represent a milestone on the way to clinical applications. Furthermore, manufacturing processes could be specifically designed to develop individual treatments for each patient, bringing personalized medicine within reach [103].

7 Bioreactor Systems for MSC-sEV Production

As discussed above, bioreactors are required to control the microenvironment of MSCs *in vitro*, enabling the regulation of DO, pH, temperature, metabolite levels, and the concentration of viable MSCs. The production of MSC-sEVs is strongly dependent on the microenvironment, so the manufacture of MSC-sEVs for clinical applications requires robust and reproducible processes that comply with good manufacturing practice (GMP). The therapeutic potential of MSCs *in vivo* is determined by external triggers that arise following infection or injury. Similar triggers must be provided to produce MSC-sEVs *in vitro* by exploiting bioreactor design and equipment-related parameters. The STRs used to produce MSCs can also be used to manufacture MSC-sEVs, but hollow-fiber and fixed-bed systems are suitable too because there is no requirement for cell harvesting [6, 38]. The bioreactor types used for MSC-sEVs therefore include many commercially available disposable bioreactors (Table 1).

Hollow-fiber and fixed-bed bioreactors allow the continuous production and harvesting of EVs from the culture medium. MSCs grow densely on the fibers and macrocarriers because the 3D structure better represents the physiological cell niche. The benefits of 3D cultivation have been demonstrated by the aggregation of MSCs into spheroids, but the same advantages also allow the efficient production of MSC-sEVs and other EVs [104]. For example, HEK293 cells in a hollow-fiber bioreactor achieved a 40-fold increase in sEV production compared to static cultures [105], whereas ad-MSCs in a hollow-fiber bioreactor achieved a ten-fold increase in MSC-sEV production compared to static cultures [106]. The cultivation of bm-MSCs in a FiberCell Systems hollow-fiber bioreactor, a smaller version of the C2018 (Table 1) with a surface area of 0.4 m^2, led to a decrease in MSC numbers due to the use of a specific EV-collection medium, but continual EV production was confirmed by the detection of specific markers [107]. Fixed-bed bioreactors combine the advantages of hollow-fiber bioreactors (3D growth) with increased metabolite availability and exposure to moderate shear stress as a trigger for MSC-sEV production [36]. Large-scale disposable hollow-fiber and fixed-bed bioreactors are currently available with surface areas of up to 600 m^2 (Table 1). However, few studies have been published about the production of MSC-sEVs and further investigation is required.

Although hollow-fiber and fixed-bed bioreactors appear suitable for large-scale EV production, cell density cannot be controlled, leading to heterogeneous cell distribution and zones with metabolite limitations and/or waste accumulation. The high cell densities in 3D-like structures combined with low diffusion rates can also lead to a general state of nutrient limitation. Although starvation can improve EV production and low metabolite concentrations/metabolite gradients are also found *in vivo*, the heterogeneous microenvironments in hollow-fiber and fixed-bed reactors hamper process standardization.

As stated above, suspension bioreactors such as STRs lack these disadvantages because they are homogenous systems that allow the online control of cell density, viability and size by dielectric spectroscopy [47]. STRs therefore provide an interesting alternative for the production of MSC-sEVs. Because the cells are not harvested, it is also possible to use porous microcarriers, which offer a larger growth surface for the cells and a 3D-like growth environment even in a STR. The cells on porous microcarriers are also protected from destructive shear effects. On the other hand, rationally designed shear forces can be used to trigger sEV production. The company EVerZom has developed a method that triggers massive EV release by applying turbulence/shear (www.everzom.com, data from 2021). The benefits of dynamic suspension cultivation have also been demonstrated by comparing uc-MSCs in static culture to those in spinner flasks on Star-Plus microcarriers, with the latter producing 20-fold more MSC-sEVs while maintaining the characteristic EV phenotype and size distribution [48]. Another dynamic culture system based on a vertical-wheel bioreactor was used to produce MSC-sEVs derived from three different MSC types. Compared to static cultures in T-flasks, dynamic cultivation resulted in a ~ three-fold increase of MSC-sEVs yields regardless of the cell type [108]. Although these small-scale processes using suspension bioreactors are promising, the hydrodynamic parameters affecting MSC-sEV production are unknown and detailed investigations are required. Once these aspects are understood, process development will be facilitated by the compatibility of suspension bioreactors with PAT and hence process standardization. Additionally, many single-use bioreactors are currently available for the analysis of process comparability. These bioreactors have a working volume of up to 2000 L providing 1080 m^2 of cultivation area with typical microcarriers.

8 CPPs Affecting the Production of MSC-sEVs

8.1 Cell-Related Parameters Influencing MSC-sEVs

Many process parameters that are critical for the production of MSCs are also critical for the production of MSC-sEVs. For the standardized and high-yield production of MSC-sEVs, an appropriate donor cell is required and the cell-related parameters must be characterized. For example, uc-MSCs not only proliferate faster than ad-MSCs as discussed above, but also produce four times as many MSC-sEVs per cell, and the EVs differ in size suggesting a difference in functionality

[48]. Although standardized MSC isolation methods are now available, this process is considered a bottleneck because the enzymatic treatment causes cell stress and affects the mechanotranscription profile [109]. Rather than processing their own MSCs, many groups working on EVs use commercial primary cells or develop immortalized MSC cell lines. However, as stated earlier, the properties of MSCs are highly dependent on age, and the functionality of MSC-sEVs is also age-dependent [110]. This was determined by comparing the gap closure ability of MSC-sEVs obtained from MSCs at various passage numbers (P2–P5), revealing that all MSC-sEVs promoted vascularization but the activity of the EVs from P5 was the weakest [110]. Cell passaging and the resulting increase in cell age is associated with the modulation of gene expression with effects on the cell cycle, protein ubiquitination, and senescence [111]. The comparison of MSC-sEVs secreted by primary bm-MSCs and the immortalized cell line hMSC-TERT (expressing the telomerase reverse transcriptase gene, and also originating from bone marrow) revealed that immortalization resulted in a slightly higher yield of $CD63^+$ $CD81^+$ sEVs [112]. All EVs were similar in morphology and size, as confirmed by phase-contrast transmission electron microscopy, but functionality was not evaluated. The effect of immortalization (transfection with lentivirus) has also been tested on human embryonic stem cell-derived MSCs (hESC-MSC) and cord-derived MSCs. The morphology of the hESC-MSCs changed and they were no longer MSCs according to the ISCT classification, but these cells produced a larger quantity of sEVs that significantly reduced the size of infarcts in mice. In contrast, cord-derived MSCs produced fewer sEVs post-immortalization but the cells retained their therapeutic efficacy [113]. These results clearly show that immortalization cannot serve as a universal strategy to enhance MSC-sEV production because of the diverse effects on different donor cells. Each cell and immortalization method must be evaluated to generate well-characterized cell lines that produce high yields of potent sEVs, representing an important step towards the standardized production of sEVs and more comparability in EV research.

8.2 *Biochemical Parameters Influencing MSC-sEVs*

Several triggers are already known that enhance sEV production *in vivo*, such as injury and infection, and the corresponding molecular signals must be provided *in vitro* to achieve the same therapeutic effect. The production of sEVs *in vitro* is often induced by adding cytokines such as IFN-γ and TNF-α, by starving the cells of serum, or depleting essential nutrients.

IFN-γ and TNF-α are pro-inflammatory mediators and thus mimic the behavior of damaged or infected tissues. MSCs and their sEVs therefore upregulate class I/II major histocompatibility complex (MHC) and stimulatory molecules to boost proliferation, enhance immunomodulatory and immunosuppressive functions, and increase the production of sEVs [114]. In the absence of IFN-γ, ad-MSCs released 281 sEVs per cell and hour, whereas those exposed to IFN-γ released 463 sEVs per cell and hour, a 1.7-fold increase without changing the size distribution of expression of specific markers [115]. Additionally, a priming approach, using IFN-γ and TNF-α simultaneously, increased the production of ad-MSC-sEVs compared to the non-primed control group. These findings were confirmed by differences in protein expression, especially the upregulation of Rab27b, which represents a regulator for the release of exosomes [116]. However, the same combined treatment reduced the number of sEVs produced by bm-MSCs, indicating that cytokine treatment is not a universal solution for the production of MSC-sEVs [117]. Another challenge associated with the use of cytokines to stimulate MSC-sEV production is the impact on purification and the resulting safety concerns. GMP compliance requires that manufacturing processes must include steps to eliminate putative immunogenic and allergenic ingredients, which in this case would include steps to remove the cytokines that were deliberately introduced into the process, thus increasing process costs [118].

The production of EVs is also triggered by serum deprivation. FBS provides growth factors that support MSC proliferation, and these are often present in the form of FBS-EVs. Such EVs contribute to cell expansion and proliferation, but they are considered as impurities [119]. The starvation of MSCs by the depletion of FBS-EVs (or the complete removal of FBS) therefore prevents the isolation of FBS-EVs along with the target product. Serum depletion affects the three main MSC types in different ways, with limited impact on the abundance of uc-MSC-sEVs but a significant depletion of ad-MSC-sEVs and bm-MSC-sEVs [120]. The exosome fraction of the uc-MSC-sEVs also showed increased functionality (interacting with target neurons), whereas the functionality of the microvesicle fraction was reduced [120]. Nevertheless, starving cells is controversial. It is common practice to expand cells in serum-containing medium and transfer them to serum-free medium for EV production, but this approach may not be compatible with therapeutic applications. The transfer to serum-free medium triggers phenotypic changes in the donor cells, mirrored by changes in the protein and RNA content of the EVs [121], as well as growth inhibition and the induction of apoptosis [122]. Given that EVs represent their donor cell, it is important to keep these cells in an active and proliferative state so that the therapeutic potential is not affected. This does not mean that serum-free medium should be avoided. Indeed, serum-free or chemically defined media are recommended when cells do not change their characteristics in terms of proliferation and sEV production. It may be necessary to optimize the proliferation of MSCs by adding specific growth factors to the medium and selecting an appropriate growth surface in order to determine sEV characteristics under these culture conditions, thus taking a step toward standardized production [6].

Other biological triggers that enhance MSC-sEV production include hypoxia, which mimics the physiological microenvironment of MSCs (typically 2–7% pO_2) and provides appropriate conditions for the investigation of MSC proliferation, metabolism, and EV release. Hypoxic conditions of 1–10% pO_2 increase the proliferative capacity and survival of cells by limiting the generation of ROS [123]. Accordingly, the same approaches have been applied to MSC-sEVs. There was no difference in the production of ad-MSC-sEVs when switching from normoxic

(21% pO_2) to hypoxic (5% pO_2) conditions. However, the hypoxic sEVs showed significantly enhanced functionality in a tube formation assay [124]. Another study confirmed the enhanced functionality of bm-MSC-sEVs produced under hypoxic conditions in cell proliferation, cell migration and tube formation assays, and the simultaneous use of serum-free medium also significantly increased MSC-sEV yields [125–127]. Hypoxic conditions lead to the production of potent MSC-sEVs and no additional purification steps are required, but strict control of O_2 is necessary, which can only be achieved in bioreactors.

9 Conclusions

The development of MSCs and MSC-sEVs as novel APIs still involves many challenges. Both are complex products with unique manufacturing processes, in which the microenvironment needs to be strictly controlled because it has a huge influence on the final product quality. MSCs and MSC-sEVs are strongly dependent on cell culture parameters such as the origin and handling of the cells, the composition of the medium, the nature of the growth surface/matrix and the hydrodynamics in the bioreactor. A standardized environment is essential for the manufacture of clinical products. It is necessary to define this environment in order to determine the CPPs for individual MSC and sEV products. Large-scale biomanufacturing processes are needed and bioreactors facilitate MSC expansion *in vitro* (STRs) and the production of sEVs (STRs, hollow-fiber reactors and fixed-bed systems). We are only just beginning to understand the influence of the microenvironment on MSCs and MSC-sEVs, and further investigation is required to establish CPPs that will enable standardized GMP-compliant production.

References

1. Alliance RM Quarterly regenerative medicine global data report, vol 2019
2. Iglesias-López C, Agustí A, Obach M et al (2019) Regulatory framework for advanced therapy medicinal products in Europe and United States. Front Pharmacol 10:921. https://doi.org/10.3389/fphar.2019.00921
3. Maciulaitis R, D'Apote L, Buchanan A et al (2012) Clinical development of advanced therapy medicinal products in Europe: evidence that regulators must be proactive. Mol Ther 20:479–482. https://doi.org/10.1038/mt.2012.13
4. Olsen TR, Ng KS, Lock LT et al (2018) Peak MSC-are we there yet? Front Med (Lausanne) 5:178. https://doi.org/10.3389/fmed.2018.00178
5. Dominici M, Le Blanc K, Mueller I et al (2006) Minimal criteria for defining multipotent mesenchymal stromal cells. Int Soc Cellular Therapy Position Statement Cytotherapy 8:315–317. https://doi.org/10.1080/14653240600855905
6. Barekzai J, Petry F, Zitzmann J et al (2020) Bioprocess development for human mesenchymal stem cell therapy products. In: María Martínez-Espinosa R (ed) New advances on fermentation processes. IntechOpen

7. Heathman TRJ, Nienow AW, McCall MJ et al (2015) The translation of cell-based therapies: clinical landscape and manufacturing challenges. Regen Med 10:49–64. https://doi.org/10.2217/rme.14.73
8. Caplan AI (2017) Mesenchymal stem cells: time to change the name! Stem Cells Transl Med 6:1445–1451. https://doi.org/10.1002/sctm.17-0051
9. Murray IR, West CC, Hardy WR et al (2014) Natural history of mesenchymal stem cells, from vessel walls to culture vessels. Cell Mol Life Sci 71:1353–1374. https://doi.org/10.1007/s00018-013-1462-6
10. Andrzejewska A, Lukomska B, Janowski M (2019) Concise review: mesenchymal stem cells: from roots to boost. Stem Cells 37:855–864. https://doi.org/10.1002/stem.3016
11. Rennerfeldt DA, van Vliet KJ (2016) Concise review: when colonies are not clones: evidence and implications of Intracolony heterogeneity in mesenchymal stem cells. Stem Cells 34:1135–1141. https://doi.org/10.1002/stem.2296
12. Calle A, Barrajón-Masa C, Gómez-Fidalgo E et al (2018) Iberian pig mesenchymal stem/stromal cells from dermal skin, abdominal and subcutaneous adipose tissues, and peripheral blood: in vitro characterization and migratory properties in inflammation. Stem Cell Res Ther 9:178. https://doi.org/10.1186/s13287-018-0933-y
13. Petrenko Y, Vackova I, Kekulova K et al (2020) A comparative analysis of multipotent mesenchymal stromal cells derived from different sources, with a focus on Neuroregenerative potential. Sci Rep 10:4290. https://doi.org/10.1038/s41598-020-61167-z
14. Donders R, Bogie JFJ, Ravanidis S et al (2018) Human Wharton's jelly-derived stem cells display a distinct immunomodulatory and Proregenerative transcriptional signature compared to bone marrow-derived stem cells. Stem Cells Dev 27:65–84. https://doi.org/10.1089/scd.2017.0029
15. Ménard C, Dulong J, Roulois D et al (2020) Integrated transcriptomic, phenotypic, and functional study reveals tissue-specific immune properties of mesenchymal stromal cells. Stem Cells 38:146–159. https://doi.org/10.1002/stem.3077
16. Vieira Paladino F, de Moraes Rodrigues J, da Silva A et al (2019) The immunomodulatory potential of Wharton's jelly mesenchymal stem/stromal cells. Stem Cells Int 2019:3548917. https://doi.org/10.1155/2019/3548917
17. Pinto DS, Ahsan T, Serra J et al (2020) Modulation of the in vitro angiogenic potential of human mesenchymal stromal cells from different tissue sources. J Cell Physiol 235:7224–7238. https://doi.org/10.1002/jcp.29622
18. Dostert G, Mesure B, Menu P et al (2017) How do mesenchymal stem cells influence or are influenced by microenvironment through extracellular vesicles communication? Front Cell Dev Biol 5:6. https://doi.org/10.3389/fcell.2017.00006
19. Takeda YS, Xu Q (2015) Neuronal differentiation of human mesenchymal stem cells using exosomes derived from differentiating neuronal cells. PLoS One 10:e0135111. https://doi.org/10.1371/journal.pone.0135111
20. Lozito TP, Tuan RS (2014) Endothelial and cancer cells interact with mesenchymal stem cells via both microparticles and secreted factors. J Cell Mol Med 18:2372–2384. https://doi.org/10.1111/jcmm.12391
21. Omar OM, Granéli C, Ekström K et al (2011) The stimulation of an osteogenic response by classical monocyte activation. Biomaterials 32:8190–8204. https://doi.org/10.1016/j.biomaterials.2011.07.055
22. Kim N, Cho S-G (2016) Overcoming immunoregulatory plasticity of mesenchymal stem cells for accelerated clinical applications. Int J Hematol 103:129–137. https://doi.org/10.1007/s12185-015-1918-6
23. Li X, Wang S, Zhu R et al (2016) Lung tumor exosomes induce a pro-inflammatory phenotype in mesenchymal stem cells via NFκB-TLR signaling pathway. J Hematol Oncol 9:42. https://doi.org/10.1186/s13045-016-0269-y
24. Lindoso RS, Collino F, Camussi G (2015) Extracellular vesicles derived from renal cancer stem cells induce a pro-tumorigenic phenotype in mesenchymal stromal cells. Oncotarget 6:7959–7969. https://doi.org/10.18632/oncotarget.3503

25. Pramotton FM, Robotti F, Giampietro C et al (2019) Optimized topological and topographical expansion of epithelia. ACS Biomater Sci Eng 5:3922–3934. https://doi.org/10.1021/acsbiomaterials.8b01346
26. Ragelle H, Naba A, Larson BL et al (2017) Comprehensive proteomic characterization of stem cell-derived extracellular matrices. Biomaterials 128:147–159. https://doi.org/10.1016/j.biomaterials.2017.03.008
27. Marinkovic M, Block TJ, Rakian R et al (2016) One size does not fit all: developing a cell-specific niche for in vitro study of cell behavior. Matrix Biol 52-54:426–441. https://doi.org/10.1016/j.matbio.2016.01.004
28. Donnelly H, Salmeron-Sanchez M, Dalby MJ (2018) Designing stem cell niches for differentiation and self-renewal. J R Soc Interface 15. https://doi.org/10.1098/rsif.2018.0388
29. Pittenger MF, Discher DE, Péault BM et al (2019) Mesenchymal stem cell perspective: cell biology to clinical progress. NPJ Regen Med 4:22. https://doi.org/10.1038/s41536-019-0083-6
30. Akhmanova M, Osidak E, Domogatsky S et al (2015) Physical, spatial, and molecular aspects of extracellular matrix of in vivo niches and artificial scaffolds relevant to stem cells research. Stem Cells Int 2015:167025. https://doi.org/10.1155/2015/167025
31. Argentati C, Morena F, Tortorella I et al (2019) Insight into Mechanobiology: how stem cells feel mechanical forces and orchestrate biological functions. Int J Mol Sci 20. https://doi.org/10.3390/ijms20215337
32. Heo S-J, Szczesny SE, Kim DH et al (2018) Expansion of mesenchymal stem cells on electrospun scaffolds maintains stemness, mechano-responsivity, and differentiation potential. J Orthop Res 36:808–815. https://doi.org/10.1002/jor.23772
33. Yang C, Tibbitt MW, Basta L et al (2014) Mechanical memory and dosing influence stem cell fate. Nat Mater 13:645–652. https://doi.org/10.1038/nmat3889
34. Werner M, Blanquer SBG, Haimi SP et al (2017) Surface curvature differentially regulates stem cell migration and differentiation via altered attachment morphology and nuclear deformation. Adv Sci (Weinh) 4:1600347. https://doi.org/10.1002/advs.201600347
35. Mannello F, Tonti GA (2007) Concise review: no breakthroughs for human mesenchymal and embryonic stem cell culture: conditioned medium, feeder layer, or feeder-free; medium with fetal calf serum, human serum, or enriched plasma; serum-free, serum replacement nonconditioned medium, or ad hoc formula? All that glitters is not gold! Stem Cells 25:1603–1609. https://doi.org/10.1634/stemcells.2007-0127
36. Weber C, Freimark D, Pörtner R et al (2010) Expansion of human mesenchymal stem cells in a fixed-bed bioreactor system based on non-porous glass carrier – part B: modeling and scale-up of the system. Int J Artif Organs 33:782–795. https://doi.org/10.1177/039139881003301103
37. Hanley PJ, Mei Z, Durett AG et al (2014) Efficient manufacturing of therapeutic mesenchymal stromal cells with the use of the quantum cell expansion system. Cytotherapy 16:1048–1058. https://doi.org/10.1016/j.jcyt.2014.01.417
38. Catapano G, Czermak P, Eibl R et al (2009) Bioreactor design and scale-up. In: Eibl R, Eibl D, Pörtner R et al (eds) Cell and tissue reaction engineering, vol 14. Springer, Berlin, Heidelberg, pp 173–259
39. Meuwly F, Ruffieux P-A, Kadouri A et al (2007) Packed-bed bioreactors for mammalian cell culture: bioprocess and biomedical applications. Biotechnol Adv 25:45–56. https://doi.org/10.1016/j.biotechadv.2006.08.004
40. Barckhausen C, Rice B, Baila S et al (2016) GMP-compliant expansion of clinical-grade human mesenchymal stromal/stem cells using a closed hollow Fiber bioreactor. Methods Mol Biol 1416:389–412. https://doi.org/10.1007/978-1-4939-3584-0_23
41. Salzig D, Schmiermund A, P Grace P et al. (2013) Enzymatic detachment of therapeutic mesenchymal stromal cells grown on glass carriers in a bioreactor. Open Biomed Eng J 7:147–158. https://doi.org/10.2174/1874120701307010147
42. Petry F, Weidner T, Czermak P et al (2018) Three-dimensional bioreactor Technologies for the Cocultivation of human mesenchymal stem/stromal cells and Beta cells. Stem Cells Int 2018:2547098. https://doi.org/10.1155/2018/2547098

43. Elseberg C, Leber J, Weidner T et al. (2017) The challenge of human mesenchymal stromal cell expansion: current and prospective answers. In: Gowder SJT (ed) New insights into cell culture technology. INTECH
44. Weber C, Pohl S, Pörtner R et al (2007) Expansion and harvesting of hMSC-TERT. Open Biomed Eng J 1:38–46. https://doi.org/10.2174/1874120700701010038
45. Leber J, Barekzai J, Blumenstock M et al (2017) Microcarrier choice and bead-to-bead transfer for human mesenchymal stem cells in serum-containing and chemically defined media. Process Biochem 59:255–265. https://doi.org/10.1016/j.procbio.2017.03.017
46. Cierpka K, Elseberg CL, Niss K et al (2013) hMSC production in disposable bioreactors with regards to GMP and PAT. Chem-Ing-Tech 85:67–75. https://doi.org/10.1002/cite.201200151
47. Zitzmann J, Weidner T, Eichner G et al (2018) Dielectric spectroscopy and optical density measurement for the online monitoring and control of recombinant protein production in stably transformed Drosophila melanogaster S2 cells. Sensors (Basel) 18. https://doi.org/10.3390/s18030900
48. Haraszti RA, Miller R, Stoppato M et al (2018) Exosomes produced from 3D cultures of MSCs by tangential flow filtration show higher yield and improved activity. Mol Ther 26:2838–2847. https://doi.org/10.1016/j.ymthe.2018.09.015
49. Block TJ, Marinkovic M, Tran ON et al (2017) Restoring the quantity and quality of elderly human mesenchymal stem cells for autologous cell-based therapies. Stem Cell Res Ther 8:239. https://doi.org/10.1186/s13287-017-0688-x
50. Ng TT, Mak KH-M, Popp C et al. (2020) Murine mesenchymal stromal cells retain biased differentiation plasticity towards their tissue of origin cells 9. https://doi.org/10.3390/cells9030756
51. Ko E, Lee KY, Hwang DS (2012) Human umbilical cord blood-derived mesenchymal stem cells undergo cellular senescence in response to oxidative stress. Stem Cells Dev 21:1877–1886. https://doi.org/10.1089/scd.2011.0284
52. Stab BR, Martinez L, Grismaldo A et al (2016) Mitochondrial functional changes characterization in young and senescent human adipose derived MSCs. Front Aging Neurosci 8:299. https://doi.org/10.3389/fnagi.2016.00299
53. Sotiropoulou PA, Perez SA, Salagianni M et al (2006) Characterization of the optimal culture conditions for clinical scale production of human mesenchymal stem cells. Stem Cells 24:462–471. https://doi.org/10.1634/stemcells.2004-0331
54. Mareschi K, Rustichelli D, Calabrese R et al (2012) Multipotent mesenchymal stromal stem cell expansion by plating whole bone marrow at a low cellular density: a more advantageous method for clinical use. Stem Cells Int 2012:920581. https://doi.org/10.1155/2012/920581
55. Silveira GP, Ishimura ME, Teixeira D et al (2018) Improvement of mesenchymal stem cell immunomodulatory properties by heat-killed Propionibacterium acnes via TLR2. Front Mol Neurosci 11:489. https://doi.org/10.3389/fnmol.2018.00489
56. Boeri L, Albani D, Raimondi MT et al (2019) Mechanical regulation of nucleocytoplasmic translocation in mesenchymal stem cells: characterization and methods for investigation. Biophys Rev 11:817–831. https://doi.org/10.1007/s12551-019-00594-3
57. Dalloul A (2013) Hypoxia and visualization of the stem cell niche. Methods Mol Biol 1035:199–205. https://doi.org/10.1007/978-1-62703-508-8_17
58. Li J, Liu X, Zuo B et al (2016) The role of bone marrow microenvironment in governing the balance between Osteoblastogenesis and Adipogenesis. Aging Dis 7:514–525. https://doi.org/10.14336/AD.2015.1206
59. Boregowda SV, Krishnappa V, Chambers JW et al (2012) Atmospheric oxygen inhibits growth and differentiation of marrow-derived mouse mesenchymal stem cells via a p53-dependent mechanism: implications for long-term culture expansion. Stem Cells 30:975–987. https://doi.org/10.1002/stem.1069
60. Bader AM, Klose K, Bieback K et al (2015) Hypoxic preconditioning increases survival and pro-Angiogenic capacity of human cord blood mesenchymal stromal cells in vitro. PLoS One 10:e0138477. https://doi.org/10.1371/journal.pone.0138477

61. Ciria M, García NA, Ontoria-Oviedo I et al (2017) Mesenchymal stem cell migration and proliferation are mediated by hypoxia-inducible factor-1α upstream of notch and SUMO pathways. Stem Cells Dev 26:973–985. https://doi.org/10.1089/scd.2016.0331
62. Lv B, Li F, Fang J et al (2017) Hypoxia inducible factor 1α promotes survival of mesenchymal stem cells under hypoxia. Am J Transl Res 9:1521–1529
63. Estrada JC, Albo C, Benguría A et al (2012) Culture of human mesenchymal stem cells at low oxygen tension improves growth and genetic stability by activating glycolysis. Cell Death Differ 19:743–755. https://doi.org/10.1038/cdd.2011.172
64. Vertelov G, Kharazi L, Muralidhar MG et al (2013) High targeted migration of human mesenchymal stem cells grown in hypoxia is associated with enhanced activation of RhoA. Stem Cell Res Ther 4:5. https://doi.org/10.1186/scrt153
65. Antebi B, Rodriguez LA, Walker KP et al (2018) Short-term physiological hypoxia potentiates the therapeutic function of mesenchymal stem cells. Stem Cell Res Ther 9:265. https://doi.org/10.1186/s13287-018-1007-x
66. Grayson WL, Zhao F, Izadpanah R et al (2006) Effects of hypoxia on human mesenchymal stem cell expansion and plasticity in 3D constructs. J Cell Physiol 207:331–339. https://doi.org/10.1002/jcp.20571
67. Potier E, Ferreira E, Andriamanalijaona R et al (2007) Hypoxia affects mesenchymal stromal cell osteogenic differentiation and angiogenic factor expression. Bone 40:1078–1087. https://doi.org/10.1016/j.bone.2006.11.024
68. Felka T, Schäfer R, Schewe B et al (2009) Hypoxia reduces the inhibitory effect of IL-1beta on chondrogenic differentiation of FCS-free expanded MSC. Osteoarthr Cartil 17:1368–1376. https://doi.org/10.1016/j.joca.2009.04.023
69. Tsai C-C, Chen Y-J, Yew T-L et al (2011) Hypoxia inhibits senescence and maintains mesenchymal stem cell properties through down-regulation of E2A-p21 by HIF-TWIST. Blood 117:459–469. https://doi.org/10.1182/blood-2010-05-287508
70. Liu W, Li L, Rong Y et al (2020) Hypoxic mesenchymal stem cell-derived exosomes promote bone fracture healing by the transfer of miR-126. Acta Biomater 103:196–212. https://doi.org/10.1016/j.actbio.2019.12.020
71. Ikeda Y, Keese SM, Tanaka K (1985) Molecular heterogeneity of variant isovaleryl-CoA dehydrogenase from cultured isovaleric acidemia fibroblasts. Proc Natl Acad Sci U S A 82:7081–7085
72. Li Y-M, Schilling T, Benisch P et al (2007) Effects of high glucose on mesenchymal stem cell proliferation and differentiation. Biochem Biophys Res Commun 363:209–215. https://doi.org/10.1016/j.bbrc.2007.08.161
73. Weil BR, Abarbanell AM, Herrmann JL et al (2009) High glucose concentration in cell culture medium does not acutely affect human mesenchymal stem cell growth factor production or proliferation. Am J Physiol Regul Integr Comp Physiol 296:R1735–R1743. https://doi.org/10.1152/ajpregu.90876.2008
74. Deorosan B, Nauman EA (2011) The role of glucose, serum, and three-dimensional cell culture on the metabolism of bone marrow-derived mesenchymal stem cells. Stem Cells Int 2011:429187. https://doi.org/10.4061/2011/429187
75. Liang C, Li H, Tao Y et al (2012) Responses of human adipose-derived mesenchymal stem cells to chemical microenvironment of the intervertebral disc. J Transl Med 10:49. https://doi.org/10.1186/1479-5876-10-49
76. Wuertz K, Godburn K, Neidlinger-Wilke C et al (2008) Behavior of mesenchymal stem cells in the chemical microenvironment of the intervertebral disc. Spine (Phila Pa 1976) 33:1843–1849. https://doi.org/10.1097/BRS.0b013e31817b8f53
77. Wuertz K, Godburn K, Iatridis JC (2009) MSC response to pH levels found in degenerating intervertebral discs. Biochem Biophys Res Commun 379:824–829. https://doi.org/10.1016/j.bbrc.2008.12.145
78. Tan L, Liu X, Dou H et al (2020) Characteristics and regulation of mesenchymal stem cell plasticity by the microenvironment —specific factors involved in the regulation of MSC plasticity. Genes & Diseases. https://doi.org/10.1016/j.gendis.2020.10.006

79. Tsuji K, Kitamura S, Wada J (2018) Secretomes from mesenchymal stem cells against acute kidney injury: possible heterogeneity. Stem Cells Int 2018:8693137. https://doi.org/10.1155/2018/8693137
80. García-Sánchez D, Fernández D, Rodríguez-Rey JC et al (2019) Enhancing survival, engraftment, and osteogenic potential of mesenchymal stem cells. World J Stem Cells 11:748–763. https://doi.org/10.4252/wjsc.v11.i10.748
81. Heng BC, Li J, Chen AK-L et al (2012) Translating human embryonic stem cells from 2-dimensional to 3-dimensional cultures in a defined medium on laminin- and vitronectin-coated surfaces. Stem Cells Dev 21:1701–1715. https://doi.org/10.1089/scd.2011.0509
82. Buxboim A, Rajagopal K, Brown AEX et al (2010) How deeply cells feel: methods for thin gels. J Phys Condens Matter 22:194116. https://doi.org/10.1088/0953-8984/22/19/194116
83. Tsai A-C, Jeske R, Chen X et al (2020) Influence of microenvironment on mesenchymal stem cell therapeutic potency: from planar culture to microcarriers. Front Bioeng Biotechnol 8:640. https://doi.org/10.3389/fbioe.2020.00640
84. Penna V, Lipay MV, Duailibi MT et al (2015) The likely role of proteolytic enzymes in unwanted differentiation of stem cells in culture. Future Sci OA 1:FSO28. https://doi.org/10.4155/fso.15.26
85. Brindley D, Moorthy K, Lee J-H et al (2011) Bioprocess forces and their impact on cell behavior: implications for bone regeneration therapy. J Tissue Eng 2011:620247. https://doi.org/10.4061/2011/620247
86. Cherry RS, Papoutsakis ET (1986) Hydrodynamic effects on cells in agitated tissue culture reactors. Bioprocess Eng 1:29–41. https://doi.org/10.1007/BF00369462
87. Nienow AW, Rafiq QA, Coopman K et al (2014) A potentially scalable method for the harvesting of hMSCs from microcarriers. Biochem Eng J 85:79–88. https://doi.org/10.1016/j.bej.2014.02.005
88. Lawson T, Kehoe DE, Schnitzler AC et al (2017) Process development for expansion of human mesenchymal stromal cells in a 50L single-use stirred tank bioreactor. Biochem Eng J 120:49–62. https://doi.org/10.1016/j.bej.2016.11.020
89. Zwietering T (1958) Suspending of solid particles in liquid by agitators. Chem Eng Sci 8:244–253. https://doi.org/10.1016/0009-2509(58)85031-9
90. Kraume M (2003) Mischen und Rühren: Grundlagen und moderne Verfahren. Wiley-VCH, Weinheim
91. Liepe F, Sperling R, Jembere S (eds) (1998) Rührwerke: Theoretische Grundlagen, Auslegung und Bewertung, 1. Aufl. Fachhochschule Anhalt, Köthen
92. Heathman TR, Nienow AW, Rafiq QA et al (2018) Agitation and aeration of stirred-bioreactors for the microcarrier culture of human mesenchymal stem cells and potential implications for large-scale bioprocess development. Biochem Eng J 136:9–17. https://doi.org/10.1016/j.bej.2018.04.011
93. Czermak P, Pörtner R, Brix A (2009) Special engineering aspects. In: Cell and tissue reaction engineering: with a contribution by Martin Fussenegger and Wilfried Weber. Springer, Berlin, Heidelberg, pp 83–172
94. Kanda Y, Hinata T, Kang SW et al (2011) Reactive oxygen species mediate adipocyte differentiation in mesenchymal stem cells. Life Sci 89:250–258. https://doi.org/10.1016/j.lfs.2011.06.007
95. Pattappa G, Heywood HK, de Bruijn JD et al (2011) The metabolism of human mesenchymal stem cells during proliferation and differentiation. J Cell Physiol 226:2562–2570. https://doi.org/10.1002/jcp.22605
96. Phelps J, Sanati-Nezhad A, Ungrin M et al (2018) Bioprocessing of mesenchymal stem cells and their derivatives: toward cell-free therapeutics. Stem Cells Int 2018:9415367. https://doi.org/10.1155/2018/9415367
97. Abbasi-Malati Z, Roushandeh AM, Kuwahara Y et al (2018) Mesenchymal stem cells on horizon: a new arsenal of therapeutic agents. Stem Cell Rev Rep 14:484–499. https://doi.org/10.1007/s12015-018-9817-x

98. Piffoux M, Nicolás-Boluda A, Mulens-Arias V et al (2019) Extracellular vesicles for personalized medicine: the input of physically triggered production, loading and theranostic properties. Adv Drug Deliv Rev 138:247–258. https://doi.org/10.1016/j.addr.2018.12.009
99. van Niel G, D'Angelo G, Raposo G (2018) Shedding light on the cell biology of extracellular vesicles. Nat Rev Mol Cell Biol 19:213–228. https://doi.org/10.1038/nrm.2017.125
100. Théry C, Witwer KW, Aikawa E et al (2018) Minimal information for studies of extracellular vesicles 2018 (MISEV2018): a position statement of the International Society for Extracellular Vesicles and update of the MISEV2014 guidelines. J Extracell Vesicles 7:1535750. https://doi.org/10.1080/20013078.2018.1535750
101. Varderidou-Minasian S, Lorenowicz MJ (2020) Mesenchymal stromal/stem cell-derived extracellular vesicles in tissue repair: challenges and opportunities. Theranostics 10:5979–5997. https://doi.org/10.7150/thno.40122
102. Gudbergsson JM, Johnsen KB, Skov MN et al (2016) Systematic review of factors influencing extracellular vesicle yield from cell cultures. Cytotechnology 68:579–592. https://doi.org/10.1007/s10616-015-9913-6
103. Patel DB, Santoro M, Born LJ et al (2018) Towards rationally designed biomanufacturing of therapeutic extracellular vesicles: impact of the bioproduction microenvironment. Biotechnol Adv 36:2051–2059. https://doi.org/10.1016/j.biotechadv.2018.09.001
104. Xie L, Mao M, Zhou L et al (2016) Spheroid mesenchymal stem cells and mesenchymal stem cell-derived microvesicles: two potential therapeutic strategies. Stem Cells Dev 25:203–213. https://doi.org/10.1089/scd.2015.0278
105. Watson DC, Bayik D, Srivatsan A et al (2016) Efficient production and enhanced tumor delivery of engineered extracellular vesicles. Biomaterials 105:195–205. https://doi.org/10.1016/j.biomaterials.2016.07.003
106. Whitford W, Ludlow JW, Cadwell JJ (2015) Continuous production of exosomes. Genetic Eng Biotechnol News 35:34. https://doi.org/10.1089/gen.35.16.15
107. Gobin J, Muradia G, Mehic J et al (2021) Hollow-fiber bioreactor production of extracellular vesicles from human bone marrow mesenchymal stromal cells yields nanovesicles that mirrors the immuno-modulatory antigenic signature of the producer cell. Stem Cell Res Ther 12:127. https://doi.org/10.1186/s13287-021-02190-3
108. de Almeida Fuzeta M, Bernardes N, Oliveira FD et al (2020) Scalable production of human mesenchymal stromal cell-derived extracellular vesicles under serum-/Xeno-free conditions in a microcarrier-based bioreactor culture system. Front Cell Dev Biol 8:553444. https://doi.org/10.3389/fcell.2020.553444
109. Secunda R, Vennila R, Mohanashankar AM et al (2015) Isolation, expansion and characterisation of mesenchymal stem cells from human bone marrow, adipose tissue, umbilical cord blood and matrix: a comparative study. Cytotechnology 67:793–807. https://doi.org/10.1007/s10616-014-9718-z
110. Patel DB, Gray KM, Santharam Y et al (2017) Impact of cell culture parameters on production and vascularization bioactivity of mesenchymal stem cell-derived extracellular vesicles. Bioeng Transl Med 2:170–179. https://doi.org/10.1002/btm2.10065
111. Izadpanah R, Kaushal D, Kriedt C et al (2008) Long-term in vitro expansion alters the biology of adult mesenchymal stem cells. Cancer Res 68:4229–4238. https://doi.org/10.1158/0008-5472.CAN-07-5272
112. Ramos L, Sánchez-Abarca LI, Muntión S et al (2016) MSC surface markers (CD44, CD73, and CD90) can identify human MSC-derived extracellular vesicles by conventional flow cytometry. Cell Commun Signal 14:2. https://doi.org/10.1186/s12964-015-0124-8
113. Chen TS, Arslan F, Yin Y et al (2011) Enabling a robust scalable manufacturing process for therapeutic exosomes through oncogenic immortalization of human ESC-derived MSCs. J Transl Med 9:47. https://doi.org/10.1186/1479-5876-9-47
114. Noronha NC, Mizukami A, Caliári-Oliveira C et al (2019) Priming approaches to improve the efficacy of mesenchymal stromal cell-based therapies. Stem Cell Res Ther 10:131. https://doi.org/10.1186/s13287-019-1224-y

115. Ragni E, Perucca Orfei C, de Luca P et al (2020) Inflammatory priming enhances mesenchymal stromal cell secretome potential as a clinical product for regenerative medicine approaches through secreted factors and EV-miRNAs: the example of joint disease. Stem Cell Res Ther 11:165. https://doi.org/10.1186/s13287-020-01677-9
116. Cheng A, Choi D, Lora M et al (2020) Human multipotent mesenchymal stromal cells cytokine priming promotes RAB27B-regulated secretion of small extracellular vesicles with immunomodulatory cargo. Stem Cell Res Ther 11:539. https://doi.org/10.1186/s13287-020-02050-6
117. Di Trapani M, Bassi G, Midolo M et al (2016) Differential and transferable modulatory effects of mesenchymal stromal cell-derived extracellular vesicles on T, B and NK cell functions. Sci Rep 6:24120. https://doi.org/10.1038/srep24120
118. Sundin M, Ringdén O, Sundberg B et al (2007) No alloantibodies against mesenchymal stromal cells, but presence of anti-fetal calf serum antibodies, after transplantation in allogeneic hematopoietic stem cell recipients. Haematologica 92:1208–1215. https://doi.org/10.3324/haematol.11446
119. Shelke GV, Lässer C, Gho YS et al (2014) Importance of exosome depletion protocols to eliminate functional and RNA-containing extracellular vesicles from fetal bovine serum. J Extracell Vesicles 3. https://doi.org/10.3402/jev.v3.24783
120. Haraszti RA, Miller R, Dubuke ML et al (2019) Serum deprivation of mesenchymal stem cells improves exosome activity and alters lipid and protein composition. iScience 16:230–241. https://doi.org/10.1016/j.isci.2019.05.029
121. de Jong OG, Verhaar MC, Chen Y et al (2012) Cellular stress conditions are reflected in the protein and RNA content of endothelial cell-derived exosomes. J Extracell Vesicles:1. https://doi.org/10.3402/jev.v1i0.18396
122. Pirkmajer S, Chibalin AV (2011) Serum starvation: caveat emptor. Am J Physiol Cell Physiol 301:C272–C279. https://doi.org/10.1152/ajpcell.00091.2011
123. Lavrentieva A, Majore I, Kasper C et al (2010) Effects of hypoxic culture conditions on umbilical cord-derived human mesenchymal stem cells. Cell Commun Signal 8:18. https://doi.org/10.1186/1478-811X-8-18
124. Almeria C, Weiss R, Roy M et al (2019) Hypoxia conditioned mesenchymal stem cell-derived extracellular vesicles induce increased vascular tube formation in vitro. Front Bioeng Biotechnol 7:292. https://doi.org/10.3389/fbioe.2019.00292
125. Bian S, Zhang L, Duan L et al (2014) Extracellular vesicles derived from human bone marrow mesenchymal stem cells promote angiogenesis in a rat myocardial infarction model. J Mol Med (Berl) 92:387–397. https://doi.org/10.1007/s00109-013-1110-5
126. Salomon C, Ryan J, Sobrevia L et al (2013) Exosomal signaling during hypoxia mediates microvascular endothelial cell migration and vasculogenesis. PLoS One 8:e68451. https://doi.org/10.1371/journal.pone.0068451
127. Lo Sicco C, Reverberi D, Balbi C et al (2017) Mesenchymal stem cell-derived extracellular vesicles as mediators of anti-inflammatory effects: endorsement of macrophage polarization. Stem Cells Transl Med 6:1018–1028. https://doi.org/10.1002/sctm.16-0363

Bio-Production of Adeno-Associated Virus for Gene Therapy

Nicholas Donohue, Niamh Keogh, Stefano Boi, and Niall Barron

Abbreviations

2DE	2-dimensional gel electrophoresis
AAP	Assembly activating protein
AAV	Adeno-associated virus
AAVR	Adeno-associated virus receptor
AAVS1	Adeno-associated virus integration site 1
BHK	Baby hamster kidney cells
CFTR	Cystic fibrosis transmembrane conductance regulator
E1/2/4	Adenovirus early region 1/2/4
FGFR	Fibroblast growth factor receptor
FVIII	Coagulation factor VIII
GeLC-MS	Gel electrophoresis liquid chromatography-mass spectrometry
GFP	Green fluorescent protein
GPN3	GPN-loop GTPase 3
GPR78	G protein-coupled receptor 78
HBoV1	Human bocavirus 1
HEK293	Human embryonic kidney 293 cells
HeLa	Henrietta Lacks cells

N. Donohue · S. Boi
National Institute of Bioprocessing, Research and Training, Dublin, Ireland
e-mail: nicholas.donohue@nibrt.ie; stefano.boi@nibrt.ie

N. Keogh · N. Barron (✉)
National Institute of Bioprocessing, Research and Training, Dublin, Ireland

UCD School of Chemical and Bioprocess Engineering, University College Dublin, Belfield, Ireland
e-mail: niamh.keogh@nibrt.ie; niall.barron@nibrt.ie

R. Pörtner (ed.), *Cell Culture Engineering and Technology*, Cell Engineering 10,
https://doi.org/10.1007/978-3-030-79871-0_11

HEM4	Uroporphyrinogen-III synthase
HGFR	Hepatocyte growth factor receptor
HSV	Herpes simplex viruses
ITR	Inverted terminal repeat
LPLD	Lipoprotein lipase deficiency
MAGv	Multiplex AAV Genotyping
miRNA	microRNA
ORF	Open reading frame
PDGFR	Platelet-derived growth factor receptor
PEI	polyethyleneimine
PR E4	Proteinase yscE4
QC	Quality control
qPCR	quantitative polymerase chain reaction
rAAV	recombinant Adeno-associated virus
RBE	Rep binding element
Sf9/Sf21	*Spodoptera frugiperda* 9/21 cells
shRNA	short hairpin RNA
SMN1	Survival motor neuron 1
SMRT	Single molecule real-time
SOD1	Superoxide dismutase
TOP2	DNA topoisomerase 2
TRAP	Tryptophan RNA-binding attenuation protein
trs	Terminal resolution site
UTR	Untranslated terminal region
VA-RNA	Virus-associated RNA
VP1/2/3	Viral protein 1/2/3
ZF5	Zinc-finger 5 protein

1 Introduction

1.1 Adeno-Associated Virus General Overview

Genetic diseases caused by absent or defective genes can inhibit normal cell functions and severely impact a patient's health. These conditions can be treated through gene therapy, by delivering a functional copy of a particular gene to affected cells. A common strategy to achieve this involves using recombinant viruses as gene therapy vectors. Adeno-Associated Virus (AAV) has emerged as a highly promising option due to its low pathogenicity, safety, and ability to infect multiple different tissue types. Currently, three AAV-based therapies have been approved for human treatment: Glybera, Luxturna and Zolgensma.

Given that genetic defects are usually present in every cell in a patient's body, gene therapy must be applied to the majority of cells in the body, or within an organ to achieve a therapeutic effect. As a result, AAV-based treatments require the administration of very high amounts of virus. Producing sufficient AAV, without compromising safety is a major challenge in bio-production. This chapter focuses on the different platforms used to produce AAV and potential optimization strategies.

1.1.1 AAV Genome and Capsid Structure

AAV is a small (20–25 nm), non-enveloped, single-stranded DNA virus that belongs to the *Parvoviridae* family, genus *Dependoparvovirus*. The AAV genome size is 4.7 kb. In wild-type AAV, the genome encodes the *rep* and *cap* genes, flanked by inverted terminal repeats (ITRs). *rep* codes for proteins involved in genome replication, whereas *cap* provides the structural elements of the viral capsid. The *cap* sequence determines the AAV serotype. At least 12 AAV serotypes occur naturally [1]. The ITRs serve to initiate genome replication and are further required for packaging the viral genome into the capsid [2]. ITRs contain inverted repeat sequences that are necessary to form T-shaped hairpin loops, leading to formation of the Rep Binding element (RBE). As the name suggests, the RBE binds the Rep protein to initiate genome replication and packaging [2, 3].

Both *rep* and *cap* make use of alternative splicing to produce different proteins from the same gene: *rep* mRNA can be spliced to generate Rep78, Rep68, Rep52 and Rep40 (named for their weights in kDa), while *cap* translation generates VP1, VP2 and VP3.

Rep68 and Rep78 bind the RBE of the AAV ITR region and then cleave the DNA at the terminal resolution site (*trs*). All four Rep proteins display helicase activity and can bind ATP [4, 5].

The viral coat is estimated to be composed of 60 proteins arranged in an icosahedron. The capsid proteins are present in a ratio of 1:1:10 for VP1, VP2 and VP3, respectively [3].

In addition, AAP (assembly-activating protein) is produced from an in-frameshift ORF in the *cap* gene [2, 6, 7]. AAP may act as a scaffolding protein for capsid assembly [8]. Interestingly, AAP is required for nuclear localization of the VP proteins as well as capsid assembly, but only in AAV serotype 2. In other AAV serotypes, the subnuclear localization of AAP is variable and AAP is indeed non-essential in AAV4, AAV5 and AAV11 [9].

1.1.2 AAV Infection and Replication

AAV was first identified in the 1960s, as a frequent contaminant in purifications of Adenovirus, hence the name. AAV requires a helper virus (e.g. Adenovirus) for replication, as its own genome does not encode the required factors to force a host cell to produce viral proteins. Without a helper virus, AAV establishes a latent infection as the viral genome persists in the host cell nucleus [3, 10].

AAV infects cells by binding to primary receptors (e.g. heparin, sialic acid or galactose) and co-receptors (e.g. integrin, FGFR, HGFR, AAVR) on the cell surface, leading to endocytosis into endosomes [2, 6, 11]. A structural change in VP1 and VP2 leads to exposure of their N-terminus and escape from the endosome, then accumulation at the perinuclear region of the host cell [12, 13]. Following entry into the nucleus, AAV is uncoated and releases the single-stranded DNA genome. The genome is then replicated to form dsDNA, which can then be transcribed and translated [6, 7].

Wild-type AAV genomes can integrate into a specific location (termed the AAVS1 site) on the long arm of chromosome 19 (19q13-qter) in human host cells [14–16]. The AAVS1 site contains a RBE, which in conjunction with the AAV Rep protein is required for integration of the viral genome [17]. Only approximately 0.1% of wild-type AAV genomes are integrated into AAVS1 during the course of an infection [15].

Wild-type AAV genomes have also been observed to integrate into other sites in the human genome, presumably in a Rep-independent manner. AAV proviruses have been found on chromosome 1q31.1 in human tonsillar tissue, while in vitro experiments showed integration into multiple different chromosomes [18–20].

If the Rep proteins are not present, as is the case in target cells during AAV-mediated gene therapy, the ITR-flanked transgene cassettes can form circularized concatemers, which then persist as episomal DNA in the host cell's nucleus [21]. Episomal DNA will be gradually diluted due to host cell replication, meaning that the beneficial effect of gene therapy will be lost over time. Therefore, the duration of the treatment is dependent on the replication turnover rate of the target cells.

1.1.3 AAV Serotypes and Tissue Tropism

A variety of AAV serotypes have been identified that display different tissue tropism, making them highly useful for targeting specific organs. Wild-type AAV DNA has been discovered in lung tissue, which is not unusual given that the virus infects via the respiratory route [18]. In addition, AAV DNA has also been found in the following tissues: brain, heart, liver, bone marrow, spleen, colon, lymph node and kidney, which suggests that the virus spreads throughout the body during infection [22].

At least 12 natural serotypes and more than 100 mutant AAVs have been discovered thus far [1]. The preferred serotypes used in pharmaceutical production are AAV2, 4, 5, 6, 8, although mutant versions have also been tried to increase tissue tropism or reduce immunogenicity (e.g. AAV7m.8, mutant AAV Spark100, etc.) [2]. The choice of serotype depends on the intended tissue tropism of the transgene, i.e. where the virus is going to be applied in the body. AAV2 is often preferred as it has such a broad tissue tropism: heparin sulfate, FGFR/HGFR (fibroblast growth factor receptor/hepatocyte growth factor receptor), lamin receptor and a5b1 are all considered receptors for AAV2 [23–25]. AAV5 on the other hand displays tropism for neuronal, retinal pigmented epithelia and photoreceptors with its preferred receptors being sialic acid and platelet-derived growth factor receptor (PDGFR) [26–28].

1.2 *AAV as a Gene Therapy Agent*

Compared to other viruses that have been explored as gene therapy agents, AAV has the advantages of low immunogenicity as well as broad tissue tropism. As a result, AAVs can be safely injected in large amounts and can treat a wide variety of organs and tissues.

A notable feature of recombinant AAV (rAAV) is that almost the entire AAV genome can be replaced with a transgene (or cargo) cassette containing the required gene. Typically, this approach involves a promoter (often specific to the intended target tissue), the cargo gene itself (e.g. coagulation factor IX for the treatment of hemophilia) and a polyA tail [2, 29]. As a result, the DNA delivered to a patient's cells contains almost no viral genome sequences and is incapable of replicating. To produce AAVs for gene therapy applications, the viral genes required for AAV replication are supplied in trans, on separate vectors.

Following delivery to a target cell, rAAV genomes persist as episomal DNA in the cell nucleus. Transcription of the cargo gene can then commence, which compensates for the gene defect in a patient's genome. rAAV-based gene therapy agents are intentionally designed to not integrate their genome into the host cell's DNA. This approach was chosen to limit the chance of off-target insertions, which could lead to dangerous mutations. Therefore, the therapeutic benefit is only maintained for as long as the episomal rAAV genome persists in the nucleus. Cell division will lead to gradual dilution or removal of episomal rAAV genomes, causing the therapeutic benefit to be lost over time. Therefore, rAAV therapy is a transient treatment, rather than a permanent cure.

1.2.1 Licensed rAAV Gene Therapy Treatments

The first approved AAV gene therapy treatment was Glybera in 2012, intended for lipoprotein lipase deficiency (LPLD), which manifests as pancreatitis [30]. AAV1 capsids deliver an intact copy of the LPL gene to muscle cells. Doses of 3×10^{12} vg/ml solution per injection are used, although multiple injections are required. In clinical trials, reductions in the incidence of pancreatitis were observed up to 6 years after administration [31]. However, this drug has been removed from the market due to profitability concerns, including the high cost (about 1.1 million USD per treatment), as well as LPLD being a very rare condition [31, 32].

Zolgensma, for the treatment of spinal muscular atrophy was approved for use by the U.S. Food and Drug Administration (FDA) in 2019 [2, 33]. Zolgensma is administered as a one-time intravenous injection for children under 2 years old, at a cost of 2.1 million USD per treatment. Corticosteroids are usually also given to prevent immune reactions. The drug uses AAV8 capsids to deliver a functional copy of the SMN1 gene to motor neurons. Since motor neurons do not divide, it is possible that the treatment could last life-long, although this is currently unclear. The recommended dosage is 1.1×10^{14} vg/kg [34].

Finally, Luxturna is an approved rAAV therapy for the eye disease Leber congenital amaurosis which affects retinal cells, leading to vision loss [35]. Administered as sub-retinal injection, AAV2 capsids deliver intact copies of the RPE65 gene to the target cells. Doses of 1.5×10^{11} vg are injected, at a cost of 425,000 USD per eye [36]. Treated patient's showed improved vision up to 3 years after administration [36].

The above numbers emphasize the quantities of virus needed for rAAV gene therapy and the associated costs. These factors are significant challenges in bio-production of AAV gene therapy agents. rAAV development continues and as of November 2018 there were 145 AAV-based drugs in clinical trials [37].

2 Challenges and Opportunities in rAAV Production

2.1 rAAV Production Challenges

Currently, rAAV production can be performed using HEK293, Sf9/Sf21, HeLa, BHK or yeast cells [2, 38]. A summary of the various platforms and their potential yields is shown in Fig. 1. While every rAAV production platform has specific drawbacks, there are some common issues. These include high production costs, sub-optimal yields and contamination with empty capsids, or mis-packaged DNA.

High production costs can push the price of a single rAAV treatment up to 2.1 million USD (as in the case of Zolgensma). The commercial failure of the drug Glybera is a stark reminder of this problem. AAV production methods must either become cheaper or increase yields.

As an example, spinal muscular dystrophy in children under 2 years old can be treated with Zolgensma, which requires a dose of 1.1×10^{14} vg/kg body weight [34, 39]. Assuming the average weight for a 2 -year old child is 12 kg, this would equate to a dosage of 1.32×10^{15} vg per patient.

The HEK293 system can produce up to 2.1×10^{5} vg/cell in a WAVE reactor with a maximum volume of 50 L. Cell density before transfection is typically around 5×10^{6} cells/ml. A 50 L WAVE reactor would therefore contain at most 2.5×10^{11} cells. If each cell produces 2.1×10^{5} vg, the final yield would be 5.25×10^{16} vg. This yield would be sufficient to treat 39 children. However this calculation is based on overly optimistic assumptions, i.e. perfectly efficient transfection, production of 2.1×10^{5} vg/cell (without empty capsids) and rAAV purification without any losses.

Spinal muscular atrophy is estimated to affect 1 in 10,000 births [40]. Based on an estimate of 140 million births per year worldwide, up to 14,000 children per year may require this treatment [41]. To make enough Zolgensma for 14,000 patients would require 359 WAVE reactors with 50 L volumes. This number would be substantially higher for more common conditions such as hemophilia, which affects up to 1 in 5000 births [42]. Treating hemophilia in adult patients would also require larger doses, due to the higher body weight of adults.

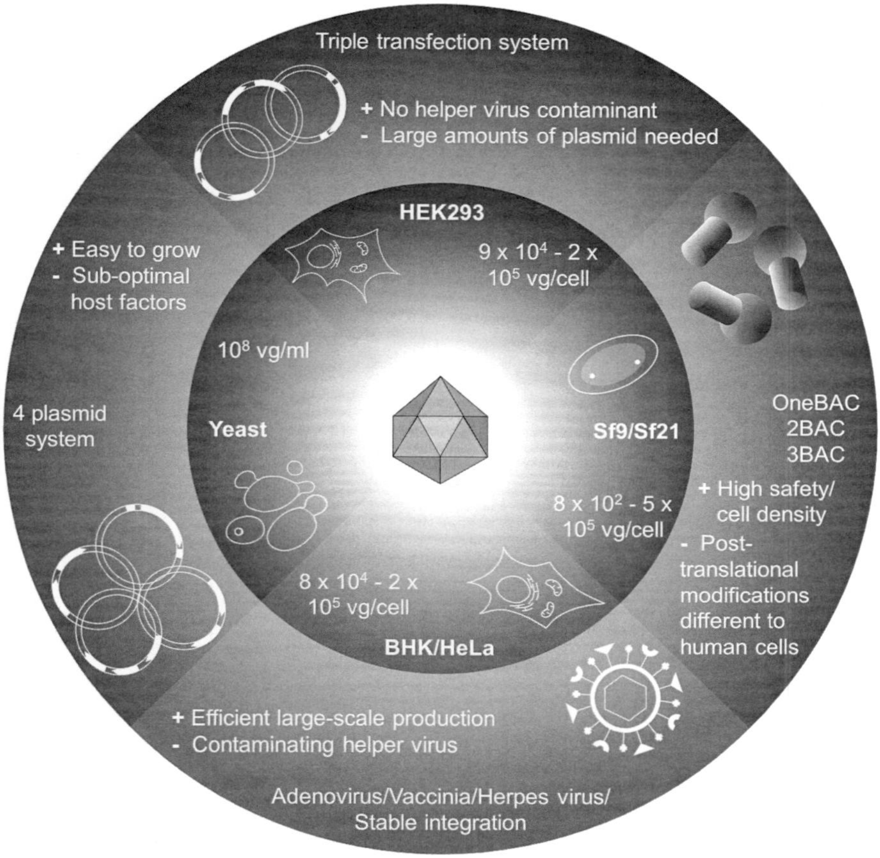

Fig. 1 Overview of AAV production systems. The inner circle shows the possible production cell lines and reported potential yields. The outer circle indicates the plasmids/viruses employed in AAV production, as well as advantages and disadvantages of the various systems [3, 45–51]

A further issue concerns the packaging of undesired sequences (such as antibiotic resistance markers or viral sequences), or production of empty capsids, as these contaminants must be removed in later steps. Mis-packaged sequences are often derived from the backbones of plasmids used to deliver the transgene into the producer cell. Although the exact cause is unclear, it has been hypothesized that replication of the transgene could continue beyond the ITRs and include DNA from the backbone, which is then packaged.

Alternatively, transgene replication may initiate at an ITR and proceed in the wrong direction, causing only backbone DNA to be packaged [43]. Inserting non-dangerous 'stuffer' DNA adjacent to the ITRs can prevent the packaging of harmful plasmid backbone elements, as the AAV capsid is limited to packaging 4.7 kb of

DNA. Although this approach is an improvement, mis-packaging can still occur, reducing the safety and effectiveness of rAAV drugs.

rAAV coat proteins self-assemble into capsids, into which DNA is subsequently packaged. The presence of empty capsids in rAAV purifications indicates that the packaging step is the limitation in this system, i.e. more capsids are produced than can be filled with DNA. Manipulating the expression levels of *rep* and *cap* genes could offer a solution, although such efforts would need to avoid producing excess capsid proteins which can also contaminate purifications and provoke an immune response [44].

An ideal rAAV production platform would need to satisfy the following criteria:

- Cost-effective, high-yield production and purification, by improving the cell line, plasmids, or production methods.
- Absent or very low contamination by mis-packaged or empty capsids. Alternatively, a cheap and reliable method to remove these contaminants is needed.
- Versatility, i.e. the ability to produce rAAVs to treat any condition using the same system, by simply swapping the transgene. This would allow for a higher return on the investment of setting up a manufacturing system.
- Safety, including low immunogenicity and toxicity. This may require the avoidance of helper viruses, especially those of human origin.
- Efficient delivery to target cells, for example through capsid modification or improving delivery methods. This would allow for a smaller amount of rAAVs to be used, which would reduce cost of treatment and improve safety.
- A solution to the transgene size limitation, as this would allow for treatment of a wider array of conditions. While splitting large transgenes across different rAAV vectors is a potential solution, it also increases the amount of rAAVs that need to be manufactured and administered to patients, which increases the cost and risk involved.

2.2 Current rAAV Production Platforms

2.2.1 HEK293 Triple Transfection System

HEK293 cells (derived from human embryonic kidney) are a popular and well-established cell line used to produce rAAVs for gene therapy. HEK293 were originally generated through transfection with Adenovirus 5 DNA [52]. As a result, HEK293 contains the Adenovirus helper genes E1a and E1b [53]. The remaining helper genes E4, E2a and VA-RNA must be supplied through plasmid transfection, along with the AAV *rep/cap* genes and the ITR-flanked transgene [54]. These elements are typically delivered on separate plasmids (i.e. one plasmid for the helper genes, one for *rep/cap* genes and one for the ITR-transgene), in a process termed the triple-transfection system. Although it would be possible to combine the helper genes, *rep/cap* and the ITR-transgene on a single plasmid, this may be unsafe as recombination events could lead to production of infectious rAAV particles carrying *rep/cap* instead of the ITR-transgene.

The triple transfection system has a disadvantage, in that delivering three different plasmids into a single cell is difficult. The currently preferred protocols rely on polyethyleneimine (PEI) for transfection, as it is well-tolerated and efficiently transfects HEK293 cells. In addition, it does not necessitate a media change after transfection and is comparatively cheap, compared to similar reagents on the market [46]. However, the plasmid DNA must still be supplied in excess quantities and one of the difficulties lies in acquiring enough purified plasmid DNA for this process.

A further issue is the potential for cells to be transfected with the *rep/cap* and helper plasmids but not the ITR-transgene vector, which results in assembly of empty viral capsids which contaminate the resulting product. This complicates downstream purification and quality control [55, 56].

HEK293 cells grown in suspension using WAVE bioreactors (with volumes in the tens of liters) and shaker flasks yield more virus than by growing the same cells in adherent culture [3]. For this reason, this method is generally preferred in industry [46]. This method yields between 9×10^4 to 2.1×10^5 vg/cell [3].

2.2.2 Sf9/Baculovirus System

The Baculovirus system uses Sf9 or Sf21 insect cells, derived from ovarian tissue of the fall armyworm moth (*Spodoptera frugiperda*). The cells can be grown attached or in suspension without serum in the growth media [57]. For the purpose of rAAV production, Sf9 cells are grown in suspension, then co-infected with recombinant Baculoviruses carrying the rAAV *rep/cap* genes and the ITR-transgene [58, 59]. Vectors generated in this system typically have lower transduction efficiency, compared to those generated with the HEK293 system, possibly due to a lower amount of VP1 packaged into the capsids [51]. Fortunately, this can be addressed by altering the Kozak sequence to express VP1/VP2/VP3 in a ratio that is similar to wildtype AAV, which results in higher transduction efficiency [60, 61].

Various set-ups have been tried for the Baculovirus system, using between one and three different Baculoviruses:

- In an approach using three different Baculoviruses, one virus contains *rep*, another *cap* and a final virus carries the ITR-transgene [3, 62].
- A further improvement was to combine the *rep* and *cap* genes into a single Baculovirus [63]. Co-infection with a second Baculovirus carrying the ITR-transgene initiates expression of rAAVs.
- An additional approach includes all of the elements (*rep*, *cap* and ITR-transgene) in one baculovirus [64].

Alternatively, the Sf9 cells can be altered by stably integrating the ITR-transgene into the cell genome. rAAV production is then initiated by infection with the above-mentioned *rep* and *cap* Baculoviruses [54, 55].

A further option is to generate a Sf9 cell line with the *rep/cap* genes stably integrated into the genome. Expression of the genes is then initiated from upstream promoters that are induced by a later infection with Baculovirus carrying the ITR-transgene. This approach (termed OneBac) is considered much simpler [51, 65]. Expression of the *rep* gene is toxic to mammalian and insect cells, so unless the inducible promoter used in this system is capable of preventing all transcription, this approach could lead to growth inhibition [66].

Sf9 cells used in bio-production are typically grown in 200 L stirred tank reactors [3]. rAAV yields from the Sf9/Baculovirus system typically range around 8×10^2 to 5×10^5 vg/cell. Sf9 cells can grow to very high densities, which makes them very suitable for producing rAAVs. In addition, given that Baculovirus is highly unlikely to infect mammalian cells, the system is very safe compared to other virus-based systems discussed below [3, 48]. Currently, the post-translational modifications of rAAV capsids produced in Sf9 cells are being investigated, as these may differ from mammalian cells [3].

2.2.3 HeLa/Adenovirus System

An alternative approach to the HEK293/triple transfection system that somewhat reduces the requirement on manufacturing large amounts of plasmid DNA is to use a cell line with a stable integration of the helper genes.

For instance, HeLa cells have been engineered to carry the ITR-flanked transgene as well as the *rep/cap* genes [55]. The *rep* proteins are not expressed until infection with a wild-type adenovirus occurs. A downside of this approach is that it relies on using a replication competent Adenovirus, which is a safety concern, due to the potential of an infectious virus contaminating a pharmaceutical product [55, 56]. The system has also been adapted to allow for co-transduction with an AAV capsid carrying the ITR-transgene to be packaged, rather than having this sequence stably integrated. Production typically involves a 250 L stirred tank reactor. Yields of 5×10^4 vg/cell have been reported [3, 67].

2.2.4 HeLa/Vaccinia Virus System

Vaccinia virus has also been used to create rAAV. The *rep/cap* genes were integrated into the Vaccinia virus genome. The resulting virus was used in co-infection of HeLa or QW158–7 cells with replication-deficient Adenovirus to generate rAAV. The resulting rAAVs showed higher transduction efficiency than rAAVs made using the HEK293/triple transfection system. This may have been the result of higher VP1 expression in the HeLa/Vaccinia virus system, due to placing a codon-optimized VP1 gene under control of a dedicated promoter, rather than generating VP1 along with VP2/VP3/AAP via *cap* transcription. This approach yielded 2×10^5 vg/cell [68]. A safety concern with this system is the use of Vaccinia/Adenovirus as these could infect human cells if recombination allowed them to become active.

2.2.5 HEK293/Triple Transfection/Bocavirus System

rAAV production can potentially be enhanced by including genes from other viruses on the helper plasmid. An example of such a virus is human Bocavirus (HBoV1), which can assist with AAV replication [69, 70]. Co-transfection of HEK293 cells with an ITR-transgene plasmid, a *rep/cap* plasmid and a plasmid containing the HBoV1 genes NP1, NS2 and BocaSR produced rAAVs with similar transduction efficiency compared to the HEK293/triple transfection system using Adenovirus helper genes. Inclusion of the adenovirus-derived E2A gene in the HBoV1 helper plasmid lead to a 3–7 fold increase in rAAV production, compared to using the standard Adenovirus helper plasmid [69].

2.2.6 HeLa/BHK/Herpesvirus System

Herpesvirus can also provide helper genes in rAAV production [71]. The current approach is to use two replication-deficient Herpes simplex viruses (HSV), one containing the AAV *rep/cap* genes, and another for the ITR-transgene. In HeLa cells, this approach yielded 1×10^5 vg/cell. A further improvement was made by infecting suspension-adapted BHK cells with these viruses. Producing between 8×10^4 and 2×10^5 vg/cell [49, 59, 72]. Production is usually performed in a 10 L WAVE reactor [3]. HSV have also been used in the HEK293 system, where optimizing the multiplicity of infection of the HSV-transgene and HSV-*rep/cap* at a ratio of 2:12, respectively, achieved a yield of up to 1×10^5 vg/cell [73].

2.2.7 Yeast System

Yeast cells (*Saccharomyces cerevisiae*) have been used successfully to generate rAAVs. Yeast cells transfected with plasmids expressing *cap* genes were used to demonstrate assembly of rAAV [74]. Later work succeeded in producing fully assembled and infectious rAAVs in yeast, using 4 plasmids in total. In this case, yeast-specific, galactose-inducible promoters were used to drive expression of *rep* and *cap* genes [75]. A titer yield of 10^8 vg/ml was reported. Although these results are promising, unfortunately this approach shows low rates of DNA encapsidation. Compared to other systems, yeast lacks the host cell factors or ability to use helper viruses that would enable optimal rAAV production [59].

Specific host proteins in yeast that are related to replication and positively impacted rAAV production have been identified. A high through-put screen utilizing a yTHC screening library consisting of 800 essential yeast genes was constructed. From this study, 22 gene candidates were discovered that improved rAAV DNA replication and vector yield as high as sixfold and 15-fold, respectively. The most prominent genes identified were PRE4, HEM4, TOP2, GPN3, SDO1. As this is the first published screen of its kind, further validation studies are needed to assess the potential application of this type of library screen in rAAV production in yeast models [76].

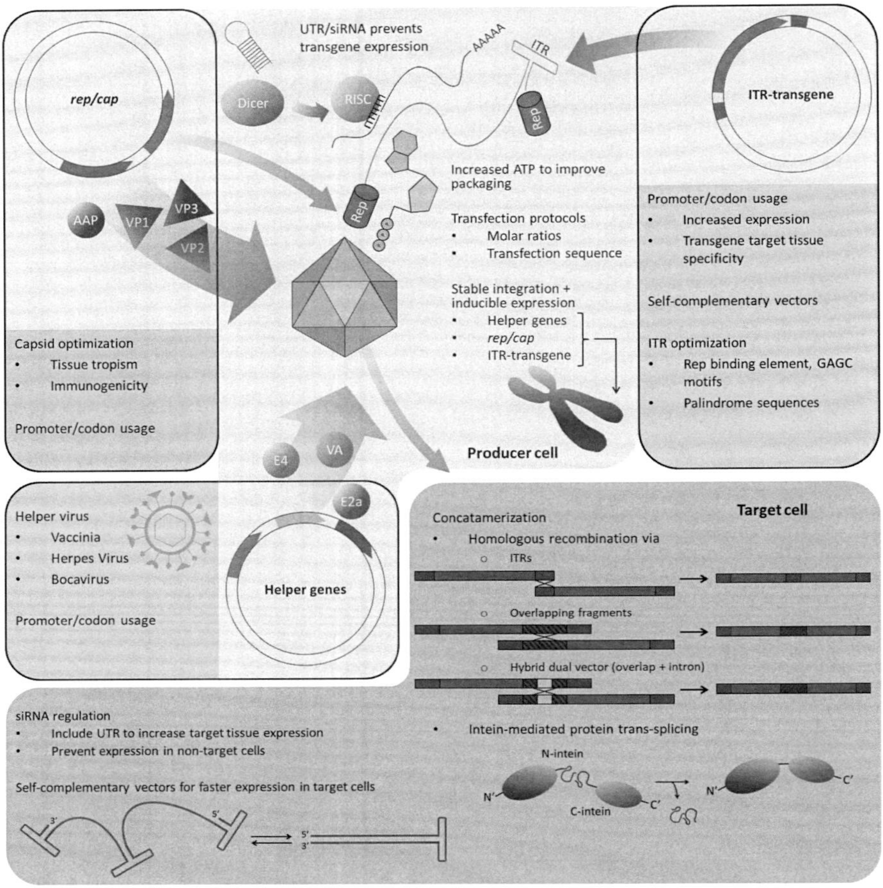

Fig. 2 Potential optimization areas in AAV production, using the HEK293/triple transfection system as an example. The AAV-assembling cell and the cell targeted in gene therapy are labeled 'Producer cell' and 'Target cell', respectively. Black edged boxes indicate the plasmids used in production (labeled '*rep/cap*', 'Helper genes' and 'ITR-transgene') and their respective optimization possibilities.

2.3 *Optimizing rAAV Production*

Extensive research has been devoted to improving various aspects of rAAV, including production optimization. Areas of potential improvement have been identified at the level of viral DNA, RNA and protein. The various components involved in rAAV production and potential optimization strategies are shown in Fig. 2.

2.4 *Improving AAV Production Through ITR-Transgene Optimization*

2.4.1 Transient Transfection vs Stable Integration of AAV Plasmids

Transient transfection requires large amounts of plasmid DNA. If 1 μg of plasmid DNA is transfected per ml of cell culture, then a 50 L reactor would require 50 mg of plasmid DNA [46]. However, this approach is currently favored in manufacturing, as it minimizes the risk of recombination events generating infectious viruses.

Aside from commercial production, rAAVs are also used in pre-clinical research, where safety requirements are less stringent. In this case, a stable insertion of rAAV plasmids into a producer cell's genome may be preferable due to lower cost. One such attempt created HeLa cells with a stable integration of both the transgene (human placental alkaline phosphatase) and the *rep/cap* sequences. rAAV production was initiated through infection by an Adenovirus helper [77, 78]. Yields in the range of 5×10^4 to 2×10^5 DNase resistant particles per cell were achieved.

Aside from the safety concerns, cell lines with stable rAAV integrations are time-consuming to establish and less versatile, as changes to the transgene or promoter sequences are harder to perform.

2.4.2 ITR-Transgene Plasmid

The rAAV ITR-flanked transgene plasmid is generally a recombinant version of the wtAAV virus genome. The *rep* and *cap* genes are replaced with the transgene (with a maximum size of 4.7 kb), leaving only the AAV ITR regions from the original virus. The AAV ITRs are the only elements required for replication and packaging of the transgene, which occurs once *rep* and *cap* are supplied. As a result, the ITR-transgene plasmid is only capable of delivering the transgene to the cell, but is completely dependent on another plasmid to actually replicate [2].

The transgene is usually inserted with a promoter (often specific to the target tissue), as well as a polyA signal. The backbone of these plasmids generally contains sequences that allow propagation in bacterial strains to amplify the vector (i.e. an origin of replication) as well as antibiotic resistance markers. ITR regions are usually flanked by multi-cloning sites.

Most ITR-transgene plasmids are derived from pSUB201, originally described in 1987 by Samulski [79]. pSUB201 and its derivatives contain a truncated AAV2 derived ITR, compared to the wild-type sequence: the ITRs are duplicated, with the second ITR being an inverted repeat of the first ITR, with a 14 bp deletion at the extremity [80]. The deleted portion corresponds to a palindromic sequence at the end of the ITR hairpin structure in wtAAV2. As a result, the pSUB201 ITR probably does not form a complete wtAAV2 ITR hairpin.

In the Baculovirus system, replacing the pSUB201 ITRs with wtAAV2 derived ITRs led to a reduction in empty capsids, as well as a tenfold reduction in non-rAAV DNA in the capsids [80].

The ITR sequences could be explored in terms of optimization: given that Rep78 covalently attaches to the ITR during genome replication, this appears to be a key requirement for packaging the viral DNA into the capsid. Rep78 binds a triplet of GAGC motifs in the Rep Binding Element. Serotype AAV2 has three GAGC motifs, but other AAV serotypes have only 2 [81]. Removal of GAGC motifs reduces viral replication in wtAAV2, especially when only a single GAGC is left intact [82].

A further option for ITR-transgene plasmids is pHpa-trs-SK, which allows for more rapid onset of gene expression in target tissues after transduction. This is achieved through deleting one of the terminal resolution sites (*trs*), where the Rep protein normally nicks the DNA. This leaves the vector with one functional ITR with a *trs*, as well as a mutant ITR that cannot be nicked by Rep [83].

Replication starts at the functional ITR through Rep nicking at the *trs*, followed by ITR replication (initially in a direction away from the transgene). After the ITR is formed, it loops back by forming a hairpin, which allows replication of the transgene. After the transgene, the mutant ITR is replicated, which also loops back and folds into an ITR hairpin. However, because this ITR cannot be nicked by Rep, DNA replication continues back through the transgene, using the recently synthesized DNA strand as a template, until the *trs* site the process started from is reached. This results in a single-stranded, inverted repeat genome, with functional ITRs at each end and the mutated ITR in the middle [83].

This approach allows more rapid expression after transduction, probably because second- strand synthesis is avoided [83]. However, the sequence available for transgene incorporation is smaller at about 2 kb, vs the 4.7 kb size of standard rAAV vectors.

On a further note, optimizing codons of the ITR-transgene leads to increased transcription and translation. Codon optimization is also beneficial to enabling tissue-specific gene expression and plays a role in reducing the innate immune response towards rAAV [84, 85].

2.4.3 ITR-Transgene Promoter Optimization

Due to the limited size of rAAV transgenes, promoter sequences must be as concise as possible, while allowing maximum expression. Various promoter and enhancer sequences from liver-specific genes have been combined into smaller promoters that allow transgenes to be larger, while still strongly inducing expression after transduction [86]. An alternative approach involved screening 100-bp long synthetic enhancer elements, which yielded an 83 bp promoter that enhanced expression of the CFTR gene in human airway cells [87].

An *in silico* rational design approach resulted in synthetic enhancer elements between 41 and 551 bp in length for use in liver-specific rAAV transgenes [2]. By including these elements along with a liver-specific promoter, rAAV transduction caused between 10 and 100-fold higher transgene expression, compared to the same construct without the synthetic elements [88]. The same approach has been used to generate cardiac specific regulatory modules [89].

2.4.4 Concatemerization

A significant problem of rAAV gene therapy is the inability to package transgenes larger than 4.7 kb. Multiple strategies have been developed to overcome this restriction.

Primarily, homologous recombination between the ITR sequences occurs naturally after delivery of the rAAV genome [90]. This procedure can be exploited by splitting a large transgene into two (or more) subunits and packaging them into individual rAAVs. If rAAVs carrying the various subunits are delivered to the same cell's nucleus, the subunits will be assembled into the desired transgene [2, 7]. This method has been successfully applied in animal models to assemble the dystrophin gene from two or three rAAVs [91].

In a similar approach truncated fragments of the transgene can be packaged into rAAVs in a randomized fashion, creating a mixed population of rAAVs. After delivery to the nucleus, homologous recombination between overlapping areas of the truncated fragments assembles the transgene [92].

A further development intends to solve the issue of non-functional products being generated if the wrong fragments are recombined, or if truncated overlapping fragments introduce undesirable ITR structures within the transgene. In a hybrid dual-vector strategy, an overlapping region is combined with intron splice sites in the split transgene [93]. Specifically, the concatemerization of rAAVs through recombination is enhanced by introducing $5'$ and $3'$ splicing elements to ensure correct assembly of the transgene.

Cross-packaging of the oversize ITR-transgenes into the capsids of other parvoviruses is another option, although this could eliminate the advantages of AAVs broad tissue tropism and low immunogenicity [94].

A final method is to use intein-mediated protein trans-splicing: intein is a catalyst for protein splicing and can ligate two distinct polypeptides together. In this way, peptides expressed from separate subunits of a transgene can be flanked by short split inteins, which are then assembled into a full protein via protein trans-splicing. This method has been used successfully in animals to deliver e.g. dystrophin, FVIII, CFTR and CRISPR-Cas9 [95].

2.4.5 Transgene Regulation During rAAV Packaging

One of the main factors limiting high production yields, especially in mammalian cells, is the expression of toxic or apoptotic transgenes in the packaging cell line. Therefore, it would be ideal to prevent transgene expression during the packaging process.

Insertion of an artificial riboswitch (termed GuaM8HDV) on the 3′-UTR of a toxic transgene, significantly downregulated transgene expression in the packaging cell line, while still allowing expression after transduction both in vitro and in vivo. The riboswitch efficiently downregulated expression of ITR-flanked GFP by 75% at a guanidine concentration of 200 μmol/L and rAAV yield was increased between 2.3 and 23-fold [96, 97].

Another approach relies on the bacterial tryptophan RNA-binding attenuation protein (TRAP). In the presence of L-tryptophan, TRAP binds a 55 bp region located on the 5′UTR and suppresses mRNA transcription. This system has been successfully employed to raise rAAV yields [98].

MicroRNAs have also been used to regulate expression. miR-373 and miR-122a have been shown to have strong silencing activity. A toxic transgene upstream of a quadruplet repeat miRNA-373 sequence was repressed in the HEK293 packaging cells, improving rAAV yields up to 22-fold [99].

Alternatively, the 3′UTR can be engineered to allow miRNA binding, leading to post-transcriptional silencing of the transgene. This can be used to prevent transgene expression in non-target cells. For example, including a synthetic 3′UTR that allows binding of miRNA-122, which is almost exclusively expressed in the liver, leads to transgene suppression [100].

A final method to prevent expression of toxic transgenes in producer cells involves shRNAs. For instance, expression of G protein-coupled receptor 78 (GPR78), was found to be toxic to HEK93 cells, reducing cell growth and vector yield. By introducing an additional plasmid expressing a shRNA targeting GPR78, AAV vector yield increased from 1 particle to >10,000 particles per producer cell [101].

2.5 *Improving rAAV Production Through Capsid Optimization*

2.5.1 Expanding rAAV Capsid Functions

Transduction can be made more selective by inserting certain proteins into different regions of the rAAV capsid. For instance, portions of protein A, cytokines and so-called DARPins (designed ankyrin repeat proteins) have all been used for this purpose and were successfully inserted into the rAAV capsid [102, 103]. This approach was also used to target rAAVs towards CD4+ T-cells as well as cancer cells [104, 105].

Another development concerns the presence of surface-exposed serine and tyrosine residues on the rAAV capsid: it has been hypothesized that these are phosphorylated upon cell entry, which leads to ubiquitinylation and subsequent degradation by the proteasome [106]. Preventing phosphorylation by mutating tyrosine to phenylalanine, lead to substantially increased transduction efficiency [107].

Of further interest in this context are pXR1-5 plasmids, which can be used to package any AAV2 derived ITR-transgene into a different AAV serotype [108].

2.6 *Improving rAAV Production Through Omics-Based Approaches*

2.6.1 High Throughput Methods in AAV Research

In gene therapy research, omics approaches are used to better understand the complex biology behind AAV, in the hopes of providing useful insights that may not have been otherwise deciphered from individual target analysis. The integration of multiple omics layers has the potential to provide a useful framework, capturing an overall image of what is being expressed, translated and produced during the AAV life cycle. As a leading platform in gene delivery for the treatment of a variety of human diseases [37], it is crucial that a comprehensive understanding of the biological system behind AAV is achieved, not only for production optimization but also treatment efficacy and patient safety.

2.6.2 Promoter, Transgene and ITR Studies

As discussed elsewhere in this chapter, there have been considerable efforts made over the past two decades to improve and optimize AAV capsid diversification and selection. Although these efforts have led to many key successes and improvements in the production of AAV, it is evident there is still room for further discovery using complementary strategies that also act on the level of vector DNA but are not solely focused on the capsid. Up until now, in comparison to capsid engineering, there has been relatively little emphasis placed on improvements to the engineering of the second critical AAV vector component, i.e. the recombinant genome and its multiple components. Omics is now playing a critical role in a set of new innovative strategies focused on the AAV expression vector. This includes the engineering of synthetic enhancers and promoters using high through-put screens, the identification and characterization of proteins that recognize the RBE within AAV ITRs and also the use of controlled genetic circuits utilizing miRNA to regulate AAV gene expression [109].

A major challenge in furthering AAV expression vector optimization hinges on deciphering which elements in the promoter, the distal enhancer and/or in the untranslated regions actually govern the strength and the specificity of transgene expression. The road to the next generation of synthetic enhancers and promoters for gene therapy application could lie in the use of high-throughput screening methods to characterize these promoters and enhancers. In 2017, a high-throughput enhancer reporter assay unravelled a set of mammalian promoters displaying enhancer activity. These promoters had distinct genomic and epigenomic features [110]. In a similar screen, the enhancer and promoter activities of thousands of DNA fragments transduced in vivo were measured and results suggested that gene promoters are distinguished from distal enhancers by specific complements of transcriptional activators [111]. Although there have not been extensive AAV studies published in this area thus far, these initial studies show the potential of high throughput screening and how it could be the way forward for transgene promoter/enhancer optimization.

As mentioned elsewhere in this chapter (see Sect. 2.4.5.), as a complementary strategy, the cell selectivity of AAV transgene expression can be honed by exploiting RNAi machinery and the tissue/cell specific expression pattern of the relevant microRNAs. miRNA based strategies to control gene expression incorporate an "on/off switch" mechanism.

The 'on/off switch' mechanism itself is not an omics approach, however it would not be possible without the use of transcriptomics and miRNA profiling identifying specific miRNA that would be suitable to fit into the switch. Genome-wide miRNA profiling is a well-established approach, utilized in many cell lines and systems to identify useful targets for bioengineering. The results of these studies are now being harnessed for the benefit of gene therapy. Particularly useful miRNA targets are those that are cell or tissue specific. For example, miR-122 is highly abundant and active in the liver. 2016 saw the first exploitation of these findings to use tissue and cell-specific miRNA signatures as a means to regulate transgene expression in the context of a viral vector [112].

Finally, genetic screens aimed at the ITRs have provided a powerful approach to identify relevant cellular factors in AAV production and transduction. A hybrid screen designed to identify cellular proteins that recognize the RBE within the AAV ITR of AAV2 identified cellular zinc-finger 5 protein (ZF5) as a negative regulator of AAV replication. It was shown to bind specifically to RRS motifs in vitro and in vivo [113]. Identification and characterization of proteins that bind to the ITR provide new insights into the AAV life cycle and could have utility in imprving aspects of AAV production as well as its performance as a gene therapy vector. Figure 3 displays elements of the AAV vector that omics studies have been applied to.

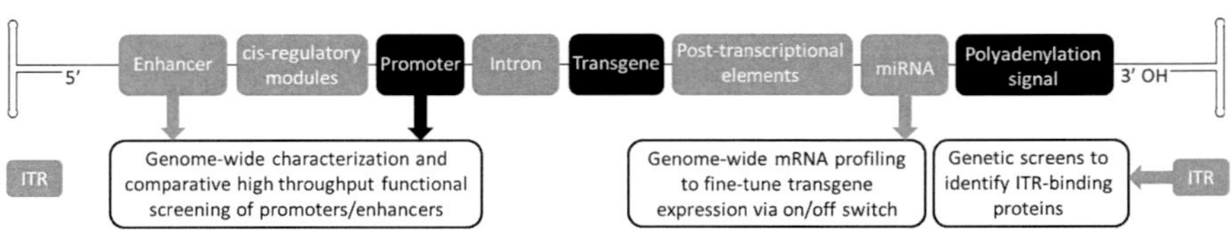

Fig. 3 Overview of a typical AAV expression vector, highlighting the components currently analyzed in published omics studies. The 3 elements detailed in black constitute the basis of any standard expression cassette; the 3 white boxes indicate components of the AAV expression vector that omics studies have been applied to. Adapted from [109].

2.6.3 *rep/cap* Plasmid Studies

A comprehensive understanding of the fundamental biology behind AAV and its host interactions is crucial in the ongoing optimization of AAV capsid selection. High-throughput dissection of AAV- host interactions has been accomplished in recent years through AAV capsid library screens [114]. These include techniques such as DNA shuffling, barcoded AAV vector evolution screening [115–117] and peptide display library screening [118–121]. These studies typically result in large libraries of synthetic AAV capsids that are then subjected to a combination of positive and negative selection pressures based on the desired outcome [122].

In one such approach, an AAV2 capsid fitness landscape was generated, to determine the effects of every single-codon substitution, insertion, and deletion in terms of in vivo delivery. This study demonstrated the importance of both surface-exposed and buried residues within the capsid proteins [123].

Although capsid engineering has been the main focus of recent optimization efforts, Rep interactions also impact the cellular proteins associated with the viral genome and so interrogation of these Rep protein interactions are being investigated through genomic [124, 125] and proteomic [126] screens in order to better understand the mechanism by which Rep proteins regulate AAV replication. Such rep/cap screening approaches facilitate the elimination of poor vectors and the optimization of promising ones, which is of particular importance in developing rAAV vectors for tailored gene therapy applications.

2.6.4 Isolation/Purification

Omics studies have a wide application and can be utilized at multiple stages of AAV production including that of QC. Standards of quality control during AAV vector production have advanced with the development of a Multiplex AAV Genotyping (MAG) assay. MAG is a powerful high-throughput tool being used to assess the purity and identity of AAV DNA plasmids and other starting materials used in production. The aim of MAG is to confirm the absence of cross-contaminating AAV sequences in plasmid stocks as well as end products using a high throughput screen. The assay is highly specific and has been used in serotype studies to specifically identify DNA from AAV serotypes 1–12 [127].

Another application of omic screening strategies is the comprehensive characterization of process and product related impurities found in rAAV stocks using an SSV-Seq technique. Measurement of impurities is imperative in any production system in order to assess the potential risks to future patients. As previously mentioned, during production rAAV capsids are known to internalize illegitimate DNA fragments in addition to their recombinant genome. These contaminants can come from plasmid or helper virus DNA as well as from the producer host cell itself and can lead to unwanted end results. A new method based on high-throughput sequencing identification and quantification of residual DNA in rAAV vector lots has been developed [128]. Contrary to qPCR, SSV-Seq (Single-Stranded DNA Virus

Sequencing) offers a nonselective approach to determine the percentage of each DNA contaminant and analyse rAAV vector genome identity.

The identification and characterization of the co-purifying cellular proteins in vector preparations can also be done using a combination of two proteomics approaches, GeLC-MS (gel electrophoresis liquid chromatography-mass spectrometry) and 2DE (two-dimensional gel electrophoresis). A 2014 proteomic study showed the detection of posttranslational modifications of capsid proteins using this method. A total of 13 cellular proteins were identified in the rAAV vectors, 9 by the GeLC-MS analysis and 4 by the 2DE analysis. Selected cellular proteins were verified and could be consistently found associated with different AAV serotypes carrying different transgenes [129].

The current standard QC methods often do not characterize the prevalence, compositions or structures of fragmented genomes and therefore preparations that have seemingly shown high genome homogeneity by gel electrophoresis have been later revealed to consist of less than 50% full-length species. Using single molecule real-time (SMRT) sequencing, researchers can now comprehensively profile packaged genomes as a single intact molecule and directly assess vector integrity without extensive preparation. This approach has been named AAV-genome population sequencing or AAV-GPseq. AAV-GPseq can reveal the accurate relative distribution of truncated genomes versus full-length genomes in vector preparations. This method used in AAV studies showed that vector populations can contain between 1.3% and 2.3% of this type of undesirable genome [130].These innovative screening techniques redefine quality control standards for viral vector preparation.

2.6.5 Patient Studies

The success of any gene therapy relies not only on a robust production system but also in vivo transduction efficiency, treatment safety, antibody evasion and cell specificity [122]. Despite applications for AAV vectors advancing fairly rapidly, in vivo performance continues to be a significant restriction to widespread rAAV applicability. Factors like transduction efficacy are proving major bottlenecks in the effort to translate new AAV approaches from the lab to a clinical setting. With the increasing number of clinical trials using rAAV vectors and a need for lower and safer vector doses, there is now a call from regulatory agencies for a more comprehensive characterization of process and product in order to generate effective cGMP compatible vectors [128]. Systems biology has the potential to identify the key drivers of therapeutic success, indicating which correlations between AAV and its host cell are most relevant. High throughput genomic screening and other omics approaches are being used to navigate the hazards and hurdles of in vivo treatment, allowing a better comprehension of AAV- host interaction that will directly influence bioreactor volume for production, therapy dosage and overall cost of a therapy [131].

2.6.6 'Back to the Future' Approach

While the power and potential of high throughput forward genetic screening approaches are undisputed, they are also inherently challenging, depending on library quality, fidelity and specificity. Strategies that look back in time, rather than forward, could alleviate some of the concerns associated with the creation of enhanced vectors for clinical production. Ancestral AAV sequence reconstruction is a reverse genetics approach that pursues the same goal as forward genetics in the race for the next generation of AAV vectors for production. This "back to the future" approach is a versatile innovative strategy being used to explore historical avenues in order to engineer proteins with novel or enhanced properties [132]. To gain insights into AAV's evolutionary history, a computationally designed putative ancestral AAV library was constructed. This study hypothesized that the divergent AAV phenotypes and structural elements underling AAV biology and pharmacology could be mapped by recreating the evolutionary lineage of the virus. Using ancestral sequence reconstruction, the amino acid sequence of ancestral AAV capsid monomers using a 'maximum likelihood' method was predicted. Evolutionary intermediates of the viral capsid using bioinformatics were designed and characterized by biological properties relevant to clinical applications. This effort led to the generation of nine functional putative ancestral AAVs. A highly potent in vivo gene therapy vector particle named Anc80 was also identified. This is believed to be a predicted ancestor of the widely studied AAV serotypes 1, 2, 8 and 9 [133].

3 Concluding Remarks

rAAV gene therapy is a very promising approach to treating genetic illnesses, particularly due to its safety. This safety is partially due to the rAAV genome persisting episomally, rather than integrating into the host cell's DNA. As a result, rAAV gene therapy can only provide temporary treatment for a condition, and patients will require multiple treatments throughout their lifetime. Given the enormous cost of these drugs and the difficulty in producing them, the current systems are inadequate for rAAV gene therapy to become widely used. In order for the field to advance and enable treatment of any genetic illness, rAAV production must become much more efficient and cost-effective.

A balance must be found between achieving high rAAV production levels while maintaining safety standards. For this reason, most helper virus-based methods are unlikely to be acceptable in pharmaceutical production (unless inactivation/removal of the helper virus can be guaranteed). Similarly, stable integration of *rep/cap* or helper genes could be dangerous due to the higher potential for recombination events to produce wild-type virus.

Many options have been tested to find an ideal production system for rAAVs. However, the HEK293/triple transfection system and the Sf9/Baculovirus system are still the favored methods due to their production output and safety. Further research should focus on optimizing production in these platforms.

In this regard, every aspect of the virus, the transgene and the producing cell must be examined to find potential avenues of improvement. New methods will be required in terms of transgene and vector optimization, cell line engineering, transfection improvement and downstream purification and processing. Ideally this would include a solution to the limited transgene size. Systems Biology and omics approaches, whose arsenal includes high-throughput experimental and computational studies that can account for the complexities of AAV-host interactions, holds significant promise in aiding the development and optimization of gene therapy production in the future.

While improvements in these areas would be very welcome, this is certainly not the only challenge in the field. Even if it were possible to cheaply produce very large amounts of rAAVs, it would be unwise to then administer large doses to patients, as this would increase the chances of an immune reaction. Therefore, the other side of the problem is to improve the administration of rAAV gene therapy, so that either a smaller amount is needed to achieve a therapeutic effect, or large amounts can be delivered safely.

Competing Interests All authors declare they have no conflict of interest.

Ethical Approval This article does not contain any studies with animals or human participants performed by any of the authors.

References

1. Kotterman MA, Schaffer DV (2014) Engineering adeno-associated viruses for clinical gene therapy. Nat Rev Genet 15:445–451. https://doi.org/10.1038/nrg3742
2. Li C, Samulski RJ (2020) Engineering adeno-associated virus vectors for gene therapy. Nat Rev Genet 21:255–272. https://doi.org/10.1038/s41576-019-0205-4
3. Naso MF, Tomkowicz B, Perry WL, Strohl WR (2017) Adeno-associated virus (AAV) as a vector for gene therapy. BioDrugs 31:317–334. https://doi.org/10.1007/s40259-017-0234-5
4. Im D-S, Muzyczka N (1990) The AAV origin binding protein Rep68 is an ATP-dependent site-specific endonuclease with DNA helicase activity. Cell 61:447–457. https://doi.org/10.1016/0092-8674(90)90526-K
5. Im DS, Muzyczka N (1992) Partial purification of adeno-associated virus Rep78, Rep52, and Rep40 and their biochemical characterization. J Virol 66:1119–1128
6. Wu Z, Asokan A, Samulski RJ (2006) Adeno-associated virus serotypes: vector toolkit for human gene therapy. Mol Ther J Am Soc Gene Ther 14:316–327. https://doi.org/10.1016/j.ymthe.2006.05.009
7. Hastie E, Samulski RJ (2015) Adeno-associated virus at 50: a golden anniversary of discovery, research, and gene therapy success–a personal perspective. Hum Gene Ther 26:257–265. https://doi.org/10.1089/hum.2015.025
8. Naumer M, Sonntag F, Schmidt K, Nieto K, Panke C, Davey NE, Popa-Wagner R, Kleinschmidt JA (2012) Properties of the adeno-associated virus assembly-activating protein. J Virol 86:13038–13048. https://doi.org/10.1128/JVI.01675-12

9. Earley LF, Powers JM, Adachi K, Baumgart JT, Meyer NL, Xie Q, Chapman MS, Nakai H (2017) Adeno-associated virus (AAV) assembly-activating protein is not an essential requirement for capsid assembly of AAV serotypes 4, 5, and 11. J Virol 91:e01980-16. https://doi.org/10.1128/JVI.01980-16
10. Rose JA, Hoggan MD, Shatkin AJ (1966) Nucleic acid from an adeno-associated virus: chemical and physical studies. Proc Natl Acad Sci U S A 56:86–92. https://doi.org/10.1073/pnas.56.1.86
11. Pillay S, Carette JE (2017) Host determinants of adeno-associated viral vector entry. Curr Opin Virol 24:124–131. https://doi.org/10.1016/j.coviro.2017.06.003
12. Bleker S, Sonntag F, Kleinschmidt JA (2005) Mutational analysis of narrow pores at the fivefold symmetry axes of adeno-associated virus type 2 capsids reveals a dual role in genome packaging and activation of phospholipase A2 activity. J Virol 79:2528–2540. https://doi.org/10.1128/JVI.79.4.2528-2540.2005
13. Kronenberg S, Böttcher B, von der Lieth CW, Bleker S, Kleinschmidt JA (2005) A conformational change in the adeno-associated virus type 2 capsid leads to the exposure of hidden VP1 N termini. J Virol 79:5296–5303. https://doi.org/10.1128/JVI.79.9.5296-5303.2005
14. Kotin RM, Siniscalco M, Samulski RJ, Zhu XD, Hunter L, Laughlin CA, McLaughlin S, Muzyczka N, Rocchi M, Berns KI (1990) Site-specific integration by adeno-associated virus. Proc Natl Acad Sci U S A 87:2211–2215. https://doi.org/10.1073/pnas.87.6.2211
15. McCarty DM, Young SM, Samulski RJ (2004) Integration of adeno-associated virus (AAV) and recombinant AAV vectors. Annu Rev Genet 38:819–845. https://doi.org/10.1146/annurev.genet.37.110801.143717
16. Samulski RJ, Zhu X, Xiao X, Brook JD, Housman DE, Epstein N, Hunter LA (1991) Targeted integration of adeno-associated virus (AAV) into human chromosome 19. EMBO J 10:3941–3950
17. Kotin RM, Linden RM, Berns KI (1992) Characterization of a preferred site on human chromosome 19q for integration of adeno-associated virus DNA by non-homologous recombination. EMBO J 11:5071–5078
18. Schnepp BC, Jensen RL, Chen C-L, Johnson PR, Clark KR (2005) Characterization of adeno-associated virus genomes isolated from human tissues. J Virol 79:14793–14803. https://doi.org/10.1128/JVI.79.23.14793-14803.2005
19. Drew HR, Lockett LJ, Both GW (2007) Increased complexity of wild-type adeno-associated virus-chromosomal junctions as determined by analysis of unselected cellular genomes. J Gen Virol 88:1722–1732. https://doi.org/10.1099/vir.0.82880-0
20. Deyle DR, Russell DW (2009) Adeno-associated virus vector integration. Curr Opin Mol Ther 11:442–447
21. Choi VW, McCarty DM, Samulski RJ (2006) Host cell DNA repair pathways in adeno-associated viral genome processing. J Virol 80:10346–10356. https://doi.org/10.1128/JVI.00841-06
22. Gao G, Vandenberghe LH, Alvira MR, Lu Y, Calcedo R, Zhou X, Wilson JM (2004) Clades of adeno-associated viruses are widely disseminated in human tissues. J Virol 78:6381–6388. https://doi.org/10.1128/JVI.78.12.6381-6388.2004
23. Summerford C, Samulski RJ (1998) Membrane-associated heparan sulfate proteoglycan is a receptor for adeno-associated virus type 2 virions. J Virol 72:1438–1445
24. Kashiwakura Y, Tamayose K, Iwabuchi K, Hirai Y, Shimada T, Matsumoto K, Nakamura T, Watanabe M, Oshimi K, Daida H (2005) Hepatocyte growth factor receptor is a coreceptor for adeno-associated virus type 2 infection. J Virol 79:609–614. https://doi.org/10.1128/JVI.79.1.609-614.2005
25. Asokan A, Hamra JB, Govindasamy L, Agbandje-McKenna M, Samulski RJ (2006) Adeno-associated virus type 2 contains an integrin alpha5beta1 binding domain essential for viral cell entry. J Virol 80:8961–8969. https://doi.org/10.1128/JVI.00843-06
26. Burger C, Gorbatyuk OS, Velardo MJ, Peden CS, Williams P, Zolotukhin S, Reier PJ, Mandel RJ, Muzyczka N (2004) Recombinant AAV viral vectors pseudotyped with viral capsids from serotypes 1, 2, and 5 display differential efficiency and cell tropism after delivery to different

regions of the central nervous system. Mol Ther J Am Soc Gene Ther 10:302–317. https://doi.org/10.1016/j.ymthe.2004.05.024
27. Walters RW, Yi SM, Keshavjee S, Brown KE, Welsh MJ, Chiorini JA, Zabner J (2001) Binding of adeno-associated virus type 5 to 2,3-linked sialic acid is required for gene transfer. J Biol Chem 276:20610–20616. https://doi.org/10.1074/jbc.M101559200
28. Davidson BL, Stein CS, Heth JA, Martins I, Kotin RM, Derksen TA, Zabner J, Ghodsi A, Chiorini JA (2000) Recombinant adeno-associated virus type 2, 4, and 5 vectors: transduction of variant cell types and regions in the mammalian central nervous system. Proc Natl Acad Sci U S A 97:3428–3432. https://doi.org/10.1073/pnas.050581197
29. Wu Z, Sun J, Zhang T, Yin C, Yin F, Van Dyke T, Samulski RJ, Monahan PE (2008) Optimization of self-complementary AAV vectors for liver-directed expression results in sustained correction of hemophilia B at low vector dose. Mol Ther J Am Soc Gene Ther 16:280–289. https://doi.org/10.1038/sj.mt.6300355
30. Ylä-Herttuala S (2012) Endgame: glybera finally recommended for approval as the first gene therapy drug in the European union. Mol Ther J Am Soc Gene Ther 20:1831–1832. https://doi.org/10.1038/mt.2012.194
31. Scott LJ (2015) Alipogene tiparvovec: a review of its use in adults with familial lipoprotein lipase deficiency. Drugs 75:175–182. https://doi.org/10.1007/s40265-014-0339-9
32. Stroes ES, Nierman MC, Meulenberg JJ, Remco F, Jaap T, Pieter HC, Maas MM, Zwinderman AH, Colin R, Eleonora A, High KA, Levi MM, Hayden MR, Kastelein JJ, Albert KJ (2008) Intramuscular administration of AAV1-lipoprotein LipaseS447X lowers triglycerides in lipoprotein lipase–deficient patients. Arterioscler Thromb Vasc Biol 28:2303–2304. https://doi.org/10.1161/ATVBAHA.108.175620
33. Keeler AM, Flotte TR (2019) Recombinant adeno-associated virus gene therapy in light of Luxturna (and Zolgensma and Glybera): where are we, and how did we get here? Annu Rev Virol 6:601–621. https://doi.org/10.1146/annurev-virology-092818-015530
34. DailyMed – ZOLGENSMA- onasemnogene abeparvovec-xioi kit. https://dailymed.nlm.nih.gov/dailymed/drugInfo.cfm?setid=68cd4f06-70e1-40d8-bedb-609ec0afa471. Accessed 21 Apr 2020
35. Dunbar CE, High KA, Joung JK, Kohn DB, Ozawa K, Sadelain M (2018) Gene therapy comes of age. Science 359:eaan4672. https://doi.org/10.1126/science.aan4672
36. EMA (2018) Luxturna. European Medcine Agency. https://www.ema.europa.eu/en/medicines/human/EPAR/luxturna. Accessed 21 Apr 2020
37. Wang D, Tai PWL, Gao G (2019) Adeno-associated virus vector as a platform for gene therapy delivery. Nat Rev Drug Discov 18:358–378. https://doi.org/10.1038/s41573-019-0012-9
38. Penaud-Budloo M, François A, Clément N, Ayuso E (2018) Pharmacology of recombinant adeno-associated virus production. Mol Ther Methods Clin Dev 8:166–180. https://doi.org/10.1016/j.omtm.2018.01.002
39. Kaemmerer WF (2018) How will the field of gene therapy survive its success? Bioeng Transl Med 3:166–177. https://doi.org/10.1002/btm2.10090
40. Spinal Muscular Atrophy. NORD National Organization for Rare Disorders. https://rarediseases.org/rare-diseases/spinal-muscular-atrophy/. Accessed 23 Apr 2020
41. Number of births and deaths per year. Our World Data. https://ourworldindata.org/grapher/births-and-deaths-projected-to-2100. Accessed 23 Apr 2020
42. CDC (2018) What is hemophilia? | CDC. Centers for Disease Control and Prevention. https://www.cdc.gov/ncbddd/hemophilia/facts.html. Accessed 1 May 2020
43. Brimble M, Zhou J, Morton C, Meagher M, Nathwani A, Gray J, Davidoff A (2016) AAV preparations contain contamination from DNA sequences in production plasmids directly outside of the ITRs. Mol Ther 24:S218
44. Hauck B, Murphy SL, Smith PH, Qu G, Liu X, Zelenaia O, Mingozzi F, Sommer JM, High KA, Wright JF (2009) Undetectable transcription of cap in a clinical AAV vector: implications for preformed capsid in immune responses. Mol Ther J Am Soc Gene Ther 17:144–152. https://doi.org/10.1038/mt.2008.227

45. Wright JF (2009) Transient transfection methods for clinical adeno-associated viral vector production. Hum Gene Ther 20:698–706. https://doi.org/10.1089/hum.2009.064
46. Grieger JC, Soltys SM, Samulski RJ (2016) Production of recombinant adeno-associated virus vectors using suspension HEK293 cells and continuous harvest of vector from the culture media for GMP FIX and FLT1 clinical vector. Mol Ther J Am Soc Gene Ther 24:287–297. https://doi.org/10.1038/mt.2015.187
47. Aslanidi G, Lamb K, Zolotukhin S (2009) An inducible system for highly efficient production of recombinant adeno-associated virus (rAAV) vectors in insect Sf9 cells. Proc Natl Acad Sci U S A 106:5059–5064. https://doi.org/10.1073/pnas.0810614106
48. Cecchini S, Virag T, Kotin RM (2011) Reproducible high yields of recombinant adeno-associated virus produced using invertebrate cells in 0.02- to 200-liter cultures. Hum Gene Ther 22:1021–1030. https://doi.org/10.1089/hum.2010.250
49. Thomas DL, Wang L, Niamke J, Liu J, Kang W, Scotti MM, Ye G, Veres G, Knop DR (2009) Scalable recombinant adeno-associated virus production using recombinant herpes simplex virus type 1 coinfection of suspension-adapted mammalian cells. Hum Gene Ther 20:861–870. https://doi.org/10.1089/hum.2009.004
50. Clément N, Knop DR, Byrne BJ (2009) Large-scale adeno-associated viral vector production using a herpesvirus-based system enables manufacturing for clinical studies. Hum Gene Ther 20:796–806. https://doi.org/10.1089/hum.2009.094
51. Mietzsch M, Grasse S, Zurawski C, Weger S, Bennett A, Agbandje-McKenna M, Muzyczka N, Zolotukhin S, Heilbronn R (2014) OneBac: platform for scalable and high-titer production of adeno-associated virus serotype 1–12 vectors for gene therapy. Hum Gene Ther 25:212–222. https://doi.org/10.1089/hum.2013.184
52. Graham FL, Smiley J, Russell WC, Nairn R (1977) Characteristics of a human cell line transformed by DNA from human adenovirus type 5. J Gen Virol 36:59–74. https://doi.org/10.1099/0022-1317-36-1-59
53. Robert M-A, Chahal PS, Audy A, Kamen A, Gilbert R, Gaillet B (2017) Manufacturing of recombinant adeno-associated viruses using mammalian expression platforms. Biotechnol J 12:1600193. https://doi.org/10.1002/biot.201600193
54. Kotin RM (2011) Large-scale recombinant adeno-associated virus production. Hum Mol Genet 20:R2–R6. https://doi.org/10.1093/hmg/ddr141
55. Clément N, Grieger JC (2016) Manufacturing of recombinant adeno-associated viral vectors for clinical trials. Mol Ther Methods Clin Dev 3:16002. https://doi.org/10.1038/mtm.2016.2
56. Wright JF (2008) Manufacturing and characterizing AAV-based vectors for use in clinical studies. Gene Ther 15:840–848. https://doi.org/10.1038/gt.2008.65
57. Invitrogen (2017) Growth and maintenance of insect cell lines: user guide. https://tools.thermofisher.com/content/sfs/manuals/Insect_Cell_Lines_UG.pdf. Accessed 21 Apr 2020
58. Kotin RM, Snyder RO (2017) Manufacturing clinical grade recombinant adeno-associated virus using invertebrate cell lines. Hum Gene Ther 28:350–360. https://doi.org/10.1089/hum.2017.042
59. Aponte-Ubillus JJ, Barajas D, Peltier J, Bardliving C, Shamlou P, Gold D (2018) Molecular design for recombinant adeno-associated virus (rAAV) vector production. Appl Microbiol Biotechnol 102:1045–1054. https://doi.org/10.1007/s00253-017-8670-1
60. Kondratov O, Marsic D, Crosson SM, Mendez-Gomez HR, Moskalenko O, Mietzsch M, Heilbronn R, Allison JR, Green KB, Agbandje-McKenna M, Zolotukhin S (2017) Direct head-to-head evaluation of recombinant adeno-associated viral vectors manufactured in human versus insect cells. Mol Ther 25:2661–2675. https://doi.org/10.1016/j.ymthe.2017.08.003
61. Bosma B, du Plessis F, Ehlert E, Nijmeijer B, de Haan M, Petry H, Lubelski J (2018) Optimization of viral protein ratios for production of rAAV serotype 5 in the baculovirus system. Gene Ther 25:415–424. https://doi.org/10.1038/s41434-018-0034-7
62. Urabe M, Ding C, Kotin RM (2002) Insect cells as a factory to produce adeno-associated virus type 2 vectors. Hum Gene Ther 13:1935–1943. https://doi.org/10.1089/10430340260355347

63. Smith RH, Levy JR, Kotin RM (2009) A simplified baculovirus-AAV expression vector system coupled with one-step affinity purification yields high-titer rAAV stocks from insect cells. Mol Ther J Am Soc Gene Ther 17:1888–1896. https://doi.org/10.1038/mt.2009.128
64. Wu Y, Jiang L, Geng H, Yang T, Han Z, He X, Lin K, Xu F (2018) A recombinant baculovirus efficiently generates recombinant adeno-associated virus vectors in cultured insect cells and larvae. Mol Ther Methods Clin Dev 10:38–47. https://doi.org/10.1016/j.omtm.2018.05.005
65. Wu Y, Mei T, Jiang L, Han Z, Dong R, Yang T, Xu F (2019) Development of versatile and flexible Sf9 packaging cell line-dependent OneBac system for large-scale recombinant adeno-associated virus production. Hum Gene Ther Methods 30:172–183. https://doi.org/10.1089/hgtb.2019.123
66. Schmidt M, Afione S, Kotin RM (2000) Adeno-associated virus type 2 Rep78 induces apoptosis through caspase activation independently of p53. J Virol 74:9441–9450. https://doi.org/10.1128/jvi.74.20.9441-9450.2000
67. Thorne BA, Takeya RK, Peluso RW (2009) Manufacturing recombinant adeno-associated viral vectors from producer cell clones. Hum Gene Ther 20:707–714. https://doi.org/10.1089/hum.2009.070
68. Wang Q, Wu Z, Zhang J, Firrman J, Wei H, Zhuang Z, Liu L, Miao L, Hu Y, Li D, Diao Y, Xiao W (2017) A robust system for production of superabundant VP1 recombinant AAV vectors. Mol Ther Methods Clin Dev 7:146–156. https://doi.org/10.1016/j.omtm.2017.11.002
69. Wang Z, Cheng F, Engelhardt JF, Yan Z, Qiu J (2018) Development of a novel recombinant adeno-associated virus production system using human Bocavirus 1 helper genes. Mol Ther Methods Clin Dev 11:40–51. https://doi.org/10.1016/j.omtm.2018.09.005
70. Wang Z, Deng X, Zou W, Engelhardt JF, Yan Z, Qiu J (2017) Human Bocavirus 1 is a novel helper for adeno-associated virus replication. J Virol 91:e00710-17. https://doi.org/10.1128/JVI.00710-17
71. Conway JE, Zolotukhin S, Muzyczka N, Hayward GS, Byrne BJ (1997) Recombinant adeno-associated virus type 2 replication and packaging is entirely supported by a herpes simplex virus type 1 amplicon expressing rep and cap. J Virol 71:8780–8789
72. Knop DR et al (2011) 593. SFM rHSV-based production of rAAV vectors in suspension BHK cells. Mol Ther 19:S227. https://doi.org/10.1016/S1525-0016(16)37166-0
73. Adamson-Small L, Potter M, Falk DJ, Cleaver B, Byrne BJ, Clément N (2016) A scalable method for the production of high-titer and high-quality adeno-associated type 9 vectors using the HSV platform. Mol Ther Methods Clin Dev 3:16031. https://doi.org/10.1038/mtm.2016.31
74. Backovic A, Cervelli T, Salvetti A, Zentilin L, Giacca M, Galli A (2012) Capsid protein expression and adeno-associated virus like particles assembly in Saccharomyces cerevisiae. Microb Cell Factories 11:124. https://doi.org/10.1186/1475-2859-11-124
75. Barajas D, Aponte-Ubillus JJ, Akeefe H, Cinek T, Peltier J, Gold D (2017) Generation of infectious recombinant adeno-associated virus in Saccharomyces cerevisiae. PLoS One 12:e0173010. https://doi.org/10.1371/journal.pone.0173010
76. Aponte J (2018) Optimization of recombinant adeno-associated virus (aav) vector production in saccharomyces cerevisiae. KGI Theses Diss
77. Martin J, Frederick A, Luo Y, Jackson R, Joubert M, Sol B, Poulin F, Pastor E, Armentano D, Wadsworth S, Vincent K (2013) Generation and characterization of adeno-associated virus producer cell lines for research and preclinical vector production. Hum Gene Ther Methods 24:253–269. https://doi.org/10.1089/hgtb.2013.046
78. Qiao C, Li J, Skold A, Zhang X, Xiao X (2002) Feasibility of generating adeno-associated virus packaging cell lines containing inducible adenovirus helper genes. J Virol 76:1904–1913. https://doi.org/10.1128/jvi.76.4.1904-1913.2002
79. Samulski RJ, Chang LS, Shenk T (1987) A recombinant plasmid from which an infectious adeno-associated virus genome can be excised in vitro and its use to study viral replication. J Virol 61:3096–3101. https://doi.org/10.1128/JVI.61.10.3096-3101.1987
80. Savy A, Dickx Y, Nauwynck L, Bonnin D, Merten O-W, Galibert L (2017) Impact of inverted terminal repeat integrity on rAAV8 production using the Baculovirus/Sf9 cells system. Hum Gene Ther Methods 28:277–289. https://doi.org/10.1089/hgtb.2016.133

81. Earley LF, Conatser LM, Lue VM, Dobbins AL, Li C, Hirsch ML, Samulski RJ (2020) Adeno-associated virus serotype-specific inverted terminal repeat sequence role in vector transgene expression. Hum Gene Ther 31:151–162. https://doi.org/10.1089/hum.2019.274
82. Bishop BM, Santin AD, Quirk JG, Hermonat PL (1996) Role of the terminal repeat GAGC trimer, the major Rep78 binding site, in adeno-associated virus DNA replication. FEBS Lett 397:97–100. https://doi.org/10.1016/s0014-5793(96)01149-0
83. McCarty DM, Fu H, Monahan PE, Toulson CE, Naik P, Samulski RJ (2003) Adeno-associated virus terminal repeat (TR) mutant generates self-complementary vectors to overcome the rate-limiting step to transduction in vivo. Gene Ther 10:2112–2118. https://doi.org/10.1038/sj.gt.3302134
84. Plotkin JB, Robins H, Levine AJ (2004) Tissue-specific codon usage and the expression of human genes. Proc Natl Acad Sci U S A 101:12588–12591. https://doi.org/10.1073/pnas.0404957101
85. Quax TEF, Claassens NJ, Söll D, van der Oost J (2015) Codon bias as a means to fine-tune gene expression. Mol Cell 59:149–161. https://doi.org/10.1016/j.molcel.2015.05.035
86. Greig JA, Wang Q, Reicherter AL, Chen S-J, Hanlon AL, Tipper CH, Clark KR, Wadsworth S, Wang L, Wilson JM (2017) Characterization of adeno-associated viral vector-mediated human factor VIII gene therapy in hemophilia A mice. Hum Gene Ther 28:392–402. https://doi.org/10.1089/hum.2016.128
87. Yan Z, Sun X, Feng Z, Li G, Fisher JT, Stewart ZA, Engelhardt JF (2015) Optimization of recombinant adeno-associated virus-mediated expression for large transgenes, using a synthetic promoter and tandem array enhancers. Hum Gene Ther 26:334–346. https://doi.org/10.1089/hum.2015.001
88. Chuah MK, Petrus I, De Bleser P, Le Guiner C, Gernoux G, Adjali O, Nair N, Willems J, Evens H, Rincon MY, Matrai J, Di Matteo M, Samara-Kuko E, Yan B, Acosta-Sanchez A, Meliani A, Cherel G, Blouin V, Christophe O, Moullier P, Mingozzi F, VandenDriessche T (2014) Liver-specific transcriptional modules identified by genome-wide in silico analysis enable efficient gene therapy in mice and non-human primates. Mol Ther J Am Soc Gene Ther 22:1605–1613. https://doi.org/10.1038/mt.2014.114
89. Rincon MY, Sarcar S, Danso-Abeam D, Keyaerts M, Matrai J, Samara-Kuko E, Acosta-Sanchez A, Athanasopoulos T, Dickson G, Lahoutte T, De Bleser P, VandenDriessche T, Chuah MK (2015) Genome-wide computational analysis reveals cardiomyocyte-specific transcriptional Cis-regulatory motifs that enable efficient cardiac gene therapy. Mol Ther J Am Soc Gene Ther 23:43–52. https://doi.org/10.1038/mt.2014.178
90. Duan D, Yue Y, Engelhardt JF (2001) Expanding AAV packaging capacity with trans-splicing or overlapping vectors: a quantitative comparison. Mol Ther J Am Soc Gene Ther 4:383–391. https://doi.org/10.1006/mthe.2001.0456
91. Hirsch ML, Wolf SJ, Samulski RJ (2016) Delivering transgenic DNA exceeding the carrying capacity of AAV vectors. Methods Mol Biol Clifton NJ 1382:21–39. https://doi.org/10.1007/978-1-4939-3271-9_2
92. Hirsch ML, Li C, Bellon I, Yin C, Chavala S, Pryadkina M, Richard I, Samulski RJ (2013) Oversized AAV transductifon is mediated via a DNA-PKcs-independent, Rad51C-dependent repair pathway. Mol Ther J Am Soc Gene Ther 21:2205–2216. https://doi.org/10.1038/mt.2013.184
93. Ghosh A, Yue Y, Lai Y, Duan D (2008) A hybrid vector system expands adeno-associated viral vector packaging capacity in a transgene-independent manner. Mol Ther J Am Soc Gene Ther 16:124–130. https://doi.org/10.1038/sj.mt.6300322
94. Fakhiri J, Schneider MA, Puschhof J, Stanifer M, Schildgen V, Holderbach S, Voss Y, El Andari J, Schildgen O, Boulant S, Meister M, Clevers H, Yan Z, Qiu J, Grimm D (2019) Novel chimeric gene therapy vectors based on adeno-associated virus and four different mammalian bocaviruses. Mol Ther Methods Clin Dev 12:202–222. https://doi.org/10.1016/j.omtm.2019.01.003
95. Tornabene P, Trapani I, Minopoli R, Centrulo M, Lupo M, de Simone S, Tiberi P, Dell'Aquila F, Marrocco E, Iodice C, Iuliano A, Gesualdo C, Rossi S, Giaquinto L, Albert S, Hoyng

CB, Polishchuk E, Cremers FPM, Surace EM, Simonelli F, De Matteis MA, Polishchuk R, Auricchio A (2019) Intein-mediated protein trans-splicing expands adeno-associated virus transfer capacity in the retina. Sci Transl Med 11:eaav4523. https://doi.org/10.1126/scitranslmed.aav4523

96. Strobel B, Klauser B, Hartig JS, Lamla T, Gantner F, Kreuz S (2015) Riboswitch-mediated attenuation of transgene cytotoxicity increases adeno-associated virus vector yields in HEK-293 cells. Mol Ther J Am Soc Gene Ther 23:1582–1591. https://doi.org/10.1038/mt.2015.123
97. Nomura Y, Zhou L, Miu A, Yokobayashi Y (2013) Controlling mammalian gene expression by allosteric hepatitis delta virus ribozymes. ACS Synth Biol 2:684–689. https://doi.org/10.1021/sb400037a
98. Maunder HE, Wright J, Kolli BR, Vieira CR, Mkandawire TT, Tatoris S, Kennedy V, Iqball S, Devarajan G, Ellis S, Lad Y, Clarkson NG, Mitrophanous KA, Farley DC (2017) Enhancing titres of therapeutic viral vectors using the transgene repression in vector production (TRiP) system. Nat Commun 8:14834. https://doi.org/10.1038/ncomms14834
99. Reid CA, Boye SL, Hauswirth WW, Lipinski DM (2017) miRNA-mediated post-transcriptional silencing of transgenes leads to increased adeno-associated viral vector yield and targeting specificity. Gene Ther 24:462–469. https://doi.org/10.1038/gt.2017.50
100. Geisler A, Fechner H (2016) MicroRNA-regulated viral vectors for gene therapy. World J Exp Med 6:37–54. https://doi.org/10.5493/wjem.v6.i2.37
101. Guimaro MC, Afione S, Tanaka T, Chiorini J (2020) Rescue of AAV production by shRNA cotransfection. Hum Gene Ther 31:1068. https://doi.org/10.1089/hum.2019.249
102. Warrington KH, Gorbatyuk OS, Harrison JK, Opie SR, Zolotukhin S, Muzyczka N (2004) Adeno-associated virus type 2 VP2 capsid protein is nonessential and can tolerate large peptide insertions at its N terminus. J Virol 78:6595–6609. https://doi.org/10.1128/JVI.78.12.6595-6609.2004
103. Ried MU, Girod A, Leike K, Büning H, Hallek M (2002) Adeno-associated virus capsids displaying immunoglobulin-binding domains permit antibody-mediated vector retargeting to specific cell surface receptors. J Virol 76:4559–4566. https://doi.org/10.1128/JVI.76.9.4559-4566.2002
104. Münch RC, Muth A, Muik A, Friedel T, Schmatz J, Dreier B, Trkola A, Plückthun A, Büning H, Buchholz CJ (2015) Off-target-free gene delivery by affinity-purified receptor-targeted viral vectors. Nat Commun 6:6246. https://doi.org/10.1038/ncomms7246
105. Münch RC, Janicki H, Völker I, Rasbach A, Hallek M, Büning H, Buchholz CJ (2013) Displaying high-affinity ligands on adeno-associated viral vectors enables tumor cell-specific and safe gene transfer. Mol Ther J Am Soc Gene Ther 21:109–118. https://doi.org/10.1038/mt.2012.186
106. Zhong L, Li B, Jayandharan G, Mah CS, Govindasamy L, Agbandje-McKenna M, Herzog RW, Weigel-Van Aken KA, Hobbs JA, Zolotukhin S, Muzyczka N, Srivastava A (2008) Tyrosine phosphorylation of AAV2 vectors and its consequences on viral intracellular trafficking and transgene expression. Virology 381:194–202. https://doi.org/10.1016/j.virol.2008.08.027
107. Zhong L, Li B, Mah CS, Govindasamy L, Agbandje-McKenna M, Cooper M, Herzog RW, Zolotukhin I, Warrington KH, Weigel-Van Aken KA, Hobbs JA, Zolotukhin S, Muzyczka N, Srivastava A (2008) Next generation of adeno-associated virus 2 vectors: point mutations in tyrosines lead to high-efficiency transduction at lower doses. Proc Natl Acad Sci U S A 105:7827–7832. https://doi.org/10.1073/pnas.0802866105
108. Rabinowitz JE, Rolling F, Li C, Conrath H, Xiao W, Xiao X, Samulski RJ (2002) Cross-packaging of a single adeno-associated virus (AAV) type 2 vector genome into multiple AAV serotypes enables transduction with broad specificity. J Virol 76:791–801. https://doi.org/10.1128/JVI.76.2.791-801.2002
109. Domenger C, Grimm D (2019) Next-generation AAV vectors – don't judge a virus (only) by its cover. Hum Mol Genet 28:R3. https://doi.org/10.1093/hmg/ddz148
110. Dao LTM, Galindo-Albarrán AO, Castro-Mondragon JA, Andrieu-Soler C, Medina-Rivera A, Souaid C, Charbonnier G, Griffon A, Vanhille L, Stephen T, Alomairi J, Martin D,

Torres M, Fernandez N, Soler E, van Helden J, Puthier D, Spicuglia S (2017) Genome-wide characterization of mammalian promoters with distal enhancer functions. Nat Genet 49:1073–1081. https://doi.org/10.1038/ng.3884

111. Nguyen TA, Jones RD, Snavely AR, Pfenning AR, Kirchner R, Hemberg M, Gray JM (2016) High-throughput functional comparison of promoter and enhancer activities. Genome Res 26:1023–1033. https://doi.org/10.1101/gr.204834.116
112. Brown BD, Venneri MA, Zingale A, Sergi Sergi L, Naldini L (2006) Endogenous microRNA regulation suppresses transgene expression in hematopoietic lineages and enables stable gene transfer. Nat Med 12:585–591. https://doi.org/10.1038/nm1398
113. Cathomen T, Stracker TH, Gilbert LB, Weitzman MD (2001) A genetic screen identifies a cellular regulator of adeno-associated virus. Proc Natl Acad Sci U S A 98:14991–14996. https://doi.org/10.1073/pnas.261567198
114. Herrmann A-K, Grimm D (2018) High-throughput dissection of AAV-host interactions: the fast and the curious. J Mol Biol 430:2626–2640. https://doi.org/10.1016/j.jmb.2018.05.022
115. Davidsson M, Wang G, Aldrin-Kirk P, Cardoso T, Nolbrant S, Hartnor M, Mudannayake J, Parmar M, Björklund T (2019) A systematic capsid evolution approach performed in vivo for the design of AAV vectors with tailored properties and tropism. Proc Natl Acad Sci 116:27053–27062. https://doi.org/10.1073/pnas.1910061116
116. de Alencastro G, Pekrun K, Valdmanis P, Tiffany M, Xu J, Kay MA (2020) Tracking adeno-associated virus capsid evolution by high-throughput sequencing. Hum Gene Ther 31:553. https://doi.org/10.1089/hum.2019.339
117. Pekrun K, Alencastro GD, Luo Q-J, Liu J, Kim Y, Nygaard S, Galivo F, Zhang F, Song R, Tiffany MR, Xu J, Hebrok M, Grompe M, Kay MA (2019) Using a barcoded AAV capsid library to select for clinically relevant gene therapy vectors. JCI Insight 4:e131610. https://doi.org/10.1172/jci.insight.131610
118. Körbelin J, Hunger A, Alawi M, Sieber T, Binder M, Trepel M (2017) Optimization of design and production strategies for novel adeno-associated viral display peptide libraries. Gene Ther 24:470–481. https://doi.org/10.1038/gt.2017.51
119. Märsch S, Huber A, Hallek M, Büning H, Perabo L (2010) A novel directed evolution method to enhance cell-type specificity of adeno-associated virus vectors. Comb Chem High Throughput Screen 13:807–812. https://doi.org/10.2174/138620710792927385
120. Varadi K, Michelfelder S, Korff T, Hecker M, Trepel M, Katus HA, Kleinschmidt JA, Müller OJ (2012) Novel random peptide libraries displayed on AAV serotype 9 for selection of endothelial cell-directed gene transfer vectors. Gene Ther 19:800–809. https://doi.org/10.1038/gt.2011.143
121. Zhang L, Rossi A, Lange L, Meumann N, Koitzsch U, Christie K, Nesbit MA, Moore CBT, Hacker UT, Morgan M, Hoffmann D, Zengel J, Carette JE, Schambach A, Salvetti A, Odenthal M, Büning H (2019) Capsid engineering overcomes barriers toward adeno-associated virus vector-mediated transduction of endothelial cells. Hum Gene Ther 30:1284–1296. https://doi.org/10.1089/hum.2019.027
122. Grimm D, Büning H (2017) Small but increasingly mighty: latest advances in AAV vector research, design, and evolution. Hum Gene Ther 28:1075–1086. https://doi.org/10.1089/hum.2017.172
123. Ogden PJ, Kelsic ED, Sinai S, Church GM (2019) Comprehensive AAV capsid fitness landscape reveals a viral gene and enables machine-guided design. Science 366:1139–1143. https://doi.org/10.1126/science.aaw2900
124. Brown KE, Muramatsu S, Handa A, Kajigaya S (1998) Transcription-positive cofactor 4 enhances rescue of adeno-associated virus genome from an infectious clone. J Gen Virol 79:2157–2161. https://doi.org/10.1099/0022-1317-79-9-2157
125. Weger S, Wendland M, Kleinschmidt JA, Heilbronn R (1999) The adeno-associated virus type 2 regulatory proteins Rep78 and Rep68 interact with the transcriptional coactivator PC4. J Virol 73:260–269
126. Smith-Moore S, Neil SJD, Fraefel C, Linden RM, Bollen M, Rowe HM, Henckaerts E (2018) Adeno-associated virus rep proteins antagonize phosphatase PP1 to counteract KAP1

repression of the latent viral genome. Proc Natl Acad Sci U S A 115:E3529–E3538. https://doi.org/10.1073/pnas.1721883115

127. Schmitt M, Pawlita M, Kleinschmidt J (2010) Genotyping of AAV plasmid stocks: quality control in adeno-associated virus vector production. Microb Physiol 19:204–212. https://doi.org/10.1159/000321449
128. Lecomte E, Leger A, Penaud-Budloo M, Ayuso E (2019) Single-stranded DNA virus sequencing (SSV-Seq) for characterization of residual DNA and AAV vector genomes. Methods Mol Biol Clifton NJ 1950:85–106. https://doi.org/10.1007/978-1-4939-9139-6_5
129. Dong B, Duan X, Chow HY, Chen L, Lu H, Wu W, Hauck B, Wright F, Kapranov P, Xiao W (2014) Proteomics analysis of co-purifying cellular proteins associated with rAAV vectors. PLoS One 9:e86453. https://doi.org/10.1371/journal.pone.0086453
130. Tai PWL, Xie J, Fong K, Seetin M, Heiner C, Su Q, Weiand M, Wilmot D, Zapp ML, Gao G (2018) Adeno-associated virus genome population sequencing achieves full vector genome resolution and reveals human-vector chimeras. Mol Ther Methods Clin Dev 9:130–141. https://doi.org/10.1016/j.omtm.2018.02.002
131. Wallen AJ, Barker GA, Fein DE, Jing H, Diamond SL (2011) Enhancers of adeno-associated virus AAV2 transduction via high throughput siRNA screening. Mol Ther 19:1152–1160. https://doi.org/10.1038/mt.2011.4
132. Weinmann J, Grimm D (2017) Next-generation AAV vectors for clinical use: an ever-accelerating race. Virus Genes 53:707–713. https://doi.org/10.1007/s11262-017-1502-7
133. Zinn E, Pacouret S, Khaychuk V, Turunen HT, Carvalho LS, Andres-Mateos E, Shah S, Shelke R, Maurer AC, Plovie E, Xiao R, Vandenberghe LH (2015) In silico reconstruction of the viral evolutionary lineage yields a potent gene therapy vector. Cell Rep 12:1056–1068. https://doi.org/10.1016/j.celrep.2015.07.019

Part IV
Cell culture process engineering, advanced bioreactor concepts, modelling, monitoring and control

Practical Considerations for the Scale-Up of Chinese Hamster Ovary (CHO) Cell Cultures

Lucas Lemire, Phuong Lan Pham, Yves Durocher, and Olivier Henry

Nomenclature

Symbol	Description	Units
ε	Eddy dissipation rate	m^2/s^3
ν	Kinetic viscosity	m/s
α	Empirical coefficient	–
β	Empirical coefficient	–
ϕ_g	Gas hold-up	–
α_G	Gas volume fraction	–
υ_s	Superficial gas velocity	m/s
a	Specific bubble surface area	m^2/m^3
A_c	Tank cross sectional area	m^2
CER	Carbon evolution rate	mmol/L/h
CTR	Carbon transfer rate	mmol/L/h
D_i	Impeller diameter	m
D_L	Gas diffusivity coefficient	m^2/s
d_b	Bubble diameter	m
D_t	Tank diameter	m
H/D	Height to diameter aspect ratio	–
K	Proportionality coefficient	–

(continued)

L. Lemire · O. Henry (✉)
Department of Chemical Engineering, Polytechnique Montréal, Montréal, QC, Canada
e-mail: olivier.henry@polymtl.ca

P. L. Pham · Y. Durocher
Human Health Therapeutics Research Center, National Research Council Canada, Montreal, QC, Canada

R. Pörtner (ed.), *Cell Culture Engineering and Technology*, Cell Engineering 10,
https://doi.org/10.1007/978-3-030-79871-0_12

Symbol	Description	Units
k_L	Film mass transfer coefficient	m/s
k_La	Gas mass transfer coefficient	s^{-1}
N	Mixing speed	RPM
N_P	Power number	–
OTR	Oxygen transfer rate	mmol/L/h
OUR	Oxygen uptake rate	mmol/L/h
P	Power input	W
P/V	Volumetric power input	W/m^3
pCO_2	Carbon dioxide partial pressure	–
qCO_2	Specific carbon dioxide production rate	mmol/cell/h
qO_2	Specific oxygen consumption rate	mmol/cell/h
Qg	Gas flow rate	m^3/s
Re	Reynolds number	–
t_m	Mixing time	s
t_r	Gas residence time	s
V	Volume	m^3
VCD	Viable cell density	cell/mL
V_t	Impeller tip speed	m/s
vvm	Volumetric gas flow rate per volume of liquid	min^{-1}

1 Introduction

It is expected that the biopharmaceutical market, which includes monoclonal antibodies (mAbs), vaccines, interferons, hormones and growth factors, will have a compound annual growth rate of 13.8% between the years 2018 and 2025 allowing markets to increase from $186 billion in 2017 to $526 billion in 2025. Monoclonal antibodies continue to dominate the global biopharmaceutical market [1]. In 2018, six monoclonal antibody products (adalimumab, nivolumab, pembrolizumab, trastuzumab, bevacizumab, and rituximab) had sales exceeding 6 billion USD [2]. Furthermore, new drugs are continually being approved by the FDA, with 18 new mAb products approved in the US in 2018–2019 [2]. These monoclonal antibodies have been used to treat osteoporosis (Romosozumab), complications in sickle-cell anemia (Crizanlizumab), cancer (Polatuzumab), and many other medical conditions [2] (Table 1). More recently, novel monoclonal antibodies have been discovered which could be used to block SARS-CoV-2 infections (human 47D11 mAb) [3]. Many antibodies which could potentially be used for treatment of SARS-Cov-2 Infection have entered clinical trials, including Regeneron's REGN-COV2 dual-antibody cocktail [4], AstraZeneca's AZD7442 dual-antibody cocktail [5], and Eli Lily, Amgen, and AbCellera's bamlanivimab (LY-CoV-555) entering clinical trials. The expiration of blockbuster biotherapeutics is also spurring the emergence of biosimilars, which will continue to drive the demand for product development and manufacturing capacity.

Table -1 Recently FDA-approved monoclonal antibody therapeutics

Monoclonal antibody	Brand name	Developer	Targeted disease	Approval year
Crizanlizumab	Adakveo	Novartis Pharmaceutical Corp.	Sickle cell disease	2019
Brolucizumab	Beovu	Novartis Pharmaceutical Corp.	Macular degeneration	2019
Polatuzumab vedotin	Polivy	Roche, F. Hoffmann-La Roche, Ltd.	Diffuse large B-cell lymphoma	2019
Risankizumab	Skyrizi	Boehringer Ingelheim Pharmaceuticals/ AbbVie Inc.	Plaque psoriasis	2019
Romosozumab	Evenity	Amgen/UCB	Osteoporosis in postmenopausal women at increased risk of fracture	2019
Caplacizumab	Cablivi	Ablynx	Acquired thrombotic thrombocytopenic purpura	2019
Rvulizumab	Ultomiris	Alexion Pharmaceuticals Inc.	Paroxysmal nocturnal hemoglobinuria	2018
Moxetumomab pasudodox	Lumoxiti	MedImmune/AstraZeneca	Hairy cell leukemia	2018
Ibalizumab	Trogarzo	Taimed Biologics Inc./ Theratechnologies Inc.	HIV infection	2018
Fremanezumab	Ajovy	Teva Pharmaceutical Industries, Ltd.	Migraine prevention	2018
Emapalumab	Gamifant	NovImmmune	Primary hemophagocytic lymphohistiocytosis	2018
Cemiplimab	Libtayo	Regeneron Pharmaceuticals Inc.	Cutaneous squamous cell carcinoma	2018
Tildrakizumab	Ilumaya	Merck & Co. Inc./Sun Pharmaceutical Industries, Ltd.	Plaque psoriasis	2018
Galcanezumab	Emgality	Eli Lilly	Migraine prevention	2018
Erenumab	Aimovig	Novartis	Migraine prevention	2018
Mogamulizumab	Poteligeo	Kyowa Hakko Kirin	Mycosis fungoides or Sézary syndrome	2018
Lanadelumab	Takhzyro	Dyax Corp.	Hereditary angioedema attacks	2018
Burosumab	Crysvita	Kyowa Hakko Kirin/Ultragenyx Pharmaceutical Inc.	X-linked hypophosphatemia	2018

Table adapted from Lu et al. [2]

1.1 CHO Cells: The Workhorse of Industrial Protein Production

Various expression systems can be used to produce recombinant proteins: bacteria, yeasts, filamentous fungi, insect cells, plant cells and mammalian cells. However, mammalian cell cultures, more specifically Chinese Hamster Ovary (CHO) cells, remain the dominant platform for the large-scale production of recombinant proteins used in clinical applications. Mammalian cells are favoured over other expression systems mainly for their unmatched ability to produce gram quantities of biologics with humanlike post-translational modifications [6, 7]. CHO cell cultures have been extensively optimized and can now routinely achieve yields of ~3 g/L [8] and sometimes up to 10 g/L [9, 10]. Another factor contributing to the popularity of CHO cells is their ability to grow well in suspension, which allows better control of the culture environment.

Proper mixing to obtain homogeneous culture conditions is of importance to maintain a desirable productive metabolic state. However, cell culture homogeneity cannot always be achieved by increasing mixing speed alone, due to concerns related to the level of shear stress imposed on the cells. The lack of cell walls makes animal cells much more sensitive to shear stress than their microbial counterparts. CHO cells nonetheless exhibit a relatively high tolerance to variations in temperature, pH, DO, and pressure, making them suitable for large-scale biomanufacturing applications.

1.2 Scale-Up, Scale-Out and Scale-Down in Process Development

The establishment of a cell culture process starts with cell line development, during which CHO cells are transfected with the desired gene to form a stable transfected cell pool. Individual cells are isolated from the stable cell pool, from which a production clone is selected based on growth and productivity characteristics, as well as product quality attributes. These clone screening steps have traditionally been performed in simple small-scale culture systems such as shake flasks, static flasks, microtiter plates or deep-well plates at a volume typically lower than 20 mL. However, fully instrumented mini-bioreactors are becoming increasingly available and have recently gained in popularity, such that plates and flasks are rapidly being replaced in many steps of upstream development [11]. Once a suitable candidate production clone is selected, process development from lab- to industrial-scale productions is typically done stepwise (Fig. 1). The cell culture process is first transferred to benchtop bioreactors (1 L–10 L) for optimization of the process conditions. Benchtop bioreactors are used as a process prototype to further optimize various culture conditions, including pH, temperature, dissolved oxygen tension (DO), medium, feeding regime, culture additives (e.g. antifoam

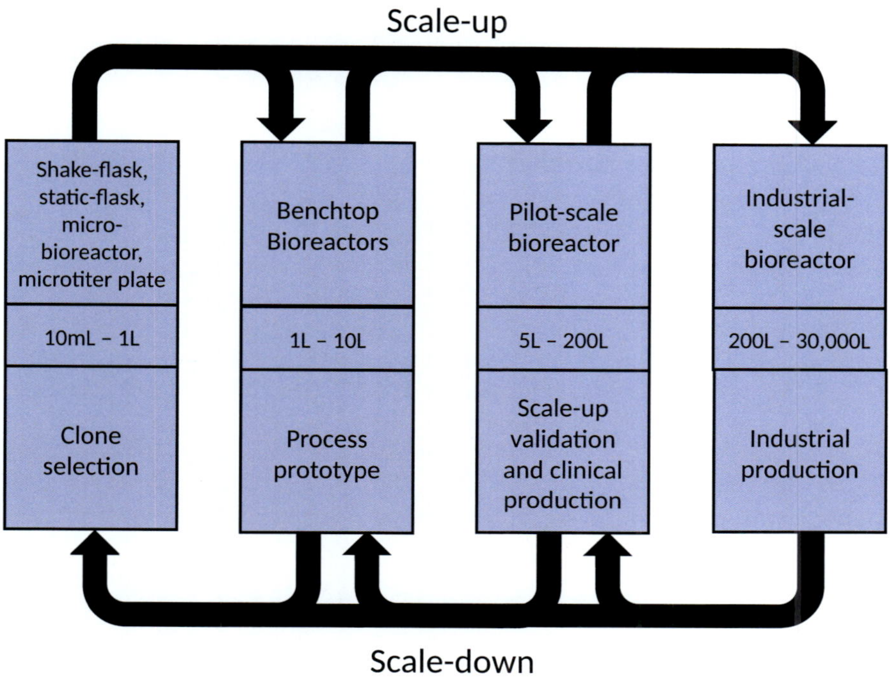

Fig. 1 Cell culture process development and scale-up

and surfactants), agitation rate and the aeration strategy. The cell culture process is then transferred to pilot-scale bioreactors (50 L–200 L) before finally being scaled-up to commercial-scale stainless-steel bioreactors (250 L–30,000 L) or single-use bioreactors (250 L–2000 L) for mass production. Process scale-up is a crucial non-trivial step, due to the many issues that may be encountered at manufacturing-scale, including culture environment heterogeneity, dissolved CO_2 accumulation, as well as excessive shear stress related to aeration and mixing conditions. Most of these problems should ideally be addressed and resolved at the pilot-scale before translating to commercial-scale production. Pilot-scale bioreactors are also typically used to supply the biological materials needed for preclinical and clinical trials. Throughout culture scale-up, scale-down models, typically 2 L–5 L benchtop reactors or mini-bioreactors, are used for troubleshooting, as well as for process understanding and optimization.

Another method of increasing production in lieu of scale-up is the scale-out approach. Scale-out consists of using multiple bioreactors rather than increasing the bioreactor volume (e.g. using five 2000 L bioreactors rather than a single 10,000 L bioreactor). Scale-out reduces risk associated with process scale-up and provides greater flexibility to cope with changes in product demand. However, the use of multiple culture vessels inherently introduces a greater source of variability which may negatively impact culture performance and, in turn, product consistency [12].

1.3 The Scale-Up Challenge

Although benchtop bioreactors and mini-bioreactors are abundantly used in upstream process development, cell culture behavior and performance (cell density and viability, final titer and product quality, cellular metabolism) observed in these systems may not always be predictive of large-scale processes. This may be attributed, on the one hand, to key differences in shear stress levels and gas transfer capabilities, and on the other hand, to the presence of undesired concentration gradients (nutrients, gases and pH) in large vessels. In bioreactor process design, operating parameters may be broadly divided into two distinct groups: scale-independent and scale-dependent variables. Scale-independent parameters include temperature, dissolved oxygen, pH, nutrient concentrations and feed regimens. In principle, these parameters can be matched across different production scales. In contrast, scale-dependent parameters may vary considerably with bioreactor size. These include, for instance, impeller tip speed, mass transfer coefficient (k_La), surface area to volume aspect ratio, gas residence time, mixing time, mixing power input, Reynolds number (Re) and superficial gas velocity. Bioreactor scale-up can be thought to consist in keeping all scale-independent parameters constant, while prioritizing which scale-dependent parameters to maintain constant. Due to their interdependence, keeping all scale-dependent parameters constant throughout scale-up is virtually impossible. For example, if agitation speeds were to be kept unchanged throughout scales in an attempt to maintain mixing times, impeller tip speed would increase with size, potentially causing cell damage. On the other hand, if impeller tip speed was maintained constant across scales, large bioreactor vessels would have drastically decreased agitation speeds and thus increased mixing times. In addition to differences in mixing during scale-up, gas transfer characteristics will also be significantly impacted. With an increase in culture size, the ratio of surface area to volume decreases, thereby reducing the contribution of the headspace on the overall gas transfer. A decrease in gas transfer in the headspace implies greater sparged gas flow rates to meet the oxygen demand, which may also result in cell damage due to shear stress caused by bubbles. Thus, there is no single obvious method to perform the scale-up of bioreactors, as critical trade-offs must be made. This is further complicated by the fact that the best scale-up practices used in industry are often kept secret. As such, scale-up of cell culture systems remains an art that often relies on trial and error to determine optimal process parameters and operating strategies.

In this review, we will focus on CHO cell culture scale-up, and the procedures and tools that can be used to tackle this challenging task. First, the most common cultivation systems, from uninstrumented flasks to large-scale stirred-tanks, and their role in the scaling-up of biopharmaceutical production will be described.

Second, the implications of culture scale-up on key process parameters will be discussed. Third, basic principles and recent strategies employed in scale-up across scales will be described. Finally, useful tools in scale-up such as computational fluid dynamics (CFD) and scale-down models will be discussed.

2 Cultivation Systems for CHO Cells

2.1 Uninstrumented Systems

Many different uninstrumented cultivation systems are available for mammalian cell culture, including Erlenmeyer shake flasks, static flasks, roller bottles, multi-well plates, TubeSpin and spinner flasks. Uninstrumented systems are mainly used for their ease of use, low costs, and high-throughput capabilities. Their major downside lies in the inability to monitor and control culture conditions such as pH and dissolved oxygen levels, although new sensor technologies adaptable to these cultivations systems are now increasingly available. Shake flasks, the most widely used uninstrumented cultivation system, come in sizes typically ranging from 25 mL to 5 L. Smaller flasks can be used for clone selection and seed train, while larger flasks of up to 5 L are routinely used for rapid productions. These containers can be made of glass or plastic, and can be with or without baffles to improve aeration and mixing. They can be agitated through orbital or linear movements, and are usually kept in a temperature and CO_2 controlled environment. Operating parameters such as vessel size, fill volume, material of construction, geometry of baffles, filter cap design, shaker orbit, and agitation speed can all potentially impact cell culture performance due to their effect on mixing and gas transfer [13, 14]. Their major limitation is the reliance on surface aeration, leading to reduced oxygen transfer rates when compared to stirred-tank bioreactors. They have thus a limited capacity to support high cell density cultures. The presence of baffles increases oxygen transfer, however high mixing speeds may cause excess splashing resulting in blocking the gas-permeable cap of the flask. Spinner flasks, which resemble shake flasks but are agitated with a magnetic stirring impeller, are easier to scale-up due to their geometry which more closely resembles bioreactors.

Plates with numbers of wells ranging from 4 (25 mL/well) to 3456 (1–2.2 μL/well) may also be used for cell culture. This allows for high and even ultra-high throughput miniaturized cultures that are particularly useful for screening campaigns. These plates can be employed for fluorescence and phosphorescence measurements, and allow for stacking to maximize space utilization in incubators. It is worthy to mention that comparable growth rates, viability, and protein production between 24 deep well plates and shake flasks were noted [15].

Static flasks and roller bottles function similarly to shake flasks; however, they provide a surface onto which adherent cells attach. Roller bottles allow for greater surface area in order to increase cell density. Expanded surface roller bottles have the same dimensions as standard bottles, but further increase the surface area available for cells by the addition of a pleated design inside the bottle. Roller bottles are advantageous over static flasks not only for their higher surface area, but also enhanced gas transfer capabilities and increased homogeneity of the culture medium. Roller bottles can be stacked and incorporated within a fully automated system for culture seeding, feeding and harvesting.

In order to allow for closer cell culture monitoring and control, new sensor technologies compatible with small culture devices have been developed. For instance, the Shake Flask Reader® by PreSens can monitor pH, dissolved oxygen levels and oxygen uptake rate (OUR), permitting to adjust the shaking speed to avoid oxygen limitations. Additionally, modern incubators may have tri-gas (carbon dioxide, oxygen, and nitrogen) control such that cell culture may be performed under hypoxia or hyperoxia conditions. RAMOS® (Respiration Activity Monitoring System) and TOM® (Transfer-Rate Online Measurement) by Kuhner allow to measure dissolved oxygen and carbon dioxide levels in the headspace. This can provide real-time measurements of oxygen uptake rate (OUR), carbon evolution rate (CER), and the respiration quotient (RQ). Until recently, these parameters could only be reliably measured within a stirred-tank bioreactor. While uninstrumented flasks used to be abundantly employed for upstream process development, they are increasingly being replaced with mini-bioreactors that better mimic large-scale conditions.

2.2 Micro- and Mini-bioreactors

Miniature bioreactor systems provide a powerful tool for the rapid optimization of cell culture parameters, including pH, DO, temperature, nutrient concentrations, pre-culture age, seeding density, aeration, mixing, etc. Miniature bioreactor systems allow for high-throughput cell cultures while monitoring and controlling DO, pH, agitation and gas sparging. Owing to ease of set-up and short turnaround times, such systems are ideal for design of experiments (DoE), resulting in a higher resolution and accurate definition of the design space, as well as shorten clone selection timelines and accelerate process development. Many different micro- and mini-bioreactor systems are commercially available (Table 2) and can be operated in either batch or fed-batch modes, and sometimes even in perfusion. These micro- and mini-bioreactor systems range in volume from 10 mL to 1 L, with 1–48 cultures ran in parallel (depending on the manufacturer) and all have pH, DO, and temperature controls. Two of the most frequently used micro- and parallel mini-bioreactor systems are the Ambr® 15 and the Ambr® 250. The latter has temperature and mixing capabilities for individual units, can be operated in both fed-batch and perfusion modes and is geometrically more similar to large-scale systems, while temperature and agitation control in the former is limited to a block of 12 micro-bioreactor vessels.

Table 2 Automated micro- and mini-bioreactor systems available for mammalian cell culture

System	Manufacturer	Volume	Units
Ambr® 15	Sartorius	10 mL–15 mL	24 or 48
Ambr® 250	Sartorius	100 mL–250 mL	12 or 24
Biostat® Qplus	Sartorius	500 mL–1000 mL	12
DASbox®	Eppendorf	60 mL–250 mL	Up to 24[a]
DASGIP®	Eppendorf	200 mL–3.8 L	Up to 12[a]
MiniBio	Applikon®	250 mL, 500 mL, 1000 mL	1
MicroMatrix	Applikon®	1 mL–7 mL	24
Multifors 2	Infors HT	300 mL–750 mL	Up to 6[b]
Micro24	Pall Life Sciences	7 mL	24
Medicel cultivation unit	Medicel Oy	100 mL–200 mL, 200 mL–500 mL	15
Clone screen	Biospectra	400 mL	32
Biopod f800	Fogale Biotech	80 mL	8
CellStation	Fluorometrix	35 mL	12

[a]In modules of 4
[b]In modules of 2

Micro and mini-bioreactor systems are invaluable tools not only for use as scale-down models of large-scale bioreactors, but are also abundantly used for clone selection. Cell culture performances in Ambr®15 were shown to more closely resemble that of bench-scale or larger bioreactors compared to shake flasks [16]. For example, cell viability, cell concentration, and product titer were reported to be comparable between Ambr® 15 and 2 L or 5 L bioreactors [17, 18]. Similarly, cell cultures performed in Ambr® 250 bioreactors have been shown to exhibit comparable final product titer, metabolite profiles, and product quality attributes to those obtained in a 5 L bioreactor [19].

2.3 *Stirred-Tank Bioreactors (Glass and Stainless-Steel)*

The use of stirred-tank bioreactors is the favoured method of producing recombinant proteins in the biopharmaceutical industry [20]. For cell culture applications, glass and stainless-steel bioreactors are available in sizes ranging from 1 L benchtop to 20,000 L industrial bioreactors. Typically, bioreactors are equipped with dissolved oxygen and pH probes and temperature sensors. Oxygen levels are controlled by sparging air, pure oxygen and sometimes nitrogen. pH setpoints are controlled by base addition and carbon dioxide sparging. Temperature is controlled either with a heating blanket and cooling coil combination, or via the use of a jacketed vessel. Mixing is achieved through the rotation of an impeller attached to a shaft, and its rotation is powered by a mechanical or magnetic motor found at either the top or the bottom of the stirred-tank.

2.4 Single-Use Bioreactors

Single-use bioreactors (SUBs) constitute an alternative to reusable glass or stainless-steel bioreactors and are increasingly adopted by the biomanufacturing industry due do their numerous advantages. SUBs allow for faster facility set-up, reduction of downtime, reduced risk of contamination and also have lower initial capital costs when compared to reusable bioreactors. Process validation and operation are also greatly simplified, as SUBs are pre-sterilized. However, these systems are limited in size (up to 2000 L) and the operating costs per batch can be higher than that of reusable bioreactors. Nonetheless, single-use bioreactor performances have been shown to be comparable to those of reusable bioreactors at various scales [21, 22]. Two main types of SUBs are commercially available: stirred-tank and Wave bioreactors. Stirred-tank SUBs consist of a plastic bag placed inside a more permanent bioreactor vessel or a rigid plastic vessel containing an integrated stirrer. A driver is connected to the stirrer mechanically or magnetically. Stirred SUBs can achieve power inputs of up to 20 W/m^3 and oxygen mass transfer coefficients of 3–8 h^{-1} with drilled hole spargers [23], which are values comparable to those typically seen in conventional bioreactors. Wave SUBs are bags placed on a rocking agitator and thus do not contain stirrers. They are often used in perfusion mode during seed train expansion of large-scale productions [24, 25]. Wave SUBs have more uniform energy dissipation and reduced foam formation, and were also shown to yield comparable cell culture performance as stirred SUBs [26]. Single-use bioreactors range in size from 15 mL (e.g. Ambr® 15) up to 2000 L for stirred-tank and 50 mL to 500 L for Wave. All sensors and control loops present in reusable bioreactors can be found in stirred-tank SUBs and come in the form of *in situ* disposable sensors or non-invasive optical sensors or ultrasonic sensors [27].

3 Key Process Conditions Impacted by Scale-Up

3.1 Aeration

In bioreactors for mammalian cells, aeration is a critical component, particularly when operating at high cell density, as sufficient oxygen needs to be supplied to the cells, while carbon dioxide must be stripped to prevent its accumulation in the culture broth. This is usually achieved using a combination of gas spargers and headspace aeration. In large vessels, oxygen is supplied by being dispersed close to the bottom using a gas sparger. Gas diffusers employed in bioreactors can be either porous-frit spargers (made of ceramic, Teflon, or stainless steel) or drilled hole spargers (DHS). The former creates smaller bubbles, thereby increasing both the interfacial surface area and gas hold-up, but is also more prone to induce foaming and not as effective for CO_2 stripping. Hence, dual-sparger designs are frequently encountered. Table 3 presents an overview of the different design and aeration configurations commonly employed. In most cases, oxygen and carbon dioxide are

Table 3 Common aeration strategies in mammalian cell cultures

Reference source	Bioreactor type & scale	Gas sparging			Headspace gas	Mixing
		Type	On demand	Continuous		
[34]	Ambr® 250 mL	Open pipe	O_2 & CO_2	Air	No gas	2-PB
	500 L SUB	0.4 mm DHS for air, frit for O_2 & CO_2	O_2 & CO_2	Air	Air	1-PB
	2000 L SUB	2 mm DHS for air, frit for O_2 & CO_2	O_2 & CO_2	Air	Air	1-PB
[19]	Ambr® 250 mL	Open pipe	O_2 & CO_2	Air	No gas	2-PB
	5 L benchtop	DHS	O_2 & CO_2	Air	Air	2-PB
	250 L SUB	DHS and frit	O_2 & CO_2	Air	Air	1-PB
[35]	Ambr® 250 mL	Open pipe	O_2 & CO_2	N/A	No gas	2-PB
	3 L benchtop	0.5 mm DHS or 20 μm frit	O_2 & CO_2	Air	Air	Marine
	2000 L SUB	2 mm DHS air/N_2, 20 μm frit O_2 & CO_2	O_2 & CO_2	Air OR N_2	Air	1-PB
[36]	5 L (glass)	15 μm frit	O_2 & CO_2	Air	Air	Marine
	20 L (glass)	15 μm frit	O_2 & CO_2	Air	Air	Marine
	5000 L (stainless steel)	1.8 mm DHS	O_2 & CO_2	Air	Air	Marine
[37]	Ambr® 250 mL	Open pipe	O_2 & CO_2	Air	No gas	2-PB
	200 L	1 mm DHS air, 20 μm frit O_2 & CO_2	O_2 & CO_2	Air	Air	1-PB
[38]	2 L benchtop	N/A	O_2 & CO_2	Air	Air	Marine
	1500 L	N/A	O_2 & CO_2	Air	Air	N/A
[39]	5000 L	N/A	O_2 & CO_2	Air	N/A	3-Marine
	25,000 L	N/A	O_2 & CO_2	Air	N/A	2-HF

DHS Drilled hole sparger, *Frit* Frit sparger, *PB* Pitched blade, *HF* Hydrofoil sparged on demand in order to control dissolved oxygen levels and pH, while air is continuously sparged and sometimes also supplied to the headspace.

3.1.1 Oxygen

In CHO cell cultures, dissolved oxygen levels are typically maintained between 10% and 80% of air saturation [17]. Excessive oxygen concentrations result in the undesired accumulation of reactive oxygen species (ROS), which can cause alterations in the mitochondrial respiratory chain and intracellular redox, ultimately leading to cell growth inhibition and decreased specific productivity [28, 29]. On the other hand, hypoxic conditions are also known to induce ROS accumulation [30]. A prolonged exposure of 10 days at 20% DO has been shown to elicit the same hypoxic response observed after 1–3 days exposure at 0.5%–5% DO. Hypoxia conditions were also associated with reduced oxygen uptake and increased lactate production [29]. ROS accumulation was correlated with a significant decrease in productivity at large-scale (5000 L), while such problem was not encountered at small-scale (20 L) [30]. Hence, DO set points must be carefully selected during scale-up, especially due to hydrostatic pressure effect; at a similar DO level (as % of air saturation), the actual oxygen concentration will be significantly higher in a large-scale vessel compared to a benchtop bioreactor [29]. And although surface aeration reduces shear stress caused by bubbles, its contribution to the overall gas transfer becomes negligible at larger scales [31]. In order to reduce shear stress caused by bubbles bursting at the gas-liquid interface [32], surfactants such as Pluronic F-68 (or Kolliphor® P188) are added to the cell culture broth. Pluronic concentrations in the range of 1 g/L (and up to 5 g/L) are typically used to confer shear protection [31, 33].

The oxygen mass transfer coefficient (k_La) is used as a measure of a system's oxygen transfer capabilities. For a given bioreactor configuration, k_La is mainly correlated to the volumetric power input (P/V) and to the superficial gas velocity (v_s) according to the relationship given below, where K is a constant and exponents α and β are approximately 0.4 and 0.5, respectively [40].

$$k_L a = K\left(\frac{P}{V}\right)^{\alpha} \upsilon_s^{\beta} \tag{1}$$

Process parameters such as gas bubble size, mixing speed, gas flow rate, volume and bioreactor geometry all impact the k_La. Various medium components, including salts, antifoam and surfactants, have also been shown to influence the mass transfer coefficient. Particularly, antifoam was found to significantly decrease k_La [18, 41]. Pluronic F-68 also decreases k_La, but to a lesser extent [18], while the addition of salts increases the oxygen mass transfer coefficient [16, 18]. Another factor found to influence the k_La is the number of impellers. The use of a single impeller was demonstrated to increase k_La compared to a 3-impeller configuration when keeping volumetric power and gassing flow rate constant in coalescence-repressing media, such as most culture broths. This indicates that concentrating the power input where bubbles are formed increases the overall mass transfer [20]. Generally speaking, an increase in bioreactor scale of similar geometry will result in an augmentation of the k_La, as noted in both conventional stirred-tank [42, 43] and single-use bioreactors [19, 44]. The increase in liquid height results in longer gas residence time allowing for more efficient gas transfer.

3.1.2 Carbon Dioxide

In mammalian cell cultures, dissolved carbon dioxide (pCO_2) levels are typically maintained at 4–10% CO_2 saturation (30–70 mmHg) [45, 46]. High levels of dissolved carbon dioxide (150–200 mmHg) are known to have detrimental effects on cell growth and productivity for many types of mammalian cells. In CHO cell cultures, high pCO_2 levels were shown to inhibit cell growth [46–49], cell productivity [46, 49], and alter product quality [50]. Dissolved carbon dioxide in water causes acidification by reacting with water to form carbonate, which dissociates into carbonic acid. The acidification of the culture broth causes base addition in pH-controlled cultures, which leads to an increase in osmolarity. Elevated levels of osmolarity (460–500 mOsm/kg) have been shown to reduce viable cell density and viability [48] when compared to normal levels (260–320 mOsm/kg) [51]. This is further exacerbated in fed-batch cultures, where feeding of concentrated nutrients can lead to osmolarity up to 540 mOsm/kg [52]. Although carbon dioxide may have detrimental effects on cell culture performance, it is nonetheless needed for the metabolic synthesis of nucleic acids so its concentration must not fall too low [20, 53]. Ultra-low pCO_2 levels (12.5–24.5 mmHg) were found to decrease viable cell density and cell viability after 6 days of cell culture compared to normal pCO_2 levels (28–54 mmHg) [54].

Large-scale animal cell cultures are faced with the issue of withdrawing (or stripping) carbon dioxide produced by cells at roughly the same molar rate as oxygen is consumed. In large vessels, headspace aeration is usually insufficient for carbon dioxide removal [36, 41, 42]. Consequently, sparge flow rates and agitation speed must be adjusted to ensure efficient carbon dioxide stripping [36]. It should be noted that carbon dioxide stripping was reported to be independent of gas mass transfer coefficients in large-scale bioreactors [39]. Instead, carbon dioxide accumulates in the culture media due to long gas residence time relative to gas bubble saturation time; gas bubbles become saturated in carbon dioxide before leaving the culture media such that mass transfer is no longer limiting carbon dioxide stripping.

3.2 Mixing

Mixing serves not only in ensuring homogeneous conditions, but also plays an important role in improving mass transfer, which is important for supplying oxygen to the cells and stripping carbon dioxide. Mixing in large-scale mammalian cell cultures is typically provided by axially pumping impellers, such as pitched blade or hydrofoil impellers. Axial flow impellers allow for decreased mixing times when compared to radial flow impellers such as Rushton impellers, and result in lower shear stress applied to cells [20]. In large-scale bioreactors, baffles are typically used around the circumference of the tank in order to induce turbulence. Without the presence of baffles, the circular energy provided by the impeller does not sufficiently contribute to the turbulent motion required for proper mixing. The absence of baffles can also cause the formation of a central vortex which may result in uncontrolled gas transfers between the headspace and culture broth. Another strategy to increase flow turbulence consists in mounting the impeller shaft eccentrically. This is mainly used in situations where baffles are inconvenient, such as in the case of single-use bioreactors. However, the use of eccentrically mounted impeller is known to require roughly double the power input in order to achieve the same mixing times as concentric impeller shafts [20].

Mixing is related to aeration through the volumetric power input (P/V), as seen in Eq. (1). The volumetric power input is an important engineering parameter often used as a criterion in scale-up to ensure comparable culture homogeneity. The volumetric power input can be measured or calculated using Eq. (2), where N_p is the dimensionless power number, N is the mixing speed, D_i is the impeller diameter, and V is the volume. Depending on the type of impeller used, various correlations are available to evaluate the power number for agitated vessels, usually as a function of the flow regime (Reynolds number) and aspect ratios. When the flow is turbulent, the power number becomes independent of the Reynolds number in baffled vessels.

$$\frac{P}{V} = \frac{N_p N^3 D_i^5}{V} \tag{2}$$

For CHO cell cultures in bioreactors, mixing speeds are typically adjusted in order to obtain a P/V in the range of 10–80 W/m^3 [19, 34, 55]. Another crucial factor that must be taken into account when setting the mixing conditions is the impeller tip speed. As high impeller tip speeds (V_t) create zones of high shear stress, it is favourable to minimize V_t.

Under increased air or oxygen sparged flow rates, lower mixing speeds are required in order to maintain the gas mass transfer coefficient (k_La) constant. This, in turn, increases the mixing time and may create heterogeneities in the culture broth [32], including the presence of significant dissolved oxygen gradients resulting from different localized gas mass transfer coefficients [56]. This can be further exacerbated by the fact that mixing times are already significantly longer in larger vessels. For instance, mixing times were found to increase from ~10 s for benchtop 3 L bioreactors to ~120 s for industrial-scale bioreactor [57]. An increase in mixing time results in pH, dissolved oxygen, carbon dioxide and nutrient concentration gradients. These gradients can have a critical impact, since cells may pass through a region of non-optimal operating conditions (the so-called micro-environment) resulting in an overall reduced culture performance. In particular, glycosylation patterns have been found to be impacted by varying dissolved oxygen levels [58]. Viable cell densities, and thus product titers, were also negatively affected due to pH inhomogeneities [59]. In bioreactors, a higher gas mass transfer coefficient is found at the bottom near the impeller due to the well-mixed region and the increased hydrostatic pressure. In order to improve mixing, Xing et al. [36] suggest minimizing culture volumes and performing alkali additions near the impeller to minimize pH peaks.

3.3 Hydrodynamic Shear Stress

Gas sparging is known to be a significant source of cell damage by shear stress. The damage to the cells can occur at the formation of bubbles, at the sparger, at impeller-bubble interactions, when bubbles rise, and during the bursting of bubbles at the liquid surface. The later was found to be the predominant source of cell damage [32]. Smaller bubbles cause more cell damage due to higher specific energy dissipation rates when bursting at the surface [60]. Due to the fact that mammalian cells are sensitive to shear forces, mixing and aeration speeds must be maintained below certain thresholds. When reducing mixing and aeration, the most important consideration is meeting the oxygen demand of the cells. Hence, pure oxygen can be used in order to reach the cells oxygen requirements at lower gas flow rates, although care has to be taken since gas sparging is essential to strip carbon dioxide. Shear stress may have impacts on cell metabolism at sublytic levels. For CHO cells, a shear stress of 0.8 N/m^2 (60,000 W/m^3) was found to drastically decrease the cell specific productivity when compared to a shear stress in the order of 0.005 N/m^2 (2 W/m^3) [61].

For the same volumetric power input, the volume average hydrodynamic stress caused by mixing was shown to be independent of scale [62]. Rather, hydrodynamic stress is mainly influenced by the bioreactor configuration (number of baffles and impellers), although small differences in vessel aspect ratios and large changes in impeller shaft angle do not have a significant impact. In most bioreactors, the average energy dissipation rate is typically in the range of 10–80 W/m^3, while the maximum energy dissipation rate is approximately 100-fold greater. CHO cells in suspension may support energy dissipations rates of up to $10^7 W/m^3$ [63]. When cells were exposed to power dissipation rates much higher than those encountered in large-scale bioreactor, only product quality was shown to be affected, hinting that cell damage due to mixing had previously been overestimated in the case of CHO cells [20]. Pluronic® F-68, a surfactant, can be added to culture media to minimize cell damage and death [32, 36, 64]. Pluronic® F-68 reduces surface tension and decreases the specific energy dissipation associated with a bubble bursting at the liquid interface [32]. Cells incorporate Pluronic® F-68 in their membrane, which increases the membrane strength [65, 66]. Additionally, high concentrations of Pluronic® F-68 in intracellular vesicles stiffen the cells mechanical properties [67]. But the largest contribution of Pluronic® F-68 to reduce cell death might be due to the fact that it reduces the hydrophobicity of the membrane, resulting in less cells attaching to bubbles and experiencing the highly localised energy dissipation generated during bursting [32].

4 Basic Principles for Bioreactor Scale-Up

The purpose of bioreactor scale-up is to increase the production volume while keeping similar product yield and quality. This implies that comparable cell specific productivity, cell density and viability, and cell metabolism must be preserved across scales. In order to accomplish this, as many operating parameters as possible must be kept constant. Scale-independent variables such as temperature, pH, dissolved oxygen setpoints, and nutrient feed strategy can easily be kept constant during process scale-up (Table 4). However, scale-dependent parameters, such as agitation, impeller tip speed, mixing time, the Reynolds number, and aeration flow rates, cannot all be simultaneously kept constant throughout scale-up. This is due to the fact that they have varying dependencies on agitation speed, impeller diameter and vessel diameter. Ultimately, these parameters can impact operating costs, culture heterogeneity, gas transfer characteristics and the shear stress applied to the cells. In essence, bioreactor scale-up involves trade-offs and compromises.

Although the volumetric power input is generally maintained within the range of 10–80 W/m^3 in stirred-tank bioreactors, other factors, namely mixing speed, mixing time, impeller tip speed, and Reynolds number have different values depending on the production scale. As seen in Table 5, agitation speeds decrease with an increase

Table 4 Scale-independent and scale-dependent variables

Scale-independent variables	Scale-dependent variables
• Temperature	• Impeller tip speed: $V_t = \pi N D_i$
• pH	• Mass transfer coefficient: $k_L a = K\left(\frac{P}{V}\right)^{\alpha}(v_s)^{\beta}$
• Dissolved oxygen	• Surface area to volume ratio: $\left(\frac{A_c}{V}\right) = \frac{\pi (D_T/2)^2}{V}$
• Nutrient concentration	• Gas residence time: $t_r = \frac{V\phi_G}{Q_g(1-\phi_G)}$
• Feed regimen	• Mixing time
• Seeding density	• Mixing power input: $P = N_p N^3 D_i^5$
	• Reynolds number: $\mathrm{Re} = \frac{N {D_i}^2 \rho}{\mu}$
	• Superficial gas velocity: $v_s = Q_g/A_c$

Table 5 Typical values of scale-dependent parameters at various production scales

Parameter	Micro and mini-bioreactor (<1 L)	Benchtop (1–10 L)	Pilot-scale (10–200 L)	Industrial-scale (>200 L)
Mixing time (s)	<10	10	20	50–200
Mixing speed (RPM)	400–1400	200–300	100–150	50–100
Impeller tip speed (m/s)	<1	0.75–1	1–2	1.5–2.5
Reynolds number	2000–15,000	20,000–50,000	50,000–150,000	>150,000

in scale. However, impeller tip speed and Reynolds number follow the opposite trends due to increased impeller size. Finally, mixing times also increase with scale due to the increase in vessel diameter.

4.1 Geometrical Similarity

Geometric similarity is typically the first criterion applied for scaling-up bioreactors. If tank diameter is increased, all other lengths (tank height, impeller diameter, and impeller width) are increased by the same scale factor. Generally, bioreactor vessels used for cell culture maintain a height-to-diameter ratio (H/D) of 1–2 for benchtop bioreactors and 2–3 for pilot and industrial scale bioreactors [68, 69]. However, preserving H/D impacts factors related to surface and volume such as heat transfer, gas transfer and mixing. Heat transfer per unit volume decreases as volume increases, due to heat exchange occurring at the walls of the vessel. A constant H/D aspect ratio will also significantly decrease the surface area to volume ratio (A_C/V), resulting in a decreased contribution of surface aeration for oxygenation and carbon dioxide stripping [42]. This is critical for shear sensitive cells, because of the importance of gas transfer rates and restrictions on mixing speeds and gas sparging flow rates.

4.2 Dynamic Similarity and Scale-Up Criteria

Dynamic similarity exists when the ratio of all relevant forces is preserved across different scales, leading to, for example, similar flow fields. During scale-up, a criterion must be selected based on which factors have the highest impact on cell culture performance. Due to the interdependencies, when one critical parameter is held constant, other factors may exhibit a significant variation following an increase in volume, potentially leading to problems that were not encountered at smaller scale. Key interdependent variations can be seen in Table 6 for some of the most common scale-up criteria employed for agitated vessels. For instance, maintaining a constant volumetric power input during scale-up will translate into an increase in maximum shear rate (from the greater tip speed) and a decrease in agitation speed. In turn, a reduced agitation speed will result in an increased mixing time, potentially leading to undesired environmental heterogeneities. Constant impeller tip speed or constant Reynolds number (i.e. similar hydrodynamic regime) during scale-up also imply decreased mixing speed.

Constant impeller tip speed may be used in the case of shear sensitive cells; however, the decrease in volumetric power must be compensated by higher gas flow rates in order to maintain acceptable oxygen transfer rates, which may also contribute to cell damage [32]. Additionally, the lower mixing speeds required to keep a constant impeller tip speed will result in poorer mixing at larger scales. Thus, impeller tip speed may not be a suitable scale-up criterion for large differences in production volume. A constant mixing time during scale-up will result in an increase in impeller tip speed, which may cause cell damage, as well as an extremely large increase in power. Oxygen transfer [70, 71], bulk mixing [72] and proper CO_2 stripping [34, 38, 41, 43, 73] are generally recognized as the most important factors during bioreactor scale-up. Hence, criteria which minimize the impacts on gas transfer and mixing are most often selected for the scale-up of CHO cell bioreactor cultures.

5 Common Scale-Up Strategies for CHO Cell Cultures

In this section, the most common scale-up strategies employed for scaling-up CHO cell cultures in bioreactors will be reviewed and their implications will be discussed. Table 7 presents an overview of recent studies that have assessed cell culture performance at different scales, and the scaling criteria that were employed.

5.1 Constant Volumetric Power Input

Using a constant impeller power input per liquid volume (P/V) is one of the most commonly used scale-up criteria for agitated and aerated vessels, where mechanical power from the impeller impacts both gas transfer characteristics and culture mixing. The P/V ratio can be set by adapting the impeller type, size and speed to the different working volumes. This scale-up criterion is either used alone or in combination with other factors (e.g. constant volumetric gas flow rate per unit volume of liquid or vvm) as shown in Table 7. Constant P/V has been used to successfully transfer from an Ambr®250 mini-bioreactor system to a 200 L pilot-scale single-used bioreactor to demonstrate the effectiveness of an air sparge-

Table 6 Interdependence of scale-up parameters for geometrically similar bioreactors when the vessel diameter is increased by a factor of 5. The factor 1 in each column indicates the criterion that is preserved at large-scale

Parameter	Small-scale (1 L)	Large-scale (125 L)			
		Constant vol. power input (P/V)	Constant impeller tip speed (V_t)	Constant Reynolds (Re)	Constant mixing time (t_m)
Power input	1	125	25	0.2	3125
Volumetric power input	1	1	0.2	0.0016	25
Mixing speed	1	0.34	0.2	0.04	1
Impeller diameter	1	5	5	5	5
Impeller tip speed	1	1.7	1	0.2	5
Reynolds number	1	8.5	5	1	25
Mixing time	1	2.92	5	25	1

Table 7 Scale-up criteria used for CHO cell cultures

References	Scale-up criteria	Scales	Results
[78]	Tip speed or P/V	10 L to Ambr®15	• Comparable cell viability and specific productivity for both criteria • Lower pCO_2 in Ambr®15 when using P/V.
[77]	P/V or $k_La_{CO_2}/k_La_{O_2}$	200 L, 200 L SUB, 1000 L SUB, and 2000 L	• P/V: Comparable VCD, viability, and titer. Higher CO_2 levels in 2000 L • k_La ratio: Comparable VCD, viability, titer and CO_2 levels
[83]	OTR	2 L / 10 L to 80 L	• Comparable cell viability and final titer • Differences in VCD
[38]	OTR and CER	2 L to 1500 L	• Consistent VCD, titer, and product glycosylation
[37]	P/V	Ambr®250 to 200 L SUB	• Comparable viability, volumetric productivity, and antibody quality attributes • Decrease of 10% in peak VCD in the 200 L SUB
[17]	P/V or Tip speed	Ambr®15 to 2 L	• Comparable cell viability, VCD, and titer for both scale-up criteria
[73]	vvm	Ambr®15, 5 L, and 15,000 L	• Comparable CO_2 evolution profiles
[76]	P/V or vvm	Ambr®250 and 18,000 L	• P/V or vvm: Comparable VCD, viability, and titer
[79]	k_La	50 L, 200 L, 500 L, 1000 L, and 2000 L	• Comparable VCD
[75]	P/V	30–50 mL SF, Ambr®250, 2 L, 50 L	• Comparable specific growth rate and mAb production

[74]	P/V	3 L and 500 L	• Comparable titer, integrated VCD, and viability
[18]	Tip speed	Ambr®250 and 5 L	• Comparable VCD and titer
[71]	P/V or k_La or OTR	2000 L to 3 L	• P/V and kLa: Significantly higher VCD and final titers in the 2000 L bioreactor • OTR: Comparable growth, productivity and protein quality
[19]	P/V and vvm or k_La	Ambr®250, 5 L, and 250 L SUB	• P/V and vvm: Comparable VCD and viability profiles. Titer was lower in the Ambr®250 due to insufficient aeration • k_La: Comparable growth, productivity, and final titer
[43]	P/V and minimum air sparge vvm	3 L, 500 L, and 2000 L SUB	• Comparable VCD, cell viability, and titer
[34]	vvm	Ambr®250, 500 L SUB, 2000 L SUB	• Comparable peak VCD, final viability, titer, and product quality

based pH control strategy [37]. Applying the same criterion resulted in comparable growth, productivity and product quality attributes when comparing 500 L pilot-scale to 3 L benchtop bioreactors [74], or when scaling up from Ambr®250 to 3 L benchtop and then 50 L pilot-scale vessels [75]. Constant P/V also proved successful as a scale-down criteria to match the conditions and performance of a 18,000 L industrial scale bioreactor in an Ambr®250 [76]. Comparable cell culture performances at 200 L and 2000 L scales were also reported using constant P/V, although pCO_2 levels were found to be higher in the larger bioreactor [77].

However, using constant P/V may translate into significant differences between small- and large-scale k_La values. In order to compensate for the change in volumetric oxygen transfer coefficient, the gassing strategy may need to be adjusted for maintaining the desired dissolved oxygen levels [34, 71]. This is particularly true when upstream process development steps are done using miniature bioreactors. However, the required flow rate adjustments may have detrimental impacts on culture performance, as exemplified by the study of Tescione et al. [71]; when scaling down from 2000 L industrial-scale vessels to 3 L lab-scale bioreactors using P/V as the sole criterion, significantly reduced viable cell densities and final product titers were observed. The normalized gas volumetric flow rate (vvm) was threefold higher in the small-scale bioreactor, resulting in increased bubble-associated cell damage.

Scale-up from a miniature (Ambr®15) to a 10 L bioreactor using either constant impeller tip speed or constant P/V were compared and evaluated by transcriptome analysis to assess the impact on gene expression [78]. Comparable cell growth and specific productivity were obtained for both scale-up criteria used, but a decrease in pCO_2 levels throughout the culture was observed in the Ambr®15 compared to the 10 L bioreactor, due to the larger oxygen flow rate required in the micro-bioreactor resulting in an increased CO_2 stripping. The transcriptome analysis revealed only minor differences (below 6%) over time in gene expression between both scales. Although 2455 and 1601 genes were uniquely regulated in the Ambr®15 and 10 L bioreactor respectively, no functional correlation was made to scale or cellular behaviour.

5.2 *Constant Oxygen Mass Transfer Coefficient (K_La)*

Maintaining proper oxygen transfer is of utmost importance in CHO cell cultures, particularly when bioreactors are operated at high cell densities. Thus, keeping constant k_La during scale-up/scale-down is also a common strategy in process development. It has successfully been employed as a criterion when using Ambr®250 as a scale-down model for 5 L lab-scale and 250 L SUB systems. A k_La of 2–3 h^{-1} satisfied oxygen requirements in the scale-down model and led to comparable growth, productivity, and final product titer for 11 out of 13 GS-CHO and DG44-CHO cell lines [19]. Maintaining a constant k_La (7.9 h^{-1}) via gas flow rate and stir speed adjustments was also shown to yield similar growth profiles in single-used bioreactors ranging from 50 L to 2000 L [79]. However, a change of production scale may require the use of different types of gas sparger, further complicating the attainment of similar process conditions. To scale-down from a 2000 L vessel equipped with a drilled-hole sparger to a 3 L bioreactor using a constant k_La, Tescione et al. [71] had to resort to a frit sparger at small-scale in order to avoid excessive mixing speeds or high gas flow rates, which had been shown to be detrimental to culture performance. But changing the sparger type led to different cell growth and metabolism when compared to the large-scale process.

To attain proper oxygen transfer, He et al. [38] developed a mathematical mass-transfer model to describe the gas exchange within a bioreactor. Mass balance equations involving both cell respiration rates (qO_2, qCO_2) and mass-transfer characteristics (k_La_{O2}, k_La_{CO2}) were used to calculate the gas flow rates required to meet the oxygen demand and predict pCO_2 levels. This scale-up method resulted in matching viable cell density and antibody production at 2 L and 1500 L. In a similar vein, the required k_La and oxygen enrichment levels required in a 3 L scale-down model were determined by assuming the same oxygen demand as that measured in a 2000 L bioreactor [71]. This allowed to set appropriate sparge rates that resulted in comparable cell culture performances with no impact on protein quality.

5.3 Constant Volumetric Gas Flow Rate per Volume (vvm)

Dissolved CO_2 accumulation being a major concern during scale-up of CHO cell cultures in bioreactors, a constant vvm is often used as a scale-up criterion to ensure sufficient carbon dioxide stripping, while adjusting the mixing speed to provide proper oxygenation. The miniature bioreactor Ambr®15 was used to develop a scale-down model of industrial and benchtop scale bioreactors using total sparge gas flow rate as a primary scale-down parameter (vvm of 0.01–0.02). This scale-down method allowed for adequate stripping and kept CO_2 evolution profiles constant across scales. Optimal operating conditions identified by DoE (growth temperature, production temperature and pH) were shown to be consistent for Ambr®15 and 5 L bioreactor [73]. A different study attempted to use constant vvm to generate a representative scale-down model for 500 L and 2000 L SUBs using the Ambr®250 platform [34]. But since surface gas transfer generally accounts for a significant portion of CO_2 stripping at small-scale, a lower normalized aeration rate was required in the mini-bioreactor system to match the pCO_2 gas evolution profiles between scales. In a different process where pCO_2 had been identified as having a major impact on culture performance, constant vvm was used as a criterion in order to match dissolved gas profiles between industrial scale processes and an Ambr®250 system. This method resulted in comparable viable cell density and product titer across scales, emphasizing that the scale-up procedure should be selected based on the specific requirements of the process [76].

Constant minimum air sparge vvm and equivalent P/V have been used in combination as scale-up criteria to ensure that cell growth, gas transfer, and mAb productivity were all matched across scales in 3 L, 500 L, and 2000 L SUBs [43]. The minimum air flow rate was selected to achieve proper stripping of CO_2. In a different study, the use of constant P/V and vvm as criteria to create a scale-down model using the Ambr®250 resulted in low DO levels (<5%) and could not mimic the performance at large-scale [19]. This was attributed to the lower gas residence time at small-scale and the ensuing decrease in the oxygen transfer coefficient.

Another important factor to take into account during bioreactor scale-up is the impact of the gas entrance velocity. This is particularly a concern in industrial-scale bioreactors where higher gas flow rates are required for oxygen supply and to ensure proper CO_2 stripping. A high gas entrance velocity (>60 m/s) is considered detrimental for culture performance, due to the sharp increase in turbulent energy transfer in the sparger region [80]. Additionally, a high sparge rate has been shown to trigger an oxidative stress response and was also associated with an increase in amino acid consumption to restore redox balance [81]. Entrance velocities of less than 20 m/s [80], and in some cases up to 50 m/s [43], were shown to not cause significant cell damage in small to large-scale bioreactors.

5.4 Constant Impeller Tip Speed

Maintaining an equivalent impeller tip speed is used as a scale-up criterion in order to maintain comparable maximum shear stress levels at different production scales. For instance, the use of this scale-up method resulted in similar viable cell densities and final product titers between Ambr® bioreactor systems and benchtop bioreactors when using tip speeds in the range of 0.7–0.8 m/s [18, 78]. Impeller tip speeds greater than 1.5 m/s are suspected to cause cell damage [82]. While comparable growth and productivity could be achieved using constant impeller tip speed to scale-up from Ambr®15 to 2 L benchtop bioreactor, dissolved oxygen levels were reduced in the 2 L vessel [17]. For large changes in production volumes, the use of constant impeller tip speed may result in insufficient oxygen mass transfer rates and poor mixing, as the volumetric power input is significantly reduced (see Table 6).

6 Tools to Assist Culture Scale-Up

6.1 Computational Fluid Dynamics (CFD)

Computational fluid dynamics (CFD) is a powerful numerical tool that is widely used for the analysis of fluid flows and can be invaluable to assist with bioreactor scale-up. Many commercial or open-source CFD packages are available to characterize flow conditions and mass transfer in bioreactors, based on the vessel geometry and operating conditions. On an existing vessel, CFD offers a way to optimize process conditions without conducting extensive and costly experimentation at large-scale. For the design and scale-up of bioreactors, CFD simulations can be employed to predict the behaviour in a large vessel. Of particular interest, the volumetric oxygen mass transfer coefficient (k_La) can be derived from the CFD simulation outputs. Most often, this is accomplished based on Higbie's penetration theory [84–88]. In this model, the film mass transfer coefficient (k_L) is expressed as a function of the energy dissipation rate (ε), the oxygen diffusivity (D_L) and the fluid properties (kinetic viscosity, ν) [89].

$$k_L = \frac{2}{\sqrt{\pi}}\sqrt{D_L\sqrt{\frac{\varepsilon}{\nu}}} \tag{3}$$

The k_La can then be estimated by multiplying the mass transfer coefficient by the specific area (a), given by:

$$a = \frac{6\alpha_G}{d_b} \tag{4}$$

where α_G is the gas volume fraction and d_b is the average bubble diameter. As reviewed by Maltby et al. [89], other expressions for the mass transfer coefficients have also been shown to give consistent estimations of k_La, some of which explicitly include the superficial velocity of the gas phase.

Computational fluid dynamics has been used to characterize and improve cell culture systems ranging from microtiter plates to industrial scale stirred-tank bioreactors. Estimates of the volumetric mass transfer coefficient can be obtained for cell culture systems in which its experimental measurement is not readily feasible, such as in microtiter plates [85]. With CFD, contour plots of k_La values can be generated [88], which allow to conveniently visualize mass transfer rate gradients within a bioreactor vessel under various operating conditions. This can provide invaluable information to optimize the process conditions so as to minimize oxygen gradients within the culture environment. For instance, Villiger et al. have determined that up to 6% of DO variation can be expected within a 15,000 L bioreactor maintained at 50% air saturation, which was in good agreement with experimental results [87]. In the same vein, CFD has also been employed to generate fluid velocity and energy dissipation contour plots [84, 86, 90–94]. Such data can help identify the presence of mixing dead zones, as well as differences in mixing patterns between bioreactor systems. Another useful application of CFD for bioreactors consists in evaluating hydrodynamic stress during cell culture [85, 87, 88, 91–94], guiding the establishment of culture conditions to minimize shear damage to cells. A 2000 L process was successfully transferred to the 5000 L scale with the aid of CFD to determine proper operating conditions based on k_La and fluid velocity contour plots [88]. First, agitation speeds providing adequate mixing times were determined, then flow rates resulting in minimal k_La were established. Other key variables, such as gas entrance velocity and shear rate, were also optimized through CFD

Table 8 Recent CFD-based characterizations of bioreactors

References	Characterizations performed by CFD
[84]	• k_La estimation • Fluid velocity, kinetic energy dissipation rate, and bubble size distribution
[85]	• k_La estimation • Hydrodynamic stress
[86]	• k_La estimation • Velocity flow field • Bubble and turbulent energy dissipation rate distribution
[87]	• k_La estimation • Hydrodynamic stress, gas volume fraction, bubble size • Mixing time • Oxygen gradient
[90]	• k_La estimation • Fluid velocity • Gas volume fraction and bubble size
[88]	• Gas exit velocity, mean shear rate near impeller, mean bubble residence time, gas hold-up, mean bubble rise velocity, mean bubble diameter • k_La contour plots
[91]	• Fluid velocity • Impeller, bulk zone, average, and maximum shear strain rate estimation
[92]	• Radial, tangential and axial fluid velocity, shear stress, and turbulent intensity contour plots
[94]	• k_La estimation • Fluid velocity • Shear stress
[93]	• Average eddy size estimation • Velocity, shear stress, and energy dissipation rate

simulations. Table 8 presents an overview of bioreactor characterization studies based on computational fluid dynamics.

6.2 *Scale-Down Models*

Scale-down models attempt to reproduce large-scale cell culture performances and operating conditions in smaller cultivation devices. Reduced size models offer the opportunity to optimize operating parameters in smaller units, often in parallel, which reduces costs and time needed for optimization. Micro- or mini-bioreactor systems (e.g. Ambr, Dasgip, Minifors) are being extensively used as scale-down models of large-scale processes. These cell cultivation platforms are favoured over plates and flasks as they have been shown to more accurately mimic cultures in large-scale bioreactors [16–19, 76]. Process conditions in scale-down models are established using the same criteria discussed above: constant k_La, constant P/V, constant vvm, or combinations of these.

Scale-down models can also be specifically designed to generate and study the impact of undesired or extreme conditions that can prevail in large-scale vessels (i.e. "worst-case" scenarios), such as poor mixing or pH/DO heterogeneities. For instance, a recent study was done at various agitations rates in benchtop bioreactors in order to mimic the impact of hydrodynamic stress encountered in various regions of large-scale bioreactors [95]. This allowed to assess the extent of hydrodynamic stress that CHO cells can tolerate before exhibiting a reduced specific productivity. Other studies have aimed at reproducing heterogeneities found in large-scale bioreactors to characterize their impact on growth and productivity. A small-scale 2-compartment bioreactor system was employed to reproduce impact of pH heterogeneities encountered at large scales [59, 96]. Such heterogeneities were found to affect cell growth, particularly towards the beginning of the cell culture and differences in pH as small as 0.4 units were found to significantly impact viable cell density.

7 Process Intensification and Scale-Up

Process intensification aims to increase the production efficiency by increasing volumetric productivity, lowering manufacturing costs, and reducing physical footprint. Since volumetric productivity is largely dependent on the cell density achieved in bioreactor, the highest values are obtained in highly optimized fed-batch, perfusion, and concentrated fed-batch processes [97, 98]. A recently popular approach to enhancing volumetric productivity in fed-batch bioreactors is to increase the inoculation viable cell density in the production vessel (N stage bioreactor) from ~0.5×10^6 cells/mL to $2–10 \times 10^6$ cells/mL through the use of a pre-stage ($N - 1$) perfusion bioreactor. High-seeding fed-batch bioreactors were found to have almost doubled final titers when compared with conventional fed-batch [99]. Seeding cell densities in the range of $10–20 \times 10^6$ cells/mL were also shown to translate into significant improvements of productivity and this strategy was successfully applied at the 500 L scale [100]. An alternate method to increase the seeding density using batch or fed-batch cultures with concentrated media at the $N - 1$ stage resulted in comparable final titers and protein quality [101]. This approach has the advantage of being simpler than $N - 1$ perfusion and the required associated equipment (i.e. retention device and storage vessels).

Although process intensification through continuous process operation is largely applied in the chemical industry, the biopharmaceutical industry is still largely dominated by fed-batch bioprocessing for recombinant protein production. Since higher productivity can be achieved with a perfusion mode, it is an interesting avenue for process intensification. In perfusion cell cultures, alternative tangential flow filtration (ATF) and tangential flow filtration (TFF) are the preferred methods of cell retention. ATF filtration benefits of reduced filter fouling and higher cell densities as a result of a self-cleaning effect [102, 103]. Perfusion cell cultures using ATF cell retention was shown to be economically attractive over fed-batch when cell concentrations in perfusion are at least threefold greater [102]. The concept of integrated continuous production is also gaining increased interest for cost-effective biomanufacturing [102, 104–108]. A proof of concept for integrated continuous processes has been done using perfusion cell culture and semicontinuous chromatography purification [108].

Process intensification has many implications for the scale-up of cell cultures. Increasing production efficiency and volumetric productivity can translate into a significant reduction in bioreactor size. Thus, intensified processes benefit from reduced risk associated with scale-up. However, operation at really high cell densities (up to 100×10^6 cells/mL) [109] results in increased oxygen transfer and carbon dioxide stripping demands. These requirements must be met through increased sparged gas flow rates and mixing speeds. This may potentially accentuate issues with mixing and mass transfer limitations which are of high concern in conventional bioreactor scale-up. Furthermore, additional considerations must be taken into account when performing scale-up of intensified processes. For example, in perfusion cell cultures, a successful scale-up is highly dependent on cell retention efficiency, fouling prevention efficiency and the similarity of equipment used at different scales [110]. Representative scale-down models will need to be developed for the rational optimization of these processes.

8 Conclusion

Predictable scale-up of biologics production is critical due to high costs associated with manufacturing downtime and failure. For example, a biotherapeutic product with expected annual sales of $1 billion would cost $80 million for every month loss during start-up, technology transfer, or scale-up [111]. However, there exists no single method of performing culture scale-up and, as such, translating optimized process conditions to larger production bioreactors remains an art. In light of this review, the following general considerations should be taken into account for the scale-up of CHO cell cultures:

- Constant P/V may be best suited for scale-up between small differences in production volumes. For large changes in scale, scale-up based on maintaining constant oxygen transfer rate may be more appropriate. In any case, cell culture processes may have specific prioritized requirements (oxygen mass transfer, CO_2 accumulation, mixing, etc.) such that scale-up criteria should be selected based on the needs of a particular process.
- Although miniature bioreactor systems are increasingly available and being used for high-throughput process optimization, they may not always constitute a representative scale-down model due mainly to mass transfer limitations. Mini-bioreactors or benchtop bioreactors, which have more flexibility in terms of mixing and sparging conditions, may more closely resemble large-scale bioreactors in their operation.

- During small-scale process optimization, it is of high importance to evaluate the impact of additives that will be employed at large-scale, such as surfactants (e.g. Pluronic F-68 or Kolliphor® P188) or anti-foaming agents, as these additives may have profound impacts on gas transfer and cell viability.
- Different equipment may be required when changing scales (e.g. gas sparger, retention device). Such modifications may greatly affect cell culture performance and gas evolution profiles. When transitioning to a vessel with different equipment, other process adjustments (i.e. agitation speed or gas flow rates) may be required to match cell culture performance across scales.
- Computational fluid dynamics (CFD) can be a powerful aid to predict important parameters like oxygen mass transfer coefficient and hydrodynamic stress at large-scale under various process conditions, thereby avoiding costly experimental trials.
- Scale-down models also constitute valuable tools to assist in process optimization and scale-up, particularly to study the effects of non-optimal conditions encountered in large production vessels. Scale-down models can also be useful during upstream process development to assess the robustness of the highest producer clones and their suitability for large-scale manufacturing.

References

1. Allied Market Research (2018) Biopharmaceuticals market by type and application-global opportunity analysis and industry forecast, 2018–2025, Market research report, July 2018
2. Lu R-M et al (2020) Development of therapeutic antibodies for the treatment of diseases. J Biomed Sci 27(1):1
3. Wang C et al (2020) A human monoclonal antibody blocking SARS-CoV-2 infection. Nat Commun 11(1):2251
4. Baum A et al (2020) Antibody cocktail to SARS-CoV-2 spike protein prevents rapid mutational escape seen with individual antibodies. Science 369(6506):1014–1018
5. Zost SJ et al (2020) Potently neutralizing and protective human antibodies against SARS-CoV-2. Nature 584(7821):443–449
6. Shirahata H et al (2019) Dynamic modelling, simulation and economic evaluation of two CHO cell-based production modes towards developing biopharmaceutical manufacturing processes. Chem Eng Res Des 150:218–233
7. Becker M et al (2019) The less the better: how suppressed base addition boosts production of monoclonal antibodies with Chinese hamster ovary cells. Front Bioeng Biotechnol 7:76
8. Rader RA, Langer ES (2015) Years of upstream productivity improvements. BioProcess Int 13(2):2015
9. Takagi Y et al (2017) The enhancement of antibody concentration and achievement of high cell density CHO cell cultivation by adding nucleoside. Cytotechnology 69(3):511–521
10. Huang YM et al (2010) Maximizing productivity of CHO cell-based fed-batch culture using chemically defined media conditions and typical manufacturing equipment. Biotechnol Prog 26(5):1400–1410
11. Carpio M (2020) Current challenges with cell culture scale-up for biologics production. BioPharm Int 33(10):23–27
12. Macdonald GJ (2019) Scale-out plus single-use can multiply yields: single-use technology reduces some biomanufacturing equations to scale-out> scale-up. Genet Eng Biotechnol News 39(11):46–50

13. Klöckner W, Büchs J (2012) Advances in shaking technologies. Trends Biotechnol 30(6):307–314
14. Klöckner W et al (2013) Correlation between mass transfer coefficient kLa and relevant operating parameters in cylindrical disposable shaken bioreactors on a bench-to-pilot scale. J Biol Eng 7(1):28
15. Chaturvedi K et al (2014) Comparison of the behavior of CHO cells during cultivation in 24-square deep well microplates and conventional shake flask systems. Biotechnol Rep 1–2:22–26
16. Nienow AW (2015) Aeration in biotechnology. In: Kirk-Othmer encyclopedia of chemical technology, pp 1–23
17. Hsu WT et al (2012) Advanced microscale bioreactor system: a representative scale-down model for bench-top bioreactors. Cytotechnology 64(6):667–678
18. Nienow AW et al (2013) The physical characterisation of a microscale parallel bioreactor platform with an industrial CHO cell line expressing an IgG4. Biochem Eng J 76:25–36
19. Xu P et al (2017) Characterization of TAP Ambr 250 disposable bioreactors, as a reliable scale-down model for biologics process development. Biotechnol Prog 33(2):478–489
20. Sieblist C et al (2011) Insights into large-scale cell-culture reactors: I. Liquid mixing and oxygen supply. Biotechnol J 6(12):1532–1546
21. Paul Smelko J et al (2011) Performance of high intensity fed-batch mammalian cell cultures in disposable bioreactor systems. Biotechnol Prog 27(5):1358–1364
22. Sherman M et al (2016) Continuous cell culture operation at 2,000-L scale. BioProcess Int 14(10):22
23. Dorceus M (2018) Cell culture scale-up in stirred-tank single-use bioreactors. BioProcess Int 16(15):2–6
24. Kaisermayer C, Yang J (2013) Highly efficient inoculum propagation in perfusion culture using WAVE Bioreactor™ systems. BMC Proc 7(6):P7
25. Wright B et al (2015) A novel seed-train process: using high-density cell banking, a disposable bioreactor, and perfusion technologies. BioProcess Int 13. https://bioprocess intl.com/upstream-processing/upstream-single-use-technologies/novel-seed-train-process-using-high-density-cell-banking-disposable-bioreactor-perfusion-technologies/
26. Imseng N et al (2014) Single-use wave-mixed versus stirred bioreactors for insect-cell/BEVS-based protein expression at benchtop scale. Eng Life Sci 14:264
27. Busse C et al (2017) Sensors for disposable bioreactors. Eng Life Sci 17(8):940–952
28. Baez A, Shiloach J (2014) Effect of elevated oxygen concentration on bacteria, yeasts, and cells propagated for production of biological compounds. Microb Cell Factories 13:181–181
29. Handlogten MW, Zhu M, Ahuja S (2018) Intracellular response of CHO cells to oxidative stress and its influence on metabolism and antibody production. Biochem Eng J 133:12–20
30. Gao Y et al (2016) Combined metabolomics and proteomics reveals hypoxia as a cause of lower productivity on scale-up to a 5000-liter CHO bioprocess. Biotechnol J 11(9):1190–1200
31. Velez-Suberbie ML et al (2013) Impact of aeration strategy on CHO cell performance during antibody production. Biotechnol Prog 29(1):116–126
32. Nienow AW (2006) Reactor engineering in large scale animal cell culture. Cytotechnology 50(1–3):9–33
33. Tharmalingam T, Goudar CT (2015) Evaluating the impact of high Pluronic® F68 concentrations on antibody producing CHO cell lines. Biotechnol Bioeng 112(4):832–837
34. Zhang X et al (2019) Systematic evaluation of high-throughput scale-down models for single-use bioreactors (SUB) using volumetric gas flow rate as the criterion. Biochem Eng J 151:107307
35. Xu S et al (2018) Probing lactate metabolism variations in large-scale bioreactors. Biotechnol Prog 34(3):756–766
36. Xing Z et al (2009) Scale-up analysis for a CHO cell culture process in large-scale bioreactors. Biotechnol Bioeng 103(4):733–746

37. Hoshan L et al (2019) Effective bioreactor pH control using only sparging gases. Biotechnol Prog 35(1):e2743
38. He C et al (2019) A systematic mass-transfer modeling approach for mammalian cell culture bioreactor scale-up. Biochem Eng J 141:173–181
39. Xing Z et al (2017) A carbon dioxide stripping model for mammalian cell culture in manufacturing scale bioreactors. Biotechnol Bioeng 114(6):1184–1194
40. Van't Riet K (1979) Review of measuring methods and results in nonviscous gas-liquid mass transfer in stirred vessels. Ind Eng Chem Process Des Dev 18(3):357–364
41. Mostafa SS, Gu X (2003) Strategies for improved dCO2 removal in large-scale fed-batch cultures. Biotechnol Prog 19(1):45–51
42. Matsunaga N et al (2009) Culture scale-up studies as seen from the viewpoint of oxygen supply and dissolved carbon dioxide stripping. J Biosci Bioeng 107(4):412–418
43. Xu S et al (2017) A practical approach in bioreactor scale-up and process transfer using a combination of constant P/V and vvm as the criterion. Biotechnol Prog 33(4):1146–1159
44. Dreher T et al (2014) Design space definition for a stirred single-use bioreactor family from 50 to 2000 L scale. Eng Life Sci 14(3):304–310
45. Blombach B, Takors R (2015) CO2 - intrinsic product, essential substrate, and regulatory trigger of microbial and mammalian production processes. Front Bioeng Biotechnol 3:108–108
46. Pattison RN et al (2000) Measurement and control of dissolved carbon dioxide in mammalian cell culture processes using an in situ fiber optic chemical sensor. Biotechnol Prog 16(5):769–774
47. Takuma S, Hirashima C, Piret JM (2003) Effects of glucose and CO2 concentrations on CHO cell physiology. Springer Netherlands, Dordrecht
48. Zhu MM et al (2005) Effects of elevated pCO2 and osmolality on growth of CHO cells and production of antibody-fusion protein B1: a case study. Biotechnol Prog 21(1):70–77
49. Matanguihan R et al (2001) Solution to the high dissolved CO2 problem in high-density perfusion culture of mammalian cells. In: Lindner-Olsson E, Chatzissavidou N, Lüllau E (eds) Animal cell technology: from target to market: proceedings of the 17th ESACT meeting Tylösand, Sweden, June 10–14, 2001. Springer Netherlands, Dordrecht, pp 399–402
50. Hossler P, Khattak SF, Li ZJ (2009) Optimal and consistent protein glycosylation in mammalian cell culture. Glycobiology 19(9):936–949
51. Min Lee G, Koo J (2009) Osmolarity effects, Chinese hamster ovary cell culture. In: Encyclopedia of industrial biotechnology: bioprocess, bioseparation, and cell technology, pp 1–8
52. Ryu JS, Lee GM (1999) Application of hypoosmolar medium to fed-batch culture of hybridoma cells for improvement of culture longevity. Biotechnol Bioeng 62(1):120–123
53. Sieblist C et al (2011) Insights into large-scale cell-culture reactors: II. Gas-phase mixing and CO(2) stripping. Biotechnol J 6(12):1547–1556
54. Wang C et al (2018) Ultra-low carbon dioxide partial pressure improves the galactosylation of a monoclonal antibody produced in Chinese hamster ovary cells in a bioreactor. Biotechnol Lett 40(8):1201–1208
55. Xu S et al (2018) Bioreactor scale-up. In: Kirk-Othmer encyclopedia of chemical technology, pp 1–35
56. Sieblist C, Jenzsch M, Pohlscheidt M (2016) Equipment characterization to mitigate risks during transfers of cell culture manufacturing processes. Cytotechnology 68(4):1381–1401
57. Paul K, Herwig C (2020) Scale-down simulators for mammalian cell culture as tools to access the impact of inhomogeneities occurring in large-scale bioreactors. Eng Life Sci 20(5–6):197–204
58. Restelli V et al (2006) The effect of dissolved oxygen on the production and the glycosylation profile of recombinant human erythropoietin produced from CHO cells. Biotechnol Bioeng 94(3):481–494
59. Brunner M et al (2017) The impact of pH inhomogeneities on CHO cell physiology and fed-batch process performance – two-compartment scale-down modelling and intracellular pH excursion. Biotechnol J 12:7

60. Boulton-Stone JM, Blake JR (1993) Gas bubbles bursting at a free surface. J Fluid Mech 254:437–466
61. Keane JT, Ryan D, Gray PP (2003) Effect of shear stress on expression of a recombinant protein by Chinese hamster ovary cells. Biotechnol Bioeng 81(2):211–220
62. Johnson C, Natarajan V, Antoniou C (2014) Verification of energy dissipation rate scalability in pilot and production scale bioreactors using computational fluid dynamics. Biotechnol Prog 30(3):760–764
63. Hu W, Berdugo C, Chalmers JJ (2011) The potential of hydrodynamic damage to animal cells of industrial relevance: current understanding. Cytotechnology 63(5):445–460
64. Sabourin M et al (2011) Increasing antibody yield and modulating final product quality using the freedom(TM) CHO-S(TM) production platform. BMC Proc 5 Suppl 8(Suppl 8):P102–P102
65. Zhang Z, Al-Rubeai M, Thomas CR (1992) Effect of pluronic F-68 on the mechanical properties of mammalian cells. Enzym Microb Technol 14(12):980–983
66. Ramirez OT, Mutharasan R (1990) The role of the plasma membrane fluidity on the shear sensitivity of hybridomas grown under hydrodynamic stress. Biotechnol Bioeng 36(9):911–920
67. Gigout A, Buschmann MD, Jolicoeur M (2008) The fate of pluronic F-68 in chondrocytes and CHO cells. Biotechnol Bioeng 100(5):975–987
68. Shuler ML, Kargi F (2001) Bioprocess engineering: basic concepts, 2nd edn. Prentice Hall, Hoboken
69. Platas Barradas O et al (2012) Evaluation of criteria for bioreactor comparison and operation standardization for mammalian cell culture. Eng Life Sci 12(5):518–528
70. Serrato JA et al (2004) Heterogeneous conditions in dissolved oxygen affect N-glycosylation but not productivity of a monoclonal antibody in hybridoma cultures. Biotechnol Bioeng 88(2):176–188
71. Tescione L et al (2015) Application of bioreactor design principles and multivariate analysis for development of cell culture scale down models. Biotechnol Bioeng 112(1):84–97
72. Marks DM (2003) Equipment design considerations for large scale cell culture. Cytotechnology 42(1):21–33
73. Janakiraman V et al (2015) Application of high-throughput mini-bioreactor system for systematic scale-down modeling, process characterization, and control strategy development. Biotechnol Prog 31(6):1623–1632
74. Nie L et al (2019) Development and qualification of a scale-down mammalian cell culture model and application in design space development by definitive screening design. AAPS PharmSciTech 20(6):246
75. Möller J et al (2020) Model uncertainty-based evaluation of process strategies during scale-up of biopharmaceutical processes. Comput Chem Eng 134:106693
76. Manahan M et al (2019) Scale-down model qualification of ambr® 250 high-throughput mini-bioreactor system for two commercial-scale mAb processes. Biotechnol Prog 35(6):e2870
77. Doi T et al (2020) Development of a scale-up strategy for Chinese hamster ovary cell culture processes using the kL a ratio as a direct indicator of gas stripping conditions. Biotechnol Prog n/a(n/a):e3000
78. Alsayyari AA et al (2018) Transcriptome analysis for the scale-down of a CHO cell fed-batch process. J Biotechnol 279:61–72
79. Metze S et al (2020) Monitoring online biomass with a capacitance sensor during scale-up of industrially relevant CHO cell culture fed-batch processes in single-use bioreactors. Bioprocess Biosyst Eng 43(2):193–205
80. Chaudhary G et al (2020) Understanding the effect of high gas entrance velocity on Chinese hamster ovary (CHO) cell culture performance and its implications on bioreactor scale-up and sparger design. Biotechnol Bioeng 117(6):1684–1695
81. Hong JK et al (2020) In silico model-based characterization of metabolic response to harsh sparging stress in fed-batch CHO cell cultures. J Biotechnol 308:10–20

82. Varley J, Birch J (1999) Reactor design for large scale suspension animal cell culture. Cytotechnology 29(3):177
83. Gimenez L, Simonet C, Malphettes L (2013) Scale-up considerations for monoclonal antibody production process: an oxygen transfer flux approach. BMC Proc 7(Suppl 6):P49
84. Li X et al (2018) Development of a computational fluid dynamics model for scaling-up ambr bioreactors. Biotechnol Bioprocess Eng 23(6):710–725
85. Wutz J et al (2018) Establishment of a CFD-based kLa model in microtiter plates to support CHO cell culture scale-up during clone selection. Biotechnol Prog 34(5):1120–1128
86. Wutz J et al (2016) Predictability of kLa in stirred tank reactors under multiple operating conditions using an Euler–Lagrange approach. Eng Life Sci 16(7):633–642
87. Villiger TK et al (2018) Experimental and CFD physical characterization of animal cell bioreactors: from micro- to production scale. Biochem Eng J 131:84–94
88. Scully J et al (2020) Beyond heuristics: CFD-based novel multiparameter scale-up for geometrically disparate bioreactors demonstrated at industrial 2kL–10kL scales. Biotechnol Bioeng 117(6):1710–1723
89. Maltby R et al (2016) Multiphase CFD modelling of single-use-technology bioreactors for industrial biotechnology applications. In: 3rd international conference on fluid flow, heat and mass transfer (FFHMT'16). Emerald Group Publishing Ltd, Dublin
90. Amer M, Feng Y, Ramsey JD (2019) Using CFD simulations and statistical analysis to correlate oxygen mass transfer coefficient to both geometrical parameters and operating conditions in a stirred-tank bioreactor. Biotechnol Prog 35(3):e2785
91. Li C et al (2020) Novel scale-up strategy based on three-dimensional shear space for animal cell culture. Chem Eng Sci 212:115329
92. Delafosse A et al (2018) Comparison of hydrodynamics in standard stainless steel and single-use bioreactors by means of an Euler-Lagrange approach. Chem Eng Sci 188:52–64
93. Borys BS et al (2019) Using computational fluid dynamics (CFD) modeling to understand murine embryonic stem cell aggregate size and pluripotency distributions in stirred suspension bioreactors. J Biotechnol 304:16–27
94. Zhu L, Song B, Wang Z (2019) Developing an orbitally shaken bioreactor featuring a hollow cylinder vessel wall. J Chem Technol Biotechnol 94(7):2212–2218
95. Sieck JB et al (2013) Development of a scale-down model of hydrodynamic stress to study the performance of an industrial CHO cell line under simulated production scale bioreactor conditions. J Biotechnol 164(1):41–49
96. Paul K et al (2020) Development, characterization, and application of a 2-compartment system to investigate the impact of pH inhomogeneities in large-scale CHO-based processes. Eng Life Sci 20(8):368–378
97. Xu S et al (2017) Bioreactor productivity and media cost comparison for different intensified cell culture processes. Biotechnol Prog 33(4):867–878
98. Yang WC et al (2016) Concentrated fed-batch cell culture increases manufacturing capacity without additional volumetric capacity. J Biotechnol 217:1–11
99. Stepper L et al (2020) Pre-stage perfusion and ultra-high seeding cell density in CHO fed-batch culture: a case study for process intensification guided by systems biotechnology. Bioprocess Biosyst Eng 43(8):1431–1443
100. Xu J et al (2020) Development of an intensified fed-batch production platform with doubled titers using N-1 perfusion seed for cell culture manufacturing. Bioresour Bioprocess 7(1):17
101. Yongky A et al (2019) Process intensification in fed-batch production bioreactors using non-perfusion seed cultures. MAbs 11(8):1502–1514
102. Pollock J et al (2017) Integrated continuous bioprocessing: economic, operational, and environmental feasibility for clinical and commercial antibody manufacture. Biotechnol Prog 33(4):854–866
103. Bielser J-M et al (2018) Perfusion mammalian cell culture for recombinant protein manufacturing – a critical review. Biotechnol Adv 36(4):1328–1340
104. Gagnon M et al (2018) Shift to high-intensity, low-volume perfusion cell culture enabling a continuous, integrated bioprocess. Biotechnol Prog 34(6):1472–1481

105. Arnold L et al (2019) Implementation of fully integrated continuous antibody processing: effects on productivity and COGm. Biotechnol J 14(2):e1800061
106. Kamga MH, Cattaneo M, Yoon S (2018) Integrated continuous biomanufacturing platform with ATF perfusion and one column chromatography operation for optimum resin utilization and productivity. Prep Biochem Biotechnol 48(5):383–390
107. Karst DJ et al (2017) Process performance and product quality in an integrated continuous antibody production process. Biotechnol Bioeng 114(2):298–307
108. Warikoo V et al (2012) Integrated continuous production of recombinant therapeutic proteins. Biotechnol Bioeng 109(12):3018–3029
109. Lim J et al (2011) An economic comparison of three cell culture techniques. BioPharm Int 24(2):54–60
110. Raja V et al (2011) Challenges in scaling up a perfusion process. BMC Proc 5 Suppl 8(Suppl 8):P122–P122
111. Matzmorr W (2016) Integrating technology transfer and facilities startup for biologics. BioPharm Int 29:46–48

Intensified and Continuous mAb Production with Single-Use Systems

Jan Müller, Misha Teale, Sandra Steiner, Stefan Junne, Peter Neubauer, Dieter Eibl, and Regine Eibl

Abbrevations

ATF	Alternating tangential flow filtration
CHO	Chinese hamster ovary
COG	Cost of goods
CSPR	Cell specific perfusion rate
DNA	Desoxyribonucleic acid
DSP	Downstream processing
GMP	Good Manufacturing Practice
HCP(s)	Host cell protein(s)
HCD(s)	High cell density/densities
L&E	Leachables and extractables
MWCO	Molecular weight cut off
mAb(s)	Monoclonal antibody/antibodies
NPCs	Net present costs
SPTFF	Single-pass tangential flow filtration
TFF	Tangential flow filtration
USP	Upstream processing
VCD(s)	Viable cell density/densities

J. Müller (✉) · M. Teale · S. Steiner · D. Eibl · R. Eibl
Institute of Chemistry and Biotechnology, School of Life Sciences and Facility Management, Zurich University of Applied Sciences, Wädenswil, Switzerland
e-mail: jan.mueller@zhaw.ch

S. Junne · P. Neubauer
Institute for Biotechnology, Technische Universität Berlin, Berlin, Germany

R. Pörtner (ed.), *Cell Culture Engineering and Technology*, Cell Engineering 10,
https://doi.org/10.1007/978-3-030-79871-0_13

1 Introduction

With the introduction of hybridoma technology in 1975 [1], Georges Köhler and César Milstein laid the basis for the *in vitro* production of therapeutic mAbs in large quantities. They created a hybridoma cell, i.e. a stable, hybrid cell capable of producing a single type of antibody directed against a specific epitope present in the antigen. To achieve this, both the German biologist and the Argentinian molecular biologist, isolated the activated B-lymphocytes of an immunized mouse and fused them with immortalized myeloma cells. Following this achievement, various new technologies (such as the phage display technique or the transgenic animal approach) have been developed for mAb generation [2]. These technologies contributed to the approval of approximately 115 therapeutic mAbs in the European Union and/or the United States of America by 2018 [3]. Alongside the murine and chimeric antibodies, increasingly humanized and even human antibodies have been made commercially available, many of which have achieved blockbuster drug status. Prominent examples of such blockbuster drugs are Adalimumab (Humira) from AbbVie, Pembrolizumab (Keytruda) from Merck & Co., and Trastuzumab (Herceptin) and Bevacizumab (Avastin) from Roche.

The vast majority of therapeutic mAbs have multiple disease indications and are produced using CHO suspension cells [3, 4] in fed-batch processes of between 12 and 20 days in combination with serum-free or chemically defined culture media [5, 6]. In this manner, medium to high cell densities (HCDs) of between $5 \cdot 10^6$ cells mL^{-1} to $3 \cdot 10^7$ cells mL^{-1} are achieved, with target mAb titers between 2 g L^{-1} and 5 g L^{-1}. In some cases, a temperature shift to values between 28 and 34 °C, either during the middle or late exponential growth phase or even at the beginning of the stationary phase, is performed to further increase the mAb titer [7–9]. The dominant 10–20 m^3 bioreactors, used for the production of mAbs, are often stirred systems. They are generally inoculated with a low viable cell density (VCD) of $2–5 \cdot 10^5$ cells mL^{-1} and are therefore referred to as low-seed fed-batch cultures.

The potential to increase the efficiency of mAb production, alongside a simultaneous reduction of production costs, accelerated the development of new methods for process intensification around unit operations, e.g. inoculum production, fermentation, clarification, capture, polishing and formulation [10]. If applied to both upstream and downstream processing, an extremely intensified overall process is achieved, with unit operations often conducted continuously. Konstantinov and Cooney have defined a continuous unit operation as a subunit of a process, in which a continuous flow input for a prolonged period of time and minimal internal hold volume is present. The output can then either be continuous or discretized into small batches, which are then produced in a cyclic manner [11]. Based on this definition of integrated unit operations with no or minimal internal hold volume, Konstantinov and Cooney define two continuous process types: (1) the hybrid continuous process, in which batch and continuous unit operations are combined, and (2) the fully integrated unit operation, also known as the fully continuous or end-to-end process. Although the fully continuous process offers the greatest advantages between the

Table 1 Trade name, therapeutic indication, manufacturer and year of approval of mAbs produced in perfusion (modified from [12, 13])

Trade name	Therapeutic indication	Manufacturer	Approval
ReoPro	High-risk angioplasty	Janssen Biotech	1994
Remicade	Crohn's disease; ulcerative colitis; rheumatoid arthritis; ankylosing spondylitis; psoriatic arthritis; plaque psoriasis	Janssen Biotech	1998
Simulect	Reversal of transplantation rejection	Novartis	1998
Campath	B-cell chronic lymphocyte leukemia	Genzyme (Sanofi)	2001
Simponi	Rheumatoid arthritis; psoriatic arthritis; ankylosing spondylitis	Janssen Biotech	2009
Stelara	Psoriasis, psoriatic arthritis, Crohn's disease, plaque psoriasis	Janssen Biotech	2009

continuous process types, it has yet to be used for commercial biotherapeutic production (see also Sect. 4).

Hybrid continuous processes, however, are already a reality in biotherapeutics production. According to Konstantinov and Cooney [11], a distinction can be made between three hybrid continuous process types:

(i) Continuous upstream followed by batch downstream
(ii) Continuous upstream and capture with batch downstream
(iii) Batch upstream followed by continuous downstream

The first of the three process types, namely *processes combining a continuous upstream process followed by a batch downstream*, has been used for more than 20 years and is, therefore, the oldest and most well established [14], resulting in the commercial production of about 20 products (therapeutic enzymes, blood factors and mAbs). An overview of mAbs produced in perfusion mode is given in Table 1. This process type, which uses smaller equipment during the USP and larger equipment for DSP, was originally chosen to facilitate the production of unstable products such as Xigris, Kogenate or Cerezyme (see also Sect. 4.1).

The extension of the continuous processing (case 2 from above) to encompass the initial chromatography step, namely the capture step (see also Sect. 4.2), is considered advantageous for three reasons [11]: (i) large holding tanks between the production bioreactor and the antibody capture step (including the clarification unit operations) are no longer required, (ii) the implementation of continuous capture leads to a 1–2 order of magnitude reduction in the protein A column size and (iii) buffer utilization can be reduced. This is important, as downstream mAb production process costs amount to between 50% and 80% of the total production costs [15, 16]. The capture step alone accounts for approximately 25% of these costs, due to the application of columns with a large diameter [17].

Finally, when applying the third process type, namely *batch upstream followed by continuous downstream*, cell cultivation is performed using the traditional fed-batch mode, while one or two downstream unit operations (e.g. capture, polishing) are operated continuously (see also Sect. 4.2). This means smaller continuous chromatography columns can be used, which reduces the amount of expensive column resins needed and leads to a reduction of operational costs [18, 19]. This approach shows how an increase of operating hours, e.g. through the application of continuous processes, can lead to the overall reduction of production costs while maintaining unit operations and product quality. The role of single-use equipment in this context will be described in the following section.

2 Single-Use Technology as a Driver of Modern mAb Production Processes

It is undisputed that the availability of a wide range of single-use (disposable) systems has driven the development of more intensive and continuous mAb production processes in recent years [20, 21]. The surfaces of these disposable systems, which are in direct contact with either the media, cells or product are generally made of polymer films (e.g. polyethylene, polycarbonate, ethylene vinyl acetate) and, as the name implies, are designed for single use only. This may also mean that such a unit is used over an entire campaign, i.e. a production cycle, and only then replaced for a new production campaign.

Single-use systems may be either be rigid or flexible and are usually purchased following gamma sterilization. This eliminates the need for equipment sterilization, which is mandatory for their reusable glass or stainless-steel counterparts. This makes single-use systems advantageous, as they can be installed and operated quickly, after which they can easily be decontaminated and disposed of. As a result, lengthy and expensive cleaning procedures and the cleaning validations required by Good Manufacturing Practice (GMP) are no longer necessary for these technologies. This means that new production processes and product changes can be undertaken immediately following process interruption.

Furthermore, compared to the cleaning and sterilization of stainless-steel equipment, the bioburden of such single-use systems, which only require gamma irradiation prior to use, is far less. Therefore, if single-use systems are correctly selected and handled, they are not only considered greener and more flexible but are also safer, cheaper and smaller, due to the decreased risk of microbial and cross-contamination, the lower capital investment and the reduced facility footprint [22, 23].

There are, however, also limitations to the application of single-use systems, resulting from the properties and processing of the polymers used for their manufacture. In addition to leakage/breakage, the majority of users see the primary risk associated with the use of single-use systems in the possible occurrence of unacceptable amounts of leachables and extractables (L&E) [24]. L&E are chemical substances that are released from the plastic either under worst-case conditions (extractables, e.g. at high temperature) or process conditions (leachables), after which they migrate into medium if in contact with the surface. These substances (e.g. *bis*(2,4-di-*tert*-butylphenyl)phosphate, 2,4-di-*tert*-butylphenol) may either inhibit cell growth, may have a negative impact on mAb quantity and quality or, in the worst case, may even be cytotoxic [25–27]. Therefore, unit operations closest to the final product and patient are considered exceptionally critical in the context of L&E [28].

2.1 Single-Use Technology: Main Developmental Milestones

The improvements made by the manufacturers of single-use systems in recent years, in terms of film materials, design, sensors and automation technology, have led to greater user confidence in them. Therefore, it is unsurprising, that more than 85% of the equipment used in biopharmaceutical processes for the production of preclinical and clinical samples is currently based on single-use systems [24]. Single-use technology is also increasingly used for commercial production. While in the 1980s, 2D (pillow-like) bags were used for the storage of medium, serum and buffer, and the first single-use filter capsules were launched, 3D bags found application in the 1990s [29]. Another important milestone in the development of single-use technology was reached at the end of the 1990s when the first scalable single-use bioreactor, the Wave Reactor [30], was launched. The success of this innovative bioreactor led to the development of further single-use bioreactor systems. Since the launch of the first stirred single-use bioreactors (Hyclone's S.U.B. and Xcellerex's XDR-Disposable Stirred Tank Bioreactor) in 2006, the majority of new bioreactors have been stirred bag systems [31]. The largest single-use bioreactor currently available is ABEC's 7,500 L Custom Single Run Bioreactor with a working volume of 6 m^3 [32]. Apart from stirred bioreactors, several wave-mixed and shaken single-use bioreactors were introduced to the market at about the same time, e.g. the 2-dimensional rocking motion bioreactor CELL-tainer and the orbitally shaken bioreactor SBX, among others [33].

In addition to the single-use bioreactors, there are now also single-use bioprocess containers, mixers, centrifuges, chromatography systems, membrane adsorbers, various filtration devices, pumps, freeze-thaw systems and isolators, as well as sampling and coupling systems. Some disposable product combinations are even offered as process platforms by key single-use technology manufacturers, with platforms available for medium and buffer production, inoculum production and fermentation or concentration and purification. A few examples of such platforms are Pall's Allegro, Sartorius's FlexAct, Cytiva's FlexFactory and ReadyToProcess, and ThermoFisher's HyPerforma series. In principle, the commercially available single-use technologies currently offered by various manufacturers already allow for the design of complete single-use facilities (from media production to filling) for the production of mAbs on a medium volume scale.

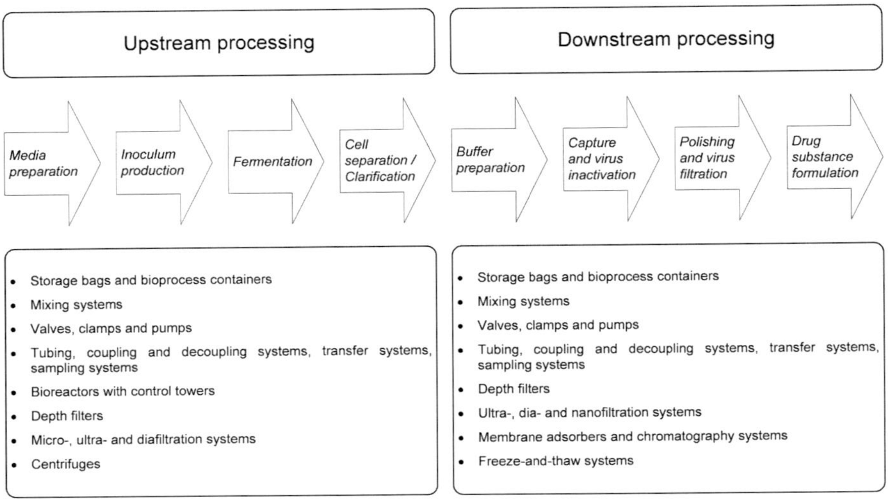

Fig. 1 Commercially available single-use systems for USP and DSP (without fill & finish)

2.2 *Single-Use USP vs. DSP*

An overview of the single-use systems available on the market for USP and DSP is shown in Fig. 1. Modern upstream mAb production processes, where the bioreactors used for inoculum or seed generation and production play an important role, can already be completely realized with single-use technology. While the inoculum preparation for modern mAb production processes is often carried out in a 10 L/20 L and 50 L/200 L wave-mixed bioreactor, stirred single-use bioreactors with a 1 m^3 and 2 m^3 working volume dominate regarding product expression [34].

In general, the scalable, dynamic single-use bioreactors available on the market can be categorized according to their mixing principle, namely into those with mechanical power input (stirred, wave-mixed, orbitally shaken), pneumatic power input and hydraulic power input [31]. Compared to their reusable counterparts, the range of mechanically driven single-use bioreactors is more diverse, the largest part of which constitutes stirred single-use bioreactors. An overview of commercially available stirred versions of single-use bioreactors, including existing bioengineering parameters and instrumentation, is given by Jossen et al. and Oosterhuis in their respective book chapters [31, 35]. Moreover, certain stirred single-use bioreactors are offered by various manufacturers with the same height-to-diameter ratio, stirring and aeration devices as their reusable counterparts have. This facilitates the transfer of a process developed in reusable bioreactors to a single-use technology platform when adapting the infrastructure of traditional production facilities [36]. For more information on the most important points which should be considered when designing such "Facilities of the Future", the interested reader should refer to the DECHEMA status paper by the same name [37].

Single-use DSP, however, currently still represents a bottleneck. On one hand, it should be noted that various single-use systems in the DSP do not support a production scale above 4,000 L [38], while on the other hand, chromatography unit operations, especially the previously mentioned expensive Protein A chromatography, still presents a fundamental weak point in DSP [24].

3 Intensified mAb Production with Single-Use Systems

It is generally accepted that existing single-use technology facilitates the implementation of unit operation intensification during mAb production. It increases productivity and reduces the bottleneck presented by specific process steps with the primary goal of implementing such intensification steps in existing fed-batch process platforms. The main approaches described in literature and explained below distinguish between USP and DSP intensification. While USP has generally been the focus thus far, in recent years DSP intensification has also become increasingly popular.

3.1 USP Intensification

Upstream intensification usually necessitates the application of a perfusion mode. Perfusion means that cells are retained within the bioreactor and that part of the culture medium is continuously exchanged. The highest cell densities ever achieved using an optimized perfusion mode were approximately $2{\cdot}10^8$ cells mL^{-1} [39], with process optimization achieved through the application of the cell-specific perfusion rate (CSPR) introduced by Ozturk [40]. The CSPR describes the available volume of media per cell and day, can be used as an indicator for the nutrient supply per cell and is therefore often used to control the perfusion process [41]. The CSPR is calculated from the quotient of the perfusion rate (the rate at which fresh medium is added to the bioreactor, usually between 1 and 3 vvd = volume fresh medium per bioreactor volume per day) and the cell density.

Perfusions, capable of lasting for months, require the use of cell retention devices [42, 43]. Here, a distinction is made between internal perfusion (where the cell retention device is located inside the bioreactor) and external perfusion (where the cell retention device is located outside the bioreactor). On a pilot and industrial scale, externally attached cell retention devices are used [44], as this facilitates device access for cleaning or replacement purposes. Ideally, such a cell retention device should have the following characteristics: (1) robustness, (2) high cell retention efficiency without cell damage, (3) high product yield, (4) scalable to a volumetric perfusion rate of at least 1000 L d^{-1}, (5) low running costs and (6) the possibility of continuous harvesting [45]. A comparison of commonly used single-use cell retention devices can be found in Table 2.

Table 2 Comparison of currently used single-use cell retention devices (partially based on [44])

Characteristics	External				Internal
	TFF system	ATF system (0.2 μm pore size)[a]	Acoustic settler	Centrifuge	Filter in bag (1.2 μm pore size)
Working principle	Hollow fiber media exchange	Hollow fiber media exchange with alternating cell broth direction	Sonic wave media exchange	Centrifugal media exchange	Membrane filter media exchange
Advantages	Low residence time of cells	Low residence time of cells, reduced membrane fouling and shear stress	Robust long-term operation	Separation of living cells from dead cells possible	Easy to operate
Disadvantages	Membrane fouling and blockage, no elimination of dead cells	No elimination of dead cells, problems at high viable cell densities due to viscosity	Generated power and heat may cause cell damage when exceeding pilot scale	Cell damage increases with rotor speed, continuous long-term operation may be challenging	Filter clogging at HCD, no elimination of dead cells, only available for wave-mixed bioreactors
Recirculation loop and (low shear-force) pump	Yes	No	Yes	Yes	No
Cell retention efficiency [%]	100	100	85–99	99	100
Maximum cell density [$\cdot 10^7$ cells mL^{-1}]	20	20	4.2	2.0	20
Simplicity of operation	Very easy	Very easy	Easy	Difficult	Very easy
Scalability to industrial scale	Yes	Yes	Yes (parallel use)	Yes	Limited to 100 L working volume
Running cost	Very expensive	Very expensive	Less expensive	Expensive	Expensive

[a]Reusable versions with different ultra- und microfiltration membranes available

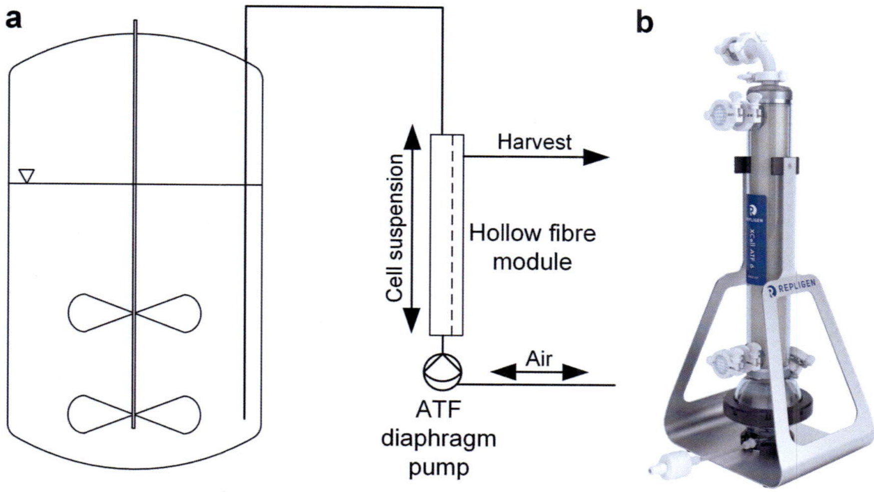

Fig. 2 Repligen's ATF system: (**a**) Schematic diagram of the working principle, (**b**) the XCell ATF 6 SU. (Image provided courtesy of Repligen Corporation)

These cell retention devices include tangential flow filtration (TFF) systems, Repligens alternating tangential flow filtration (ATF) systems, perfusion filters in wave-mixed bags, acoustic separators (e.g. AWS, Cytoperf) and centrifuges (e.g. Centritech, kSep). They are either based on the principles of filtration or acceleration/gravity [46–51].

The ATF system, as it is shown schematically in Fig. 2a, is available for laboratory scale (XCell ATF 2), pilot scale (XCell ATF 6, Fig. 2b) and industrial scale (XCell ATF 10) as a single-use version. It is currently the most frequently used external cell retention device alongside the hollow fiber-based TFF systems [52]. Characterized by an alternating direction of culture broth flow, its main features are the hollow fiber unit and the diaphragm pump. Shear stress [53] and fouling [54, 55] are reduced and only one single connection to the bioreactor is required. The main advantages of the ATF system are its ability to generate ultra-HCDs, around $2{\cdot}10^8$ cells mL^{-1}, and its proven reliable operation in processes, which are operated for longer than 30 days. For efficient ATF operation, installation in the vicinity of the bioreactor is recommended.

In recent years, the availability of single-use perfusion systems has contributed to the increased development of N-1 (i.e. seed stage) perfusion processes, high-seed fed-batch production and concentrated fed-batch processes. These three intensification approaches are aimed at increasing the viable cell density, improving cell specific productivity and improving bioreactor utilization.

3.1.1 N-1 Seed Perfusion Process

It takes approximately 20 days to produce enough biomass to inoculate a CHO-cell based mAb batch process carried out in a 1 m^3 working volume production bioreactor [56]. In order to prevent nutrient limitations, the cell suspension is therefore preferably diluted with new medium during the seed generation process, which takes place over several stages in N-x bioreactors, where x describes the number of necessary bioreactor transfer steps prior to the production scale. If the number of cells is too low in the final stage, a long lag phase leads to a delay in production and, finally, reduced process performances if the equipment is not suitable for prolonged production phases. Alternatively, perfusion allows for the generation of HCDs in the seed train step prior to production [57] and results in a reduction of the production time at the final scale with equal or even higher product titer [5, 57–61]. It might even reduce the required transfer steps [35]. Usually, such N-1 seed perfusion processes take 4–7 days. Furthermore, as a consequence of biomass generation in one perfusion step instead of in several sequential batch precursors, the number and size of the seed bioreactors can be significantly reduced [51].

When using single-use perfusion devices for N-1 processes, different users prefer implementing the internal filters listed in Table 2 in combination with the wave-mixed bioreactors. A long-term goal of N-1 perfusion processes is to be able to inoculate a stirred production bioreactor with a 2 m^3 working volume directly with cells from such a perfusion bag (10 L) and to reduce the total process time (including the seed production) to less than 2 weeks. This, however, demands the direct inoculation of the perfusion-operated wave-mixed bioreactor with cells from a high density working cell bank with 4.5 mL vials containing approximately $50 \cdot 10^6$ cells mL^{-1} [62].

In Fig. 3, the experimental set-up and growth curve of an exemplary N-1 perfusion process (unpublished; 1 L working volume) performed in a wave-mixed BIOSTAT RM 20/50 (Sartorius) is shown [63]. A maximum VCD of $2 \cdot 10^8$ cells mL^{-1} with viability over 95% was achieved after 11 d. Following the termination of the media exchange process after 11 d, the cells were discarded on day 12 and not used for further cultures.

3.1.2 High-Seed Fed-Batch Production

An N-1 seed perfusion process allows for higher seeding of the production bioreactor. Through the application of this process, it becomes possible to seed a fed-batch production bioreactor with an initial viable cell density of between $2 \cdot 10^6$ cells mL^{-1} to $1 \cdot 10^7$ cells mL^{-1}, meaning that high-seed fed-batch production (VCD = $1 \cdot 10^7$ cells mL^{-1}) could become a reality [64]. Furthermore, as it is possible to inoculate the production bioreactor with a higher cell density, the cultivation time required until the target product titer is reached can be reduced by 13–43%, compared to traditional fed-batch processes [5, 60]. In other words, higher product titers can

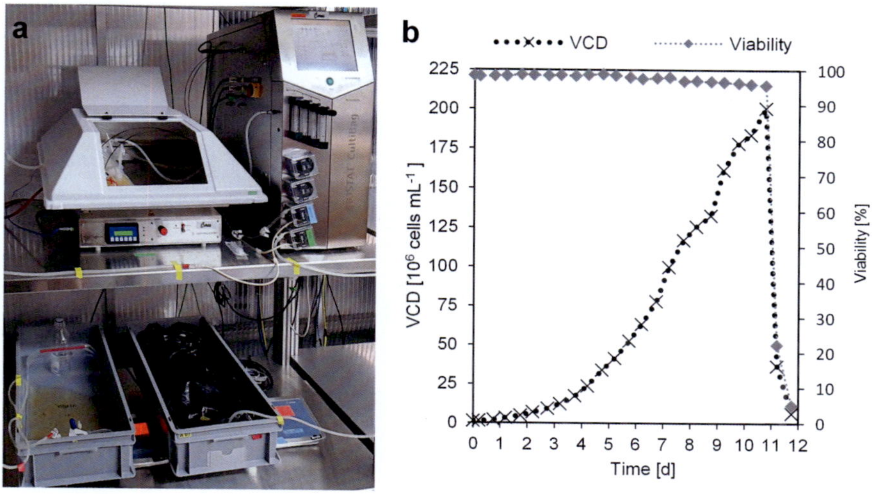

Fig. 3 Single-use N-1 seed perfusion process: (**a**) Photo of the experimental set-up with control unit, rocking platform and feed and harvest bags on balances (**b**) exemplary growth curve. A Flexsafe RM 2L Perfusion bag was used

be achieved using the high-seed fed-batch approach after the same process time, compared to the low-seed fed-batch approach [5].

Studies have shown that high-seed fed-batch production can improve productivity by about 30% [60, 65]. Consequently, more bioreactor runs can be performed per year and thus more mAbs can be produced in the same time period after switching to high-seed fed-batch production. Moreover, no negative influences on product quality have been described thus far for high-seed fed-batch production. Nevertheless, a critical point of the high-seeding approach is to keep the cells in the exponential growth phase while in the N-1 bioreactor until the desired cell densities are reached [66].

Figure 4b shows the results (unpublished) of a high-seeded fed-batch cultivation over a period of 4 d following an N-1 perfusion process [67]. A stirred ambr 250 modular bioreactor (Sartorius) served as the production bioreactor (Fig. 4a). The high-seed fed-batch process was able to reach the target mAb titer 4.2 d sooner than the low-seed fed-batch, resulting in a reduction of the production process time by over 50%.

3.1.3 Concentrated Fed-Batch Process

The concentrated fed-batch process is a perfusion process in which both the cells and the product are retained via an ultrafiltration membrane and the product is harvested in batch mode [68]. This requires the Molecular Weight Cut Off

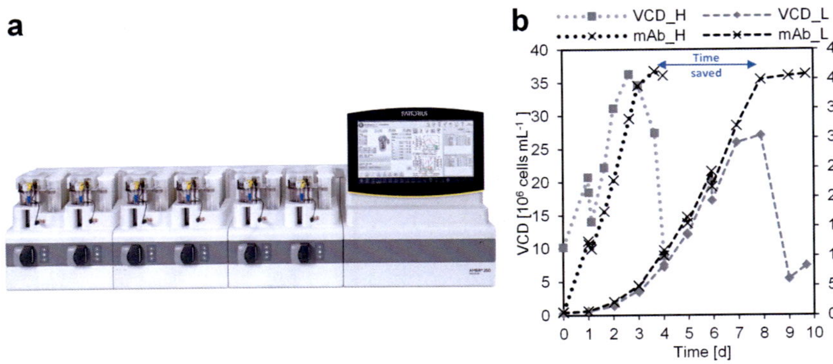

Fig. 4 N-1 perfusion-based high-seed and low-seed fed-batch production process of a mAb: (**a**) The ambr 250 modular system (Image provided courtesy of Sartorius), (**b**) course of mAb concentration and VCD over the duration of the high-seed (mAb/VCD_H) and low-seed (mAb/VCD_L) fed-batch processes

(MWCO) of the perfusion device membrane to be 2–3 times smaller than the target product. Both ATF and TFF systems have proven themselves for mAb production in concentrated fed-batch processes [68, 69]. However, only a reusable version of the ATF system (XCell ATF 2, 4, 6, 10 Stainless Steel) can be currently implemented for this purpose, as there were no single-use versions with ultrafiltration membranes available during the writing of this paper. According to the manufacturer, the ultrafiltration modules with MWCOs between 30 and 50 kDa are the most popular for concentrated fed-batch processes and have been successfully implemented in combination with stirred single-use bioreactors and working volumes of between 1 L and 2 m^3.

According to the literature, the concentrated fed-batch approach, which generally takes between 2 and 3 weeks, was able to generate not only the highest VCDs (> $2 \cdot 10^8$ cells mL^{-1}) but also the highest mAb titers (25 g L^{-1}), a factor 5–15 times higher than those achieved in traditional fed-batch processes [68, 70–73]. This process approach, which is also patented as XD (extreme density) technology [74], is therefore recommended for improving low titers of stable products in fed-batch mode.

3.2 *DSP Intensification*

An intensified USP and subsequent HCDs result not only in a higher concentration of solid and cell impurities such as host cell proteins (HCPs) and deoxyribonucleic acid (DNA) but also in protein aggregates. These make harvesting and subsequent processing more difficult than in traditional fed-batch processes. Currently, two main approaches are being pursued to intensify DSP. (1) Binding capacities of the chromatography columns are increased or used more effectively [75, 76] and (2) higher productivity and a reduction in buffer consumption are achieved through the continuous operation of individual process steps [38, 77]. Here the capture step, performed with Protein A columns [75, 78] and taking place directly after cell separation and clarification in mAb productions, is the focus of DSP intensification [38, 79].

Protein A, a protein originally derived from *Staphylococcus aureus* and now commonly used in DSP, is capable of binding to the Fc region of various immunoglobulins with high affinity [76]. Since its discovery, the protein has been optimized through numerous modifications: (i) by increasing its binding capacity and improving the lifetime of the affinity columns and, (ii) by reducing the immunoglobulin G binding strength of the protein allowing for the implementation of less acidic elution techniques and (iii) by improving the stability of the columns to allow for cleaning under alkaline conditions. A detailed overview of the studies concerning the modification of protein A was published by Zarrineh et al. [76].

Alternative intensified capture steps, that dispense completely with protein A resins, such as performing several cheaper chromatography steps instead of affinity chromatography [80] or performing the capture step exclusively using cation exchange chromatography [81], were also described. Other purification options include precipitation [82–85], crystallization of the target product [86] and two-phase extractions [87, 88].

Capture and polishing steps may now also be performed using membrane chromatography instead of the classic packed bed columns [89], as these membranes have the advantage of operating at flow rates which are 10–50 fold higher with less operating pressure than packed beds. In addition, time-consuming steps such as packing and validating the bed are no longer necessary. These columns are offered as single-use capsules by various manufacturers (e.g. Pall Life Sciences, Sartorius, Natrix Separations) and can be used both in bind-elute mode, i.e. the target product binds to the membrane and must be eluted, and in flow-through mode, where the product passes the membrane while impurities are retained [89–92]. This is performed with anion exchange chromatography membranes that are suitable for virus, DNA and HCP removal. Membrane technology is also cheaper: Lim et al. [93] determined cost savings of at least 65% when switching to disposable anion exchange membranes compared to conventional stainless steel columns for mAb purification. Nevertheless, some limitations of this technology do exist: the membrane processes have the disadvantage that the binding capacity is lower than with packed bed columns. Therefore, membrane technology as a capture step is not widely used [89]. An alternative means of reducing labor and validation costs in connection with packed bed chromatography is the use of pre-packed columns, which are commercially available from various suppliers (e.g. Cytiva, Repligen, Merck/Millipore). The first pre-packed single-use chromatography system with disposable flow paths and columns, namely the Äkta ready, was launched in 2008 by GE Healthcare, today known as Cytiva [94].

Another intensification approach in the field of polishing is the use of mixed-mode chromatography, its suitability has been investigated since the end of the

1990s [95–97]. Instead of performing several chromatography steps in succession, several components with different affinities are used as the stationary phase of the chromatography column.

Casey et al. proposed the addition of a fivefold single-pass TFF (SPTFF, see also Sect. 4) to the capture step in order to reduce the size of the capture column in fed-batch production systems [98]. This approach saves 80% of the resin and buffer, which would correspond to a cost reduction of $500,000 per 1000 L of antibody broth. It is therefore crucial, regardless of the intensification approach chosen, to determine whether a continuous process platform would not ultimately succeed in greater cost savings.

4 Continuous mAb Production with Single-Use Systems

The continuous production of mAbs with single-use systems results in the production of higher amounts of mAbs in less time and with smaller equipment, more precisely with a factor 5–10 reduction in volume [18, 99]. This is achieved by increasing productivity and reducing non-value adding operations, such as hold steps, which simultaneously reduces capital and operational expenses. This stands in contrast to the higher demands on process control and equipment, high media consumption and often low product titers [12]. Typical specific productivity rates of this method in connection with single-use systems are around 1 g L^{-1} d^{-1}, with product titers ranging between 0.3 and 1 g L^{-1} [94]. Recently, however, WuXi Biologics reported an unprecedented productivity of 2.5 g L^{-1} d^{-1} using its laboratory-scale continuous USP platform for the production of a fusion protein [100].

The increasing interest in continuous processes, which has been observed for a few years, especially for mAb production, is also due to the positive influence of continuous operation on product quality and product supply [101, 102]. Therefore, it is possible to: (1) achieve consistent product quality when operating under steady state conditions (which can be problematic in fed-batch mode depending on the cultivation time), (2) reduce the risk of human error and (3) achieve a quick response to changing supply requirements. Furthermore, the level of impurities (e.g. HCPs and cell debris) in the product is kept low through consistently high cell viability.

In 2015, Godawat et al. described the first fully continuous mAb production at laboratory scale with 12 L working volume [18]. The process was operated for 31 days with fully automated product purification by ATF, Protein A chromatography, viral inactivation, cation exchange chromatography and final membrane filtration. In 2019, Arnold et al. published results of the first fully continuous pilot scale mAb production (50 L) [103]. The authors calculated that by replacing the fed-batch facility (4 × 12.5 m^3 bioreactors) with a continuous one (5 × 2 m^3 continuously operated single-use bioreactors and purification train) a cost saving by 15% could be achieved [103]. In the same year, the Dutch-Australian company Biosana received approval for phase 1 clinical trials of a therapeutic biosimilar mAb (Omalizumab) with a fully continuous manufacturing approach (bioreactor with 50 L working volume), which gave promising results in March 2020 [104, 105]. However, as mentioned in Sect. 1 of the book chapter, commercial mAbs produced in fully continuous facilities are still in clinical development.

4.1 Continuous Single-Use USP

Conversely, hybrid process approaches, which implement continuous USP [106], have been in use since 1994 for the production of commercial biotherapeutic products (even if they were not declared as continuous single-use USP). A current trend to produce increasingly stable molecules has also been reported [24]. The goal of such continuous perfusion processes, in which spent medium and product are discharged continuously, is to achieve a steady state in which productivity and product quality can be maintained over a long period of time with minimal deviation. In order to maintain the necessary constant VCD with high cell viability over the process operation period of between 30 and 90 days, the bleed has to be adjusted according to the growth rate. The longer the continuous USP runs under stable conditions, the higher the space-time yield of the bioreactor. Critical components of continuous USP are the single-use bioreactors, including their sensors and automation technology, as well as the cell retention devices presented in Sect. 3.1. ATF modules in particular have been proven to support such continuous upstream processes where targeted specific productivities exceed 1 g L^{-1} d^{-1} [14]. However, while pilot and industrial scale processes favor the use of single-use ATF systems, single-use TFF systems are still favored for laboratory scale processes.

A typical representation of a modern mAb production process using continuous USP is shown in Fig. 5. Inoculum production is performed in a wave-mixed bioreactor with a perfusion bag, which is inoculated directly with an HCD cryovial. The stirred single-use production bioreactor operates continuously with an ATF system. While multiple media feeds can be added to the production bioreactor, the volume is kept constant by control of the harvest- and bleed-stream. Their ratio depends on the growth rate of the production cell line. In this context, the use of an optimized perfusion medium is an important step towards increasing process stability and process costs reduction.

Long-term operation with high VCDs ($> 1 \cdot 10^8$ cells mL^{-1}), respectively, leads to high demands on both the production bioreactor and its internal and external peripherals (see also Sect. 5) [107]. An example of such a next generation stirred single-use bioreactor, specifically designed for continuous as well as intensified fed-batch operation at very HCDs, is the 50 L bioreactor of the HyPerforma DynaDrive series from Thermo Fisher Scientific, as shown in Fig. 6a. The Hyperforma DynaDrive series includes a 50 L, 500 L and 5 m^3 version of the bioreactor with a cube-shaped bag and special turndown ratios, namely 10:1 (50 L) and 20:1 (500 L, 5 m^3) instead of the usual 5:1. This means that this series of bioreactors can already be operated at 10%, or even 5%, of their maximum working volume. They

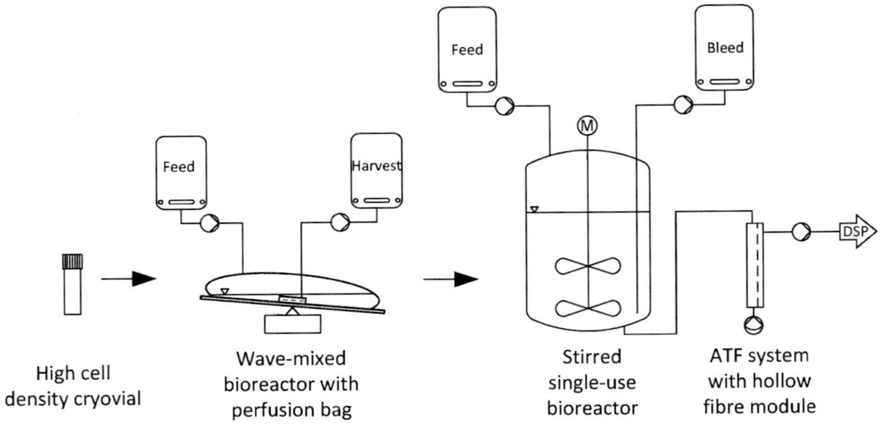

Fig. 5 Flowchart of a typical continuous USP of a modern mAb production process

are equipped with a low-shear drilled hole sparger (capable of producing sufficient oxygen transfer and CO_2 stripping) and a flexible drive train with multiple two-bladed impellers (Fig. 6b), capable of achieving mixing times of < 60 s at moderate revolutions and tip speeds, according to the supplier [108].

Other single-use bioreactors suitable for perfusion at production scale are distributed by ABEC, Cytiva, Merck Millipore, PALL and Sartorius. The product stream, which is harvested from the bioreactor e.g. with the ATF system (Fig. 5), is forwarded to the DSP, where the mAbs are purified.

4.2 Continuous Single-Use DSP

The use of continuously operated single-use devices enables savings to be made on buffers, resins and consumables. More specifically, findings show that savings in the range of 36–55% are possible when using disposable continuous DSP, in comparison to a classic stainless-steel batch process [109, 110].

Although the capture step is usually still performed with Protein A columns in bind-elute mode, several columns are connected in parallel and run cyclically, meaning that significantly smaller columns can be used. This method is known as multi-column or counter-current chromatography and varies in design. Here, the hold-up step after the production bioreactor is eliminated and the consumption of buffer and resins is reduced [111, 112]. Examples of continuous single-use multi-column chromatography systems are the BIOSMB platform from Sartorius (Fig. 7) and the systems from ARTeSYN Biosolutions, which can be used for both bind-elute chromatography in the capture step and subsequent flow-through polishing steps. They are GMP-compatible multi-column chromatography systems with a

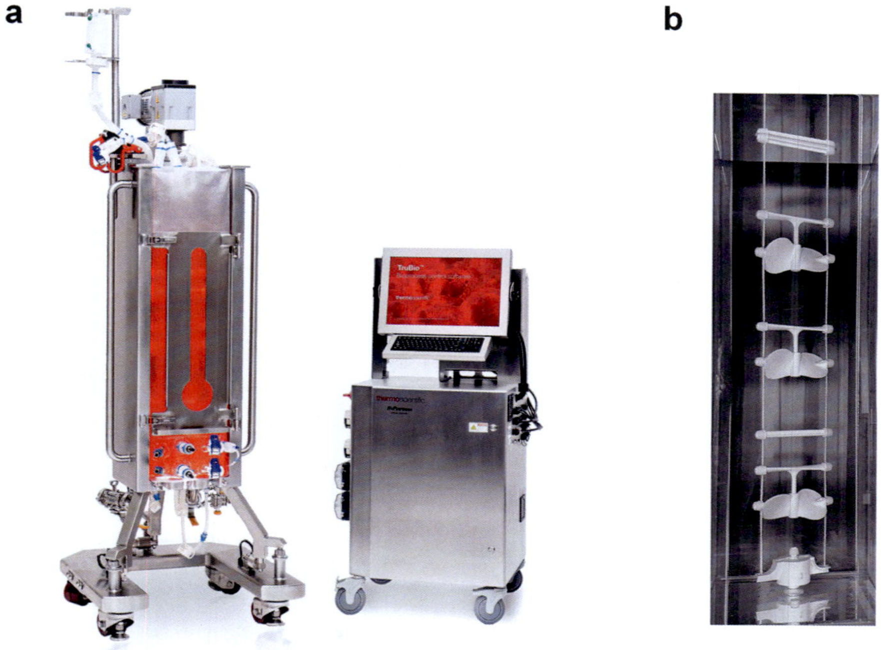

Fig. 6 Thermo Fisher Scientific's HyPerforma DynaDrive series: (**a**) 50 L bioreactor with control unit, (**b**) Plexiglas model showing the drive train with impellers. (Images provided courtesy of Thermo Fisher Scientific)

fully single-use product flow path. Both Merck and Bayer use the BioSMB system in continuous mAb production processes for the capture step and cation exchange or mixed-mode chromatography [94].

Until now, polishing steps have mostly been performed in bind-elute mode. Generally, polishing is performed using ion-exchange chromatography, hydrophobic-interaction chromatography or mixed-mode chromatography to further purify the target product [113]. In this case, a change to continuous use may have the same positive effects as in capture chromatography. Continuous operation is made feasible by binding the contaminants instead of the target product. The target product then flows through the column. Once the column is fully loaded with contaminants, contaminant break-through occurs and the column has to be replaced. While there is no true continuous mode, this flow-through mode is considered pseudo-continuous [114]. Gupte et al. achieved a more than 90% removal of HCPs in flow-through mode using a salt-tolerant single-use anion exchange membrane, compared to less than 70% when using a conventional resin [115].

A report describes a multi-column chromatography, in which both the capture step and two polishing steps were connected in series. Three parallel flow paths were alternately loaded, and disposable membrane absorbers were used. This mode

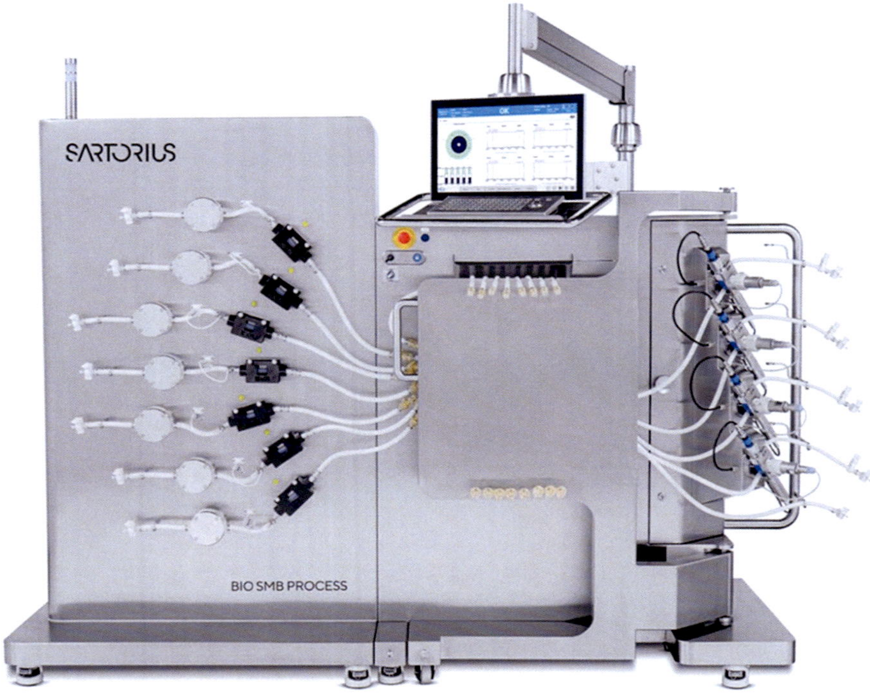

Fig. 7 BioSMB Process for continuous protein A capture and polishing steps. (Image provided courtesy of Sartorius)

of operation, implemented in a 500 L process, resulted in a 66% reduction in column size with a tenfold productivity increase of an antibody purification process at Sanofi [90].

Currently, it is considered common practice to perform a virus inactivation step following the capture step and a virus filtration step usually preceding the final ultra-/diafiltration [116]. Virus inactivation is typically performed by a pH shift to the acidic range. If, for the virus inactivation step in the context of a traditional (batch) DSP, a mixing tank was recently used, and acid was added and then neutralized with base, two approaches exist for the implementation of continuous operation: as a first option, the classic approach can be adopted with two single-use mixers which are alternately operated [38]. The Cadence Virus Inactivation System from Pall (Fig. 8) uses this method, for example. A second possibility is the implementation of plug-flow reactors, where acid is added to the product flow at the reactor inlet. The low pH value is maintained for the product residence time in the plug-flow reactor, after which it is neutralized at the outlet by adding base. In order to converge on optimal plug-flow conditions, back-mixing should be avoided and the variance of product residence time kept to a minimum [38, 117, 118]. In addition to inactivation by low pH, it is also possible to inactivate viruses using UV-C radiation in plug-

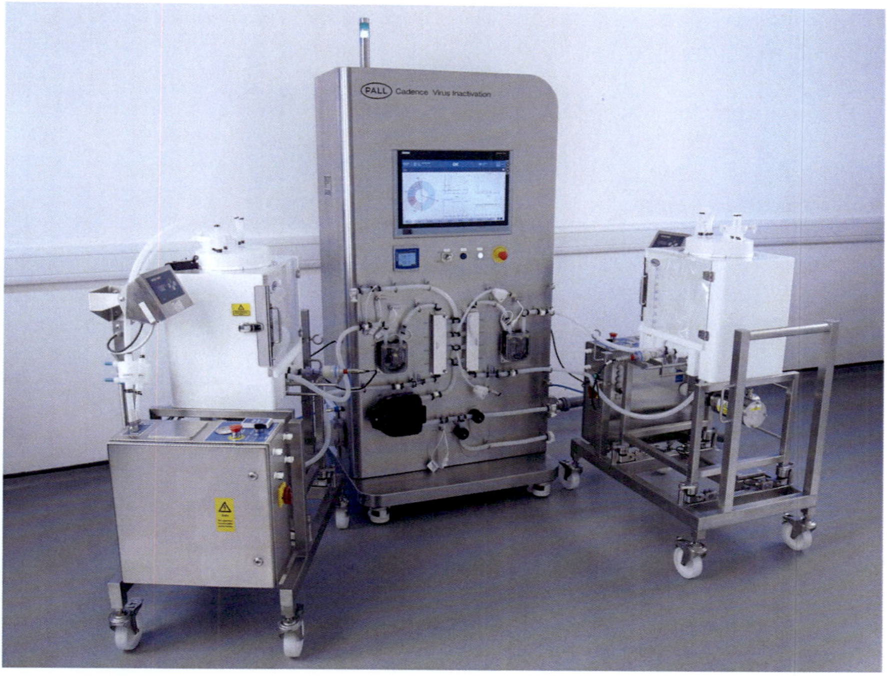

Fig. 8 Cadence^{TM} Virus Inactivation System. (Image provided courtesy of Pall Corporation)

flow reactors. However, radiation can lead to protein damage. There are currently no commercial single-use devices for UV virus inactivation on the market [119].

The second step in virus reduction, namely the virus filtration, follows the polishing steps. This usually involves dead-end nanofiltration. The successful use of these filters for continuous DSP, i.e. over a long period of time and with low flow, has already been described in various studies [120, 121]. In continuous operation, two individual filters are connected in parallel. As soon as one filter reaches its capacity limit, e.g. due to a pressure limit, the system automatically switches to the other filter. The first filter is then replaced with the help of a tube welder [122].

Recently the performance of four different filters used for viral filtration in connection with a mAb production process was investigated for batch and continuous cultivation [123]. During continuous filtration for 72 h at a flow rate of 0.3 L m^{-2} h^{-1}, start-stop scenarios were applied to identify which filters were suitable for continuous operation in a low-pressure environment. The two tested second-generation filters, namely Pall Pegasus Prime and Sartorius Virosart HF, were able to maintain a decadic logarithm reduction value of above 4 for at least 20 L m^{-2} during the entire filtration time. Additional experiments with short and long-term stops to investigate the impact of pressure fluctuation revealed a steady state after a certain period, making continuous operation feasible. Nevertheless, the pressure reduction of up to 1 bar, as is necessary for the single-use based continuous operation mode, might allow the flow of phages through larger filter pores at the onset of operation. This stands in contrast to batch processing with higher operating pressures where drag forces are assumed to prevent the crossing of phages [123].

Usually, a mAb formulation requires several TFF steps, for which single-use equipment is available [119]. The product is then concentrated by ultrafiltration, transferred to the appropriate buffer system via diafiltration and brought to the final concentration through a second ultrafiltration step. Here, SPTFF (see also Sect. 3.2) can be used for continuous applications. The filtration time can be increased through higher membrane areas and lower flow rates to such an extent, that one continuous filtration step is sufficient. This method can be used for both protein concentration and diafiltration [119, 124, 125]. For both steps, single-use device versions are already available [125, 126]. Buffer consumption in SPTFF diafiltration processes is comparatively high, but can be reduced by using several SPTFF modules and buffer recycling from late to early steps in a method known as counter current diafiltration [119, 127].

5 Overall Cost Comparison

The consideration of costs in dependence of the process mode is as important as the technological feasibility while comparing batch and continuous processing. While several reports compare the costs between batch and continuous process mode, most of them do not consider fully single-use based production, but rather hybrid processes [128]. It is common sense, however, that single-use based manufacturing is favorable at smaller production scales, where the limited size of single-use equipment, especially the bioreactor working volumes, does not present a challenge. Furthermore, smaller facility sizes are usually achievable through the use of continuous production processes. A study recently reported that a fed-batch process resulted in operating cost of goods (COG) of \$99 g^{-1} for mAb production, as opposed to \$51 g^{-1} for the continuous process. This difference in cost is also partly due to the smaller footprint and fewer storage tanks, which resulted in the reduction of facility costs by 66% [129]. Broken down to single unit operations, lower resin costs in the downstream part led to a reduction in consumables cost by 68% in the continuous process. Single-use based continuous upstream processes have proven to be cost-efficient options as well, however, larger benefits are seen if both, up- and downstream operations are performed with single-use equipment, as shown by several COG simulations [130]. Cost benefits were also seen in parallelization and flexibilization, which may be supported through the application of single-use equipment. The elimination of clean- and sterilize-in-place procedures reduces equipment complexity, which itself lowers capital cost and reduces downtimes. It was calculated that from an annual production scale of 1300 kg and at 3 g L^{-1} of upstream titer, a fully single-use mAb production with 2000 L bioreactors becomes the preferable choice over the stainless steel option, with either the same or higher upstream working volume capacity. At higher upstream titers, single-use variations were always favorable. If broken down to unit operations, cost savings can be large for several steps when applying continuous instead of batch processing, especially when coupled with a single-use based approach where possible. More specifically, savings of over 20% were achieved dependent on titer and scale for the protein A capture step during antibody production [131]. Pollard et al. went on to estimate the costs for three mAb production scenarios using the net present costs (NPCs), which constitute the net present value of the product without the profits from sales [132]. In one scenario, conventional stainless-steel equipment was compared with single-use equipment, for an intensified process with titers above 10 g L^{-1}, and a fully continuous process, respectively. The initial process was considered as operated with 6 × 15,000 L bioreactors with a titer of 3 g L^{-1}, the single-use bioreactor process was considered as operated with 6 × 2000 L bioreactors. The continuous perfusion process was set at a productivity of 2 g L^{-1} d^{-1}. At an annual production of 1 t, a reduction of the NPCs of approximately 60% for both the intensified single-use and the continuous production process was determined.

Further emphasis to simplify operation and improve process robustness, as well as cost-efficiency, should also focus on process analytical techniques for monitoring and control strategies, especially for single-use equipment [133]. This will remove one of the drawbacks, as a similar degree of sensor application cannot yet be fully reached in single-use based processing.

6 Summary and Conclusions

The global market for therapeutic mAbs is said to continue its growth in the near future, with revenues expected to reach $300 billion in 2025 [4]. In this framework, intensified and continuous USP and DSP unit operations with single-use systems are well-suited to reduce the time and costs for mAb development and production.

Regarding continuous processes, operational times between 30 and 120 days are being actively pursued. Continuous USP is already used for some approved mAbs, while the first mAb produced fully continuously is currently already in clinical trials. There are, however, still challenges to be overcome. For USP, these are attributed to the robustness and mechanical stability of the bioreactor bag (with its components such as impellers, sealings, spargers etc.) and the plastic used for the perfusion device. The demands on the design of the bag, especially regarding the stirred tank reactors' impeller, the sparger, the design of the exhaust gas filters and the foam control, as a result of the long-term usage are high. To this end continuous-use grade disposable plastics are indispensable. An oxygen transfer coefficient > 15 h^{-1} and adequate CO_2 stripping need to be guaranteed, as well as sufficient specific power input at moderate stirrer tip speeds [107]. Moreover, there is a need for: (1) improved tubing and connectors, (2) improved and new disposable sensors and probes, (3) installation of redundant sensors as well as filters and corresponding exchange strategies, and (4) single-use pumps operating at low flow rates ($0.1\ L\ min^{-1} - 2\ L\ min^{-1}$) with high accuracy. These four points also apply to DSP.

As previously mentioned, a further challenge, which concerns both continuous USP and DSP unit operations, is the effect of continuous process solution exposure to L&E profiles for process times of up to 4 months. An L&E characterization for all USP and DSP unit operations and corresponding mitigation strategies (e.g. water or buffer flushing, leachable clearance, single-use device exchange) have, therefore, to be developed and implemented [64, 107]. If successful, the currently dominating classic and intensified fed-batch processes using single-use systems for mAb production will increasingly be replaced by a continuous operation approach.

Acknowledgment We would like to thank Nina Steffen and Maren Grün for their support in generating the data of our intensified mAb productions at laboratory scale.

Conflict of Interest The authors declare that they have no conflict of interest.

References

1. Köhler G, Milstein C (1975) Continuous cultures of fused cells secreting antibody of predefined specificity. Nature 256:495–497. https://doi.org/10.1038/256495a0
2. Carvalho LS, da Silva OB, da Almeida GC et al (2017) Production processes for monoclonal antibodies. In: Jozola A (ed) Fermentation processes. InTech, www.intechopen.com, pp 181–198
3. Walsh G (2018) Biopharmaceutical benchmarks 2018. Nat Biotechnol 36:1136–1145. https://doi.org/10.1038/nbt.4305
4. Lu RM, Hwang YC, Liu IJ et al (2020) Development of therapeutic antibodies for the treatment of diseases. J Biomed Sci 27:1–30. https://doi.org/10.1186/s12929-019-0592-z
5. Bielser JM, Wolf M, Souquet J et al (2018) Perfusion mammalian cell culture for recombinant protein manufacturing – a critical review. Biotechnol Adv 36:1328–1340. https://doi.org/10.1016/j.biotechadv.2018.04.011
6. Kelley B (2009) Industrialization of mAb production technology: the bioprocessing industry at a crossroads. MAbs 1:443–452. https://doi.org/10.4161/mabs.1.5.9448
7. Becerra S, Berrios J, Osses N, Altamirano C (2012) Exploring the effect of mild hypothermia on CHO cell productivity. Biochem Eng J 60:1–8. https://doi.org/10.1016/j.bej.2011.10.003
8. Masterton RJ, Smales CM (2014) The impact of process temperature on mammalian cell lines and the implications for the production of recombinant proteins in CHO cells. Pharm Bioprocess 2:49–61. https://doi.org/10.4155/pbp.14.3
9. Torres M, Zúñiga R, Gutierrez M et al (2018) Mild hypothermia upregulates myc and xbp1s expression and improves anti-TNFα production in CHO cells. PLoS One 13:e0194510. https://doi.org/10.1371/journal.pone.0194510
10. Xu J, Xu X, Huang C et al (2020) Biomanufacturing evolution from conventional to intensified processes for productivity improvement: a case study. MAbs 12:1770669. https://doi.org/10.1080/19420862.2020.1770669
11. Konstantinov KB, Cooney CL (2015) White paper on continuous bioprocessing May 20–21 2014 continuous manufacturing symposium. J Pharm Sci 104:813–820. https://doi.org/10.1002/jps.24268
12. Pollock J, Ho SV, Farid SS (2013) Fed-batch and perfusion culture processes: economic, environmental, and operational feasibility under uncertainty. Biotechnol Bioeng 110:206–219. https://doi.org/10.1002/bit.24608

13. Lindskog EK (2017) The upstream process: principal modes of operation. In: Jagschies G, Lindskog E, Lacki K, Galliher P (eds) Biopharmaceutical processing: development, design, and implementation of manufacturing processes. Elsevier, pp 625–635
14. Gupta SK (2017) Upstream continuous process development. In: Subramanian G (ed) Continuous biomanufacturing – innovative technologies and methods. Wiley-VCH Verlag GmbH & Co. KGaA, Weinheim, pp 201–232
15. Farid SS (2007) Process economics of industrial monoclonal antibody manufacture. J Chromatogr B 848:8–18. https://doi.org/10.1016/j.jchromb.2006.07.037
16. Gavara P, Bibi N, Sanchez M et al (2015) Chromatographic characterization and process performance of column-packed anion exchange fibrous adsorbents for high throughput and high capacity bioseparations. PRO 3:204–221. https://doi.org/10.3390/pr3010204
17. Broly H, Costioli MD, Guillemot-Potelle C, Mitchell-Logean C (2010) Cost of goods modeling and quality by design for developing cost-effective processes. BioPharm Int 23:26–35
18. Godawat R, Konstantinov K, Rohani M, Warikoo V (2015) End-to-end integrated fully continuous production of recombinant monoclonal antibodies. J Biotechnol 213:13–19. https://doi.org/10.1016/j.jbiotec.2015.06.393
19. Somasundaram B, Pleitt K, Shave E et al (2018) Progression of continuous downstream processing of monoclonal antibodies: current trends and challenges. Biotechnol Bioeng 115:2893–2907. https://doi.org/10.1002/bit.26812
20. Whitford WG (2013) Single-use technology supporting the comeback of continuous bioprocessing. Pharm Bioprocess 1:249–253. https://doi.org/10.4155/pbp.13.28
21. Langer ES, Rader RA (2014) Single-use technologies in biopharmaceutical manufacturing: a 10-year review of trends and the future. Eng Life Sci 14:238–243. https://doi.org/10.1002/elsc.201300090
22. Chulkova TY, Kurbanova EK, Novikov YN, Gusarov DA (2013) Single-use technologies in biopharmaceutical production: advantages and disadvantages. Upstream process solutions (mini-review). Russ J Biopharm 5:3–12
23. Rogge P, Müller D, Schmidt SR (2015) The single-use or stainless steel decision process: a CDMO perspective. Bioprocess Int 13:10–15
24. BioPlan Associates Inc. (2020) Seventeenth annual report and survey of biopharmaceutical manufacturing capacity and production. www.bioplanassociates.com
25. Kunas K, Horvath B, Frank G et al (2013) A generic growth test method for improving quality control of disposables in industrial cell culture. BioPharm Int 26:34–41
26. Sobańtka A, Weiner C (2019) Extractables/leachables from single-use equipment. In: Eibl R, Eibl D (eds) Single-use technology in biopharmaceutical manufacture, 2nd edn. Wiley, Hoboken, pp 143–158
27. Dorival-García N, Carillo S, Ta C et al (2018) Large-scale assessment of extractables and leachables in single-use bags for biomanufacturing. Anal Chem 90:9006–9015. https://doi.org/10.1021/acs.analchem.8b01208
28. Pahl I, Hauk A, Schosser L, Orlikowski S (2019) Considerations on performing quality risk analysis for production processes with single-use systems. In: Eibl R, Eibl D (eds) Single-use technology in biopharmaceutical manufacture, 2nd edn. Wiley, Hoboken, pp 211–218
29. Martin JM (2011) A brief history of single-use manufacturing. BioPharm Int 24(Suppl):5–7
30. Singh V (1999) Disposable bioreactor for cell culture using wave-induced agitation. Cytotechnology 30:149–158. https://doi.org/10.1023/a:1008025016272
31. Jossen V, Eibl R, Eibl D (2019) Single-use bioreactors – an overview. In: Eibl R, Eibl D (eds) Single-use technology in biopharmaceutical manufacture, 2nd edn. Wiley, Hoboken, pp 37–52
32. Stanton D (2019) ABEC breaks plastic ceiling again with 6,000 L single-use bioreactor. In: Bioprocess Int. https://bioprocessintl.com/bioprocess-insider/upstream-downstream-processing/abec-breaks-plastic-ceiling-again-with-6000-l-single-use-bioreactor/. Accessed 10 Nov 2020

33. Oosterhuis NMG, Junne S (2016) Design, applications, and development of single-use bioreactors. In: Mandenius C-F (ed) Bioreactors: design, operation and novel applications. Wiley-VCH Verlag GmbH & Co. KGaA, Weinheim, pp 261–294
34. Kaiser SC, Kraume M, Eibl D, Eibl R (2015) Single-use bioreactors for animal and human cells. In: Al-Rubeai M (ed) Animal cell culture. Cell engineering, vol 9. Springer, Cham, pp 445–500
35. Oosterhuis NMG (2017) Single-use bioreactors for continuous bioprocessing: challenges and outlook. In: Subramanian G (ed) Continuous biomanufacturing – innovative technologies and methods. Wiley-VCH Verlag GmbH & Co. KGaA, Weinheim, pp 131–148
36. Schirmer C, Müller J, Steffen N et al (2020) How to produce mAbs in a cube-shaped stirred single-use bioreactor at 200 L scale. In: Pörtner R (ed) Animal cell biotechnology: methods and protocols, 4th edn. Humana Press, New York, pp 169–186
37. Anderlei T, Eibl D, Eibl R et al (2017) Facility of the future. DECHEMA, Frankfurt a.M. ISBN 978-3-89746-199-4
38. Manser B, Glenz M, Bisschops M (2019) Single-use downstream processing for biopharmaceuticals. In: Eibl R, Eibl D (eds) Single-use technology in biopharmaceutical manufacture, 2nd edn. Wiley, Hoboken, pp 117–126
39. Clincke MF, Mölleryd C, Samani PK et al (2013) Very high density of Chinese hamster ovary cells in perfusion by alternating tangential flow or tangential flow filtration in WAVE bioreactor™-part II: applications for antibody production and cryopreservation. Biotechnol Prog 29:768–777. https://doi.org/10.1002/btpr.1703
40. Ozturk SS (1996) Engineering challenges in high density cell culture systems. Cytotechnology 22:3–16. https://doi.org/10.1007/BF00353919
41. Konstantinov K, Goudar C, Ng M et al (2006) The "push-to-low" approach for optimization of high-density perfusion cultures of animal cells. Adv Biochem Eng Biotechnol 101:75–98. https://doi.org/10.1007/10_016
42. Castilho LR, Medronho RA (2002) Cell retention devices for suspended-cell perfusion cultures. In: Schügerl K, Zeng A-P (eds) Advances in biochemical engineering/biotechnology, 74th edn. Springer, Berlin, Heidelberg, pp 129–169
43. Warnock JN, Al-Rubeai M (2006) Bioreactor systems for the production of biopharmaceuticals from animal cells. Biotechnol Appl Biochem 45:1–12. https://doi.org/10.1042/BA20050233
44. Chotteau V (2015) Perfusion processes. In: Al-Rubeai M (ed) Animal cell culture. Cell engineering, vol 9. Springer, Cham, pp 407–443
45. Voisard D, Meuwly F, Ruffieux P-A et al (2003) Potential of cell retention techniques for large-scale high-density perfusion culture of suspended mammalian cells. Biotechnol Bioeng 82:751–765. https://doi.org/10.1002/bit.10629
46. Clincke MF, Mölleryd C, Zhang Y et al (2013) Very high density of CHO cells in perfusion by ATF or TFF in WAVE bioreactor™: part I: effect of the cell density on the process. Biotechnol Prog 29:754–767. https://doi.org/10.1002/btpr.1704
47. Oosterhuis NMG (2014) Perfusion process design in a 2D rocking single-use bioreactor. In: Subramanian G (ed) Continuous processing in pharmaceutical manufacturing. Wiley-VCH Verlag GmbH & Co. KGaA, Weinheim, pp 155–164
48. Whitford WG (2015) Single-use perfusion bioreactors support continuous biomanufacturing. Pharm Bioprocess 3:75–93. https://doi.org/10.4155/pbp.14.58
49. Karst DJ, Serra E, Villiger TK et al (2016) Characterization and comparison of ATF and TFF in stirred bioreactors for continuous mammalian cell culture processes. Biochem Eng J 110:17–26. https://doi.org/10.1016/j.bej.2016.02.003
50. Madsen B, Cobia J, Jones N (2019) S.U.B. enhancements for high-density perfusion cultures. Appl. Note. https://assets.thermofisher.com/TFS-Assets/BPD/Application-Notes/sub-enhancements-perfusion-cultures-app-note.pdf. Accessed 22 Nov 2020
51. Stepper L, Filser FA, Fischer S et al (2020) Pre-stage perfusion and ultra-high seeding cell density in CHO fed-batch culture: a case study for process intensification guided by systems biotechnology. Bioprocess Biosyst Eng 43:1431–1443. https://doi.org/10.1007/s00449-020-02337-1

52. Sieck JB, Schild C, von Hagen J (2017) Perfusion formats and their specific medium requirements. In: Subramanian G (ed) Continuous biomanufacturing – innovative technologies and methods. Wiley-VCH Verlag GmbH & Co. KGaA, Weinheim, pp 171–200
53. Zhan C, Bidkhori G, Schwarz H et al (2020) Low shear stress increases recombinant protein production and high shear stress increases apoptosis in human cells. iScience 23:–101653. https://doi.org/10.1016/j.isci.2020.101653
54. Radoniqi F, Zhang H, Bardliving CL et al (2018) Computational fluid dynamic modeling of alternating tangential flow filtration for perfusion cell culture. Biotechnol Bioeng 115:2751–2759. https://doi.org/10.1002/bit.26813
55. Walther J, McLarty J, Johnson T (2019) The effects of alternating tangential flow (ATF) residence time, hydrodynamic stress, and filtration flux on high-density perfusion cell culture. Biotechnol Bioeng 116:320–332. https://doi.org/10.1002/bit.26811
56. Woodgate JM (2018) Perfusion N-1 culture—opportunities for process intensification. In: Jagschies G, Lindskog E, Łącki K, Galliher P (eds) Biopharmaceutical processing - development, design, and implementation of manufacturing processes. Elsevier, Amsterdam/Oxford/Cambridge, pp 755–768
57. Zamani L, Lundqvist M, Zhang Y et al (2018) High cell density perfusion culture has a maintained exoproteome and metabolome. Biotechnol J 13:1–11. https://doi.org/10.1002/biot.201800036
58. Padawer I, Ling WLW, Bai Y (2013) Case study: an accelerated 8-day monoclonal antibody production process based on high seeding densities. Biotechnol Prog 29:829–832. https://doi.org/10.1002/btpr.1719
59. Pohlscheidt M, Jacobs M, Wolf S et al (2013) Optimizing capacity utilization by large scale 3000 L perfusion in seed train bioreactors. Biotechnol Prog 29:222–229. https://doi.org/10.1002/btpr.1672
60. Yang WC, Lu J, Kwiatkowski C et al (2014) Perfusion seed cultures improve biopharmaceutical fed-batch production capacity and product quality. Biotechnol Prog 30:616–625. https://doi.org/10.1002/btpr.1884
61. Xu J, Rehmann MS, Xu M et al (2020) Development of an intensified fed-batch production platform with doubled titers using N-1 perfusion seed for cell culture manufacturing. Bioresour Bioprocess 7:17. https://doi.org/10.1186/s40643-020-00304-y
62. GE Healthcare (2016) One-step seed culture expansion from one vial of high-density cell bank to 2000 L production bioreactor. Appl. Note 29160932. https://www.cytivalifesciences.co.kr/wp-content/uploads/2020/04/One-step-seed-culture-expansion-from-one-vial-of-high-density-cell-bank-to-2000-L-production-bioreactor.pdf. Accessed 10 Nov 2020
63. Steffen N (2020) Neue Ansätze zur CHO-Zell-basierten Antikörperproduktion. Master thesis, Zürcher Hochschule für Angewandte Wissenschaften
64. BioPhorum Operations Group (2017) Continuous downstream processing for biomanufacturing: an industry review. https://www.biophorum.com/download/continuous-downstream-processing-for-biomanufacturing-an-industry-review/. Accessed 8 Nov 2020
65. Hwang I (2019) Samsung biologics implements large scale N-1 perfusion for commercial application. Press release. https://www.prnewswire.com/news-releases/samsung-biologics-implements-large-scale-n-1-perfusion-for-commercial-application-300899326.html. Accessed 10 Nov 2020
66. Jordan M, Mac Kinnon N, Monchois V et al (2018) Intensification of large-scale cell culture processes. Curr Opin Chem Eng 22:253–257. https://doi.org/10.1016/j.coche.2018.11.008
67. Grün M (2018) Scale-up strategie für eine CHO-Zell-basierte IgG-Produktion (Fed-Batch) in gerührten Single-Use Bioreaktoren. Master thesis, Zürcher Hochschule für Angewandte Wissenschaften
68. Yang WC, Minkler DF, Kshirsagar R et al (2016) Concentrated fed-batch cell culture increases manufacturing capacity without additional volumetric capacity. J Biotechnol 217:1–11. https://doi.org/10.1016/j.jbiotec.2015.10.009
69. Fenge C, Weyand J, Greller G, Adams T (2018) Large-scale perfusion and concentrated fed-batch operation of BIOSTAT ® STR single-use bioreactor. Appl. Note. https://

www.sartorius.com/resource/blob/11984/8e3d506edce9939b03efd4e2352d7e6b/appl-large-scale-perfusion-sbt1018-e-data.pdf. Accessed 10 Nov 2020
70. Warikoo V, Godawat R, Brower K et al (2012) Integrated continuous production of recombinant therapeutic proteins. Biotechnol Bioeng 109:3018–3029. https://doi.org/10.1002/bit.24584
71. Zijlstra G, Touw K, Koch M, Monge M (2019) Design considerations towards an intensified single-use facility. In: Eibl R, Eibl D (eds) Single-use technology in biopharmaceutical manufacture, 2nd edn. Wiley, Hoboken, pp 181–192
72. Chon JH, Zarbis-Papastoitsis G (2011) Advances in the production and downstream processing of antibodies. New Biotechnol 28:458–463. https://doi.org/10.1016/j.nbt.2011.03.015
73. Butler M, Meneses-Acosta A (2012) Recent advances in technology supporting biopharmaceutical production from mammalian cells. Appl Microbiol Biotechnol 96:885–894. https://doi.org/10.1007/s00253-012-4451-z
74. Zijlstra GM, Hof RP, Schilder J (2006) Improved process for the culturing of cells. Patent WO2008006494A1, 14 July 2006
75. Ramos-de-la-Peña AM, González-Valdez J, Aguilar O (2019) Protein a chromatography: challenges and progress in the purification of monoclonal antibodies. J Sep Sci 42:1816–1827. https://doi.org/10.1002/jssc.201800963
76. Zarrineh M, Mashhadi IS, Farhadpour M, Ghassempour A (2020) Mechanism of antibodies purification by protein a. Anal Biochem 609:113909. https://doi.org/10.1016/j.ab.2020.113909
77. Gao Z-Y, Zhang Q-L, Shi C et al (2020) Antibody capture with twin-column continuous chromatography: effects of residence time, protein concentration and resin. Sep Purif Technol 253:117554. https://doi.org/10.1016/j.seppur.2020.117554
78. Hilbold N-J, Le Saoût X, Valery E et al (2017) Evaluation of several protein a resins for application to multicolumn chromatography for the rapid purification of fed-batch bioreactors. Biotechnol Prog 33:941–953. https://doi.org/10.1002/btpr.2465
79. Nadar S, Shooter G, Somasundaram B et al (2020) Intensified downstream processing of monoclonal antibodies using membrane technology. Biotechnol J:2000309. https://doi.org/10.1002/biot.202000309
80. Follman DK, Fahrner RL (2004) Factorial screening of antibody purification processes using three chromatography steps without protein A. J Chromatogr A 1024:79–85. https://doi.org/10.1016/j.chroma.2003.10.060
81. Lain B, Cacciuttolo M, Zarbis-Papastoitsis G (2009) Development of a high-capacity MAb capture step based on cation-exchange chromatography. Bioprocess Int 7:26–34
82. Hammerschmidt N, Tscheliessnig A, Sommer R et al (2014) Economics of recombinant antibody production processes at various scales: industry-standard compared to continuous precipitation. Biotechnol J 9:766–775. https://doi.org/10.1002/biot.201300480
83. Gronemeyer P, Ditz R, Strube J (2014) Trends in upstream and downstream process development for antibody manufacturing. Bioengineering 1:188–212. https://doi.org/10.3390/bioengineering1040188
84. Kateja N, Agarwal H, Saraswat A et al (2016) Continuous precipitation of process related impurities from clarified cell culture supernatant using a novel coiled flow inversion reactor (CFIR). Biotechnol J 11:1320–1331. https://doi.org/10.1002/biot.201600271
85. Martinez M, Spitali M, Norrant EL, Bracewell DG (2019) Precipitation as an enabling technology for the intensification of biopharmaceutical manufacture. Trends Biotechnol 37:237–241. https://doi.org/10.1016/j.tibtech.2018.09.001
86. Smejkal B, Agrawal NJ, Helk B et al (2013) Fast and scalable purification of a therapeutic full-length antibody based on process crystallization. Biotechnol Bioeng 110:2452–2461. https://doi.org/10.1002/bit.24908
87. Rosa PAJ, Azevedo AM, Ferreira IF et al (2007) Affinity partitioning of human antibodies in aqueous two-phase systems. J Chromatogr A 1162:103–113. https://doi.org/10.1016/j.chroma.2007.03.067

88. Kruse T, Kampmann M, Rüddel I, Greller G (2020) An alternative downstream process based on aqueous two-phase extraction for the purification of monoclonal antibodies. Biochem Eng J 161:107703. https://doi.org/10.1016/j.bej.2020.107703
89. Boi C (2019) Membrane chromatography for biomolecule purification. In: Basile A, Charcosset C (eds) Current trends and future developments on (Bio-) membranes. Elsevier, Amsterdam/Oxford/Cambridge, pp 151–166
90. Mothes B, Pezzini J, Schroeder-Tittmann K, Villain L (2016) Accelerated, seamless antibody purification: process intensification with continuous disposable technology. BioProcess Int 14:34–58
91. Weaver J, Husson SM, Murphy L, Wickramasinghe SR (2013) Anion exchange membrane adsorbers for flow-through polishing steps: part II. Virus, host cell protein, DNA clearance, and antibody recovery. Biotechnol Bioeng 110:500–510. https://doi.org/10.1002/bit.24724
92. Jacquemart R, Stout J (2017) Membrane adsorbers, columns: single-use alternatives to resin chromatography. Bioprocess Int 14:18–19
93. Lim JAC, Sinclair A, Kim DS, Gottschalk U (2007) Economic benefits of single-use membrane chromatography in polishing – a cost of goods model. Bioprocess Int 5:60–64
94. Bisschops M, Schofield M, Grace J (2017) Two mutually enabling trends: continuous bioprocessing and single-use technologies. In: Subramanian G (ed) Continuous biomanufacturing – innovative technologies and methods. Wiley-VCH Verlag GmbH & Co. KGaA, Weinheim, pp 149–170
95. Zhang K, Liu X (2016) Mixed-mode chromatography in pharmaceutical and biopharmaceutical applications. J Pharm Biomed Anal 128:73–88. https://doi.org/10.1016/j.jpba.2016.05.007
96. Saufi SM, Fee CJ (2012) Preparation of multiple interaction membrane chromatography using mixed matrix membrane preparation concept. Procedia Eng 44:133–135. https://doi.org/10.1016/j.proeng.2012.08.335
97. Freitag R, Splitt H, Reif OW (1996) Controlled mixed-mode interaction chromatography on membrane adsorbers. J Chromatogr A 728:129–137. https://doi.org/10.1016/0021-9673(95)01024-6
98. Casey C, Rogler K, Gjoka X et al (2016) Cadence™ Single-pass TFF Coupled with Chromatography Steps Enables Continuous Bioprocessing while Reducing Processing Times and Volumes. In: Am Pharm Rev. https://www.americanpharmaceuticalreview.com/Featured-Articles/239951-Cadence-Single-pass-TFF-Coupled-with-Chromatography-Steps-Enables-Continuous-Bioprocessing-while-Reducing-Processing-Times-and-Volumes/. Accessed 12 Nov 2020
99. Kossik J (2002) Think small: pharmaceutical facilities could boost capacity and slash costs by trading in certain batch operations for continuous versions. In: Pharma Manuf. https://www.pharmamanufacturing.com/articles/2002/6/. Accessed 13 Nov 2020
100. Stanton D (2019) Up titer: WuXi breaks 50g/L with continuous CHO process. In: Bioprocess Int. https://bioprocessintl.com/bioprocess-insider/upstream-downstream-processing/up-titer-wuxi-breaks-50g-l-with-continuous-cho-process/. Accessed 2 Nov 2020
101. Yu L (2016) Continuous manufacturing has a strong impact on drug quality. In: FDA voice. https://www.pharmaceuticalprocessingworld.com/continuous-manufacturing-has-a-strong-impact-on-drug-quality/. Accessed 13 Nov 2020
102. U.S. Food and Drug Administration (2004) PAT—a framework for innovative pharmaceutical development, manufacturing, and quality assurance. https://www.fda.gov/node/379301. Accessed 13 Nov 2020
103. Arnold L, Lee K, Rucker-Pezzini J, Lee JH (2019) Implementation of fully integrated continuous antibody processing: effects on productivity and COGm. Biotechnol J 14:1800061. https://doi.org/10.1002/biot.201800061
104. Genengnews (2020) First mAb produced via fully continuous biomanufacturing. https://www.genengnews.com/news/first-mab-produced-via-fully-continuous-biomanufacturing/. Accessed 7 Nov 2020

105. Biosimilar Development (2020) BiosanaPharma announces successful outcome of comparative Phase I Study of BP001 A Biosimilar Candidate To Xolair (omalizumab). https://www.biosimilardevelopment.com/doc/biosanapharma-announces-successful-phase-i-study-of-bp-a-biosimilar-xolair-omalizumab-0001. Accessed 13 Nov 2020
106. Bonham-Carter J, Shevitz J (2011) A brief history of perfusion biomanufacturing. Bioprocess Int 9:24–28
107. Bayer M, Castan A, Eibl R et al (2020) Technical state-of-the-art and risk analysis on single-use equipment in continuous processing steps. DECHEMA, Frankfurt a.M. ISBN 978-3-89746-226-7
108. Thermo Fisher Scientific Inc. (2020) HyPerforma DynaDrive single-use bioreactor (S.U.B.). https://assets.thermofisher.com/TFS-Assets/BPD/brochures/dynadrive-sub-brochure.pdf. Accessed 10 Nov 2020
109. Walther J, Godawat R, Hwang C et al (2015) The business impact of an integrated continuous biomanufacturing platform for recombinant protein production. J Biotechnol 213:3–12. https://doi.org/10.1016/j.jbiotec.2015.05.010
110. Hummel J, Pagkaliwangan M, Gjoka X et al (2019) Modeling the downstream processing of monoclonal antibodies reveals cost advantages for continuous methods for a broad range of manufacturing scales. Biotechnol J 14:1700665. https://doi.org/10.1002/biot.201700665
111. Steinebach F, Müller-Späth T, Morbidelli M (2016) Continuous counter-current chromatography for capture and polishing steps in biopharmaceutical production. Biotechnol J 11:1126–1141. https://doi.org/10.1002/biot.201500354
112. Ötes O, Flato H, Vazquez Ramirez D et al (2018) Scale-up of continuous multicolumn chromatography for the protein a capture step: from bench to clinical manufacturing. J Biotechnol 281:168–174. https://doi.org/10.1016/j.jbiotec.2018.07.022
113. Ichihara T, Ito T, Kurisu Y et al (2018) Integrated flow-through purification for therapeutic monoclonal antibodies processing. MAbs 10:325–334. https://doi.org/10.1080/19420862.2017.1417717
114. Yoshimoto N, Hasegawa S, Yamamoto S (2019) A method for designing flow-through chromatography processes. MATEC Web Conf 268:01004. https://doi.org/10.1051/matecconf/201926801004
115. Gupte P, Gavasane M, Samagod A et al (2018) Establishing effective high-throughput contaminant removal with membrane chromatography. Bioprocess Int 16:60–63
116. International Conference on Harmonization of Technical Requirements for Registration of Pharmaceuticals for Human Use (1999) Viral Safety Evaluation of Bbiotechnology Products Derived from Cell Lines of Human or Animal Origin Q5A(R1). www.ich.org
117. Parker SA, Amarikwa L, Vehar K et al (2018) Design of a novel continuous flow reactor for low pH viral inactivation. Biotechnol Bioeng 115:606–616. https://doi.org/10.1002/bit.26497
118. Klutz S, Lobedann M, Bramsiepe C, Schembecker G (2016) Continuous viral inactivation at low pH value in antibody manufacturing. Chem Eng Process Process Intensif 102:88–101. https://doi.org/10.1016/j.cep.2016.01.002
119. Kavara A, Sokolowski D, Collins M, Schofield M (2020) Recent advances in continuous downstream processing of antibodies and related products. In: Matte A (ed) Approaches to the purification, analysis and characterization of antibody-based therapeutics. Elsevier, Amsterdam/Oxford/Cambridge, pp 81–103
120. Lute S, Kozaili J, Johnson S et al (2020) Development of small-scale models to understand the impact of continuous downstream bioprocessing on integrated virus filtration. Biotechnol Prog 36:e2962. https://doi.org/10.1002/btpr.2962
121. Bohonak D, Mehta U, Weiss ER, Voyta G (2020) Adapting virus filtration to enable intensified and continuous mAb processing. Biotechnol Prog. https://doi.org/10.1002/btpr.3088
122. Klutz S, Magnus J, Lobedann M et al (2015) Developing the biofacility of the future based on continuous processing and single-use technology. J Biotechnol 213:120–130. https://doi.org/10.1016/j.jbiotec.2015.06.388

123. David L, Niklas J, Budde B et al (2019) Continuous viral filtration for the production of monoclonal antibodies. Chem Eng Res Des 152:336–347. https://doi.org/10.1016/j.cherd.2019.09.040
124. Yehl CJ, Zydney AL (2020) Single-use, single-pass tangential flow filtration using low-cost hollow fiber modules. J Memb Sci 595:117517. https://doi.org/10.1016/j.memsci.2019.117517
125. Casey C, Gallos T, Alekseev Y et al (2011) Protein concentration with single-pass tangential flow filtration (SPTFF). J Memb Sci 384:82–88. https://doi.org/10.1016/j.memsci.2011.09.004
126. BioPharm International Editors (2017) Pall debuts new inline diafiltration modules for continuous bioprocessing. In: BioPharm Int. https://www.biopharminternational.com/view/pall-debuts-new-inline-diafiltration-modules-continuous-bioprocessing-0. Accessed 8 Nov 2020
127. Nambiar AMK, Li Y, Zydney AL (2018) Countercurrent staged diafiltration for formulation of high value proteins. Biotechnol Bioeng 115:139–144. https://doi.org/10.1002/bit.26441
128. Yang O, Qadan M, Ierapetritou M (2020) Economic analysis of batch and continuous biopharmaceutical antibody production: a review. J Pharm Innov 15:182–200
129. Yang O, Prabhu S, Ierapetritou M (2019) Comparison between batch and continuous monoclonal antibody production and economic analysis. Ind Eng Chem Res 58:5851–5863. https://doi.org/10.1021/acs.iecr.8b04717
130. Jacquemart R, Vandersluis M, Zhao M et al (2016) A single-use strategy to enable manufacturing of affordable biologics. Comput Struct Biotechnol J 14:309–318. https://doi.org/10.1016/j.csbj.2016.06.007
131. Brower M, Hou Y, Pollard D (2014) Monoclonal antibody continuous processing enabled by single use. In: Subramanian G (ed) Continuous processing in pharmaceutical manufacturing. Wiley-VCH Verlag GmbH & Co. KGaA, Weinheim, pp 255–296
132. Pollard D, Brower M, Abe Y et al (2016) Standardized economic cost modeling for next-generation MAb production. Bioprocess Int 14:14–23
133. Kornecki M, Schmidt A, Lohmann L et al (2019) Accelerating biomanufacturing by modeling of continuous bioprocessing—piloting case study of monoclonal antibody manufacturing. PRO 7:495. https://doi.org/10.3390/pr7080495

Mathematical Modelling of Cell Culture Processes

Veronique Chotteau, Erika Hagrot, Liang Zhang, and Meeri E. L. Mäkinen

Abbreviations

BFMs	basic flux modes
CG	column generation
CHO	Chinese Hamster Ovary
DoE	design of experiment
ECs	extreme currents
EFMs	elementary flux modes called as well elementary modes
EM	elementary modes
EPs	extreme pathways
FBA	flux balance analysis
FVA	flux variability analysis
GFA	glycosylation flux analysis
MB-DoE	model-based design of experiments
MFA	metabolic flux analysis
MGs	minimal generators

V. Chotteau (✉) · M. E. L. Mäkinen
Department of Industrial Biotechnology, School of Engineering Sciences in Chemistry, Biotechnology and Health, KTH-Royal Institute of Technology, Stockholm, Sweden

AdBIOPRO, VINNOVA Competence Centre for Advanced Bioproduction by Continuous Processing, KTH- Royal Institute of Technology, Stockholm, Sweden

Digital Futures – KTH-Royal Institute of Technology, Stockholm, Sweden
e-mail: chotteau@kth.se; meerim@kth.se

E. Hagrot · L. Zhang
Department of Industrial Biotechnology, School of Engineering Sciences in Chemistry, Biotechnology and Health, KTH-Royal Institute of Technology, Stockholm, Sweden

AdBIOPRO, VINNOVA Competence Centre for Advanced Bioproduction by Continuous Processing, KTH- Royal Institute of Technology, Stockholm, Sweden
e-mail: hagrot@kth.se; lianz@kth.se

R. Pörtner (ed.), *Cell Culture Engineering and Technology*, Cell Engineering 10,
https://doi.org/10.1007/978-3-030-79871-0_14

ODE	ordinary differential equations
POI	product of interest

Mathematical Terms

$\mathbf{A}$	stoichiometric matrix of the reaction network
$\mathbf{A}_{ext}$	stoichiometric matrix related to the extracellular metabolites
$\mathbf{A}_{int}$	stoichiometric matrix related to the intracellular metabolites
$\mathbf{A}_{mac}$	stoichiometric matrix of the macroscopic reaction network
$\mathbf{C}$	vector of the metabolite concentrations
$\mathbf{C}_{\mathbf{ext}}$	vector of the extracellular metabolite concentrations
$\mathbf{C}_{int}$	vector of the intracellular metabolite concentrations
$C_{I,\,i}$	concentration of inhibiting metabolite in the kinetic model
$C_{S,\,i}$	concentration of substrate in the kinetic model
J	set of all the reactions of the network
$K_{C_{I,i}}^{\text{Monod}}$	kinetic parameter of inhibition for generalized Monod model
$K_{C_{S,i}}^{\text{Monod}}$	kinetic parameter of saturation for generalized Monod model
m	number of metabolites
n	number of reactions
$\mathbf{q}_{ext}$	vector of the cell specific rates of uptake or secretion respective to the extracellular metabolites $\mathbf{C}_{ext}$, i.e. cell specific external fluxes
$\mathbf{E}$	matrix of the EFMs
$K_{C_{S,i,l}}$	kinetic parameter of saturation for generalized Michaelis-Menten model
$K_{C_{I,i,l}}$	kinetic parameter of inhibition for generalized Michaelis-Menten model
$\mathbf{r}$	vector of reaction rates r_i
S_i	substrate concentration of substrate $\mathrm{S_i}$
$\mathbf{v}$	vector of the fluxes v_j in a reaction network
$\mathbf{w}$	vector of the macroscopic reaction fluxes w_i
$w_{max,\,i}$	maximum flux of the macroscopic reaction rate w_i
X	biomass
μ_{max}	apparent maximum cell specific growth rate
ν_0	rate of an enzymatic reaction
ν_{max}	maximum rate of an enzymatic reaction

1 Modelling of Mammalian Cell Processes

Due to their ability to produce glycoproteins, mammalian cells are nowadays used for the manufacturing of many biopharmaceuticals. Biologics bioproduction processes are complex and costly. Their development is time-consuming and costly [1]. It is mostly driven by trial-and-error approach supported by statistically based Design of Experiment such as factorial design. The field is though seeking as well for more rational approach based on mathematical models [2]. These can provide better understanding, storage of generated knowledge, tools for simulation, prediction, optimization and control support [3–8]. Furthermore, software sensors are mathematical models, which can support process monitoring [9]. Finally, *in silico* process twin, i.e. digital twin, are also ideal tools for operator training and trajectory tracking. From these multiple facets, the interest in mathematical modelling from large industrial actors has significantly increased these last years [4, 10, 11]. The complexity of the cells, in particular of mammalian cells, combined with their agility to adapt to different environment, render bioprocess modelling a challenge task. Modelling of bioprocesses is highly multi-disciplinary as it includes the knowledge of cell biology and bioprocessing, as well as, ideally, requires tools of system identification, automatic control, data processing, statistics, optimization theory, bioinformatics and data-driven methods.

Mammalian cells have a flexible metabolism, i.e. they can use different biochemical pathways depending of the received stimuli. This can be regarded as an advantage but increases the challenge of mechanistic understanding of these systems [12–15]. Furthermore, the genetic transformation applied to express a product of interest (POI), can affect the cell in multiple ways, e.g. metabolism, nutritional requirements, production of by-products, post-translational modifications, due to altered genome [1]. This implies that a process optimal for a given cell line might not be optimal for another line. The behavior of the cell, i.e. cell growth rate and productivity of a recombinant product are largely influenced by the cell metabolism. These "living factories" have also an inherent variability and heterogeneity. These are triggered by the macro- but also micro-variations of their complex environment, constituted of the culture medium including 50 to >100 components and their own productions of metabolites and large molecules [16, 17]. The metabolism is determined by a network of intracellular metabolic reaction catalyzed by enzymes. Information about the metabolite secretion, depletion or uptake is analyzed for medium improvement [18–20], however *why* certain changes in the medium composition lead to improvements of the process is often not understood [21, 22]. Mathematical modelling can be instrumental to understand the metabolic behavior and cell needs from the study of these uptakes and secretions [15, 20, 23–28]. Besides the metabolism, the field is also interested to model the secretion of the POI and its post-translational modifications, in particular glycosylation. The present chapter will first introduce mathematical modelling of mammalian cells bioprocesses. It will then review the present state of the art of the field, in particular for the cell metabolism and the production of the product of interest including its glycosylation.

2 Overview of the Modelling Approach

The whole modelling approach is not trivial. In general, the model setup can be divided into several steps: *(i)* identification of the modelling project and goal, *(ii)* creation of the model, *(iii)* evaluation of the model [29]. Notice that this section presents the general concepts for modelling approach however several of these concepts have not been applied for mammalian cell bioprocesses.

(i) ***Identifying the modelling project*** begins by determining the model goal. The model goal includes defining the application of the model, target values, acceptance criteria and boundary conditions [29]. The model goal will aid in choosing the type of model and act as a guideline when weighing modelling tradeoffs. For instance, a goal can be to achieve a better understanding of a process; or another one can be to create a model-based feedback control of a given output parameter in a manufacturing process. In the first instance, the model should be mechanistic and comprehensive, while, in the second case, the model should be robust and simple, focusing on the parameters having an influence on the controlled output.

(ii) ***Setting up the model*** requires expert knowledge [29]. Ideally the model structure is assumed, experiments are adequately designed for the model identification and performed to collect the data. Although this seems obvious, it is good to recall that the model will potentially describe the information carried by the data but that it will not be able to render phenomena not present in the experimental data in a reliable way. Therefore, it is important to generate a data set rich in information, in response to various stimuli and covering the space where the model is intended. If necessary, iterations of generating experimental data, developing the model, performing experiments again to improve the model, re-modelling, and so on, should be performed. This iterative process can be difficult to carry over due to time or resource constraints, a criticality typical for mammalian cell culture work. Furthermore, in many cases, models are created from data sets generated for another purpose, and might not be fully adequate for the envisioned modelling purpose. During process development, DoE can typically be carried out to characterize a process in a given parameter design space. The DoE exercise generates a data-driven model of the process in this space [30]. This model is used to catch the influence of the parameters on the process and for optimization but do not include a mechanistic description. Such data could also be used for mechanistic modelling however this is rarely performed [31]. Choosing the experiments could also be approached by relying on information theory. Model-based design of experiments (MB-DoE) provides experimental design maximizing the possibility to estimate the model parameters according to different criteria, i.e. D-optimal, A-optimal, E-optimal [32, 33]. MB-DoE has been proposed in theoretical work simulating microorganism process in case of continuous and very frequent measurements [34]. For this, the continuous dynamic measurement function of the Fisher information matrix was optimized with the objectives of estimating the model parameters in the easiest and quickest way.

(iii) ***Evaluation of the model*** is performed with collected data. During this phase the unknown model parameters are estimated by optimizing the fit of the data simulated by the model to the experimental data, often by minimizing the error between these [29]. This fit can be accompanied by identifiability analysis of the model parameters and their deviations. Structural identifiability ensures that the model parameters are not compensating each other's due to cross correlations in the data [29, 35, 36], while practical identifiability evaluates if the confidence intervals of the parameters allow their estimation with the available data [36]. In addition to identifiability, the model robustness can be evaluated. A simple test is to check the model behavior with extreme input values [29]. A more sophisticated approach is to employ a sensitivity analysis, which describes the impact of the model parameters, process parameters and model inputs on the model outputs [37, 38]. In the following text, the focus is on the model setup for modelling mammalian cell metabolism in the context of biopharmaceutical production.

3 Mathematical Models for Biopharmaceutical Production Processes

Different model types have been used for the biopharmaceutical production processes. These models can be classified based on their ideals and approaches, to highlight the purpose they are suited for [39].

3.1 Modelling Ideals

A model is only a simplified description of the reality. The choice of simplifications and assumptions depends on model objectives. The modeler has the task of choosing which tradeoffs to make between different ideals. Weisberg listed 5 ideals to guide the model construction process: *completeness*, *simplicity*, *1-causal*, *Maxout*, and *generality* [40].

1. The *completeness* ideal is to represent the complexity of the target by including every known property of the target.
2. In contrast, the *simplicity* ideal aims to include as few properties as possible to achieve predictability.
3. As furthest from completeness, the *1-causal* ideal aims to include only the most important causal factors.
4. The *Maxout* ideal aims to maximize the fit to the experimental data, as well as precision and accuracy of the prediction [41].
5. The *generality* ideal aims to maximize the generalizability of the model to different targets.

Along these ideals, the compromises to be made can be segregated by complexity *vs.* simplicity as well as fit to specific data *vs.* generality.

3.2 Model Classifications

Fredrickson *et al.* established a systematic classification for mathematical models of cell cultivation processes into structured *vs.* unstructured, and segregated *vs.* unsegregated [42].

- In unstructured models, the intracellular description of the cell is omitted while the structured models aim to describe the intracellular structure to some detail, i.e. by dividing the cell model into an internal structure of multiple compartments, components and reactions [43]. Data driven models are an example of unstructured models, in which the structure of the model is irrelevant and hidden while mechanistic models are an example of structured models, which rely on the underlying biological mechanisms [44].
- Unsegregated models consider the cells as identical, while segregated models consider the heterogeneity of the cells in the cultivation, i.e. the different phases of cell cycle [45, 46]. Some probabilistic models consider the behavior of individual cells or molecules [47] whereas deterministic models consider the average behavior of the population as a continuous process [39, 47].

  The *unstructured unsegregated* models aim at simplicity ideal while *structured segregated* aim at completeness. S*tructured unsegregated* models reflect the generality ideal while *unstructured segregated* can correspond to the Maxout ideal.

Beside this classification, different features, potentially non-exclusive, can be distinguished as presented below;

3.2.1 Mechanistic Models

Mechanistic models rely on the underlying biological mechanisms and provide an accurate and exhaustive description; however a drawback is that the underlying theory must be correct. These models have the potential to improve our understanding of cellular and process behaviors, and allow for optimization tasks by changes in the internal structure of the cells as well as by the inputs given to the cells [43, 48]. Mechanistic models can be classified as structured models, where the level of structure detail depends on the modelling decisions. These models can be further divided into probabilistic or deterministic models. Mechanistic probabilistic models are often structured segregated models, while mechanistic deterministic models are structured unsegregated or segregated models.

Probabilistic Models

Probabilistic models are based on statistics and widely used for modelling gene regulation and metabolism. The statistical basis of the model allows the representation of the population variability, including random variables, and modelling results as a probability distribution. These models can take into account the uncertainties of the experimental data. Due to their statistical nature, the same set of parameter values and initial conditions can lead to different outputs. Models can be built to avoid overfitting. From a hierarchical assembly of simple modules, more complex models can be created. They can provide coherent relationships despite the data complexity.

Deterministic Models

Opposite to probabilistic models, deterministic models include no randomness factor. They provide a quantitative description of the bioprocess phenomena, which can include a dynamical evolution. Their output prediction is always the same for given inputs and initial conditions. They are based on continuous variables using ordinary differential equations (ODE) of the law of mass action describing these phenomena. They allow the description of the system at a level of detail decided according to the needs and the available experimental information, from very simple to genome-scale. A drawback is that they include a large to very large number of parameters depending of the model size. These have to be determined either through model identification, by dedicated experiments or from the literature. A large number of parameters can lead to overfitting, where the model is fitted to anomalies in the data rather than the actual information. An example of deterministic models valuable for bioprocesses are ***kinetic models***, which include kinetic equations reflecting the enzymatic reactions, e.g. the Michaelis-Menten equation [49]. These models are non-linear, which may be challenging for the identification of their parameters.

3.2.2 Black-Box Models

Black-box models consist of suitable mathematical relationships that allow connecting the inputs and outputs while hiding the biologically irrelevant structure [44, 50]. Black-box models are pure data-driven and sacrifice the explanatory power in favor of fitting the data. Due to lack of structure within the cells or the cell population, black-box models can be classified as unstructured unsegregated models. Construction of these models does not require extensive knowledge of the underlying biological mechanisms [51]. However compared to mechanistic models, they have the drawback of limitation of predictability, in particular outside of the space covered by the experimental data. Many models are so called ***grey-box models***, where black-box modelling is combined with mechanistic information or observation.

The present Chapter is mainly focusing on mechanistic models for mammalian cell culture bioprocesses, and refers to the above-mentioned reviews for black-box models.

3.3 Mechanistic Models of the Cell Metabolism

Mechanistic models rely on the underlying biological mechanisms to describe the observed phenomena. Accurate mechanistic modeling of the mammalian cell metabolism is a challenging task due to the complexity and flexibility of these cells. Two key components of the cell metabolism are the ***network(s)*** of the ***metabolic reactions*** and the ***fluxes*** of molecules through these reactions. A flux is the turnover rate of molecules through one reaction or a series of reactions constituting a metabolic pathway and can be understood as the movement of matter through metabolic networks [12]. In most cases, a flux is the cell specific rate of a reaction or a pathway, expressed as mass per time unit and cell. To describe a culture process, a model needs to include the ***kinetics*** of the fluxes, expressing the influence of components and factors on the reaction rates. Furthermore for processes, which are not in steady-state such as batch or fed-batch modes, these models have also to be dynamical, i.e. expressing a time dependency.

3.3.1 Metabolic Reaction Networks

A structured model for the cell metabolism is generated from the knowledge of the metabolic reactions network. This network consists of a set of reactions interconnecting the metabolites. The information about the reactions can be obtained from e.g. biochemistry textbooks, databases such as KEGG pathways [52, 53], or literature. Many of the reactions in living cells are known. Detailed information from large databases provides these reactions, organized in metabolic systems and pathways [53]. Extensive network models have been established for Chinese Hamster Ovary (CHO) cells, the workhorse of biopharmaceutical manufacturing, both based on mouse models [26] and on CHO genome [54–56]. These models provide detailed information at the molecular level and can be powerful tools for cell engineering. The identification of the parameters of these models is however very tedious and difficult since it requires a lot of information at the intracellular and extracellular levels that has to be obtained from literature or from experimental data. This together with the complexity of the reaction network renders the task of determining as well the flux kinetics even more challenging, although attempts have been proposed [26].

Metabolic reactions and reaction networks can be translated into mathematical representation expressing their stoichiometry. Stoichiometry is based on the law of conservation of matter: the number of atoms consumed by a reaction matches the number of atoms produced by this reaction. The information of a metabolic network can be represented as a stoichiometric matrix $\mathbf{A}$ of the stoichiometric factors of all the reactions. In $\mathbf{A}$, each column corresponds to a reaction while each row corresponds to a metabolite. The stoichiometric factors are negative for the consumed metabolites and positive for the products.

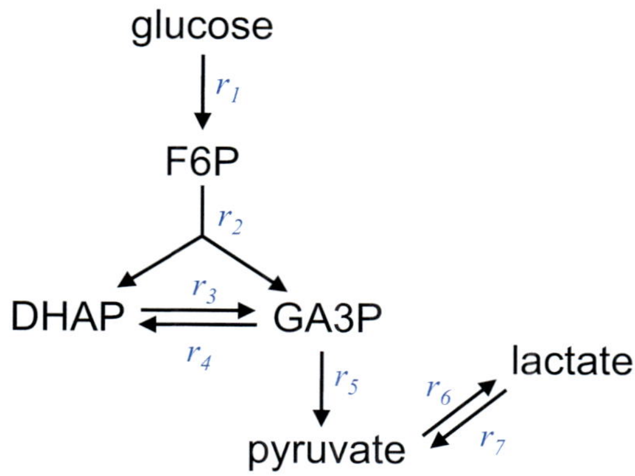

Fig. 1 Example 1; reaction network, consisting of simplified glycolysis where glucose is transformed into pyruvate, followed by pyruvate reversibly transformed into lactate where the co-factors are omitted for clarity

$$A = \overbrace{\begin{pmatrix} a_{1,1} & \cdots & a_{1,n} \\ \vdots & \ddots & \vdots \\ a_{m,1} & \cdots & a_{m,n} \end{pmatrix}}^{n \text{ reactions}} \Bigg] \; m \text{ metabolites}$$

An illustration is given below in Example 1 of a reaction network, consisting of simplified glycolysis where glucose is transformed into pyruvate, followed by pyruvate reversibly transformed into lactate (Fig. 1). This is expressed by the following reactions where H_2O is omitted for clarity and reversible reactions are expressed by two reactions in opposite directions;

Reaction r_1	glucose + ATP → F6P + ADP + P_i
Reaction r_2	F6P + ATP → DHAP + GA3P + ADP + P_i
Reaction r_3	DHAP → GA3P
Reaction r_4	GA3P → DHAP
Reaction r_5	GA3P+2 ADP+2P_i+NAD^+ → pyruvate+2 ATP + NADH + H^+
Reaction r_6	pyruvate + NADH + H^+ → lactate + NAD^+
Reaction r_7	lactate + NAD^+ → pyruvate + NADH + H^+

where DHAP = Dihydroxyacetone phosphate; F6P = Fructose-6-phosphate; GA3P = Glyceraldehyde-3-phosphate

The mass balances of the fluxes through these 7 reactions are given by the derivative with time of the 6 metabolite masses equal to the formation and disappearance of these in the reactions according to their stoichiometric factors (positive for formation and negative for disappearance), as follows, where the co-factors are omitted for clarity;

$$\frac{d\text{ glucose}}{dt} = -r_1 \tag{1a}$$

$$\frac{d\,F6P}{dt} = r_1 - r_2 \tag{1b}$$

$$\frac{d\text{ DHAP}}{dt} = r_2 - r_3 + r_4 \tag{1c}$$

$$\frac{d\text{ GA3P}}{dt} = r_2 + r_3 - r_4 \tag{1d}$$

$$\frac{d\text{ pyruvate}}{dt} = r_5 - r_6 + r_7 \tag{1e}$$

$$\frac{d\text{ lactate}}{dt} = r_6 - r_7 \tag{1f}$$

where $r_1, \ldots, r_7$ are the reaction rates.

Introducing the vector of metabolite concentrations, **C**, the vector of rates reactions, **r**, and expressing the stoichiometric matrix according to (Eqs. 1a–1f);

$$\underset{\text{vector of metabolites,}\,\mathbf{C}}{\begin{bmatrix} \text{glucose} \\ \text{F6P} \\ \text{DHAP} \\ GA3P \\ \text{pyruvate} \\ \text{lactate} \end{bmatrix}} \quad \underset{\text{vector of reactions,}\,\mathbf{r}}{\begin{bmatrix} r_1 \\ r_2 \\ r_3 \\ r_4 \\ r_5 \\ r_6 \\ r_7 \end{bmatrix}} \quad \underset{\text{stoichiometric matrix,}\,\mathbf{A}}{\begin{bmatrix} -1 & 0 & 0 & 0 & 0 & 0 & 0 \\ 1 & -1 & 0 & 0 & 0 & 0 & 0 \\ 0 & 1 & -1 & 1 & 0 & 0 & 0 \\ 0 & 1 & 1 & -1 & 0 & 0 & 0 \\ 0 & 0 & 0 & 0 & 1 & -1 & 1 \\ 0 & 0 & 0 & 0 & 0 & 1 & -1 \end{bmatrix}}$$

the mass balances of the fluxes (Eqs. 1a–1f) are now rewritten as

$$\frac{d}{dt}\begin{bmatrix} \text{glucose} \\ \text{F6P} \\ \text{DHAP} \\ \text{GA3P} \\ \text{pyruvate} \\ \text{lactate} \end{bmatrix} = \begin{bmatrix} -1 & 0 & 0 & 0 & 0 & 0 & 0 \\ 1 & -1 & 0 & 0 & 0 & 0 & 0 \\ 0 & 1 & -1 & 1 & 0 & 0 & 0 \\ 0 & 1 & 1 & -1 & 0 & 0 & 0 \\ 0 & 0 & 0 & 0 & 1 & -1 & 1 \\ 0 & 0 & 0 & 0 & 0 & 1 & -1 \end{bmatrix} \begin{bmatrix} r_1 \\ r_2 \\ r_3 \\ r_4 \\ r_5 \\ r_6 \\ r_7 \end{bmatrix} \tag{2a}$$

or

$$\frac{d\mathbf{C}}{dt} = \mathbf{A}\ \mathbf{r} \tag{2b}$$

where $\mathbf{C}$ (m x 1); $\mathbf{A}$ (m x n); $\mathbf{r}$ (n x 1)

The number of reactions n is usually higher than the number of metabolites m; for instance Example 1 includes 7 reactions and 6 metabolites.

Let us define v_j as the flux through a reaction r_j, expressed as mass per time unit and cell, and $\mathbf{v}$ (n x 1) as the vector of the all the fluxes v_j in a reaction network. The reaction rate r_j can be expressed as the product of the cell specific flux of mass through this reaction, v_j, times the biomass, X, i.e. $\mathrm{r}_j = v_j\ X$, where X is a scalar. Applied to Eq. 2b, it follows that

$$\frac{d\mathbf{C}}{dt} = \mathbf{A}\ \mathbf{r} = \mathbf{A}\ \mathbf{v}\ X \tag{3}$$

The stoichiometric matrix can be given further structural details of compartmentalization by defining metabolites as internal, external or belonging to compartments such as mitochondria. Transport reactions are added in case compartments are included. In particular, $\mathbf{A}$ can be split into the reactions related only to the intracellular metabolites and the ones related to the extracellular metabolites, into two matrices $\mathbf{A}_{int}$ and $\mathbf{A}_{ext}$ respectively.

$$\mathbf{A} = \begin{bmatrix} \mathbf{A}_{ext} \\ \mathbf{A}_{int} \end{bmatrix} = \begin{bmatrix} a_{ext\,1,1} & \cdots & a_{ext\,1,n} \\ \vdots & \ddots & \vdots \\ a_{ext\,k,1} & \cdots & a_{ext\,k,n} \\ a_{int\,k+1,1} & \cdots & a_{int\,k+1,n} \\ \vdots & \ddots & \vdots \\ a_{int\,m,1} & \cdots & a_{int\,m,n} \end{bmatrix} \tag{4}$$

Accordingly Eq. 3 can be rewritten as

$$\frac{d}{dt}\begin{bmatrix} \mathbf{C}_{ext} \\ \mathbf{C}_{int} \end{bmatrix} = \begin{bmatrix} \mathbf{A}_{ext} \\ \mathbf{A}_{int} \end{bmatrix} \mathbf{v}\ X \tag{5}$$

where suffix *ext* corresponds to the extracellular elements, and suffix *int* to the intracellular elements, i.e. $\mathbf{C}_{\mathbf{ext}}$ and $\mathbf{C}_{int}$ are respectively the vectors of the extracellular and intracellular metabolite concentrations. A further constraint is that $\mathbf{v} \geq 0$, which indicates that $v_j \geq 0, \forall j \in J$ (for all j belonging to J) where J is the set of all the reactions of the network. In case a reaction is reversible, it is split into two reactions, one in each direction, and thus $\mathbf{A}$ includes two columns for a reversible reaction, see Example 1 for instance for the reversible transformation of pyruvate into lactate.

3.3.2 Metabolic Pathway and Macroscopic Reactions

A ***metabolic pathway*** is a chain of interconnected metabolic reactions, i.e. a defined series of reactions where the product of a reaction acts as substrate for the next one. A ***macroscopic reaction*** or ***macro-reaction*** describes the path between entry substrates in the cell and final products into a single reaction without considering the intracellular metabolites. It can be expressed by the stoichiometry relating the metabolites taken up by the cells to the secreted ones. It is characterized by the flux of the turnover of the extracellular substrates into extracellular products. A macroscopic view of a system is obtained by a network of macroscopic reactions.

In Example 1, the following macro-reaction is a simplification of the system of reactions r_1 to r_7;

$$\text{glucose} + 2\ \text{ADP} + 2\ \text{P}_\text{i} \rightarrow 2\ \text{lactate} + 2\ \text{ATP}$$

Unlike metabolic reactions, macroscopic reactions lack a complete underlying biochemical description, which is lost when reactions are lumped into a macroscopic reaction. However, experimental data of metabolic fluxes can be used to aid the creation of the model. The network of macroscopic reactions can be expressed by a stoichiometric matrix in a similar way as a network of single reactions. The concept of metabolic pathways or macroscopic reactions is often used to reduce the model size.

Macroscopic models are based on macroscopic reactions. They provide simpler models for the simulation of the reaction rates and metabolite concentrations in a cell culture where the intracellular metabolism is not detailed [14, 44, 57–63]. ***Macroscopic kinetic models*** include also the kinetic equations to describe a simplified metabolism of multiple pathways running through the metabolic network into reactions linking observable extracellular substrates to products [57–60, 62].

3.3.3 Introduction to Flux Analysis

As previously mentioned, a flux is the turnover rate of molecules through a reaction or a series of reactions. Flux analysis addresses the movement of matter through a network of metabolic reactions and can be divided into two approaches: *(i)* kinetic analysis, which provides a dynamical description and can be variable with the process components and with time, or *(ii)* static analysis, which is constraint-based steady-state and do not consider kinetics.

Kinetic analysis enables the evolution of the concentrations of metabolites over time and is thus the most appropriate to describe a culture process. However the kinetics are complex non-linear functions including a large number of parameters, and thus challenging for the model identification. Inherent to this challenge, the number of reactions of these models needs to be limited, leading to a compromise of the detail level of the reaction network.

In contrast, static analysis does not require the identification of the kinetics and is thus compatible with larger reaction networks, where constraint(s) can be instrumental for the model identification. This approach is suitable for cell engineering and synthetic biology but does not provide suitable tools for process modelling.

The measurement of the extracellular component fluxes is typically carried out by e.g. high-pressure liquid chromatography or mass spectrometry. The quantification of the fluxes of the intracellular metabolites is a highly valuable information for the model but is much more challenging. These fluxes can be studied by isotope labeling, typically ^{13}C. The cells are cultured (or pulsed) with an isotope labelled substrate, and the distribution of the ^{13}C labelled metabolites can be followed by mass spectrometry or nuclear magnetic resonance ^{13}C-NMR [64]. The high cost of the labelling substrates, the instruments and critical requirement for cell quenching makes this approach less feasible [21], and is hardly compatible with kinetic analysis both from economical and methodological point-of-views.

3.3.4 Determination of the Relevant Metabolic Pathways

In the analysis of the metabolic pathways, these are identified by combining sets of metabolic reactions of the network to describe how the cells produce and consume the metabolites corresponding to the measurements of the reaction fluxes. The identification of the pathways is a central problem in mechanistic modelling of cell metabolism. There are several methods for the systematic identification of the pathways: extreme currents, ECs [65], elementary flux modes, EFMs [66], extreme pathways, EPs [67], and minimal generators, MGs [68]; for a review, see [65, 69].

All four methods are related to extreme rays: If all the possible combinations of reactions that form pathways through a metabolic network are contained within the feasible solution space, then the extreme rays are the pathways that can be used to represent all the other pathways within this space. The extreme rays cannot be decomposed in simpler vectors and are the edges of a convex polyhedral pointed cone representing the feasible solution space, see Fig. 2 [65]. The extreme rays are all the non-decomposable vectors in the flux space, where a non-decomposable vector is composed of reactions such that if any of the constituent reaction does not transfer a flux, the other vector reactions cannot operate. Any flux distribution in the flux space can be represented as a non-negative combination of the extreme rays.

Elementary flux modes, called as well elementary modes (EM), are the most commonly used among the methods listed above. The majority of the models based on pathways for mammalian cells are based on EFMs [14, 57–63, 70]. A set of EFMs in a flux space can be defined as all the non-decomposable vectors in this space such that there are no cancellations of reversible fluxes, i.e. a backward flux cannot be cancelled by a forward flux [71]. EFMs are also not systemically

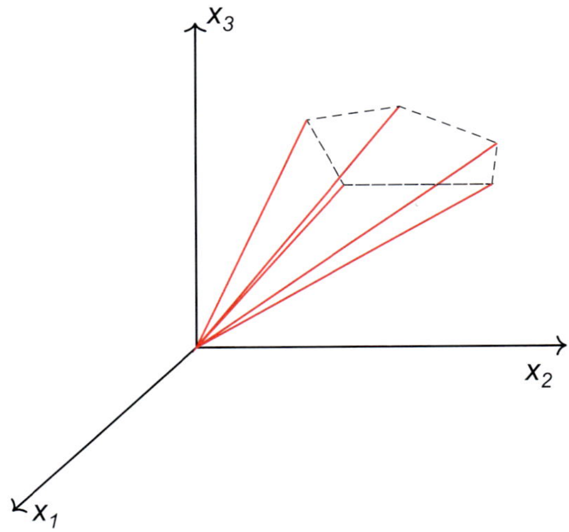

Fig. 2 Illustration of the extreme rays for a three-dimensional system. The convex polyhedron contains the feasible solutions. The extreme rays are the edge vectors of the cone (represented as red continuous line). These are a basis for all the feasible solutions

independent, thus an EFM can be a non-negative combination of other EFMs [72]. EFMs allow the identification of important and physiologically meaningful pathways not found via other pathway identification methods [73]. A set of EFMs for a given reaction network are *all* the non-decomposable pathway vectors that can generate the flux space.

3.3.5 Identification of Relevant Elementary Flux Modes, EFMs

A standard approach for the identification of the EFMs, is based on the double description method [74], generating a systematic enumeration of all the EFMs. EFMs are generated by pairwise combination of existing EFMs while ensuring that the EFM has not already been identified. The method can be computed using the software's Efmtool [75] or Metatool [76]. The main challenge of systematic enumeration of all the EMFs is that the number of possible combination of reactions into pathways grows exponentially with the size of the metabolic network [73, 77, 78], while most of the enumerated EFMs are biologically irrelevant. Due to the exponential increase of the EFMs leading to computational limitation, the size of the network is practically limited to $\approx$ 36 reactions [14, 57–61], see Table 1 for examples. A reduced set of EFMs can be obtained by using experimental data to guide the simplification of the network by selecting the largest measured macroscopic fluxes prior to pathway enumeration [59, 61, 79]. Unfortunately, the reaction mechanisms removed by this simplification are lost.

An alternative is to keep the original reaction network without simplification and to use a method generating only subsets of EFMs instead of their complete enumeration. Several methods apply this principle; selection of only the *shortest EFMs* [80], feasible *elementary flux patterns* of sub-networks [81] and *decompo-*

Table 1 Illustration of kinetic models of mammalian cell bioprocesses

Cell line	Mode	Network	Macro-reactions	Kinetics	Reference
Structured models					
CHO	FB	34 *n*, 24 *m*	/	Convenience rate laws	[115]
CHO	Batch	30 *n*	/	Monod/MiMe	[119]
CHO	Batch	35 *n*	/	Monod/MiMe	[120]
GS-CHO	FB/Batch	24 *n*	/	MiMe	[123]
CHO-DXB11	FB/Batch	34 *n*	/	Monod/MiMe	[107]
Reaction network-based macroscopic models					
CHO	Batch	18 *n*, 20 *m*	7 EFMs [a]	Monod/MiMe	[57]
CHO	Batch	25 *n*	3-11 EFMs [a]	Monod/MiMe	[58]
Hybridoma	Batch	32 *n*	9 EFMs [b]	Monod/MiMe	[59]
Hybridoma	FB/Batch	32 *n*, 30 *m*	10 EFMs [b]	Monod/MiMe	[60]
CHO	FB/Batch	34 *n*, 30 *m*	8 EFMs [b]	Monod/MiMe	[61]
CHO	Batch	100 *n*, 72 *m*	19 EFMs [c]	Monod/MiMe	[62]
Hybridoma/ GS-CHO	FB/perfusion	35 *n*	9 EFMs [a, b]	No kinetics	[79]
CHO-K1SV	P-perfusion	34 *n*, 38 *m*	379 EFMs [a]	Monod/MiMe	[14]
CHO-K1SV	P-perfusion	126 *n*, 118 *m*	125 EFMs [d]	Monod/MiMe	[63]
CHO-K1SV	P-perfusion	131 *n*, 128 *m*	25 BFMs	Monod/MiMe	[86]

FB fed-batch, P-Perfusion pseudo-perfusion, *n* Number of reactions, *m* number of metabolites, MiMe Michealis-Menten; reproduced from [85]
[a]systematic enumeration of EFMs by Metatool
[b]simplification of the network by selecting the largest measured macroscopic fluxes prior to pathway enumeration
[c]subset of EFMs by decomposition of the flux vector
[d]subset of EFMs by column generation

sition of the flux vector obtained from measurements [82]. Another approach is to utilize the experimental data via flux analysis to systematically identify an EFM subset directly from the detailed network [62, 70, 82, 83]. This approach can be implemented by utilizing the mathematical technique *column generation* (CG), to identify the EFMs relevant for the experimental data from the reaction network [70]. CG, is a dynamic identification method, which provides the EFMs relevant to the information of the experimental data, without their systematic enumeration [63, 70]. CG has been successfully applied to mAb production by CHO cells in different media mammalian cell cultivations for identifying EFM subsets in metabolic networks of 125 reactions [63, 70, 84], and larger networks, ≥ 1000 reactions, can be addressed by CG [85].

A recent alternative to the EFMs is the basic flux modes (BFMs), which allow a reduction of the number of macroscopic fluxes to the same order of magnitude as the number of extracellular metabolites [86]. The stoichiometries of BFMs are constructed as linear combinations of the stoichiometries of independent reactions of the reaction network. BFMs have been demonstrated in mammalian pseudo-perfusion process of CHO cells [86].

3.3.6 Metabolic Flux Analysis, MFA

Metabolic flux analysis deals with the distribution of the intracellular fluxes in metabolic networks based on the application of material balances, and including coupling with the extracellular metabolites. It addresses variations in the flux distribution and intensity caused by perturbations, e.g. in the cell environment, or in the genome. The vector of the fluxes over the reaction network, v, gives a quantitative characterization of the studied phenotype, e.g. cell metabolism. Metabolic flux analysis (MFA) and flux balance analysis (FBA) are based on the application of material balances as introduced here and are the most popular constraint-based approaches [87]. In both cases, the ***pseudo steady state assumption*** is commonly used; implying that there is no net accumulation or reduction of the intracellular metabolites [88, 89], in other words

$$\frac{d\mathbf{C}_{int}}{dt} = 0 \tag{6}$$

where $\mathbf{C}_{int}$ is the vector of the concentrations of all the intracellular metabolites.

Constant environmental conditions, such as achieved in chemostat and perfusion operations, provides this state to the cells. On the contrary, batch or fed-batch cultures have a dynamic evolution. However, for these modes, a common assumption is that the cells adapt quickly to changing environment so that the pseudo-steady state assumption is valid for phase(s) of the culture, where the cell behavior is consistent such as exponential growth.

Let us segregate Eq. 5 into the extracellular and the intracellular elements, and take Eq. 6 into account

$$\frac{d\mathbf{C}_{ext}}{dt} = \mathbf{A}_{ext}\,\mathbf{v}X = \mathbf{q}_{ext}\;X \quad \text{or} \quad \mathbf{A}_{ext}\,\mathbf{v} = \mathbf{q}_{ext} \tag{7a}$$

$$\frac{d\mathbf{C}_{int}}{dt} = \mathbf{A}_{int}\,\mathbf{v}X = 0 \quad \text{or} \quad \mathbf{A}_{int}\,\mathbf{v} = 0 \tag{7b}$$

where $\mathbf{q}_{ext}$ (m x 1) is the vector of the cell specific rates of uptake or secretion respective to the extracellular metabolites $\mathbf{C}_{ext}$, i.e. cell specific external fluxes, and can potentially be measured extracellularly. System Eqs. 7a–7b is rewritten as

$$\begin{bmatrix} \mathbf{A}_{ext} \\ \mathbf{A}_{int} \end{bmatrix} \mathbf{v} = \begin{bmatrix} \mathbf{q}_{\mathbf{ext}} \\ 0 \end{bmatrix} \quad \text{and} \quad v_j \geq 0, \; v_j \in \mathbf{v} \tag{8}$$

where $\mathbf{A}$ (m x n); $\mathbf{v}$ (n x 1); $\mathbf{q}_{ext}$ (k x 1)

Equation 8 is a system of equations where $\mathbf{A}$ is the stoichiometric matrix and is known, $\mathbf{v}$ is the vector of the fluxes v_j and is unknown, $\mathbf{q}_{ext}$ can be measured. Each row of Eq. 8 corresponds to a metabolite. Equation 8 includes m equations corresponding to the number of metabolites and n unknown variables, which are the fluxes. Importantly, usually the number of metabolites is smaller than the number of reactions and thus fluxes. In other words, $m < n$ and Eq. 8 has fewer equations than variables, and thus this problem is ***underdetermined***.

Due to the complexity of the biological system, for bioprocess modelling, metabolic networks of MFA are usually simplified, and the analysis only focuses on a set of related metabolic reactions, thus non-important reactions are ignored to reduce the number of unknown fluxes. This leads to a loss of some metabolic information. Another option is to increase the amount of detail in metabolic network, however $\mathbf{q}_{\mathbf{ext}}$ is limited by the available measurements, which are often limited to only a few metabolites. Consequently, several vectors of fluxes can satisfy the Eq. 8 and the solution is not unique. A possible approach is to have intervals on each reaction flux. This is known as *interval metabolic flux analysis* [83], and the *flux spectrum approach* [90] or *flux variability analysis*, FVA. Another option is to narrow the flux intervals by introducing additional constraints [83]. MFA has been applied to CHO and hybridoma cell cultures to investigate the effects of different stimuli such as different amino acid supplementations [14, 20], sugars [24], temperature shifts [91] and butyrate addition [92]. Alternatively, the system is complemented by intracellular flux information, however this is hardly adequate to model processes.

3.3.7 Flux Balance Analysis, FBA

In flux balance analysis (FBA), a biologically relevant objective function is optimized to solve the distribution of the fluxes given by Eq. 8 system [93–95]. For a review we refer to [96] and [21]. For micro-organisms, the objective can be to maximize the growth rate or minimize the energy consumption within the space of possible fluxes, determined by the reaction network [21, 93, 97]. Linear programming is used to solve this optimization problem, and is applied for genome-scale networks. The optimization problem is constrained by *(i)* the stoichiometry of the metabolic network given by $\mathbf{A}_{ext}$ and $\mathbf{A}_{int}$; *(ii)* the uptake/secretion rates $\mathbf{q}_{ext}$, which can be replaced by upper and lower limits in case they are not measured, and *(iii)* constraints of lower and upper bounds on the fluxes v_j. In case of minimization, the FBA problem can be expressed as

$$\min_{\mathbf{v}} f_{obj}\ (\mathbf{v}) = b_1\ v_1 + b_2\ v_2 + \cdots + b_l\ v_l \tag{9a}$$

$$\text{subject to } \mathbf{A}_{ext}\ \ \mathbf{v} \ = \ \mathbf{q}_{ext} \tag{9b}$$

$$\mathbf{A}_{int} \quad \mathbf{v} \quad = \quad 0 \tag{9c}$$

$$v_{j,\min} \quad \leq \quad v_j \quad \leq \quad v_{j,\max} \qquad \forall j \tag{9d}$$

Similar to MFA, the solution is not necessarily unique as several different flux combinations may give an optimal solution to the problem. The feasible space can be further constricted to rule out alternative solutions by adding constraints to the optimization problem, i.e. the rates $\mathbf{q}_{ext}$ [71]. A subtype of FBA is flux variability analysis (FVA), in which each flux is systematically minimized and maximized over the entire reaction network providing an interval on each reaction flux [26].

FBA is widely used for microorganisms, which grow in conditions where constraints on the nutrient uptake can be readily applied. Furthermore, it is reasonable to assume a driving force of the cells to grow, why growth maximization is typically selected as objective function. Consequently, FBA has been extensively used in genome-scale modelling, for cell engineering and synthetic biology, e.g. *in silico* knock-out, metabolic engineering, and bacterial evolution [98, 99]. The application of FBA to mammalian cells systems is more challenging. Their metabolism is more complex and flexible, not amenable to constraints on the nutrient uptake in a similar way as microorganisms.

FBA has been applied to CHO cells systems [21, 91, 100, 101], to hybridoma cell fed-batch cultures [102], complemented with additional information from isotope labeling [103] or omics data [21, 22, 26]. In addition to modelling the metabolism, the steady state assumption can also be applied to model the glycosylation in cell culture. Based on pseudo steady state assumption, glycosylation flux analysis (GFA) has been proposed in a model used in fed-batch culture [104].

3.3.8 EFM-Based MFA

As previously introduced, a macroscopic reaction is an overall reaction between extracellular metabolites, where intracellular reactions are lumped and intracellular metabolites are not considered. A flux through a macroscopic reaction, w_i, is issued from a combination of fluxes v_j of individual reactions. EFMs-based MFA is based on the MFA concept and EFM principle to obtain the flux pathways of macroscopic reactions and provide a link between the experimental rates of uptake and secretion $\mathbf{q}_{ext}$ and $\mathbf{w}$, where $\mathbf{w}$ is the vector of $w_{i\,(i\,=\,1,\,\ldots,\,k)}$ elements. EFMs-based MFA describes the cell behavior using the extracellular measurements by determining macroscopic reactions. The vector of fluxes $\mathbf{v}$ can be expressed as the set of EFMs, given by the matrix $\mathbf{E}$, and the vector of the macroscopic reaction fluxes $\mathbf{w}$ as follows;

$$\mathbf{v} = \mathbf{E} \ \mathbf{w} \tag{10}$$

Considering Eqs. 7a and 10 becomes

$$\mathbf{q}_{ext} \quad = \quad \mathbf{A}_{ext} \ \mathbf{v} \quad = \quad \mathbf{A}_{ext} \ \mathbf{E} \quad \mathbf{w} \tag{11}$$

where $\mathbf{q}_{ext}$ (k x 1); $\mathbf{A}_{ext}$ (k x n); $\mathbf{E}$ (n x k); $\mathbf{w}$ (k x 1)

In Eq. 11, $\mathbf{q}_{ext}$ is the vector of the extracellular fluxes of uptake and secretion, which is measured, and the stoichiometric matrix $\mathbf{A}_{ext}$ is known. Considering that $\mathbf{E}$ is known, see Sect. 3.3.5 Identification of relevant EFMs, the EFMs-based MFA problem consists in identifying an estimate of the pathway flux vector $\mathbf{w}$ that minimizes the difference between the measured fluxes $\mathbf{q}_{ext}$ and $\mathbf{A}_{ext}$ $\mathbf{E}$ $\mathbf{w}$, i.e. linear least-squares data fitting problem, as follows;

$$\min_{\mathbf{w}} \ \frac{1}{2} \sum \|\mathbf{q}_{ext} \ - \ \mathbf{A}_{ext} \ \mathbf{E} \ \ \mathbf{w}\|_2^2, \qquad \mathbf{w} \geq \ 0 \tag{12}$$

which is equivalent to $\min_{\mathbf{w}} \quad \sum \|\mathbf{q}_{ext} \ - \ \mathbf{A}_{ext} \ \mathbf{E} \ \ \mathbf{w}\|_2^2, \qquad \mathbf{w} \geq \ 0$

Notice that a macroscopic stoichiometric matrix $\mathbf{A}_{mac}$ such as

$$\mathbf{A}_{mac} = \ \mathbf{A}_{ext} \ \mathbf{E} \tag{13}$$

can be defined. Considering Eqs. 13 and 11 becomes

$$\mathbf{q}_{ext} \quad = \quad \mathbf{A}_{mac} \ \mathbf{w} \tag{14}$$

As previously mentioned, the systematic enumeration of all the EFMs is practically limited since their number increases exponentially with the number of reactions. Only rather small reaction networks, typically including less then 36 reactions, can be addressed by complete enumeration to identify the macro-reactions, as illustrated in Table 1. Instead of systematic enumeration of the EFMs, considering a relevant subset of EFMs is possible as described in Sect. 3.3.5 Identification of relevant elementary flux modes, EFMs. Linear least-squares data fitting problem regression can be used to identify EFMs that explain the variations of the experimental fluxes [105].

3.4 Mechanistic Kinetic Models of the Cell Metabolism

As presented above, in order to simulate a culture process, the description of the kinetics is required, however this adds to the modelling complexity. The kinetics should be able to describe the changes in the cell metabolism in response to variations in the culture to support dynamic simulations and predictions. A kinetic model is made of *(i)* the relevant reactions, and *(ii)* the kinetic equations, including the estimation of the kinetic parameters. Due to the complexity of the kinetics generating challenging modelling with risk of over-fitting, and the fact that the measurements are extracellular and not intracellular for economical and methodological reasons, most of the times, the model are based on macro-reactions, i.e. macroscopic kinetic model, and not extremely large models. The macroscopic kinetic model structure consists thus of two parts: *(i)* the macroscopic reactions involving the extracellular metabolites; and *(ii)* the kinetic equations expressing how the culture components and conditions influence the flux rates of the macroscopic reactions [44].

The macroscopic reactions are determined with a mechanistic approach by utilizing information from a known metabolic network, typically using an EFM-based approach [14, 58, 63]. The macro-reactions linking the extracellular metabolites are built by elimination of the intracellular elements as presented above.

The kinetic equations are mathematical expressions of the impact of the metabolites or physical parameters on the reaction rates [49, 106]. Modelling the dynamic conditions is a difficult task [14, 58, 59, 62, 63, 107]. The identification of the kinetic equations is quite challenging in dynamic cultures, such as batch, or fed-batch modes, where the cell environment is constantly changing with time. In continuous perfusion mode where a culture steady-state is established, the kinetic equations are not function of time, which alleviates their identification.

3.4.1 Kinetic Equations

The Monod equation models the experimentally observed growth rate of a cell population as a function of a limiting nutrient concentration [108]. This equation can be derived from data alone, without structure. The Monod equation and its generalized forms have been successfully applied to describe the growth of cultured cells as function of limiting substrates [44, 109], with the following generalization

$$\mu = \mu_{\max} \prod_i \frac{S_i}{S_i + K_{C_{S,i}}^{\text{Monod}}} = \mu_{\max} \frac{S_1}{S_1 + K_{C_{S,1}}^{\text{Monod}}} \frac{S_2}{S_2 + K_{C_{S,2}}^{\text{Monod}}} \cdots \tag{15}$$

where the kinetic parameters are the apparent maximum cell specific growth rate μ_{max}, and the half-saturation constants $K_{C_{S,i}}^{\text{Monod}}$ associated with each limiting substrate $S_{i\,(i=1,\ldots,l)}$. Effects of inhibition by by-products or other metabolites can also be included in this kinetic equation, which becomes

$$\mu = \mu_{max} \underbrace{\prod_{i \in M_S} \frac{C_{S,i}}{C_{S,i} + K_{C_{S,i}}^{\text{Monod}}}}_{saturation\ by\ substrates} \underbrace{\prod_{i \in M_I} \frac{K_{C_{I,i}}^{\text{Monod}}}{C_{I,i} + K_{C_{I,i}}^{\text{Monod}}}}_{inhibition\ by\ products\ or\ metabolites} \tag{16}$$

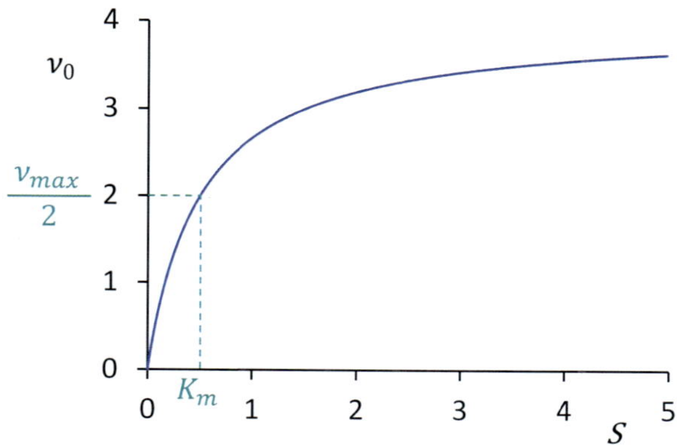

Fig. 3 Michaelis-Menten equation $\nu_0 = \nu_{\max} \frac{S}{S+K_m}$ where $\nu_{max} = 4$ and $K_m = 0.5$, reproduced from [85]

where $C_{S,i}$ and $C_{I,i}$ are substrates and inhibiting metabolites concentrations, and $K_{C_{S,i}}^{\text{Monod}}$ and $K_{C_{I,i}}^{\text{Monod}}$ their respective kinetic parameters of saturation and inhibition. For instance, saturation by substrates glucose and glutamine, and inhibition by by-products lactate and ammonium in hybridoma cell culture were included by Jang and Barford [110], as follows

$$\mu = \mu_{max} \underbrace{\frac{Glc}{Glc + K_{Glc}} \frac{Gln}{Gln + K_{Gln}}}_{saturation\ by\ substrates} \underbrace{\frac{K_{Lac}}{Lac + K_{Lac}} \frac{K_{Amm}}{Amm + K_{Amm}}}_{inhibition\ by\ products}$$

In a reaction, the enzyme kinetics describes the influence of the metabolites, i.e. substrates, products or other regulating metabolites, on the rate of the reaction. The Michaelis-Menten rate model was evaluated by Leonor Michaelis and Maud Menten, from an equation proposed by Victor Henri [111, 112]. It is based on the assumption of steady-state, expressing that the concentration of the intermediate complex substrate-enzyme, ES, does not vary with time, in the reaction $E + S \rightleftharpoons ES \rightleftharpoons P + S$ transforming a substrate S into a product P with the catalyzation of enzyme E. The rate ν_0 of this reaction is

$$\nu_0 = \nu_{\max} \frac{S}{S + K_m} \tag{17}$$

where ν_{max} stands for the maximum reaction rate, S denotes the substrate concentration, K_m is the Michaelis constant, a kinetic parameter, corresponding to the S concentration when the reaction is at half its maximum, see Fig. 3. The mathematical form of the Monod equation is similar to the Michaelis-Menten equation.

Modelling the kinetic behaviour of cell cultures related to a given metabolic reaction network has been approached by convenience kinetics, which are generalised Michaelis-Menten equation accounting for effects of saturation, activation and inhibition [113, 114]. Other constraints, such as thermodynamic constraints, can also be added to the equations. Nolan and Lee [115] applied this approach to model CHO cell cultures by manually tuning the equations, including regulation factors and optimizing the fitting of the simulation to the experimental data. The kinetic parameters for the enzymes are available in databases such as BRENDA, however these rates have mostly been obtained *in vitro*, i.e. in conditions artificial compared to live cell systems [116, 117]. Furthermore, although very important in this context, the allosteric regulation is challenging to include in cell culture modelling [118]. On another hand, the kinetic equations can be constructed by considering all the intracellular fluxes of the macro-reactions with a kinetic term per intracellular reaction using generalized Michaelis-Menten equations and parameter estimation based on intracellular measurements, e.g. nucleotides, as proposed for CHO cells by Ghorbaniaghdam, Robitaille and co-workers [107, 119, 120]. An alternative way is to adopt a network of macroscopic reactions and model the kinetics of the macro-reactions at a macroscopic level. In other words, instead of creating models from the detailed kinetics of individual reactions, the models describing culture behaviour are based on generalized kinetic equations, providing a global behaviour of the system [118]. This approach is also potentially amenable to automation [117]. Table 1 lists examples of macroscopic models, where generalized kinetic functions have been used for a large majority of the models.

A major problem in modelling metabolic kinetics is the lack of kinetic information of the metabolism [117]. Simplified kinetic models can be used to reduce the number of unknown kinetic parameters to simulate the dynamics of cell behavior [58, 59, 62]. The simplification can be a reduction of the number of inputs and outputs, or a reduction of the reaction network [62, 82]. Some argue that pathways contain bottleneck regulator enzyme(s). Others that the flux are controlled by multiple steps and not by single enzymes [121].

In macroscopic kinetic EFM-based models, each element of **w**, the vector of macroscopic reaction fluxes, is replaced with a kinetic function of the substrate, product and influencing metabolites.

$$\mathbf{w} = \mathbf{F}(C_1, C_2, \ldots, k_1, k_2, \ldots) \quad \mathbf{w}_{max} \tag{18}$$

So that Eq. 11 becomes

$$\mathbf{q}_{ext} = \mathbf{A}_{ext}\ \mathbf{v} = \mathbf{A}_{ext}\ \mathbf{E}\quad \mathbf{F}(C_1, C_2, \ldots, k_1, k_2, \ldots) \quad \mathbf{w}_{max} \tag{19}$$

This approach has been applied by [14, 57–63]. Models can be developed from approximative and generalized kinetic equations, e.g. power-laws and linearizable logarithmic equations [44, 117] generalized Michaelis-Menten/Monod equations [14, 57–59, 62, 63]. Adopting generalized Michaelis-Menten equations, a macroscopic flux rate $w_l$is given by

$$w_l = w_{\max,l} \prod_i \frac{S_i}{S_i + K_{S_i}} = w_{\max,l} \frac{S_1}{S_1 + K_{S_1}} \frac{S_2}{S_2 + K_{S_2}} \cdots \quad (20)$$

where $w_{max,\,l}$ is the maximum flux of the macroscopic reaction rate w_l; S_i is the concentration of substrate S_i and K_{S_i} is the half-saturation constant for this substrate.

Due to the constraint of positivity of the flux rates, inhibition effects from by-products or other metabolites can be necessary to account for to obtain a satisfying model fitting. Such effects were included in the detailed enzyme-mechanistic based equations in the model of CHO cell metabolism [107, 120]. An improvement of Eq. 20 to include inhibition in the generalized Michaelis-Menten can be adopted [14, 63], providing the following equation for the macroscopic flux rate w_l

$$w_l = w_{\max,l} \underbrace{\prod_{i \in M_{S,l}} \frac{C_{S,i}}{C_{S,i} + K_{C_{S,i,l}}}}_{substrate\ saturation\ or\ enhancement\ by\ a\ metabolite} \underbrace{\prod_{i \in M_{I,l}} \frac{K_{C_{I,i,l}}}{C_{I,i} + K_{C_{I,i,l}}}}_{inhibition\ by\ a\ product\ or\ a\ metabolite} \quad (21)$$

where $K_{C_{S,i,l}}$ and $K_{C_{I,i,l}}$ are the kinetic parameters of saturation and inhibition for respectively $C_{S,\,i}$ and $C_{I,\,i}$.

3.4.2 Modelling Approaches for Processes with Multiple Phases or in Diverse Conditions

In a process, variations in the culture behavior can occur due to multiple factors. Batch and fed- batch cultures are dynamical processes where the conditions are continuously changing. Shifts in temperature, pH or medium/feed media, as well as addition of a chemical inducer trigger different phases in a process. Variations can be introduced by working with different clones. Different approaches have been used to address these changes.

To model the dynamical evolution in batch or fed- batch modes, a strategy is to model the process by generating several consecutive phase-specific models [59, 115, 120]. Typically, the process is divided in growth phase, stationary phase and decay phase and the whole process is modelled with two or three consecutive models, each associated to a different phase. To deal with the consecutive models and merge them for the process simulation, switching actuators, which are functions of time, are decided from the experimental data of the culture [58, 62]. A drawback of this approach is that the switch is decided before hand, so the power of prediction is limited. Single model able to describe several phases without switch, can therefore be preferred. Robitaille *et al.* [107] included information from biological mechanism and quantification of intracellular metabolites to model batch and fed-batch cultures in two different media, which was achieved with a single model without switching functions. Flexible kinetics describing multiple metabolic states was developed by

Hagrot *et al.* This allowed capturing diverse metabolic behaviors in one single model [14] and even to replace uncertain measurements or information with estimations of intervals [63]. In this latter work, the model described the behavior of CHO cells when submitted to 16 different media, for which the relevant EFMs were determined by CG from an extensive reaction network of 126 reactions, and the model consisted of the union of all the EFMs and their kinetics [63].

Models including a culture change caused by the addition of a chemical agent have been developed as well using bi-phasic models with a switching approach. In this case, the prediction ability is not affected by the switching approach since the timing of the agent addition is always decided before hand by the operator. Models with production enhancement by sodium butyrate addition or with induction with the cumate system for the production of recombinant protein by CHO cells have been reported [119, 122].

3.4.3 Estimation of the Kinetic Parameters

Very few large-scale kinetic models of CHO cells have been built [21]. One prohibiting aspect is the estimation of the kinetic parameters. These can be taken from literature and/or estimated during the modelling exercise based on data fitting by minimizing the difference between simulation and experimental data, with e.g. least squares regression or maximum likelihood. However, estimation of multiple kinetic parameters in several simultaneous non-linear equations is difficult due to risk of over-fitting and local minima [44].

The number of parameters can be reduced by transforming the non-linear problem into a linear one. In practice this can be done by providing fixed half-saturation parameters and only estimating the maximum flux rates from the measurements [14, 57–60, 62]. This simplification is based on the assumption that the reactions rates are maximal since the substrates are not limiting and thus the half-saturation parameters are set to a low value in relation to the initial concentrations of the substrates [44, 59, 62]. The kinetic parameters can then be adjusted by trial and error based on the obtained model fitting [14] possibly with automatized selection of the parameters providing the best model fitting from a systematic screening of values [63]. A drawback of all these methods is that the kinetics models rely on a manual selection.

A more objective and automated identification of kinetic parameters is desirable and can be obtained by utilizing the rational structure of the model in a re-parameterization [86, 124]. The reparameterization consists in a mathematical transformation such that the kinetics equation becomes linear while preserving the nonlinearity of the parameters, and is then followed by the parameter estimation by least-squares regression or maximum-likelihood [124].

Furthermore, the selection of the macroscopic kinetics structures can also be objectively and systematically approached based on a multilinear Gaussian process [125]. This method generates information about the correlation between the metabolites and the measured fluxes, where tailoring to the kinetic effects of activation, inhibition and their combinations, has been included. It determines thus which metabolites are influent for each flux and the nature of this effect, i.e. activation and/or inhibition. Furthermore the parameter estimation is simplified as each kinetic effect is modeled separately, as opposed to attempting simultaneous estimation of all the parameters [125].

3.5 Models of Glycosylation

Glycosylation modelling has been recently reviewed [126, 127]. The present section focusses mainly on models of glycosylation for bioprocesses.

Several approaches have been developed to control the glycosylation by cell engineering and medium design supported by mathematical modelling. Shelikoff *et al* developed a model where the complex glycosylation process in the Golgi was viewed as a plug flow reactor [128]. This model considered the influences of factors, such as enzyme activity, the rates of mRNA elongation and protein synthesis, intracellular nucleotide sugars and competition with protein folding. Umaña *et al* built a model of the influence of the glycosylation enzymes on the glycoforms [129]. This structured model was based on a reaction network with 33 glycan structures and 8 glycosylation enzymes. It assumed a glycoprotein transport through four compartments of the Golgi with different distributions of the glycosyltransferase. The model parameters for the kinetics and enzyme distributions, glycoprotein residence time, volume of the Golgi compartments, *etc.* were estimated from the literature.

Umaña's model was extended to 7565 glycans with 22871 reactions by considering a larger number of enzymes and using numerical methods to enable parameters adjustment to fit the experimental data [130]. In 2009, Krambeck *et al* further expanded their model with human specific glycans by adding 8 enzymes and including experimental data for human cells. This was increased to glycoforms 50605 later [131] and utilized for CHO cell lines. Based on the Golgi maturation theory, Hossler *et al* compared a representation of the Golgi as four continuous stirred tanks in series or as four plug-flow reactors and concluded in favor of the latter [132].

To simulate the glycosylation of the product of interest in a culture, kinetic models based on metabolic reaction network are the most appropriated. del Val *et al.* linked the glycosylation process in the Golgi to the cellular metabolism by introducing the impact on the glycosylation of the nucleotide sugar donors (NSDs) transported into the Golgi [133]. The kinetic model of the glycosylation included the NSDs metabolic synthesis network followed by the glycosylation maturation. They also considered the Golgi as one plug-flow reactor instead of four and introduced transport proteins recycling between compartments. In another model, they linked extracellular nutrients, i.e. depending on the culture conditions, with intracellular NSDs metabolism and cell growth, and simulated the glycosylation with time in culture [134]. The influences of the Golgi volume and the antibody specific productivity in batch culture on the glycosylation were later included in their model [135]. The glycosylation process is complex, and the kinetic models contain partial differentiation equations with a high number of parameters, e.g. related to enzyme activities, to be determined, representing a challenge for their applications. One of the approaches to reduce the reaction network complexity and the number of parameters has been to use the EFM concept creating macro-reactions. A challenge is however that the proportion of sugar(s) used for the glycosylation is extremely small compared to their total consumption, mainly channeled into the cell energy. The identification of the mass balance of the sugar(s) used for the glycosylation is therefore very difficult and even more challenging in case of feeding different sugars. This can be solved by linking the concentration of sugars in the medium to the distribution of glycosylation patterns and utilizing macro-reactions to simulate the process of glycosylation [136, 137]. A constraint-based model named glycosylation flux analysis, GFA, was developed by Hutter *et al.* using pseudo steady state assumption for fed-batch cultures, based on the fact that the residence time of a protein in the Golgi is 20–40 min [104]. The influence on the glycosylation of manganese addition during a fed-batch culture was investigated with this model.

Alternatively, non-kinetic models can present the advantage to include a lower number of parameters to identify, [7, 104, 138]. A model based on flux balance and Markov chain of states of glycosylation maturation, where the transition between states is a probabilistic event has been used for cell engineering, e.g GnT-IV and FucT knocked-out cells producing erythropoietin or antibody, and for biosimilars [138, 139]. A Bayesian network-based model for the prediction of antibody glycosylation was utilized in fed-batch, pseudo-perfusion and perfusion cultures [7].

4 Models as Support to Process Monitoring and Optimization

Process monitoring, is necessary to compensate for disturbances affecting quality and productivity. Monitoring includes the collection of information via repeated measurements and subsequent data processing to determine control actions [140]. The control actions rely on the availability of suitable of models. Automation and *in silico* processes have commonly been utilized in chemical engineering, but are much less developed for applications in biopharmaceutical production. This can be attributed to the complexity of the studied system used in the production processes, i.e. live cells. The data processing prior to control actions to perform tasks such as outlier detection, data conversion and estimation of states and parameters can be performed with model-based methods [29] containing prior knowledge of the process [141]. The model-based methods in combination with continuous measurements are often referred to as soft-sensors, which instead of measuring a variable, are able to provide an estimate of the variable [29]. The ability of soft-sensors to estimate a desired variable (parameters of the process) is referred to as observability; it ensures that the information is sufficient to reconstruct the value of the target parameter and can be assessed by observability analysis [142]. Furthermore, the observability can be used to define the measurement accuracy and frequency needed for applying the particular soft-sensor in a control scheme [29]. The algorithms used in soft-sensors can be classified either as observers or filters [143]. Observers reconstruct current estimates of parameter values from the collected information based on the process model. They can also take measurement errors and process noise into consideration [144] allowing even unmeasured, noisy or infrequently measured parameters to be reconstructed [145]. Filtering algorithms, i.e. Kalman filters, allow the soft-sensors to cope with nonlinearities and biological complexity [89]. Monitoring can be performed with simple mass balance models [146–148] or more detailed kinetic models [29]. Complex kinetic models have the risk of being limited by observability, however, this can be solved by simplifying the model [149, 150].

The ***process optimization***, and ***continuous improvement of the process*** include estimation of the optimal process inputs (optimal design vector) and identification of critical process limitations, such as physiological and technical constraints as well as product and system rationales [29]. If a reliable process model exists, it can be used to determine the optimal process inputs and to ideally control the process to achieve an optimal process performance. Mathematically, optimization problems are typically arranged as minimization problems of an objective function. The optimization objective depends on the aim of the model and is generally related to optimizing the *(i)* information content, *(ii)* productivity [37, 151, 152] or *(iii)* robustness and reproducibility [29, 150, 153, 154]. The optimization space is frequently constrained by what is physically possible or feasible and where the model is reliable to reduce the size of the optimization space. The size of the optimization space is often related to the computation time of the solution [29]. The optimization problems are frequently solved with numerical algorithms (detailed descriptions: [155, 156]; common in bioprocessing [150–152, 157–159]. The choice of optimization algorithm depends on the number of variables to be optimized, the complexity of the model and the desired duration of the optimization calculation [29]. In case the optimal design vector is time-dependent it might need to be parametrized. This can be done by discretizing the input signal via partially constant, linear or parabolic functions (also termed as zero, first or second order hold) [29]. Simulations are a valuable tool for investigating how to parameterize the design vector [37].

5 Perspectives

Modelling of bioprocesses for mammalian cell cultures has received a renewal of attention, including from industry. The perspective of creating *in silico* tools to simulate these processes is nowadays becoming a reality. For instance, their usage for process and medium development is offered as commercial service, or large biopharma companies develop their own modelling tools. For this, hybrid models allying mechanistic description and black-box approach for the kinetics, i.e. machine learning, are potentially used. It is important though that the field continues to investigate new methods supporting a more complete mechanistic description and many issues are unsatisfactory solved today. Among others, the inclusion of omics information can be instrumental however this creates new challenges, requiring novel technical approaches. Another challenge is the generation of the experimental data, both in terms of resources needed, i.e. cultures and analyses, but also in terms of appropriately designing the experiments to gather the measurements the most critical to catch the cell and process behavior in the model. There, smart experimental design, high-through-put solutions for cultures and analytical methods are needed. Process simulation can also be used for the training of staff or students, as performed at KTH-Royal Institute of Technology nowadays. Finally, model-based feed-back control of a process to track a given trajectory is also explored in the field. In this latter case, this can guide how the model should be designed, potentially favoring robustness to completeness. Here process analytical technology is of course instrumental.

Modelling of biopharmaceutical processes using mammalian cells has been investigated since several decades, however the field is starting to harvest these efforts nowadays. From the actual interest and current efforts, high quality *in silico* possibilities can be expected to grow significantly in the future.

Acknowledgments The authors were supported by the Competence Centre for Advanced Bio-Production by Continuous Processing, AdBIOPRO, funded by the Sweden's Innovation Agency VINNOVA (diaries nr. 2016-05181), by Digital Futures – KTH Royal Institute of Technology, by the Chinese Scholarship Council, and by the Innovative Medicines Initiative 2 Joint Undertaking [grant agreement No 777397]. This Joint Undertaking, project iConsensus, receives support from the European Union's Horizon 2020 research and innovation programme and EFPIA Partners Sanofi, GSK, Bayer, Rentschler, UCB, Byondis and Pfizer.

References

1. Carrillo-Cocom LM, Genel-Rey T, Araíz-Hernández D et al (2015) Amino acid consumption in naïve and recombinant CHO cell cultures: producers of a monoclonal antibody. Cytotechnology 67:809–820. https://doi.org/10.1007/s10616-014-9720-5
2. Rajendra Y, Kiseljak D, Baldi L et al (2012) Reduced glutamine concentration improves protein production in growth-arrested CHO-DG44 and HEK-293E cells. Biotechnol Lett 34:619–626. https://doi.org/10.1007/s10529-011-0809-z
3. Mears L, Stocks SM, Albaek MO et al (2017) Mechanistic fermentation models for process design, monitoring, and control. Trends Biotechnol 35:914–924. https://doi.org/10.1016/j.tibtech.2017.07.002
4. Portela RMC, Varsakelis C, Richelle A et al (2020) When is an in silico representation a digital twin? A biopharmaceutical industry approach to the digital twin concept. Adv Biochem Eng Biotechnol 176:1–33

5. Roush D, Asthagiri D, Babi DK et al (2020) Toward in silico CMC: an industrial collaborative approach to model-based process development. Biotechnol Bioeng 117:3986–4000. https://doi.org/10.1002/bit.27520
6. Shirsat NP, English NJ, Glennon B, Al-Rubeai M (2015) Modelling of mammalian cell cultures. In: Animal cell culture, cell engineering, vol 9, Springer, Cham, pp 259–326. https://doi.org/10.1007/978-3-319-10320-4
7. Zhang L, Wang M, Castan A et al (2021) Probabilistic model by Bayesian network for the prediction of antibody glycosylation in perfusion and fed-batch cell cultures. Biotechnol Bioeng 27769. https://doi.org/10.1002/bit.27769
8. Zhang L, Schwarz H, Wang M et al (2021) Control of IgG glycosylation in CHO cell perfusion cultures by GReBA mathematical model supported by a novel targeted feed, TAFE. Metab Eng 65:135–145. https://doi.org/10.1016/j.ymben.2020.11.004
9. Randek J, Mandenius CF (2018) On-line soft sensing in upstream bioprocessing. Crit Rev Biotechnol 38:106–121
10. Ben Yahia B, Gourevitch B, Malphettes L, Heinzle E (2017) Segmented linear modeling of CHO fed-batch culture and its application to large scale production. Biotechnol Bioeng 114:785–797. https://doi.org/10.1002/bit.26214
11. Traustason B, Cheeks M, Dikicioglu D (2019) Computer-aided strategies for determining the amino acid composition of medium for Chinese hamster ovary cell-based biomanufacturing platforms. Int J Mol Sci 20:5464-1–13. https://doi.org/10.3390/ijms20215464
12. Altamirano C, Berrios J, Vergara M, Becerra S (2013) Advances in improving mammalian cells metabolism for recombinant protein production. Electron J Biotechnol 16:1–14. https://doi.org/10.2225/vol16-issue3-fulltext-2
13. Feichtinger J, Hernández I, Fischer C et al (2016) Comprehensive genome and epigenome characterization of CHO cells in response to evolutionary pressures and over time. Biotechnol Bioeng 113:2241–2253. https://doi.org/10.1002/bit.25990
14. Hagrot E, Oddsdóttir HÆ, Hosta JG et al (2017) Poly-pathway model, a novel approach to simulate multiple metabolic states by reaction network-based model – application to amino acid depletion in CHO cell culture. J Biotechnol 259:235–247. https://doi.org/10.1016/j.jbiotec.2017.05.026
15. Sidorenko Y, Wahl A, Dauner M et al (2008) Comparison of metabolic flux distributions for MDCK cell growth in glutamine- and pyruvate-containing media. Biotechnol Prog 24:311–320. https://doi.org/10.1021/bp0702673
16. Zhang J (2014) Mammalian cell culture for biopharmaceutical production. In: Manual of industrial microbiology and biotechnology. ASM Press, Washington, DC, pp 157–178
17. Zamani L, Lundqvist M, Zhang Y et al (2018) High cell density perfusion culture has a maintained exoproteome and metabolome. Biotechnol J 13:1800036. https://doi.org/10.1002/biot.201800036
18. Wahrheit J, Nicolae A, Heinzle E (2014) Dynamics of growth and metabolism controlled by glutamine availability in Chinese hamster ovary cells. Appl Microbiol Biotechnol 98:1771–1783. https://doi.org/10.1007/s00253-013-5452-2
19. Sellick CA, Croxford AS, Maqsood AR et al (2011) Metabolite profiling of recombinant CHO cells: designing tailored feeding regimes that enhance recombinant antibody production. Biotechnol Bioeng 108:3025–3031. https://doi.org/10.1002/bit.23269
20. Xing Z, Kenty B, Koyrakh I et al (2011) Optimizing amino acid composition of CHO cell culture media for a fusion protein production. Process Biochem 46:1423–1429. https://doi.org/10.1016/j.procbio.2011.03.014
21. Huang Z, Lee D-Y, Yoon S (2017) Quantitative intracellular flux modeling and applications in biotherapeutic development and production using CHO cell cultures. Biotechnol Bioeng 114:2717–2728. https://doi.org/10.1002/bit.26384
22. Lewis AM, Abu-Absi NR, Borys MC, Li ZJ (2016) The use of 'omics technology to rationally improve industrial mammalian cell line performance. Biotechnol Bioeng 113:26–38. https://doi.org/10.1002/bit.25673

23. Ahn WS, Antoniewicz MR (2012) Towards dynamic metabolic flux analysis in CHO cell cultures. Biotechnol J 7:61–74. https://doi.org/10.1002/biot.201100052
24. Altamirano C, Illanes A, Becerra S et al (2006) Considerations on the lactate consumption by CHO cells in the presence of galactose. J Biotechnol 125:547–556. https://doi.org/10.1016/j.jbiotec.2006.03.023
25. Quek LE, Dietmair S, Krömer JO, Nielsen LK (2010) Metabolic flux analysis in mammalian cell culture. Metab Eng 12:161–171. https://doi.org/10.1016/j.ymben.2009.09.002
26. Selvarasu S, Ho YS, Chong WPK et al (2012) Combined in silico modeling and metabolomics analysis to characterize fed-batch CHO cell culture. Biotechnol Bioeng 109:1415–1429. https://doi.org/10.1002/bit.24445
27. Xie L, Wang DIC (1994) Stoichiometric analysis of animal cell growth and its application in medium design. Biotechnol Bioeng 43:1164–1174. https://doi.org/10.1002/bit.260431122
28. Xie L, Wang DIC (1996) Material balance studies on animal cell metabolism using a stoichiometrically based reaction network. Biotechnol Bioeng 52:579–590. https://doi.org/10.1002/(SICI)1097-0290(19961205)52:5<579::AID-BIT5>3.0.CO;2-G
29. Kroll P, Hofer A, Ulonska S et al (2017) Model-based methods in the biopharmaceutical process lifecycle. Pharm Res 34:2596–2613. https://doi.org/10.1007/s11095-017-2308-y
30. Mandenius C-F, Brundin A (2008) Bioprocess optimization using design-of-experiments methodology. Biotechnol Prog 24:1191–1203. https://doi.org/10.1002/btpr.67
31. Möller J, Pörtner R (2017) Model-based design of process strategies for cell culture bioprocesses: state of the art and new perspectives. In: New insights into cell culture technology. InTech, Rijeka, pp 157–172
32. Franceschini G, Macchietto S (2008) Model-based design of experiments for parameter precision: state of the art. Chem Eng Sci 63:4846–4872. https://doi.org/10.1016/j.ces.2007.11.034
33. Telen D, Logist F, Van Derlinden E et al (2012) Optimal experiment design for dynamic bioprocesses: a multi-objective approach. Chem Eng Sci 78:82–97. https://doi.org/10.1016/j.ces.2012.05.002
34. Galvanin F, Barolo M, Bezzo F (2010) A framework for model-based design of experiments in the presence of continuous measurement systems. In: IFAC proceedings volumes. IFAC, Geneva, pp 571–576
35. Chis OT, Banga JR, Balsa-Canto E (2011) Structural identifiability of systems biology models: a critical comparison of methods. PLoS One 6:e27755-1–16. https://doi.org/10.1371/journal.pone.0027755
36. Raue A, Kreutz C, Maiwald T et al (2009) Structural and practical identifiability analysis of partially observed dynamical models by exploiting the profile likelihood. Bioinformatics 25:1923–1929. https://doi.org/10.1093/bioinformatics/btp358
37. Kawohl M, Heine T, King R (2007) Model based estimation and optimal control of fed-batch fermentation processes for the production of antibiotics. Chem Eng Process Process Intensif 46:1223–1241. https://doi.org/10.1016/j.cep.2006.06.023
38. Lemaire C, Schoefs O, Lamy E et al (2014) Modeling of an aerobic bioprocess based on gas exchange and dynamics: a novel approach. Bioprocess Biosyst Eng 37:1809–1816. https://doi.org/10.1007/s00449-014-1157-x
39. Tsuchiya HM, Fredrickson AG, Aris R (1966) Dynamics of microbial cell populations. Adv Chem Eng 6:125–206. https://doi.org/10.1016/S0065-2377(08)60275-6
40. Weisberg M (2007) Three kinds of idealization. J Philos 104:639–659. https://doi.org/10.5840/jphil20071041240
41. Cawley GC, Talbot NLC (2010) On over-fitting in model selection and subsequent selection bias in performance evaluation. J Mach Learn Res 11:2079–2107
42. Fredrickson AG, Megee RD, Tsuchiya HM (1970) Mathematical models for fermentation processes. In: Advances in applied microbiology. Elsevier Science & Technology, San Diego, pp 419–465
43. Sanderson CS, Phillips PJ, Barford JP (1996) Structured modelling of animal cells. Cytotechnology 21:149–153. https://doi.org/10.1007/BF02215664

44. Ben Yahia B, Malphettes L, Heinzle E (2015) Macroscopic modeling of mammalian cell growth and metabolism. Appl Microbiol Biotechnol 99:7009–7024. https://doi.org/10.1007/s00253-015-6743-6
45. Bailey JE (1998) Mathematical modeling and analysis in biochemical engineering: past accomplishments and future opportunities. Biotechnol Prog 14:8–20. https://doi.org/10.1021/bp9701269
46. Kyriakopoulos S, Ang KS, Lakshmanan M et al (2018) Kinetic modeling of mammalian cell culture bioprocessing: the quest to advance biomanufacturing. Biotechnol J 13:1700229. https://doi.org/10.1002/biot.201700229
47. Székely T, Burrage K (2014) Stochastic simulation in systems biology. Comput Struct Biotechnol J 12:14–25. https://doi.org/10.1016/j.csbj.2014.10.003
48. Hodgson BJ, Taylor CN, Ushio M et al (2004) Intelligent modelling of bioprocesses: a comparison of structured and unstructured approaches. Bioprocess Biosyst Eng 26:353–359. https://doi.org/10.1007/s00449-004-0382-0
49. Gombert AK, Nielsen J (2000) Mathematical modelling of metabolism. Curr Opin Biotechnol 11:180–186. https://doi.org/10.1016/S0958-1669(00)00079-3
50. Bunge M (1963) A general black box theory. Philos Sci 30:346–358
51. Ferentinos KP (2005) Biological engineering applications of feedforward neural networks designed and parameterized by genetic algorithms. Neural Netw 18:934–950. https://doi.org/10.1016/j.neunet.2005.03.010
52. Kanehisa M, Goto S, Sato Y et al (2014) Data, information, knowledge and principle: Back to metabolism in KEGG. Nucleic Acids Res 42:199–205. https://doi.org/10.1093/nar/gkt1076
53. Kanehisa M (2000) KEGG: Kyoto encyclopedia of genes and genomes. Nucleic Acids Res 28:27–30. https://doi.org/10.1093/nar/28.1.27
54. Harcum SW, Lee KH (2016) CHO cells can make more protein. Cell Syst 3:412–413. https://doi.org/10.1016/j.cels.2016.11.007
55. Hefzi H, Ang KS, Hanscho M et al (2016) A consensus genome-scale reconstruction of Chinese hamster ovary cell metabolism. Cell Syst 3:434–443.e8. https://doi.org/10.1016/j.cels.2016.10.020
56. Rejc Ž, Magdevska L, Tršelič T et al (2017) Computational modelling of genome-scale metabolic networks and its application to CHO cell cultures. Comput Biol Med 88:150–160. https://doi.org/10.1016/j.compbiomed.2017.07.005
57. Provost A, Bastin G (2004) Dynamic metabolic modelling under the balanced growth condition. J Process Control 14:717–728. https://doi.org/10.1016/j.jprocont.2003.12.004
58. Provost A, Bastin G, Agathos SN, Schneider YJ (2005) Metabolic design of macroscopic models: Application to CHO cells. In: Proceedings of the 44th IEEE Conference on Decision and Control, and the European Control Conference, CDC-ECC'05. IEEE, pp 2982–2989
59. Gao J, Gorenflo VM, Scharer JM, Budman HM (2007) Dynamic metabolic modeling for a MAb bioprocess. Biotechnol Prog 23:168–181. https://doi.org/10.1021/bp060089y
60. Dorka P, Fischer C, Budman H, Scharer JM (2009) Metabolic flux-based modeling of mAb production during batch and fed-batch operations. Bioprocess Biosyst Eng 32:183–196. https://doi.org/10.1007/s00449-008-0236-2
61. Naderi S, Meshram M, Wei C et al (2011) Development of a mathematical model for evaluating the dynamics of normal and apoptotic Chinese hamster ovary cells. Biotechnol Prog 27:1197–1205. https://doi.org/10.1002/btpr.647
62. Zamorano F, Vande Wouwer A, Jungers RM, Bastin G (2013) Dynamic metabolic models of CHO cell cultures through minimal sets of elementary flux modes. J Biotechnol 164:409–422. https://doi.org/10.1016/j.jbiotec.2012.05.005
63. Hagrot E, Oddsdóttir HÆ, Mäkinen M et al (2019) Novel column generation-based optimization approach for poly-pathway kinetic model applied to CHO cell culture. Metab Eng Commun 8. https://doi.org/10.1016/j.mec.2018.e00083
64. Goudar C, Biener R, Boisart C et al (2010) Metabolic flux analysis of CHO cells in perfusion culture by metabolite balancing and 2D [13C, 1H] COSY NMR spectroscopy. Metab Eng 12:138–149. https://doi.org/10.1016/j.ymben.2009.10.007

65. Llaneras F, Picó J (2010) Which metabolic pathways generate and characterize the flux space? A comparison among elementary modes, extreme pathways and minimal generators. J Biomed Biotechnol 2010. https://doi.org/10.1155/2010/753904
66. Schuster S, Hilgetag C (1994) On elementary flux modes in biochemical reaction systems at steady state. J Biol Syst 02:165–182. https://doi.org/10.1142/S0218339094000131
67. Papin JA (2002) Extreme pathway lengths and reaction participation in genome-scale metabolic networks. Genome Res 12:1889–1900. https://doi.org/10.1101/gr.327702
68. Urbanczik R, Wagner C (2005) An improved algorithm for stoichiometric network analysis: theory and applications. Bioinformatics 21:1203–1210. https://doi.org/10.1093/bioinformatics/bti127
69. Zanghellini J, Ruckerbauer DE, Hanscho M, Jungreuthmayer C (2013) Elementary flux modes in a nutshell: properties, calculation and applications. Biotechnol J 8:1009–1016. https://doi.org/10.1002/biot.201200269
70. Oddsdóttir HÆ, Hagrot E, Chotteau V, Forsgren A (2015) On dynamically generating relevant elementary flux modes in a metabolic network using optimization. J Math Biol 71:903–920. https://doi.org/10.1007/s00285-014-0844-1
71. Klamt S, Regensburger G, Gerstl MP et al (2017) From elementary flux modes to elementary flux vectors: metabolic pathway analysis with arbitrary linear flux constraints. PLoS Comput Biol 13:1–22. https://doi.org/10.1371/journal.pcbi.1005409
72. Papin JA, Stelling J, Price ND et al (2004) Comparison of network-based pathway analysis methods. Trends Biotechnol 22:400–405. https://doi.org/10.1016/j.tibtech.2004.06.010
73. Klamt S, Stelling J (2003) Two approaches for metabolic pathway analysis? Trends Biotechnol 21:64–69
74. Fukuda K, Prodon A (1996) Double description method revisited. Lect Notes Comput Sci (including Subser Lect Notes Artif Intell Lect Notes Bioinformatics) 1120:91–111. https://doi.org/10.1007/3-540-61576-8_77
75. Terzer M, Stelling J (2008) Large-scale computation of elementary flux modes with bit pattern trees. Bioinformatics 24:2229–2235. https://doi.org/10.1093/bioinformatics/btn401
76. von Kamp A, Schuster S (2006) Metatool 5.0: fast and flexible elementary modes analysis. Bioinformatics 22:1930–1931. https://doi.org/10.1093/bioinformatics/btl267
77. Abbate T, de Sousa S, Dewasme L, Bastin G (2019) Inference of dynamic macroscopic models of cell metabolism based on elementary flux modes analysis. Biochem Eng J 151:107325
78. Hebing L, Neymann T, Thüte T et al (2016) Efficient generation of models of fed-batch fermentations for process design and control. IFAC-PapersOnLine 49:621–626. https://doi.org/10.1016/j.ifacol.2016.07.237
79. Niu H, Amribt Z, Fickers P et al (2013) Metabolic pathway analysis and reduction for mammalian cell cultures-towards macroscopic modeling. Chem Eng Sci 102:461–473. https://doi.org/10.1016/j.ces.2013.07.034
80. de Figueiredo LF, Podhorski A, Rubio A et al (2009) Computing the shortest elementary flux modes in genome-scale metabolic networks. Bioinformatics 25:3158–3165. https://doi.org/10.1093/bioinformatics/btp564
81. Kaleta C, de Figueiredo LF, Schuster S (2009) Can the whole be less than the sum of its parts? Pathway analysis in genome-scale metabolic networks using elementary flux patterns. Genome Res 19:1872–1883. https://doi.org/10.1101/gr.090639.108
82. Jungers RM, Zamorano F, Blondel VD et al (2011) Fast computation of minimal elementary decompositions of metabolic flux vectors. Automatica 47:1255–1259. https://doi.org/10.1016/j.automatica.2011.01.011
83. Zamorano F, Wouwer AV, Bastin G (2010) A detailed metabolic flux analysis of an underdetermined network of CHO cells. J Biotechnol 150:497–508. https://doi.org/10.1016/j.jbiotec.2010.09.944
84. Oddsdóttir HÆ, Hagrot E, Chotteau V, Forsgren A (2016) Robustness analysis of elementary flux modes generated by column generation. Math Biosci 273:45–56. https://doi.org/10.1016/j.mbs.2015.12.009

85. Hagrot E (2019) Macroscopic models of Chinese hamster ovary cell cultures. Doctoral dissertation, KTH Royal Institute of Technology, Stockholm, Sweden
86. Rodrigues D, Abdalmoaty MR, Jacobsen EW et al (2021) An integrated approach for modeling and identification of perfusion bioreactors via basis flux modes. Comput Chem Eng 149:107238. https://doi.org/10.1016/j.compchemeng.2021.107238
87. Antoniewicz MR (2013) Dynamic metabolic flux analysis-tools for probing transient states of metabolic networks. Curr Opin Biotechnol 24:973–978. https://doi.org/10.1016/j.copbio.2013.03.018
88. Zupke C, Stephanopoulos G (1995) Intracellular flux analysis in hybridomas using mass balances and in vitro 13C nmr. Biotechnol Bioeng 45:292–303. https://doi.org/10.1002/bit.260450403
89. Nyberg GB, Balcarcel RR, Follstad BD et al (1999) Metabolism of peptide amino acids by Chinese hamster ovary cells grown in a complex medium. Biotechnol Bioeng 62:324–335. https://doi.org/10.1002/(SICI)1097-0290(19990205)62:3<324::AID-BIT9>3.0.CO;2-C
90. Llaneras F, Picó J (2008) Stoichiometric modelling of cell metabolism. J Biosci Bioeng 105:1–11. https://doi.org/10.1263/jbb.105.1
91. Martínez VS, Dietmair S, Quek LE et al (2013) Flux balance analysis of CHO cells before and after a metabolic switch from lactate production to consumption. Biotechnol Bioeng 110:660–666. https://doi.org/10.1002/bit.24728
92. Carinhas N, Duarte TM, Barreiro LC et al (2013) Metabolic signatures of GS-CHO cell clones associated with butyrate treatment and culture phase transition. Biotechnol Bioeng 110:3244–3257. https://doi.org/10.1002/bit.24983
93. Feist AM, Palsson BO (2010) The biomass objective function. Curr Opin Microbiol 13:344–349. https://doi.org/10.1016/j.mib.2010.03.003
94. Orman MA, Berthiaume F, Androulakis IP, Ierapetritou MG (2011) Advanced stoichiometric analysis of metabolic networks of mammalian systems. Crit Rev Biomed Eng 39:511–534. https://doi.org/10.1615/CritRevBiomedEng.v39.i6.30
95. Schuetz R, Kuepfer L, Sauer U (2007) Systematic evaluation of objective functions for predicting intracellular fluxes in Escherichia coli. Mol Syst Biol 3:319. https://doi.org/10.1038/msb4100162
96. Antoniewicz MR (2015) Methods and advances in metabolic flux analysis: a mini-review. J Ind Microbiol Biotechnol 42:317–325. https://doi.org/10.1007/s10295-015-1585-x
97. Orth J, Thiele I, Palsson B (2010) What is flux balance? Nat Biotechnol 28:245–248. https://doi.org/10.1038/nbt.1614.What
98. Alper H, Miyaoku K, Stephanopoulos G (2005) Construction of lycopene-overproducing E. coli strains by combining systematic and combinatorial gene knockout targets. Nat Biotechnol 23:612–616
99. Pál C, Papp B, Lercher M (2005) Adaptive evolution of bacterial metabolic networks by horizontal gene transfer. Nat Genet 37:1372–1375
100. Ivarsson M, Noh H, Morbidelli M, Soos M (2015) Insights into pH-induced metabolic switch by flux balance analysis. Biotechnol Prog 31:347–357. https://doi.org/10.1002/btpr.2043
101. Sou SN, Sellick C, Lee K et al (2015) How does mild hypothermia affect monoclonal antibody glycosylation? Biotechnol Bioeng 112:1165–1176. https://doi.org/10.1002/bit.25524
102. Richelle A, Gziri K, Bogaerts P (2016) A methodology for building a macroscopic FBA-based dynamical simulator of cell cultures through flux variability analysis. Biochem Eng J 114:50–64. https://doi.org/10.1016/j.bej.2016.06.017
103. Sengupta N, Rose ST, Morgan JA (2011) Metabolic flux analysis of CHO cell metabolism in the late non-growth phase. Biotechnol Bioeng 108:82–92. https://doi.org/10.1002/bit.22890
104. Hutter S, Villiger TK, Brühlmann D et al (2017) Glycosylation flux analysis reveals dynamic changes of intracellular glycosylation flux distribution in Chinese hamster ovary fed-batch cultures. Metab Eng 43:9–20. https://doi.org/10.1016/j.ymben.2017.07.005
105. Ferreira AR, Dias JML, Teixeira AP et al (2011) Projection to latent pathways (PLP): a constrained projection to latent variables (PLS) method for elementary flux modes discrimination. BMC Syst Biol 5:1–13. https://doi.org/10.1186/1752-0509-5-181

106. Kim OD, Rocha M, Maia P (2018) A review of dynamic modeling approaches and their application in computational strain optimization for metabolic engineering. Front Microbiol 9:1–22. https://doi.org/10.3389/fmicb.2018.01690
107. Robitaille J, Chen J, Jolicoeur M (2015) A single dynamic metabolic model can describe mAb producing CHO cell batch and fed-batch cultures on different culture media. PLoS One 10:e0136815. https://doi.org/10.1371/journal.pone.0136815
108. Monod J (1949) The growth of bacterial cultures. Annu Rev Microbiol 3:371–394. https://doi.org/10.1146/annurev.mi.03.100149.002103
109. Liu YH, Bi JX, Zeng AP, Yuan JQ (2008) A simple kinetic model for myeloma cell culture with consideration of lysine limitation. Bioprocess Biosyst Eng 31:569–577. https://doi.org/10.1007/s00449-008-0204-x
110. Jang JD, Barford JP (2000) An unstructured kinetic model of macromolecular metabolism in batch and fed-batch cultures of hybridoma cells producing monoclonal antibody. Biochem Eng J 4:153–168. https://doi.org/10.1016/S1369-703X(99)00041-8
111. Henri V (1903) General Laws for the Action of Diastases 75
112. Michaelis VL, Menten ML (1913) The kinetics of Invertase action (Die Kinetik der Invertinwirkung). Eschenbach, Easton
113. Liebermeister W, Klipp E (2006) Bringing metabolic networks to life: integration of kinetic, metabolic, and proteomic data. Theor Biol Med Model 3:1–11. https://doi.org/10.1186/1742-4682-3-42
114. Liebermeister W, Klipp E (2006) Bringing metabolic networks to life: convenience rate law and thermodynamic constraints. Theor Biol Med Model 3:1–13. https://doi.org/10.1186/1742-4682-3-41
115. Nolan R, Lee K (2011) Dynamic model of CHO cell metabolism. Metab Eng 13:108–124. https://doi.org/10.1016/j.ymben.2010.09.003
116. Barthelmes J, Ebeling C, Chang A et al (2007) BRENDA, AMENDA and FRENDA: the enzyme information system in 2007. Nucleic Acids Res 35(suppl_1):D511–D514
117. Almquist J, Cvijovic M, Hatzimanikatis V et al (2014) Kinetic models in industrial biotechnology – improving cell factory performance. Metab Eng 24:38–60. https://doi.org/10.1016/j.ymben.2014.03.007
118. Steuer R, Junker B (2009) Computational models of metabolism: stability and regulation in metabolic networks. In: Advances in chemical physics. Wiley, Newark, pp 105–251
119. Ghorbaniaghdam A, Henry O, Jolicoeur M (2013) A kinetic-metabolic model based on cell energetic state: study of CHO cell behavior under Na-butyrate stimulation. Bioprocess Biosyst Eng 36:469–487. https://doi.org/10.1007/s00449-012-0804-3
120. Ghorbaniaghdam A, Chen J, Henry O, Jolicoeur M (2014) Analyzing clonal variation of monoclonal antibody-producing CHO cell lines using an in silico metabolomic platform. PLoS One 9. https://doi.org/10.1371/journal.pone.0090832
121. Moreno-Sánchez R, Saavedra E, Rodríguez-Enríquez S, Olín-Sandoval V (2008) Metabolic control analysis: a tool for designing strategies to manipulate metabolic pathways. J Biomed Biotechnol 2008. https://doi.org/10.1155/2008/597913
122. McMurray-Beaulieu V, Hisiger S, Durand C et al (2009) Na-butyrate sustains energetic states of metabolism and t-PA productivity of CHO cells. J Biosci Bioeng 108:160–167
123. Chen N, Bennett M, Kontoravdi C (2014) Analysis of Chinese hamster ovary cell metabolism through a combined computational and experimental approach. Cytotechnology 66:945–966
124. Wang M, Jacobsen E, Chotteau V, Hjalmarsson H (2019) A multi-step least-squares method for nonlinear rational models. In: American control conference (ACC) IEEE. IEEE, Piscataway, pp 4509–4514
125. Wang M, Risuleo RS, Jacobsen EW et al (2020) Identification of nonlinear kinetics of macroscopic bio-reactions using multilinear Gaussian processes. Comput Chem Eng 133:106671. https://doi.org/10.1016/j.compchemeng.2019.106671
126. Tejwani V, Andersen MR, Nam JH, Sharfstein ST (2018) Glycoengineering in CHO cells: advances in systems biology. Biotechnol J 13:1700234. https://doi.org/10.1002/biot.201700234

127. Spahn PN, Lewis NE (2014) Systems glycobiology for glycoengineering. Curr Opin Biotechnol 30:218–224. https://doi.org/10.1016/j.copbio.2014.08.004
128. Shelikoff M, Sinskey AJ, Stephanopoulos G (1996) A modeling framework for the study of protein glycosylation. Biotechnol Bioeng 50:73–90. https://doi.org/10.1002/(SICI)1097-0290(19960405)50:1<73::AID-BIT9>3.0.CO;2-Z
129. Umaña P, Bailey JE (1997) A mathematical model of N-linked glycoform biosynthesis. Biotechnol Bioeng 55:890–908. https://doi.org/10.1002/(SICI)1097-0290(19970920)55:6<890::AID-BIT7>3.0.CO;2-B
130. Krambeck FJ, Betenbaugh MJ (2005) A mathematical model of N-linked glycosylation. Biotechnol Bioeng 92:711–728. https://doi.org/10.1002/bit.20645
131. Krambeck FJ, Bennun SV, Andersen MR, Betenbaugh MJ (2017) Model-based analysis of N-glycosylation in Chinese hamster ovary cells. PLoS One 12:e0175376. https://doi.org/10.1371/journal.pone.0175376
132. Hossler P, Mulukutla BC, Hu WS (2007) Systems analysis of N-glycan processing in mammalian cells. PLoS One 2. https://doi.org/10.1371/journal.pone.0000713
133. Jimenez del Val I, Nagy JM, Kontoravdi C (2011) A dynamic mathematical model for monoclonal antibody N-linked glycosylation and nucleotide sugar donor transport within a maturing Golgi apparatus. Biotechnol Prog 27:1730–1743. https://doi.org/10.1002/btpr.688
134. Jedrzejewski PM, Jimenez Del Val I, Constantinou A et al (2014) Towards controlling the Glycoform: a model framework linking extracellular metabolites to antibody glycosylation. Int J Mol Sci 15:4492–4522. https://doi.org/10.3390/ijms15034492
135. del Val JI, Fan Y, Weilguny D (2016) Dynamics of immature mAb glycoform secretion during CHO cell culture: an integrated modelling framework. Biotechnol J 11:610–623. https://doi.org/10.1002/biot.201400663
136. Zhang L, Wang M, Castan A et al (2020) Glycan residues balance analysis - GReBA: a novel model for the N-linked glycosylation of IgG produced by CHO cells. Metab Eng 57:118–128. https://doi.org/10.1016/j.ymben.2019.08.016
137. Zhang L, Schwarz H, Castan A, et al (2021) Prediction of IgG glycosylation in CHO cell perfusion cultures by GReBA mathematical model supported by a novel targeted feed, TAFE. Biotechnol Bioeng. https://doi.org/10.1002/bit.27769
138. Spahn PN, Hansen AH, Hansen HG et al (2016) A Markov chain model for N-linked protein glycosylation – towards a low-parameter tool for model-driven glycoengineering. Metab Eng 33:52–66. https://doi.org/10.1016/j.ymben.2015.10.007
139. Spahn PN, Hansen AH, Kol S et al (2017) Predictive glycoengineering of biosimilars using a Markov chain glycosylation model. Biotechnol J 12:1600489. https://doi.org/10.1002/biot.201600489
140. Glassey J, Gernaey KV, Clemens C et al (2011) Process analytical technology (PAT) for biopharmaceuticals. Biotechnol J 6:369–377. https://doi.org/10.1002/biot.201000356
141. Glacken MW, Fleischaker RJ, Sinskey AJ (1986) Serum in cultures of mammalian cells. Biotechnology XXVIII:1376–1389
142. Nakhaeinejad M, Bryant MD (2011) Observability analysis for model-based fault detection and sensor selection in induction motors. Meas Sci Technol 22. https://doi.org/10.1088/0957-0233/22/7/075202
143. Mohd Ali J, Ha Hoang N, Hussain MA, Dochain D (2015) Review and classification of recent observers applied in chemical process systems. Comput Chem Eng 76:27–41. https://doi.org/10.1016/j.compchemeng.2015.01.019
144. Dochain D (2003) State and parameter estimation in chemical and biochemical processes: a tutorial. J Process Control 13:801–818. https://doi.org/10.1016/S0959-1524(03)00026-X
145. Goffaux G, Vande Wouwer A (2005) Bioprocess state estimation: some classical and less classical approaches. In: Lecture notes in control and information sciences. Springer, Berlin, pp 111–128
146. Jobé AM, Herwig C, Surzyn M et al (2003) Generally applicable fed-batch culture concept based on the detection of metabolic state by on-line balancing. Biotechnol Bioeng 82:627–639. https://doi.org/10.1002/bit.10610

147. Mou D-G, Cooney CL (1983) Growth monitoring and control in complex medium: a case study employing fed-batch penicillin fermentation and computer-aided on-line mass balancing. Biotechnol Bioeng 25:257–269. https://doi.org/10.1002/bit.260250119
148. Wechselberger P, Sagmeister P, Herwig C (2013) Real-time estimation of biomass and specific growth rate in physiologically variable recombinant fed-batch processes. Bioprocess Biosyst Eng 36:1205–1218. https://doi.org/10.1007/s00449-012-0848-4
149. Aehle M, Kuprijanov A, Schaepe S et al (2011) Simplified off-gas analyses in animal cell cultures for process monitoring and control purposes. Biotechnol Lett 33:2103–2110. https://doi.org/10.1007/s10529-011-0686-5
150. Aehle M, Kuprijanov A, Schaepe S et al (2011) Increasing batch-to-batch reproducibility of CHO cultures by robust open-loop control. Cytotechnology 63:41–47. https://doi.org/10.1007/s10616-010-9320-y
151. Amribt Z, Niu H, Bogaerts P (2013) Macroscopic modelling of overflow metabolism and model based optimization of hybridoma cell fed-batch cultures. Biochem Eng J 70:196–209. https://doi.org/10.1016/j.bej.2012.11.005
152. del Rio-Chanona EA, Zhang D, Vassiliadis VS (2016) Model-based real-time optimisation of a fed-batch cyanobacterial hydrogen production process using economic model predictive control strategy. Chem Eng Sci 142:289–298. https://doi.org/10.1016/j.ces.2015.11.043
153. Craven S, Whelan J, Glennon B (2014) Glucose concentration control of a fed-batch mammalian cell bioprocess using a nonlinear model predictive controller. J Process Control 24:344–357. https://doi.org/10.1016/j.jprocont.2014.02.007
154. Goudar CT, Biener R, Konstantinov KB, Piret JM (2009) Error propagation from prime variables into specific rates and metabolic fluxes for mammalian cells in perfusion culture. Biotechnol Prog 25:986–998. https://doi.org/10.1002/btpr.155
155. Biegler LT (2007) An overview of simultaneous strategies for dynamic optimization. Chem Eng Process Process Intensif 46:1043–1053. https://doi.org/10.1016/j.cep.2006.06.021
156. Nocedal J, Wright S (2006) Numerical Optimization, 2nd edn. Springer, New York
157. Craven S, Shirsat N, Whelan J, Glennon B (2013) Process model comparison and transferability across bioreactor scales and modes of operation for a mammalian cell bioprocess. Biotechnol Prog 29:186–196. https://doi.org/10.1002/btpr.1664
158. Das S, Suganthan PN (2011) Differential evolution: a survey of the state-of-the-art. IEEE Trans Evol Comput 15:4–31. https://doi.org/10.1109/TEVC.2010.2059031
159. Lagarias JC, Reeds JA, Wright MH, Wright PE (1998) Convergence properties of the Nelder-Mead simplex method in low dimensions. SIAM J Optim 9:112–147. https://doi.org/10.1137/S1052623496303470

Computational Efforts for the Development and Scale-up of Antibody-Producing Cell Culture Processes

Johannes Möller and Ralf Pörtner

List of Abbreviations

Abbreviation	Explanation
CFD	Computational fluid dynamics
cGMP	Current good manufacturing practice
CHO	Chinese hamster ovary
DoE	Design of Experiments
mDoE	Model-assisted Design of Experiments
RS	Response surface
RQ	Respiratory quotient
QbD	Quality by Design

1 Introduction

The demand for highly potent next-generation pharmaceuticals has continuously increased in the last decades [1, 2]. Besides small molecules, biopharmaceuticals are used as a medication for difficult-to-treat diseases and constitute a novel class of therapeutics. Trends for the future indicate a growing market share of up to 50% of the top 100 pharmaceuticals to be bio-based [3]. At the same time, the

J. Möller (✉) · R. Pörtner
Institute of Bioprocess and Biosystems Engineering, Hamburg University of Technology, Hamburg, Germany
e-mail: johannes.moeller@tuhh.de; poertner@tuhh.de

R. Pörtner (ed.), *Cell Culture Engineering and Technology*, Cell Engineering 10,
https://doi.org/10.1007/978-3-030-79871-0_15

development costs of biopharmaceuticals have increased drastically (620% from 1980 to 2013) [4]. Between 2015–2018, 129 different biopharmaceuticals were approved by the EU and US government, which makes this the highest 4 year period since the first biopharmaceuticals entered the market in the mid-twentieth century [5]. In 2018, 374 approved biopharmaceuticals were available in total, of which 316 are with distinct individual ingredients and a current marketing license. Furthermore, 47% of the newly approved pharmaceuticals were bio-based. Most of the biopharmaceuticals are glycoproteins, and their efficacy is highly dependent on human-like post-translational modifications (i.e., glycosylation), which can be performed human-like using CHO cells as expression systems [6].

Additionally, there is a rising demand for accelerated process development and increased efficiency and economics of biopharmaceutical production processes [7, 8]. Especially for the development of processes for biopharmaceutical products, requirements for increased process understanding have evolved from the Process Analytical Technology (PAT) and Quality by Design (QbD) philosophy. Processes have become more complex and sophisticated, e.g., by switching from simple batch to more complex fed-batch or continuous perfusion processes [9, 10]. The number of process variables that have to be monitored and their complexity has increased. Furthermore, the demands related to quality management and documentation (current good manufacturing practice, cGMP) have increased dramatically. A benefit of these higher efforts for development is the increasing process knowledge [11]. This can speed up technology transfer from development into manufacturing, deliver a more optimized, robust process with higher titers and greater reproducibility, and aid in troubleshooting and root-cause analysis of deviations during production [12, 13].

Even though CHO cells have been used for more than two decades in the biopharmaceutical industry, the general understanding of the dynamics of the growth, metabolism, and product expression in mammalian cell culture processes is still insufficient. To capture the process dynamics, mathematical process modeling is an efficient tool because it includes the mechanistics of the biological system and the interactions of critical process parameters and critical quality attributes [14–17]. A mathematical model mainly summarizes the so far available understanding and works as an initial starting point to obtain a deeper insight that is required for knowledge-driven process design and optimization. Although the application of mathematical process models for the development of sophisticated processes has many advantages, it is still not commonly applied in bioprocess development. Reasons for this include the variety and complexity of mathematical models including different assumed mechanistics and quality of predictions (recently reviewed in [18]). In the past years, we developed novel model-assisted concepts for the design and scale-up of antibody-producing cell culture processes, which are reviewed in this chapter. Beforehand, a short review about the history of CHO cell culture is presented.

Short History of CHO Cells

CHO cells are spontaneously immortalized mammalian cells initially isolated from the ovary of a Chinese hamster in 1957 [19]. Historically, these cells became essential to study mammalian genetics and cell physiology due to their large chromosome (observable using a light microscope) and their robustness and fast growth rates in cultures. The cells were used and distributed in different laboratories, and a variety of cell lines were generated over time (e.g., CHO DG-44, CHO K-1, CHO-S, reviewed in [20]). Commercialization of CHO cell culture processes was first achieved in the mid-1980 by Genentech, who started a 10 m^3 production process of a recombinant tissue plasminogen activator with suspension-adapted CHO cells [21]. Today, a collection of methods for genetic modification and selection of suitable clones have been established [22]. Modern highly productive CHO processes achieve product concentrations of over 4–10 l^{-1} and are optimized regarding process control, media composition, and feeding strategies [23]. Furthermore, CHO cells grow relatively fast compared to other mammalian cell expression systems and can be cultivated in chemically defined, serum-free media without antibiotics and serum. In 2011, [24] presented the first draft genomic sequence of a CHO-K1 ancestral cell line. They investigated the genes responsible for the glycosylation and viral entry to optimize biopharmaceutical production further [24]. To utilize CHO cells, the genetic information of the expressed protein needs to be included in the cells. Commonly, the DHFR-MTX selection system is applied during clone development [25]. Nevertheless, clonal stability is not ensured, even if the DHFR-MTX system is applied and genetic losses were described in long-term studies with [26] and without [27] MTX selection pressure. Currently, novel gene-editing methods (e.g., CRISPR/Cas9) are applied to CHO cells to develop presumed next-generation expression cell lines with improved productivity and product quality [28, 29].

2 Model-Assisted Design of Experiments

In this section, an introduction to Design of Experiments (DoE) is presented first. Then, a new model-assisted Design of Experiments (mDoE) approach is shown followed by its extension and transformation into a software toolbox. Please notice that this section is based on our recent publications, in which further details can be found [30–32].

2.1 Design of Experiments

In contrast to evaluating just one factor at a time (i.e., changing only glucose concentration in growth medium), DoE tools enable to systematically evaluate multiple variables at the same time. DoE methods are recommended within the QbD methodology to describe the interdependency of process variables on the final biopharmaceutical [33]. In the beginning, a screening design is used to statistically identify relevant process variables (i.e., factors), e.g., medium compositions [34, 35] or process parameters [36, 37]. Secondly, the experimental space, consisting of significant variables and user-defined boundaries, is experimentally evaluated concerning their impact on the targeted outcomes (responses, e.g., product concentration, yields) [38, 39]. For the manufacturing of biopharmaceuticals, a stable and valid setpoint (e.g., medium concentration) of the process variables is aimed in order to guarantee the product quality and defining boundaries for stable operation, referred to as design space [40]. By using DoE, experiments can be efficiently planned and the interactions on experimental variables (i.e., factors) and the investigated outcome (i.e., quality attributes) are described by statistic equations [41, 42]. In the field of the design of biopharmaceutical manufacturing processes (upstream, as well as downstream) they are mostly applied and DoE methods are used to show the process understanding in regulatory dossiers.

Although DoE can be used to identify correlations between process parameters and their influence on the final product, the complex bioprocess is reduced to a few key numbers (e.g., final product concentration), and the dynamics of growth and metabolism are not sufficiently taken into account [30, 31]. DoE methods are based on user-defined choices of the experimental design and the definition of factor boundary values, including the definition of variables and their evaluated levels [33, 43]. Expert knowledge is required to define suitable boundary values for process development and optimization [44–46]. The heuristic conception of a DoE by choosing the limits of the parameter space poses a particular challenge [47]. Thus, there is a high risk that the experiments carried out were wrongly chosen and have only insufficient validity, which results in further costs and time delays [30, 32]. In Table 1, common examples of the application of DoE methods for the design and optimization of biopharmaceutical production processes are listed.

Table 1 Exemplary applications of DoE in bioprocess engineering studies with respect to the applied experimental design

Product	Design and factors	Improvement	
Ab	Rotable central composite design, valproic acid and time of adding	Optimal valproic acid concentration	[48]
Ab G8.8	D-optimal (18 runs), time of feed start, temperature and osmolality shift and feed volume	Optimal feed volume determined	[44]
Ab 4A1	Factorial design, initial cell and glutamine concentration and feed rate	Successfully adaption of a fed-batch protocol	[49]
EPO	Rotable central composite design (16 runs, γ=1.68, 2 center points), 3 nucleotide sugar precursors, time of adding (lag or exponential phase)	Optimal concentration in exponential phase	[50]

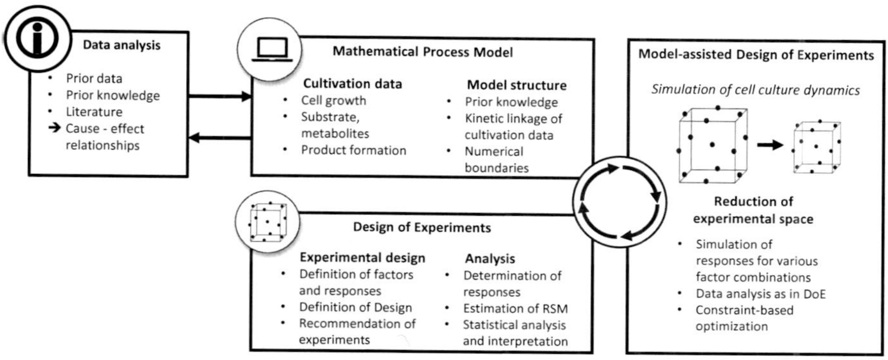

Fig. 1 Workflow of mDoE consisting of the combination of mathematical process models and classical Design of Experiments, please see [30] for further details

2.2 *Model-Assisted DoE: Approach*

To enable a more knowledge-driven bioprocess development and optimization, a model-assisted DoE approach (mDoE) was introduced [30, 51]. Early process development data from cultivation experiments based on literature or prior knowledge/literature are generated (see Fig. 1).

The first data are evaluated in a data analysis step and the data and prior knowledge is used to model the growth, metabolic rates and productivity of the bioprocess. Prior and expert knowledge is incorporated in the mathematical process model since it consists of mechanistic links describing the interactions of the culture dynamics [52]. For mDoE, typically a low amount of data is available (i.e. in early process development) and the used model structures are simple and describing known mechanistic effects, but the model parameters can be estimated based on few data. After modeling, a classical DoE is planned including the definition of factors and responses and the choice of an appropriate experimental design. The factor combinations are exported and the dynamics of the cell culture process are simulated based on the process model. The responses are treated like experimental data and the DoE is evaluated in the used statistic program including a response-surface model and the constraint-based optimization of the experimental space. This loop can be repeated several times to reduce the boundary values for an experimental DoE and the number of experiments during process development. As an example for the application of the mDoE approach, the optimization of a fed-batch strategy in laboratory scale is explained in the following. This example is based on the previous publication of [30].

Initially, experiments were designed using standard DoE software resulting in a I-optimal design with 29 recommended experiments (see Fig. 2a). The boundary values of this DoE were defined based on the available feeding solutions and vendor data, as typically in the development of cell cultivation processes. Simulations were used instead of laboratory experiments to predict the responses of each experiment

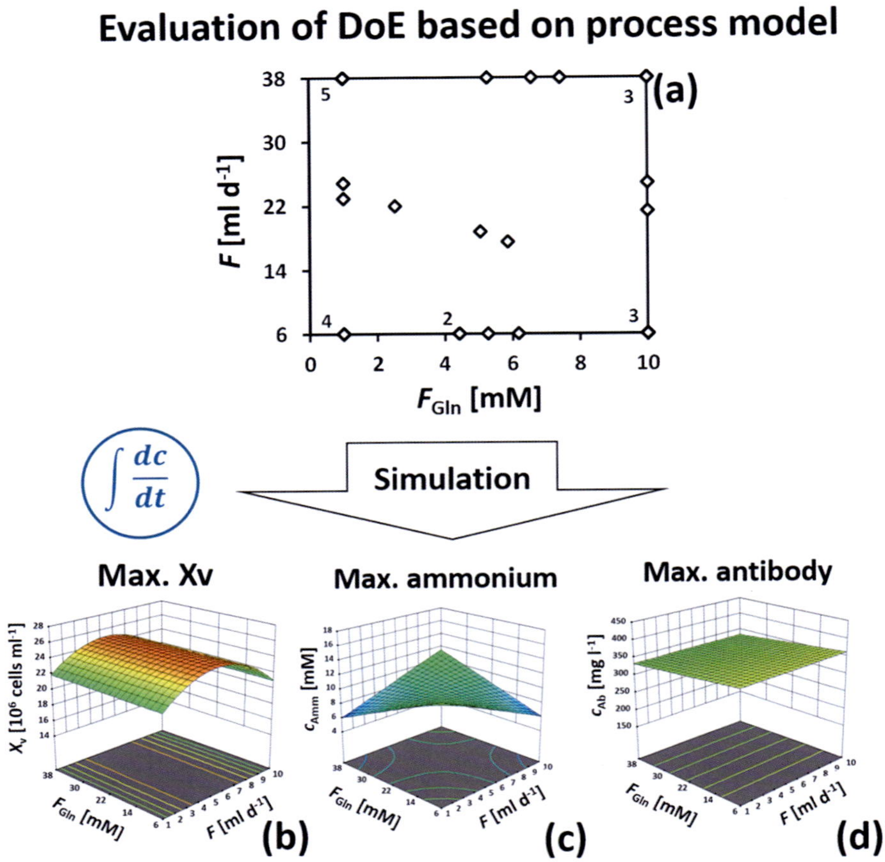

Fig. 2 Application of the mDoE approach for the optimization of a fed-batch process, (**a**) initially planned experiments in a DoE ($n = 29$ experiments), Response surfaces are based on the model-based simulations of each experiment and the following response surfaces were predicted: (**b**) maximal viable cell density, (**c**) maximal ammonium concentration, (**d**) maximal antibody concentration. See [30] for further details. F-feeding rate, F_{Gln}-glutamine concentration in feed, F_{Glc}-glucose concentration in feed

suggested in this DoE plan. The responses of each DoE is exported out of the simulations and these data are used in the DoE software to predict response surfaces as in a traditional evaluation, as can be seen in Fig. 2b–d.

Based on these simulations, the initially defined boundary conditions can be questioned and challenged, leading to knowledge-based DoE designs. In the presented example, different surfaces were generated and these can hardly be predicted beforehand without an appropriate mathematical model. No heuristic restrictions with several iterative rounds were necessary, because the mathematical process model incorporates the known factors, their dynamics and interactions in

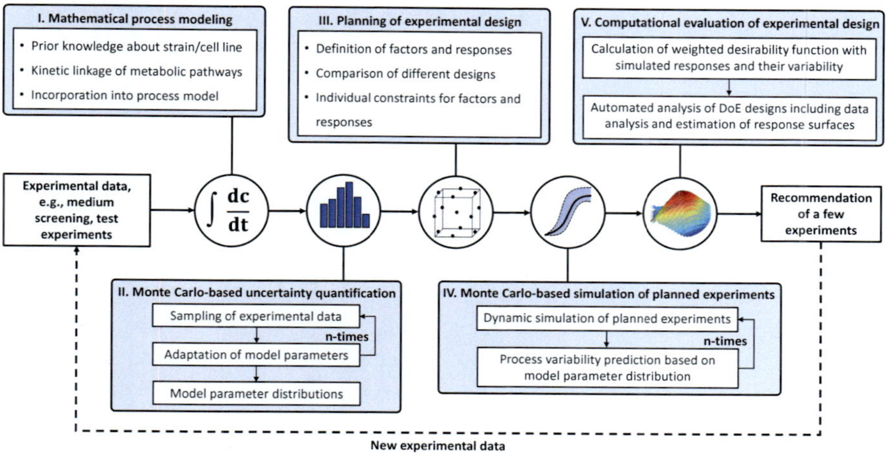

Fig. 3 Structural workflow of mDoE-toolbox consisting of the combination of mathematical process models and classical DoE under the consideration of model parametric uncertainty based on experimental variability, adapted from page 685 in [32]

DoE. In general, process modeling itself is a rather undefined work, and a variety of models and modeling approaches of different considered complexity exist in literature [53, 54]. mDoE is designed to be applied in the initial phase of process development for which very little data is available. Therefore, structurally simple [31] or generalized models [55] are applied, for which model parameters can be adapted based on few data points typically generated in medium tests or first cultivations.

2.3 *Model-Assisted DoE: Software Toolbox*

The formerly shown mDoE approach was extended and incorporated into a software toolbox [32], implemented in the programming software MATLAB and R. The description of the toolbox in the following is summarized from [32]. The main parts of the mDoE-toolbox are the combination of a mathematical process model, including model-parametric uncertainties with the computational planning and evaluation of DoE designs, as can be seen in Fig. 3.

In the beginning, the objective of the study (i.e., maximization of product concentration, minimization of inhibitory component) is defined. Then, the biotechnological system is modeled first (Fig. 3 box I). Thus, prior knowledge (e.g., pre-experiments, literature) about the strain is used to define mathematical expressions for cell growth, metabolism, and productivity [30]. After defining a mathematical model, the model-parametric uncertainties are derived with Monte Carlo sampling based on the experimental uncertainty (i.e., measurement error, Fig. 3 box II).

Therefore, the expected process variability based on the measurement errors is simulated and later used in the DoE evaluation [51]. Next, the experimental factors and responses are defined (Fig. 3 box III) with individual boundary values, e.g., a tolerated concentration of an inhibitory component or a minimum required product concentration. A DoE design, such as an optimal design or Box-Behnken design is subsequently planned (see [31] for a review on different designs). For each recommended factor combination, the time courses of the modeled state variables (e.g., cell weight, substrate, and product concentration) are simulated multiple times (Monte Carlo simulations, Fig. 3 box IV) taking into account the previously determined parameter probability functions (box II). From these simulations, the average expected response (e.g., average maximal cell dry weight) and its variability (difference in Monte-Carlos simulations) are calculated. These simulations are used for the computational evaluation (box V) of the former planned experimental design (box III). This enables an evaluation of each planned experiment with respect to its simulated average and its expected variability with the aim of simultaneously maximizing the response and minimizing its variability. After simulating the experimental designs, response surface plots are generated automatically for visualization. Only a few (e.g., 2–4) experiments with the highest response are recommended to be performed, while others can be neglected. Generally, uncertainty-based modeling techniques have been widely used in chemical systems or systems biology, but not often in bioprocess simulation studies [16].

Using the mDoE-toolbox, the available knowledge can be captured in the mathematical model, which can serve as a basis for advanced process understanding and digital twins [56]. In this way, the new data obtained from the recommended experiments can be used to re-adapt the model parameters and their probability distribution or to modify the model structure if so far unknown effects are identified [51]. In summary, the model-assisted DoE toolbox can be used to:

- Model the bioprocess and the interaction of critical process parameters to critical quality attributes,
- Enable knowledge-guided planning of experiments within a risk-based approach,
- Reduce the number of experiments needed for bioprocess development,
- Evaluate new experimental data and identify new cellular effects,
- Understand the definition of in-process controls, and the expected deviation of the manufacturing process,
- Comparison of different DoE designs,
- Transfer the process model during technology transfer and scale-up.

So far, successful applications of the mDoE approach were in the field of mammalian cell culture and bio-economy [30–32, 51].

3 Model-Assisted Evaluation of Scale-up

After process development, the bioprocess including its process strategy needs to be scaled up to pilot and production scale, for which mostly data-driven approaches are currently used. This is conventionally done by keeping a hydrodynamic state constant, for example, volumetric power input (P/V_L) [57, 58], mixing time [59, 60], impeller tip speed [61, 62] or the volumetric mass transfer coefficient k_La [63, 64], and hydrodynamical characterization is recommended at each scale (experimental recommendation see [65]). As an alternative to the experimental characterization of the bioreactor, computational fluid dynamics (CFD) has gained rising importance to obtain an improved understanding of the bioreactor hydrodynamics from small to large scale [66–69]. However, the so far applied purely data-based scale-up procedures do not consider the dynamics of the bioprocess itself, and it is not ensured that the previously developed process strategy is scaled up sufficiently and that the process dynamics stay constant during scale-up.

To enable more knowledge-driven decision-making during scale-up and scale-down, mathematical process models can be combined with statistical tools to understand the dynamics of bioprocesses in silico [16, 32, 70]. Therefore, the mechanistic of a biological process are modeled in different bioreactor scales and with varying process strategies (e.g., fed-batch, perfusion). Additional process knowledge is gained if effects of input uncertainties (e.g., experimental variations) on model outcomes are considered and quantified (Monte-Carlo methods) [51, 71]. By this, biological variability and fluctuations inherent in bioprocesses are incorporated into the model and model parameter distributions [32, 51]. Subsequently, individual model parameter distributions can be statistically compared to identify changes in the process dynamics between the modeled scales [51, 72].

This section summarizes our recent efforts in developing a new model uncertainty-based workflow for the evaluation of process strategies in different bioreactor scales. Please notice that this section is based on our recent publications, for which further details can be found in [51, 72].

3.1 *Computational Workflow*

As can be seen at the top of Fig. 4, experimental data sets at two different bioprocess scales are used as input (exemplary Scale A and Scale B), e.g., process development and process implementation scale (typically using different bioreactor systems). It should be noticed that this workflow does not focus on how the scale-up needs to be performed hydrodynamically. The aim was to develop a method to statistically evaluate if the process dynamics are comparable at both scales and if the targeted process optimum (i.e., process strategy) is still met [51].

The basis of the introduced concept is the quantification of model-parametric uncertainties under consideration of experimental uncertainty due to variability in

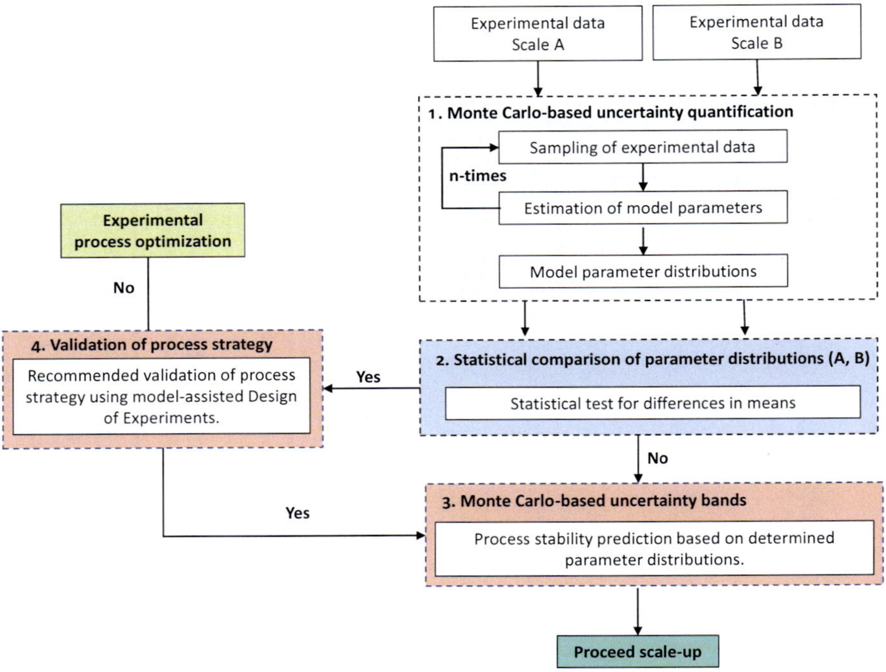

Fig. 4 Uncertainty-based workflow for the evaluation of dynamics of bioprocesses at different process scales, please see [51] for further details

measurements (Fig. 4, box 1). Please notice that the same approach is also used in the formerly introduced mDoE toolbox and the process understanding, captured in the same mathematical model, can be transferred between both approaches. The model parameters are estimated multiple times (Monte Carlo sampling) for each investigated scale considering measurement errors. Then, the parameter distributions and the prediction quantiles are used to visualize the process variability based on the model parameter distributions (Fig. 4, box 2). In the next step (Fig. 4, box 3), a statistical comparison of the parameter distributions is performed to evaluate if there are statistically significant differences in the dynamics between both scales. The same process dynamics and targeted process strategy could be ensured if no changes in the parameter distributions are identified. Otherwise, if the parameters differ significantly, a validation of the process strategy is recommended (Fig. 4, box 4, e.g., adjusting the feed composition). In this validation step, the previously introduced mDoE toolbox can be used to re-adjust the process strategy with a reduced number of experiments.

As can be seen as exemplary in Fig. 5a,b for two different single-use reactors at the laboratory scale, the same mathematical process model can be used to describe the process dynamics. The probability functions derived at different scales (e.g.,

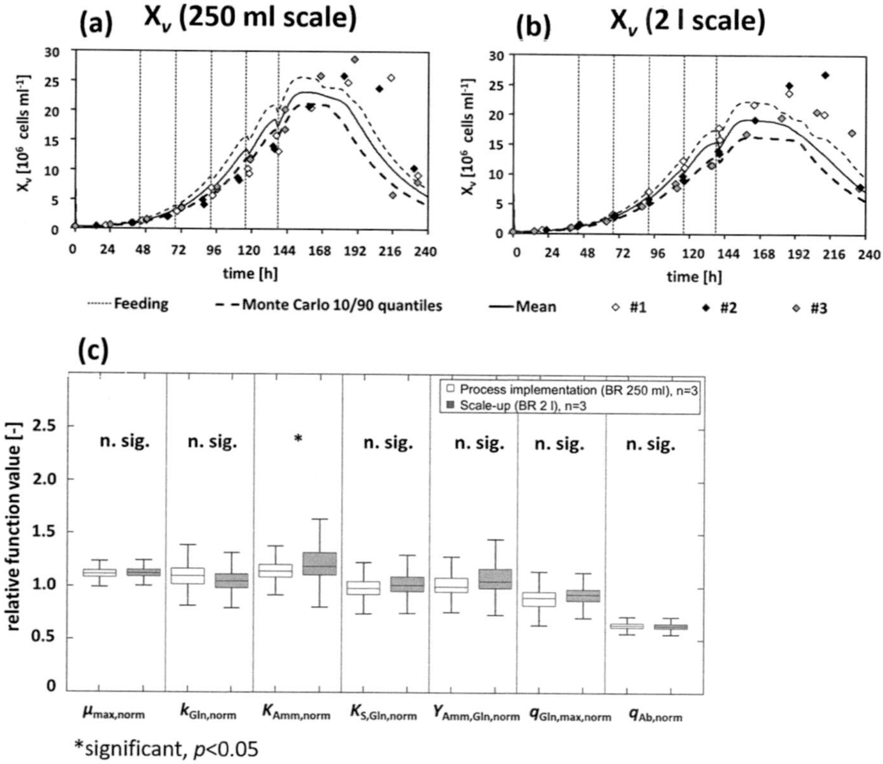

Fig. 5 Viable cell density of fed-batch cultures at 250 ml (**a**) and 2 l (**b**) bioreactor scale. Solid line is the median of 1000 simulations based on the Monte Carlo-based method; dashed line represents the 10% and 90% quantiles of the simulations. (**c**) shows the Box-plots of the normalized parameters of these runs, * = significant; n. sig. = not significant, please see [51] for further details

250 ml and 2 l scale) were then statistically compared to identify changes in the bioprocess dynamics (Fig. 5c).

μ_{max}, norm, k_{Gln}, norm, K_{s}, Gln, norm, Y_{Amm}, Gln, norm, q_{Gln}, max, norm, and q_{Ab}, norm were identified to be not significantly different. However, k_{Amm}, norm was slightly higher in the 2 l scale than in the 250 ml scale, but no differences were present in the maximal ammonium concentration and this change was therefore neglected. In summary, the process dynamics remain comparable in both scales, proven using the above-explained workflow. In conventional scale-up studies, the pure cultivation data of both scales (250 ml and 2 l, respectively) would have been compared, and a heuristic decision of the goodness of scale-up would have been drawn (e.g., same maximal titer, trends) [73, 74]. In the proposed workflow, the model uncertainty is quantified based on the available experimental variability and measurement error. Therefore, the process variability is determined on a timely axis

(10% and 90% quantiles of simulation, Fig. 5a, b) and in the parameter distributions (Fig. 5c).

Overall, the knowledge-driven computational evaluation of the process strategy during scale-up can be used to:

- Evaluate the interactions of critical process parameters to critical quality attributes during technology transfer and scale-up,
- Evaluate the right number of growth data and sampling,
- Understand the effect of different scales on the bioprocess performance,
- Validate that scale-up was sufficient,
- Guarantee the same process performance during scale-up from a few ml to pilot and production scale,
- Mathematicall adaptation of process strategies if a changing process performance was identified.

The workflow was also adapted for the scale-up of process strategy developed in microbioreactors (e.g., 24 well plates) to a larger scale, please see [72] for more details.

4 Future Steps Towards Digital Twins

Bioprocess design and optimization have undergone significant changes during the last decades moving from heuristically experimental approaches into twenty-first-century digitalization of the bioprocess industry [75]. Moreover, the high demand for biotechnological products calls for smart and efficient methods during research and development, as well as during tech transfer and routine manufacturing [1]. A promising tool is the usage of comprehensive mathematical process models as a virtual representation of the bioprocess, called a digital twin. The term digital twin is still not clearly defined and has different meanings in different parts of the industry. Historically, it is a model of a machine tool or a mechanical manufacturing site to handle the increased complexity [76, 77]. In the bioprocess industry, it progressively includes multiple parts of the manufacturing steps and their interaction. The complexity of digital twins highly depends on the desired task, and they can include mathematical process models of different complexity to reflect the mechanistics of the biological system and the interactions between process parameters, key performance indicators and product quality attributes. Therefore, they enable the use of computer-aided methods to test and plan novel bioprocesses, to efficiently monitor and control them, to predict the process variability as well as to gain an improved process understanding. One currently discussed approach is the implementation of machine learning tools (i.e., artificial intelligence) to enable self-learning tools to improve bioprocess [78–80]. A digital twin-based framework can be seen as a knowledge-based process development and engineering strategy, for which the main advantages are a comprehensive summary of knowledge of the investigated system, increased understanding of the process, usage of next-generation computational tools, a decrease of the development cost for experimental design, support of regulatory documentation, evolving of the digital twin during the lifecycle of the product, as well as evaluation, screening and virtual testing of new configurations/settings, among others [53]. The benefits of the application of digital twins heavily rely on their intended use and that the above-described points are only valid for the design, characterization and operation of the culture system and the respective cultivation process. More details regarding the current efforts towards digital twins can be found in [81, 82].

5 Conclusion

Biopharmaceutical cell culture processes require novel model-assisted methods for the development of process strategies and scale-up. Here a model-assisted workflow for the combination of mathematical process models as part of a digital twin with DoE was introduced and applied to design and scale-up of a manufacturing process for therapeutic antibodies. Compared to conventional DoE, mDoE supplies a more knowledge-based development of bioprocesses. Furthermore, domain knowledge is required and can be captured as additional constraints to the system, leading to a focused screening or optimization of bioprocesses using the mathematical model as a digital twin in mDoE. The introduced approaches provides novel knowledge-driven decision-making tools for bioprocess development and implementation.

References

1. Walsh G (2014) Biopharmaceutical benchmarks 2014. Nat Biotechnol 32(10):992–1000. https://doi.org/10.1038/nbt.3040
2. Nelson AL, Dhimolea E, Reichert JM (2010) Development trends for human monoclonal antibody therapeutics. Nat Rev Drug Discov 9(10):767–774. https://doi.org/10.1038/nrd3229
3. Chen C, Le H, Goudar CT (2016) Integration of systems biology in cell line and process development for biopharmaceutical manufacturing. Biochem Eng J 107:11–17. https://doi.org/10.1016/j.bej.2015.11.013
4. DiMasi JA, Grabowski HG, Hansen RW (2016) Innovation in the pharmaceutical industry: new estimates of r and d costs. J Health Econ 47:20–33. https://doi.org/10.1016/j.jhealeco.2016.01.012. http://www.sciencedirect.com/science/article/pii/S0167629616000291
5. Kretzmer G (2002) Industrial processes with animal cells. Appl Microbiol Biotechnol 59(2):135–142. https://doi.org/10.1007/s00253-002-0991-y
6. Lingg N, Zhang P, Song Z, Bardor M (2012) The sweet tooth of biopharmaceuticals: importance of recombinant protein glycosylation analysis. Biotechnol J 7(12):1462–1472. https://doi.org/10.1002/biot.201200078. https://onlinelibrary.wiley.com/doi/abs/10.1002/biot.201200078
7. Baldea M, Edgar TF (2018) Dynamic process intensification. Curr Opin Chem Eng 22:48–53. https://doi.org/10.1016/j.coche.2018.08.003. https://www.sciencedirect.com/science/article/pii/S2211339818300479. Biotechnology and bioprocess engineering

8. Stepper L, Filser FA, Fischer S, Schaub J, Gorr I, Voges R (2020) Pre-stage perfusion and ultra-high seeding cell density in cho fed-batch culture: a case study for process intensification guided by systems biotechnology. Bioprocess Biosyst Eng 43(8):1431–1443
9. Clincke MF, Mölleryd C, Zhang Y, Lindskog E, Walsh K, Chotteau V (2013) Very high density of cho cells in perfusion by ATF or TFF in wave bioreactor. Part I. effect of the cell density on the process. Biotechnol Prog 29(3):754–767. https://doi.org/10.1002/btpr.1704. https://aiche.onlinelibrary.wiley.com/doi/abs/10.1002/btpr.1704
10. Wolf MKF, Müller A, Souquet J, Broly H, Morbidelli M (2019) Process design and development of a mammalian cell perfusion culture in shake-tube and benchtop bioreactors. Biotechnol Bioeng 116(8):1973–1985. https://doi.org/10.1002/bit.26999
11. Smiatek J, Jung A, Bluhmki E (2020) Towards a digital bioprocess replica: computational approaches in biopharmaceutical development and manufacturing. Trends Biotechnol 38(10):1141–1153. https://doi.org/10.1016/j.tibtech.2020.05.008. https://www.sciencedirect.com/science/article/pii/S0167779920301426. Special Issue: Therapeutic Biomanufacturing
12. Gargalo CL, Udugama I, Pontius K, Lopez PC, Nielsen RF, Hasanzadeh A, Mansouri SS, Bayer C, Junicke H, Gernaey KV (2020) Towards smart biomanufacturing: a perspective on recent developments in industrial measurement and monitoring technologies for bio-based production processes. J Ind Microbiol Biotechnol 47(11):947–964. https://doi.org/10.1007/s10295-020-02308-1. https://academic.oup.com/jimb/article-pdf/47/11/947/36796286/jimb0947.pdf
13. Zobel-Roos S, Schmidt A, Uhlenbrock L, Ditz R, Köster D, Strube J (2021) Digital Twins in biomanufacturing. Springer, Cham, pp 181–262. https://doi.org/10.1007/10_2020_146
14. Möhler L, Flockerzi D, Sann H, Reichl U (2005) Mathematical model of influenza a virus production in large-scale microcarrier culture. Biotechnol Bioeng 90(1):46–58. https://doi.org/10.1002/bit.20363
15. López-Meza J, Araiz D, Carrillo L, López-Pacheco F, Rocha M, Alvarez M (2015) Using simple models to describe the kinetics of growth, glucose consumption, and monoclonal antibody formation in naive and infliximab producer cho cells. Cytotechnology 68:1287–1300. https://doi.org/10.1007/s10616-015-9889-2
16. Hernández Rodríguez T, Posch C, Schmutzhard J, Stettner J, Weihs C, Pörtner R, Frahm B (2019) Predicting industrial scale cell culture seed trains—a bayesian framework for model fitting and parameter estimation, dealing with uncertainty in measurements and model parameters, applied to a nonlinear kinetic cell culture model, using a MCMC method. Biotechnol Bioeng 116(11):2944–2959. https://doi.org/10.1002/bit.27125. https://onlinelibrary.wiley.com/doi/abs/10.1002/bit.27125
17. Zalai D, Kopp J, Kozma B, Küchler M, Herwig C, Kager J (2021) Microbial technologies for biotherapeutics production: key tools for advanced biopharmaceutical process development and control. Drug Discov Today Technol. https://doi.org/10.1016/j.ddtec.2021.04.001. https://www.sciencedirect.com/science/article/pii/S174067492100007X
18. Shirsat N, English N, Glennon B, Al-Rubeai M (2015) Modelling of mammalian cell cultures. Anim Cell Cult Springer 9:259–326. https://doi.org/10.1007/978-3-319-10320-4_10
19. Puck TT (1957) The genetics of somatic mammalian cells. Adv Biol Med Phys 5:40–47. https://doi.org/10.1016/B978-1-4832-3111-2.50006-7
20. Wurm F, Wurm M (2017) Cloning of cho cells, productivity and genetic stability-a discussion. Processes 5(2):20. https://doi.org/10.3390/pr5020020
21. Wurm FM, Hacker D (2011) First cho genome. Nat Biotechnol 29(8):718–720. https://doi.org/10.1038/nbt.1943
22. Stolfa G, Smonskey MT, Boniface R, Hachmann AB, Gulde P, Joshi AD, Pierce AP, Jacobia SJ, Campbell A (2018) Cho-omics review: the impact of current and emerging technologies on Chinese hamster ovary based bioproduction. Biotechnol J 13(3):1700227. https://doi.org/10.1002/biot.201700227. https://onlinelibrary.wiley.com/doi/abs/10.1002/biot.201700227

23. Handlogten MW, Lee-O'Brien A, Roy G, Levitskaya SV, Venkat R, Singh S, Ahuja S (2018) Intracellular response to process optimization and impact on productivity and product aggregates for a high-titer cho cell process. Biotechnol Bioeng 115(1):126–138. https://doi.org/10.1002/bit.26460. https://onlinelibrary.wiley.com/doi/abs/10.1002/bit.26460
24. Xu X, Nagarajan H, Lewis NE, Pan S, Cai Z, Liu X, Chen W, Xie M, Wang W, Hammond S (2011) The genomic sequence of the chinese hamster ovary (cho)-k1 cell line. Nat Biotechnol 29(8):735–741. https://doi.org/10.1038/nbt.1932
25. Kaufman RJ, Sharp PA (1982) Amplification and expression of sequences cotransfected with a modular dihydrofolate reductase complementary DNA gene. J Mol Biol 159(4):601–621. https://doi.org/10.1016/0022-2836(82)90103-6. http://www.sciencedirect.com/science/article/pii/0022283682901036
26. Beckmann TF, Krämer O, Klausing S, Heinrich C, Thüte T, Büntemeyer H, Hoffrogge R, Noll T (2012) Effects of high passage cultivation on cho cells: a global analysis. Appl Microbiol Biotechnol 94(3):659–671. https://doi.org/10.1007/s00253-011-3806-1
27. Kim NS, Kim SJ, Lee GM (1998) Clonal variability within dihydrofolate reduct-ase-mediated gene amplified chinese hamster ovary cells: stability in the absence of selective pressure. Biotechnol Bioeng 60(6):679–688. https://onlinelibrary.wiley.com/doi/10.1002/(SICI)1097-0290(19981220)60:6%3C679::AID-BIT5%3E3.0.CO;2-Q
28. Lee JS, Grav LM, Lewis NE, Faustrup Kildegaard H (2015) Crispr/cas9-mediated genome engineering of cho cell factories: application and perspectives. Biotechnol J 10(7):979–994. https://doi.org/10.1002/biot.201500082. https://onlinelibrary.wiley.com/doi/abs/10.1002/biot.201500082
29. Dangi AK, Sinha R, Dwivedi S, Gupta SK, Shukla P (2018) Cell line techniques and gene editing tools for antibody production: a review. Front Pharmacol 9:630. https://doi.org/10.3389/fphar.2018.00630
30. Möller J, Kuchemüller KB, Steinmetz T, Koopmann KS, Pörtner R (2019) Model-assisted design of experiments as a concept for knowledge-based bioprocess development. Bioprocess Biosyst Eng 42:867–882. https://doi.org/10.1007/s00449-019-02089-7
31. Kuchemüller KB, Pörtner R, Möller J (2021) Digital Twins and their role in model-assisted design of experiments. Springer, Berlin, pp 1–33. https://doi.org/10.1007/10_2020_136
32. Moser A, Kuchemüller KB, Deppe S, Rodríguez TH, Frahm B, Pörtner R, Hass VC, Möller J (2021) Model-assisted doe software: optimization of growth and biocatalysis in saccharomyces cerevisiae bioprocesses. Bioprocess Biosyst Eng 44(4):683–700. https://doi.org/10.1007/s00449-020-02478-3
33. Mandenius CF, Graumann K, Schultz TW, Premstaller A, Olsson IM, Petiot E, Clemens C, Welin M (2009) Quality-by-design for biotechnology-related pharmaceuticals. Biotechnol J 4(5):600–609. https://doi.org/10.1002/biot.200800333
34. Torkashvand F, Vaziri B, Maleknia S, Heydari A, Vossoughi M, Davami F, Mahboudi F (2015) Designed amino acid feed in improvement of production and quality targets of a therapeutic monoclonal antibody. PLoS One 10:1–21. https://doi.org/10.1371/journal.pone.0140597
35. Esteban PP, Patel H, Veraitch F, Khalife R (2021) Optimization of the nutritional environment for differentiation of human-induced pluripotent stem cells using design of experiments—a proof of concept. Biotechnol Prog :e3143. https://doi.org/10.1002/btpr.3143. https://aiche.onlinelibrary.wiley.com/doi/pdf/10.1002/btpr.3143
36. Gmeiner C, Saadati A, Maresch D, Stanimira K, Frank M, Altmann F, Herwig C, Spadiut O (2015) Development of a fed-batch process for a recombinant *Pichia pastoris* OCH1 strain expressing a plant peroxidase. Microb Cell Factories 14:1. https://doi.org/10.1186/s12934-014-0183-3
37. Rowland-Jones RC, Jaques C (2019) At-line raman spectroscopy and design of experiments for robust monitoring and control of miniature bioreactor cultures. Biotechnol Prog 35(2):e2740. https://doi.org/10.1002/btpr.2740. https://aiche.onlinelibrary.wiley.com/doi/pdf/10.1002/btpr.2740

38. Galbraith SC, Bhatia H, Liu H, Yoon S (2018) Media formulation optimization: current and future opportunities. Curr Opin Chem Eng 22:42–47. https://doi.org/10.1016/j.coche.2018.08.004. https://www.sciencedirect.com/science/article/pii/S221133981830025X. Biotechnology and bioprocess engineering
39. Lam L, Yunling B, Cheng C, Ishai P, Changjian W (2015) Development and manufacturability assessment of chemically-defined medium for the production of protein therapeutics in cho cells. Biotechnol Prog 31(5):1163–1171. https://doi.org/10.1002/btpr.2108
40. Sokolov M, Ritscher J, MacKinnon N, Souquet J, Broly H, Morbidelli M, Butté A (2017) Enhanced process understanding and multivariate prediction of the relationship between cell culture process and monoclonal antibody quality. Biotechnol Prog 33(5):1368–1380. https://doi.org/10.1002/btpr.2502. https://aiche.onlinelibrary.wiley.com/doi/pdf/10.1002/btpr.2502
41. Montgomery D (2017) Design and analysis of experiments. https://www.wiley.com/en-us/Design+and+Analysis+of+Experiments%2C+10th+Edition-p-9781119492443
42. Nasrabadi MRN, Razavi SH (2010) Use of response surface methodology in a fed-batch process for optimization of tricarboxylic acid cycle intermediates to achieve high levels of canthaxanthin from dietzianatronolimnaea hs-1. J Biosci Bioeng 109(4):361–368. https://doi.org/10.1016/j.jbiosc.2009.10.013
43. Mandenius CF, Brundin A (2008) Bioprocess optimization using design-of-experiments methodology. Biotechnol Prog 24(6):1191–1203. https://doi.org/10.1002/btpr.67
44. Duvar S, Hecht V, Finger J, Gullans M, Ziehr H (2013) Developing an upstream process for a monoclonal antibody including medium optimization. BMC Proc 7:P34. https://doi.org/10.1186/1753-6561-7-S6-P34
45. Legmann R, Schreyer HB, Combs RG, McCormick EL, Russo AP, Rodgers ST (2009) A predictive high-throughput scale-down model of monoclonal antibody production in cho cells. Biotechnol Bioeng 104(6):1107–1120. https://doi.org/10.1002/bit.22474
46. Moran EB, McGowan ST, McGuire JM, Frankland JE, Oyebade IA, Waller W, Archerand LC, Morris L, Pandya J, Nathan SR, Smith L, Cadette ML, Michal-owski JT (2000) A systematic approach to the validation of process control parameters for monoclonal antibody production in fed-batch culture of a murine myeloma. Biotechnol Bioeng 69(3):242–255. https://onlinelibrary.wiley.com/doi/10.1002/1097-0290(20000805)69:3%3C242::AID-BIT2%3E3.0.CO;2-I
47. Brunner M, Fricke J, Kroll P, Herwig C (2017) Investigation of the interactions of critical scale-up parameters (ph, po_2 and pco_2) on cho batch performance and critical quality attributes. Bioproc Biosyst Eng 40(2):251–263. https://doi.org/10.1007/s00449-016-1693-7
48. Yang WC, Lu J, Nguyen NB, Zhang A, Healy NV, Kshirsagar R, Ryll T, Huang YM (2014) Addition of valproic acid to cho cell fed-batch cultures improves monoclonal antibody titers. Mol Biotechnol 56(5):421–428. https://doi.org/10.1007/s12033-013-9725-x
49. Nasseri SS, Ghaffari N, Braasch K, Jardon MA, Butler M, Kennard M, Gopaluni B, Piret JM (2014) Increased cho cell fed-batch monoclonal antibody production using the autophagy inhibitor 3-ma or gradually increasing osmolality. Biochem Eng J 91(Supplement C):37–45. https://doi.org/10.1016/j.bej.2014.06.027
50. Cha HM, Lim JH, Lee KS, Kim DI (2018) Nucleotide sugar precursor feeding strategy to enhance sialylation of albumin-erythropoietin in cho cell cultures. Process Biochem 66:197–204. https://doi.org/10.1016/j.procbio.2017.12.014. http://www.sciencedirect.com/science/article/pii/S135951131630993X
51. Möller J, Hernández Rodríguez T, Müller J, Arndt L, Kuchemüller KB, Frahm B, Eibl R, Eibl D, Pörtner R (2020) Model uncertainty-based evaluation of process strategies during scale-up of biopharmaceutical processes. Comput Chem Eng 134:106693. https://doi.org/10.1016/j.compchemeng.2019.106693
52. Glassey J, Gernaey KV, Clemens C, Schulz TW, Oliveira R, Striedner G, Mandenius CF (2011) Process analytical technology (pat) for biopharmaceuticals. Biotechnol J 6(4):369–377. https://doi.org/10.1002/biot.201000356

53. Möller J, Pörtner R (2021) Digital twins for tissue culture techniques—concepts, expectations, and state of the art. Processes 9(3):447. https://doi.org/10.3390/pr9030447. https://www.mdpi.com/2227-9717/9/3/447
54. Bailey JE (1998) Mathematical modeling and analysis in biochemical engineering: past accomplishments and future opportunities. Biotechnol Prog 14(1):8–20. https://doi.org/10.1021/bp9701269
55. Brüning S, Gerlach I, Pörtner R, Mandenius CF, Hass VC (2017) Modeling suspension cultures of microbial and mammalian cells with an adaptable six-compartment model. Chem Eng Technol 40(5):956–966. https://doi.org/10.1002/ceat.201600639. https://onlinelibrary.wiley.com/doi/pdf/10.1002/ceat.201600639
56. Narayanan H, Luna MF, von Stosch M, Cruz Bournazou MN, Polotti G, Morbidelli M, Butte A, Sokolov M (2020) Bioprocessing in the digital age: the role of process models. Biotechnol J 15(1):1900172. https://doi.org/10.1002/biot.201900172
57. Klöckner W, Tissot S, Wurm F, Büchs J (2012) Power input correlation to characterize the hydrodynamics of cylindrical orbitally shaken bioreactors. Biochem Eng J 65:63–69. https://doi.org/10.1016/j.bej.2012.04.007. http://www.sciencedirect.com/science/article/pii/S1369703X12000915
58. Catapano G, Czermak P, Eibl R, Eibl D, Pörtner R (2009) Bioreactor design and scale-up. Cell and tissue reaction engineering springer, pp 173–259. https://doi.org/10.1007/978-3-540-68182-3_5
59. Varley J, Birch J (1999) Reactor design for large scale suspension animal cell culture. Cytotechnology 29(3):177–205. https://doi.org/10.1023/A:1008008021481
60. Rosseburg A, Fitschen J, Wutz J, Wucherpfennig T, Schlüter M (2018) Hydrodynamic inhomogeneities in large scale stirred tanks—influence on mixing time. Chem Eng Sci 188:208–220. https://doi.org/10.1016/j.ces.2018.05.008. http://www.sciencedirect.com/science/article/pii/S0009250918302999
61. Ju LK, Chase G (1992) Improved scale-up strategies of bioreactors. Bioprocess Eng 8(1–2):49–53. https://doi.org/10.1007/BF00369263
62. Alsayyari AA, Pan X, Dalm C, van der Veen JW, Vriezen N, Hageman JA, Wijffels RH, Martens DE (2018) Transcriptome analysis for the scale-down of a cho cell fed-batch process. J Biotechnol 279:61–72. https://doi.org/10.1016/j.jbiotec.2018.05.012
63. Xing Z, Kenty BM, Li ZJ, Lee SS (2009) Scale-up analysis for a cho cell culture process in large-scale bioreactors. Biotechnol Bioeng 103(4):733–746. https://doi.org/10.1002/bit.22287
64. Nienow AW, Langheinrich C, Stevenson NC, Emery AN, Clayton TM, Slater NKH (1996) Homogenisation and oxygen transfer rates in large agitated and sparged animal cell bioreactors: some implications for growth and production. Cytotechnology 22(1):87–94. https://doi.org/10.1007/BF00353927
65. Meusel W, Löffelholz C, Husemann U, Dreher T, Greller G, Kauling J (2016) Recommendations for process engineering characterisation of single-use bioreactors and mixing systems by using experimental methods. https://dechema.de/studien.html
66. Sharma C, Malhotra D, Rathore AS (2011) Review of computational fluid dynamics applications in biotechnology processes. Biotechnol Prog 27(6):1497–1510. https://doi.org/10.1002/btpr.689. https://aiche.onlinelibrary.wiley.com/doi/pdf/10.1002/btpr.689
67. Werner S, Kaiser SC, Kraume M, Eibl D (2014) Computational fluid dynamics as a modern tool for engineering characterization of bioreactors. Pharm Biol 2(1):85–99. https://doi.org/10.4155/PBP.13.60
68. Kaiser SC, Eibl R, Eibl D (2011) Engineering characteristics of a single-use stirred bioreactor at bench-scale: the mobius cellready 3l bioreactor as a case study. Eng Life Sci 11(4):359–368. https://doi.org/10.1002/elsc.201000171. https://onlinelibrary.wiley.com/doi/abs/10.1002/elsc.201000171
69. Nienow AW, Rielly CD, Brosnan K, Bargh N, Lee K, Coopman K, Hewitt CJ (2013) The physical characterisation of a microscale parallel bioreactor platform with an industrial cho cell line expressing an igg4. Biochem Eng J 76:25–36. https://doi.org/10.1016/j.bej.2013.04.011. http://www.sciencedirect.com/science/article/pii/S1369703X13001150

70. Kroll P, Hofer A, Ulonska S, Kager J, Herwig C (2017) Model-based methods in the biopharmaceutical process lifecycle. Pharm Res 34(12):2596–2613. https://doi.org/10.1007/s11095-017-2308-y
71. Anane E, García AC, Haby B, Hans S, Krausch N, Krewinkel M, Hauptmann P, Neubauer P, Cruz Bournazou MN (2019) A model-based framework for parallel scale-down fed-batch cultivations in mini-bioreactors for accelerated phenotyping. Biotechnol Bioeng 116(11):2906–2918. https://doi.org/10.1002/bit.27116. https://onlinelibrary.wiley.com/doi/pdf/10.1002/bit.27116
72. Arndt L, Wiegmann V, Kuchemüller KB, Baganz F, Pörtner R, Möller J (2021) Model-based workflow for scale-up of process strategies developed in miniaturized bioreactor systems. Biotechnol Prog 37(3):e3122. https://doi.org/10.1002/btpr.3122. https://aiche.onlinelibrary.wiley.com/doi/pdf/10.1002/btpr.3122
73. Rameez S, Mostafa SS, Miller C, Shukla AA (2014) High-throughput miniaturized bioreactors for cell culture process development: reproducibility, scalability, and control. Biotechnol Prog 30(3):718–727. https://doi.org/10.1002/btpr.1874. https://aiche.onlinelibrary.wiley.com/doi/pdf/10.1002/btpr.1874
74. Li F, Shen A, Amanullah A (2013) Cell culture processes in monoclonal antibody production. Pharm Sci Encycl 1:1–38. https://doi.org/10.1002/9780470571224.pse506
75. Steinwandter V, Borchert D, Herwig C (2019) Data science tools and applications on the way to pharma 4.0. Drug Discov Today 24(9):1795–1805. https://doi.org/10.1016/j.drudis.2019.06.005. https://www.sciencedirect.com/science/article/pii/S1359644618305324
76. Geris L, Lambrechts T, Carlier A, Papantoniou I (2018) The future is digital: in silico tissue engineering. Curr Opin Biomed Eng 6:92–98. https://doi.org/10.1016/j.cobme.2018.04.001. http://www.sciencedirect.com/science/article/pii/S2468451118300102. Tissue Engineering and Regenerative Medicine/Biomaterials
77. Negri E, Fumagalli L, Macchi M (2017) A review of the roles of digital twin in CPS-based production systems. Proc Manuf 11:939–948. https://doi.org/10.1016/j.promfg.2017.07.198. http://www.sciencedirect.com/science/article/pii/S2351978917304067. 27th International Conference on Flexible Automation and Intelligent Manufacturing, FAIM2017, 27-30 June 2017, Modena, Italy
78. Nargund S, Guenther K, Mauch K (2019) The move toward biopharma 4.0. Genet Eng Biotechnol News 39(6):53–55. https://doi.org/10.1089/gen.39.06.18
79. Narayanan H, Luna MF, von Stosch M, Cruz Bournazou MN, Polotti G, Morbidelli M, Butté A, Sokolov M (2020) Bioprocessing in the digital age: the role of process models. Biotechnol J 15(1):1900172. https://doi.org/10.1002/biot.201900172. https://onlinelibrary.wiley.com/doi/pdf/10.1002/biot.201900172
80. Narayanan H, Sokolov M, Morbidelli M, Butté A (2019) A new generation of predictive models: the added value of hybrid models for manufacturing processes of therapeutic proteins. Biotechnol Bioeng 116(10):2540–2549. https://doi.org/10.1002/bit.27097. https://onlinelibrary.wiley.com/doi/pdf/10.1002/bit.27097
81. Herwig C, Pörtner R, Möller J (2021) Digital Twins: applications to the design and optimization of bioprocesses, vol 177. Springer, Berlin
82. Herwig C, Pörtner R, Möller J (2021) Digital Twins: tools and concepts for smart biomanufacturing, vol 176. Springer, New York

Monitoring Tools for the Development of High Cell Density Culture Strategies

Martí Lecina, Pere Comas, Ivan Martínez-Monge, and Jordi J. Cairó

Abbreviations

CER	Carbon Dioxide Evolution Rate
CHO	Chinese Hamster Ovary cells
CPP	Critical Process Parameters
CQA	Critical Quality Attributes
C_L	Dissolved oxygen concentration (mM)
C_{L*}	Dissolved oxygen concentration in the gas-liquid interphase (mM)
DNA	Deoxyribonucleic acid
DO	Dissolved Oxygen (%)
GOI	Gene of Interest
HCDC	High Cell Density Cultures
Gm	Gas flow (L/h)
HEK293	Human Embryonic Kidney 293 cells
ICH	International Council for Harmonisation
ICV	Integral of Viable Cells density

V. Chotteau (✉) · M. E. L. Mäkinen
Department of Industrial Biotechnology, School of Engineering Sciences in Chemistry, Biotechnology and Health, KTH-Royal Institute of Technology, Stockholm, Sweden

AdBIOPRO, VINNOVA Competence Centre for Advanced Bioproduction by Continuous Processing, KTH- Royal Institute of Technology, Stockholm, Sweden

Digital Futures – KTH-Royal Institute of Technology, Stockholm, Sweden
e-mail: chotteau@kth.se; meerim@kth.se

E. Hagrot · L. Zhang
Department of Industrial Biotechnology, School of Engineering Sciences in Chemistry, Biotechnology and Health, KTH-Royal Institute of Technology, Stockholm, Sweden

AdBIOPRO, VINNOVA Competence Centre for Advanced Bioproduction by Continuous Processing, KTH- Royal Institute of Technology, Stockholm, Sweden
e-mail: hagrot@kth.se; lianz@kth.se

R. Pörtner (ed.), *Cell Culture Engineering and Technology*, Cell Engineering 10,
https://doi.org/10.1007/978-3-030-79871-0_16

K_{des}	Constant of oxygen desorption (h^{-1})
K_{La}	global oxygen transport volumetric coefficient (h^{-1})
MOI	Multiplicity of Infection [pfu/cell]
OTR	Oxygen Transfer Rate (mM $O_2 \cdot h^{-1}$)
OUR	Oxygen Uptake Rate (mM $O_2 \cdot h^{-1}$)
PAT	Process Analytical Technologies
PFU	Plaque Forming Units
PTM	Post Translational Modifications
QbD	Quality by Design
QPA	Quality Product Attributes
q_i	Specific production/consumption rate of metabolite *i*, [g·(106 cells·h)$^{-1}$]
r_i	Volumetric production/consumption rate of metabolite *i*, [g·(L·h)$^{-1}$]
RTRT	Real-Time Release Testing
TNF	Tumour Necrosis Factor
TOA	Time of Action, [h]
TOF	Time of Infection [h]
TOH	Time of Harvesting [h]
Xv	Viable cell density, (10^6 Cells·mL^{-1})
vvd	volume media per volume bioreactor vessel per day
V_L	volume of liquid in the bioreactor
V_p	Volumetric productivity, [g_p·(L·h)$^{-1}$]
y_{O2}	oxygen concentration in the gas phase
μ	specific growth rate (h^{-1})

1 Introduction

The twenty-first century has been the era of biopharmaceuticals, this is evidenced by the number of biopharmaceuticals approved by Regulatory Agencies (about 300 products). Approvals have been increasing yearly along the last decade. In 2014 sales of biotherapeutics represented about 18% of drug sales worldwide and reached a market volume about 195 billion USD [1]. The best performing market segment in terms of sales is the monoclonal antibodies sector. Only during the period from 2010 to 2016 monoclonal antibodies sales have doubled [2, 3]. The bestselling molecule with a total sale of almost 19 billion USD was Humira (adalimumab, a TNF inhibitor) manufactured by AbbVie (USA) and Eisai Co., Ltd. (Japan) [2]. This fact, together with the irruption of biosimilars due to several patents expiration have boosted the industry in areas such as developing and manufacturing tools.

Furthermore, biotech products comprise seven of the top ten drugs based on worldwide sales not a long time ago (2014). In addition, 52 biotechnological products were blockbusters, defined as a product with $1 billion sales per year; and considering that at present there are only 67 blockbusters drugs in total, the strength of Biotechnology industry is absolutely proved [1, 4].

Even though modern bioprocessing and biopharmaceuticals manufacturing can be already considered as mature fields, still biomanufacturing challenges remain. For this reason, new technologies are constantly being developed and applied, and such advances themselves often bring again new challenges. Besides the development of new biological drugs, main biomanufacturing aims are focused on increasing process and product quality attributes, and on improving productivity and process efficiency [5] to reduce bioprocessing time and costs [6].

For example, a critical QPA defined for biotherapeutics is the glycosylation pattern of glycoproteins. Many therapeutics in the pipeline are glycoproteins with at least one or more N-glycan/ O-glycan chains. The role of N-glycans is related to the half-life and biological activity of protein products, as well as immune response and cellular homeostasis, among others [7]. These glycoproteins are mainly produced in Chinese Hamster Ovary (CHO) cells [8]. CHO cells are well-characterized and its continued usage over several decades, without any clear adverse effects, have allowed regulatory approval over 100 biopharmaceuticals [9]. However, CHO cells generate diverse N-glycan structures leading to protein product heterogeneity and they neither cannot perform all types of human glycosylation [10]. Therefore, the final product consists in a mixture of glycoproteins with different glycan chains, and this becomes a drawback when the desired product consists of only one specific N-glycan structure. Thus, glycosylation still remains as a current challenge for the biotherapeutic manufacturing industry, and much effort has been dedicated to improve the production of proteins with specific glycosylation patterns [11]

For these reasons, in some cases, it is necessary to produce certain recombinant proteins in human cells, such as human embryonic kidney (HEK293) or human-retina-derived (PERC6) [12]. One example is Xigris (activated protein C), which is produced in HEK293 cells, as the post-transitional modifications performed by CHO cells were found to be inadequate [13].

The huge increase in the biopharmaceuticals produced in mammalian cells has placed in the first line the optimization and intensification of bioprocess as mandatory factors to design competitive processes. In this respect, improvements focused on achieving higher cell densities cultures and productivities in bioreactors have attracted a great deal of interest for the biomanufacturing industry [14]. The success of a bioprocess is based in its technical viability, but also on its economic feasibility. The interaction and well-definition of these two parameters leads the process to be successfully sustained for a long time.

In this regard, although the fed-batch has dominated the production of biopharmaceuticals in the last years, the pressure to obtain more and more competitive processes has led the industry to look for better strategies, such as perfusion systems, due to the possibility to obtain higher cell densities in a continuous mode [15]. As an example, [16] published a study in which more than $200{\cdot}10^6$ cells$\cdot$mL^{-1} were obtained in CHO perfusion system, which proves the huge improvements achieved in this area. Another example of the huge interest in perfusion systems, is the recent book published by Wolf et al. [17], describing the current state of the art in the design and operation of continuous bioreactors focused on mammalian cells. However, it must be said that the efficient application of these processes requires the availability of reliable and real-time monitoring systems for substrates, cell density and cell metabolic activity estimation, in order to achieve the process and product quality desired.

2 Bioprocess Parameters Monitoring as a Tool for Quality Assurance in Biopharmaceuticals Manufacturing. Quality by Design (QbD) Built on Process Analytical Technologies (PAT)

With the fast development of the biotechnology industry over the past 30 years, many efforts have been done by biopharmaceutic companies to ensure the quality and safety of their products. For this, new regulations have been addressed to ensure the quality requirements of the product mainly associated with viral safety, product characterisation and reproducibility [18]. From the first guideline on process validation made by the Food and Drug Administration in 1987, there has been much progress to these days. In 2004, the Food and Drug Administration (FDA) published a new initiative called "Pharmaceutical Current Good Manufacturing Practices (cGMPs) for the 21st century – a Risk Based Approach", with the intention of modernizing the regulation quality for human and veterinary drugs.

Biopharmaceuticals quality in terms of guarantee product quality, efficacy and safety is shifting from the final product attributes control testing, to the new paradigm based on Quality by Design (QbD). The International Council for Harmonisation (ICH), composed by the main Regulatory Agencies, details in ICH Q8 (R2) guidelines [19] that quality should be "built in" by bioprocess design. The aim of developers of biopharmaceuticals is to design a quality product and its manufacturing process to consistently deliver the intended performance of the desired product. A deep knowledge of the bioprocess generated along the biopharmaceutical development studies, as well as from manufacturing experience provide scientific understanding to define the design space, specifications, and manufacturing control parameters. For this reason, robust bioprocesses should be carefully developed. The QbD aims to achieve consistent and robust production of biopharmaceuticals with their desired attributes, and therefore their intended performance, being these attributes directly linked to the design space of the bioprocess parameters. The success of a design space defined lies on a better understanding of the bioprocess and their critical process parameters (CPP), which directly affect critical quality attributes (CQA) of the manufactured product, thus product performance. In the specific case of biopharmaceuticals produced in mammalian cells, mainly those large and complex molecules with post translational modifications such as glycosylation, CPP controls should go beyond the basic culture parameters like pH, temperature or Dissolved Oxygen Concentration (pO_2).

As an example, it has been described that carbon dioxide concentration affects product titter [20] as well as post-translational modifications in the final product [21].

Process Analytical Technology (PAT) initiative promoted by Food and Drug Administration is called to be the basis for a wider implementation of QbD methodologies in the manufacturing industry. Although the concept of PAT is not completely new, as process analysis/control is well established in important areas of chemical engineering, PAT introduced the idea of real-time process control for real-time quality assurance (QA) implementation in modern pharmaceutical manufacturing.

As previously defined [22] "from a PAT standpoint, a process is considered well understood when: (1) all critical sources of variability are identified and explained; (2) variability is managed by the process and (3) product quality attributes can be accurately and reliably predicted". In conclusion, the final goal has to be the on-line identification of the physiological state of cells in order to detect any deviation from the predefined parameters, and thus to take the corresponding control actions to redirect the process therein the boundaries [23]. By combining such an approach with measured product attributes (final product analysis needs will be reduced but not completely eliminated) Real-Time Release Testing (RTRT) will be a reality in the near future [24]. Currently, still exists a gap between the technological advances related to PAT and their implementation by the manufacturing industry of biopharmaceuticals for a host of reasons that include regulatory and cost issues, however, but it is expected that "PAT will soon become a widespread reality for real-time release testing".

3 Bioprocess Intensification: Increasing Productivity of Bioprocesses

One of the main current interests and challenges of the manufacturing industry of biopharmaceuticals is the increase of their competitiveness, what involves developing more cost-effective manufacturing bioprocesses and reducing the manufacturing time. To be able to reach such targets, manufacturing companies should move on to intensified bioprocesses. In other words, to achieve the desired total production rate using the smallest possible volume vessel, the volumetric productivity of the bioreactor must be increased as much as possible.

Volumetric productivity is defined as the amount of the biopharmaceutical of interest that can be produced per bioreactor volume unit and per time unit. Therefore, if the volumetric productivity is increased it has a direct impact on the volume of bioreactor needed to manufacture a given amount of the product of interest (reducing the bioreactor's volume), or alternatively, the production capacity increases. Consequently, the volume of other unit operations of the process, including upstream and downstream equipment, can be also reduced. Eventually, by reducing the volume of the manufacturing plant while maintaining the same amount of product of interest manufactured, the cost of goods decreases accordingly. The lowering of the cost of goods is not only due to the decrease of the plant capital investment for the manufacturing plant construction and start up, but also for the reduction of the operating cost (raw materials needs, manpower and plant maintenance costs).

Volumetric productivity (V) is described by Eq. 1. The volumetric productivity depends on two parameters: the specific productivity (v) and the amount of biocatalyst accumulated into the bioreactor (i.e. Biomass concentration, Xv).

$$\mathrm{Vp}\left[\frac{\text{mass of product}}{\text{volume}\cdot\text{ time}}\right]=\mathrm{Xv}\left[\frac{\text{cells}}{\text{volume}}\right]\cdot\mathrm{qp}\left[\frac{\text{mass of product}}{\text{cells}\cdot\text{ ime}}\right] \quad (1)$$

The intensification of a given bioprocess should be technically and economically addressed. From the technical perspective, the researcher is in pursuit of the maximal increment of Vp, which is directly proportional to the biocatalyst concentration and to the biocatalyst specific production rate of the product of interest. As can be deduced from Eq. 1, there are two potential strategies to increase Vp: (1) by increasing the amount of active biocatalyst in the bioreactor or (2) by improving the specific productivity of the biocatalyst. The most effective and smart strategy is two spread the endeavours in enhancing both parameters, instead of focusing on only one of them, since the improvements achieved in each of them has a multiplicative effect in the general increase of Vp. So, a good balance between the two viewpoints would lead to a more efficient optimization of the bioprocess.

The specific productivity of the biocatalyst can be enhanced mainly by cell engineering, by means of recombinant DNA technology, and by setting optimal culture conditions. Cell engineering includes the expression systems and cell machinery involved in the DNA transcription and protein synthesis [25]. The emerge of CRISPR/Cas 9 technology has enabled the modification of target DNA sites in a precise manner, allowing to perform gene knock-outs and knock-ins, as well as the expression control of target genes [26]. In addition, many efforts have been done in targeted integration methods in order to shorten the traditional timelines of random integration during bioprocess development [27]. Although is a very fascinating field, is out of the scope of this chapter.

As said, Vp also depends on the amount of biocatalyst (i.e. cell concentration) present in the bioreactor where the product of interest is generated. In this sense, not only the cell concentration in the bioreactor, but also for how long such biomass is kept actively producing into the bioreactor. The former parameter is directly represented by the Xv (viable cell density), while the later can be expressed as the Integral Viable Cell density (IVC), understood as the concentration of viable cells that were working during the time the cell culture lasted. In other words, ICV introduces the time parameter offering an overall bioprocess information beyond the instantaneous idea of the value of cell concentration. It can be determined as the area under the cell growth time profile curve by applying the trapezium rule. For example, what makes a difference in fed-batch or perfusion systems, is not only the cell density reached at the end of the growth phase, but also for how long producing cells can be maintained during the stationary phase, prolonging the process and then increasing its productivity.

First attempts of bioprocess intensification were based on cell culture media optimization and the development of many nutrients supplements to support and extend culture development [28]. Also, the paradigm of one-media-fits-all approach has been changing towards the development of complex formulations optimised for a particular cell line and even for a particular application or bioprocess [29].

The current proposed alternatives for increasing bioprocess productivity are bioprocess intensification through culture optimization based on fed-batch operation, and even more promising, the adoption of continuous bioprocessing technologies [30]. While fed batch was well adopted by the industry, and it represents yet a good alternative in many bioprocesses, perfusion cultures are slowly adapted by the manufacturing industries. In the following section, Bioprocess intensification through High Cell Density Cultures (HCDC) implementation is described in detail.

4 Cell Culture Strategies for Bioprocess Intensification

Current industrial mammalian cell-based processes for large-scale production are mostly produced using suspension cultures in stirred-tank bioreactors (about 70% of licensed processes for therapeutic recombinant proteins production) [31]. The ability to adapt many cell types to suspension culture and the use of polymeric additives to reduce shear damage have enabled the widespread application of suspension cell culture [32]. Furthermore, stirred-tank bioreactors ease monitoring and control of the process (pH, dissolved oxygen, temperature and others), provide homogeneous cell cultures due to the stirring and feasible scale-up.

Three different operation strategies can be used in stirred-tank for mammalian cell-based cultures depending on the application. The choice must be done according a compromise between the final productivity desired, the scale of the process, the investment in both cost and time, and on product quality required [33], that could facilitate the selection of a particular mode of production during drug development or lifecycle management depending on the project objectives.

Figure 1 shows the most common culture strategies used by the biotechnological industry [34]. The main features of the main culture strategies are as follows:

- *Batch*: the simplest strategy in which cells are cultured in a finite media until they stop growing. The main disadvantages of this system are the nutrient limitation, the low cell densities achieved, the low productivity and the high toxic accumulation by-products from the cell metabolism. Nowadays is only used in few specific bioprocesses or for screening purposes.
- *Fed batch*: consists in a batch culture with a gradual addition of a concentrated nutrient solution is performed during the culture. This operation mode avoids the

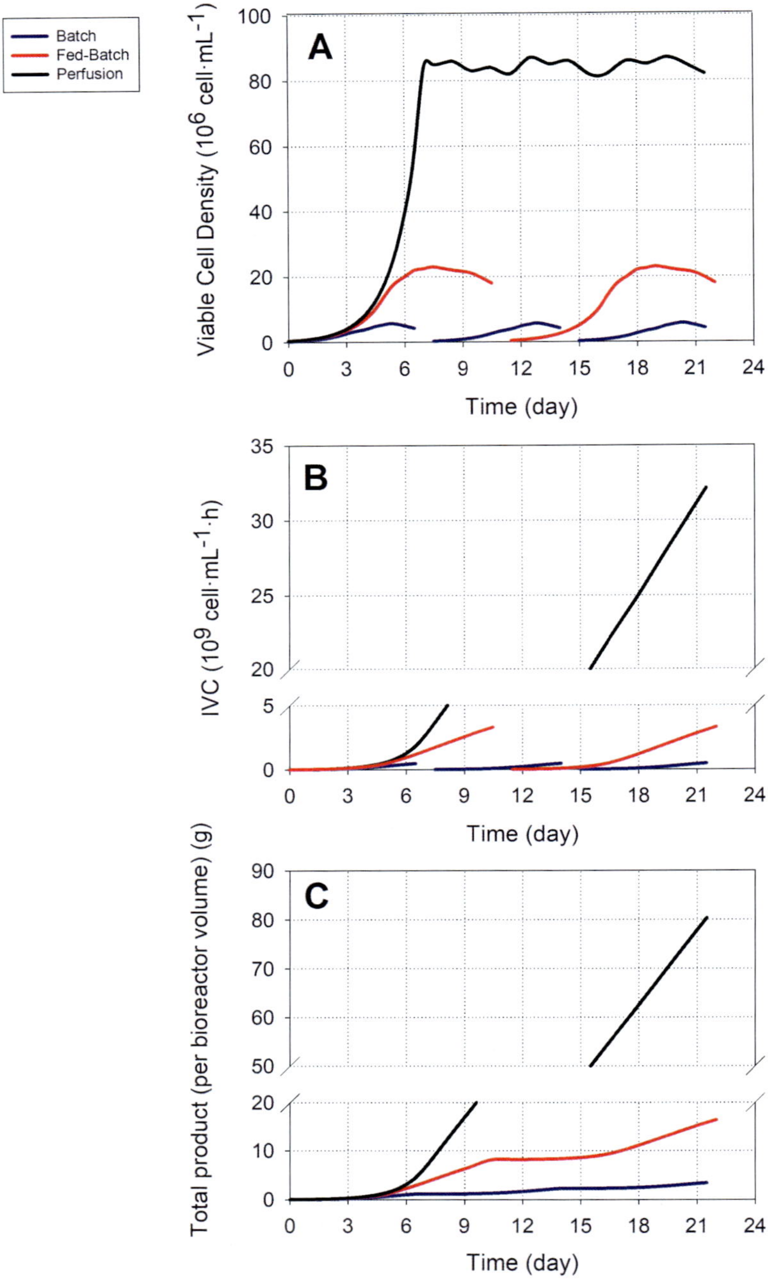

Fig. 1 Simulated time profiles comparison between different culture strategies: batch (blue), fed batch (red), and perfusion (black). (**a**) Viable cell density in [10^6 cells·mL^{-1}], (**b**) Integral of Viable Cells density (IVC) in [10^9 cells·mL^{-1}·h], and (**c**) Total amount of product generated in a 1-Litre bioreactor in [g]. The physiologic parameters used for the simulation are representative of experimental data: $\mu = 0{,}03$ h^{-1} for all simulations while the specific productivity qp (mg·(106 cells·h)$^{-1}$) depends on the strategy simulated being as follows: 2,5·10_3 for the batch culture, 2,25·10_3 for fed-batch and 2,1·10_3 for the perfusion culture

nutrient limitation, obtaining both higher cell densities and final product concentration compared with batch system. As in the batch mode, the accumulation of toxic metabolites in the culture broth leads cells to stop growing at some point. It normally has two phases: (1) the growth phase, starting with a short batch phase followed by feeding, the cells are growing until a maximum VCD is reached; (2) stationary phase: VCD is maintain constant during some days while the cells are producing, the culture ends when the viability drops.

- *Perfusion*: a continuous culture equipped with a cell retention system at the outlet stream that permits removal of the growth-inhibitory by-products while maintaining the cells in the bioreactor. It can be operated in different modes, but normally also two different phases can be differentiated: (1) the growth phase, starting with a short batch phase which is followed by initial stages of the medium exchange. Cell growth and culture expansion is sustained until the maximum perfusion rate is reached, normally limited to 1–2 volume of medium per reactor volume per day (vvd) in large-scale volumes; (2) stationary phase: perfusion rate is maintained constant, and a purge stream has to be included to remove dead cells from the bioreactor. VCD is maintain constant for long periods of time, where a harvest line is continuously obtained with the desired product. There exists an increasing interest in the use of perfusion culture attributed to the higher product output in a smaller bioreactor volume.

Culture media composition commonly used in batch cultures of mammalian cells contain all the nutrients needed for cell growth up to a certain level. Some of them are at very high concentrations, as glucose and glutamine, that lead to a metabolic deregulation in their uptake rates [35, 36], generating high levels of lactate and ammonium, that are toxic at certain concentration and negatively affect both cell growth and product generation. However, batch operations are normally limited by the depletion of an essential nutrient.

To overcome the previously described batch process disadvantages, alternative cell culture strategies, such as fed-batch [31, 37] and perfusion processes have been developed and successfully implemented in bioreactor [38–40].

Fed-batch bioreactors has been the most utilized industry standard for biological protein production, and the developments performed over the past few decades has led to steadily increasing productivity in this platform [41]. Fed-batch cultivation allows controlling cell growth environment by the addition of nutrients in order to maintain the nutrient concentrations at levels in which cells can regulate more efficiently their metabolism, reducing the generation and accumulation of by-products as lactate or ammonium [42]. However, fed-batch cultures still end up by a drop in cell viability due to toxic metabolites accumulation [43, 44].

Perfusion has been the preferred mode of operation to overcome these problems related to by-product accumulation in fed-batch cultures. Although many critical parameters (such as dissolved oxygen, pH, and temperature) are well controlled in a regular fed-batch bioreactor, the control of other parameters is more challenging, such as nutrient and waste concentrations and osmolality. The combination of a constant perfusion rate with a cell density control loop allows even these factors to be tightly regulated, ultimately resulting in a consistent microenvironment for cells across the bioreactor duration, which in turn promotes consistent metabolism, productivity and product quality. Perfusion also provides product of interest with narrower residence time distribution, limiting it exposure to degradation by extracellular enzymes (e.g., proteases, glycosidases, sialidases) as well as by the extracellular environment conditions (i.e. pH, temperature).

As depicted in Fig. 1, perfusion cell culture systems presented significant increased productivities due to the achievement of higher cell densities and long culturing times [45, 46]. Integral of Viable Cell density (ICV) is a parameter that gather together the concentration of producing cells and how long the cells have been active, and as it can be observed from Fig. 1 B, the amount of product generated is directly proportional to ICV. Volumetric productivity of perfusion cultures can range 2–4 g/L-day, being about 5–10 times higher than those obtained in fed-batch [39]. In theory, perfusion operation at constant perfusion rate combined with a daily purge of a percentage of the total bioreactor volume (25–50%, Fig. 2) should allow for better control of product quality. This is due to the fact that death cells are purged from the system, and cell lysis and the concomitant release of intracellular proteins like proteases, is significantly reduced. It is well recognized that perfusion bioreactors are the best solution for labile bioproducts that are affected by the longer operation time in fed-batch mode. Interesting is the last tendency of using perfusion cultures for seed train intensification, prior to production bioreactor

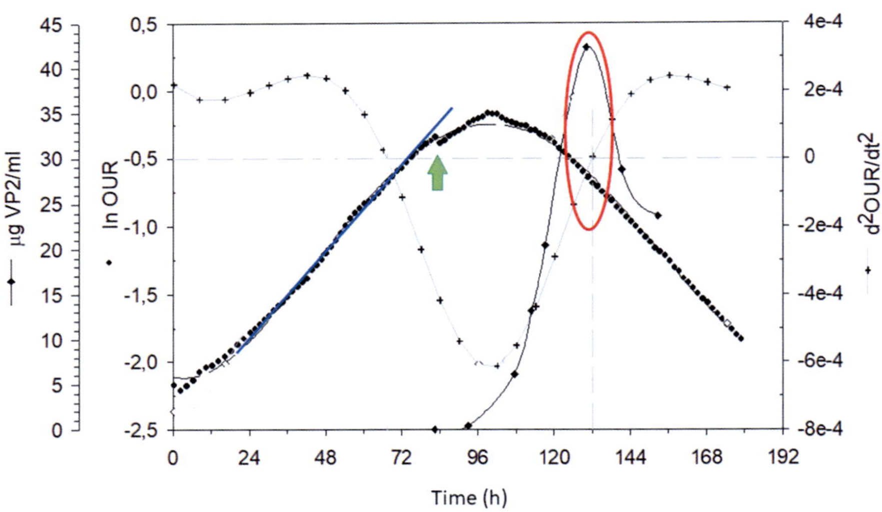

Fig. 2 Time profile of Ln(OUR) (black dots), VP2 protein concentration (black diamonds) and d2OUR/dt2 (+ symbol and grey line) of Sf9 insect cells culture infected with Baculovirus (AcVP2.IBDV). Blue line represents the linear regression of Ln(OUR) during the exponential growth phase. Time of Infection (TOI) is marked with a green arrow, upon the linearity of Ln(OUR) ends. Red ellipse highlights the coincidence of the intracellular VP2 protein peak with the zero root for the second derivative of Ln(OUR) profile, that can be defined as the Time of Harvest (TOH)

inoculation, for its capacity to obtain high cell density inoculum for either fed batch or continuous modes. This allows to start the process with higher cell density, shortening dramatically the production bioreactor running time [47].

One of the main challenges of perfusion cultures is the huge amounts of media needed, that limits its implementation (limited to 1–2 vvd in large scale bioreactors). The minimum perfusion rate of a specific medium needed to sustain a given cell density in perfusion is defined as the medium depth. In this direction, several strategies have been proposed to increase medium depth by increasing limiting nutrients concentration while maintaining the baseline components concentration (i.e. other buffers, Pluronic and salts) [40, 48]. The increase of medium depth allows reducing the perfusion rate and thus, the amount of medium that have to be prepared, eventually easing the handling of perfusion operations.

In any case, these culture systems, fed batch and perfusion, can be designed to provide an environment closer to the *in vivo* physiological state: constant supply of nutrients at the right concentration levels, avoiding the metabolic deregulation observed when glucose and glutamine are present at high concentrations. Under these conditions a proper cell growth is promoted, as well as, final cell concentration and product titter are higher, and the most important product expected quality is achieved.

The main challenge of both fed batch and perfusion systems is that their efficient implementation requires the availability of reliable and robust on-line measuring systems for cell density, cell metabolic activity (i.e. specific consumption and production rates) and main nutrient concentrations (i.e. glucose, glutamine or oxygen concentrations), enabling control of their concentrations at the desired level. The different set of monitoring tools developed and implemented to this end are exposed in further down in the coming sections.

The development of bioprocesses is not a trivial task and designing an appropriate flowsheet, including every step of the process, is essential to detect the possible bottlenecks. Understanding the interactions between operations and how the process can change as function of key operating variables should lead to design an optimal process for the desired product.

5 PAT: Monitoring Tools and Strategies

Currently there is a myriad of different monitoring tools, sensors or probes as well as monitoring strategies. The selection of the most suitable monitoring tools for each purpose should be done carefully, fulfilling the requirements defined *a priori* for each specific use. The monitoring of CPP can be classified depending on the strategy followed to measure a parameter (direct or indirect measurements) or depending on where the measuring device is placed in relation to the bioprocess. The later classification criterium indirectly involves the time needed for the analysis to be done. The time needed for obtaining the value of the parameter analysed is very relevant when selecting the proper monitoring strategies for the bioprocess control.

Short analytical times, if not real-time, are the most suitable monitoring tools for bioprocess control and automatization.

Regarding the nature of the measuring method, direct and indirect measurements can be distinguished:

- *Direct measurements*: the parameter of interest is directly measured with a specific probe or analysis. Some examples could be the pH, dissolved oxygen or analysis of metabolites concentration by means of High-Pressure Liquid Chromatography, etc.
- *Indirect measurements*: the parameter of interest is determined or estimated through the measurement of other related parameters. For example, biomass concentration can be directly measured using an impedance probe, or indirectly estimated by means of the Oxygen Uptake Rate (OUR), which is a parameter that offers information about the physiological state of the cells.

A second classification of the monitoring methods considers the placement of the sensor or measuring device in regards the bioreactor [49] and the time needed for the measurement or analysis:

- *Off-line measurement:* samples are manually or automatically collected from the bioreactor or bioprocess stream and taken to a specialized and centralized laboratory of analysis that is not located in the bioprocessing area. Samples are measured posteriori, and there is a gap between sampling and analysis time, so off-line measurements cannot be considered as a real-time monitoring system. For this reason, off-line monitoring is not the preferred solution for bioprocess control, but it is worth for high specialized measurement methods and for product CQA testing.
- *At-line measurements*: this method also requires a manual or automatic sampling, and the sample is taken to the analytical instrument, that in this case is located into the bioprocessing area. In at-line measurements sample are discarded and not returned to the bioprocessing stream. Normally, the time gap between sampling and the measurement is reduced in comparison to Off-line techniques. In some cases, in which the time required for the analysis is considerable (more than 0.5–1 h), the technique can be considered as off-line even though the measurement is performed in the bioprocessing area.
- *Real-time measurements*: real-time measurements, normally also known as on-line measurements, offer almost instantaneous measurements systems. This category is also divided into two other sub-categories.
  - *In-line measurements*: the measurement is performed without the need of extracting a sample from the bioprocess, since the sensor head is directly inserted in the bioreactor and in contact with the bioprocess stream. In-line measurements are the truly real-time techniques. Some examples are the pH, pressure or gas-flow measurements.
  - *On-line:* the main difference compared with *In-line* measurements, is that the sensor is not directly placed into the bioprocess mainstream or bioreactor. It requires an automatic sampling loop to aseptically transport the sample to

the sensor connected to the bioreactor. Often the sample can be recirculated to the bioreactor once the measurement is performed. Measurements can be real-time or carried out in minutes.

6 PAT Tools Used to Implement and Control High Cell Density Cultures

As described before, the huge market pressure for being competitive has led the biotechnological industry to intensify its processes to obtain even better productivities. Proof of that is the amount of publications focused on fed-batch and perfusion optimization released in the last years, with the main goal of increasing the final cell density and product titters in the processes [16, 50]. Nevertheless, although a lot of effort has been done in other areas as media optimization and cell line engineering, the use of robust monitoring and controlling tools to automate the control of the culture conditions and nutrients feedings is still a workhorse for the industry.

Over the recent years, various methods for measuring some of the most important variables related to cell culture have been proposed. One of the most crucial decisions to be made is what variable/s are the chosen ones to be monitored, as the control strategy will be affected by this decision. One monitoring strategy is focused on the measurement of the main nutrients from the culture broth considered as essential to maintain cell growth and cell activity. Normally essential nutrients (carbon and nitrogen sources) are present in the culture composition at higher concentration in respect to non-essential. Some examples of monitoring techniques that have been used for nutrient concentration monitoring are compiled in Table 1.

The monitoring of main nutrients concentration and its evolution along the culture allows controlling their concentration. Also, cell density can be determined from the evolution of the nutrients consumption rate (r_i).

Other analytical techniques are centred on the direct measurement of the cell density. Cell density can be measured using different approaches as summarized in Table 2. Commercial probes to measure total cell density, what includes life and death cells, and viable cell density (can measure live cells despite the presence of death cells) are commercially available.

Even more attractive can be other viable cell density monitoring strategies directed to measure cell activity. In these cases, viable cell density is estimated indirectly using other molecules like oxygen consumption, NADPH/NADH concentration, or volatile organic compounds derived from the lipid metabolism. The main advantage of measuring cell activity is that these techniques are very sensitive to metabolic changes which normally anticipates the start of the stationary or death phase of the culture profile.

Finally, the monitoring target can be the different products generated by the cells (Techniques complied in Table 3). Products monitoring englobe not only the generation and accumulation of by-products from the cell metabolism, but also the

Table 1 Different monitoring techniques used for the monitoring of main nutrients of the culture

Parameter measured	Parameter estimated	Technique	Type of measurement	Cell culture strategy	Reference
Glucose concentration	r_{glc}/Xv	FIA (glucose oxidase)	At-line	Fed-batch/Perfusion	[51]
		NIR spectrometry	In-line	Batch	[52]
		MIR spectrometry	In-line	Batch	[53]
		RAMAN spectrometry	In-line	Fed-batch	[54]
		HPLC	At-line	Batch	[55]
Glutamine concentration	r_{glu}/Xv	NIR spectrometry	In-line	Batch	[52]
		RAMAN spectrometry	In-line	Fed-batch	[54]
		MIR spectrometry	In-line	Batch	[53]
		FIA (L-glutaminase)	At-line	Batch	[56]
		HPLC	At-line	Batch	[55]
Glutamate concentration	r_{glut}	MIR spectrometry	In-line	Batch	[53]

concentration of the recombinant protein of interest, and even some of their quality attributes, as mAb glycosylation pattern measured by NIR spectroscopy.

In this context, it must be mentioned the promising technology based on different spectrometric methods, as Raman or IR spectroscopy for cell culture monitoring. Spectroscopic methods have been used to chemically characterize pure compounds or mixtures of compounds [23]. They allow the real-line measurement of viable cell density, glucose, lactate, glutamine and other amino acids among others components altogether [54, 69, 70]. The fact of offering the simultaneous measurements of mixtures of compounds represents a breakthrough in regard the monitoring tools for a single component measurement listed above. Nevertheless, although Raman seems to be the most promising tool for online monitoring cultures, its application relies on chemometric models that need an extensive calibration for each specific process, although some recent publications proposed novel methods based on machine-learning and real-time-just-in-time-learning for automatic calibration [71, 72].

However, the complexity in the implementation of these techniques, together with the fact that some of them do not offer a real-time measurement, have limited their use. Interestingly, easy measurable parameters, as the oxygen uptake rate (OUR) or the CO_2 production rate (CER), can be used to indirectly estimate the viable cell density along the culture, and consequently, have been extensively applied for a wide variety of mammalian cells-based processes [64, 73, 74]. In the next sections, the different methods for the measurement of OUR and its application for cell culture monitoring and automatization are detailed.

Table 2 Summary of the different monitoring techniques used for the monitoring of cell density

Parameter measured	Parameter estimated	Technique	Type of measurement	Cell culture strategy	Reference
		Image analysis	Inline/on-line	–	[57]
Viable Cell density	μ/Xv	NIR spectrometry	In-line	Batch	[58]
		RAMAN spectrometry	In-line	Fed-batch	[54]
		Fluorescence spectroscopy	Off-line	–	[59]
		RF Impedance	In-line	Fed-batch/Perfusion	[60,61]
		Acoustic resonance densitometry	On-line	Fed-batch/Perfusion	[62]
O_2 conc. (gas phase)	OUR/Xv	Mass spectrometry	At-line	Batch	[63]
O_2 conc. (liquid phase)	OUR/ Xv	Amperometric or optical probe	In-line	Fed-batch	[51, 64]
NADPH/NADH concentration	q_{NADH}	UV/VIS spectrometry	Off-line	–	[65]

7 OUR Measurement as Example of a Robust Tool for Cell Activity Monitoring and Bioprocess Control

Indirect methods, commonly called "soft sensors", are very promising tools to indirectly estimate the viable cell density and substrates/products concentration along the culture. Examples of soft sensors can be the oxygen uptake rate (OUR) or the alkali addition to control the pH as an indirect measurement of viable cells, glucose and glutamine consumptions [64, 75, 76]. The huge amount of studies published in the last years evidence that OUR behave become a reliable alternative for the monitoring and controlling of high cell density culturing strategies with very high productivities [77]. Regarding the cell metabolism, OUR determines the capacity of the cells to generate energy. This is why OUR is one of the most informative parameters for the estimation of cell density and cell activity, and can be also very informative to detect the transition phases commonly observed in mammalian cell cultures, as for example from the end of the exponential to the stationary phase in fed-batch [78].

Table 3 Summary of the different monitoring techniques used for the monitoring of by-products of the metabolism or to measure product of interest concentration or product quality attributes.

Parameter measured	Parameter estimated	Technique	Type of measurement	Cell culture strategy	Reference
Ammonium concentration	q_{amm}/Xv	Fluorescence spectroscopy	Off-line	–	[59]
		NIR spectrometry	In-line	Batch	[52]
		FIA (ammonium-selective electrode)	At-line	Batch	[56]
		MIR spectrometry	In-line	Batch	[53]
Lactate concentration	q_{lac}/X_v	RAMAN spectrometry	In-line	Fed-batch	[54]
		HPLC	At-line	Batch	[55]
		FIA (lactate oxidase)	At-line	Perfusion	[66]
		NIR spectrometry	In-line	Batch	[52]
mAb Glycosylation	Product quality	NIR spectrometry	In-line	Batch	[67]
mAb concentration	Productivity	RAMAN spectrometry	Off-line	–	[68]

As previously described [75], three different methods for the measurement of OUR have been developed up to date: dynamic method (invasive), and global mass balance or stationary liquid mass balance methods (non-invasive):

(a) The dynamic technique is based on the periodic measurement of the dissolved oxygen concentration (C_L) extinction profile in the liquid phase when air supply is discontinued with a nitrogen inlet flow into the headspace. The nitrogen atmosphere in the headspace of the bioreactor, forces the oxygen desorption characterized by the desorption constant (K_{des}). The dissolved oxygen concentration decreases due to the respiratory activity of cells and OUR is proportional to the slope of the decay curve. OUR is calculated using the Equation 2.

$$\text{O.U.R.} = \frac{C_L(t_0) - C_L(t_f)}{t_f - t_0} + \frac{\int_{t_0}^{t_f} (-K_{des} \cdot C_L(t)) \cdot dt}{t_f - t_0} \tag{2}$$

(b) The global mass balance method (Eq. 3) consists on analysing the differential oxygen concentration (y_{O2}) between the bioreactor's gas inlet and outlet (G_m), keeping the DO concentration constant.

$$\text{O.U.R.} = \frac{G_m \cdot \left(yO_{2,in} - yO_{2,out}\right)}{V_L} \tag{3}$$

(c) Finally, the stationary liquid mass balance method is based on the estimation of the oxygen uptake rate (OUR) by means of measurement of the oxygen transfer rate (OTR) as described by Eq. 4

$$\text{O.U.R.} = \text{O.T.R.} = k_L a \cdot \left(C_L{}^* - C_L\right) \tag{4}$$

The cell stress produced by DO and pH variations related to the dynamic method have promoted the development of the non-invasive alternative methods for OUR measurement. However, the low oxygen consumption rates of mammalian cells limit the use of the global mass balance. In contrast, stationary liquid mass balance has successfully been implemented, being a realistic alternative to the dynamic method.

The incorporation of on-line measurements for cell culture monitoring have played an important role on the successful implementation of high cell density cultures, especially regarding fed-batch and perfusion. In addition, the complexity of this processes regarding the possible metabolic changes during the culture made them to require adaptative feeding strategies depending on the stage of the culture. The first nutrient feeding phase in fed-batch operation aims to increase viable cell density concentration, while the second phase starts once the maximum cell density is reached being the objective to maintain the cells actively producing for as long as possible. In this direction, many studies have reported the promising applicability of OUR as a monitoring tool, from small culture platforms to big scale bioreactors. OUR can be corelated with the state of the culture in terms of VCD and viability, but also to key metabolites in the nutrient definition as glucose or glutamine. Stoichiometric ratios between OUR and nutrient consumption have been extensively defined [79].

The next step is the application of OUR as a monitoring tool to control the processes to further implement the optimal feeding strategies. In the case of fed-batch, having an optimal feeding strategy to feed only the nutrients required by the cells at every phase of the culture is crucial for not depleting any essential nutrient, as well as to not overfed the culture. In this regard, feeding media composition optimization is crucial for the successful of the process, but this must go hand in hand with having an optimal feeding strategy that requires to know the state of the culture in real time.

The feeding strategy will be different in the two main phases of a fed-batch culture, focusing on obtaining the maximum growth and final VCD in the growth phase, and extending the viability and culture long in the stationary phase. Regarding the perfusion, one of the main drawbacks is the huge amounts of media required for sustaining the high number of cells in the bioreactor. The possibility of having a tool to online monitor the state of the culture is crucial in the growth phase to not waste media and becomes even more important in the stationary phase to sustain high cell densities during long periods of time, obtaining high quantities of product in a continuous mode.

We have a lot of examples in which OUR has been used to control mammalian cell bioprocesses, as stated in the last published review by [77]. From fed-batch to perfusion, at different scales, OUR has been very useful to monitor the physiological state of cells in bioreactor and then to adjust the nutrient feeding according to that. With that, many goals have been achieved, as the possibility of keeping the main nutrients concentration at low values avoiding the nutrients accumulation in fed-batch [79], controlling the feeding rates in fed-batch, increasing the final cell density and productivity [51], as well as controlling the feeding and time of infection in HEK293 perfusion cultures for adenoviral vectors production [80] and in baculovirus-insect cell cultures (Sf9) [81].

In this context of bioprocess optimization, an interesting application for OUR arise recently with the emerge of the multi-parallel bioreactor systems. The huge necessity from the industry for cell line screening and media optimization had emerge the use of small-scale platforms for high throughput screening purposes or for the analysis of the metabolism/physiology of hundreds of clones in a very short time. Nowadays, systems that are able to measure the oxygen consumption in extremely small volumes (from 2 μL on) are commercially available [82]. However, these systems were often not scalable in terms of big bioreactors production. Interestingly, bench scale parallel bioreactors systems have considerably attracted the interest of the industry for its scalability and its capacity to test multiple conditions in very small-scale volumes, with the possibility to have up to 48 controlled bioreactors running at the same time. The complicated part of using these high throughput parallel systems is the difficulty to adapt the feeding strategies, either in fed-batch or "simulated" perfusions, when different cell lines with different nutrients requirements are running at the same time. Therefore, OUR becomes very interesting as it is an easy measurement that can be performed in these systems, adapting the feeding strategy to each bioreactor demands.

Recently, [83] demonstrated the applicability of the OUR measurement to control the feeding of glucose at different bioreactor's scales with CHO cells (micro-scale at 15 mL, laboratory-scale at 7 L and pilot-scale at 50 L). Comparing this method with just daily punctual nutrient additions, the new method allows maintaining lower set points of glucose concentration during the fed-batch phase while avoiding the glucose depletion in the culture broth. Interestingly, these multi-parallel bioreactors have been shown to be a good system for developing scale-down models for fed-batch [84], and even perfusion models, using for example SAM (sedimentation in an automated bioreactor) strategy presented by [85].

8 OUR Measurement as Example for Bioprocess Automatization: Time of Action (TOA)

OUR is an excellent indicator of the metabolic activity of cells, and it is highly sensitive to physiologic or/and metabolic changes along culture progression. As explained in the previous section, OUR measurements correlate linearly with cell density during the exponential growth phase. As most cell culture bioprocesses are run in fed-batch or perfusion, OUR becomes a very useful tool for estimating the nutrient feeding rate with the aim of keeping the main carbon and nitrogen sources at low concentration. This correlation can be observed meanwhile cells grow with no nutrient limitation nor inhibitory concentration of by-products is reached (mainly lactate and ammonia).

A deviation of the linear correlation between OUR and cell density indicates a change in metabolism of cell activity, what can be used for the automation of bioprocesses beyond the command of fed-batch or perfusion nutrient addition (marked with a green arrow in Fig. 2. Such metabolic changes are reflected rapidly in the OUR profile, before any change in cell growth rate or in main metabolites concentration (i.e. glucose, glutamine, lactate or ammonia) is observable. The plot of OUR time profile (using log y-axis scale) draws the end of the linear behaviour upon reaching the end of the exponential growth phase faster than cell density plot [86]. So, OUR is the easiest measurable parameter that better anticipates the change from the exponential growing phase to the deceleration phase during a cell culture progression as shown in Fig. 2 (blue linear regression). The detection of the end of the exponential growth phase, is named as Time of Action (TOA), being a key cell culture parameter because if no action is taken, the culture will enter in the stationary and eventually the death phase.

OUR is also a good parameter for monitoring more complex bioprocesses in which two biocatalysts are involved: host cells and viruses. Virus can be used for different purposes like vectors for heterologous protein expression, production of virus themselves for prophylactic applications, or also recombinant virus or VLP manufacturing for gene therapy. In all these examples, the viral infection should be done in fully metabolic active cells in order to guarantee the availability of biosynthesis precursor metabolites and thus the proper production of the desired product. TOA can be used as Time of Infection (TOF) of the culture, by simply adding some fresh media at TOF. Such methodology allows to extend cell activity for a time window comparable to the viral cycle, ensuring the efficient virus infection and replication. A scheme of a bioprocess automation based on OUR measurements in Sf9 insect cell cultures infected with recombinant baculovirus is depicted in Fig. 3.

Once the cells are infected, the evolution of the OUR profile is again worth monitoring since it offers useful information of the viral infection progress. However, there is not a single behaviour of the OUR profile. In some cases, depending on the virus/host cell used, OUR increases at the first stages of the viral infection [87]. The increase of OUR could be related to an increase of energy demand, triggered

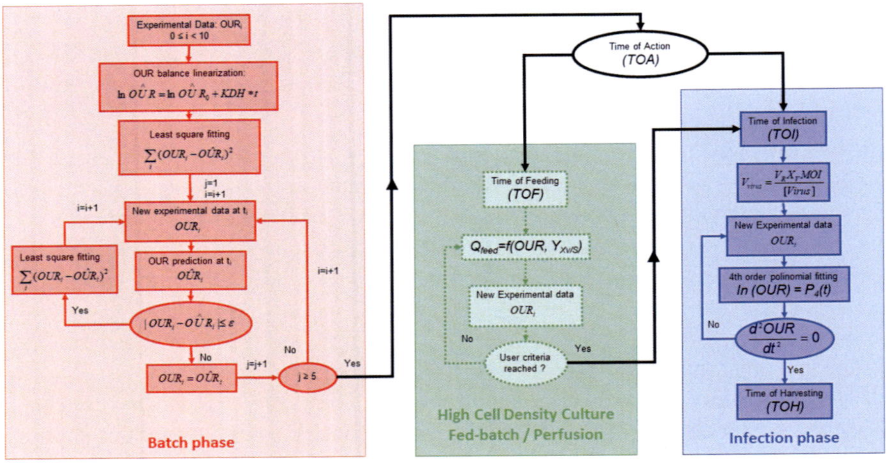

Fig. 3 Flow diagram for bioprocess automation. In red the batch phase where OUR measurements are used for the detection of the end of the exponential growth phase (TOA). TOA corresponds at the end of the linearized exponential growth phase, and is detected by the deviation of the measured OUR values and the linear regression of the previous OUR measurements. Then, TOA can be used for starting the nutrients feeding for the development of High Cell Density Cultures (both Fed batch or perfusion cultures) depicted in green. In HCDC, OUR is used for biomass and nutrient feeding estimation by means of the specific oxygen consumption rate. Eventually, TOA can be used for cell culture baculovirus infection for VP2 protein production (Blue section). In the infection phase both virus amount (given a specific MOI) and the detection of the TOH are determined by means of OUR measurements

by viral DNA replication, protein synthesis and budded virus release. Wong et al. [88] observed that the increase of oxygen consumption coincides with high rates of DNA synthesis during viral replication. In insect cells cultures infected with baculovirus, the evolution of OUR profiles was correlated with the Multiplicity of Infection (MOI) [81]. When a low MOI value is used (i.e. less than 0.1 PFU/cell) cell density increases clearly even after the infection due to the fact that at this low MOI value, the viruses only infect at the maximum less than 10% of cells at the beginning of the infection. Then, non-infected cells remain dividing during the initial stages of infection, what is reflected in the increase of the OUR value. Afterwards, viruses replicate and spread around the culture and eventually cell growth is completely arrested. It is unclear whether cell-cell communication also has an effect on arresting cell growth upon infection. On the contrary, if MOI is higher than 1 the infection process is very effective. The entire cell culture is infected from the very beginning of the infection phase. Cell machinery of infected cells is used for the virus replication and no longer for cell growth, what is reflected in the OUR profile. OUR value is quite stable during the infection and only decreases when cell lysis occurs, due to virions and protein release into the culture broth. Therefore, once the biological system (host cells and virus) and the bioprocess conditions are defined and fixed, OUR measurements are useful for the monitoring of the infection progress. The evolution of the OUR time profile correlates with the efficiency of the culture infection.

Another important parameter in bioprocesses for recombinant heterologous protein production based on virus/host cells is the Time of Harvesting. In order to ease the downstream of the heterologous protein, it could be advantageous to harvest the protein intracellularly. Using this strategy, the downstream volume is reduced in comparison to the total bioreactor volume, and also protein is preserved from degradation. Thus, culture harvesting should be done before cell lysis occurs. Further analysing the OUR profile along the entire infection process, as viral infection progresses the OUR time profile decreases softly. After approximately the time needed for the virus replication cycle, the OUR decays sharply coinciding with cell lysis. If the transition from the initial infection phases and cell lysis can be anticipated, it would be possible to harvest intact cells with the intracellular protein. Regarding the OUR profile, such transition is characterized by a change in the slope of the decay profile [81]. A methodology proposed to detect the TOH is based on fitting the profile of the natural logarithm of OUR values to a fourth-grade polynomial function. First and second derivatives of the fourth-grade polynomial function in respect to time can be deduced analytically from their coefficients. Finally, the maximum peak of intracellular protein concentration matches with the minimum for the first derivative, or what is even easier to estimate with a zero root for the second derivative. It can be deduced from this match that the maximum of protein concentration accumulated into the cells corresponds to the second inflexion point of the ln(OUR) profile (declining phase of the culture). Interestingly, TOH detection can be performed in an anticipated manner with good accuracy due to the high parameter sensitivity and the large number of OUR data generated during the culture. Both features, together with the low cost, make this methodology a potential candidate for bioprocess automatization, since it displays the wishful requirements of a suitable monitoring system.

9 Concluding Remarks

The manufacturing industry of biopharmaceuticals is evolving quite fast, pushed by the large number of new biological molecules that have been approved during the last years. The market of biopharmaceuticals in terms of both, the amount of sales and number of products is expected to keep growing for the futures years. In order to front the expected increased demand of production capacity of biopharmaceuticals, the industry has put lots of interest to increase bioprocesses volumetric productivity, by means of intensifying the bioprocesses, and to ensure and increase the product quality. The most promising mode of bioreactor's operation, also known as culture strategies, to overcome both challenges is the continuous bioprocessing mode, what in terms of bioprocessing corresponds to perfusions cultures. Perfusion cultures allow bioprocess intensification while controlling cell culture parameters and the environmental conditions. Moreover, the continuous product removal increases the chance to avoid product degradation. The successful implementation of high cell density perfusion bioprocesses producing high quality products is strongly linked to the availability of reliable real-time monitoring tools.

The challenge of the monitoring system in the future comes with the necessity of ensuring the robustness of the process as well as the quality of the product. This can only be achieved through the entire automatization process, and this has to be implemented in almost all the steps of the biopharmaceutical manufacturing. Especially important will be monitoring of the cell physiology during the cell culture, and this will lead to increasingly keep improving the capabilities of the sensors used for the measurement of the culture parameters, as well as to replace methods with certain delay between sampling and analysis to real-time in-line measurements that allow to obtain the information required even faster.

Disclosure of Interests All authors declare they have no conflict of interest.

References

1. Evens RP (2016) Pharma success in product development—does biotechnology change the paradigm in product development and attrition. AAPS J 18:281–285
2. Walsh G (2018) Biopharmaceutical benchmarks 2018. Nat Biotechnol 36:1136–11455
3. Estes S, Melville M (2014) Mammalian cell line developments in speed and efficiency. Adv Biochem Eng Biotechnol 139:11–33
4. Walsh G (2014) Biopharmaceutical benchmarks 2014. Nat Biotechnol 32:992–1000
5. Langer ES, Rader RA, Gillespie DE (2018) 15th Annual report and survey of biopharmaceutical manufacturing capacity and production: a study of biotherapeutic developers and contract manufacturing organizations. Rockville
6. Zijlstra G, Gupta P (2017) Moving toward continuous processing. Gen Eng News
7. Jennewein MF, Alter G (2017) The Immunoregulatory roles of antibody glycosylation. Trends Immunol 38:358–372
8. Zhu J (2012) Mammalian cell protein expression for biopharmaceutical production. Biotechnol Adv 30:1158–1170. https://doi.org/10.1016/j.biotechadv.2011.08.022
9. Butler M, Meneses-Acosta A (2012) Recent advances in technology supporting biopharmaceutical production from mammalian cells. Appl Microbiol Biotechnol 96:885–894
10. Grabenhorst E, Schlenke P, Pohl S et al (1999) Genetic engineering of recombinant glycoproteins and the glycosylation pathway in mammalian host cells. In: Glycotechnology. Springer, Boston, pp 1–17
11. Amann T, Hansen AH, Kol S et al (2019) Glyco-engineered CHO cell lines producing alpha-1-antitrypsin and C1 esterase inhibitor with fully humanized N-glycosylation profiles. Metab Eng 52:143–152. https://doi.org/10.1016/j.ymben.2018.11.014
12. Butler M (2005) Animal cell cultures: recent achievements and perspectives in the production of biopharmaceuticals. Appl Microbiol Biotechnol 68:283–291
13. Durocher Y, Butler M (2009) Expression systems for therapeutic glycoprotein production. Curr Opin Biotechnol 20:700–707
14. Xu S, Gavin J, Jiang R, Chen H (2017) Bioreactor productivity and media cost comparison for different intensified cell culture processes. Biotechnol Prog 33:867–878
15. Challener (2016) A Look At Perfusion: The Upstream Continuous Process – BioProcess InternationalBioProcess International. Bioprocess Int.

16. Clincke MF, Mölleryd C, Zhang Y et al (2013) Very high density of CHO cells in perfusion by ATF or TFF in WAVE bioreactor™: part I: effect of the cell density on the process. Biotechnol Prog 29:754–767
17. Wolf M, Bielser J-M, Morbidelli M (2020) Perfusion cell culture processes for biopharmaceuticals. Cambridge University Press, New York
18. Haigney S (2018) A look into the future of biopharmaceutical quality | BioPharm international. BioPharm Int 31
19. ICH_Q8 (2017) ICH guideline Q8 (R2) on pharmaceutical development
20. Román R, Farràs M, Camps M et al (2018) Effect of continuous feeding of CO2 and pH in cell concentration and product titers in hIFNγ producing HEK293 cells: induced metabolic shift for concomitant consumption of glucose and lactate. J Biotechnol 287:68–73. https://doi.org/10.1016/j.jbiotec.2018.10.005
21. Linthwaite VL, Janus JM, Brown AP et al (2018) The identification of carbon dioxide mediated protein post-translational modifications. Nat Commun:9. https://doi.org/10.1038/s41467-018-05475-z
22. Rao G, Moreira A, Brorson K (2009) Disposable bioprocessing: the future has arrived. Biotechnol Bioeng 102:348–356
23. Zhao L, Fu H-Y, Zhou W, Hu W-S (2015) Advances in process monitoring tools for cell culture bioprocesses. Eng Life Sci 15:459–468. https://doi.org/10.1002/elsc.201500006
24. Jiang M, Severson KA, Love JC et al (2017) Opportunities and challenges of real-time release testing in biopharmaceutical manufacturing. Biotechnol Bioeng 114:2445–2456. https://doi.org/10.1002/bit.26383
25. Román R, Miret J, Scalia F et al (2016) Enhancing heterologous protein expression and secretion in HEK293 cells by means of combination of CMV promoter and IFNα2 signal peptide. J Biotechnol 239:57–60. https://doi.org/10.1016/j.jbiotec.2016.10.005
26. Lee JS, Kallehauge TB, Pedersen LE, Kildegaard HF (2015) Site-specific integration in CHO cells mediated by CRISPR/Cas9 and homology-directed DNA repair pathway. Sci Rep 5:1–11
27. Sergeeva D, Lee GM, Nielsen LK, Grav LM (2020) Multicopy targeted integration for accelerated development of high-producing Chinese hamster ovary cells. ACS Synth Biol 9:2546–2561
28. Liste-Calleja L, Lecina M, Cairó JJ (2014) HEK293 cell culture media study towards bioprocess optimization: animal derived component free and animal derived component containing platforms. J Biosci Bioeng 117:471–477. https://doi.org/10.1016/j.jbiosc.2013.09.014
29. Grossvenor S (2012) A new era in cell culture media development. BioPharm Int 25:20–23
30. Ultee M E (2018) The top 5 trends to watch in bioprocessing. In: Bioprocess online. https://www.bioprocessonline.com/doc/the-top-trends-to-watch-in-bioprocessing-0001
31. Chu L, Robinson DK (2001) Industrial choices for protein production by large-scale cell culture. Curr Opin Biotechnol 12:180–187
32. van der Pol L, Tramper J (1998) Shear sensitivity of animal cells from a culture-medium perspective. Trends Biotechnol 16:323–328
33. Kadic E, Heindel TJ (2014) An introduction to bioreactor hydrodynamics and gas-liquid mass transfer. Wiley, Hoboken
34. Hu W-S, Zhou W (2012) Cell culture bioprocess engineering, 2nd edn. CRC Press, Taylor and Francis Group
35. Martínez-Monge I, Comas P, Triquell J et al (2019) Concomitant consumption of glucose and lactate: a novel batch production process for CHO cells. Biochem Eng J 151:107358
36. Martínez-Monge I, Albiol J, Lecina M et al (2019) Metabolic flux balance analysis during lactate and glucose concomitant consumption in HEK293 cell cultures. Biotechnol Bioeng 116:388–404
37. Yang J-D, Lu C, Stasny B et al (2007) Fed-batch bioreactor process scale-up from 3-L to 2,500-L scale for monoclonal antibody production from cell culture. Biotechnol Bioeng 98:141–154. https://doi.org/10.1002/bit.21413
38. Lecina M, Tintó A, Gálvez J et al (2011) Continuous perfusion culture of encapsulated hybridoma cells. J Chem Technol Biotechnol 86:1555–1564. https://doi.org/10.1002/jctb.2680

39. Bielser JM, Wolf M, Souquet J et al (2018) Perfusion mammalian cell culture for recombinant protein manufacturing – a critical review. Biotechnol Adv 36:1328–1340
40. Lin H, Leighty RW, Godfrey S, Wang SB (2017) Principles and approach to developing mammalian cell culture media for high cell density perfusion process leveraging established fed-batch media. Biotechnol Prog 33:891–901. https://doi.org/10.1002/btpr.2472
41. Kelley B (2009) Industrialization of mAb production technology: the bioprocessing industry at a crossroads. MAbs 1:443–452
42. Gagnon M, Hiller G, Luan Y-T et al (2011) High-end pH-controlled delivery of glucose effectively suppresses lactate accumulation in CHO fed-batch cultures. Biotechnol Bioeng 108:1328–1337
43. Pereira S, Kildegaard HF, Andersen MR (2018) Impact of CHO metabolism on cell growth and protein production: an overview of toxic and inhibiting metabolites and nutrients. Biotechnol J 13:1700499
44. Mulukutla BC, Kale J, Kalomeris T et al (2017) Identification and control of novel growth inhibitors in fed-batch cultures of Chinese hamster ovary cells. Biotechnol Bioeng 114:1779–1790
45. Advances in Biochemical Engineering/Biotechnology. https://www.springer.com/series/10. Accessed 20 Jan 2021
46. Godawat R, Konstantinov K, Rohani M, Warikoo V (2015) End-to-end integrated fully continuous production of recombinant monoclonal antibodies. J Biotechnol 213:13–19. https://doi.org/10.1016/j.jbiotec.2015.06.393
47. Woodgate JM (2018) Perfusion N-1 culture-opportunities for process intensification. In: Biopharmaceutical processing: development, design, and implementation of manufacturing processes. Elsevier, Amsterdam, pp 755–768
48. Bielser J, Kraus L, Burgos-Morales O et al (2020) Reduction of medium consumption in perfusion mammalian cell cultures using a perfusion rate equivalent concentrated nutrient feed. Biotechnol Prog 36. https://doi.org/10.1002/btpr.3026
49. Rantanen J, Khinast J (2015) The future of pharmaceutical manufacturing sciences. J Pharm Sci 104:3612–3638. https://doi.org/10.1002/jps.24594
50. Zhan C, Schwarz H, Lundqvist M, et al (2017) Development and application of screening scale bioreactor systems for very high cell density perfusion of mammalian cells. Integr Contin Biomanufacturing III
51. Casablancas A, Gámez X, Lecina M et al (2013) Comparison of control strategies for fed-batch culture of hybridoma cells based on on-line monitoring of oxygen uptake rate, optical cell density and glucose concentration. J Chem Technol Biotechnol 88:1680–1689. https://doi.org/10.1002/jctb.4019
52. Arnold SA, Crowley J, Woods N et al (2003) In-situ near infrared spectroscopy to monitor key analytes in mammalian cell cultivation. Biotechnol Bioeng 84:13–19
53. Sandor M, Rüdinger F, Bienert R et al (2013) Comparative study of non-invasive monitoring via infrared spectroscopy for mammalian cell cultivations. J Biotechnol 168:636–645
54. Abu-Absi NR, Kenty BM, Cuellar ME et al (2011) Real time monitoring of multiple parameters in mammalian cell culture bioreactors using an in-line Raman spectroscopy probe. Biotechnol Bioeng 108:1215–1221
55. Favre E, Pugeaud P, Péringer P (1990) Automated HPLC monitoring of glucose, glutamine, lactate and alanine on suspended mammalian cell reactors. Biotechnol Tech 4:315–320. https://doi.org/10.1007/BF00157428
56. Campmajó C, Cairó JJ, Sanfeliu A et al (1994) Determination of ammonium and L-Glutamine in hybridoma cell cultures by sequential flow injection analysis. Cytotechnology 14:177–182
57. Höpfner T, Bluma A, Rudolph G et al (2010) A review of non-invasive optical-based image analysis systems for continuous bioprocess monitoring. Bioprocess Biosyst Eng 33:247–256
58. Roychoudhury P, O'Kennedy R, McNeil B, Harvey LM (2007) Multiplexing fibre optic near infrared (NIR) spectroscopy as an emerging technology to monitor industrial bioprocesses. Anal Chim Acta 590:110–117. https://doi.org/10.1016/j.aca.2007.03.011

59. Ohadi K, Aghamohseni H, Legge RL, Budman HM (2014) Fluorescence-based soft sensor for at situ monitoring of chinese hamster ovary cell cultures. Biotechnol Bioeng 111:1577–1586
60. Carvell JP, Dowd JE (2006) On-line measurements and control of viable cell density in cell culture manufacturing processes using radio-frequency impedance. Cytotechnology 50:35–48. https://doi.org/10.1007/s10616-005-3974-x
61. Metze S, Ruhl S, Greller G et al (2020) Monitoring online biomass with a capacitance sensor during scale-up of industrially relevant CHO cell culture fed-batch processes in single-use bioreactors. Bioprocess Biosyst Eng 43:193–205. https://doi.org/10.1007/s00449-019-02216-4
62. Kilburn DG, Fitzpatrick P, Blake-Coleman BC et al (1989) On-line monitoring of cell mass in mammalian cell cultures by acoustic densitometry. Biotechnol Bioeng 33:1379–1384. https://doi.org/10.1002/bit.260331103
63. Behrendt U, Koch S, Gooch DD et al (1994) Mass spectrometry: a tool for on-line monitoring of animal cell cultures. Cytotechnology 14:157–165
64. Martínez-Monge I, Comas P, Triquell J et al (2018) A new strategy for fed-batch process control of HEK293 cell cultures based on alkali buffer addition monitoring: comparison with O.U.R. dynamic method. Appl Microbiol Biotechnol 102:10469–10483
65. Niklas J, Melnyk A, Yuan Y, Heinzle E (2011) Selective permeabilization for the high-throughput measurement of compartmented enzyme activities in mammalian cells. Anal Biochem 416:218–227
66. Dremel BAA, Li S-Y, Schmid RD (1992) On-line determination of glucose and lactate concentrations in animal cell culture based on fibre optic detection of oxygen in flow-injection analysis. Biosens Bioelectron 7:133–139. https://doi.org/10.1016/0956-5663(92)90018-I
67. Zavala-Ortiz DA, Ebel B, Li M et al (2020) Interest of locally weighted regression to overcome nonlinear effects during in situ NIR monitoring of CHO cell culture parameters and antibody glycosylation. Biotechnol Prog 36:e2924
68. Ashton L, Xu Y, Brewster VL et al (2013) The challenge of applying Raman spectroscopy to monitor recombinant antibody production. Analyst 138:6977–6985
69. Whelan J, Craven S, Glennon B (2012) In situ Raman spectroscopy for simultaneous monitoring of multiple process parameters in mammalian cell culture bioreactors. Biotechnol Prog 28:1355–1362
70. Bhatia H, Mehdizadeh H, Drapeau D, Yoon S (2018) In-line monitoring of amino acids in mammalian cell cultures using raman spectroscopy and multivariate chemometrics models. Eng Life Sci 18:55–61
71. Tulsyan A, Schorner G, Khodabandehlou H et al (2019) A machine-learning approach to calibrate generic Raman models for real-time monitoring of cell culture processes. Biotechnol Bioeng 116:2575–2586
72. Tulsyan A, Wang T, Schorner G et al (2020) Automatic real-time calibration, assessment, and maintenance of generic Raman models for online monitoring of cell culture processes. Biotechnol Bioeng 117:406–416. https://doi.org/10.1002/bit.27205
73. Ruffieux P-A, Von Stockar U, Marison IW (1998) Measurement of volumetric (OUR) and determination of specific (q O 2) oxygen uptake rates in animal cell cultures. J Biotechnol 63:85–95
74. Kussow CM, Zhou W, Gryte DM, Hu W-S (1995) Monitoring of mammalian cell growth and virus production process using on-line oxygen uptake rate measurement. Enzym Microb Technol 17:779–783. https://doi.org/10.1016/0141-0229(94)00035-P
75. Fontova A, Lecina M, López-Repullo J et al (2018) A simplified implementation of the stationary liquid mass balance method for on-line OUR monitoring in animal cell cultures. J Chem Technol Biotechnol 93:1757–1766. https://doi.org/10.1002/jctb.5551
76. Lin J, Takagi M, Qu Y, Yoshida T (2002) Possible strategy for on-line monitoring and control of hybridoma cell culture. In: Biochemical engineering journal. Elsevier, New York, pp 205–209
77. Martínez-Monge I, Roman R, Comas P et al (2019) New developments in online OUR monitoring and its application to animal cell cultures. Appl Microbiol Biotechnol 103:6903–6917

78. Ducommun P, Ruffieux PA, Furter MP et al (2000) A new method for on-line measurement of the volumetric oxygen uptake rate in membrane aerated animal cell cultures. J Biotechnol 78:139–147
79. Wlaschin KF, Hu WS (2007) Engineering cell metabolism for high-density cell culture via manipulation of sugar transport. J Biotechnol 131:168–176
80. Gálvez J, Lecina M, Solà C et al (2012) Optimization of HEK-293S cell cultures for the production of adenoviral vectors in bioreactors using on-line OUR measurements. J Biotechnol 157:214–222
81. Lecina M, Soley A, Gràcia J et al (2006) Application of on-line OUR measurements to detect actions points to improve baculovirus-insect cell cultures in bioreactors. J Biotechnol 125:385–394. https://doi.org/10.1016/j.jbiotec.2006.03.014
82. de Moura MB, Van Houten B (2014) Bioenergetic analysis of intact mammalian cells using the seahorse XF24 extracellular flux analyzer and a luciferase ATP assay. Methods Mol Biol 1105:589–602
83. Goldrick S, Lee K, Spencer C et al (2018) On-line control of glucose concentration in high-yielding mammalian cell cultures enabled through oxygen transfer rate measurements. Biotechnol J 13:1700607
84. Janakiraman V, Kwiatkowski C, Kshirsagar R et al (2015) Application of high-throughput mini-bioreactor system for systematic scale-down modeling, process characterization, and control strategy development. Biotechnol Prog 31:1623–1632
85. Kreye S, Stahn R, Nawrath K et al (2019) A novel scale-down mimic of perfusion cell culture using sedimentation in an automated microbioreactor (SAM). Biotechnol Prog 35:e2832
86. Zhou W, Hu W-S (1994) On-line characterization of a hybridoma cell culture process. Biotechnol Bioeng 44:170–177. https://doi.org/10.1002/bit.260440205
87. Reuveny S, Kim YJ, Kemp CW, Shiloach J (1993) Effect of temperature and oxygen on cell growth and recombinant protein production in insect cell cultures. Appl Microbiol Biotechnol 38:619–623. https://doi.org/10.1007/BF00182800
88. Wong TKK, Nielsen LK, Greenfield PF, Reid S (1994) Relationship between oxygen uptake rate and time of infection of Sf9 insect cells infected with a recombinant baculovirus. Cytotechnology 15:157–167. https://doi.org/10.1007/BF00762390

Industrial Monitoring of Cell Culture

Sebastian Schwamb, Markus Engel, Tobias Werner, and Philipp Wiedemann

Abbreviations

AAV	adeno-associated-virus
CER	carbon dioxide emission rate
CIP	cleaning in place
CoG	cost of goods
CPP	critical process parameter
CQA	critical quality attribute
CSPR	cell specific perfusion rate
DO	dissolved oxygen
EDTA	Ethylendiamintetraacetate
FDA	Food and Drug Administration
FIA	flow injection analysis
GC	gas chromatography
GMP	good manufacturing practice
GOD	glucose oxidase

V. Chotteau (✉) · M. E. L. Mäkinen
Department of Industrial Biotechnology, School of Engineering Sciences in Chemistry, Biotechnology and Health, KTH-Royal Institute of Technology, Stockholm, Sweden

AdBIOPRO, VINNOVA Competence Centre for Advanced Bioproduction by Continuous Processing, KTH- Royal Institute of Technology, Stockholm, Sweden

Digital Futures – KTH-Royal Institute of Technology, Stockholm, Sweden
e-mail: chotteau@kth.se; meerim@kth.se

E. Hagrot · L. Zhang
Department of Industrial Biotechnology, School of Engineering Sciences in Chemistry, Biotechnology and Health, KTH-Royal Institute of Technology, Stockholm, Sweden

AdBIOPRO, VINNOVA Competence Centre for Advanced Bioproduction by Continuous Processing, KTH- Royal Institute of Technology, Stockholm, Sweden
e-mail: hagrot@kth.se; lianz@kth.se

R. Pörtner (ed.), *Cell Culture Engineering and Technology*, Cell Engineering 10,
https://doi.org/10.1007/978-3-030-79871-0_17

HCA	hierarchical cluster analysis
HEK	human embryonic kidney
HPLC	high performance liquid chromatography
HSE	health and safety executive
IEC	ion exchange chromatography
KPP	key process parameter
LC	liquid chromatography
LED	light emitting diode
MAB	monoclonal antibody
MIR	mid-infrared
MS	mass spectrometry
NIR	near-infrared
ORP	oxidation-reduction-potential
OUR	oxygen uptake rate
PAT	process analytical technology
PCA	principle component analysis
PerC6	a human continuous cell line
PI	propidium iodide
PLS	partial least square
PQ	product quality
QbD	quality by design
RQ	respiratory quotient
SEC	size exclusion chromatography
SF9/Sf11	*Spodoptera frugiperda* cells
SIP	sterilization in place
TCD	total cell density
TTM	time to market
UPLC	ultra-performance liquid chromatography
VCD	viable cell density

1 Introduction

Every cell cultivation is accompanied by suitable monitoring of the culture. In its most basic form, this could be a rough assessment of cell concentration over cultivation time by simple microscopy of culture vessels, i.e. assessing the degree of confluency (when working with adherent cells) or applying counting chambers (for adherent and suspension cells). At the other extreme, monitoring could comprise cell counts as well as a number of metabolic and physicochemical parameters – and all these could be monitored not only off-line but also, at best, in-line and in real-time.

Independent of the extent of monitoring and of the cell cultivation environment itself – academia or industry, laboratory, process development units, large scale production sites – the key reasons for cell cultivation and process monitoring are always the same: Allowing robust and repeatable cultivation processes and ensuring reproducible product quality; whereby the product can be the cells themselves or a protein which is expressed for functional studies or subsequent purification.

This volume primarily covers the industrial application of cell cultivation processes, i.e. the production of biopharmaceuticals. In monitoring such cultures, the emphasis was initially on cell concentration, viability and recombinant product titer as well as basic physicochemical parameters like pH and DO [1]. Quickly, physiological parameters like specific uptake (substrates) or production (by-products and products) rates, as well as metabolites, began to come into focus [2]. Automated methods for off-line metabolite monitoring and automated cell counters developed the field further. Nowadays, monitoring technology development efforts are directed particularly towards methods that provide data in-line and in real-time, e. g. infrared (IR-) and Raman-spectroscopy [3].

What are the key drivers for such developments, i.e. new strategies and technologies in cell culture monitoring?

Firstly, there is a regulatory dimension: The "Process Analytical Technology" (PAT) initiative of the FDA, the US Food and Drug Administration (e. g. [1, 4]) demands process understanding and control based on monitoring raw materials and quality attributes, at best in real-time, necessitating a paradigm shift in process monitoring and sensor technology [5]. Additionally, efficient monitoring is fundamental in the context of Quality by Design (QbD; [6–10]). In brief, in QbD the task is to identify the desired product quality and to design processes capable of maintaining this quality. In order to do that, the critical process parameters (CPP) which influence the critical quality attributes (CQAs) of the product must be found. The QbD approach relies on several pillars like the use of qualified scale-down models, application of risk assessment methodologies and sound experimental planning based on the design of experiments – and importantly, the analytical capabilities to monitor process parameters and quality attributes satisfactorily in a validated environment.

Lastly, automation and digitalization of biomanufacturing as envisioned by Industry 4.0 require thorough process understanding, data acquisition and subsequent analysis and modelling (see e. g. [11]) – the basis of all this being thorough process monitoring.

If we combine the above-mentioned concepts and intentions with the very down-to-earth demands of the biopharmaceutical industry, together with the growing competition of biosimilars, the main drivers for monitoring-technology development are obvious. They are product quality (PQ), costs and speed. In other words:

Only if one thoroughly understands and monitors the process, can one improve process control, thereby reducing process variability, raising product quality and ultimately reducing failed production batches and therefore costs.

Secondly, improved process monitoring, results in better process understanding and robustness which in turn enables more efficient process development – leading to shorter development timelines, lower costs and ultimately shorter time to market (TTM).

Thirdly, in-line monitoring is, in particular, a potent means to gain process understanding, raise data in real-time and is key to "building quality into the product rather than testing quality into the product". Furthermore, if one understands the interconnections between CPPs and CQAs and monitors those CPPs (and CQAs) in or near real-time by in-line monitoring, real-time release of production batches could be possible (this, as well, is one idea of PAT) which would, in turn, reduce costs.

Despite the strong focus of this contribution on monitoring of the cell cultivation itself, one should keep in mind that a prerequisite of appropriate process monitoring and control is sound knowledge of potential process-disrupting factors such as critical raw materials. Even the most advanced process control cannot lead to the desired PQ if – to name but one example – raw materials are contaminated with trace elements like copper or manganese [12].

Furthermore, monitoring of process parameters is sometimes done without the intention of controlling them. Sampling from cell culture vessels for microbial sterility testing would be an example of such a situation. Nevertheless, in the majority of cases, monitoring of a process parameter is connected to controlling the respective parameter. These aspects are elucidated further in chapter 16 of this volume (Lecina et al. Monitoring tools for the development of high cell density culture strategies).

2 Principles of Monitoring

2.1 *Scope of Monitoring*

In the context of biopharmaceutical cell cultivation processes, monitoring comprises CPPs (see above), key process parameters (KPPs) which are defined as non-CPPs that, while not affecting product quality, still may affect process performance such as yield, and non-KPPs which are those parameters that have no effect on process performance or product quality [13] (Castillo et al. 2016). The goal of monitoring is to ensure a robust process performance and the desired PQ as defined by CQAs.

For cell culture ("upstream") processes, to date, monitored process input parameters (CPPs/KPPs) are typically physiological ones such as VCD, viability and metabolites as well as physicochemical ones such as osmotic pressure, pH, temperature or dissolved gas concentrations e. g., Oxygen (usually given as dissolved oxygen, DO, i.e., as a percentage with reference to pO_2) and dissolved Carbon Dioxide (typically given as partial gas pressure, pCO_2).

Relevant process outputs (CQAs) are defined by an appropriate risk assessment and are mainly determined by the recombinant protein itself. Typically, they comprise physical, chemical and biological properties of the protein, alongside with, for example, the absence of any sort of contamination.

2.2 Monitoring Categories

General monitoring procedures can be divided into two categories: Offline monitoring and real-time – or at least, near real-time – monitoring.

The first category, offline monitoring, always requires sampling. When sampling, the sterile barrier of the culture vessel is breached and so here, great care has to be taken. Since microbial or viral contamination is a major risk of all cell culture operations, suitable sampling devices that maintain sterility need to be in place. These range from properly operated laminar flow hoods if a culture vessel has to be physically opened over steam sterilised sampling containers connected to a bioreactor to disposable sterile sampling devices. Furthermore, the sample has to be correctly treated in order not to obscure the original conditions inside the culture vessel.

Such samples, usually taken for the analysis of the cells themselves or the cell-free supernatant, are then handed over to laboratories close to or outside the production area. There, often extensive sample preparation is necessary which does not facilitate real-time process control/feedback loop control.

In the second category, real-time or near real-time monitoring, several different setups are possible:

Ideally, process parameters can be monitored in-line (or "*in situ*") without sampling. An advantage of this method is that no sterile barrier has to be breached to obtain monitoring data. Also, this method has the advantage of speed – no time is lost through sample extraction, sample preparation and external analysis. Physicochemical parameters like pH, partial gas pressure and temperature are often measured in such a setting, providing a constant stream of measurement data and thereby the basis of control of the parameters. Also conductivity measurement and the more advanced techniques like Dielectric, Infrared (IR), and Raman spectroscopy fall in this category of real-time monitoring. With regard to PAT, in-line spectroscopic methods have become the main route of development as they allow multivariate data analysis to build calibration models for multiple CPPs/KPPs. In-line monitoring requires capable sensors, which means they have to be operational inside a cultivation vessel (e. g. dye-based cell monitoring methods and devices usually are not so) and they must also withstand cleaning and sterilisation-in-place operations (CIP/SIP) of the vessel [1].

For parameters that cannot be measured in-line, one option is online monitoring. This is characterised by either continuous removal of the process liquid, guiding it past the sensor and feeding the unaltered liquid back into the cultivation vessel (i.e., bypass configuration) or by automated removal and preparation of the sample (e. g. dilution, addition of dyes), subsequent analysis and discarding of the sample. A prominent example of the latter is FIA (flow injection analysis, e. g. [14]).

Alternatively, in at-line monitoring, samples are removed (e. g. by an operator) but the analyser is located in close proximity to the process vessel. In at-line monitoring, usually little or no sample preparation is necessary. This enables fast turnaround times – i.e. near real-time monitoring. Cell counting, external

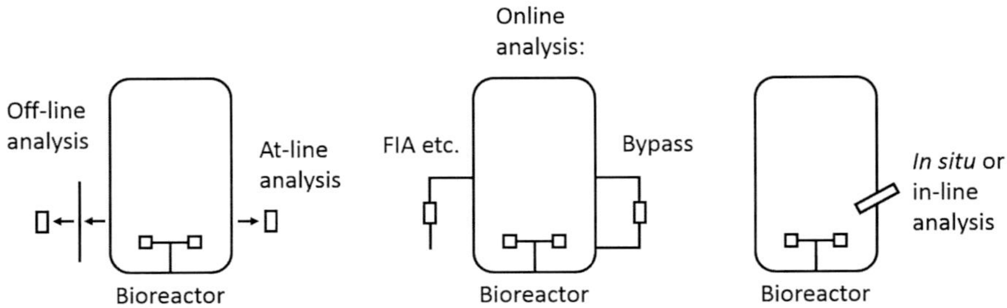

Fig. 1 Monitoring principles. *FIA* flow injection analysis. (Taken from [1])

pH measurements and at-line glucose measurements, for example, fall into this category. As in offline monitoring, care has to be taken to not obscure the samples. For example, if the monitoring parameter is pCO_2, pH or DO, care has to be taken to prevent outgassing of CO_2 or O_2 between taking the sample and analysing it. Figure 1 summarizes the different monitoring principles.

Whether installed inside the cultivation vessel or used outside of it, monitoring devices must fulfil certain criteria to be usable in a laboratory and even more so if they are to be used in the tightly regulated area of industrial cell culture technology. These would include adequate sensitivity and specificity, (mechanical) robustness and ease of use, suitability for cleaning and sterilising, and ease of validation.

3 Intention and Scope of this Contribution

In the previous issue of this series, we concentrated on monitoring of cell and metabolite-related parameters [1]. In the present contribution, we carry on from there and focus on solutions and trends that are fit for industrial practice, development departments and laboratories.

Additionally, emphasis is placed on the growing progress made in real-time monitoring of key upstream input and output parameters. Despite the significant progress that has been made over the past five years, there is still a substantial lack of real-time or near real-time methods to measure a broad range of CQAs, except very basic physiological ones such as product titer of the desired molecule.

We would like to stress that in the field of industrial cell cultivation requirements are substantially different from those in scientific cell culture laboratories:

In large scale manufacturing, companies will only use monitoring devices that operate in an extremely robust fashion, and can therefore be easily operated by personnel on the production floor and also have the necessary mechanical rigidity. Such systems must be fit for use in a Good Manufacturing Practice (GMP) – regulated environment and its functions easy to validate within the scope of a production process. These demands should not be underestimated. Regulatory validation requirements are constantly growing and even without these, the implementation of new monitoring systems involves monetary investment, time and risks. In the end, only those new technologies will be implemented that are worth the challenges – that means they must offer a proven and substantial (monitoring) advantage over the technologies used so far. Which technology is fit for use in a GMP environment, which is not and which has the potential to get there? For example, although technologically more advanced and although the value of monitoring these parameters is undisputed, it took years for digital probes (pO_2, pH) to prove their robustness and be accepted in large scale manufacturing.

Also, technologies that are useful during process development might not be necessary in routine production. For example, a pCO_2 probe can offer a distinct added value in process scale-up but once the process is fixed might not be necessary in large-scale production i.e. the added value of monitoring pCO_2 does not outweigh the cost and risk to operate the system in a validated process under GMP.

In process development and scale-up or R&D in general, emphasis is on gaining a deep understanding of the cell culture and process. Here, many different methods and parameters might be assessed for process and product characterisation, as well as those that aid process scale-up. As an example, NIR or Raman-spectroscopy can provide very relevant data and are increasingly employed in the field. Unfortunately, extensive chemometric operations are necessary to interpret original measurement results. This is time-consuming and costly. If platform processes are developed that are theoretically similar to each other, this effort could be reduced since a chemometric model that is developed for one process could be extrapolated to another. This would alleviate the situation. However, if you work with such platform processes, they are well understood from the outset, so that sophisticated monitoring systems are not necessary in the first place.

Concluding from the above, it comes as no real surprise that, although new technologies are constantly being developed and evaluated on a scientific level and also used in process development (and some emerge that have the necessary "regulated environment potential"), particularly large scale operations to date operate with the same set of proven monitoring parameters and robust technologies as they did for quite some time in the past.

A list of typical monitoring parameters with emphasis on those being employed in the biopharmaceutical industry on or close to the production floor can be found in Table 1. It is non-exclusive and certainly partly subjective. In Sect. 4 of this contribution, we will briefly highlight a subset of those monitoring parameters.

We have omitted many monitoring techniques – even though they might provide very valuable process information and have many published reports of successful investigational use – if they seem not yet advanced enough to us to fit in the scope of this article. Examples are UV/VIS-spectroscopy, MIR-spectroscopy, fluorescence-spectroscopy, and – from our own work – at-line mass spectrometry [15] and *in situ* microscopy [16]. One exception is in-line glucose monitoring (once again our own work, see Sect. 5).

Yet another aspect of high importance within the concepts of Industry 4.0, PAT and QbD which we do not have space here to elaborate on, is data science

Table 1 Typical process parameters for animal cell cultures. Additional parameters might be of value for different organisms or applications. Underlined are those parameters that are most often taken into account in large scale manufacturing, and are to date, still sufficient for such operations

Parameter	Typical monitoring device	Typical monitoring method
Physicochemical parameters		
Temperature	Resistance thermometer (PT 100 probe)	In-line
pH value	Electrochemical Ingold pH glass electrode/optical sensor	In-line
Oxygen partial pressure (pO_2)	Amperometric Clark electrode/optical sensor	In-line
Pressure	Membrane sensor	In-line
Stirrer speed	Revolution counter	In-line (on stirrer motor)
Foam Level	Conductivity probe	In-line
Redox potential	ORP probe	In-line
Gas Flow	Mass flow meters	In-line (gas lines)
Carbon dioxide partial pressure (pCO_2)	pCO_2 probe (Severinghaus probe/optical sensor)	In-line
Osmolality	NIR/Raman, Freezing point depression	In-line, online, off-line
Physiological/biological parameters		
Cell number and Viability; Biomass	Dye-based (mostly trypan blue, PI) and dye-free (Capacitance, NIR, Raman) methods; Light scattering/absorbance probes	Off-line, online or in-line
Substrates and metabolites	Enzymatic, spectroscopic (Raman, NIR) and other methods	Off-line, online or in-line
Lactate dehydrogenase	Enzymatic/photometric methods	Off-line or online
Product concentration	HPLC, photometric, spectroscopic and other methods	Off-line, online or in-line
Product quality	Various methods, e. g. SEC, IEC, LC-MS, HPLC, Raman*, light scattering*	Off-line (first attempts of in-line aggregate detection*)
Apoptosis	Flow cytometry/Mass spectrometry	Off-line
Sterility	Cultivation assays	Off-line

Altered, from [1]

(including soft sensors and machine learning). The best monitoring device is useless without the integration of monitoring data into a capable data management and analysis platform. To this end, several companies have developed such platforms, e. g. BioPAT MFCS (Sartorius Stedim Biotech, Göttingen, Germany) or Lucullus (Securecell AG, Urdorf, Switzerland). Ideally, these systems are meant to design, execute and analyse bioprocesses on one integrative software platform. As an additional step, a combination of sensor signals, data analysis and modelling can be performed for attribute estimation – often referred to as "soft sensors" (e. g. [11, 17, 18]). These will certainly be important in the future even though not extensively being applied to platform development or in GMP production momentarily.

4 Monitoring Parameter and Technologies

It is important to keep in mind that the choice of monitoring methods, to a large extent, depends on the environment these methods will be used in: The less regulated the environment, the broader the set of potential methods.

During (early and late stage) process development, the interconnection between process inputs (CPP/KPP) and process outputs (CQA) is typically going to be developed. According to the QbD paradigm, which calls for smarter experimental design, there is a substantial demand for real-time or near real-time monitoring methods to analyse not only key input but also key output parameters.

During scale-up and process validation, the potential range of monitoring methods is already limited to those of critical importance to the process, i.e. those to run the process within acceptable operational ranges in order to meet predefined acceptance criteria.

Lastly, in a fully regulated, commercial manufacturing environment, any attempt to introduce a new process-analytical technology is stringently linked to the ability to demonstrate a strong advantage of this technology with respect to process reproducibility while, at the same time, avoiding any additional process risks or impacting the costs of the process (CoGs; costs of goods).

In the following, we will give details on a subset of the monitoring techniques shown in Table 1. As stated earlier, this contribution places emphasis on the growing progress in real-time or near-real-time monitoring of key upstream input and – where available – output parameters. Therefore, we will not go into detail on off-line methods, even if now widely employed, as, for example, mass spectrometry for product analysis (e. g. [19]). Furthermore, we will not elaborate on off-line or in-line methods that have been proven for decades (e. g. trypan blue cell counting, pH and pO_2 monitoring via electrochemical probes) ~~or~~ but deal with them only briefly.

For each technology described, a – partly subjective – "experience matrix" is given that outlines its application in process development (non-GMP) or routine monitoring (GMP).

4.1 Physiological/Biological Parameters

4.1.1 Viable Cell Density, Total Cell Density and Cell Viability

Two of the most important KPPs in mammalian cell culture are viable cell density (VCD) and viability, i.e. the percentage of living cells with respect to the total cell density (TCD). All stages of cell cultivation, from establishing a simple sub-cultivation routine in basic cell culture to a defined process control strategy in commercial production, are based on an accurate determination of cell density (e. g. [2]).

For cell density determination, the gold standard, not only for commercial GMP manufacturing, but also in most development cell culture applications, is still at-line analysis employing live/dead staining (trypan blue, propidium iodide PI). Nevertheless, there are industry trends that are leading to a (partial) paradigm shift away from at-line monitoring approaches, e. g.:

- Increased application of perfusion processes (e. g. intensified n-1 or perfusion production bioreactor) and thus an increasing demand for continuous biomass/cell density determination to control e. g. the cell-specific perfusion rate (CSPR).
- Increased application of anchorage-dependent cells and cells tending to form cell clusters, e. g. HEK cells for AAV production, leading to an increased demand for alternative cell density determination approaches without the need for sample preparation.
- Control of fed-batch processes: Adjustment of feeding rates based on continuous cell density measurement helps to determine optimal feed rates and therefore reduces CoGs.

Ideally, cell density would be determined in-line to avoid contamination from breaching the sterile barrier when sampling [20] and to enable a constant stream of real-time data. Techniques that have reached the threshold to industrial application (at least in development and partly in large-scale) are measurements of optical density or capacitance as well as (mostly NIR and Raman) spectroscopy. A challenge when applying these "dye-free" methods is often the discrimination between live and dead cells (e. g. [21], and sometimes combinations of two probes are used to determine both, e. g. the Incyte (permittivity/capacitance – VCD) and Dencytee (optical density – TCD) probes, both from Hamilton (Bonaduz, Switzerland). It is worthwhile mentioning that the benchmark for determining live and dead cells by any alternative method is just about always its validation against trypan blue staining, simply because trypan blue is the gold standard. This method does not take into account staining artefacts, or cells that are en route to death but still possess a functional membrane. Neither does it discriminate between different modes of death (apoptosis, necrosis, autophagy). It could be useful to get a more detailed picture of the state of the cells, particularly in development, like when assessing cultivation hardware (e. g. pumps for perfusion) or feeding strategies. At-line MS

[15], offline methods like Flow Cytometry or advanced data analysis (e. g. of dielectric properties, [21]) are exploratory possibilities here.

Cell density	Monitoring Method	Application in Development	Application in GMP
Capacitance	in-line	yes	yes
Raman	in-line	yes (e. g. Sanofi[2])	presumably yes; no reference available
NIR	in-line	yes	yes
Cell counter (Trypan blue) integration	FIA (online)	yes	no application references available

Cell Density: Capacitance Measurement

For commercial bioprocesses, in particular, capacitance probes are available from Aber Instruments (Aberystwyth, UK) and Hamilton (Bonaduz, Switzerland). Both manufacturers provide stainless steel reusable probes (CIP/SIPable) as well as single-use patches embedded in single-use bag systems, e. g. for Hyclone STR (Thermo Scientific HyClone, South Logan, USA), Sartorius RM and STR (Sartorius Stedim Biotech, Göttingen, Germany) and Mobius STR (Millipore/Merck, Darmstadt, Germany).

Probes enable in-line monitoring of VCD and therefore feedback control loops for VCD-based feeding, bleeding and osmolality control. The technology is comparatively robust and allows for successful qualification in an industrial environment. Also, in the case of anchorage-dependent cells and cells tending to form cell clusters, capacitance measurement might be a valuable alternative to trypan blue counting and time-intensive and cell-disrupting sample preparation. The technology has found considerable acceptance in industry [22].

Capacitance measurement (radio-frequency impedance; dielectric spectroscopy) exploits the fact that cells with an intact cellular membrane act as small capacitors when subjected to a periodically alternating electrical field [23]. Ions in the medium and the cytoplasm of cells move in this field, their mobility being constrained by the cellular membrane. This results in a charge separation across the membrane at the poles of the cell, also termed polarization [24]. The magnitude of that polarization is measured by the resulting capacitance in picoFarad (pF). The polarizability of the cell suspension corresponds to the permittivity (pF per cm) of the cell suspension, i.e. at higher cell concentrations, more cells contribute to the polarization, leading to a higher permittivity [25]. Frequencies from 0.1 to 10 MHz can be applied to the cells, resulting in additional information e. g., compensating measurements for changing diameter of the cells over the time of a culture or measurement of apoptosis [26]. Cell death is accompanied characterized by a loss of membrane integrity, which results in a free diffusive exchange of intra and extracellular compounds. Therefore, this method does not allow for the determination of the total cell density.

Cell Density: Raman Spectroscopy

Raman spectroscopy is probably the most advanced PAT in-line monitoring method [27–30]. It enables the multivariate analysis of a broad range of process parameters like VCD and TCD; also – see Metabolites: Raman Spectroscopy section – glucose, glutamine, lactate, ammonium [31, 32], and even antibody concentration [33]. The disadvantage of this method is that for spectra evaluation, advanced mathematical and statistical chemometric methods are required, such as Principle Component Analysis, PCA; Hierarchical Cluster Analysis, HCA; or Partial Least Square, PLS [32, 34, 35]. Furthermore, the technology is cost-intensive due to the employment of class 3B lasers [34] and potentially critical from an HSE regulation point of view.

Raman probes are available as in-line probes (standard port size, various lengths, CIP/SIP-able; predominantly Kaiser Optical Systems Inc., Ann Arbor, USA) as well as online systems, e. g., via the BioPAT Spectro (Sartorius Stedim Biotech), applicable to single-use systems such as Ambr15 and Ambr250 or Biostat STR (all: Sartorius Stedim Biotech). Kaiser Optical Systems offers additionally, disposable adaptors that enable a reusable Kaiser probe to be inserted into a SUB and collect in situ measurements without making product contact.

Compared to near and mid-infrared (NIR and MIR) spectroscopy (see below) which are based on detecting absorption of electromagnetic radiation, Raman spectroscopy is based on light scattering [36]. Raman spectroscopy relies on the inelastic scattering of monochromatic light (usually from an adjustable laser source in the spectral range between near-infrared and near-ultraviolet light) due to the interaction of photons with the sample [37]. Spectra that can be detected after irradiation and light scattering consist mostly of unaltered frequencies from the excitation light (Rayleigh scattering), and of shifted frequencies (Raman scattering). The frequency shift between ~~of~~ excitation light and scattered light is molecule-specific. As such, Raman spectroscopy can provide information regarding the composition of the analysed sample. Nevertheless, challenges of the method include the strong fluorescence of many biological molecules that may overlay Raman scattering signals and the generally weak Raman scattering of cell culture broths [38].

Cell Density: Near-Infrared (NIR) Spectroscopy

Hamilton (Bonaduz, Switzerland) offers probes enabling in-line determination of both viable cell density and total cell density (Dencytee and Incyte Arc). While the viable cell density is analysed via capacitance (Incyte; see above), the total density or turbidity is measured via near-infrared (NIR) spectroscopy (Dencytee). Sartorius' BioPAT Fundalux probe operates on a transmittance principle, using IR light to track total biomass, albeit predominantly for microbial cultivations (Sartorius Stedim Biotech, Göttingen, Germany). Other than for determination of cell density, NIR is seen as a potent PAT tool for at-line raw material testing (e. g. [39–41]; online glucose monitoring via NIR has also been reported [42]. Moreover, there are several reports on using NIR for bioprocess control [43–46] by determining metabolite concentration (e. g. glucose or lactate) and less often cell density during mammalian cell cultivation processes. Regulatory agencies like the FDA have even published guidelines on the implementation of NIR in the pharmaceutical industry [47].

NIR spectroscopy determines absorption in the near-infrared region (wavenumbers 13,000–4,000 cm^{-1}) of the electromagnetic spectrum. NIR relevant functional groups of molecules in the sample are predominantly those containing a hydrogen atom (like O-H, C-H or N-H). It is actually the chemical bond whose "natural" vibration is moved from one energy level to another, which causes so-called overtones and combination-bands in the NIR spectra [48].

An advantage of the NIR technique is lesser interference with water than in the case of MIR. However, both show higher water absorbance than Raman spectroscopy. A disadvantage is that spectral peaks are broad and overlapping, making their interpretation difficult. Therefore, complex chemometric methods – as in Raman spectroscopy – must be applied for data interpretation.

Cell Density: Integration of Automated Cell Counters

The principle of Trypan blue cell staining for live and dead cell detection, dates as far back as 1914 [49]. To overcome the disadvantages of manual hemocytometer-based cell counting – the method is subjective, prone to human error and time consuming (e. g. [50]) – automated cell counters based on trypan blue staining have been developed and, in some instances, connected to bioreactors in an online/FIA-like setting.

Those machines, such as e. g. Vi-CELL (Beckman Coulter, Brea, USA) or CedexHiRes (Roche, Basel, Switzerland) are image-based cell analysers using a fully automated liquid handling systems with an integrated flow-through ("pseudo hemocytometer") chamber and a high-resolution image scanner. Image analysis algorithms (cell line dependent adaptation is required) enable the system to classify detected objects into debris, viable and dead cells, aggregates and other undefined objects. Besides the main parameters (viable-cell density, total-cell density and viability), other parameters such as object-diameter can be examined. The cell diameter, for example, is a known indicator for culture healthiness or ongoing apoptosis [51–54].

Very often, these analysers are operated at-line. This does not enable real-time monitoring but near real-time, since an assay does not take more than a few minutes. This is often appropriate since cell culture processes are comparatively slow, provided the sampling intervals suit the monitoring needs. An alternative would be connecting the analyser to the cell cultivation platform and performing automated sampling/cell counting. Examples for such combinations, which are used in the biopharmaceutical industry, include:

- Integration of Beckman's ViCell and Roche's CedexHiRes into the Ambr250 – an automated stirred tank reactor platform of up to 24 parallel reactors at working volumes of 100 – 250 mL (Sartorius Stedim Biotech GmbH, Göttingen, Germany)
- Combination of Nova Chemicals (Waltham, MA, USA) BioProfile Flex 2 and Ambr250 or Ambr15 (Sartorius, up to 48 parallel cultivations at the 10–15 mL microbioreactor scale)
- General option for automated sampling through 12, 19 and 25 mm standard ports: Seg-Flow (Flownamics/Sentinel Process Systems, Hatboro, USA) for Cedex HiRes (Roche, Basel, Switzerland) und ViCell XR (Beckman Coulter, Brea, USA)

General disadvantages of this approach are: it is cost-intensive (initial investment, maintenance), it can be mechanically challenging (autosampling), it might have limited flexibility (length of connective tubing), and only a limited number of bioreactors can be connected to one analyser.

The advantages are: the trypan blue assay itself is cost-effective, its use in automated counting avoids variability inherent in manual sample preparation and counting, it is the long-term "gold standard" of cell counting and the assay is widely accepted by regulatory agencies.

It is worthwhile mentioning that also in the field of automated cell counters there is a constant drive for new developments. For instance, [55]; Selexis SA, Switzerland assess the Guava easyCyte HT (Luminex, Austin, USA) and the Cytoflex (Beckman), both equipped with 96-well-plate loaders, as a replacement for trypan blue counters. Both machines are "originally" flow cytometers, the Guava already conveniently equipped for direct absolute counts.

4.1.2 Metabolic Parameters and Product Titer

Since the early days of biopharmaceutical development, one key objective in process development was and still is the optimization of cell-specific productivities [56] with the overall aim to reduce the costs of goods (CoGs) while maintaining the desired product quality.

Optimizing productivity can be achieved, for example, by balancing nutrient resources in the culture, preventing a decline in specific productivity or preventing variation of product quality (e. g. glycosylation).

Therefore, the monitoring and control of key metabolic parameters such as substrates and metabolites are of utmost importance for developing and operating robust and reproducible processes. In process development, this helps to determine the needs and boundaries of the processes, while later in production, it is an indicator for culture reproducibility.

In contrast to CPPs like pH, DO or temperature, the major drawback in the case of metabolic parameters is the lack of robust and reliable in-line probes, which results in sampling from the bioreactor and subsequent at-line or off-line analysis. Fortunately, the new, mainly spectroscopic, techniques for both online and in-line measurement enable measuring metabolic parameters, too.

It should be noted here that a few companies developed alternatives to off-line measurements in the past e. g., Trace Analytics (Braunschweig, Germany) offers online glucose and lactate monitoring via standard immobilized enzyme technology connected to the bioreactor by single-use tubing systems and dialysis or filtration probes. C-CIT Sensors AG, a start-up company from Wädenswil, Switzerland, developed *in situ* enzymatic glucose and lactate sensors.

In the following section, we give a condensed overview of emerging trends and applications successfully overcoming the "historical burden" of limiting the monitoring of metabolic parameters to at-line and off-line analysis. The availability of adequate in-line and online monitoring capabilities, together with the availability of closed feedback loop control, opens up new possibilities to operate processes with the aim of avoiding accumulation of waste products or maintaining desired product quality attributes. For example, [57] report the successful use of Raman spectroscopy for online glucose and lactate monitoring to control glucose feeding. Thereby, they significantly lower the lactate level of cultures, leading to improved culture viability and increased harvest titer. In contrast, [29] report applying Raman spectroscopy to operate a fed-batch process at different low glucose levels to reduce the percentage of glycation.

Metabolites/titer	Monitoring Method	Application in Development	Application in GMP
Raman	in-line or online	yes	yes
Metabolite Analyser integration	FIA (online)	yes	no application references available

Metabolites: Raman Spectroscopy

Ever since the first reports of Raman spectroscopy in combination with multivariate regression techniques for multi-parameter-monitoring in the context of mammalian cell culture technology (e. g. [32]), the number of applications of this technology has been constantly rising. Its dominant role in the analysis of aqueous solutions such as cell culture media, compared to other spectroscopic techniques (e. g. NIR or MIR), can mainly be explained by the weak Raman scattering of water [58].

To develop a reliable model using spectroscopic techniques, spectra must be correlated with at-line measurement-values using, for example, Partial Least Square (PLS) regression. This means that, theoretically, every component of the cell culture broth can be monitored, bearing in mind the technical limit of detection of analytes.

It is a great advantage of Raman spectroscopy that multiple parameters can be monitored with only one system.

Of course, this technology also has its disadvantages and limitations, as already mentioned in 4.1.1.2. Please also refer to 4.1.1.2 for a brief overview of available probes and technical solutions.

In this subchapter, we examine monitoring via Raman spectroscopy of key substrates such as glucose or glutamine [31, 57], metabolites such as lactate and ammonia [36] and product titer [33, 59, 60]. Another recently reported application of Raman spectroscopy is the in-line monitoring of amino acids. E. g., Bhatia et al. [61] report the in-line monitoring of key amino acids, known to be essential for cell growth and productivity [62]. Typically, amino acid profiling (also known as spent media analysis), is performed as an off-line analysis, using liquid chromatography (e. g. HPLC or GC-MS), with ready- to- use kits, e. g., EZfaast (Phenomenex, Torrance, USA). Therefore, these analyses can support process optimization or troubleshooting activities, but are not suitable for routine process control purposes.

It is worth mentioning that nowadays Raman technology is applied in development units and likewise in GMP manufacturing facilities. Also, increased attention is being given to the scalability of spectroscopic models, e. g., for substrate, metabolite and product concentrations, developed using small-scale bioreactor models of manufacturing scale reactors, as recently reported by Janssen Biologics [63], or earlier by Biogen Idec [28].

Metabolites: Integration of Metabolite Analysers

Commonly used metabolite analysers, such as CedexBioHT (Roche, Basel, Switzerland), BioProfile Flex (Nova Biomedical, Waltham, USA) or YSI7100 (Yellow Spring Instruments, Yellow Springs, USA) are based on automated enzymatic (enzymatic-amperometric/enzymatic-photometric) measurement principles.

Like automated cell counters (see 4.1.1.4), metabolite analysers are mainly operated at-line, especially in "traditionally" designed development laboratories, where multiple benchtop bioreactors are spread over and operated across several hundred square meters of laboratory space. The paradigm shift towards highly automated small-scale multi-bioreactor systems (e. g. the Ambr platform, Roche, Basel, Switzerland), operating up to 48 bioreactors on a few square meters of laboratory space, combined with the availability of smart interfaces connecting analysers and bioreactors via automated online sampling systems, has successfully closed the gap between process and analyser [64].

Similar approaches can also be applied to standalone bioreactors (benchtop or technical scale), although in this case, the economic burden will be higher.

Independent of the scale of bioreactor systems connected via an autosampler to a metabolite analyser, the advantage of these solutions is the availability of data for both, monitoring and control purposes.

Examples, successfully demonstrating those combinations, applied in the biopharmaceutical industry include:

- Automated bioreactor sampling via Numera (Securecell AG, Urdorf, Switzerland) for glutamine and glutamate analysis using Roche's CedexBioHT [65].
- Connecting Roche's CedexBioHT with an Eppendorf (Hamburg, Germany) standalone 2 L benchtop bioreactor and the corresponding Eppendorf Process control software (BioComand Batch control software) via SegFlow (Flownamics, Madison, USA) for automated glucose control [66].

As the number of bioreactors that can be connected to each autosampler is limited, depending on the actual number of bioreactor systems and the layout of the laboratory, this technology requires a high initial investment.

4.1.3 Product Quality Attributes: Protein Aggregates

To date, product quality is, predominantly assessed off-line, often employing methods like HP-SEC, MS, HPLC and others. A significant gap still exists in the development of rapid in-line measurement techniques [3]; only the first concepts of real-time release-testing exist up to now. Particularly in the area of upstream processes, real-time or near real-time product-quality-attribute analysis in crude process bulk or harvest broth is still a very explorative field. To give just two examples: [67] developed an automated at-line assay to monitor monoclonal antibody (MAB) charge variants near real-time during production and [68] showed near real-time Glycosylation determination via fluorescence UPLC. Unfortunately, none of these developments seems mature enough yet to be applied in routine analysis in a development or ~~a~~ regulated production environment.

Attempts to quantify molecule aggregates seem to be more promising in this respect. Such protein aggregates are a major concern since they not only lower the effective dosing concentration of a drug but can result in adverse toxicological and immunological responses that can be life-threatening [69]. It is interesting to see that in the monitoring of biopharmaceutical protein aggregates, once again, Raman spectroscopy plays a considerable role: The method has not only been used to monitor protein degradation/aggregation in an off-line configuration (e. g. [70–72], the latter pointing towards real-time release); but a report from Bristol-Myers Squibb shows the application of in-line Raman spectroscopy for monitoring aggregation of therapeutic proteins as a function of formulation and process variables [73].

Although the state of in-line – or at least (near) real-time – aggregate monitoring is still in its infancy, it should be mentioned here that Raman spectroscopy is not the only method under investigation: In a collaboration with Biogen Idec, [74] measured light scattering simultaneously from many independent MAB solutions in order to monitor their time-dependent aggregation behaviour.

The overall challenge now is to incorporate any such technology in a robust sensor design, allowing for long-time operation in a production environment, as well as ease of handling, the ability to recalibrate in-process and incorporation in automated process control strategies.

Product aggregates	Monitoring Method	Application in Development	Application in GMP
Raman	in-line	no application references available	no application reference available
Light Scattering	in-line	no application references available	No application reference available

4.2 *Physicochemical Parameters*

4.2.1 Osmotic Pressure/Osmolality

Osmolality measures solute concentration and is defined as the number of osmoles per kilogram of a solution [18]. It is the sum of all dissolved components (salts, nutrients, waste products etc.) in the medium. For mammalian cell culture processes, the osmolality is usually in a range from 260 to 450 mOsm/kg, depending on the cell line (e. g. [75]). Freezing point depression is often used to measure osmolality [76]. Increases in osmolality can be used to increase specific productivity (e. g. [77]) but can also inhibit cell growth (e. g. [78, 79]). Therefore, controlling osmolality via the osmotic strength of the medium itself or using cheap additives like NaCl (e. g. [80]) could be of high importance for fed-batch and perfusion processes – requiring, at best, in-line monitoring instead of off-line freezing point depression measurements.

Osmolality	Monitoring Method	Application in Development	Application in GMP
Raman	In-line	yes	No reference available
NIR	In-line	yes	No reference available
Metabolite Analyser integration	FIA (online)	yes	No references available

In this respect and for the scope of this contribution, three methods are of particular interest for determining osmolality, since they use the same equipment as that used for in-line cell density (see 4.1.1.) and metabolite (see 4.1.2) monitoring.

Firstly, a collaboration of Biogen Idec and East Carolina University reported osmolality can be determined by Raman spectroscopy [28, 81]. Secondly, in a report from Hoffmann-La Roche, NIR was used to determine, among others, medium osmolality [82]. The same was shown in 2009 by [83]. Thirdly, metabolite analysers, connected to the bioreactor in an online fashion can be used to determine osmolality, as, for example, the BioProfile FLEX2 (Nova Biomedical, Waltham, MA, USA) or the CedexBio (Roche, Basel, Switzerland). For a determination with Raman and NIR-spectroscopy, osmolality needs to be included in the chemometric models that have to be developed when using these methods. Also, the CedexBio does not directly measure osmolality by freezing point determination but calculates it on the basis of Na^+, glucose and glutamine measurements since these three medium constituents contribute the most to osmolality. The influence of other compounds is included through empirically derived factors. The Nova Bioprofile FLEX2 uses freezing point depression.

4.2.2 pO_2, pCO_2, pH

Oxygen supply and an appropriate pH, as well as pCO_2 level, are key contributors to a successful bioreactor run. Traditionally, pO2 is measured with an amperometric ("Clark") electrode [84]. The pH is measured potentiometrically by a combination electrode; the general principle goes back to Arnold Beckman [85]. Standard pCO_2 probes function in a similar way to pO_2 probes and consist of a pH probe surrounded by bicarbonate buffer. CO_2 molecules diffuse from the bioreactor into the carbonate buffer (see also pCO_2: In-Line Probe section).

Even though CO_2 is of importance to stabilize the pH in the medium, dissolved CO_2 can be a harmful and undesirable by-product of cell culture processes [86]. High levels of dissolved CO_2 cause undesirable metabolic changes, inhibit growth and reduce productivity (e. g. [87]). In fact, CO_2 levels in bioreactors will fluctuate throughout the run; in the past, these were mostly monitored offline.

Additionally, in bioreactors, the pO_2 and the pH are controlled within a specific range, more or less automatically leading to a specific and hopefully appropriate pCO_2 concentration. However, especially in a large-scale environment, pCO_2 levels are often too high (as compared to small and technical-scale; and also from a cell-physiological point of view), leading to differences in the lactate profile, which in turn impacts the pH (see e. g. [88]).

Ideally, all these parameters – pCO_2, lactate/pH, pO_2 – would be monitored in-line. This is already well established for pO_2 and pH; for lactate measurements, several technical solutions exist as well (see 4.1.2); also, pCO_2 probes are available. However, both pCO_2 probes and pH probes have the disadvantage that they often drift and need to be re-calibrated.

As an alternative to the "traditional" probes, optical sensors for pO_2, pH, and pCO_2 have been developed (see pH, pO_2, pCO_2: Optical Sensor Spots section) and are – at least in the case of pO_2 – widely accepted in all scales, from micro to commercial/large-scale applications. Like pH probes, pH spots are reported to show drift-problems, as well.

Monitoring technology	Monitoring method	Application in development	Application in GMP
Probe (pCO_2)	in-line	yes	no reference available
Off-gas analyser (pO_2, pCO_2)	online	yes	no reference available
Optical sensor spots (pH, pO_2, pCO_2)	in-line	yes	yes

pH, pO_2, pCO_2: Optical Sensor Spots

Optical sensors for pH, pO_2, and, to a lesser degree, pCO_2 have become more and more important in many fields of biotechnology over the last 20 years. Commercially available sensors for pO_2 are based on lifetime detection of fluorescence quenching and sensors for pH and CO_2 are generally based on ratiometric detection using pH-sensitive fluorescent dyes [89]. A detailed description of the working principle is outside of the scope of this contribution and can easily be found in many excellent reviews [90, 91]. Here, we briefly discuss the application of these optical sensors in a cell culture environment. More specifically, we concentrate on the application of pH and pO_2 "sensor-spots" (see below) in single-use applications, since those are the two most often monitored parameters in small-scale applications. Please note, we do not elaborate here, on the widespread successful replacement of reusable amperometric pO_2 probes by their optical counterparts, as, for example, Hamilton's VisiFerm or Memosens COS81D (Endress+Hauser, Rheinach, Switzerland) – one of their major advantages being that optical oxygen probes do not require lengthy equilibration.

Already in the early 1990s, a so-called triple LED compatible sensor unit measuring pH, oxygen and carbon dioxide for the use in bioreactors was developed [92]. The sensing membranes that could be easily sterilized were placed in an external flow cell and the sample fluid was continuously passed through the sensors, in order to reduce the risk of contamination. Because almost all sensor parameters such as accuracy, resolution and reproducibility fulfilled the requirements of biotechnological applications, these results initiated the further rapid development of optical sensors for use in biotechnology.

Nowadays, it is more common to integrate planar optical "sensor-spots" into single-use reactors, microbioreactors or shake flasks. Optical sensors offer several unique advantages. Their small size, for example, allows the integration of several different sensors in a single microwell. One example is the micro-Matrix from Applikon Biotechnology (Delft, Netherlands) which operates with working volumes of 2–7 mL and yet offers monitoring and control of culture conditions. pH, DO and temperature profiles are controlled in 24 wells in parallel (e. g. [93]). Other microbioreactors such as the BioLector from m2p-labs GmbH (Baesweiler, Germany) are able to monitor pH and pO_2 online during fully-automated high-throughput cultivations of cell lines such as PerC6, Sf9 and Sf21 [94]. Sartorius Stedim Biotech equips its microbioreactor vessels Ambr15 with optical single-use pH and DO sensors; the Ambr250 comes with DO sensor spots but disposable pH electrodes.

Optical sensor technology and instrumentation to a large extent rely on PreSens GmbH (Regensburg, Germany), which develops an extensive range of sensors and signal readers for different cultivation formats. Its technology is widely applied e. g. in the Applikon MicroFlask, where a 24-channel reader for non-invasivedetection of oxygen and pH in deep-well and low-well plates is used. The PreSens SensorDish Reader can be used in incubators and on shakers and is thus also a good tool for monitoring pH or pO_2 for cultivations in combination with the MicroFlask system [95].

Other optical sensor systems are making use of incubator shaker systems for cell cultivation where measurements are performed in cultivation tubes or shake flasks of larger dimension equipped with pH and pO_2 sensor spots [96]. Several systems are in use for different purposes, for instance for process optimization and recombinant protein production (e. g. [97, 98], the latter showing results of a comparison of different non-invasive optical sensor technologies in shake flasks). Cultivation conditions and monitoring in shake flasks have been extensively described (e. g. [99]). Apart from an in-line configuration, bypass options are available; e. g. [100] describe a fully embeddable microarray system to measure pO_2 and pH non-invasively when coupled to a bioreactor.

As mentioned above, a CO_2 optical sensor spot is now available as well. As O_2 or pH spots, it is attached to the inner surface of any transparent vessel, while non-contact measurements are taken from the outside [101]. In another prototype application, this sensor spot is integrated into stainless steel bioreactor probes [102] – providing an alternative to the probes mentioned in pCO_2: In-Line Probe section.

Importantly, sensor-spot technology has also found its way into the field of wave bags and benchtop reactors. Pre-sterilized, single-use bioreactor bags equipped for non-invasive optical pH and DO cell cultivation monitoring are commercially available from different suppliers such as Cytiva (Little Chalfont, UK) or Satorius Stedim Biotech while some companies even offer optical sensing devices for the detection of dissolved oxygen (and, partly, pH) in benchtop reactors (e. g. Satorius Stedim Biotech – Univessel SU with optical pO_2 and pH sensors).

All together, optical sensors have proven to be important enablers particularly of micro and small-scale cell-cultivation applications and can be very advantageous in mid to large-scale reactors as well.

pO2, pCO2: Off-Gas Analysers

The monitoring of pCO_2 and pO_2 concentrations in the outlet gas stream allows – in combination with data from the inlet stream, e. g., the O_2 and CO_2 MFCs (mass flow controllers) – quantifying rates of gas utilization and production (oxygen uptake rate OUR, carbon dioxide emission rate (CER)) and the respiratory quotient (RQ). Additionally, off-gas analysers monitoring the partial pressure of carbon dioxide can be an alternative to in-line CO_2 probes in the development of cell culture processes [103].

There is a number of methods for off-gas analysis available for bioprocesses, among others electrochemical sensors, gas chromatography, infrared absorption and mass spectrometry [104]. Nevertheless, interesting for the scope of this chapter are those systems that are accepted for use in industry. A prominent example is the BlueInOne (BlueSens gas sensor GmbH, Herten, Germany), a combined in-line O_2/CO_2 analyser which measures pO_2 electro-galvanically or via a zirconium dioxide sensor and pCO_2 via dual wavelength IR. The system is also used by several bioreactor suppliers like Sartorius Stedim Biotech (Göttingen, Germany – "BioPAT Xgas"), Eppendorf (Hamburg, Germany), INFORS HT (Bottmingen, Switzerland) and Applikon Biotechnology (Delft, The Netherlands).

pCO_2: In-Line Probe

To date, only a few measurement principles of CO_2 in-line probes have been successfully commercialized and been applied to cell culture technology, namely:

- InPro5000 (Mettler Toledo AG, Urdorf, Switzerland)
- YSI 8500 (Yellow Spring Instruments, Yellow Springs, USA)

The measurement principles of the InPro5000 relies on the Severinghaus principle [105]. A CO_2-semipermeable membrane separates a bicarbonate buffer from the aqueous solution (e. g. culture medium). As a consequence of CO_2 diffusion across the membrane, an equilibrium will result. A glass electrode measures the resulting pH change of the bicarbonate buffer, which is converted into a pCO_2 signal. The disadvantage of this probe-setup is that the internal bicarbonate solution needs to be replaced regularly to ensure an adequate measurement accuracy.

In contrast to the traditional glass electrode design, the YSI 8500 in-line probe is based on an optical measurement. It also relies on the Severinghaus principle, but the difference is that CO_2 moves through a permeable polymer into a fluorescent dye [106]. CO_2 reversibly reacts with the dye which changes fluorescence conditions and subsequently light emission.

The availability of in-line pCO_2 signals allows, for example, the implementation of a pCO_2 control strategy using N_2 sparging [107]. It also allows the controlled addition of CO_2 in small-scale models to mimic technical or production-scale pCO_2 concentrations.

A downside of the available in-line pCO_2 probe technology that still exists is, the tendency of signals to drift, similar to pH in-situ probes and pH sensor spots. Therefore, appropriate at-line recalibration procedures are required to ensure reliable pCO2 measurement during long-lasting processes.

4.2.3 Redox Potential

The redox potential of a solution, also known as oxidation-reduction-potential (ORP), is the result of the equilibrium between oxidizing and reducing species. It plays an important role in understanding protein secretion [108], cell growth and productivity [109] as well as modulating glycosylation [110]. Additionally, these and subsequent studies [111] have shown that changes in the cell culture redox potential correlate with the disruption of disulphide-bonds in secreted molecules.

During cell culture processes, changes in the redox potential are caused by changes in the concentration of ions in the medium, pH, temperature and DO. Furthermore, the medium redox potential is changed by cellular oxidative stress which can be caused by bioreactor oxygen inhomogeneity, rich cell culture media, high productivity rates and waste product accumulation.

The loss of interchain disulphide-bonds during manufacturing processes is a phenomenon that has attracted increasing attention in the biopharmaceutical industry over the last decade [112, 113], as it leads to batch failure or requires additional adaptation of the downstream process. It is known that molecule reduction can happen either during the upstream bioreactor process, during the cell separation step or during the intermediate bulk storage time. For the latter two, several mitigation strategies have been established, such as chilling the bioreactor broth prior to cell separation, adjusting the pH, decreasing filter loads, controlling the DO during intermediate storage or adding enzyme-inhibitors such as EDTA or metal ions [112]. For molecule reduction that occurs during the bioreactor process, mitigation strategies are limited, as they must neither affect cell growth, productivity nor product quality.

Online monitoring of the ORP as an indicator of reduction-propensity was demonstrated by [111]. The study reports the development of an online bioreactor control system that prevents MAB reduction by controlling the cell culture redox potential.

Redox potential	Monitoring method	Application in development	Application in GMP
Probe	in-line	yes	no reference available

Existing in-line probes are available from Hamilton (e. g. EasyFerm® Plus) and Mettler Toledo (e. g. InPro 3253i), among others. Redox probes typically are pH/ORP probes, combined with a temperature sensor.

As a side note, it should be mentioned here that other than monitoring the propensity for a reductive reactor environment, the ORP signal can also help to monitor and solve erratic oxygen control situations: The ORP can be used as a signature or "fingerprint" for a cell culture in a certain scale or process. This signature can be used to scale-up the process or to track the batch-to-batch consistency.

5 Future Perspectives: Sensor Spots

As stated above, optical sensor spots have proven advantageous over other types of in-line sensors and even off-line measurements. This trend will be accelerated by new sensing concepts, integration of new materials and novel fabrication technologies. This will lead to smaller, faster, cheaper optical sensor systems possessing

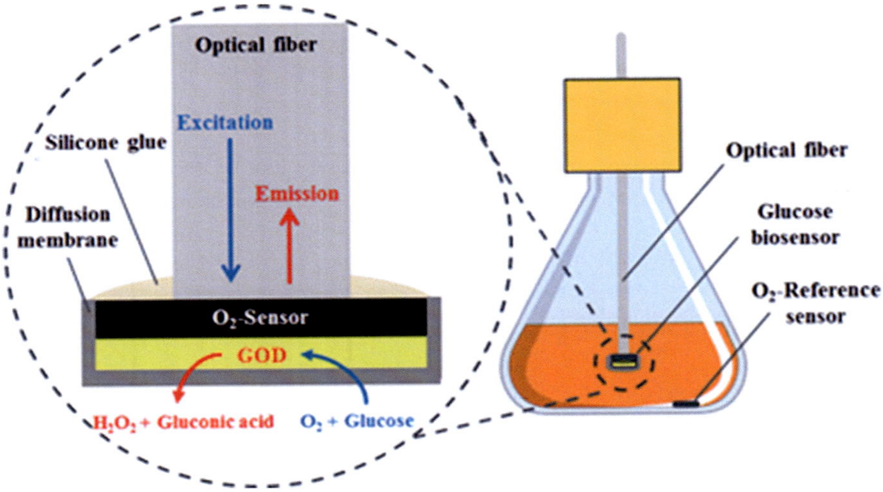

Fig. 2 Measuring principle of a glucose biosensor in a 125-mL shake flask. The glucose biosensor consists of an optical oxygen transducer that measures the oxygen consumption inside a glucose oxidase (GOD) enzyme layer due to the enzymatic conversion of glucose and oxygen, which enter the enzyme layer by diffusion through a diffusion membrane. An additional oxygen reference sensor at the bottom of the shake flask is used to compensate for oxygen fluctuations in the medium. The glucose response (ΔpO2) is attained by subtracting the oxygen partial pressure within the biosensor from the oxygen partial pressure in the medium. (Adapted from [97])

acceptable reliability for industrial application. Progress in this field will, among other things, facilitate the development of efficient application-specific optical (bio-) sensors for important metabolites such as glucose, lactate, glutamine and ammonium/ammonia. Such biosensors have already been introduced into different fields of biotechnology. As a PAT tool, they will provide valuable insights into cultivation processes and help optimizing culture conditions.

An important CPP is glucose. Not only is optimal glucose monitoring and feeding crucial for cell growth and prolonged viability, but it is also important in sensing and ensuring correct glycosylation (e. g. [60, 114]). Recently, a glucose biosensor (Fig. 2) was developed that is applicable for in-line and long-term glucose monitoring in mammalian cell bioreactors [97].

This sensor is based on an oxygen sensor that is coated with crosslinked glucose oxidase (GOD; see Fig. 2). It can be integrated with an automated glucose monitoring and feeding system capable of maintaining stable glucose levels, for example. An additional benefit is that the system is based on an easy-to-handle plug-and-play principle, which does not necessitate complex chemometric data evaluation as in Raman-based systems.

Acknowledgements The authors are indebted to Andrea McIntosh-Suhr and Tanja Bergfeld for critical comments on this manuscript.

Disclosure of interests All authors declare they have no conflict of interest.

References

1. Schwamb S, Puskeiler R, Wiedemann P (2015) Monitoring of cell culture. In: Al-Rubeai M (ed) Cell engineering: animal cell culture. Springer, Basel, pp 185–221
2. Konstantinov KB (1996) Monitoring and control of the physiological state of cell cultures. Biotechnol Bioeng 52:271–289
3. BioPhorum (2017) Biomanufacturing technology roadmap in-line monitoring and real-time release. https://www.biophorum.com/download/in-line-monitoring-and-real-time-release/. Accessed 9.12.2020
4. Food and Drug Administration (2004) Guidance for Industry: PAT – a framework for innovative pharmaceutical development, manufacturing, and quality assurance. U.S. Department of Health and Human Services, Food and Drug Administration. https://www.fda.gov/media/71012/download. Accessed 9.12.2020
5. Read EK, Shah RB, Riley BS et al (2009) Process analytical technology (PAT) for biopharmaceutical products: part II. Concepts and applications. Biotechnol Bioeng 105:285–295
6. International Conference on Harmonisation of Technical Requirements for Registration of Pharmaceuticals for Human Use (2005) ICH harmonised tripartite guideline: quality risk management (Q9). https://database.ich.org/sites/default/files/Q9_Guideline.pdf. Accessed 7.12.2020
7. International Conference on Harmonisation of Technical Requirements for Registration of Pharmaceuticals for Human Use (2008) ICH harmonised tripartite guideline: pharmaceutical quality system (Q10). https://database.ich.org/sites/default/files/Q10%20Guideline.pdf. Accessed 9.12.2020
8. International Conference on Harmonisation of Technical Requirements for Registration of Pharmaceuticals for Human Use (2009) ICH harmonised tripartite guideline: pharmaceutical development Q8(R2). https://database.ich.org/sites/default/files/Q8%28R2%29%20Guideline.pdf. Accessed 9.12.2020
9. Rathore AS (2009) Roadmap for implementation of quality by design (QbD) for biotechnology products. Trends Biotechnol 27:546–553
10. Zurdo J, Arnell A, Obrezanova O et al (2015) Early implementation of QbD in biopharmaceutical development: a practical example. Biomed Res Int 2015:605427
11. Narayanan H, Luna MF, von Stosch M et al (2020) Bioprocessing in the digital age: the role of process models. Biotechnol J 15(1):e1900172
12. Graham Ryan J, Bhatia H, Yoon S (2019) Consequences of trace metal variability and supplementation on Chinese hamster ovary (CHO) cell culture performance: a review of key mechanisms and considerations. Biotechnol Bioeng 116(12):3446–3456
13. Castillo FC, Cooney B, Levine HL (2016) Biopharmaceutical manufacturing process validation and quality risk management. Pharm Eng 36(3):82–92
14. Rhee JI, Ritzka A, Scheper T (2004) On-line monitoring and control of substrate concentrations in biological processes by flow injection analysis systems. Biotechnol Bioprocess Eng 9:156–165
15. Schwamb S, Munteanu B, Meyer B et al (2013) Monitoring CHO cell cultures: cell stress and early apoptosis assessment by mass spectrometry. J Biotechnol 168:452–461
16. Wiedemann P, Guez JS, Wiegemann HB et al (2011) In situ microscopic cytometry enables noninvasive viability assessment of animal cells by measuring entropy states. Biotechnol Bioeng 108(12):2884–2993

17. Mandenius CF, Gustavsson R (2015) Mini-review: soft sensors as means for PAT in the manufacture of bio-therapeutics. J Chem Technol Biotechnol 90:215–227
18. Pohlscheidt M, Charaniya S, Bork C et al (2013) Bioprocess and fermentation monitoring. In: Flickinger MC (ed) Upstream industrial biotechnology: equipment, process design, sensing, control, and cGMP operations, vol 2, 1st edn. Wiley, Hoboken, pp 1471–1491
19. Rogstad S, Faustino A, Ruth A et al (2017) A retrospective evaluation of the use of mass spectrometry in FDA biologics license applications. J Am Soc Mass Spectrom 28(5):786–794
20. Rehbock C, Beutel S, Bruckerhoff T et al (2008) Bioprozessanalytik. Chemie Ingenieur Technik 80(3):267–286
21. Braasch K, Nikolic-Jaric M, Cabel T et al (2013) The changing dielectric properties of CHO cells can be used to determine early apoptotic events in a bioprocess. Biotechnol Bioeng 110:2902–2914
22. Justice C, Brix A, Freimark D et al (2011) Process control in cell culture technology using dielectric spectroscopy. Biotechnol Adv 29:391–401
23. Carvell JP, Dowd JE (2006) On-line measurements and control of viable cell density in cell culture manufacturing processes using radio-frequency impedance. Cytotechnology 50:35–48
24. Arnold WM (2001) Monitoring of biological cell collection by dielectric spectroscopy. In: 2001 annual report conference on electrical insulation and dielectric phenomena. Electrical Insulation and Dielectric Phenomena, 2001 annual report, conference on, Kitchener, pp 40–44
25. Metze S, Blioch S, Matuszczyk J et al (2020) Multivariate data analysis of capacitance frequency scanning for online monitoring of viable cell concentrations in small-scale bioreactors. Anal Bioanal Chem 412(9):2089–2102
26. Ma F, Zhang A, Chang D et al (2019) Real-time monitoring and control of CHO cell apoptosis by in situ multifrequency scanning dielectric spectroscopy. Process Biochem 80:138–145
27. Wasalathanthri DP, Rehmann MS, Song Y et al (2020) Technology outlook for real-time quality attribute and process parameter monitoring in biopharmaceutical development-a review. Biotechnol Bioeng 117(10):3182–3198
28. Berry B, Moretto J, Matthews T et al (2015) Cross-scale predictive modeling of CHO cell culture growth and metabolites using Raman spectroscopy and multivariate analysis. Biotechnol Prog 31(2):566–577
29. Berry BN, Dobrowsky TM, Timson RC et al (2016) Quick generation of Raman spectroscopy based in-process glucose control to influence biopharmaceutical protein product quality during mammalian cell culture. Biotechnol Prog 32(1):224–234
30. Esmonde-White KA, Cuellar M, Uerpmann C et al (2017) Raman spectroscopy as a process analytical technology for pharmaceutical manufacturing and bioprocessing. Anal Bioanal Chem 409(3):637–649
31. Webster TA, Hadley BC, Hilliard W et al (2018) Development of generic Raman models for a GS-KOTM CHO platform process. Biotechnol Prog 34:730–737
32. Abu-Absi NR, Kenty BM, Cuellar ME et al (2011) Real time monitoring of multiple parameters in mammalian cell culture bioreactors using an in-line Raman spectroscopy probe. Biotechnol Bioeng 108:1215–1221
33. Yilmaz D, Connor MO, Mccormick P (2020) Application of raman spectroscopy in monoclonal antibody producing continuous systems for downstream process intensification. Biotechnol Prog 36:e2947
34. Lourenco ND, Lopes JA, Almeida CF et al (2012) Bioreactor monitoring with spectroscopy and chemometrics: a review. Anal Bioanal Chem 404:1211–1237
35. Rolinger L, Rüdt M, Hubbuch J (2020) A critical review of recent trends, and a future perspective of optical spectroscopy as PAT in biopharmaceutical downstream processing. Anal Bioanal Chem 412(9):2047–2064

36. Whelan J, Craven S, Glennon B (2012) In situ Raman spectroscopy for simultaneous monitoring of multiple process parameters in mammalian cell culture bioreactors. Biotechnol Prog 28(5):1355–1362
37. Jestel NL (2005) Process Raman spectroscopy. In: Bakeev KA (ed) Process analytical technology. Blackwell, Oxford
38. Ulber R, Frerichs JG, Beutel S (2003) Optical sensor systems for bioprocess monitoring. Anal Bioanal Chem 376:342–348
39. Trunfio N, Lee H, Starkey J et al (2017) Characterization of mammalian cell culture raw materials by combining spectroscopy and chemometrics. Biotechnol Prog 33(4):1127–1138
40. Rodionova OY, Sokovikov YV, Pomerantsev AL (2009) Quality control of packed raw materials in pharmaceutical industry. Anal Chim Acta 642:222–227
41. Hakemeyer C, Strauss U, Werz S (2013) Near-infrared and two-dimensional fluorescence spectroscopy monitoring of monoclonal antibody fermentation media quality: aged media decreases cell growth. Biotechnol J 8(7):835–846
42. Kambayashi T, Noguchi T, Nojima A et al (2020) Glucose monitoring in cell culture with online ultrasound-assisted near-infrared spectroscopy. Anal Chem 92(4):2946–2952
43. Arnold SA, Crowley J, Woods N et al (2003) In-situ near infrared spectroscopy to monitor key analytes in mammalian cell cultivation. Biotechnol Bioeng 84:13–19
44. Cervera AE, Petersen N, Lantz AE et al (2009) Application of near-infrared spectroscopy for monitoring and control of cell culture and fermentation. Biotechnol Prog 25:1561–1581
45. Qui J, Arnold MA, Murhammer DW (2014) On-line near infrared bioreactor monitoring of cell density and concentrations of glucose and lactate during insect cell cultivation. J Biotechnol 173:106–111
46. Scarff M, Arnold SA, Harvey LM et al (2006) Near infrared spectroscopy for bioprocess monitoring and control: current status and future trends. Crit Rev Biotechnol 26:17–39
47. Food and Drug Administration (2015) Development and submission of near infrared analytical procedures guidance for industry. U.S. Department of Health and Human Services, Food and Drug Administration. https://www.fda.gov/media/91343/download. Accessed 7.12.2020
48. Simpson MB (2005) Near-infrared spectroscopy for process analytical chemistry: theory, technology and implementation. In: Bakeev KA (ed) Process analytical technology. Blackwell, Oxford, pp 107–153
49. Evans HM, Schulemann W (1914) The action of vital stains belonging to the benzidine group. Science 39:443–454
50. Chen A, Leith M, Tu R et al (2017) Effects of diluents on cell culture viability measured by automated cell counter. PLoS One 12(3):e0173375
51. Darzynkiewicz Z, Juan G, Li X, Gorczyca W et al (1997) Cytometry in cell necrobiology: analysis of apoptosis and accidental cell death (necrosis). Cytometry 27:1–20
52. Ishaque A, Al-Rubeai M (2002) Role of vitamins in determining apoptosis and extent of suppression by bcl-2 during hybridoma cell culture. Apoptosis 7:231–239
53. Kroemer G, Galluzzi L, Vandenabeele P et al (2009) Classification of cell death: recommendations of the nomenclature committee on cell death 2009. Cell Death Differ 16:3–11
54. Krysko DV, Vanden Berghe T, D'herde K et al (2008) Apoptosis and necrosis: detection, discrimination and phagocytosis. Methods 44:205–221
55. Fagète S, Steimer C, Girod PA (2019) Comparing two automated high throughput viable-cell counting systems for cell culture applications. J Biotechnol 305:23–26
56. Cotter TG, Al-Rubeai M (1995) Cell death (apoptosis) in cell culture systems. Trends Biotechnol 13:150–155
57. Matthews TE, Berry BN, Smelko J et al (2016) Closed loop control of lactate concentration in mammalian cell culture by Raman spectroscopy leads to improved cell density, viability, and biopharmaceutical protein production. Biotechnol Bioeng 113:2416–2424
58. André S, Cristau LS, Gaillard S et al (2015) In-line and real-time prediction of recombinant antibody titer by in situ Raman spectroscopy. Anal Chim Acta 892:148–152
59. Ashton L, Xu Y, Brewster VL et al (2013) The challenge of applying Raman spectroscopy to monitor recombinant antibody production. Analyst 138:6977–6985

60. Liu B, Villacres-Barragan C, Lattova E et al (2013) Differential affects of low glucose on the macroheterogeneity and microheterogeneity of glycosylation in CHO-EG2 camelid monoclonal antibodies. BMC Proc 7(Suppl 6):P112
61. Bhatia H, Mehdizadeh H, Drapeau D, Yoon S (2017) In-line monitoring of amino acids in mammalian cell cultures using raman spectroscopy and multivariate chemometrics models. Eng Life Sci 18(1):55–61
62. Hiller GW, Mulukutla BC (2017) Method of cell culture. WO2015140708A1. Google Patents, https://patents.google.com/patent/WO2015140708A1/en. Accessed 20.11.2020
63. Douma R (2019) Monitoring the scale-up of a biopharmaceutical cell culture process using Raman Spectroscopy at 2000 L scale in a GMP environment. Presentation: BioProcess International Europe, Vienna, April 2019
64. Rees-Manley A, Carpio M (2019) Evaluating integrated analytics for single-use mini bioreactors. American Pharmaceutical Review. https://www.americanpharmaceuticalreview.com/1504-White-Papers-Application-Notes/557243-Evaluating-Integrated-Analytics-for-Single-use-Mini-Bioreactors/?catid=6262. Accessed 15.11.2020
65. Hofer A, Herwig C (2018) Automated bioreactor sampling for on-line analysis of amino acids using two approaches: HPLC and Cedex® Bio HT Analyzer. Secure Cell Application Note #004. https://securecell.ch/de/automated-bioreactor-sampling-for-on-line-analysis-of-amino-acids-using-two-approaches-hplc-and-cedex-bio-ht-analyzer/. Accessed 9.12.2020
66. Willard SS, Suttle A, Schneider FJ et al (2018) Automated Glucose Control for a Cell Culture Process. http://www.flownamics.com/docs/BioFlo320_SegFlow-Integration.pdf. Accessed 3.11.2020
67. St Amand MM, Ogunnaike BA, Robinson AS (2014) Development of at-line assay to monitor charge variants of MAbs during production. Biotechnol Prog 30(1):249–255
68. Pais DA, Carrondo MJ, Alves PM et al (2014) Towards real-time monitoring of therapeutic protein quality in mammalian cell processes. Curr Opin Biotechnol 30:161–167
69. Berkowitz SA, Engen JR, Mazzeo JR et al (2012) Analytical tools for characterizing biopharmaceuticals and the implications for biosimilars. Nat Rev Drug Discov 11(7):527–540
70. Gryniewicz CM, Kauffman JF (2008) Multivariate calibration of covalent aggregate fraction to the Raman Spectrum of regular human insulin. J Pharm Sci 97(9):3727–3734
71. McAvan BS, Bowsher LA, Powell T et al (2020) Raman spectroscopy to monitor post-translational modifications and degradation in monoclonal antibody therapeutics. Anal Chem 92(15):10381–10389
72. Zhang C, Springall JS, Wang X et al (2019) Rapid, quantitative determination of aggregation and particle formation for antibody drug conjugate therapeutics with label-free Raman spectroscopy. Anal Chim Acta 1081:138–145
73. Mungikar A, Kamat M (2010) Use of In-line Raman Spectroscopy as a Non-destructive and Rapid Analytical Technique to Monitor Aggregation of a Therapeutic Protein. American Pharmaceutical Reviews. https://www.americanpharmaceuticalreview.com/Featured-Articles/37071-Use-of-In-line-Raman-Spectroscopy-as-a-Non-destructive-and-Rapid-Analytical-Technique-to-Monitor-Aggregation-of-a-Therapeutic-Protein/. Accessed 15.10.2020
74. Drenski MF, Brader ML, Alston RW et al (2013) Monitoring protein aggregation kinetics with simultaneous multiple sample light scattering. Anal Biochem 437(2):185–197
75. Waymouth C (1970) Osmolality of mammalian blood and of media for culture of mammalian cells. In Vitro 6:109–127
76. Sweeney TE, Beuchat CA (1993) Limitations of methods of osmometry: measuring the osmolality of biological fluids. Am J Phys 264(3 Pt 2):R469–R480
77. Qin J, Wu X, Xia Z et al (2019) The effect of hyperosmolality application time on production, quality, and biopotency of monoclonal antibodies produced in CHO cell fed-batch and perfusion cultures. Appl Microbiol Biotechnol 103(3):1217–1229

78. De Zengotita VM, Schmelzer AE, Miller WM (2002) Characterization of hybridoma cell responses to elevated pCO(2) and osmolality: intracellular pH, cell size, apoptosis, and metabolism. Biotechnol Bioeng 77(4):369–380
79. Zhu MM, Goyal A, Rank DL et al (2005) Effects of elevated pCO2 and osmolality on growth of CHO cells and production of antibody-fusion protein B1: a case study. Biotechnol Prog 21(1):70–77
80. Ryu JS, Kim TK, Chung JY et al (2000) Osmoprotective effect of glycine betaine on foreign protein production in hyperosmotic recombinant chinese hamster ovary cell cultures differs among cell lines. Biotechnol Bioeng 70(2):167–175
81. Moretto J, Smelko JP, Cuellar M et al (2011) Process Raman Spectroscopy for In-Line CHO Cell Culture Monitoring. American Pharmaceutical Reviews. https://www.americanpharmaceuticalreview.com/Featured-Articles/37040-Process-Raman-Spectroscopy-for-In-Line-CHO-Cell-Culture-Monitoring/. Accessed 14.12.2020
82. Clavaud M, Roggo Y, Von Daeniken R et al (2013) Chemometrics and in-line near infrared spectroscopic monitoring of a biopharmaceutical Chinese hamster ovary cell culture: prediction of multiple cultivation variables. Talanta 111:28–38
83. Mattes R, Root D, Sugui MA et al (2009) Real-time bioreactor monitoring of osmolality and pH using near-infrared spectroscopy. Bioprocess Int 7:44–50.146
84. Clark LC Jr, Wolf R, Granger D et al (1953) Continuous recording of blood oxygen tensions by polarography. J Appl Physiol 6(3):189–193
85. Beckman AO, Fracker HE (1934) US Patent No. 2,058,761
86. Chopda VR, Holzberg T, Ge X et al (2020) Real-time dissolved carbon dioxide monitoring I: application of a novel in situ sensor for CO2 monitoring and control. Biotechnol Bioeng 117(4):981–991
87. Kimura R, Miller WM (1996) Effects of elevated pCO2 and/or osmolality on the growth and recombinant tPA production of CHO cells. Biotechnol Bioeng 52:152–160
88. Abu-Absi S, Xu S, Graham H et al (2014) Cell culture process operations for recombinant protein production. Adv Biochem Eng Biotechnol 139:35–68
89. Chen A, Chitta R, Chang D et al (2009) Twenty-four well plate miniature bioreactor system as a scale-down model for cell culture process development. Biotechnol Bioeng 102:148–160
90. Steinegger A, Wolfbeis OS, Borisov SM (2020) Optical sensing and imaging of pH values: spectroscopies, materials, and applications. Chem Rev 120(22):12357–12489
91. Wang XD, Wolfbeis OS (2014) Optical methods for sensing and imaging oxygen: materials, spectroscopies and applications. Chem Soc Rev 43(10):3666–3761
92. Weigl BH, Holobar A, Trettnak W et al (1994) Optical triple sensor for measuring pH, oxygen and carbon dioxide. J Biotechnol 32(2):127–138
93. Wiegmann V, Bernal C, Kreukniet M et al (2017) micro-Matrix: Monitored CHO cell batch in 24 parallel micro-bioreactors. https://www.applikon-biotechnology.com/files/appk-poster2a0-chocell-210617lr.pdf. Accessed 7.12.2020
94. m2p-labs, 2020.: https://www.m2p-labs.com/bioreactors/products/biolector. Accessed 7.12.2020
95. Applikon Biotechnology (2020) https://www.applikon-biotechnology.com/en/products/cultivation-systems/micro-flask/. Accessed 7.12.2020
96. Schneider K, Schütz V, John GT et al (2010) Optical device for parallel online measurement of dissolved oxygen and pH in shake flask cultures. Bioprocess Biosyst Eng 33:541–547
97. Tric M, Lederle M, Neuner L et al (2017) Optical biosensor optimized for continuous in-line glucose monitoring in animal cell culture. Anal Bioanal Chem 409(24):5711–5721
98. Tsai W-L, Autsen JL, Ma J et al (2012) Noninvasive optical sensor technology in Shake Flasks for mammalian cell cultures. BioProcess International 10(1):50–56
99. Takahashi M, Aoyagi H (2018) Practices of shake-flask culture and advances in monitoring CO2 and O2. Appl Microbiol Biotechnol 102:4279–4289

100. Lee S, Ibey BL, Coté GL et al (2008) Measurement of pH and dissolved oxygen within cell culture media using a hydrogel microarray sensor. Sensors Actuators B Chem 128:388–398
101. Couture J, Lindsay J, Baggio R (2018) Impact of dCO2 on Process Parameters: Effect of High dCO2 on Viable Cell Density and Product Titer in CHO Cell Culture. https://www.presens.de/knowledge/publications/application-note/impact-of-dco2-on-process-parameters-575. Accessed 14.12.2020
102. Hill M, Kern M, Bose S et al (2019) Evaluation of an Optical CO_2 Probe for Long-Term Monitoring in Stirred-Tank Bioreactors, Appl. Note: Prototype Autoclavable CO_2 Sensor Used in Low and High Cell Density Cultures https://www.presens.de/knowledge/publications/application-note/evaluation-of-an-optical-co2-probe-for-long-term-monitoring-in-stirred-tank-bioreactors-1715. Accessed 12.12.2020
103. Kroll P, Herwig C (2019) Control of dissolved carbon dioxide in driving force limited cell culture processes. Semantic Scholar Corpus ID: 202620766. https://www.semanticscholar.org/paper/Control-of-dissolved-carbon-dioxide-in-driving-cell-Kroll-Herwig/d82a5020f309f51b5b7bb71df0f3e704db36b3b5. Accessed 12.12.2020
104. Pollard D, Christensen J (2013) Vent gas analysis. In: Flickinger MC (ed) Upstream industrial biotechnology: equipment, process design, sensing, control, and cGMP operations, vol 2, 1st edn. Wiley, Hoboken, pp 1567–1584
105. Severinghaus J, Bradley F (1958) Electrodes for blood pO2 and pCO2 determination. J Appl Physiol 13:515–520
106. Uttamlal M, Walt D (1995) A fiber-optic carbon dioxide sensor for fermentation monitoring. Biotechnology 13:597–601
107. Pattison RN, Swamy J, Mendenhall B et al (2000) Measurement and control of dissolved carbon dioxide in mammalian cell culture processes using an in situ fiber optic chemical sensor. Biotechnol Prog 16(5):769–774
108. Pluschkell SB, Flickinger MC (1995) Improved methods for investigating the external redox potential in hybridoma cell culture. Cytotechnology 19(1):11–26
109. Meneses-Acosta A, Gómez A, Ramírez OT (2012) Control of redox potential in hybridoma cultures: effects on MAb production, metabolism, and apoptosis. J Ind Microbiol Biotechnol 39(8):1189–1198
110. Dionne B, Mishra N, Butler M (2017) A low redox potential affects monoclonal antibody assembly and glycosylation in cell culture. J Biotechnol 246:71–80
111. Handlogten MW, Wang J, Ahuja S (2020) Online control of cell culture redox potential prevents antibody interchain disulfide bond reduction. Biotechnol Bioeng 117(5):1329–1336
112. Trexler-Schmidt M, Sargis S, Chiu J et al (2010) Identification and prevention of antibody disulfide bond reduction during cell culture manufacturing. Biotechnol Bioeng 106(3):452–461
113. Hutterer KM, Hong RW, Lull J et al (2013) Monoclonal antibody disulfide reduction during manufacturing: untangling process effects from product effects. MAbs 5(4):608–613
114. Nyberg GB, Balcarcel RR, Follstad BD et al (1999) Metabolic effects on recombinant interferon-gamma glycosylation in continuous culture of Chinese hamster ovary cells. Biotechnol Bioeng 62:336–347

Index

R. Pörtner (ed.), *Cell Culture Engineering and Technology*, Cell Engineering 10,
https://doi.org/10.1007/978-3-030-79871-0

The manufacturer's authorised representative in the EU is Springer Nature Customer Service Centre GmbH, Europaplatz 3, 69115 Heidelberg, Germany. If you have any concerns regarding our products, please contact ProductSafety@springernature.com

Printed and bound by CPI Group (UK) Ltd, Croydon, CR0 4YY
30/09/2026
02224059-0010